AIDS Booklet

This booklet describes how AIDS and related diseases are commonly spread so that readers can protect themselves and their friends against this debilitating and deadly disease.

Human Biology Laboratory Manual

Known for its human emphasis and self-contained content, this laboratory manual, written by Dr. Sylvia S. Mader, with contributions by Nancy Segsworth, helps students understand how the human body works and the relationship of humans to other living things in the biosphere. With few exceptions, each chapter in the text has an accompanying laboratory exercise in the manual.

Laboratory Resource Guide

More extensive information regarding preparation is found in this helpful guide. The guide includes suggested sources for materials and supplies, directions for making up solutions and otherwise setting up the laboratory, expected results for the exercises, and suggested answers to questions in the laboratory manual.

Student Study Guide

Study Tips, Study Questions, Definitions, Chapter Tests, and Answer Key will help your students solidify *Human Biology* concepts.

200 Transparencies

A set of 200 full-color transparency acetates accompanies *Human Biology*. These acetates contain key illustrations from the text.

100 Micrograph Slides

This ancillary provides 35mm slides of many photomicrographs and all electron micrographs in the text.

Instructor's Manual with Test Item File

The *Instructor's Manual*, prepared by Dr. Steve Badger, includes features such as: Behavioral Objectives, Extended chapter lecture outlines, Teaching Strategies, Discussion Activities, Demonstration Activities, additional Critical Thinking questions, and a Technology Correlation Guide. The *Test Item File*, prepared by Dr. Thomas Pitzer, includes a classification system for difficulty level and type of question.

Classroom Testing Software (MicroTest)

This helpful testing software—available in either Macintosh or Windows format—provides well-written and researched book-specific questions featured in the Test Item File.

How to Study Science, Third Edition

This excellent workbook offers students helpful suggestions for meeting the considerable challenges of a college science course. It offers tips on how to take notes, how to get the most out of laboratories, and how to overcome science anxiety.

Schaum's Outlines: Biology

Updated to include the latest advances, *Schaum's Outlines: Biology*, features detailed illustrations of complex biologic systems and processes, ranging from the smallest elements of life to primates. Hundreds of problems with fully explained solutions cut down on study time and make important points easy to remember.

Basic Chemistry for Biology, Second Edition

Basic Chemistry for Biology is a self-paced supplement for students who need additional material to understand the basic concepts of chemistry. This text leads biology students through fundamental chemical concepts.

Critical Thinking Case Study Workbook

This ancillary provides 34 critical thinking case studies that are designed to immerse students in the "process of science" and challenge them to solve problems in the same way biologists do. An answer key accompanies this workbook.

Biology Start-up Software

This software is a five-disk Macintosh tutorial that helps nonmajors master challenging biological concepts such as basic chemistry, photosynthesis, and cellular respiration.

For My Family

Sixth Edition

HUMAN
B I O L O G Y

Test April 10.

Sylvia S. Mader

579-0290.

Boston   Burr Ridge, IL   Dubuque, IA   Madison, WI   New York   San Francisco   St. Louis
Bangkok   Bogotá   Caracas   Lisbon   London   Madrid
Mexico City   Milan   New Delhi   Seoul   Singapore   Sydney   Taipei   Toronto

McGraw-Hill Higher Education

A Division of The McGraw·Hill Companies

HUMAN BIOLOGY, SIXTH EDITION

Published by McGraw-Hill, an imprint of The McGraw-Hill Companies, Inc., 1221 Avenue of the Americas, New York, NY 10020. Copyright © 2000, 1998, 1995, 1992, 1990, 1988 by The McGraw-Hill Companies, Inc. All rights reserved. No part of this publication may be reproduced or distributed in any form or by any means, or stored in a database or retrieval system, without the prior written consent of The McGraw-Hill Companies, Inc., including, but not limited to, in any network or other electronic storage or transmission, or broadcast for distance learning.

Some ancillaries, including electronic and print components, may not be available to customers outside the United States.

 This book is printed on recycled, acid-free paper containing 10% postconsumer waste.

4 5 6 7 8 9 0 QPD/QPD 0 9 8 7 6 5 4 3 2 1 0

ISBN 0–07–290584–0

1 2 3 4 5 6 7 8 9 0 QPD/QPD 0 9 8 7 6 5 4 3 2 1 0

ISBN 0–07–117940–2 (ISE)

Vice president and editor-in-chief: *Kevin T. Kane*
Publisher: *Michael D. Lange*
Sponsoring editor: *Patrick E. Reidy*
Senior developmental editor: *Suzanne M. Guinn*
Senior marketing manager: *Lisa L. Gottschalk*
Senior project manager: *Marilyn M. Sulzer*
Production supervisor: *Laura Fuller*
Designer: *K. Wayne Harms*
Senior photo research coordinator: *Lori Hancock*
Senior supplement coordinator: *Audrey A. Reiter*
Compositor: *GTS Graphics, Inc.*
Typeface: *10/12 Palatino*
Printer: *Quebecor Printing Book Group/Dubuque, IA*

The credits section for this book begins on page C–1 and is considered an extension of the copyright page.

Library of Congress Cataloging-in-Publication Data

Mader, Sylvia S.
   Human biology / Sylvia S. Mader. — 6th ed.
       p.   cm.
   Includes index.
   ISBN 0–07–290584–0
   1. Human biology. I. Title.

QP36.M2   2000
   612—dc21   99–14988
                                                          CIP

INTERNATIONAL EDITION ISBN 0–07–117940–2
Copyright © 2001. Exclusive rights by The McGraw-Hill Companies, Inc., for manufacture and export. This book cannot be re-exported from the country to which it is sold by McGraw-Hill. The International Edition is not available in North America.

www.mhhe.com

Brief Contents

Contents

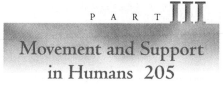

PART III

Movement and Support in Humans 205

Readings

Ecology Focus

Health Focus

Visual Focus

Preface

Human Biology is suitable for use in one-semester biology courses that emphasize human physiology and the role that humans play in the biosphere. All students should leave college with a firm grasp of how their bodies normally function and how the human population can become more fully integrated into the biosphere. This knowledge can be applied daily and helps assure our continued survival as individuals and as a species. The application of biological principles to practical human concerns is now widely accepted as a suitable approach to the study of biology because it fulfills a great need. Human beings are frequently called upon to make decisions about their bodies and their environment. Wise decisions require adequate knowledge.

In this edition, as in previous editions, each chapter presents the topic clearly, simply, and distinctly so that students will feel capable of achieving an adult level of understanding. Detailed, high-level scientific data and terminology are not included because I believe that true knowledge consists of working concepts rather than technical facility.

Homeostasis

This edition has a renewed emphasis on homeostasis. The last chapter in Part 1 is entitled "Introduction to Homeostasis". The principles of homeostasis are discussed and the contributions of the various systems to homeostasis are outlined before a feature of the text called a Working Together box is introduced. Working Together boxes throughout the text describe how each organ system works with other systems to achieve homeostasis. In each chapter, an icon calls attention to those portions of the text which discuss homeostasis. Applying Your Knowledge to the Concepts questions at the end of chapter pertain to the maintenance of homeostasis.

The urinary system chapter and the endocrine system chapter were rewritten to better emphasize the contribution of these systems to homeostasis.

Vibrant, New Illustration Program

Almost every illustration in the text is new or has been revised to better engage students in the study of human biology. Students are visually motivated, and the new art program has many features they will find helpful. Visual Focus illustrations give a conceptual overview that relates structure to function. Color coordination includes assigning colors to the various classes of organic molecules and to the different human tissues and organs.

Readings

As in the previous edition, health and ecology concerns are carried through the book by Health Focus and Ecology Focus readings. The Health Focus readings are designed to help students cope with common health problems. The Ecology Focus readings draw attention to a particular environmental problem.

In this edition, students are asked to apply the concepts to the many and varied perplexing bioethical issues that face us every day. Each chapter ends with a description of a bioethical situation that calls for a value judgement on the part of the reader. Students are challenged to develop a point of view by answering a series of questions that deal with the issue. The myriad of issues considered include genetic disease testing, modern reproductive technologies, human cloning, AIDS vaccine trials, animal rights, responsibility for one's health, and fetal research.

Technology

New to this edition, the free *Essential Study Partner* CD-ROM, accompanies the text. A CD-ROM icon has been placed throughout each chapter to remind students that this important learning tool can assist them in reviewing the concepts. *The Dynamic Human 2.0*, which offers a pictorial review of each human system, has been revised to have even more student appeal. The Mader Home Page contains interactive exercises to help students master the objective of each chapter and provides further information on most topics discussed in the text.

Pedagogical Features

As before, *Human Biology* excels in pedagogical features. Each chapter begins with an integrated chapter outline that lists the chapter's concepts according to numbered sections of the chapter. This numbering system is continued in the chapter and summary so that instructors can assign just certain portions of the chapter, if they like. The text is paged so that major sections start at the top of the page and illustrations are on the same or facing page to its reference.

The questions at the end of the chapter are of both the essay and objective type. New to this edition, the Testing Your Knowledge of the Concepts questions include multiple choice, fill in the blanks, and true-false questions. The questions called Applying Your Knowledge to the Concepts help stress the homeostasis theme of the text. All the boldfaced terms in the chapter are listed and page referenced. A matching exercise tests student comprehension of the terms.

Revised Chapters

Almost every chapter in *Human Biology* has been revised. Every systems chapter now has a major section entitled Homeostasis, which outlines how that system works with other systems to maintain homeostasis. The nervous system chapter describes new findings in the field of memory and learning. The cardiovascular system chapter has a new section which pulls together material on pulse rate, blood pressure, and blood flow. In the lymphatic system chapter, the inflammatory reaction has been rewritten. The respiratory system chapter now includes a greater number of respiratory tract infections and disorders. In the chapter on senses, the mechanism of smelling and tasting has been updated. The AIDS Supplement and the chapter on cancer include the latest information on these disorders.

Applications

Educational theory tells us that students are most interested in knowledge of immediate practical application. This text is consistent with and remains true to this approach.

Each chapter begins with a short story that applies chapter material to real-life situations. The readings stress applications and so does the running text material. This edition features expanded treatment of such topics as eating disorders, allergies, pulmonary disorders, hepatitis infections, modern reproductive technologies, the human genome project, and gene therapy. Some topics such as the cloning of animals, xenotransplantation, and gene therapy to treat cancer are new.

New to This Edition

- Homeostasis has a renewed emphasis. An icon calls attention to those portions of the text that discuss homeostasis; each systems chapter has a major section that discusses how that system works with other systems of the body to achieve homeostasis, and Applying Your Knowledge to the Concepts contains questions that pertain to homeostasis.
- Revised Working Together boxes appear in each of the system chapters. These illustrations describe how each organ system works with the other systems to achieve homeostasis, which is also discussed in a major section of the chapter.
- Technology aids are described at the end of each chapter. The *Essential Study Partner* CD-ROM tutorial which supports and enhances the concepts presented is offered free with the text. A CD-ROM icon is used throughout the chapter to remind students to consult this useful learning tool. *The Dynamic Human 2.0* CD-ROM is an interactive three-dimensional visual guide to human anatomy and

physiology. The Mader Home Page provides interactive study exercises and further information for each chapter of the text.
- Health Focus and Ecology Focus readings support the two major themes of the text. A new bioethical issue is discussed in a featured section at the end of each chapter. Challenging questions are provided that can be used as a basis for class discussion.
- A new illustration program adds vitality to the art and enhances the appeal of the text. Many new micrographs provide realism. Visual Focus illustrations give a pictorial overview of key topics. Color coding is used both for molecular structures and for human tissues and organs.
- Relevancy of the text is increased with the inclusion or expanded treatment of such topics as eating disorders, allergies, pulmonary disorders, hepatitis infections, xenotransplantation, modern reproductive technologies, human cloning, the human genome project, and gene therapy to treat cancer.

Acknowledgments

The personnel at WCB/McGraw-Hill have always lent their talents to the success of *Human Biology*. My publisher Michael Lange was always there to offer advice and my editor Patrick Reidy stepped in when needed to encourage us all. Suzanne Guinn, my developmental editor, served as a liaison between me and everyone else on the book team. Suzanne had many creative suggestions and was an inspiration to us all despite the long hours she labored.

Those in production also worked diligently toward the success of this edition. Marilyn Sulzer was the project manager, Jodi Banowetz, the visuals coordinator, and Lori Hancock was the photo research coordinator. And I especially want to thank Wayne Harms for the beautiful book he designed for all of us to enjoy. Everyone remained cheerful and helpful while going beyond the call of duty.

In my office Evelyn Jo Hebert has consistently provided support through several editions of the text, and Norma Costain's contributions have also made the success of *Human Biology*, sixth edition, possible. Kathleen Hagelston has been a wonderful resource for creative and expert input on illustrations through the editions of *Human Biology*.

The Reviewers

Many instructors have contributed not only to this edition of *Human Biology* but also to previous editions. I am extremely thankful to each one, for they have all worked diligently to remain true to our calling to provide a product that will be the most useful to our students.

It is appropriate to acknowledge the help of the following individuals for the sixth edition:

Sister Jane Anne Molinaro
Immaculate College

Joanna Borvcinska
University of Hartford

Hessel Bouma
Calvin College

Kathleen Lively
Marquette University

Steve Badger
Central Bible College

Peter Biesmeyer
North County Community College

Julia Brown
Northeast Iowa Community College North Campus

Karen Vanmeter
Des Moines Area Community College

John Sternick
Mansfield University

Surendra Singh
Kansas Newman College

Don Nabor
University of Maine

Ronald Salyx
Bloomfield College

Anthony Serino
Washburn University

Mary Louise Greeley
Salve Regina University

Maka Najaragan
Wilberforce University

Mary Catherine Cox
Wingate University

Cecilia Golnazarian
Community College of Vermont

Dr. Carl Frankel
Penn State University

Jacqueline Shepperson
Winston-Salem State University

Oian Frances Moss
Des Moines Area Community College

Patricia Klofenstein
Edison Community College

Caren Shapiro
D'Youville College

Deborah Dodson
Vincennes University

Susan Karr
Carson Newman College

Debra Zehner
Wilkes University

Lewis Lutton
Mercyhurst College

Jacquelin McLaughlin
Penn State University

Curt Walker
Dixie College

Elizabeth Lawrence
Miles Community College

G. Malcolm Amerson
Oglethorpe University

Sebastian Haskel
Nova Southeastern University

Soma Sanyal
Penn State-Altoona

Thanks also to reviewers of the previous edition:

Donald Jasper
Illinois Institute of Technology

Allan R. Stevens
Snow College

C. L. Swendson
Warren Wilson College

Don Naber
University College, University of Maine

Tom Denton
Auburn University at Montgomery

Mary King Kananen
Penn State–Altoona

Gina Erickson
Highline Community College

Diane Merlos
Grossmont College

Arlene Marian
Niagara University

Lawton Owen
Kansas Wesleyan University

Donald A. Wheeler
Edinboro University of Pennsylvania

Craig Berezowsky
The University of British Columbia

Elaine Rubenstein
Skidmore College

Ronald F. Cooper

Dr. Charles Hummel
New York Institute of Technology

Valerie Vander Vleit
Lewis University

David Wolfrom
Paducah Community College

Al Avenoso
University of Houston–Downtown

James J. Greene
The Catholic University of America

Grant M. Barkley
Kent State University

Michael Emsley
George Mason University

Felix Baerlocher
Mount Allison University

Charlene L. Forest
Brooklyn College of CUNY

Isaac Elegbe
College of New Rochelle

Pat Selelyo
College of Southern Idaho

Joseph V. Martin
Rutgers University

Loretta M. Parsons
Skidmore College

James R. Phillips
Babson College

Garry Davies
University of Alaska–Anchorage

Fritz Taylor
University of New Mexico

Lynette Rushton
South Puget Sound Community College

Alexander Varkey
Liberty University

Donald S. Emmeluth
Fulton-Montgomery Community College

Penelope ReVelle
Essex Community College

Patricia Matthews
Grand Valley State University

Joe Connell
Leeward Community College

Vaughn Rundquist
Montana State University–Northern

Arnold E. S. Gussin
St. Francis College

Robert H. Chesney
William Paterson College of New Jersey

Robert J. Ratterman
Jamestown Community College

Stephen R. Karr
Carson-Newman College

Hessell Bouma III
Calvin College

Walt Sinnamon
Southern Wesleyan University

Marirose T. Ethington
Genesee Community College

Robert S. Greene
Niagara University

Michelle A. Green
SUNY at Alfred

Stanton F. Hoegerman
College of William and Mary

Dana Demmans
Finger Lakes Community College

Douglas J. Burks
Wilmington College of Ohio

Helen Cadwallader
Black Hawk College

J. D. Brammer
North Dakota State University

Rodney Mowbray
University of Wisconsin–LaCrosse

Edward W. Carroll
Marquette University

Char A. Beeanson
St. Olaf College

Barbara Wineinger
Vincennes University–Jasper

Charles Ellison
Willmington College–Cincinnati Branch

Lee H. Lee
Montclair State University

Robert Olson
Briar Cliff College

Benjamin C. Stark
Illinois Institute of Technology

Debra J. Martin
St. Mary's University of Minnesota

Ann L. Henninger
Wartburg College

Ted Johnson
St. Olaf College

Florence M. Dusek
Des Moines Area Community College

Madeline M. Hall
Cleveland State University

Albert C. Jensen
Central Florida Community College

William E. Dunscombe
Union County College

Kathleen Lauckner
UNLV-Harry Reid Center for Environmental Studies

Penny Bernstein
Kent State University–Stark Campus

Darryl L. Daley
Snow College

Doris M. Shoemaker
Dalton College

Kim R. Finer
Kent State University–Stark Campus

Lisa Danko
Mercyhurst College

James A Gessaman
Utah State University

Debra Zehner
Wilkes University

Robert H. Tamarin
University of Massachusetts–Lowell

George A. Hudock
Indiana University

Allan Hunt
Elizabeth Community College

Dale Lambert
Tarrant Count Junior College

Caren K. Shapiro
D'Youville College

Theresa Hoffman-Till
Northern Virginia Community College

David E. Dallas
Northeastern Oklahoma Agri. & Mechanical College

Orrie O. Stenroos
Lynchburg College in Virginia

Marcus Young Owl
California State University–Long Beach

Kathleen Lauber
Catonsville Community College

Dalia Giedrimiene
Saint Joseph College

M. L. Tiell
Mercy College

Gregory L. Stewart
State University of West Georgia

William P. Ventura
Pace University

Kathryn Sergeant Brown

Stephen Smith
Johnson Bible College

Carl M. Christenson
Indiana University Southeast

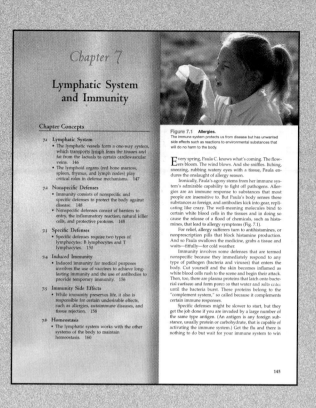

Before you begin, spend a little time looking over the next few pages. They provide a quick guide to the learning tools found throughout the text that have been designed to enhance your understanding of biology.

Concepts are Stressed

In this edition, the major topics are numbered, and the concepts listed on the chapter's opening page are grouped according to these topics. This numbering system, which is used in the text material and in the summaries, allows instructors to assign specific portions of the chapter. It also allows students to study the chapter in terms of the concepts presented.

In addition, *Human Biology* now has a further enhanced art and pedagogical system. Improved page layout, an outstanding art program, revised charts, and rewritten and reorganized chapters ensure that *Human Biology* will continue to be a winner in the classroom.

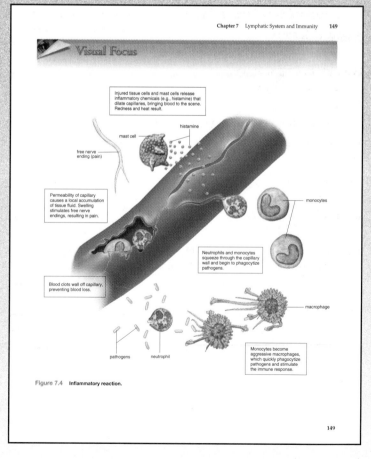

Working Together

Human Biology emphasizes human physiology and the role humans play in the biosphere. Each chapter presents topics clearly, simply, and distinctly using a concepts approach. For example, homeostasis is emphasized throughout the text (shown by an icon) and each systems chapter has its own main section explaining, in depth, how that particular system helps maintain homeostasis.

In addition, full-page illustrations entitled "Working Together" visually summarize and describe how each organ system interacts with other body systems. This is featured in every systems chapter.

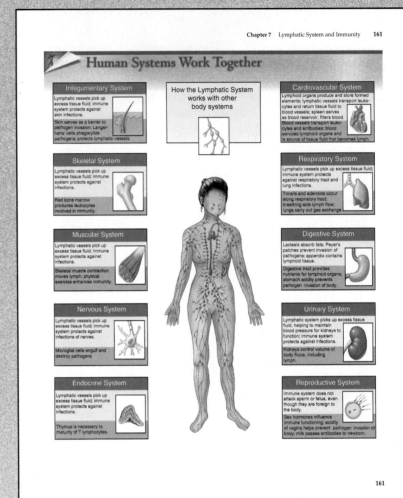

Human Systems Work Together

How the Lymphatic System works with other body systems

Integumentary System
Lymphatic vessels pick up excess tissue fluid; immune system protects against skin infections.
Skin serves as a barrier to pathogen invasion; Langerhans' cells phagocytize pathogens; protects lymphatic vessels.

Cardiovascular System
Lymphoid organs produce and store formed elements; lymphatic vessels transport leukocytes and return tissue fluid to blood vessels; spleen serves as blood reservoir, filters blood.
Blood vessels transport leukocytes and antibodies; blood services lymphoid organs and is source of tissue fluid that becomes lymph.

Skeletal System
Lymphatic vessels pick up excess tissue fluid; immune system protects against infections.
Red bone marrow produces leukocytes involved in immunity.

Respiratory System
Lymphatic vessels pick up excess tissue fluid; immune system protects against respiratory tract and lung infections.
Tonsils and adenoids occur along respiratory tract; breathing aids lymph flow; lungs carry out gas exchange.

Muscular System
Lymphatic vessels pick up excess tissue fluid; immune system protects against infections.
Skeletal muscle contraction moves lymph; physical exercise enhances immunity.

Digestive System
Lacteals absorb fats; Peyer's patches prevent invasion of pathogens; appendix contains lymphoid tissue.
Digestive tract provides nutrients for lymphoid organs; stomach acidity prevents pathogen invasion of body.

Nervous System
Lymphatic vessels pick up excess tissue fluid; immune system protects against infections of nerves.
Microglial cells engulf and destroy pathogens.

Urinary System
Lymphatic system picks up excess tissue fluid, helping to maintain blood pressure for kidneys to function; immune system protects against infections.
Kidneys control volume of body fluids, including lymph.

Endocrine System
Lymphatic vessels pick up excess tissue fluid; immune system protects against infections.
Thymus is necessary to maturity of T lymphocytes.

Reproductive System
Immune system does not attack sperm or fetus, even though they are foreign to the body.
Sex hormones influence immune functioning; acidity of vagina helps prevent pathogen invasion of body; milk passes antibodies to newborn.

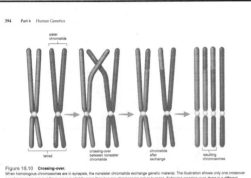

sister chromatids

tetrad | crossing-over between nonsister chromatids | chromatids after exchange | resulting chromosomes

Figure 18.10   Crossing-over.
When homologous chromosomes are in synapsis, the nonsister chromatids exchange genetic material. The illustration shows only one crossover per chromosome pair, but the average is slightly more than two per chromosome pair in humans. Following crossing-over, there is a different combination of genes on each chromatid.

18.4 Meiosis

Meiosis, which requires two cell divisions, results in *four daughter cells, each having one of each kind of chromosome and therefore half the number of chromosomes as the parent cell.*[2] The parent cell has the 2n number of chromosomes, while the daughter cells have the n number of chromosomes. Therefore, meiosis is often called reduction division. Following meiotic cell division, the daughter cells are not genetically identical, and neither is identical to the parent cell.

Overview of Meiosis: 2n → n

Meiosis results in four daughter cells because it consists of two divisions called meiosis I and meiosis II. Before meiosis I begins, each chromosome has duplicated and is composed of two sister chromatids. The parental cell is 2n. Recall that when a cell is 2n, the chromosomes occur in pairs. For example, the 46 chromosomes of humans occur in 23 pairs of chromosomes. These pairs are called **homologous chromosomes**.

During meiosis I, the homologous chromosomes of each pair come together and line up side by side due to a means of attraction still unknown. This so-called synapsis results in a **tetrad**, an association of four chromatids that stay in close proximity until they separate. During synapsis, nonsister chromatids may exchange genetic material. The exchange of genetic material between chromatids is called **crossing-over.** Crossing-over recombines the genes of the parental cell without the loss or gain of genetic material (Fig. 18.10).

Following synapsis during meiosis I, the homologous chromosomes of each pair separate. This separation means that one chromosome from each homologous pair will be found in each daughter cell. There are no restrictions as to which chromosome goes to each daughter cell, and therefore, all possible combinations of chromosomes occur within the daughter cells.

Notice that following meiosis I, the daughter cells have half the number of chromosomes and the chromosomes are still duplicated (Fig. 18.11). Again, counting the number of centromeres tells the number of chromosomes in each daughter cell.

During meiosis I, homologous chromosomes separate, and the daughter cells receive one of each pair. The daughter cells are not genetically identical. The chromosomes are still duplicated.

[2]The term meiosis technically refers only to nuclear division, but for convenience, it is used here to refer to the division of the entire cell.

The Essential Study Partner CD-ROM is an interactive student study tool referenced in the text and packed with over 100 animations and more than 200 learning activities. From quizzes to interactive diagrams, you will find that there has never been a more exciting way to study biology. A self-quizzing feature allows users to test their knowledge of a topic before moving on to a new module. Additional unit exams provide the opportunity to review an entire subject area. The quizzes and unit exams hyperlink back to tutorial sections so students can easily review coverage.

Summarizing the Concepts

These summaries help students review the important concepts and topics discussed in the chapter.

Bioethical Issue

Students are challenged to read the bioethical issue, then give a point of view by answering a series of questions that pertain to the issue. Issues include genetic disease testing, human cloning, AIDS vaccine trials, animal rights, responsibility for one's health, and fetal research.

Testing Your Knowledge of the Concepts

These objective questions allow students to test their ability to answer recall-based questions. At least one question requires that students label a diagram or fill in a table. Answers to *Testing Your Knowledge of the Concepts* appear in the appendix.

Bioethical Issue

The many types of animals in a coral reef form a complex community that is admired by both snorkelers and scuba divers. The various types of fish and shellfish in a coral reef are sources of food for millions of people. Like a tropical forest, coral reefs are most likely sources of medicines yet to be discovered. And a reef serves as a storm barrier that protects the shoreline and provides a safe harbor for ships.

Reefs around the globe are being destroyed. Tons of soil from deforested tracts of land bring nutrients that stimulate the growth of all kinds of algae. This has contributed to population explosion of the crown-of-thorn starfish that are devouring Australia's 1,200 mile-long Great Barrier Reef. Reefs are also being damaged by pollutants that seep into the sea from factories, farm fields, and sewers. Stress, combined with unusually warm seawater, has caused the corals to expel their symbiotic colorful algae, which carry on photosynthesis and help sustain them. So-called coral bleaching has been noticed in reefs of the Pacific Ocean and Caribbean. Might worldwide global warming also contribute to coral bleaching and death?

Marine scientist Eduardo Gomez estimates that 90% of coral reefs of the Philippines are dead or deteriorating due to pollution, but especially due to overfishing. The methods are sinister, including the use of dynamite to kill the fish, making it easier to scoop them up, use of cyanide to stun the fish to capture them alive, and using satellite navigation systems to home in on areas where mature fish are spawning to reproduce. If all large herbivores are killed off, seaweed overgrows and kills the coral.

Paleobiologist Jeremy Jackson of the Smithsonian Tropical Research Institute near Panama City wonders if he is doing enough to warn the public that reefs around the world are in danger. He estimates that we may lose 60% of all coral reefs by the year 2050.

Questions

1. Do you think it would be possible to make the public care about the loss of coral reefs? Explain.
2. When and under what circumstances do dire predictions help preserve the environment?
3. Considering what is causing the loss of coral reefs, would it be possible to save them? How?

Summarizing the Concepts

24.1 Human Population Growth

The human population is expanding exponentially, and it is unknown when growth will level off. Presently, each year exhibits a large increase, and the doubling time is now about 47 years. Populations have a biotic potential for increase in size. Biotic potential is normally held in check by environmental resistance, thereby producing an **S**-shaped growth curve, leveling off at the carrying capacity of the environment.

24.2 The Human Population and Pollution

An increasing human population is causing air, water, and land pollution.

Like the panes of a greenhouse, carbon dioxide, nitrous oxide, methane, and CFCs allow the sun's rays to pass through but impede the release of infrared wavelengths. It is predicted that a buildup in these "greenhouse gases" will lead to a global warming. The effects of global warming could be a rise in sea level and a change in climate patterns. An effect on agriculture could follow.

24.3 The Human Population and Biodiversity

Human activities have brought about a biodiversity crisis. Individuals and commercial hunting, habitat destruction or fragmentation, and introduction of new species and pollution are all major causes of species extinction.

Studying the Concepts

1. Draw a growth curve to represent exponential growth, and explain why a curve representing population growth usually levels off. 496
2. Calculate the growth rate and the doubling time for a population in which the birthrate is 20 per 1,000 and the death rate is 2 per 1,000. 497
3. Distinguish between MDCs and LDCs. Include a reference to age-structure diagrams. 498–99
4. Explain why the population of LDCs is expected to increase tremendously. What steps could be taken to prevent this from occurring? 498–99
5. How and why is the global climate expected to change, and what are the predicted consequences of this change? 500–1
6. What causes acid deposition, and what are its effects? 502
7. How does photochemical smog develop, and what is thermal inversion? 503
8. Of what benefit is the ozone shield? What pollutant in particular should be associated with stratospheric ozone depletion, and what are the consequences of this depletion? 504
9. What are several ways in which surface waters, aquifers, and oceans can be polluted? What is biological magnification? 505–6
10. Explain how soil erosion and desertification are related. 507
11. What are the primary ecological concerns associated with the destruction of rain forests? 508
12. Explain the primary causes of the biodiversity crisis and the goals of conservation biology. 509–10

Testing Your Knowledge of the Concepts

In questions 1–4, match the molecule to an environmental problem below.
a. sulfur dioxide
b. hydrocarbons
c. CFCs
d. carbon dioxide
_____ 1. photochemical smog
_____ 2. global warming
_____ 3. ozone shield destruction
_____ 4. acid deposition

In questions 5–7, indicate whether the statement is true (T) or false (F).
_____ 5. After a country has undergone the demographic transition, the death rate and the birthrate are both high.
_____ 6. Pesticides and radioactive wastes are both subject to biological magnification.
_____ 7. If global warming occurs, it is predicted that rising waters will threaten many coastal cities.

In questions 8 and 9, fill in the blanks.
8. Photochemical smog contains ozone and PAN, which are sometimes trapped near ground level due to a _____.
9. If the age-structure diagram has a pyramid shape, there are more women _____ the reproductive years than older women leaving them.
10. Label this **S**-shaped growth curve.

Applying Your Knowledge to the Concepts

These questions pertain to population concerns.
1. What are the two factors, for decreasing the growth rate? Explain.
2. How long would it take to stabilize the world's population if the growth rate is reduced to zero? Explain.
3. Humans, as well as other animals, have been dumping their wastes into the environment for thousands of years. What is the reason(s) that this appears to be such a problem today?
4. Some individuals believe that the carrying capacity of the earth is between 50–100 billion people; others believe that the present population of 5.5–6 billion people already exceeds the earth's carrying capacity. How is it possible for so-called "experts" to arrive at such different numbers?

Understanding the Terms

acid deposition 502
aquifer 505
biological magnification 506
biotic potential 497
carrying capacity 497
chlorofluorocarbons (CFCs) 504

environmental resistance 497
exponential growth 496
greenhouse effect 500
growth rate 496
ozone hole 504
PAN (peroxyacetylnitrate) 503
photochemical smog 503

Match the terms to these definitions.
a. _____ Water-bearing stratum of permeable rock that constitutes an underground reservoir.
b. _____ The yearly percentage of increase or decrease in the size of a population.
c. _____ Transformation of marginal lands to desert conditions.

Applying Technology to the Concepts

Your study of population concerns is supported by these available technologies:

Essential Study Partner CD-ROM
Ecology → Human Impact
Visit the Mader web site for related ESP activities.

Exploring the Internet
The Mader Home Page provides resources and tools as you study this chapter.
http://www.mhhe.com/biosci/genbio/mader

Applying Your Knowledge to the Concepts

In this section, three or four questions ask students to relate concepts they have learned to matters of practical concern. Answers to these questions appear in the appendix.

Applying Technology to the Concepts

References to McGraw-Hill technology point students to other sources for more information.

Studying the Concepts

These questions, which are page-referenced and organized according to the major sections of the chapter, review important chapter material.

Understanding the Terms

A matching exercise ensures that students understand the chapter's terms before proceeding to the next chapter.

Introduction

A Human Perspective

Chapter Concepts

Figure I.1 Living things are alike.
The bird and the boy have many characteristics in common. Their cells, tissues, and organ systems function similarly.

Young Billy Hanson was quite excited. Today, he would let loose the black-capped chickadee he had cared for after it fell from the nest several weeks before. He had kept it warm and well fed until now it was ready to take flight. He knew from hours of watching birds that black-capped chickadees commonly flit from one tree to another, even though they can fly long distances. Billy was hopeful this one would stay close to home, especially since he had a bird-feeder. He yearned for what he thought of as their "friendship" to continue.

We know that Billy, and, yes, the bird and the trees surrounding Billy's home are alive. By what criteria do we make this judgment? All living things respond to stimuli as they grow and reproduce. They take materials and energy from the environment to keep on living. And all living things are adapted to their way of life—the bird flies, Billy walks. Even so, they both have the same organ systems. Both have a heart, a liver, intestines, and so forth. There is a unity of life that goes beyond its diversity. This chapter discusses the characteristics of human beings; how they are like other living things and how they are different from other living things.

Biology is the scientific study of life. When biologists do their work they attempt to answer specific questions. How do genes control who we are, whether a human or a bird? What impact do humans have on their environment, and the environment of all living things? Biologists use the scientific method, which is explained more fully in this chapter, to come to conclusions which they share with others. This text shares their knowledge with you.

The text as a whole has two main primary goals. The first goal is to explore human anatomy and physiology so you will know how the body functions. The second goal is to look at human evolution and ecology so that you will better understand the place of humans in nature. Both the human body and the environment are self-regulating systems that can be thrown out of kilter by misuse and mismanagement. An appreciation of the delicate balance present in both systems provides the perspective from which future decisions can be made. It is hoped that adequate information will better enable you to keep your body and the environment healthy.

I.I Biologically Speaking

You are about to launch on a study of human biology. Before you begin, it is appropriate to define who humans are and how they fit into the world of living things.

Who Are We?

Certain characteristics tell us who human beings are biologically speaking.

Human beings are highly organized. A **cell** is the basic unit of life, and human beings are multicellular since they are composed of many types of cells. Like cells form tissues, and **tissues** make up organs. Each type of **organ** is a part of an organ system. The different systems perform the specific functions listed in Table I.1. Together, the **organ systems** maintain **homeostasis,** an internal environment for cells that varies only within certain limits. The text emphasizes how all the systems of the human body help maintain homeostasis. The digestive system takes in nutrients, and the circulatory system distributes these to the cells. The waste products are excreted by the excretory system. The work of the nervous and endocrine systems is critical because they coordinate the functions of the other systems.

Table I.1	Human Organ Systems
System	**Function**
Digestive	Converts food particles to nutrient molecules
Circulatory	Transports nutrients to and wastes from cells
Immune	Defends against disease
Respiratory	Exchanges gases with the environment
Excretory	Eliminates metabolic wastes
Nervous	Regulates systems and internal environment
Musculoskeletal	Supports and moves organism
Endocrine	Regulates systems and internal environment
Reproductive	Produces offspring

Human beings reproduce and grow. Reproduction and growth are fundamental characteristics of all living things. Just as cells come only from preexisting cells, so living things have parents. When living things **reproduce,** they create a copy of themselves and assure the continuance of the species. (A species is a type of living thing.) Human reproduction requires that a sperm contributed by the male fertilize an egg contributed by the female. Growth occurs as the resulting cell develops into the newborn. Development includes all the changes that occur from the fertilized egg to death and, therefore, all the changes that occur during childhood, adolescence, and adulthood.

Humans have a cultural heritage. We are born without knowledge of civilized ways of behavior, and we gradually acquire these by adult instruction and imitation of role models. It is our cultural inheritance that makes us think we are separate from nature. But actually we are a product of **evolution,** a process of change that has resulted in the diversity of life, and we are a part of the **biosphere,** a network of life that spans the surface of the earth.

Like other living things, humans are composed of cells, and when they reproduce, growth and development occur. Unlike other living things, humans have a cultural heritage.

How Do We Fit In?

Certain characteristics tell us how human beings fit into the world of living things.

Human beings are a product of an evolutionary process. Life has a history that began with the evolution of the first cell(s) about 3.5 billion years ago. It is possible to trace human ancestry from the first cell through a series of prehistoric ancestors until the evolution of modern-day humans. The presence of the same types of chemicals tells us that *human beings are related to all other living things.* DNA is the genetic material, and ATP is the energy currency in all cells, including human cells. It is even possible to do research with bacteria and have the results apply to human beings.

The classification of living things mirrors their evolutionary relationships. It is common practice to classify living things into five major groups called **kingdoms** (Fig. I.2). *Humans are vertebrates in the animal kingdom.* **Vertebrates** have a nerve cord that is protected by a vertebral column whose repeating units (the vertebrae) indicate that we and other vertebrates are segmented animals. Among the vertebrates, we are most closely related to the apes, specifically the chimpanzee, from whom we are distinguished by our highly developed brains, completely upright stance, and the power of creative language.

Human beings are a part of the biosphere. All living things are a part of the biosphere, where living things live

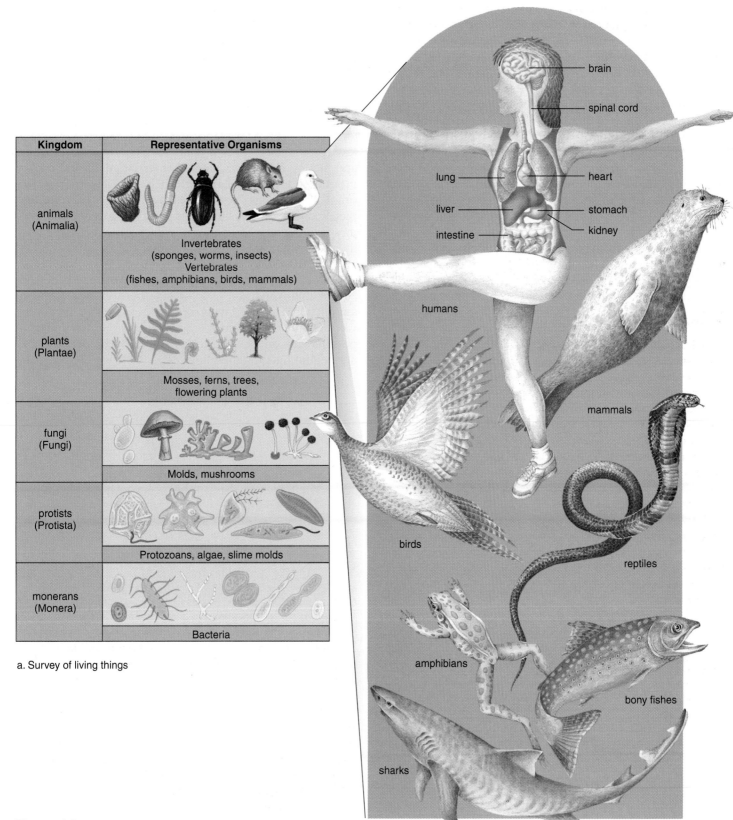

a. Survey of living things

b. Survey of vertebrates

Figure I.2   **Classification and evolution of humans.**
a. Living things are classified into five kingdoms, and humans are in the animal kingdom. **b.** Human beings are most closely related to the other vertebrates shown. The evolutionary tree of life has many branches; the vertebrate line of descent is just one of many.

in the air, in the sea, and on land. In any portion of the biosphere, such as a particular forest or pond, the various populations interact with one another, and with the physical environment to form an **ecosystem** in which chemicals cycle and energy flows. All organisms of one type in a particular ecosystem belong to a **population.** A major part of the interactions between populations pertains to who eats whom. Plants produce organic food, and animals that eat plants may be food for other animals. Both plants and animals interact with the physical environment, as when they exchange gases with the atmosphere. As Figure I.3 shows, all living things are dependent upon solar energy and upon plants, which use this energy to convert inorganic nutrients into a form that is usable by all living things, including humans.

Normally, ecosystems remain relatively stable. Although a forest or pond changes—trees fall, ducks come and go, seeds sprout—each ecosystem remains recognizable year after year. We say it is in dynamic balance. In many cases, even the extinction of species (and their replacement by new species through evolution) still allows the dynamic balance of the system to be maintained.

If the ecosystem is big enough, it needs no raw materials from the outside. A big ecosystem just keeps cycling its raw materials, like water and nitrogen. The only input it needs is energy.

Humans Threaten the Biosphere

Human populations tend to modify existing ecosystems for their own purposes. For example, humans clear forests or grasslands in order to grow crops; later, they build houses on what was once farmland; and, finally, they convert small towns into cities. Human populations ever increase in size and require greater amounts of material goods and energy input each year (Fig. I.4). With each step, fewer and fewer original organisms remain, until at last ecosystems are completely altered. If this continues, only humans and their domesticated plants and animals will largely exist where once there were many diverse populations.

More and more ecosystems are threatened as the human population increases in size. As discussed in the Ecology Focus reading on page 7, presently there is great concern among scientists and laypersons about the destruction of the world's rain forests due to logging and the large numbers of persons who are starting to live and to farm there. But we are beginning to realize how dependent we are on intact

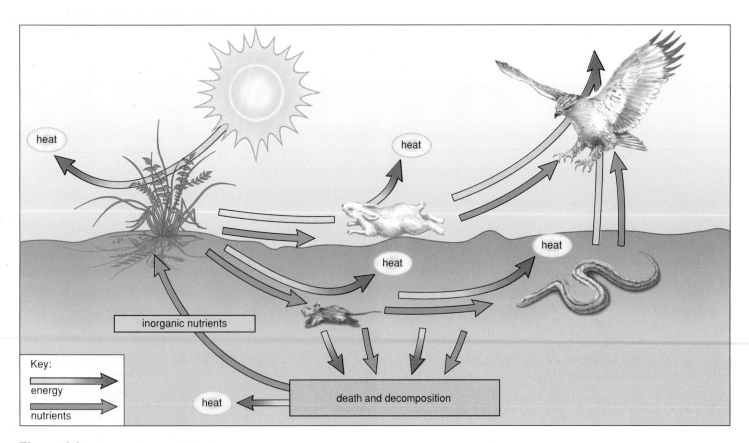

Figure I.3 Ecosystem organization.
Within an ecosystem, nutrients cycle (see blue arrows); plants make and use their own organic food, and this becomes food for several levels of animal consumers, including humans. When these organisms die and decompose, the inorganic remains are used by plants as they produce organic food. Energy flows (see yellow arrows); solar energy used by plants to produce organic food is eventually converted to heat by all members of an ecosystem (therefore, a constant supply of solar energy is required for life to exist).

ecosystems and the services they perform for us. For example, the tropical rain forests act like a giant sponge, which absorbs carbon dioxide, a pollutant that pours into the atmosphere from the burning of fossil fuels, like oil and coal. An increased amount of carbon dioxide in the atmosphere is expected to have many adverse effects, such as an increase in the average daily temperature.

An ever-increasing human population size is a threat to the continued existence of *Homo sapiens* when it means that the dynamic balance of the biosphere is upset. The recognition that the workings of the biosphere need to be preserved is one of the most important developments of our new ecological awareness.

Biodiversity: Going, Going, Gone   When humans modify existing ecosystems, they reduce biodiversity. **Biodiversity** is the total number of species, the variability of their genes, and the ecosystems in which they live. The present biodiversity of our planet has been estimated to be as high as 80 million species, and so far, under 2 million have been identified and named. Extinction, the death of a species, occurs when a species is unable to adapt to a change in environmental conditions. It's estimated that presently we are losing from 24 to even 100 species a day due to human activities. For example, the existence of the species featured in the reading on page 7 is threatened because tropical rain forests are being reduced in size. As another example,

because of seaside development, pollution, and overfishing, 14 of the most valuable finfishes are becoming commercially extinct, meaning that too few remain to justify the cost of catching them.

Most biologists are alarmed over the present rate of extinction and believe the rate may eventually rival that of the five mass extinctions that have occurred during our planet's history. The dinosaurs became extinct during the last mass extinction, 65 million years ago. Everyone needs to realize that humans are totally dependent on other species for food, clothing, medicines, and various raw materials. Therefore, it is very shortsighted of us to allow other species to become extinct. Ecosystems and the species living in them should be preserved because only then can the human species continue to exist. And it takes from 2,000 to 10,000 generations for new species to evolve and to replace the ones that have died out. Because we are dependent upon the normal function and the present biodiversity of the biosphere, the existing species should be preserved.

Humans belong to the world of living things and are vertebrates. They have modified existing ecosystems to the point that they must now be seriously concerned about the continued existence of the biosphere.

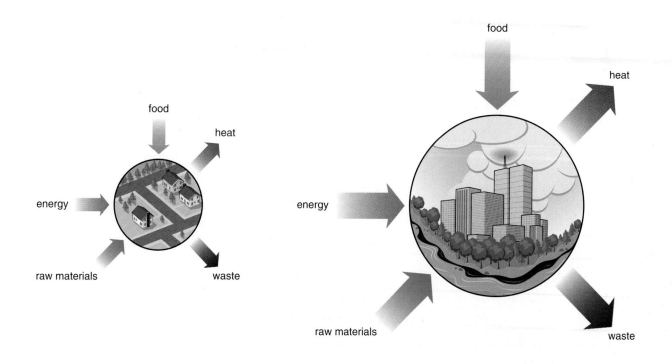

Figure I.4   Humans threaten the biosphere.
As cities grow in size, more materials, fuel, and food are taken from the environment, and more heat and waste are returned to the environment. **a.** Small town. **b.** Large city.

Ecology Focus

Tropical Rain Forests: Can We Live Without Them?

Figure IA Tropical rain forest inhabitants.

These animals and plants make their homes in the Amazon basin.

The tropics are home for 66% of the plant species, 90% of the nonhuman primates, 40% of the birds of prey, and 90% of the insects that have been identified thus far. Many more species of organisms are estimated to exist but have not yet been discovered (perhaps as many as 30 million), and these are believed to live in the tropical rain forests that occur in a green belt spanning the planet on both sides of the equator.

If tropical rain forests are preserved, the rich diversity of plants and animals will continue to exist for scientific and pharmacological study (Fig. IA). One-fourth of the medicines we currently use come from tropical rain forests. For example, the rosy periwinkle from Madagascar has produced two potent drugs for use against Hodgkin disease, leukemia, and other blood cancers. It is hoped that many of the still-unknown plants will provide medicines for other human ills.

Tropical forests cover 6–7% of the total land surface of the earth—an area roughly equivalent to our contiguous 48 states. Every year humans destroy an area of forest equivalent to the size of Oklahoma (Fig. IB). At this rate, these forests and the species they contain will disappear completely in just a few more decades. Even if the forest areas now legally protected survive, 58–72% of all tropical forest species would still be lost.

The loss of tropical rain forests results from an interplay of social, economic, and political pressures. Many people already live in the forest, and as their numbers increase, more of the land is cleared for farming. People move to the forests because internationally financed projects build roads and open up the forests for exploitation. Small-scale farming accounts for about 60% of tropical deforestation, and decreasing percentages are due to commercial logging, cattle ranching, and mining. International demand for timber promotes destructive logging of rain forests in Southeast Asia and South America. The market for low-grade beef encourages their conversion to pastures for cattle. The lure of gold draws miners to rain forests in Costa Rica and Brazil.

The destruction of tropical rain forests gives only short-term benefits but is expected to cause long-term problems. The forests act like a giant sponge, soaking up rainfall during the wet season and releasing it during the dry season. Without them, a regional yearly regime of flooding followed by drought is expected to destroy property and reduce agricultural harvests. Worldwide, there could be changes in climate that would affect the entire human race.

However, studies show that if the forests were used as a sustainable source of nonwood products, such as nuts, fruits, and latex rubber, they would generate as much or more revenue while continuing to perform their various ecological functions and biodiversity could still be preserved. Brazil is exploring the concept of "extractive reserves," in which plant and animal products are harvested, but the forest itself is not cleared. Ecologists have also proposed "forest farming" systems, which mimic the natural forest as much as possible while providing abundant yields. But for such plans to work maximally, the human population size and the resource consumption per person must be stabilized.

Preserving tropical rain forests is a wise investment. Such action promotes the survival of most of the world's species—indeed, the human species, too.

Figure IB **Burning of trees in a tropical rain forest.**
It is estimated that an area the size of Oklahoma is being lost each year. While trees ordinarily take up carbon dioxide, burning releases carbon dioxide to the atmosphere.

I.2 The Process of Science

Science helps human beings understand the natural world. Science aims to be objective rather than subjective even though it is very difficult to make objective observations and to come to objective conclusions because we are often influenced by our own particular prejudices. Still, scientists strive for objective observations and conclusions. We also keep in mind that scientific **conclusions** are subject to change whenever new findings so dictate. Quite often in science, new studies, which might utilize new techniques and equipment, tell us when previous conclusions need to be modified or changed entirely.

Scientific Theories in Biology

The ultimate goal of science is to understand the natural world in terms of **scientific theories,** concepts based on the conclusions of observations and experiments. In a movie, a detective might claim to have a theory about the crime, or you might say that you have a theory about the win-loss record of your favorite baseball team, but in science, the word *theory* is reserved for a conceptual scheme supported by a large number of observations and not yet found lacking. Some of the basic theories of biology are as follows:

Name of Theory	Explanation
Cell	All organisms are composed of cells.
Biogenesis	Life comes only from life.
Evolution	All living things have a common ancestor, but each is adapted to a particular way of life.
Gene	Organisms contain coded information that dictates their form, function, and behavior.

Evolution is the unifying concept of biology because it pertains to various aspects of living things. For example, the theory of evolution enables scientists to understand the history of life, the variety of living things, and the anatomy, physiology, and development of organisms—even their behavior. Because the theory of evolution has been supported by so many observations and experiments for over a hundred years, some biologists refer to the **principle** of evolution. They believe this is the appropriate terminology for theories that are generally accepted as valid by an overwhelming number of scientists.

The Scientific Method Has Steps

Scientists, including biologists, employ an approach to gathering information that is known as the **scientific method.** The approach of individual scientists to their

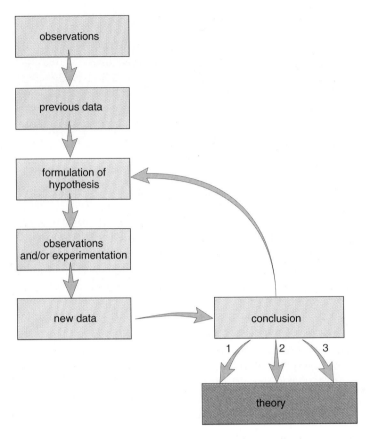

Figure I.5   Flow diagram for the scientific method.
On the basis of observations and previous data, a scientist formulates a hypothesis. The hypothesis is tested by further observations or a controlled experiment, and new data either support or falsify the hypothesis. The return arrow indicates that a scientist often chooses to retest the same hypothesis or to test a related hypothesis. Conclusions from many different but related experiments may lead to the development of a scientific theory. For example, studies in biology of development, anatomy, and fossil remains all support the theory of evolution.

work is as varied as they themselves are; still, for the sake of discussion, it is possible to speak of the scientific method as consisting of certain steps (Fig. I.5). After making initial observations, a scientist will most likely study any previous **data** which are facts pertinent to the matter at hand. Imagination and creative thinking also help a scientist formulate a **hypothesis** that becomes the basis for more observation and/or experimentation. The new data help a scientist come to a conclusion that either supports or does not support the hypothesis. Because hypotheses are always subject to modification, they can never be proven true; however, they can be proven untrue—that is, hypotheses are falsifiable. When the hypothesis is not supported by the data, it must be rejected; therefore, some think of the body of science as what is left after alternative hypotheses have been rejected.

The Discovery of Lyme Disease

In order to examine the scientific method in more detail, we will relate how scientists discovered the cause of Lyme disease, a debilitating illness that affects the whole body.

Observations

When Allen C. Steere began his work on Lyme disease in 1975, a number of adults and children in the city of Lyme, Connecticut, had been diagnosed as having rheumatoid arthritis (Fig. I.6). Steere knew that children rarely get rheumatoid arthritis, so this made him suspicious and he began to make observations. He found that (1) most victims lived in heavily wooded areas, (2) the disease was not contagious; whole groups of people did not come down with Lyme disease, (3) symptoms first appeared in the summer, and (4) several victims remembered a strange bull's-eye rash occurring several weeks before the onset of symptoms.

Formulating the Hypothesis

Inductive reasoning occurs when you generalize from assorted facts. Steere used inductive reasoning; that is, he put the pieces together to formulate the *hypothesis* that Lyme disease was caused by a pathogen most likely transmitted by the bite of an insect or a tick.

Testing the Hypothesis

Deductive reasoning helps scientists decide what further observations and experimentations they will make. *Deductive reasoning* utilizes an "if . . . then" statement: If Lyme disease is caused by the bite of a tick, it should be possible to show that a tick carries the pathogen and the pathogen is in the blood of those who have the disease. However, when Steere tested the blood of Lyme disease victims for the presence of infectious microbes, not a single test was positive. Finally, in 1977, one victim saved the tick that bit him and it was identified as *Ixodes dammini*, the deer tick. Then Willy Burgdorfer, an authority on tick-borne diseases, was able to isolate a spirochete (spiral bacterium) from deer ticks, and he also found this microbe in the blood of Lyme disease victims. The new spirochete was named *Borrelia burgdorferi*, after Burgdorfer.

The Conclusion

The new data collected by Burgdorfer supported the hypothesis and allowed scientists to conclude that Lyme disease is caused by the bacterium *Borrelia burgdorferi* transmitted by the bite of the deer tick.

Even though the scientific method is quite variable, it is possible to point out certain steps that characterize it: making observations, formulating a hypothesis, testing it, and coming to a conclusion.

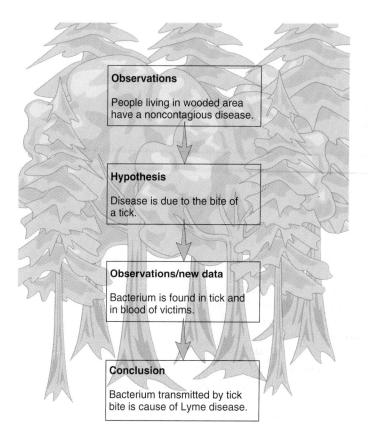

Figure I.6   **Flow diagram for Lyme disease study.**

Reporting the Findings

It is customary to report findings in a scientific journal so that the design and the results of the experiment are available to all. For example, data about tick-borne diseases are often reported in the journal *Clinical Microbiology Review*. It is necessary to give other researchers details on how experiments were conducted because results must be repeatable; that is, other scientists using the same procedures must get the same results. Otherwise, the hypothesis is no longer supported.

Often authors of a report suggest what other types of experiments might clarify or broaden the understanding of the matter under study. People reading the report may think of other experiments to do, also. In our example, the bull's-eye rash was later found to be due to the Lyme disease spirochete.

Observations and the results of experiments are published in a journal, where they can be examined. These results are expected to be repeatable; that is, they will be obtained by anyone following the same procedure.

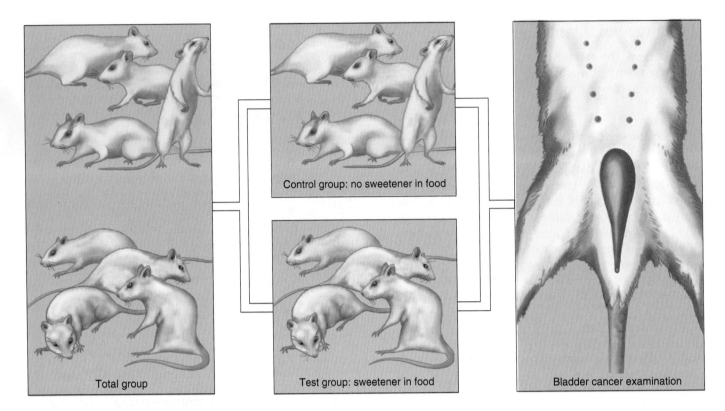

Figure I.7   **Design of a controlled experiment.**
From *left* to *right:* Genetically similar mice are randomly divided into a control group and the test groups. All groups are exposed to the same conditions, such as housing, temperature, and water supply. The control group is not subjected to the test (presence of sweetener S in the food). At the end of the experiment, all mice are examined for bladder cancer.

Scientists Use Controlled Experiments

When scientists are studying a phenomenon, they often perform **experiments** in a laboratory where extraneous variables can be eliminated. A **variable** is a factor that can cause an observable change during the progress of an experiment. Experiments are considered more rigorous when they include a control group. A **control group** goes through all the steps of the experiment except the one being tested.

Designing the Experiment

Suppose, for example, physiologists want to determine if sweetener S is a safe food additive (Fig. I.7). On the basis of available information, they formulate a hypothesis that sweetener S is a safe food additive even up to 50% of dietary intake. Next, they design the experiment described in Figure I.7 to test the hypothesis.

Test group: 50% of diet is sweetener S

Control group: diet contains no sweetener S

The researchers first place a certain number of randomly chosen inbred (genetically identical) mice into the various groups—say, 100 mice per group. If any of the mice

are different from the others, it is hoped random selection has distributed them evenly among the groups. The researchers also make sure that all conditions, such as availability of water, cage setup, and temperature of the surroundings, are the same for both groups. The food for each group is exactly the same except for the amount of sweetener S.

At the end of the experiment, both groups of mice are to be examined for bladder cancer. Let's suppose that one-third of the mice in the test group are found to have bladder cancer, while none in the control group have bladder cancer. The results of this experiment do not support the hypothesis that sweetener S is a safe food additive up to 50% of dietary intake.

Continuing the Experiment

Science is ongoing, and one experiment frequently leads to another. Physiologists might now wish to hypothesize that sweetener S is safe if the diet contains a limited amount of sweetener S. They feed sweetener S to groups of mice at ever-greater concentrations:

Group 1: diet contains no sweetener S (the control)

Group 2: 5% of diet is sweetener S

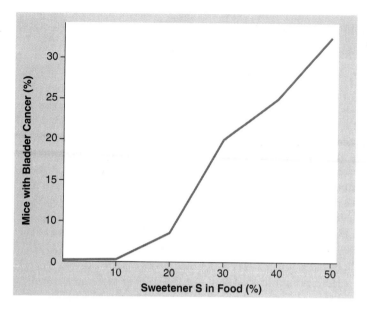

Figure I.8   Presenting the data.
Scientists often report mathematical data in the form of a table or a graph. The data in this instance suggest there is a correlation between the amount of sweetener S in food and the incidence of bladder cancer. Similar experiments will be repeated many times to test these results, and the results will be statistically analyzed to determine if they are significant or due to chance alone.

Group 3: 10% of diet is sweetener S

↓

Group 11: 50% of diet is sweetener S

Usually, data obtained from experiments such as this are presented in the form of a table or a graph (Fig. I.8). Researchers might run a statistical test to determine if the difference in the number of cases of bladder cancer between the various groups is significant. After all, if a significant number of mice in the control group develop cancer, the results are invalid. Scientists prefer *mathematical data* because such data lends itself to objectivity.

On the basis of the data, the experimenters try to develop a recommendation concerning the safety of sweetener S in the food of humans. They might caution, for example, that the intake of sweetener S beyond 10% of the diet is associated with too great a risk of bladder cancer.

Scientists ask questions and carry on investigations that pertain to the natural world. The conclusions of these investigations are tentative and subject to change. Eventually, it may be possible to arrive at a theory that is generally accepted by all.

I.3 Science and Social Responsibility

Science is objective and not subjective. It assumes that each person is capable of collecting data and seeing natural events in the same way and that the same theories and principles are applicable to past, present, and future events. Therefore, science seeks a natural cause for the origin and history of life. Doctrines of creation that have a mythical, philosophical, or theological basis are not a part of science because they are not subject to objective observations and experimentation by all. Many cultures have their own particular set of supernatural beliefs, and various religions within a culture differ as to the application of these beliefs. Such approaches to understanding the world are not within the province of science. Similarly, scientific creationism, which states that God created all species as they are today, cannot be considered science because explanations based on supernatural rather than natural causes involve faith rather than data.

There are many ways in which science has improved our lives. The discovery of antibiotics, such as penicillin, and of the polio, measles, and mumps vaccines, has increased our life span by decades. Cell biology research is helping us understand the causes of cancer. Genetic research has produced new strains of agricultural plants that have eased the burden of feeding our burgeoning world population.

Science also has effects we may find disturbing. For example, it sometimes fosters technologies that can be ecologically disastrous if not controlled properly. Too often we blame science for these developments and think that scientists are duty bound to pursue only those avenues of research that are consistent with our present system of values. But making value judgments is not a part of science. Ethical and moral decisions must be made by all people. The responsibility for how we use the fruits of science, including a given technology, must rest with people from all walks of life, not with scientists alone. Scientists should provide the public with as much information as possible when such issues as the use of atomic energy, fetal research, and genetic engineering are being debated. Then they, along with other citizens, can help make decisions about the future role of these technologies in our society. The text, while covering all aspects of biology, focuses on human biology. It is hoped that your study of biology will enable you to make wise decisions regarding your own individual well-being and also the well-being of all species, including our own.

It is the task of all persons to use scientific information as they make value judgments about their own lives and about the environment.

Bioethical Issue

The Endangered Species Act requires the federal government to identify endangered and threatened species and to protect their habitats, even to the extent of purchasing their habitats. Developers feel that the act protects wildlife at the expense of jobs for U.S. citizens. In an effort to allow development in sensitive areas, it is now possible to move forward after a Habitat Conservation Plan (HCP) is approved. An HCP permits, say, new-home construction or logging on a part of the land if wildlife habitat is conserved on another part. Conservation can also mean helping the government buy habitat some place else.

The nation's first HCP was approved in 1980. It permitted housing construction on San Bruno Mountain near San Fran-cisco, if 97% of the habitat for the endangered mission blue butterfly was preserved. That sounds pretty good, but by this time hundreds of HCPs have been approved, and the conservation requirement may have slipped a bit. Tim Cullinan, a director of the National Audubon Society, recently found that logging companies in the Pacific Northwest are proposing the exchange of habitat on public land for the right to log privately owned old forests. In other words, nothing has been given up. By now, there are so many HCPs in the works they are being rubber-stamped by government officials with no public review at all.

Do you favor development over preservation of habitat or vice versa? Do you think the federal government should be in the business of trying to preserve endangered species? Do you think that public review of HCPs should be allowed, even if it slows down the approval process? Is it the public's responsibility to remain vigilant or is governmental review of HCPs sufficient?

Questions

1. What are the concerns of developers versus environmentalists with regard to natural areas?
2. Is it short-sighted to stress the importance of jobs over the rights of wildlife? Why or why not?
3. In what ethical ways can each side make their concerns known to the general public?

Summarizing the Concepts

I.1 Biologically Speaking
Human beings, just like other organisms, are a product of the evolutionary process. They are members of the animal kingdom and are the vertebrates most closely related to other primates, including the apes. Like other living things, human beings reproduce, are highly organized, and maintain a fairly constant internal environment. They are members of the biosphere but have a cultural heritage that sometimes hinders the realization of their place in nature.

I.2 The Process of Science
When studying the world of living things, biologists and other scientists use the scientific method, which consists of the following steps:

observations, previous data, hypothesis, observation and

experimentation, new data, conclusions

I.3 Science and Social Responsibility
It is the responsibility of all to make ethical and moral decisions about how best to make use of the results of scientific investigations.

Studying the Concepts

1. Name five characteristics of human beings, and discuss each one. 2–4
2. What is homeostasis, and how is it maintained? Choose one organ system and tell how it helps maintain homeostasis. 2
3. Give evidence that human beings are related to all other living things. 2
4. Describe the five kingdom system of classification, and name types of organisms in each kingdom. 3
5. Human beings are dependent upon what services performed by plants? 4
6. Discuss the importance of scientific theory and name several theories that are basic to understanding biological principles. 8
7. Name the steps of the scientific method, and discuss each one. 8
8. How do you recognize a control group, and what is its purpose in an experiment? 10
9. What is our social responsibility in regard to scientific findings? 11

Testing Your Knowledge of the Concepts

In questions 1-4, match the human characteristics to the descriptions below.

Human beings:
  a. are organized.
  b. reproduce and grow.
  c. have a cultural heritage.
  d. are the product of evolutionary process.
  e. are a part of the biosphere.
  _____ 1. Humans are related to all other living things.
  _____ 2. The human population encroaches on natural habitats.
  _____ 3. Like cells form tissues in the human body.
  _____ 4. We learn how to behave from our elders.

In questions 5–7, indicate whether the statement is true (T) or false (F).
  _____ 5. Once a scientist formulates a hypothesis, he tests it by observation and/or experimentation.
  _____ 6. The theory of evolution is so poorly supported that many scientists feel it should be discarded.
  _____ 7. When an experiment has a control group, it lends validity to the resulting data.

In questions 8–9, fill in the blanks.
  8. To reproduce is to make a _____ of one's self.
  9. _____ has a responsibility to decide how scientific knowledge should be used.

10. An investigator spills dye on a culture plate and then notices that the bacteria live despite exposure to sunlight. He hypothesizes that the dye protects bacteria against death by ultraviolet (UV) light. To test this hypothesis, he decides to expose two hundred culture plates to UV light. One hundred plates contain bacteria and dye; the other hundred plates contain only bacteria. Result: after exposure to UV light, the bacteria on both plates die. Fill in the right-hand portion of this diagram.

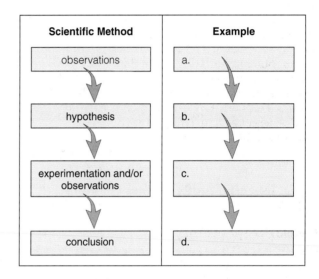

Applying Your Knowledge to the Concepts

These questions pertain to human perspectives.
  1. Homeostasis is the maintenance of a dynamic equilibrium by the body. How does a physician determine that your body is maintaining homeostasis?
  2. What is (are) the reason(s) that humans are considered to be a severe threat to the biosphere, whereas other animals are not considered to be a threat?
  3. Many industries today test various products on lower organisms, even as simple as bacteria, before they are put on the market for human consumption. What justification is there for assuming that bacteria can be used for these tests?
  4. Each of the systems listed in Table I.1 contribute to homeostasis except the reproductive system. Show that this is the case by telling what consequences you would expect if each of these systems were to malfunction.

Understanding the Terms

biodiversity 5
biosphere 2
cell 2
conclusion 8
control group 10
data 8
ecosystem 4
evolution 2
experiment 10
homeostasis 2
hypothesis 8
kingdom 2

organ 2
organ system 2
population 4
principle 8
reproduce 2
science 8
scientific method 8
scientific theory 8
tissue 2
variable 10
vertebrate 2

Match the terms to these definitions:

a. _____ Concept consistent with conclusions based on a large number of experiments and observations.

b. _____ Statement that is capable of explaining present observations and will be tested by further experimentation and observations.

c. _____ Capacity to do work and bring about change; occurs in a variety of forms.

d. _____ Suitability of an organism for its environment, enabling it to survive and produce offspring.

e. _____ Maintenance of the internal environment of an organism within narrow limits.

Applying Technology to the Concepts

Your study of biology is supported by these available technologies:

Essential Study Partner CD-ROM

Diversity → Classification

Visit the Mader web site for related ESP activities.

Exploring the Internet

The Mader Home Page provides resources and tools as you study this chapter.

http://www.mhhe.com/biosci/genbio/mader

Further Readings

Balick, M. J., and Cox, P. A. 1996. *Plants, people, and culture: The science of ethnobotany.* New York: Scientific American Library. This interesting, well-illustrated book discusses the medicinal and cultural uses of plants, and the importance of rain forest conservation.

Barnard, C., et al. 1993. *Asking questions in biology.* Essex: Longman Scientific & Technical. First-year life science students are introduced to the skills of scientific observation.

Carey, S. S. 1997. *A beginner's guide to scientific method.* 2d ed. Belmont, Calif.: Wadsworth Publishing. The basics of the scientific method are explained.

Dobson, A. P. 1996. *Conservation and biodiversity.* New York: Scientific American Library. Discusses the extent and the value of biodiversity, and describes attempts to manage endangered species.

Drewes, F. 1997. *How to study science.* 2d ed. Dubuque, Iowa: Wm. C. Brown Publishers. Supplements any introductory science text; shows students how to study and take notes and how to interpret text figures.

Frenay, A. C. F., and Mahoney, R. M. 1997. *Understanding medical terminology.* 10th ed. Dubuque, Iowa: Wm. C. Brown Publishers. A structural approach to the study of medical terminology.

Johnson, G. B. 1996. *How scientists think.* Dubuque, Iowa: Wm. C. Brown Publishers. Presents the rationale behind 21 important experiments in genetics and molecular biology that became the foundation for today's research.

Kellert, S. R. 1996. *The value of life: Biological diversity and human society.* Washington, D.C.: Island Press/Shearwater Books. The importance of biological diversity to the well-being of humanity is explored.

Marchuk, W. N. 1992. *A life science lexicon.* Dubuque, Iowa: Wm. C. Brown Publishers. Helps students master life sciences terminology.

Margulis, L., et al. 1998. *Five kingdoms: An illustrated guide to the phyla of life on earth.* New York: W. H. Freeman & Co. Introduces the kingdoms of organisms.

Minkoff, E. C., and Baker, P. J. 1996. *Biology today: An issues approach.* New York: The McGraw-Hill Companies, Inc. This introductory text emphasizes understanding of selected biological issues, and discusses each issue's social context.

Nemecek, S. August 1997. Frankly, my dear, I don't want a dam. *Scientific American* 277(2):20. The article discusses how dams affect biodiversity.

Primak, R. B. 1995. *A primer of conservation biology.* Sunderland, Mass.: Sinauer Associates. The relatively new discipline of conservation biology addresses the alarming loss of biological diversity throughout the world.

Schmidt, M. J. January 1996. Working elephants. *Scientific American* 274(1):82. In Asia, teams of elephants serve as an alternative to destructive logging equipment.

Serafini, A. 1993. *The epic history of biology.* New York: Plenum Press. This is a history of biology beginning with ancient Egyptian medicine.

Human Organization

The human body is composed of cells, the smallest units of life. An understanding of cell structure, physiology, and biochemistry serves as a foundation for understanding how the human body functions.

Principles of inorganic and organic chemistry are discussed before a study of human cell structure is undertaken. The human cell is bounded by a membrane and contains organelles, which are also membranous. Membranes regulate entrance and exit of molecules and help cellular organelles carry out their functions.

The many cells of the body are specialized into tissues that are found within the organs of the various systems of the body. All body systems help maintain homeostasis, a dynamic equilibrium of the internal environment, so that proper physical conditions exist for each cell.

Chapter 1

Chemistry of Life

Chapter Concepts

1.1 **Elements and Atoms**
- All matter is composed of elements, each having one type of atom.   16

1.2 **Molecules and Compounds**
- Atoms react with one another, forming ions, molecules, and compounds.   19

1.3 **Water and Living Things**
- The existence of living things is dependent on the characteristics of water.   21
- The hydrogen ion concentration in water changes when acids or bases are added to water.   23

1.4 **Molecules of Life**
- Macromolecules are polymers that arise when their specific monomers (unit molecules) join together.   26
- The macromolecules found in cells are carbohydrates, lipids, proteins, and nucleic acids.   26

1.5 **Carbohydrates**
- Carbohydrates function as a ready source of energy in most organisms.   27
- Glucose is a simple sugar; starch, glycogen, and cellulose are polymers of glucose.   27
- Cellulose lends structural support to plant cell walls.   28

1.6 **Lipids**
- Lipids are varied molecules.   29
- Fats and oils, which function in long-term energy storage, are composed of glycerol and three fatty acids.   29
- Sex hormones are derived from cholesterol, a complex ring compound.   30

1.7 **Proteins**
- Proteins help form structures (e.g., muscles and membranes) and function as enzymes.   31
- Proteins are polymers of amino acids.   32

1.8 **Nucleic Acids**
- Nucleic acids are polymers of nucleotides.   34
- Genes are composed of DNA (deoxyribonucleic acid). DNA specifies the correct ordering of amino acids in proteins, with RNA as an intermediary.   34

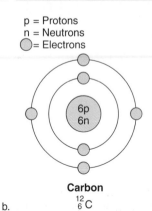

Figure 1.1   Chemicals and the body.
The human body is affected by the chemicals that we breathe and consume as nutrients. Even the sweeteners used in sodas can influence body metabolism, as those with a metabolic disorder, such as phenylketonuria, know all too well.

a.

Common Elements in Living Things				
Element	Atomic Symbol	Atomic Number	Atomic Weight	Comment
hydrogen	H	1	1	These
carbon	C	6	12	elements
nitrogen	N	7	14	make up
oxygen	O	8	16	most
phosphorus	P	15	31	biological
sulfur	S	16	32	molecules.
sodium	Na	11	23	These
magnesium	Mg	12	24	elements
chlorine	Cl	17	35	occur mainly
potassium	K	19	39	as dissolved
calcium	Ca	20	40	salts.

p = Protons
n = Neutrons
⬤ = Electrons

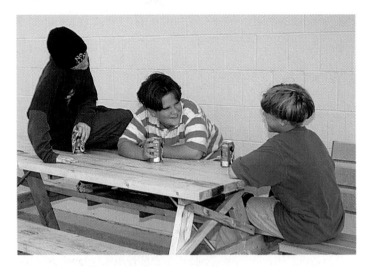

Carbon
$^{12}_{6}$C

b.

Figure 1.2   Elements and atoms.

a. The atomic symbol, atomic number, and atomic weight are given for the common elements in living things. **The atomic symbol for calcium is Ca, the atomic number is 20, and the atomic weight is 40. b.** An atom contains the subatomic particles called protons (p) and neutrons (n) in the nucleus (colored pink) and electrons (colored gray) in shells about the nucleus.

Glance at the back of any diet soda can, and you'll find an important warning: "Contains Phenylalanine." For most of us, this chemical—one of 20 naturally occurring amino acids—is harmless. That's because our bodies contain a liver enzyme that converts phenylalanine into a useful chemical for the body.

But people with PKU, or phenylketonuria, lack the necessary enzyme. In these people, phenylalanine builds up, and dangerous by-products flood the bloodstream. Brain damage or other problems follow.

Thus, doctors keep PKU sufferers on a strict diet, limiting their intake of phenylalanine. Soda companies help by labeling products containing the chemical (Fig. 1.1).

PKU is just one of hundreds of disorders caused by malfunctioning or absent enzymes. From simple inorganic molecules to complex organic macromolecules, such as enzymes, chemicals are essential to our being. As people with PKU know, lacking just one of these precious chemicals can be disastrous.

This chapter reviews the structure of atoms and how they join to form both inorganic and organic chemicals. Inorganic chemicals like salts have significant functions in the human body just as organic chemicals like proteins do. All chemicals whether inorganic or organic are composed of elements.

1.1 Elements and Atoms

Matter is anything that takes up space and has weight. All matter, both nonliving and living, is composed of certain basic substances called **elements.** Considering the variety of living and nonliving things in the world, it's quite remarkable that there are only 92 naturally occurring elements. It is even more surprising that over 90% of the human body is composed of just three elements: carbon, oxygen, and hydrogen.

Every element has a name and a symbol; for example, calcium has been assigned the atomic symbol Ca (Fig. 1.2a). Some of the symbols we use for elements are derived from Latin. For example, the symbol for sodium is Na (*natrium* in Latin means sodium).

Atoms

An **atom** is the smallest unit of an element that still retains the chemical and physical properties of an element. While it is possible to split an atom by physical means, an atom is the smallest unit to enter into chemical reactions. For our purposes, it is satisfactory to think of each atom as having a central nucleus, where subatomic particles called **protons** and **neutrons** are located, and shells, which are pathways about the nucleus where **electrons** orbit (Fig. 1.2b). Most of an atom is empty space. If we could draw an atom the size of a football stadium, the nucleus would be like a gumball in the center of the field, and the electrons would be tiny specks whirling about in the upper stands.

Two important features of protons, neutrons, and electrons are their charge and weight:

Name	Charge	Weight
Electron	One negative unit	Almost no mass
Proton	One positive unit	One atomic mass unit
Neutron	No charge	One atomic mass unit

The atomic number of an atom tells you how many protons (+) and therefore how many electrons (−) an atom has when it is electrically neutral. For example, the atomic number of calcium is 20; therefore, when calcium is neutral, it has 20 protons and 20 electrons. How many electrons are there in each shell of an atom? The inner shell has the lowest energy level and can hold only two electrons; after that, each shell for the atoms noted in Figure 1.2a can hold up to eight electrons. Using this information, calculation determines that calcium has four shells and the outer shell has two electrons. As we shall see, an atom is most stable when the outer shell has eight electrons. (Hydrogen with only one shell is an exception to this statement. Atoms with only one shell are stable when this shell contains two electrons.)

The subatomic particles are so light that their weight is indicated by special designations called atomic mass units. Notice in the chart above that protons and neutrons each have about one atomic unit of weight and electrons have almost no weight. Therefore, the atomic weight generally tells you the number of protons plus the number of neutrons. How could you calculate that carbon (C) has 6 neutrons? Carbon's atomic weight is 12, and you know from its atomic number that it has 6 protons. Therefore, carbon has 6 neutrons (Fig. 1.2b).

As shown in Figure 1.2b, the atomic number of an atom is often written as a subscript to the lower left of the atomic symbol. The atomic weight is often written as a superscript to the upper left of the atomic symbol. Therefore, carbon can be designated in this way:

$$_6^{12}C$$

All matter is composed of elements, each containing particles called atoms. Atoms have an atomic symbol, atomic number (number of protons), and atomic weight (number of protons and neutrons).

Isotopes

The atomic weights given in the periodic table are the average weight for each kind of atom. This is because atoms of the same type may differ in the number of neutrons; therefore, their weight varies. Atoms that have the same atomic number and differ only in the number of neutrons are called **isotopes**. Isotopes of carbon can be written in the following manner, where the subscript stands for the atomic number and the superscript stands for the atomic weight:

$$_6^{12}C \qquad _6^{13}C \qquad _6^{14}C$$

Carbon 12 has six neutrons, carbon 13 has seven neutrons, and carbon 14 which has eight neutrons is radioactive.

Isotopes have many uses. Each type of food has its own proportion of isotopes, and this information allows biologists to study mummified or fossilized human tissue to know what ancient peoples ate. Most isotopes are stable, but radioactive isotopes break down and emit radiation in the form of radioactive particles or radiant energy. Because carbon 14 breaks down at a known rate, the amount of carbon 14 remaining is often used to determine the age of fossils. Radioactive isotopes are widely used in biological and medical research; for example, because the thyroid gland uses iodine (I), it is possible to administer a dose of radioactive iodine and then observe later that the thyroid has taken it up (Fig. 1.3).

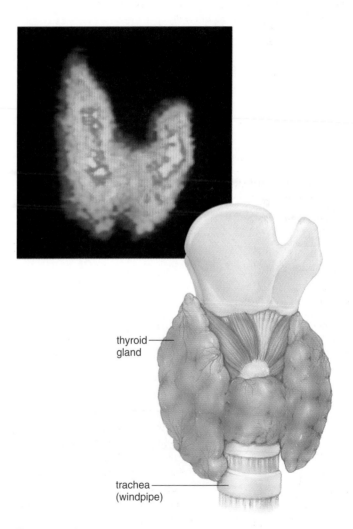

thyroid gland

trachea (windpipe)

Figure 1.3   **Use of radioactive iodine.**
A scan of the thyroid gland 24 hours after the patient was administered radioactive iodine. The thyroid gland is located at the base of the neck.

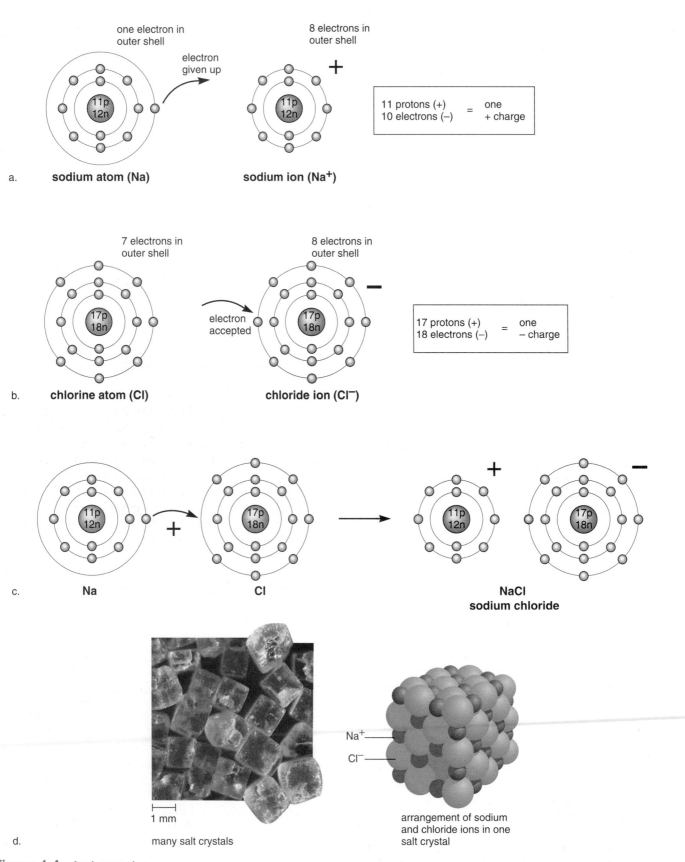

Figure 1.4   Ionic reaction.

a. When a sodium atom gives up an electron, it becomes a positive ion. **b.** When a chlorine atom gains an electron, it becomes a negative ion.
c. When sodium reacts with chlorine, the compound sodium chloride (NaCl) results. In sodium chloride, an ionic bond exists between the ions. **d.**
In a sodium chloride crystal, the ionic bonding between Na$^+$ and Cl$^-$ causes ions to form a three-dimensional lattice in which each sodium ion is
surrounded by six chlorine ions, and each chlorine ion is surrounded by six sodium ions.

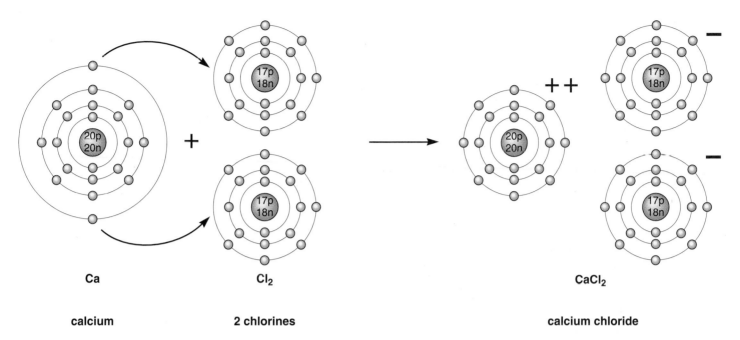

Ca   Cl₂

calcium   2 chlorines   calcium chloride

Figure 1.5   Ionic reaction.
The calcium atom gives up two electrons, one to each of two chlorine atoms. In the compound calcium chloride ($CaCl_2$), the calcium ion is attracted to two chloride ions.

1.2 Molecules and Compounds

Atoms often bond with each other to form a chemical unit called a **molecule.** A molecule can contain atoms of the same kind, as when an oxygen atom joins with another oxygen atom to form oxygen gas. Or the atoms can be different, as when an oxygen atom joins with two hydrogen atoms to form water. When the atoms are different, a compound results.

Two types of bonds join atoms: the ionic bond and the covalent bond.

Ionic Reactions

Recall that atoms (with more than one shell) are most stable when the outer shell contains eight electrons. During an ionic reaction, atoms give up or take on an electron(s) in order to achieve a stable outer shell.

Figure 1.4 depicts a reaction between a sodium (Na) and chlorine (Cl) atom in which chlorine takes an electron from sodium. **Ions** are particles that carry either a positive (+) or negative (−) charge. The sodium ion carries a positive charge because it now has one more proton than electrons, and the chloride ion carries a negative charge because it now has one fewer proton than electrons. The attraction between oppositely charged sodium ions and chloride ions forms an **ionic bond.** The resulting compound, sodium chloride, is table salt, which we use to enliven the taste of foods.

Figure 1.5 shows an ionic reaction between a calcium atom and two chlorine atoms. Notice that calcium with two electrons in the outer shell reacts with two chlorine atoms. Why? Because with seven electrons already, each chlorine requires only one more electron to have a stable outer shell. The resulting salt is called calcium chloride.

Significant ions in the human body are listed in Table 1.1. The balance of these ions in the body is important to our health. Too much sodium in the blood can cause high blood pressure; not enough calcium leads to rickets (a bowing of the legs) in children; too much or too little potassium results in heartbeat irregularities. Bicarbonate, hydrogen, and hydroxide ions are all involved in maintaining the acid-base balance of the body. If the blood is too acidic or too basic, the body's cells cannot function properly.

An ionic bond is the attraction between oppositely charged ions.

Table 1.1	Significant Ions in the Body	
Name	**Symbol**	**Special Significance**
Sodium	Na^+	Found in body fluids; important in muscle contraction and nerve conduction.
Chloride	Cl^-	Found in body fluids.
Potassium	K^+	Found primarily inside cells; important in muscle contraction and nerve conduction.
Phosphate	PO_4^{3-}	Found in bones, teeth, and the high-energy molecule ATP.
Calcium	Ca^{2+}	Found in bones and teeth; important in muscle contraction.
Bicarbonate	HCO_3^-	Important in acid-base balance.
Hydrogen	H^+	Important in acid–base balance.
Hydroxide	OH^-	Important in acid–base balance.

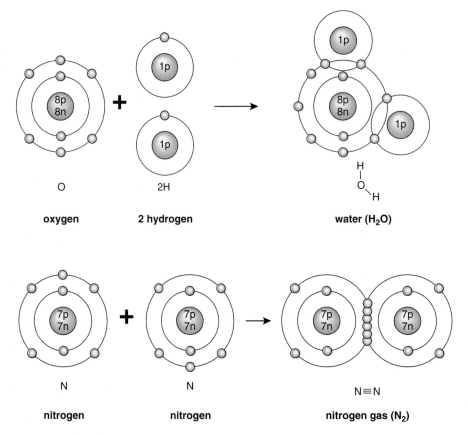

Figure 1.6   **Covalent reactions.**
After a covalent reaction, each atom will have filled its outer shell by sharing electrons. To show this, it is necessary to count the shared electrons as belonging to both bonded atoms. Oxygen and nitrogen are most stable with eight electrons in the outer shell; hydrogen is most stable with two electrons in the outer shell.

Covalent Reactions

After covalent reactions, the atoms share electrons in **covalent bonds** instead of losing or gaining them. Covalent bonds can be represented in a number of ways. The overlapping outermost shells in Figure 1.6 indicate that the atoms are sharing electrons. Just as two hands participate in a handshake, each atom contributes one electron to the pair that is shared. These electrons spend part of their time in the outer shell of each atom; therefore, they are counted as belonging to both bonded atoms.

Structural formulas use straight lines to show the covalent bonds between the atoms. Each line represents a pair of shared electrons. Molecular formulas indicate only the number of each type of atom making up a molecule.

Structural formula: Cl — Cl

Molecular formula: Cl_2

Double and Triple Bonds

Besides a single bond, in which atoms share only a pair of electrons, a double or a triple bond can form. In a double bond, atoms share two pairs of electrons, and in a triple bond, atoms share three pairs of electrons between them. For example, in Figure 1.6, each nitrogen atom (N) requires three electrons to achieve a total of eight electrons in the outer shell. Notice that six electrons are placed in the outer overlapping shells in the diagram and that three straight lines are in the structural formula for nitrogen gas (N_2).

A covalent bond arises when atoms share electrons. In double covalent bonds, atoms share two pairs of electrons, and in triple covalent bonds, atoms share three pairs of electrons.

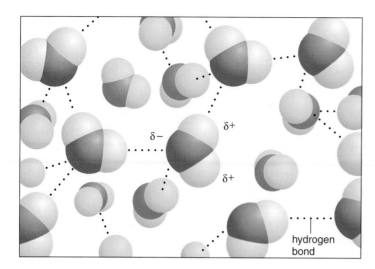

Figure 1.7 **Hydrogen bonding between water molecules.**
The polarity of the water molecules allows hydrogen bonds (dotted lines) to form between the molecules.

1.3 Water and Living Things

Water is the most abundant molecule in living organisms, and it makes up about 60 to 70% of the total body weight of most organisms. We will see that the physical and chemical properties of water make life possible as we know it.

Water is a polar molecule; the oxygen end of the molecule has a slight negative charge, and the hydrogen end has a slight positive charge:

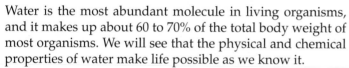

The diagram on the left shows the structural formula of the molecule and the one on the right shows the space-filling model of the molecule.

In polar molecules, covalently bonded atoms share electrons unevenly; that is, the electrons spend more time circling the nucleus of one atom than circling the other. In water, the electrons spend more time circling the larger oxygen (O) than the smaller hydrogen (H) atoms.

In water, the negative ends and positive ends of the molecules attract one another. Each oxygen forms loose bonds to hydrogen atoms of two other water molecules (Fig. 1.7). These bonds are called hydrogen bonds. A **hydrogen bond** occurs whenever a covalently bonded hydrogen is positive and attracted to a negatively charged atom some distance away. A hydrogen bond is represented by a dotted line in Figure 1.7 because it is relatively weak and can be broken rather easily.

Properties of Water

Because of their polarity and hydrogen bonding, water molecules are cohesive and cling together. Polarity and hydrogen bonding causes water to have many characteristics beneficial to life.

1. Water is a liquid at room temperature. Therefore we are able to drink it, cook with it, and bathe in it.

Compounds with low molecular weights are usually gases at room temperature. For example, oxygen (O_2) with a molecular weight of 32 is a gas, but water with a molecular weight of 18 is a liquid. The hydrogen bonding between water molecules keeps water a liquid and not a gas at room temperature. Water does not boil and become a gas until 100°C, one of the reference points for the Celsius temperature scale. (See Appendix B.) Without hydrogen bonding between water molecules, our body fluids and indeed our bodies would be gaseous!

2. Water is the universal solvent for polar (charged) molecules and thereby facilitates chemical reactions both outside of and within our bodies.

When a salt such as sodium chloride (NaCl) is put into water, the negative ends of the water molecules are attracted to the sodium ions, and the positive ends of the water molecules are attracted to the chloride ions. This causes the sodium ions and the chloride ions to separate and to dissolve in water:

The salt NaCl dissolves in water

When ions and molecules disperse in water, they move about and collide, allowing reactions to occur. Therefore, water is a solvent that facilitates chemical reactions.

Ions and molecules that interact with water are said to be **hydrophilic.** Nonionized and nonpolar molecules that do not interact with water are said to be **hydrophobic.**

3. Water molecules are cohesive and therefore liquids will fill vessels.

Water molecules cling together because of hydrogen bonding, and yet, water flows freely. This property allows dissolved and suspended molecules to be evenly distributed throughout a system. Therefore, water is an excellent transport medium. Within our bodies, blood which fills our arteries and veins is 92% water. Blood transports oxygen and nutrients to the cells and removes wastes such as carbon dioxide.

a.   b.   c.

Figure 1.8   Characteristics of water.
a. Water boils at 100°C. If it boiled and was a gas at a lower temperature, life could not exist. **b.** It takes much body heat to vaporize sweat, which is mostly liquid water, and this helps keep bodies cool when the temperature rises. **c.** Ice is less dense than water, and it forms on top of water, making skate sailing possible.

4. The temperature of liquid water rises and falls slowly, preventing sudden or drastic changes.

The many hydrogen bonds that link water molecules cause water to absorb a great deal of heat before it boils (Fig. 1.8a). A **calorie** of heat energy raises the temperature of one gram of water 1°C. This is about twice the amount of heat required for other covalently bonded liquids. On the other hand, water holds heat, and its temperature falls slowly. Therefore, water protects us and other organisms from rapid temperature changes and helps us maintain our normal internal temperature. This property also allows great bodies of water, such as oceans, to maintain a relatively constant temperature. Water is a good temperature buffer.

5. Water has a high heat of vaporization, keeping the body from overheating.

It takes a large amount of heat to change water to steam (Fig. 1.8a). (Converting one gram of the hottest water to steam requires an input of 540 calories of heat energy.) This property of water helps moderate the earth's temperature so that life can continue to exist. Also, in a hot environment, animals sweat and the body cools as body heat is used to

evaporate sweat, which is mostly liquid water (Fig. 1.8b). Body heat is also given off when certain blood vessels dilate, bringing more blood to the surface of the body. The control of body temperature is an example of homeostasis which is the maintenance of the internal environment within normal limits.

6. Frozen water is less dense than liquid water so that ice floats on water.

As water cools, the molecules come closer together. They are densest at 4°C, but they are still moving about. At temperatures below 4°C, there is only vibrational movement, and hydrogen bonding becomes more rigid but also more open. This makes ice less dense. Bodies of water always freeze from the top down, making skate sailing possible (Fig. 1.8c). When a body of water freezes on the surface, the ice acts as an insulator to prevent the water below it from freezing. Aquatic organisms are protected, and they have a better chance of surviving the winter.

Because of its polarity and hydrogen bonding, water has many characteristics that benefit life.

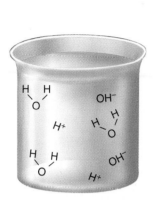

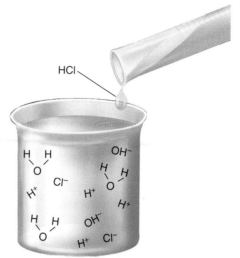

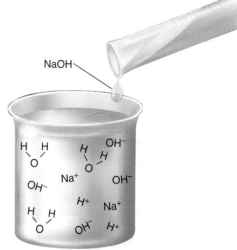

Figure 1.9 **Dissociation of water molecules.**
Dissociation produces an equal number of hydrogen ions (H^+) and hydroxide ions (OH^-). (These illustrations are not meant to be mathematically accurate.)

Figure 1.10 **Addition of hydrochloric acid (HCl).**
HCl releases hydrogen ions (H^+) as it dissociates. The addition of HCl to water results in a solution with more H^+ than OH^-.

Figure 1.11 **Addition of sodium hydroxide (NaOH), a base.**
NaOH releases OH^- as it dissociates. The addition of NaOH to water results in a solution with more OH^- than H^+.

Acidic and Basic Solutions

When water dissociates (breaks up), it releases an equal number of hydrogen ions (H^+) and hydroxide ions (OH^-).

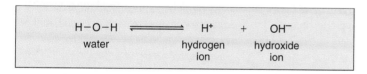

Only a few water molecules at a time are dissociated (Fig. 1.9). The actual number of ions is 10^{-7} moles/liter. A mole is a unit of scientific measurement for atoms, ions, and molecules.[1]

Acidic Solutions

Lemon juice, vinegar, tomato juice , and coffee are all familiar acidic solutions. What do they have in common? Acidic solutions have a sharp or sour taste, and therefore, we sometimes associate them with indigestion. To a chemist, **acids** are molecules that dissociate in water, releasing hydrogen ions (H^+). For example, an important acid in the laboratory is hydrochloric acid (HCl), which dissociates in this manner:

$$HCl \rightarrow H^+ + Cl^-$$

Dissociation is almost complete; therefore, this is called a strong acid. When hydrochloric acid is added to a beaker of water, the number of hydrogen ions increases (Fig. 1.10).

Basic Solutions

Milk of magnesia and ammonia are common bases that most people have heard of. Bases have a bitter taste and feel slippery when in water. To a chemist, **bases** are molecules that either take up hydrogen ions (H^+) or release hydroxide ions (OH^-). For example, an important inorganic base is sodium hydroxide (NaOH), which dissociates in this manner:

$$NaOH \rightarrow Na^+ + OH^-$$

Dissociation is almost complete; therefore sodium hydroxide is called a strong base. If sodium hydroxide is added to a beaker of water, the number of hydroxide ions increases (Fig. 1.11).

It is not recommended that you taste a strong acid or base, because they are quite destructive to cells. Any container of household cleanser like ammonia has a poison symbol and carries a strong warning not to ingest the product.

The Litmus Test

A simple laboratory test for acids and bases is called the litmus test. Litmus is a vegetable dye that changes color from blue to red in the presence of an acid and from red to blue in the presence of a base. The litmus test has become a common figure of speech, as when you hear a commentator say, "The litmus test for a Republican is . . . ."

---

[1]A mole is the same amount of atoms, molecules, ions as the number of atoms in exactly 12 grams of ^{12}C.

The pH Scale

The **pH scale**[2] is used to indicate the acidity and basicity (alkalinity) of a solution. Since there are normally few hydrogen ions (H⁺) in a solution, the pH scale was devised to eliminate the use of cumbersome numbers. For example, the possible hydrogen ion concentrations of a solution are on the left of this listing and the pH is on the right:

moles/liter

1×10^{-6} [H^+] = pH 6 (an acid)

1×10^{-7} [H^+] = pH 7 (neutral)

1×10^{-8} [H^+] = pH 8 (a base)

Pure water (HOH) has an equal number of hydrogen ions (H⁺) and hydroxide ions (OH⁻); therefore, one of each is released when water dissociates. One mole of pure water contains only 10^{-7} moles/liter of hydrogen ions; therefore, a pH of exactly 7 is neutral pH. At a pH of 7 there is an equal number of hydrogen ions and hydroxide ions. Above pH 7 there are more hydroxide ions than hydrogen ions, and below pH 7 there are more hydrogen ions than hydroxide ions. Therefore any solution with a pH below 7 is an acidic solution and any solution with a pH above 7 is a basic solution. Also, as we move down the pH scale from 14 to 0, each unit has 10 times the [H^+] of the previous unit (Fig. 1.12). As we move up the pH scale from 0 to 14, each unit has 10 times the [OH^-] of the previous unit.

As discussed in the Ecology reading on page 25, there have been detrimental environmental consequences to non-living and living things as rain and snow have become more acidic. In plants and animals, including ourselves, pH needs to be maintained within a narrow range or there are health consequences.

Buffers and pH

Buffers resist pH changes because they are chemicals or combinations of chemicals that can take up excess hydrogen ions (H⁺) or hydroxide ions (OH⁻). Many commercial products like Bufferin or shampoos or deodorants are buffered as an added incentive to have us buy them.

To maintain homeostasis, our cells and body fluids are naturally buffered so that the pH remains within normal limits. The pH of our blood is usually about 7.4, in part because it contains a combination of carbonic acid and bicarbonate ions. Carbonic acid (H_2CO_3) is a weak acid that minimally dissociates and then re-forms in the following manner:

H_2CO_3	dissociates $\rightleftharpoons$ re-forms	H^+	+	HCO_3^-
carbonic acid		hydrogen ion		bicarbonate ion

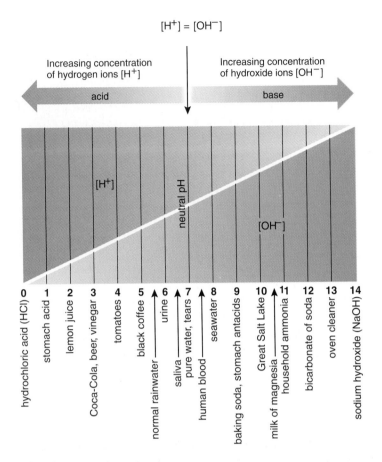

Figure 1.12   The pH scale.
The diagonal line indicates the proportionate concentration of hydrogen ions (H⁺) to hydroxide ions (OH⁻) at each pH value. Any pH value above 7 is basic, while any pH value below 7 is acidic.

When hydrogen ions (H⁺) are added to blood, the following reaction occurs:

$$H^+ + HCO_3^- \rightarrow H_2CO_3$$

When hydroxide ions (OH⁻) are added to blood, this reaction occurs:

$$OH^- + H_2CO_3 \rightarrow HCO_3^- + H_2O$$

These reactions prevent any significant change in blood pH.

Acids have a pH that is less than 7, and bases have a pH that is greater than 7. Buffers, which can combine with both hydrogen ions and hydroxide ions, resist pH changes.

---

[2]pH is defined as the negative logarithm of the molar concentration of the hydrogen ion [H^+].

Ecology Focus

The Harm Done by Acid Deposition

Normally, rainwater has a pH of about 5.6 because the carbon dioxide in the air combines with water to give a weak solution of carbonic acid. Rain falling in the northeastern United States and southeastern Canada now has a pH between 5.0 and 4.0. We have to remember that a pH of 4 is ten times more acidic than a pH of 5 to comprehend the increase in acidity this represents.

There is very strong evidence that this observed increase in rainwater acidity is a result of the burning of fossil fuels, like coal and oil, as well as gasoline derived from oil. When fossil fuels are burned, sulfur dioxide and nitrogen oxides are produced, and they combine with water vapor in the atmosphere to form the acids sulfuric acid and nitric acid. These acids return to earth contained in rain or snow, a process properly called wet deposition, but more often called acid rain. Dry particles of sulfate and nitrate salts descend from the atmosphere during dry deposition.

Unfortunately, regulations that require the use of tall smokestacks to reduce local air pollution only cause pollutants to be carried far from their place of origin. Acid deposition in southeastern Canada is due to the burning of fossil fuels in factories and power plants in the Midwest. Acid deposition adversely affects lakes, particularly in areas where the soil is thin and lacks limestone (calcium carbonate, $CaCO_3$), a buffer to acid deposition. It leaches aluminum from the soil, carries aluminum into the lakes, and converts mercury deposits in lake bottom sediments to soluble and toxic methyl mercury. Lakes not only become more acidic, but they also show accumulation of toxic substances. In Norway and Sweden, at least 16,000 lakes contain no fish, and an additional 52,000 lakes are threatened. In Canada, some 14,000 lakes are almost fishless, and an additional 150,000 are in peril because of excess acidity. In the United States, about 9,000 lakes (mostly in the Northeast and upper Midwest) are threatened, one-third of them seriously.

In forests, acid deposition weakens trees because it leaches away nutrients and releases aluminum. By 1988, most spruce, fir, and other conifers atop North Carolina's Mt. Mitchell were dead from being bathed in ozone and acid fog for years. The soil was so acidic, new seedlings could not survive. Nineteen countries in Europe have reported woodland damage, ranging from 5 to 15% of the forested area in Yugoslavia and Sweden to 50% or more in the Netherlands, Switzerland, and the former West Germany. More than one-fifth of Europe's forests are now damaged.

These aren't the only effects of acid deposition. Reduction of agricultural yields, damage to marble and limestone monuments and buildings, and even illnesses in humans have been reported. Acid deposition has been implicated in the increased incidence of lung cancer and possibly colon cancer in residents of the East Coast. Tom McMillan, Canadian Minister of the Environment, says that acid rain is "destroying our lakes, killing our fish, undermining our tourism, retarding our forests, harming our agriculture, devastating our heritage, and threatening our health."

There are, of course, things that can be done. We could

a. whenever possible use alternative energy sources, such as solar, wind, hydropower, and geothermal energy.

b. use low-sulfur coal or remove the sulfur impurities from coal before it is burned.

c. require factories and power plants to use scrubbers, which remove sulfur emissions.

d. require people to use mass transit rather than driving their own automobiles.

e. reduce our energy needs through other means of energy conservation.

a.

b.

c.

Figure 1A **Effects of acid deposition.**
Trees die and statues deteriorate due to the burning of fossil fuels. The combustion of fossil fuels results in atmospheric acids that return to the earth as acid rain.

I.4 Molecules of Life

Inorganic molecules constitute nonliving matter, but even so inorganic molecules like salts (e.g., NaCl) and water play important roles in living things. The molecules of life are organic molecules. **Organic molecules** always contain carbon (C) and hydrogen (H). The chemistry of carbon accounts for the formation of the very large variety of organic molecules found in living things. A carbon atom has four electrons in the outer shell. In order to achieve eight electrons in the outer shell, a carbon atom shares electrons covalently with as many as four other atoms. Methane is a molecule in which a carbon atom shares electrons with four hydrogen atoms.

$$H-\underset{\underset{H}{|}}{\overset{\overset{H}{|}}{C}}-H$$

A carbon atom can share with another carbon atom, and in so doing, a long hydrocarbon chain can result:

$$H-\underset{\underset{H}{|}}{\overset{\overset{H}{|}}{C}}-\underset{\underset{H}{|}}{\overset{\overset{H}{|}}{C}}-\underset{\underset{H}{|}}{\overset{\overset{H}{|}}{C}}-\underset{\underset{H}{|}}{\overset{\overset{H}{|}}{C}}-\underset{\underset{H}{|}}{\overset{\overset{H}{|}}{C}}-\underset{\underset{H}{|}}{\overset{\overset{H}{|}}{C}}-\underset{\underset{H}{|}}{\overset{\overset{H}{|}}{C}}-\underset{\underset{H}{|}}{\overset{\overset{H}{|}}{C}}-\underset{\underset{H}{|}}{\overset{\overset{H}{|}}{C}}-H$$

A hydrocarbon chain can also turn back on itself to form a ring compound:

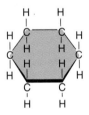

So-called functional groups can be attached to carbon chains. A **functional group** is a particular cluster of atoms that always behaves in a certain way. One functional group of interest is the acidic (carboxyl) group —COOH because it can give up a hydrogen (H⁺) and ionize to —COO⁻.

hydrocarbon
(hydrophobic)

acid in ionized form
(hydrophilic)

Whereas a hydrocarbon chain is *hydrophobic* (not attracted to water) because it is nonpolar, a hydrocarbon chain with an attached ionized group is *hydrophilic* (is attracted to water) because it is polar.

The molecules of life are divided into four classes: carbohydrates, lipids, proteins, and nucleic acids. Carbohydrates, lipids, and proteins are very familiar to you because certain foods are known to be rich in these molecules, as illustrated in Figures 1.13–1.15. The nucleic acid DNA makes up our genes which are hereditary units that control our cells and the structure of our bodies.

Many molecules of life are macromolecules. Just as atoms can join to form a molecule, so molecules can join to form a macromolecule. The smaller molecules are called monomers, and the macromolecule is called a polymer. A polymer is a chain of monomers.

Polymer	Monomer
polysaccharide	monosaccharide
protein	amino acid
nucleic acid	nucleotide

Figure 1.13   **Foods rich in carbohydrates.**
Breads, pasta, rice, corn, and oats all contain complex carbohydrates.

Figure 1.14   **Foods rich in lipids.**
Butter and oils contain fat, the most familiar of the lipids.

Figure 1.15   **Foods rich in proteins.**
Meat, eggs, cheese, and beans have a high content of protein.

1.5 Carbohydrates

Carbohydrates first and foremost function for quick and short-term energy storage in all organisms, including humans. Carbohydrate molecules are characterized by the presence of the atomic grouping H—C—OH, in which the ratio of hydrogen atoms (H) to oxygen atoms (O) is approximately 2:1. Since this ratio is the same as the ratio in water, the name—hydrates of carbon—seems appropriate.

Simple Carbohydrates

If the number of carbon atoms in a molecule is low (from three to seven), then the carbohydrate is a simple sugar, or **monosaccharide**. The designation **pentose** means a 5-carbon sugar, and the designation **hexose** means a 6-carbon sugar. Ribose and deoxyribose are two pentoses of significance because they are found respectively in the nucleic acids RNA and DNA. RNA and DNA are discussed later in the chapter. **Glucose,** a hexose, is blood sugar (Fig. 1.16); our bodies use glucose as an immediate source of energy. Other common hexoses are fructose, found in fruits, and galactose, a constituent of milk. These three hexoses (glucose, fructose,

and galactose) all occur as ring structures with the molecular formula $C_6H_{12}O_6$, but the exact shape of the ring differs, as does the arrangement of the hydrogen (—H) and the hydroxyl groups (—OH) attached to the ring.

A small organic molecule can be a unit of a larger organic molecule often called a macromolecule. A unit is called a monomer and the macromolecule is called a polymer. Glucose is a monomer for larger carbohydrates like glycogen, starch, and cellulose which are discussed on the next page. Organisms have a common way of joining monomers to build larger molecules. **Condensation synthesis** of a larger molecule is so called because synthesis means "making of" and condensation means that water has been removed as monomers are joined. Breakdown of the larger molecule is a **hydrolysis** reaction because water is used to split bonds between monomers. Polymers are synthesized and broken down in this manner:

$$\text{monomers} \rightleftharpoons \text{polymer} + \text{H}_2\text{O molecules}$$

Figure 1.17 shows how condensation synthesis results in a disaccharide called maltose and how hydrolysis of the maltose results in two glucose molecules again. A **disaccharide** (*di* means two and *saccharide* means sugar) contains two monosaccharides. Maltose is a disaccharide of interest because it is found in our digestive tract as a result of starch digestion. When glucose and fructose join, the disaccharide sucrose forms. Sucrose, which is ordinarily derived from sugarcane and sugar beets, is commonly known as table sugar. When we eat a candy bar, our digestive enzymes hydrolyze sucrose into glucose and fructose.

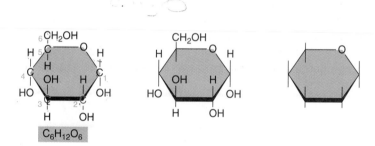

Figure 1.16   **Three ways to represent the structure of glucose.**
The *far left* structure shows the carbon atoms; $C_6H_{12}O_6$ is the molecular formula for glucose. The *far right* structure is the simplest way to represent glucose

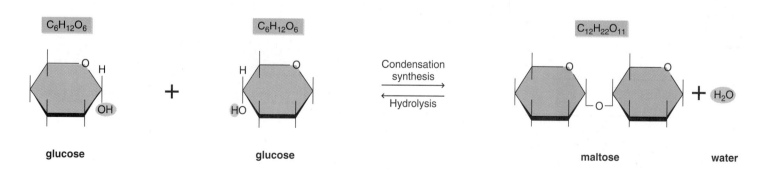

Figure 1.17   **Condensation synthesis and hydrolysis of maltose, a disaccharide.**
During condensation synthesis of maltose, a bond forms between the two glucose molecules and the components of water are removed. During hydrolysis, the components of water are added, and the bond is broken.

Starch and Glycogen

Starch and glycogen are ready storage forms of glucose in plants and animals, respectively. Starch and glycogen are **polysaccharides;** that is, they are polymers of glucose formed just as a necklace might be made using only one type of bead. The following equation shows how starch is synthesized:

| glucose molecules (monomers) | $\xrightarrow[\text{hydrolysis}]{\text{synthesis}}$ | starch + H_2O molecules (polymer) |

Some of the polymers in starch are long chains of up to 4,000 glucose units. Others are branched as is glycogen (Fig. 1.18). Starch has fewer side branches, or chains of glucose that branch off from the main chain, than does glycogen.

Starch is the storage form of glucose inside plant cells. Flour, which we usually acquire by grinding wheat and use to bake bread and rolls, is high in starch. **Glycogen** is the storage form of glucose in humans. Figure 1.18 includes a micrograph of glycogen granules inside the liver.

After we eat starchy foods like bread, potatoes, and cake, starch is hydrolyzed to glucose. Then the bloodstream carries excess glucose to the liver where it is stored as glycogen. In between eating, the liver releases glucose so that the blood glucose concentration is always about 0.1%. As mentioned in chapter 3, this function of the liver is an example of homeostasis.

Cellulose

The polysaccharide **cellulose** is found in plant cell walls, and this accounts, in part, for the strong nature of these walls. In cellulose (Fig. 1.19), the glucose units are joined by a slightly different type of linkage than that in starch or glycogen. (Observe the alternating position of the oxygen atoms in the linked glucose units.) While this might seem to be a technicality, actually it is important because we are unable to digest foods containing this type of linkage; therefore, cellulose largely passes through our digestive tract as fiber, or roughage. Recently, it has been suggested that fiber in the diet is necessary to good health and may even help to prevent colon cancer.

Cells usually use the monosaccharide glucose as an energy source. The polysaccharides starch and glycogen are storage compounds in plant and animal cells, respectively. The polysaccharide cellulose is found in plant cell walls.

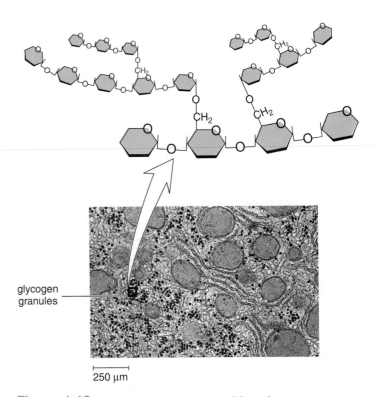

glycogen granules

250 μm

Figure 1.18   Glycogen structure and function.
Glycogen is a highly branched polymer of glucose molecules. The electron micrograph shows glycogen granules in liver cells. Glycogen is the storage form of glucose in animals.

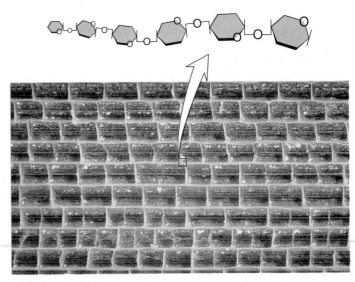

cattail leaf cell walls

Figure 1.19   Cellulose structure and function.
Cellulose contains a slightly different type of linkage between glucose molecules than that in starch or glycogen. Plant cell walls contain cellulose, and the rigidity of the cell walls permits nonwoody plants to stand upright as long as they receive an adequate supply of water.

Figure 1.20 **Condensation synthesis and hydrolysis of a fat molecule.**
Fatty acids can be saturated (no double bonds between carbon atoms) or unsaturated (have double bonds, colored yellow, between carbon atoms). When a fat molecule forms, three fatty acids combine with glycerol, and three water molecules are produced.

1.6 Lipids

Lipids are diverse in structure and function, but they have a common characteristic: they do not dissolve in water.

Fats and Oils

The most familiar lipids are those found in fats and oils. **Fats,** which are usually of animal origin (e.g., lard and butter), are solid at room temperature. **Oils,** which are usually of plant origin (e.g., corn oil and soybean oil), are liquid at room temperature. Fat has several functions in the body: it is used for long-term energy storage, it insulates against heat loss, and it forms a protective cushion around major organs.

Fats and oils form when one glycerol molecule reacts with three fatty acid molecules. A fat is sometimes called a **triglyceride** because of its three-part structure, and the term neutral fat is sometimes used because the molecule is non-polar (Fig. 1.20).

Saturated and Unsaturated Fatty Acids

A **fatty acid** is a hydrocarbon chain that ends with the acidic group —COOH (Fig. 1.20). Most of the fatty acids in cells contain 16 or 18 carbon atoms per molecule, although smaller ones with fewer carbons are also known.

Fatty acids are either saturated or unsaturated. **Saturated fatty acids** have no double bonds between carbon atoms. The carbon chain is saturated, so to speak, with all the hydrogens it can hold. Saturated fatty acids account for the solid nature at room temperature of butter and lard, which are derived from animal sources. **Unsaturated fatty acids** have double bonds between carbon atoms wherever the number of hydrogens is less than two per carbon atom. Unsaturated fatty acids account for the liquid nature of vegetable oils at room temperature. Hydrogenation of vegetable oils can convert them to margarine and products such as Crisco.

Soaps

Strictly speaking, soaps are not lipids, but they are considered here as a matter of convenience. A **soap** is a salt formed from a fatty acid and an inorganic base. For example,

NaOH	+	RCOOH	$\longrightarrow$	RCOO$^-$ Na$^+$
sodium hydroxide		fatty acid		soap

Unlike fats, soaps have a polar end that is hydrophilic in addition to the nonpolar end that is hydrophobic (the hydrocarbon chain represented by *R*). Therefore, a soap does mix with water. When soaps are added to oils, the oils, too, mix with water because a soap positions itself about an oil droplet so that its nonpolar ends project into the fat droplet, while its polar ends project outward.

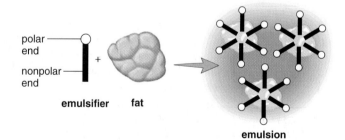

Now the droplet disperses in water, and it is said that **emulsification** has occurred. Emulsification occurs when dirty clothes are washed with soaps or detergents. Also, prior to the digestion of fatty foods, fats are emulsified by bile. A person who has had the gallbladder removed may have trouble digesting fatty foods because this organ stores bile for emulsifying fats prior to the digestive process.

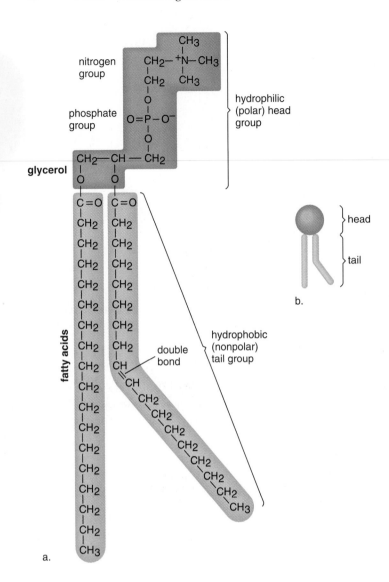

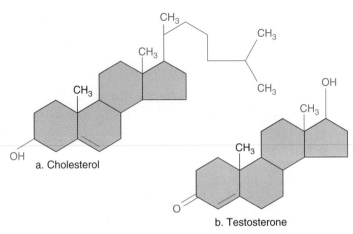

Figure 1.22   Steroid diversity.
a. Cholesterol, like all steroid molecules, has four adjacent rings, but the effects of steroids on the body largely depend on the attached groups indicated in red. **b.** Testosterone is the male sex hormone.

Phospholipids

Phospholipids, as their name implies, contain a phosphate group (Fig. 1.21). Essentially, they are constructed like fats, except that in place of the third fatty acid, there is a phosphate group or a grouping that contains both phosphate and nitrogen. These molecules are not electrically neutral as are fats because the phosphate and nitrogenous groups are ionized. It forms the so-called hydrophilic head of the molecule, while the rest of the molecule becomes the hydrophobic tails. The plasma membrane which surrounds cells is a phospholipid bilayer in which the heads face outward into a watery medium and the tails face each other because they are water repelling.

Steroids

Steroids are lipids having a structure that differs entirely from that of fats. Steroid molecules have a backbone of four fused carbon rings, but each one differs primarily by the arrangement of the atoms in the rings and the type of functional groups attached to them. Cholesterol is a component of an animal cell's plasma membrane and is the precursor of several other steroids, such as the sex hormones estrogen and testosterone (Fig. 1.22).

We know that a diet high in saturated fats and cholesterol can lead to circulatory disorders. This type of diet causes fatty material to accumulate inside the lining of blood vessels and blood flow is reduced. As discussed in the Health reading on page 36, nutrition labels are now required to list the calories from fat per serving and the percent daily value from saturated fat and cholesterol.

Lipids include fats and oils for long-term energy storage and steroids. Phospholipids, unlike other lipids, are soluble in water because they have a hydrophilic group.

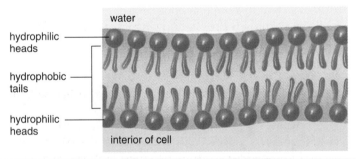

c. **Phospholipid molecules in plasma membrane**

Figure 1.21   Phospholipid structure and shape.
a. Phospholipids are constructed like fats, except that they contain a phosphate group. This phospholipid also includes an organic group that contains nitrogen. **b.** The hydrophilic portion of the phospholipid molecule (head) is soluble in water, whereas the two hydrocarbon chains (tails) are not. **c.** This causes the molecule to arrange itself as shown when exposed to water.

Name	Structural Formula	R Group

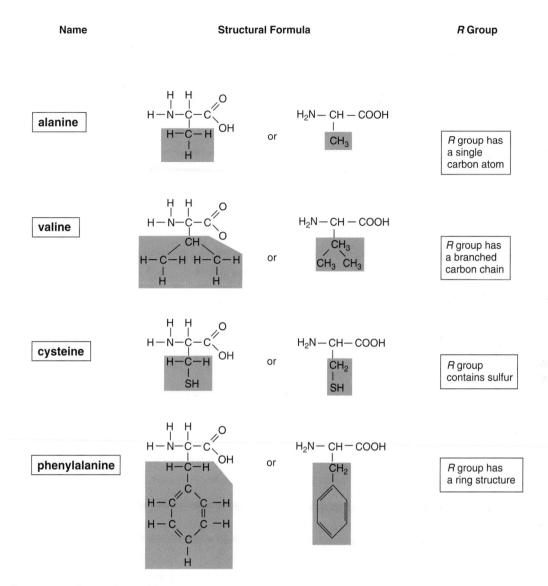

Figure 1.23 **Representative amino acids.**

Amino acids differ from one another by their R group; the simplest R group is a single hydrogen atom (H). The R groups that contain carbon vary as shown.

1.7 Proteins

Proteins sometimes have a structural function. For example, in humans, the protein keratin makes up hair and nails, whereas collagen is found in all types of connective tissue, including ligaments, cartilage, bones, and tendons. The muscles contain proteins, which account for their ability to contract.

Some proteins are enzymes, necessary contributors to the chemical workings of the cell and therefore of the body. **Enzymes** speed chemical reactions; they work so quickly that a reaction that normally takes several hours or days without an enzyme takes only a fraction of a second with an enzyme.

Proteins are macromolecules with amino acid monomers. An **amino acid** has a central carbon atom bonded to a hydrogen atom and three groups. The name of the molecule is appropriate because one of these groups is an amino group (—NH$_2$) and another is an acidic group (—COOH). The other group is called an R group because it is the *Remainder* of the molecule. Amino acids differ from one another by their R group; the R group varies from a single hydrogen (H) to a complicated ring (Fig. 1.23).

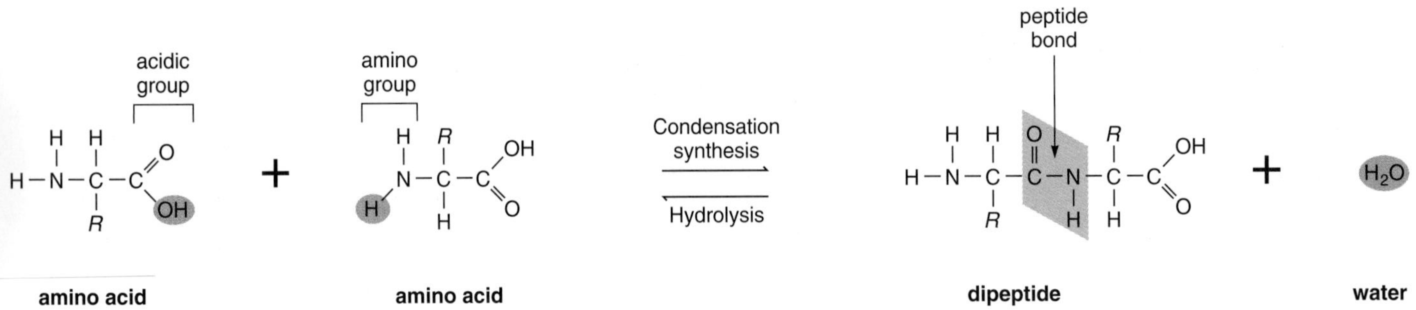

Figure 1.24   Condensation synthesis and hydrolysis of a dipeptide.
The two amino acids on the left-hand side of the equation differ by their *R*-groups. As these amino acids join, a peptide bond forms, and a water molecule is produced. During hydrolysis, water is added, and the peptide bond is broken.

Peptides

Figure 1.24 shows that a condensation synthesis reaction between two amino acids results in a dipeptide and a molecule of water. A bond that joins two amino acids is called a **peptide bond.** The atoms associated with a peptide bond—oxygen (O), carbon (C), nitrogen (N), and hydrogen (H)—share electrons in such a way that the oxygen has a partial negative charge and the hydrogen has a partial positive charge.

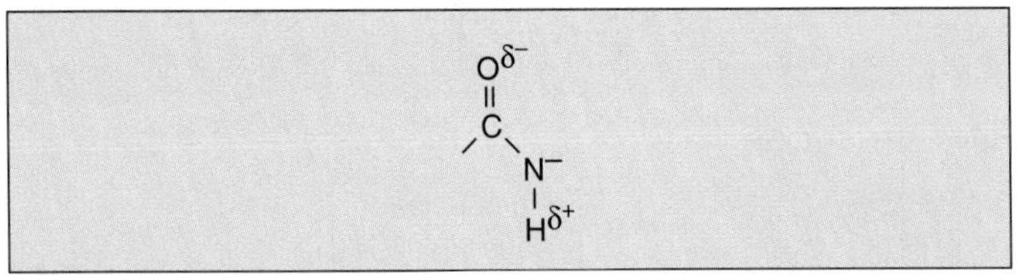

Therefore, the peptide bond is polar, and hydrogen bonding is possible between the C=O of one amino acid and the N—H of another amino acid in a polypeptide. A **polypeptide** is a single chain of amino acids.

Levels of Protein Organization

The structure of a protein has at least three levels of organization (Fig. 1.25*a*). The first level, called the *primary structure,* is the linear sequence of the amino acids joined by peptide bonds. Polypeptides can be quite different from one another. You will recall that the structure of a polysaccharide can be likened to a necklace that contains a single type "bead," namely, glucose. Polypeptides can make use of 20 different possible types of amino acids or "beads." Each particular polypeptide has its own sequence of amino acids. It can be said that each polypeptide differs by the sequence of its *R* groups and the number of amino acids in the sequence.

The *secondary structure* of a protein comes about when the polypeptide takes on a particular orientation in space. A coiling of the chain results in an alpha (α) helix, or a right-handed spiral, and a folding of the chain results in a pleated sheet. Hydrogen bonding between peptide bonds holds the shape in place.

The *tertiary structure* of a protein is its final three-dimensional shape. In muscles, the helical chains of myosin form a rod shape that ends in globular (globe-shaped) heads. In enzymes, the helix bends and twists in different ways. Invariably, the hydrophobic portions are packed mostly on the inside, and the hydrophilic portions are on the outside where they can make contact with water. The tertiary shape of a polypeptide is maintained by various types of bonding between the *R* groups; covalent, ionic, and hydrogen bonding all occur. One common form of covalent bonding between *R* groups is disulfide (S—S) linkages between two cysteine amino acids.

Some proteins have only one polypeptide, and some others have more than one polypeptide chain, each with its own primary, secondary, and tertiary structures. These separate polypeptides are arranged to give some proteins a fourth level of structure, termed the *quaternary structure* (Fig. 1.25). Hemoglobin is a complex protein having a quaternary structure; most enzymes also have a quaternary structure.

The final shape of a protein is very important to its function. As we will discuss in chapter 2, for example, enzymes cannot function unless they have their usual shape. When proteins are exposed to extremes in heat and pH, they undergo an irreversible change in shape called **denaturation.** For example, we are all aware that the addition of acid to milk causes curdling and that heating causes egg white, which contains a protein called albumin, to coagulate. Denaturation occurs because the normal bonding between the *R* groups has been disturbed. Once a protein loses its normal shape, it is no longer able to perform its usual function.

Proteins, which contain covalently linked amino acids, are important in the structure and the function of cells. Some proteins are enzymes, which speed chemical reactions.

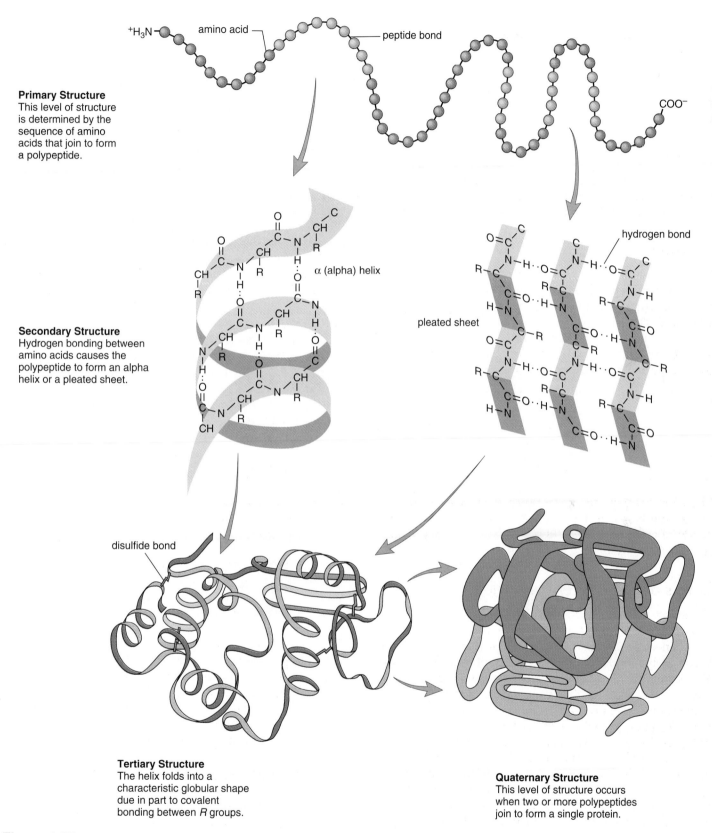

Primary Structure
This level of structure is determined by the sequence of amino acids that join to form a polypeptide.

amino acid

peptide bond

α (alpha) helix

hydrogen bond

pleated sheet

Secondary Structure
Hydrogen bonding between amino acids causes the polypeptide to form an alpha helix or a pleated sheet.

disulfide bond

Tertiary Structure
The helix folds into a characteristic globular shape due in part to covalent bonding between *R* groups.

Quaternary Structure
This level of structure occurs when two or more polypeptides join to form a single protein.

Figure 1.25   Levels of protein organization.

1.8 Nucleic Acids

Nucleic acids, such as **DNA (deoxyribonucleic acid)** and **RNA (ribonucleic acid),** are huge polymers of nucleotides. Every **nucleotide** is a molecular complex of three types of subunit molecules: phosphate (phosphoric acid), a pentose sugar, and a nitrogen-containing base:

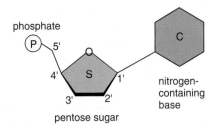

DNA makes up the genes and stores information regarding its own replication and the order in which amino acids are to be joined to form a protein. RNA is an intermediary in the process of protein synthesis, conveying information from DNA regarding the amino acid sequence in a protein.

The nucleotides in DNA contain the sugar deoxyribose, and in RNA they contain the sugar ribose; this difference accounts for their respective names (Table 1.2). As indicated in Figure 1.26, there are four different types of bases in DNA: A = adenine, T = thymine, G = guanine, and C = cytosine. The base can have two rings (adenine or guanine), or one ring (thymine or cytosine). These structures are called bases because their presence raises the pH of a solution. In RNA the base uracil replaces the base thymine.

Although the sequence can vary between molecules, any particular DNA or RNA has a definite sequence. The nucleotides form a linear molecule called a strand, in which the

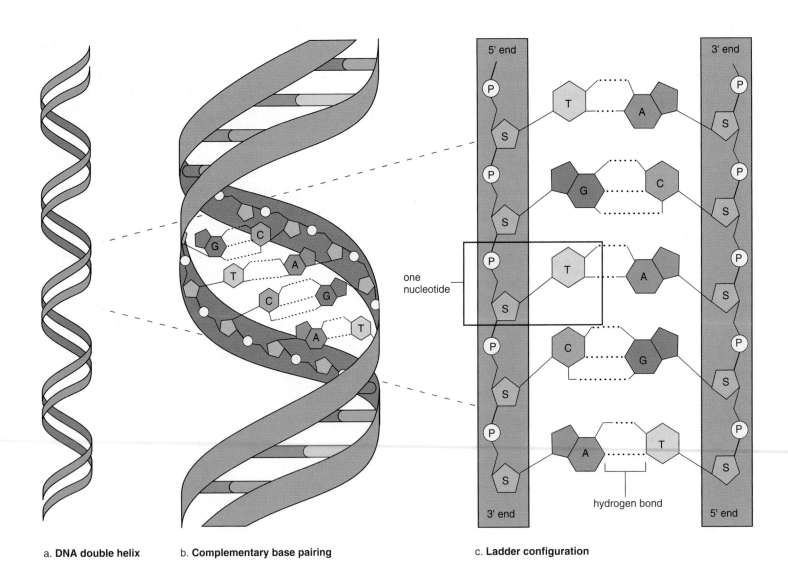

a. **DNA double helix**   b. **Complementary base pairing**   c. **Ladder configuration**

Figure 1.26   Overview of DNA structure.
a. Double helix. **b.** Complementary base pairing between strands. **c.** Ladder configuration. Notice that the uprights are composed of phosphate and sugar molecules and that the rungs are complementary paired bases.

backbone is made up of phosphate-sugar-phosphate-sugar, with the bases projecting to one side of the backbone. Since the nucleotides occur in a definite order, so do the bases.

RNA is usually single stranded, while DNA is usually double stranded, with the two strands twisted about each other in the form of a double helix. In DNA, the two strands are held together by hydrogen bonds between the bases. When unwound, DNA resembles a stepladder. The sides of the ladder are made entirely of phosphate and sugar molecules, and the rungs of the ladder are made only of complementary paired bases. Thymine (T) always pairs with adenine (A), and guanine (G) always pairs with cytosine (C) (Fig. 1.26). Complementary bases have shapes that fit together.

We shall see that complementary base pairing allows DNA to replicate in a way that assures the sequence of bases will remain the same. The sequence of the bases are the genetic information that specifies the sequence of amino acids in the proteins of the cell.

DNA has a structure like a twisted ladder: sugar and phosphate molecules make up the sides, and hydrogen-bonded bases make up the rungs of the ladder.

Table 1.2	DNA Structure Compared to RNA Structure	
	DNA	**RNA**
Sugar	Deoxyribose	Ribose
Bases	Adenine, guanine, thymine, cytosine	Adenine, guanine, uracil, cytosine
Strands	Double stranded with base pairing	Single stranded
Helix	Yes	No

ATP (Adenosine Triphosphate)

In addition to being the monomers of nucleic acids, nucleotides alone have metabolic functions in cells. When adenosine (adenine plus ribose) is modified by the addition of three phosphate groups, it becomes **ATP (adenosine triphosphate),** the primary energy carrier in cells.

Glucose contains too much energy to be used in cellular reactions and cells use the energy within a glucose molecule to build ATP molecules. The amount of energy in ATP makes it useful to supply energy for chemical reactions in cells. As an analogy, consider that twenty dollar bills (i.e., ATP) are more useful for everyday purchases than one hundred dollar bills (i.e., glucose).

ATP is called the energy currency of cells because when cells require energy, they often "spend" ATP. Cells use ATP for the synthesis of macromolecules like carbohydrates and proteins. In muscle cells, the energy within an ATP molecule is used for muscle contraction, and in nerve cells, it is used for the conduction of nerve impulses.

ATP is sometimes called a high-energy molecule because the last two phosphate bonds are unstable and are easily broken. Usually in cells the terminal phosphate bond is hydrolyzed, leaving the molecule **ADP (adenosine diphosphate)** and a molecule of inorganic phosphate Ⓟ (Fig. 1.27). The terminal bond is sometimes called a high-energy bond, symbolized by a wavy line. But this terminology is misleading—the breakdown of ATP releases energy because the products of hydrolysis (ADP and Ⓟ) are more stable than ATP. It is the entire ATP molecule that releases energy and not a particular bond.

After ATP breaks down, it is rebuilt by the addition of Ⓟ to ADP (Fig. 1.27). There is enough energy in one glucose molecule to build 36 ATP molecules in this way. Homeostasis is only possible when cells continually produce and use ATP molecules.

ATP is a high-energy molecule. When ATP breaks down to ADP + Ⓟ, releasing energy, this energy is used for all metabolic work done in a cell.

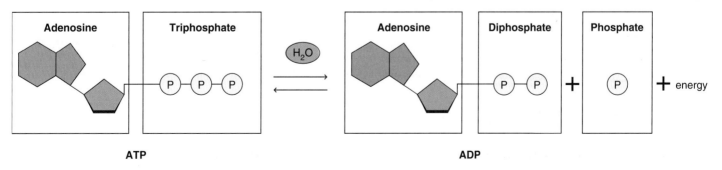

Figure 1.27 ATP reaction.
ATP, the universal energy currency of cells, is composed of adenosine and three phosphate groups. When cells require energy, ATP usually becomes ADP + Ⓟ , with the release of energy

Nutrition Labels

Packaged foods now have a nutrition label like the one depicted in Figure 1B. The nutrition information given on this label is based on the serving size (that is, 1¼ cup, 57 grams) of the cereal. A Calorie* is a measurement of energy. One serving of the cereal provides 220 Calories, of which 20 are from fat. At the bottom of the label, the recommended amounts of nutrients are based on a typical diet of 2,000 Calories for women and 2,500 Calories for men.

Fats are the nutrient with the highest energy content: 9 Cal/g compared to 4 Cal/g for carbohydrates and proteins. The body stores fat under the skin and around the organs for later use. A 2,000-Calorie diet should contain no more than 65 g (585 Calories) of fat. Dietary fat has been implicated in cancer of the colon, pancreas, ovary, prostate, and breast. Although saturated fat and cholesterol are essential nutrients, dietary consumption of saturated fats and cholesterol in particular should be controlled. Cholesterol and saturated fat contribute to the formation of deposits called plaques, which clog arteries and lead to cardiovascular disease, including high blood pressure.

For these reasons, it is important to know how a serving of the cereal will contribute to the maximum daily recommended amount of fat, saturated fat, and cholesterol. You can find this out by looking at the listing under % Daily Value: the total fat in one serving of the cereal provides 3% of the daily recommended amount of fat. How much will a serving of the cereal contribute to the maximum recommended daily amount of saturated fat? cholesterol?

Carbohydrates (sugars and polysaccharides) are the quickest, most readily available source of energy for the body. Because carbohydrates aren't usually associated with health problems, they should compose the largest proportion of the diet. Breads and cereals containing complex carbohydrates are

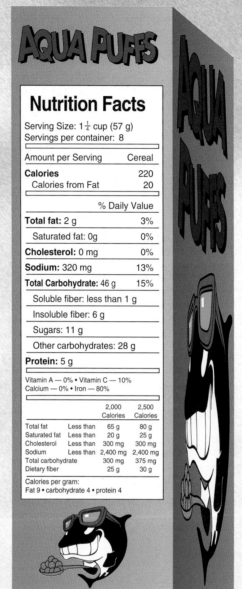

Figure 1B   Nutrition label on side panel of cereal box.

preferable to candy and ice cream containing simple carbohydrates because they are likely to contain dietary fiber (nondigestible plant material). Insoluble fiber has a laxative effect and seems to reduce the risk of colon cancer; soluble fiber combines with the cholesterol in food and prevents the cholesterol from entering the body proper.

The body does not store amino acids for the production of proteins, which are found particularly in muscles but also in all cells of the body. A woman should have about 44 g of protein per day, and a man should have about 56 g of protein a day. Red meat is rich in protein, but it is usually also high in saturated fat. Therefore, it is considered good health sense to rely on protein from plant origins (e.g., whole-grain cereals, dark breads, legumes) more than is customary in the United States. Legume is a botanical term that includes peas and beans—a combination of beans and rice can provide all of the various amino acids you need to build cellular proteins.

The amount of dietary sodium (as in table salt) is of concern because excessive sodium intake has been linked to high blood pressure in some people. It is recommended that the intake of sodium be no more than 2,400 mg per day. A serving of this cereal provides what percent of this maximum amount?

Vitamins are essential requirements needed in small amounts in the diet. Each vitamin has a recommended daily intake, and the food label tells what percent of the recommended amount is provided by a serving of this cereal.

---

*A calorie is the amount of heat required to raise the temperature of one gram of water one degree centigrade. A Calorie (capital C) is 1,000 calories.

Bioethical Issue

Eric Stevenson and more than 110,000 veterans of the Gulf War are mysteriously ill. They complain of conditions like skin rashes, breathing difficulties, fatigue, diarrhea, muscle and joint pain, headaches and loss of memory. While the Pentagon doesn't recognize what is called the Gulf War Syndrome (GWS), it does admit that soldiers were exposed to at least four categories of chemicals:

• Petroleum products such as kerosene, diesel fuel, leaded gasoline and smoke from oil-well fires.

• Pesticides and insect repellents. These were applied to clothing and sprayed into the air.

• Drugs and vaccines. Pyridostigmine bromide was given to protect against nerve gas. Vaccines against anthrax and botulism were given in case of biological warfare.

• Biological and chemical weapons. Reluctantly, the Pentagon estimates that hundreds of thousands of soldiers may have been exposed to nerve gas released into the air when Iraqi ammunition depots were bombed.

While the military says that exposure can make you sick, they cling to the idea that toxic chemicals either kill you outright or you recover completely. Therefore, they suggest that the veterans are suffering from post-traumatic stress disorder. Lingering symptoms of stress are to be expected in a certain number of soldiers that come home from war.

Epidemiologist Robert Haley, however, has studied the effects of these chemicals on chickens, and found that their combination does produce Gulf War Syndrome symptoms. He says, "The Defense Department should agree that this is a physical injury, a brain injury, aided and abetted by acute stress."

For what reasons might the Pentagon be reluctant to accept the possibility that GWS is due to exposure of troops to chemicals? In so doing are they shirking their responsibility to Gulf War veterans?

Questions

1. Do you believe that veterans with GWS should be completely supported by the government? Why or why not?

2. The Pentagon has earmarked $42 million in grants for research of GWS. Do you approve of this action? Why or why not?

3. Should veterans be allowed to sue the government for their illness, much as tobacco users have sued tobacco companies? Why or why not?

Table 1.3 Organic Compounds Associated with Living Things

Macromolecule	Monomer	Function
Proteins	Amino acids	Enzymes speed up chemical reactions; structural components (e.g., muscle and membrane proteins)
Carbohydrates		
Starch	Glucose	Energy storage in plants
Glycogen	Glucose	Energy storage in animals
Cellulose	Glucose	Plant cell walls
Lipids		
Fats and Oils	Glycerol, 3 fatty acids	Long-term energy storage
Phospholipids	Glycerol, 2 fatty acids, phosphate group	Plasma membrane structure
Nucleic Acids		
DNA	Nucleotides with deoxyribose sugar	Genetic material
RNA	Nucleotides with ribose sugar	Protein synthesis

Summarizing the Concepts

1.1 Elements and Atoms
All matter is composed of some 92 elements. Each element is made up of just one type atom. An atom has a weight, which is dependent on the number of protons and neutrons in the nucleus, and its chemical properties are dependent on the number of electrons in the outer shell.

1.2 Molecules and Compounds
Atoms react with one another by forming ionic bonds or covalent bonds. Ionic bonds are an attraction between charged ions. Atoms share electrons in covalent bonds, which can be single, double, or triple bonds.

Oxidation is the loss of electrons (hydrogen atoms), and reduction is the gain of electrons (hydrogen atoms).

1.3 Water and Living Things
Water, acids, and bases are important inorganic molecules. The polarity of water accounts for it being the universal solvent; hydrogen bonding accounts for it boiling at 100°C and freezing at 0°C. Because it is slow to heat up and slow to freeze, it is liquid at the temperature of living things.

Pure water has a neutral pH; acids increase the hydrogen ion concentration [H^+] but decrease the pH, and bases decrease the hydrogen ion concentration [H^+] but increase the pH of water.

1.4 Molecules of Life
The chemistry of carbon accounts for the chemistry of organic compounds. Carbohydrates, lipids, proteins, and nucleic acids are macromolecules with specific functions in cells (Table 1.3).

1.5 Carbohydrates
Glucose is the six-carbon sugar most utilized by cells for "quick" energy. Like the rest of the macromolecules to be studied, condensation synthesis joins two or more sugars, and a hydrolysis reaction splits the bond. Plants store glucose as starch, and animals store glucose as glycogen. Humans cannot digest cellulose, which forms plant cell walls.

1.6 Lipids
Lipids are varied in structure and function. Fats and oils, which function in long-term energy storage, contain glycerol and three fatty acids. Fatty acids can be saturated or unsaturated. Plasma membranes contain phospholipids that have a polarized end. Certain hormones are derived from cholesterol, a complex ring compound.

1.7 Proteins
The primary structure of a polypeptide is its own particular sequence of the possible 20 types of amino acids. The secondary structure is often an alpha (α) helix. The tertiary structure occurs when a polypeptide bends and twists into a three-dimensional shape. A protein can contain several polypeptides, and this accounts for a possible quaternary structure.

1.8 Nucleic Acids
Nucleic acids are polymers of nucleotides. Each nucleotide has three components: a sugar, a base, and phosphate (phosphoric acid). DNA, which contains the sugar deoxyribose, is the genetic material that stores information for its own replication and for the order in which amino acids are to be sequenced in proteins. DNA, with the help of RNA, specifies protein synthesis.

ATP, with its unstable phosphate bonds, is the energy currency of cells. Hydrolysis of ATP to ADP + Ⓟ releases energy that is used by the cell to do metabolic work.

Studying the Concepts

1. Name the subatomic particles of an atom; describe their charge, weight, and location in the atom. 16–17
2. Give an example of an ionic reaction, and explain it. 18
3. Diagram the atomic structure of calcium, and explain how it can react with two chlorine atoms. 19
4. Give an example of a covalent reaction, and explain it. 20
5. Relate the characteristics of water to its polarity and hydrogen bonding between water molecules. 21–22
6. On the pH scale, which numbers indicate a basic solution? An acidic solution? Why? 24
7. What are buffers, and why are they important to life? 24
8. Relate the variety of organic compounds to the bonding capabilities of carbon. 26
9. Name the four classes of organic molecules in cells, and relate them to macromolecules and also polymers. 26
10. Name some monosaccharides, disaccharides, and polysaccharides, and state some general functions for each. What is the most common monomer for polysaccharides? 27–28
11. How is a neutral fat synthesized? What is a saturated fatty acid? An unsaturated fatty acid? What is the function of fats? 29
12. Relate the structure of a phospholipid to that of a neutral fat. What is the function of a phospholipid? 30
13. What is the general structure and significance of cholesterol? 30
14. What are some functions of proteins? What is a peptide bond, a dipeptide, and a polypeptide? 31–34
15. Discuss the primary, secondary, and tertiary structures of globular proteins. 32
16. Discuss the structure and function of the nucleic acids, DNA and RNA. 34–35

—

Testing Your Knowledge of the Concepts

In questions 1–4, match the molecule to the functions below.
a. carbohydrates
b. lipids
c. proteins
d. nucleic acids

_____ 1. long-term energy storage
_____ 2. most enzymes
_____ 3. immediate source of energy
_____ 4. hereditary units called genes

In questions 5–7, indicate whether the statement is true (T) or false (F).

_____ 5. The higher the pH, the higher the H^+ concentration.
_____ 6. Ionic bonds share electrons and covalent bonds are an attraction between charges.
_____ 7. Protons are located in the nucleus, while electrons are located in shells about the nucleus.

In questions 8 and 9, fill in the blanks.

8. Fats and oils contain the molecules _____ and _____.

9. A nucleotide contains a _____ sugar, a _____ group, and a nitrogen-containing _____.

10. Label this diagram of condensation synthesis and hydrolysis.

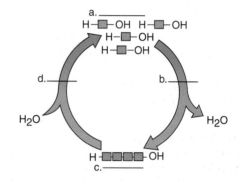

Applying Your Knowledge to the Concepts

These questions pertain to the chemistry of life.

1. Explain why you would expect the covalent bond to be stronger than an ionic bond or a hydrogen bond.

2. Many soft drinks sold today are carbonated, that is, the gas CO_2 has been added to them. Considering the equation $CO_2 + H_2O \rightleftharpoons H_2CO_3$ (carbonic acid), explain why the pH of carbonated drinks moves toward neutral once the bottle is opened.

3. Carbon atoms make up the skeleton or backbone of organic molecules. How is the carbon atom well designed for this function?

4. Heating an enzyme to a certain temperature usually destroys its enzymatic activity even though the same sequence of amino acids remains. What levels of structure have been affected by heating?

Understanding the Terms

acid 23
ADP (adenosine diphosphate) 35
amino acid 31
atom 16
ATP (adenosine triphosphate) 35
base 23
buffer 24
calorie 22
carbohydrate 27
cellulose 28
condensation synthesis 27
covalent bond 20
denaturation 32
disaccharide 27
DNA (deoxyribonucleic acid) 34
electron 16
element 16
emulsification 29
enzyme 31

fat 29
fatty acid 29
functional group 26
glucose 27
glycogen 28
hexose 27
hydrogen bond 21
hydrolysis 27
hydrophilic 21
hydrophobic 21
inorganic molecule 26
ion 19
ionic bond 19
isotope 17
lipid 29
matter 16
molecule 19
monosaccharide 27
neutron 16
nucleotide 34
oil 29
organic molecule 26

pentose 27
peptide bond 32
phospholipid 30
pH scale 24
polypeptide 32
polysaccharide 28
protein 31
proton 16

RNA (ribonucleic acid) 34
saturated fatty acid 29
soap 29
starch 28
steroid 30
triglyceride 29
unsaturated fatty acid 29

Match the terms to these definitions:

a. _____ Polymer of many amino acids linked by peptide bonds.

b. _____ Organic catalyst, usually protein, that speeds up a reaction in cells due to its particular shape.

c. _____ Polysaccharide composed of glucose molecules; the chief constituent of a plant's cell wall.

d. _____ Subatomic particle that has weight of one atomic mass unit, carries no charge, and is found in the nucleus of an atom.

e. _____ Solution in which pH is less than 7; a substance that contributes or liberates hydrogen ions in a solution.

Applying Technology to the Concepts

Your study of chemistry is supported by these available technologies:

Essential Study Partner CD-ROM

Cells → Chemistry
  → Metabolism
  → Respiration

Visit the Mader web site for related ESP activities.

Exploring the Internet

The Mader Home Page provides resources and tools as you study this chapter.

http://www.mhhe.com/biosci/genbio/mader

HealthQuest CD-ROM

3 Nutrition → Gallery → Water
              → The Food Pyramid

Life Science Animations 3D Video

1 Atomic Structure and Covalent and Ionic Bonding

Chapter 2

Cell Structure and Function

Figure 2.1   Racing cyclists.
Cycling or any human activity is dependent on the functioning of skeletal muscle cells (colored red in the insert). Oxygen reaches muscle cells by way of the capillaries (colored blue in this micrograph).

Helen is nervous. But when she hops on her bike at the start of the race, her body's harmonized network of 75 trillion cells answers the challenge. Her brain sends messages along nerves to her skeletal muscles, which are fastened to the bones of the skeletal system. When her leg muscles contract, her bones move the bike. The energy to power her muscles came from sugars absorbed by her digestive tract. Oxygen, used to release energy from sugars, was absorbed by her lungs. Sugar and oxygen molecules are delivered to muscle cells by the circulatory system. And after sugars are broken down for energy, the waste products are expelled from the body by the lungs and kidneys.

The body's organs are composed of cells, and it is at the cellular level that we must understand how the body functions. Helen can win the race because each individual muscle cell has done its job of keeping her legs moving. Because cells are small, and it takes a microscope to see them, it is sometimes hard to imagine that individual muscle cells account for the functioning of an organ like skeletal muscle.

Use of a microscope does show that muscles and all organs are composed of cells. The electron microscope, developed in this century, has revealed that cells contain organelles, little bodies that are specialized in structure to carry on a particular function. Mitochondria are the organelles in muscle cells that oxidize sugar molecules to release energy. When you are fit, your mitochondria are conditioned to start using oxygen right away so that acids don't build up and cause fatigue. Helen trained for many months to increase her endurance. Her muscle cells have more mitochondria than in those who have not trained, and her mitochondria are all set to help her win the race.

Despite specialization—muscles cells are specialized to contract—all cells have the same basic structure and metabolism. This chapter discusses the generalized structure of cells. It also describes the structure and function of the various organelles which carry on the activities of a cell. The chapter ends by describing the cellular reactions that provide energy for the workings of a cell.

2.1 Cell Size

All living things are made up of fundamental units called **cells.** Because cells are so small, the study of cells did not begin until the invention of the first microscope in the seventeenth century. Then the **cell theory,** which states that *all living things are composed of cells, and new cells arise only from preexisting cells,* was formulated.

Regardless of a cell's size and shape, it must carry on the functions associated with life—interacting with the environment, obtaining chemicals and energy, growing, and reproducing. A few cells, like a hen's egg or a frog's egg, are large enough to be seen by the naked eye, but most are not. This is the reason a microscope is needed to see cells. Why are cells so small (most are less than one cubic millimeter)? An explanation for why cells are so small and why we are multicellular is explained by considering the surface/volume ratio of cells. Nutrients enter a cell and wastes exit a cell at its surface; therefore the amount of surface represents the ability to get material in and out of the cell. A large cell requires more nutrients and produces more wastes than a small cell. In other words, the volume represents the needs of the cell. Yet, as cells get larger in volume, the proportionate amount of surface area actually decreases, as you can see by comparing these two cells:

small cell—
more surface area
per volume

large cell—
less surface area
per volume

1 × 1 × 1 cube: surface/volume ratio = 6 : 1
2 × 2 × 2 cube: surface/volume ratio = 3 : 1

We would expect, then, that there would be a limit to how large an actively metabolizing cell can become. Once a hen's egg is fertilized and starts actively metabolizing, it divides repeatedly without growth. Cell division restores the amount of surface area needed for adequate exchange of materials.

A cell needs a surface area that can adequately exchange materials with the environment. This explains why cells stay small.

Microscopy and Cell Structure

Three types of microscopes are most commonly used: the *compound light microscope, transmission electron microscope,* and *scanning electron microscope.* Figure 2.2 depicts these microscopes, along with a micrograph of red blood cells viewed with each one.

In a compound light microscope, light rays passing through a specimen are brought to a focus by a set of glass lenses, and the resulting image is then viewed by the human eye. In the transmission electron microscope, electrons passing through a specimen are brought to a focus by a set of magnetic lenses, and the resulting image is projected onto a fluorescent screen or photographic film.

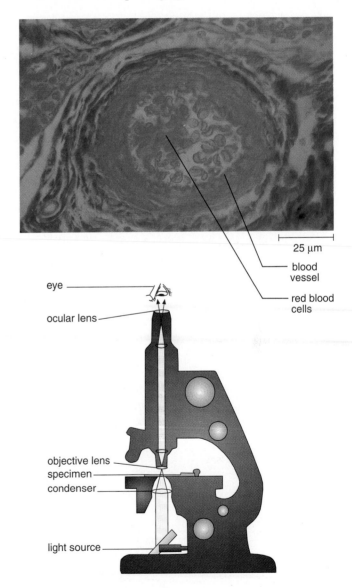

Compound light microscope

Figure 2.2   **Blood vessels and red blood cells viewed with three different types of microscopes.**

The magnification produced by an electron microscope is much higher than that of a light microscope. Also, the ability of the electron microscope to make out detail in enlarged images is much greater. In other words, the electron microscope has a higher resolving power, that is, the ability to distinguish between two adjacent points. The following shows the resolving power of the eye, light microscope, and electron microscope:

eye:	0.2 mm	=	200 µm	=	200,000 nm[1]
light microscope: (1,000×)	0.0002 µm	=	0.200 mm	=	200 nm
electron microscope (50,000×):	0.00001 mm	=	0.0001 µm	=	10 nm

[1]See Metric System in Appendix B

A scanning electron microscope provides a three-dimensional view of the surface of an object. A narrow beam of electrons is scanned over the surface of the specimen, which has been coated with a thin layer of metal. The metal gives off secondary electrons, which are collected to produce a television-type picture of the specimen's surface on a screen.

A picture obtained using a light microscope sometimes is called a photomicrograph, and a picture resulting from the use of an electron microscope is called a transmission electron micrograph (TEM) or a scanning electron micrograph (SEM), depending on the type of microscope used.

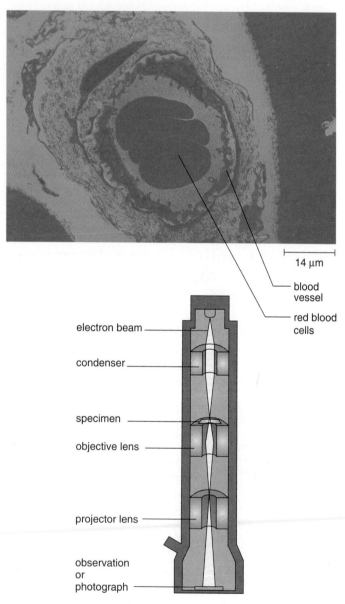

14 µm

blood vessel

red blood cells

electron beam

condenser

specimen

objective lens

projector lens

observation or photograph

Transmission electron microscope

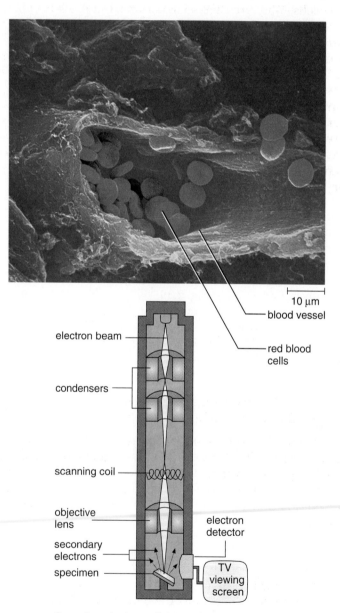

10 µm

blood vessel

red blood cells

electron beam

condensers

scanning coil

objective lens

secondary electrons

specimen

electron detector

TV viewing screen

Scanning electron microscope

2.2 Cellular Organization

The **plasma membrane** which surrounds and keeps the cell intact regulates what enters and exits a cell. The plasma membrane is a phospholipid bilayer that is said to be semipermeable because it allows certain molecules but not others to enter the cell. Proteins present in the plasma membrane play important roles in allowing substances to enter the cell.

The **nucleus** is a large, centrally located structure that can often be seen with a light microscope. The nucleus contains the chromosomes and is the control center of the cell. It controls the metabolic functioning and structural characteristics of the cell. The nucleolus is a region inside the nucleus.

The **cytoplasm** is the portion of the cell between the nucleus and the plasma membrane. The matrix of the cytoplasm is a semifluid medium that contains water and various types of molecules suspended or dissolved in the medium. The presence of proteins accounts for the semifluid nature of the cytoplasm.

The cytoplasm occurs between the **organelles,** small membranous structures that can usually only be seen with an electron microscope. Each type of organelle has a specific function. One type of organelle produces proteins, for example, and another type produces energy for the cell. Since organelles are composed of membrane, it can be seen that membrane compartmentalizes, keeping the various cellular activities separated from one another (Table 2.1 and Fig. 2.3).

Cells also have a **cytoskeleton,** a network of interconnected filaments and microtubules that occur in the cytoplasm. The name cytoskeleton is convenient in that it allows us to compare the cytoskeleton to the bones and muscles of an animal. Bones and muscle give an animal structure and produce movement. Similarly, the elements of the cytoskeleton maintain cell shape and allow the cell and its contents to move. Some cells move by using cilia and flagella which are also made up of microtubules.

The human cell has a central nucleus and an outer plasma membrane. Various organelles are found within the cytoplasm, the portion of the cell between the nucleus and the plasma membrane.

Table 2.1	**Structures in Animal Cells**	
Name	**Composition**	**Function**
Plasma membrane	Phospholipid bilayer with embedded proteins	Selective passage of molecules into and out of cell
Nucleus	Nuclear envelope surrounding nucleoplasm, chromatin, and nucleolus	Storage of genetic information
Nucleolus	Concentrated area of chromatin, RNA, and proteins	Ribosomal formation
Ribosome	Protein and RNA in two subunits	Protein synthesis
Endoplasmic reticulum (ER)	Membranous saccules and canals	Synthesis and/or modification of proteins and other substances, and transport by vesicle formation
Rough ER	Studded with ribosomes	Protein synthesis
Smooth ER	Having no ribosomes	Various; lipid synthesis in some cells
Golgi apparatus	Stack of membranous saccules	Processing, packaging, and distributing molecules
Vacuole and vesicle	Membranous sacs	Storage of substances
Lysosome	Membranous vesicle containing digestive enzymes	Intracellular digestion
Mitochondrion	Inner membrane (cristae) within outer membrane	Cellular respiration
Cytoskeleton	Microtubules, actin filaments	Shape of cell and movement of its parts
Cilia and flagella	9 + 2 pattern of microtubules	Movement of cell
Centriole	9 + 0 pattern of microtubules	Formation of basal bodies

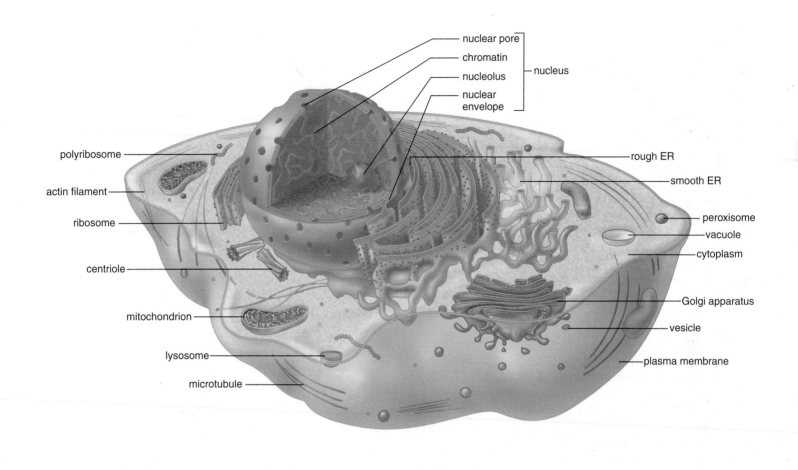

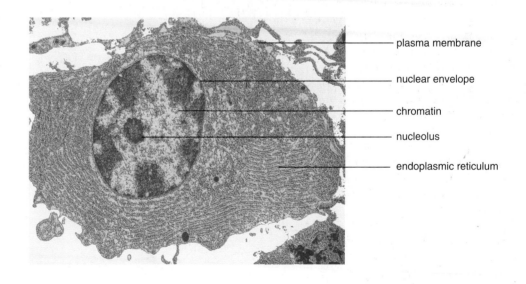

Figure 2.3 **Animal cell.**
This generalized representation is based on electron micrographs.

The Plasma Membrane

An animal cell is surrounded by an outer *plasma membrane.* The plasma membrane marks the boundary between the outside of the cell and the inside of the cell, termed the *cytoplasm.* Plasma membrane integrity and how it functions are necessary to the life of the cell.

The plasma membrane is a phospholipid bilayer with attached or embedded proteins. The structure of a phospholipid is such that the molecule has a polar head and nonpolar tails (Fig. 2.4). The polar heads, being charged, are hydrophilic (water loving) and face outward, toward the cytoplasm on one side and the tissue fluid on the other side, where they will encounter a watery environment. The nonpolar tails are hydrophobic (not attracted to water) and face inward toward each other, where there is no water. When phospholipids are placed in water, they naturally form a circular bilayer because of the chemical properties of the heads and the tails. At body temperature, the phospholipid bilayer is a liquid; it has the consistency of olive oil, and the proteins are able to change their position by moving laterally. The fluid-mosaic model, a working description of membrane structure, says that the protein molecules have a changing pattern (form a mosaic) within the fluid phospholipid bilayer (Fig. 2.4).

Short chains of sugars are attached to the outer surface of some protein and lipid molecules (called glycoproteins and glycolipids, respectively). It is believed that these carbohydrate chains, specific to each cell, mark it as belonging to a particular individual and account for such characteristics as blood type or why a patient's system sometimes rejects an organ transplant. Other glycoproteins have a special configuration that allows them to act as a receptor for a chemical messenger like a hormone. Some plasma membrane proteins form channels through which certain substances can enter cells or are carriers involved in the passage of molecules through the membrane.

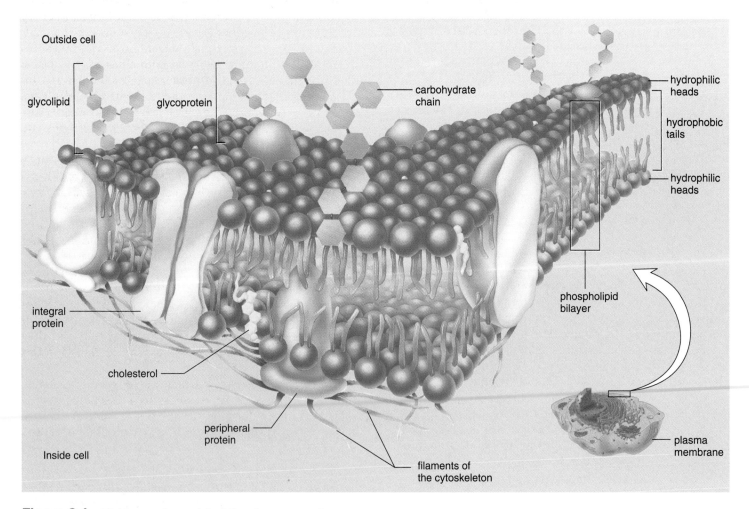

Figure 2.4   Fluid-mosaic model of the plasma membrane.
The membrane is composed of a phospholipid bilayer. The polar heads of the phospholipids are at the surfaces of the membrane; the nonpolar tails make up the interior of the membrane. Proteins are embedded in the membrane. Some of these function as receptors for chemical messengers, as conductors of molecules through the membrane, and as enzymes in metabolic reactions. Carbohydrate chains of glycolipids and glycoproteins are involved in cell-to-cell recognition.

Plasma Membrane Functions

The plasma membrane keeps a cell intact. It allows only certain molecules and ions to enter and exit the cytoplasm freely; therefore, the plasma membrane is said to be **selectively permeable.** Small molecules that are lipid soluble, such as oxygen and carbon dioxide, can pass through the membrane easily. Certain other small molecules, like water, are not lipid soluble but still cross the membrane passively by moving through a protein channel. Still other molecules and ions require the use of a carrier to enter a cell.

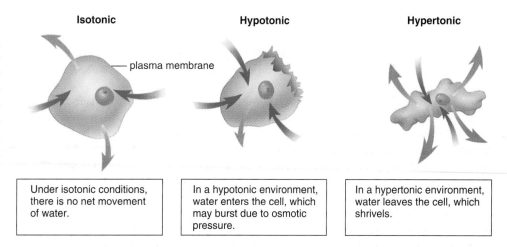

Isotonic **Hypotonic** **Hypertonic**

— plasma membrane

Under isotonic conditions, there is no net movement of water.

In a hypotonic environment, water enters the cell, which may burst due to osmotic pressure.

In a hypertonic environment, water leaves the cell, which shrivels.

Figure 2.5   Tonicity.
The arrows indicate the movement of water.

The plasma membrane, composed of phospholipid and protein molecules, is selectively permeable and regulates the entrance and exit of molecules and ions into and from the cell.

Diffusion   Diffusion is the random movement of molecules from the area of higher concentration to the area of lower concentration until they are equally distributed. To illustrate diffusion, imagine opening a perfume bottle in the corner of a room. The smell of the perfume soon permeates the room because the molecules that make up the perfume move to all parts of the room. Another example is putting a tablet of dye into water. The water eventually takes on the color of the dye as the tablet diffuses.

The chemical and physical properties of the plasma membrane allow only a few types of molecules to enter and exit a cell simply by diffusion. Lipid-soluble molecules such as alcohols can diffuse through the membrane because lipids are the membrane's main structural components. Gases can also diffuse through the lipid bilayer; this is the mechanism by which oxygen enters cells and carbon dioxide exits cells. As an example, consider the movement of oxygen from the alveoli (air sacs) of the lungs to blood in the lung capillaries. After inhalation (breathing in), the concentration of oxygen in the alveoli is higher than that in the blood; therefore, oxygen diffuses into the blood.

When molecules simply diffuse down their concentration gradients without assistance, across plasma membranes, no cellular energy is involved.

Molecules diffuse down their concentration gradients. A few types of small molecules can simply diffuse through the plasma membrane, and no carrier protein or cellular energy is involved.

Osmosis   Osmosis is the diffusion of water across a plasma membrane. It occurs whenever there is an unequal concentration of water on either side of a selectively permeable membrane. Normally, body fluids are isotonic to cells (Fig. 2.5)—there is an equal concentration of substances (solutes) and water (solvent) on both sides of the plasma membrane, and cells maintain their usual size and shape. Intravenous solutions medically administered usually have this tonicity. **Tonicity** is the degree to which a solution's concentration of solute versus water causes water to move into or out of cells.

Solutions that cause cells to swell or even to burst due to an intake of water are said to be hypotonic solutions. If red blood cells are placed in a hypotonic solution, which has a higher concentration of water (lower concentration of solute) than do the cells, water enters the cells and they swell to bursting. The term lysis is used to refer to disrupted cells; hemolysis, then, is disrupted red blood cells.

Solutions that cause cells to shrink or to shrivel due to a loss of water are said to be hypertonic solutions. If red blood cells are placed in a hypertonic solution, which has a lower concentration of water (higher concentration of solute) than do the cells, water leaves the cells and they shrink. The term crenation refers to red blood cells in this condition.

These changes have occurred due to osmotic pressure. *Osmotic pressure* is the force exerted on a selectively permeable membrane because water has moved from the area of higher to lower concentration of water (higher concentration of solute).

In an isotonic solution, a cell neither gains nor loses water. In a hypotonic solution, a cell gains water. In a hypertonic solution, a cell loses water and the cytoplasm shrinks.

Transport by Carriers   Most solutes do not simply diffuse across a plasma membrane; rather, they are transported by means of protein carriers within the membrane. During **facilitated transport,** a molecule (e.g., an amino acid or glucose) is transported across the plasma membrane from the side of higher concentration to the side of lower concentration. The cell does not need to expend energy for this type of transport because the molecules are moving down their concentration gradient.

During **active transport,** a molecule is moving contrary to the normal direction—that is, from lower to higher concentration (Fig. 2.6). For example, iodine collects in the cells of the thyroid gland; sugar is completely absorbed from the gut by cells that line the digestive tract; and sodium (Na^+) is sometimes almost completely withdrawn from urine by cells lining kidney tubules. Active transport requires a protein carrier and the use of cellular energy obtained from the breakdown of ATP. When ATP is broken down, energy is released, and in this case the energy is used by a carrier to carry out active transport. Therefore, it is not surprising that cells involved in active transport, such as kidney cells, have a large number of mitochondria near the membrane at which active transport is occurring.

Proteins involved in active transport often are called pumps because just as a water pump uses energy to move water against the force of gravity, proteins use energy to move substances against their concentration gradients. One type of pump that is active in all cells but is especially associated with nerve and muscle cells moves sodium ions (Na^+) to the outside of the cell and potassium ions (K^+) to the inside of the cell.

The passage of salt (NaCl) across a plasma membrane is of primary importance in cells. The chloride ion (Cl^-) usually crosses the plasma membrane because it is attracted by positive charged sodium ions (Na^+). First, sodium ions are pumped across a membrane; then, chloride ions simply diffuse through channels that allow their passage. Chloride ion channels malfunction in persons with cystic fibrosis, and this leads to the symptoms of this inherited (genetic) disorder.

Endocytosis and Exocytosis   During *endocytosis,* cells take in substances by vesicle formation. A portion of the plasma membrane invaginates to envelop the substance, and then the membrane pinches off to form an intracellular vesicle. During *exocytosis,* the vesicle often formed by the Golgi apparatus fuses with the plasma membrane as secretion occurs. This is the way that insulin leaves insulin-secreting cells, for instance.

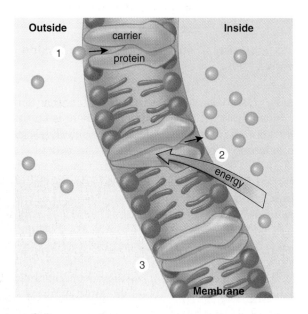

Figure 2.6   Active transport through a plasma membrane.
Active transport allows a solute to cross the membrane from lower solute concentration to higher solute concentration. (1) Molecule enters carrier. (2) Chemical energy of ATP is needed to transport the molecule which exits inside of cell. (3) Carrier returns to its former state.

Table 2.2	Passage of Molecules into and out of Cells			
	Name	**Direction**	**Requirement**	**Examples**
Passive Transport Means	DIFFUSION	Toward lower concentration	Concentration gradient	Lipid-soluble molecules, water, and gases
	FACILITATED TRANSPORT	Toward lower concentration	Carrier and concentration	Sugars and amino acids
Active Transport Means	ACTIVE TRANSPORT	Toward greater concentration	Carrier plus energy	Sugars, amino acids, and ions
	ENDOCYTOSIS	Toward inside	Vesicle formation	Macromolecules
	EXOCYTOSIS	Toward outside	Vesicle fuses with plasma membrane	Macromolecules

The Nucleus

The **nucleus,** which has a diameter of about 5 μm, is a prominent structure in the eukaryotic cell. The nucleus is of primary importance because it stores genetic information that determines the characteristics of the body's cells and their metabolic functioning. Every cell contains a complex copy of genetic information, but each cell type has certain genes, or segments of DNA, turned on, and others turned off. Activated DNA, with RNA acting as an intermediary, specifies the sequence of amino acids during protein synthesis. The proteins of a cell determine its structure and the functions it can perform.

When you look at the nucleus, even in an electron micrograph, you cannot see DNA molecules but you can see chromatin (Fig. 2.7). **Chromatin** looks grainy, but actually it is a threadlike material that undergoes coiling into rodlike structures called **chromosomes,** just before the cell divides. Chemical analysis shows that chromatin, and therefore chromosomes, contains DNA and much protein, and some RNA. Chromatin is immersed in a semifluid medium called the **nucleoplasm.** A difference in pH between the nucleoplasm and cytoplasm suggests that the nucleoplasm has a different composition.

Most likely, too, when you look at an electron micrograph of a nucleus, you will see one or more regions that look darker than the rest of the chromatin. These are nucleoli (sing., **nucleolus**) where another type of RNA, called ribosomal RNA (rRNA), is produced and where rRNA joins with proteins to form the subunits of ribosomes. (Ribosomes are small bodies in the cytoplasm that contain rRNA and proteins.)

The nucleus is separated from the cytoplasm by a double membrane known as the **nuclear envelope** which is continuous with the endoplasmic reticulum discussed on the next page. The nuclear envelope has **nuclear pores** of sufficient size (100 nm) to permit the passage of proteins into the nucleus and ribosomal subunits out of the nucleus.

The structural features of the nucleus include the following.

Chromatin:	DNA and proteins
Nucleolus:	chromatin and ribosomal subunits
Nuclear envelope:	double membrane with pores

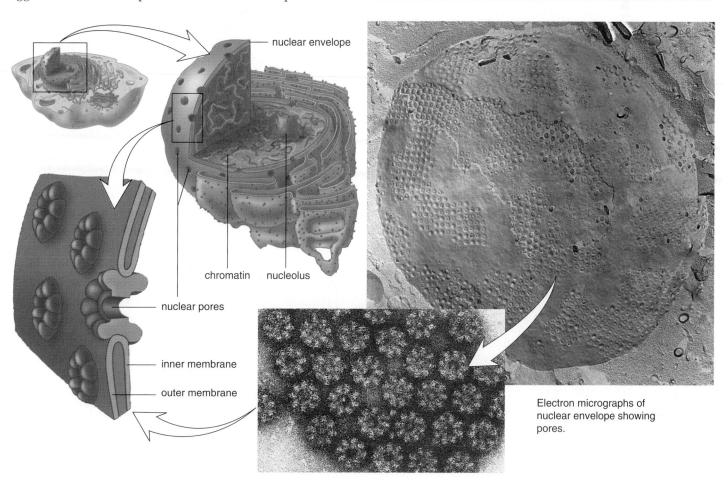

nuclear envelope

chromatin   nucleolus

nuclear pores

inner membrane

outer membrane

Electron micrographs of nuclear envelope showing pores.

Figure 2.7   The nucleus and the nuclear envelope.
The nucleoplasm contains chromatin. Chromatin has a special region called the nucleolus, which is where rRNA is produced and ribosomal subunits are assembled. The nuclear envelope, consisting of two membranes separated by a narrow space, contains pores. The electron micrographs show that the pores cover the surface of the envelope.

Ribosomes

Ribosomes are composed of two subunits, one large and one small. Each subunit has its own mix of proteins and rRNA. Protein synthesis occurs on the ribosomes. Ribosomes occur free within the cytoplasm either singly or in groups called **polyribosomes.** Ribosomes are often attached to the endoplasmic reticulum, a membranous system of saccules and channels discussed in the next section. Proteins synthesized by cytoplasmic ribosomes are used in the cell, such as in the mitochondria and chloroplasts. Those produced by ribosomes attached to endoplasmic reticulum may eventually be secreted from the cell.

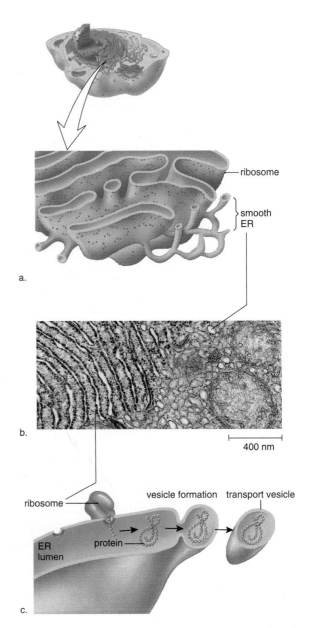

Figure 2.8   **The endoplasmic reticulum (ER).**
a. Rough ER has attached ribosomes but smooth ER does not.
b. Rough ER appears to be flattened saccules, while smooth ER is a network of interconnected tubules. **c.** A protein made on a ribosome moves into the lumen of the system and eventually is packaged in a transport vesicle for distribution inside the cell.

> Ribosomes are small organelles where protein synthesis occurs. Ribosomes occur in the cytoplasm, both singly and in groups (i.e., polyribosomes). Numerous ribosomes are attached to the endoplasmic reticulum.

Membranous Canals and Vesicles

The endomembrane system consists of the nuclear envelope, the endoplasmic reticulum, the Golgi apparatus, and several **vesicles** (tiny membranous sacs). This system compartmentalizes the cell so that particular enzymatic reactions are restricted to specific regions. Membranes that make up the endomembrane system are connected by direct physical contact and/or by the transfer of vesicles from one part to the other.

The Endoplasmic Reticulum

The **endoplasmic reticulum (ER),** a complicated system of membranous channels and saccules (flattened vesicles), is physically continuous with the outer membrane of the nuclear envelope. Rough ER is studded with ribosomes on the side of the membrane that faces the cytoplasm (Fig. 2.8). Here proteins are synthesized and enter the ER interior where processing and modification begin. Smooth ER, which is continuous with rough ER, does not have attached ribosomes. Smooth ER synthesizes the phospholipids that occur in membranes and has various other functions depending on the particular cell. In the testes, it produces testosterone, and in the liver it helps detoxify drugs. Regardless of any specialized function, smooth ER also forms vesicles in which large molecules are transported to other parts of the cell. Often these vesicles are on their way to the plasma membrane or the Golgi apparatus.

> ER is involved in protein synthesis (rough ER) and various other processes such as lipid synthesis (smooth ER). Molecules that are produced or modified in the ER are eventually enclosed in vesicles that often transport them to the Golgi apparatus.

The Golgi Apparatus

The **Golgi apparatus** is named for Camillo Golgi, who discovered its presence in cells in 1898. The Golgi apparatus consists of a stack of three to twenty slightly curved saccules whose appearance can be compared to a stack of pancakes (Fig. 2.9). In animal cells, one side of the stack (the inner face) is directed toward the ER, and the other side of the stack (the outer face) is directed toward the plasma membrane. Vesicles can frequently be seen at the edges of the saccules.

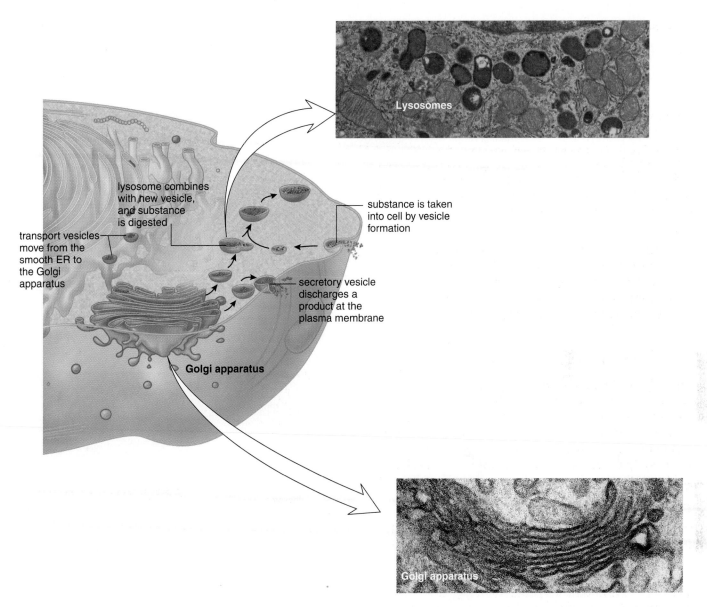

Figure 2.9   The Golgi apparatus.
The Golgi apparatus receives transport vesicles containing proteins from smooth ER. After modifying the proteins, it repackages them in either secretory vesicles or in lysosomes. When lysosomes combine with newly formed vesicles, their contents are digested. Lysosomes also break down cellular components.

The Golgi apparatus receives protein and/or lipid-filled vesicles that bud from the ER. Some biologists believe that these fuse to form a saccule at the inner face and that this saccule remains as a part of the Golgi apparatus until the molecules are repackaged in new vesicles at the outer face. Others believe that the vesicles from the ER proceed directly to the outer face of the Golgi apparatus, where processing and packaging occurs within its saccules. The Golgi apparatus contains enzymes that modify proteins and lipids. For example, it can add a chain of sugars to proteins, thereby making them glycoproteins and glycolipids, which are molecules found in the plasma membrane.

The vesicles that leave the Golgi apparatus move about the cell. Some vesicles proceed to the plasma membrane, where they discharge their contents. Because this is secretion, it is often said that the Golgi apparatus is involved in processing, packaging, and secretion. Other vesicles that leave the Golgi apparatus are lysosomes.

The Golgi apparatus processes, packages, and distributes molecules about or from the cell. It is also said to be involved in secretion.

Lysosomes

Lysosomes, vesicles produced by the Golgi apparatus, contain hydrolytic digestive enzymes. Sometimes macromolecules are brought into a cell by vesicle formation at the plasma membrane (see Fig. 2.9). When a lysosome fuses with such a vesicle, its contents are digested by lysosomal enzymes into simpler subunits that then enter the cytoplasm. Even parts of a cell are digested by its own lysosomes (called autodigestion). Normal cell rejuvenation most likely takes place in this matter, but autodigestion is also important during development. For example, when a tadpole becomes a frog, lysosomes digest away the cells of the tail. The fingers of a human embryo are at first webbed, but they are freed from one another as a result of lysosomal action.

Occasionally, a child is born with a metabolic disorder involving a missing or inactive lysosomal enzyme. In these cases, the lysosomes fill to capacity with macromolecules that cannot be broken down. The cells become so full of these lysosomes that the child dies. Someday soon it may be possible to provide the missing enzyme for these children.

Lysosomes are produced by a Golgi apparatus, and their hydrolytic enzymes digest macromolecules from various sources.

Mitochondria

Most mitochondria (sing., **mitochondrion**) are between 0.5 μm and 1.0 μm in diameter and 7 μm in length, although the size and the shape can vary. Mitochondria are bounded by a double membrane. The inner membrane is folded to form little shelves called *cristae*, which project into the *matrix*, an inner space filled with a gel-like fluid (Fig. 2.10).

Mitochondria are the site of ATP (adenosine triphosphate) production involving complex metabolic pathways. ATP molecules are the common carrier of energy in cells. A shorthand way to indicate the chemical transformation that occurs in mitochondria is as follows:

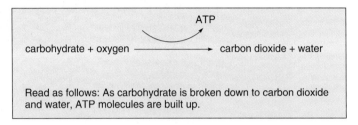

carbohydrate + oxygen ⟶ carbon dioxide + water

Read as follows: As carbohydrate is broken down to carbon dioxide and water, ATP molecules are built up.

Mitochondria are often called the powerhouses of the cell: just as a powerhouse burns fuel to produce electricity, the mitochondria convert the chemical energy of glucose products into the chemical energy of ATP molecules. In the process, mitochondria use up oxygen and give off carbon dioxide and water. The oxygen you breathe in enters cells and then mitochondria; the carbon dioxide you breathe out

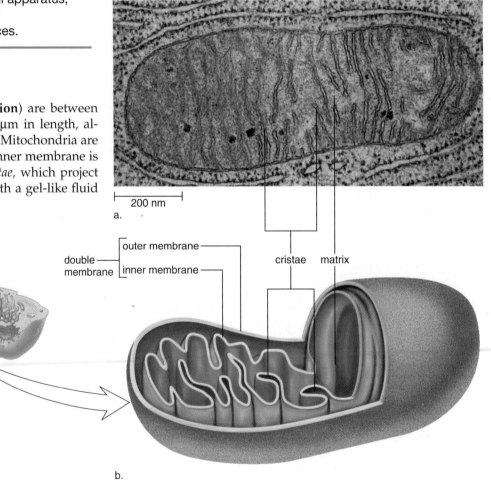

200 nm

a.

double membrane { outer membrane / inner membrane }   cristae   matrix

b.

Figure 2.10   Mitochondrion structure.
a. Electron micrograph. **b.** Generalized drawing in which the outer membrane and portions of the inner membrane have been cut away to reveal the cristae.

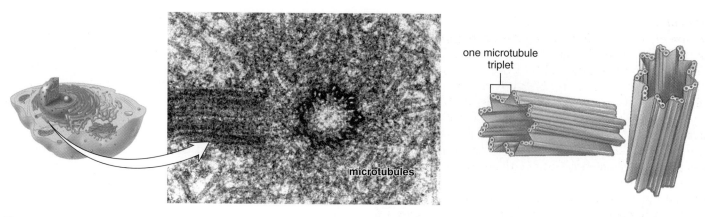

one microtubule
triplet

microtubules

Figure 2.11   Centrioles.
Centrioles are composed of nine microtubule triplets. They lie at right angles to one another within the microtubule organizing center (MTOC), which is believed to assemble microtubules at the time of cell division.

is released by mitochondria. Because oxygen is involved, it is said that mitochondria carry on aerobic cellular respiration.

The matrix of a mitochondrion contains enzymes for breaking down glucose products. ATP production then occurs at the cristae. The protein complexes that aid in the conversion of energy are located in an assembly-line fashion on these membranous shelves.

Every cell uses a certain amount of ATP energy to synthesize molecules, but many cells use ATP to carry out their specialized function. For example, muscle cells use ATP for muscle contraction, which produces movement, and nerve cells use it for the conduction of nerve impulses, which make us aware of our environment.

> Mitochondria are the sites of aerobic cellular respiration, a process that provides ATP molecules to the cell.

The Cytoskeleton

Several types of filamentous protein structures form a **cytoskeleton** that helps maintain the cell's shape and either anchors the organelles or assists their movement as appropriate. The cytoskeleton includes microtubules and actin filaments (see Fig. 2.3).

Microtubules are shaped like thin cylinders and are several times larger than actin filaments. Each cylinder contains 13 rows of tubulin, a globular protein, arranged in a helical fashion. Remarkably, microtubules can assemble and disassemble. In many cells, the regulation of microtubule assembly is under the control of a microtubule organizing center (MTOC), which lies near the nucleus. Microtubules radiate from the MTOC, helping to maintain the shape of the cell and acting as tracks along which organelles move. It is well known that during cell division, microtubules form spindle fibers, which assist the movement of chromosomes.

Actin filaments are long, extremely thin fibers that usually occur in bundles or other groupings. Actin filaments have been isolated from various types of cells, especially those in which movement occurs. Microvilli, which project from certain cells and can shorten and extend, contain actin filaments. Actin filaments, like microtubules, can assemble and disassemble.

> The cytoskeleton contains microtubules and actin filaments. Microtubules (13 rows of tubulin protein molecules arranged to form a hollow cylinder), and actin filaments (thin actin strands) maintain the shape of the cell and also direct the movement of cell parts.

Centrioles and Microtubules

In animal cells, **centrioles** are short cylinders with a 9 + 0 pattern of microtubules. There are nine outer microtubule triplets and no center microtubules (Fig. 2.11). There is always one pair of centrioles lying at right angles to one another near the nucleus. Before a cell divides, the centrioles duplicate, and the members of the new pair are also at right angles to one another. During cell division, the pairs of centrioles separate so that each daughter cell gets one pair of centrioles.

Centrioles are part of a microtubule organizing center that also includes other proteins and substances. Microtubules begin to assemble in the center, and then they grow outward, extending through the entire cytoplasm. In addition, centrioles may be involved in other cellular processes that use microtubules, such as movement of material throughout the cell or the formation of the spindle, a structure that distributes the chromosomes to daughter cells during cell division. Their exact role in these processes is uncertain, however. Centrioles also give rise to basal bodies that direct the formation of cilia and flagella.

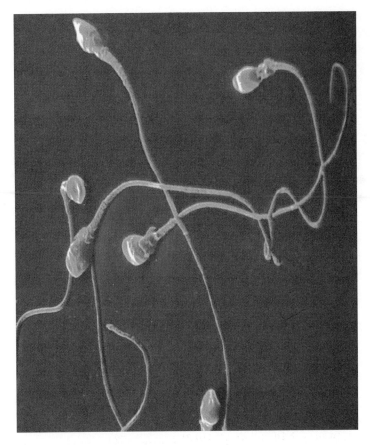

Figure 2.12   Sperm cells.
Sperm cells use long, whiplike flagella to move about.

Cilia and Flagella

Cilia and **flagella** are projections of cells that can move either in an undulating fashion, like a whip, or stiffly, like an oar. Cilia are short (2–10 μm) while flagella are longer (usually no longer than 200 μm). Cells that have these organelles are capable of self-movement or moving material along the surface of the cell. For example, sperm cells, carrying genetic material to the egg, move by means of flagella (Fig. 2.12). The cells that line our respiratory tract are ciliated. These cilia sweep debris trapped within mucus back up the throat, and this action helps keep the lungs clean.

Each cilium and flagellum has a basal body at its base, which lies in the cytoplasm. **Basal bodies,** like centrioles, have a 9 + 0 pattern of microtubule triplets. They are believed to organize the structure of cilia and flagella even though cilia and flagella have a 9 + 2 pattern of microtubules. In cilia and flagella, there are nine microtubule doublets surrounding two central microtubules. This arrangement is believed to be necessary to their ability to move.

Centrioles give rise to basal bodies that organize
the pattern of microtubules in cilia and flagella.

2.3 Cellular Metabolism

Cellular **metabolism** includes all the chemical reactions that occur in a cell. Quite often these reactions are organized into metabolic pathways.

$$\begin{array}{cccccc} 1 & 2 & 3 & 4 & 5 & 6 \end{array}$$
$$A \rightarrow B \rightarrow C \rightarrow D \rightarrow E \rightarrow F \rightarrow G$$

The letters, except A and G, are **products** of the previous reaction and the **reactants** for the next reaction. A represents the beginning reactant(s), and G represents the end product(s). The numbers in the pathway refer to different enzymes. *Every reaction in a cell requires a specific enzyme.* In effect, no reaction occurs in a cell unless its enzyme is present. For example, if enzyme number 2 in the diagram is missing, the pathway cannot function; it will stop at B. Since enzymes are so necessary in cells, their mechanism of action has been studied extensively.

Metabolic pathways contain many enzymes that
perform their reactions in a sequential order.

Enzymes and Coenzymes

When an enzyme speeds up a reaction, the reactant(s) that participate(s) in the reaction is called the enzyme's **substrate(s).** Enzymes are often named for their substrate(s) (Table 2.3). Enzymes have a specific region, called an **active site,** where the substrates are brought together so that they can react. An enzyme's specificity is caused by the shape of the active site, where the enzyme and its substrate(s) fit together in a specific way, much as the pieces of a jigsaw puzzle fit together (Fig. 2.13). After one reaction is complete, the product or products are released, and the enzyme is ready to catalyze another reaction. This can be summarized in the following manner:

$$E + S \rightarrow ES \rightarrow E + P$$

(where E = enzyme, S = substrate, ES = enzyme-substrate complex, and P = product).

Environmental conditions such as an incorrect pH or high temperature can cause an enzyme to become denatured. A denatured enzyme no longer has its usual shape and is therefore unable to speed up its reaction.

Table 2.3	Enzymes Named for their Substrates
Substrate	**Enzyme**
Lipid	Lipase
Urea	Urease
Maltose	Maltase
Ribonucleic acid	Ribonuclease
Lactose	Lactase

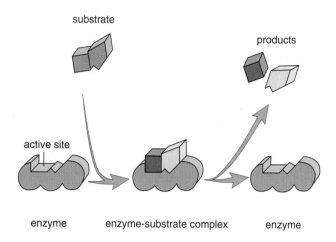

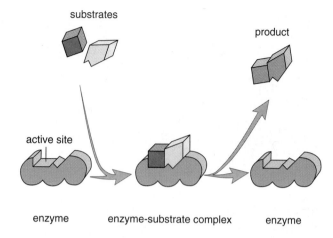

a. Degradative reaction

b. Synthetic reaction

Figure 2.13   Enzymatic action.

An enzyme has an active site where the substrates and enzyme fit together in such a way that the substrates are oriented to react. Following the reaction, the products are released.

Many enzymes require cofactors. Some cofactors are inorganic, such as copper, zinc, or iron. Some cofactors are organic, nonprotein molecules and are called **coenzymes.** These cofactors assist the enzyme and may even accept or contribute atoms to the reaction. It is interesting that vitamins are often components of coenzymes. The vitamin niacin is a part of the coenzyme NAD, which removes hydrogen (H) atoms from substrates and therefore is called a **dehydrogenase.** Hydrogen atoms are sometimes removed by NAD as molecules are broken down. NAD that is carrying hydrogen atoms is written as $NADH_2$ because NAD removes two hydrogen atoms at a time. As we shall see, the removal of hydrogen atoms releases energy that can be used for ATP buildup.

Enzymes are specific because they have an active site that accommodates their substrates. Enzymes often have organic, nonprotein helpers called coenzymes. NAD is a dehydrogenase, a coenzyme that removes hydrogen from substrates.

Cellular Respiration and Metabolic Pathways

Cellular respiration is an important part of cellular metabolism because it accounts for ATP buildup in cells. Cellular respiration includes *aerobic* (requires oxygen) cellular respiration and fermentation, an *anaerobic* (does not require oxygen) process. During both processes, the energy released as glucose breakdown occurs is used to build up ATP molecules, the common energy carrier in cells.

Aerobic Cellular Respiration

During **aerobic cellular respiration,** glucose is broken down to carbon dioxide and water. Even though it is possible to write an overall equation for the process, aerobic cellular respiration does not occur in one step. Glucose breakdown requires three subpathways: *glycolysis,* the *Krebs cycle,* and the *electron transport system.* The location of these subpathways is as follows:

Glycolysis—occurs in the cytoplasm, outside a mitochondrion

Krebs cycle—occurs in the matrix of a mitochondrion

Electron transport system—occurs on the cristae of mitochondrion

Glycolysis and the Krebs cycle are a series of reactions in which the product of the previous reaction becomes the substrate for the next reaction. Every reaction that occurs during glycolysis and the Krebs cycle requires a specific enzyme. Each pathway resembles a conveyor belt in which a beginning substrate continuously enters at the start and, after a series of reactions, end products leave at the termination of the belt. It is important to realize, too, that these two pathways and the electron transport system occur at the same time. They can be compared to the inner workings of a watch, in which all parts are synchronized.

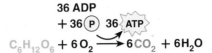

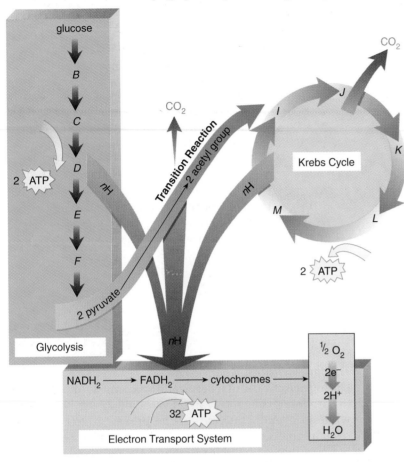

Figure 2.14   **Aerobic cellular respiration.**
The overall reaction shown at the top actually requires three subpathways: glycolysis, the Krebs cycle, and the electron transport system. As the reactions occur, a number of hydrogen (H) atoms and carbon dioxide (CO_2) molecules are removed from the various substrates. Oxygen (O_2) acts as the final acceptor for hydrogen atoms ($2e^- + 2H^+$) and becomes water (H_2O).

The reactants of aerobic cellular respiration, namely glucose and oxygen, and products, namely carbon dioxide and water, are related to the subpathways in the manner described next.

1. Glucose, a C_6 molecule, is to be associated with **glycolysis,** the breakdown of glucose to two molecules of pyruvate (pyruvic acid), a C_3 molecule. During glycolysis, energy is released as hydrogen (H) atoms are removed. This energy is used to form two ATP molecules (Fig. 2.14).
2. Carbon dioxide, CO_2, is to be associated with the transition reaction and the Krebs cycle, both of which occur in mitochondria. During the transition reaction, pyruvate is converted to a C_2 acetyl group after CO_2 comes off. Because the transition reaction occurs twice per glucose molecule, two molecules of CO_2 are released. Hydrogen (H) atoms are also removed at this time.

   The acetyl group enters the **Krebs cycle,** a cyclical series of reactions that give off two CO_2 molecules and produce one ATP molecule. Since the Krebs cycle occurs twice per glucose molecule, altogether four CO_2 and two ATP are produced per glucose molecule. Hydrogen (H) atoms are removed from the substrates and added to NAD, forming $NADH_2$ as the Krebs cycle occurs.
3. Oxygen, O_2, and water, H_2O, are to be associated with the electron transport system. The **electron transport system** begins with $NADH_2$, the coenzyme that carries most of the hydrogen (H) atoms to the system, but after that it consists of molecules that carry electrons. High energy electrons are removed from the hydrogen atoms, leaving behind hydrogen ions (H^+), and then the electrons are passed from one molecule to another until the electrons are received by an oxygen atom. At this point, $2H^+$ combine with an oxygen to give water. As the electrons are passed down the system, their energy is released to allow the buildup of ATP.
4. ATP is to be associated with glycolysis, the Krebs cycle, and the electron transport system. Altogether, 36 ATP result from the breakdown of one glucose molecule (Table 2.4).

Table 2.4	Overview of Aerobic Cellular Respiration
Name of Pathway	**Result**
Glycolysis	Removal of H from substrates produces [2 ATP]
Transition reaction	Removal of H from substrates releases **2 CO2**
Krebs cycle	Removal of H from substrates releases **4 CO2**
	Produces [2 ATP] after 2 turns
Electron transport system	Accepts H from other pathways and passes electrons on to O_2, producing **H2O**
	Produces s [32 ATP]

Aerobic cellular respiration requires glycolysis, which takes place in the cytoplasm; the Krebs cycle, which is located in the matrix of the mitochondria; and the electron transport system, which is located on the cristae of the mitochondria.

Fermentation

Fermentation is an anaerobic process. When oxygen is not available to cells, the electron transport system soon becomes inoperative because oxygen is not present to accept electrons. In this case, most cells have a safety valve so that some ATP can still be produced. Glycolysis operates as long as it is supplied with "free" NAD; that is, NAD that can pick up hydrogen atoms. Normally, $NADH_2$ takes hydrogens to the electron transport system and thereby becomes "free" of hydrogen atoms. However, if the system is not working due to lack of oxygen, $NADH_2$ passes its hydrogen atoms to pyruvate as shown in the following reaction:

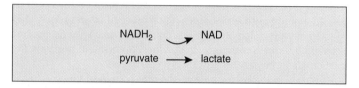

The Krebs cycle and electron transport system do not function as part of fermentation. When oxygen is available again,

lactate (lactic acid) can be converted back to pyruvate, and metabolism can proceed as usual.

Fermentation takes less time than aerobic cellular respiration, but since glycolysis alone is occurring, it produces only 2 ATP per glucose molecule. Also, fermentation results in the buildup of lactate. Lactate is toxic to cells and causes muscles to cramp and fatigue. If fermentation continues for any length of time, death follows.

It is of interest to know that fermentation takes its name from yeast fermentation. Yeast fermentation produces alcohol and carbon dioxide (instead of lactate). When yeast is used to leaven bread, it is the carbon dioxide that produces the desired effect. When yeast is used to produce alcoholic beverages, it is the alcohol that humans make use of.

> Fermentation is an anaerobic process, a process that does not require oxygen but produces very little ATP per glucose molecule and results in lactate or alcohol and carbon dioxide buildup.

Bioethical Issue

As the cell theory tells us, cells can't be manufactured. Therefore, unlike medications such as certain hormones and vaccines, which are now biotechnology products, the only way to get human cells is from previous human cells. No wonder, then, there is a market in blood cells which are easily accessible. Viacord is in the business of storing umbilical blood for possible future use by the parents of a newborn. It only costs an initial outlay of $2,500 and a yearly fee of $95.

The blood in a baby's umbilical cord is rich in stem cells, which give rise to all the other types of blood cells within an adult's red bone marrow. Should the parent's or the child's red bone marrow be destroyed by disease or cancer treatment, these stem cells will bring the red bone

marrow back to life. Viacord says that stem-cell banking is a kind of insurance, but there is no doubt it capitalizes on our fear of future illness.

The banking of umbilical cord blood is a new idea. Suppose you are an elderly cancer patient, could you buy stem cells from someone else? The U.S. National Organ Transplant Act (NOTA) bans interstate commerce in the sale of transplant organs but excludes tissues such as blood or sperm, which are replenishable. The answer, then, is most likely yes, especially if the parents now find themselves in need of cash.

Where would you draw the line in the traffic of human parts? In 1984, John Moore sued his doctor for using, without his consent, tissue from his cancerous spleen to create a commercial cell line now valued at

about $3 billion in profits. The California Supreme Court ruled against him because it might deter the work of research scientists. Was that right and proper?

Questions

1. Should there be any restrictions in the buying and selling of transplant organs, blood, or blood products? Why or why not?
2. Who should have the rights to tissues that are used to start up commercial cell lines—the person who donated the cells or the investigator who established the line, or both?
3. Who should get to use stored umbilical blood? Only family members, or also a paying customer? After all, some parents have several children.

Summarizing the Concepts

2.1 Cell Size
Cells are quite small, and it usually takes a microscope to see them. Small cubes, like cells, have a more favorable surface/volume ratio than do large cubes. Only inactive eggs are large enough to be seen by the naked eye; once development begins, cell division results in small-size cells.

2.2 Cellular Organization
A cell is surrounded by a plasma membrane, which regulates the entrance and exit of molecules and ions. Some molecules, such as water and gases, diffuse through the membrane. The direction in which water diffuses is dependent on its concentration within the cell compared to outside the cell.

Table 2.1 lists the cell organelles we have studied in the chapter. The nucleus is a large organelle of primary importance because it controls the rest of the cell. Within the nucleus lies the chromatin, which condenses to become chromosomes during cell division.

Proteins are made at the rough ER before being modified and packaged by the Golgi apparatus into vesicles for secretion. During secretion, a vesicle discharges its contents at the plasma membrane. Golgi-derived lysosomes fuse with incoming vesicles to digest any material enclosed within, and lysosomes also carry out autodigestion of old parts of cells.

Mitochondria are the powerhouses of the cell. During the process of aerobic cellular respiration, mitochondria convert carbohydrate energy to ATP energy.

Microtubules and actin filaments make up the cytoskeleton, which maintains the cell's shape and permits movement of cell parts. Centrioles are a part of the microtubule organizing center, which is associated with the formation of microtubules in general and the spindle that appears during cell division. Centrioles also produce basal bodies that give rise to cilia and flagella.

2.3 Cellular Metabolism
Cellular metabolism is the sum of all biochemical pathways of the cell. In a pathway, a series of reactions proceed in an orderly step-by-step manner. Each of these reactions requires a specific enzyme. Sometimes enzymes require coenzymes, nonprotein portions that participate in the reaction. NAD is a coenzyme.

Aerobic cellular respiration (the breakdown of glucose to carbon dioxide and water) includes three pathways: glycolysis, the Krebs cycle, and the electron transport system. If oxygen is not available in cells, the electron transport system is inoperative, and fermentation (an anaerobic process) occurs. Fermentation makes use of glycolysis only, plus one more reaction in which pyruvate is reduced to lactate.

Studying the Concepts

1. Describe the structure and biochemical makeup of a plasma membrane.   46
2. What are three mechanisms by which substances enter and exit cells? Define isotonic, hypertonic, and hypotonic solutions.   47–48
3. Describe the nucleus and its contents, including the terms DNA and RNA in your description.   49
4. Describe the structure and function of endoplasmic reticulum. Include the terms rough and smooth ER and ribosomes in your description.   50
5. Describe the structure and function of the Golgi apparatus and its relationship to vesicles and lysosomes.   50–51
6. Describe the structure of mitochondria, and relate this structure to the pathways of aerobic cellular respiration.   52–53
7. Describe the composition of the cytoskeleton.   53
8. Describe the structure and function of centrioles, cilia, and flagella.   53–54
9. Discuss and draw a diagram for a metabolic pathway. Discuss and give a reaction to describe the specificity theory of enzymatic action. Define coenzyme.   54–55
10. Name and describe the events within the three subpathways that make up aerobic cellular respiration. Why is fermentation necessary but potentially harmful to the human body?   55–57

Testing Your Knowledge of the Concepts

In questions 1–4, match the organelles to the functions below.
Key:
a. mitochondria
b. nucleus
c. Golgi apparatus
d. rough ER

_____ 1. packaging and secretion
_____ 2. powerhouses of the cell
_____ 3. protein synthesis
_____ 4. control center for cell

In questions 5–7, indicate whether the statement is true (T) or false (F).

_____ 5. Microtubules and actin filaments are a part of the cytoskeleton, the framework of the cell that provides its shape and regulates movement of organelles.

_____ 6. Water enters a cell when it is placed in a hypertonic solution.

_____ 7. Substrates react at the active site, located on the surface of their enzyme.

In questions 8 and 9, fill in the blanks.

8. During aerobic cellular respiration, most of the ATP molecules are produced at the _____, a series of carriers located on the _____ of mitochondria.

9. Fermentation of a glucose molecule produces only _____ ATP compared to the _____ ATP produced by aerobic cellular respiration.

10. Label only the parts of the cell that are involved in protein synthesis and modification. Explain your choices.

11. Complete the following diagram by labeling the pathways and by adding ATP, CO_2, O_2, and H_2O where needed.

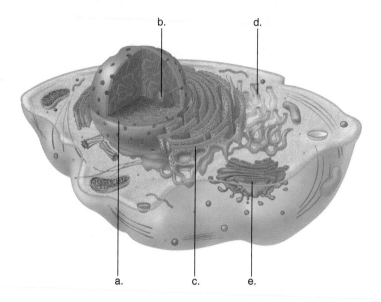

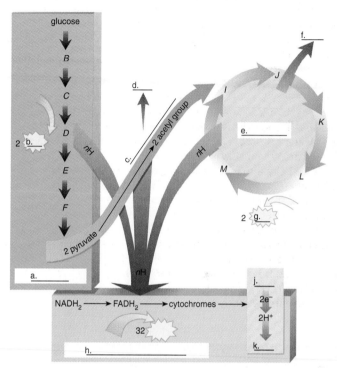

Applying Your Knowledge to the Concepts

These questions pertain to cell structure and function.

1. In examining the constituents of the plasma membrane of two individual people, which of the following constituents would probably show the greatest difference between the two: phospholipid, protein, glycoprotein, cholesterol? Give the reasoning for your answer.

2. Under certain pathological conditions, the cell digests itself; this is referred to as autolysis. Which specific cellular organelle is probably involved in this process? Explain.

3. A microtubule is a hollow cylinder composed of 13 rows of protein molecules. What evidence do you have each row is not an actin filament?

4. The process of cellular respiration results in 36 ATP, and the process of fermentation results in 2 ATP. Where is the rest of the chemical energy following fermentation?

Understanding the Terms

active site 54
active transport 48
aerobic cellular respiration 55
basal body 54
cell 42
cell theory 42
centriole 53
chromatin 49
chromosome 49
cilium 54
coenzyme 55
cytoplasm 44
cytoskeleton 44
dehydrogenase 55
diffusion 47
electron transport system 56
endoplasmic reticulum (ER) 50
facilitated transport 48
fermentation 57
flagellum 54
glycolysis 56
Golgi apparatus 50

hypotonic 47
Krebs cycle 56
lysosome 52
metabolism 54
microtubule 53
mitochondrion 52
nuclear envelope 49
nuclear pore 49
nucleolus 49
nucleoplasm 49
nucleus 44
organelle 44
osmosis 47
plasma membrane 44
polyribosome 50
product 54
reactant 54
ribosome 50
selectively permeable 47
substrate 54
tonicity 47
vesicle 50

Match the terms to these definitions:

a. _____ Metabolic pathway found in the cytoplasm that participates in aerobic cellular respiration and fermentation; it converts glucose to two molecules of pyruvate.

b. _____ Movement of molecules from a region of higher concentration to a region of lower concentration.

c. _____ Region on the surface of an enzyme where the substrate binds and where the reaction occurs.

d. _____ Organelle consisting of concentrically folded saccules that functions in the packaging, storage, and distribution of cellular products.

e. _____ Membranous system of tubules, vesicles, and sacs in cells, sometimes having attached ribosomes.

Applying Technology to the Concepts

Your study of cell structure and function is supported by these available technologies:

Essential Study Partner CD-ROM

Cells → Cell Structure → Cell Membrane

Visit the Mader web site for related ESP activities.

Exploring the Internet

The Mader Home Page provides resources and tools as you study this chapter.

http://www.mhhe.com/biosci/genbio/mader

Dynamic Human 2.0 CD-ROM

Human Body → Clinical Concepts
   → Cell Components
   → Cell Shapes

Virtual Physiology Laboratory CD-ROM

Diffusion, Osmosis, & Tonicity
Enzyme Characteristics

Life Science Animations 3D Video

4 Diffusion
5 Osmosis
7 Enzyme Action
9 Electron Transport Chain

Chapter 3

Introduction to Homeostasis

Figure 3.1 **Human organization.**
The nervous system coordinates the other systems of the body so that we respond to outside stimuli, including those from the opposite sex.

Ever wonder why a touch on the arm can send you blushing, your heart racing with swift pleasure (Fig. 3.1)? Biochemists sometimes joke that love—or the physical rush accompanying it—is really just a complex set of reactions. It's not a romantic view. Technically, though, it's true.

The coveted reactions of love begin when a light pressure on your skin is detected by receptors that send messages racing to your brain. There, a cascade of messages bounce back. The brain's emotional centers can bring about all sorts of reactions; everything from a quickened heartbeat to sweaty palms. The nervous system is in communication with the endocrine system. The look a lover gives you, the sound of a lover's voice and yes, a touch, can all start hormones racing through your body. Did you ever notice that you sometimes feel a wave of love come over you? Some investigators call one hormone, oxytocin, the love hormone and say that it is responsible for a turned-on feeling. Reactions to feelings of love by your brain can also have visible effects—your muscles can twitch or your voice can break, even as you try to act normally. When you begin to make love, oxytocin is released throughout your body, making your nerves more sensitive to pleasure and helping to account for orgasm. Some might say such a complex set of reactions caused by the nervous and endocrine systems is the best example of coordination in humans we have to offer.

In order to understand how coordination is carried out, we must first understand the organization of the body. Each system contains a number of organs and each organ is composed of different types of tissues. This chapter reviews the structure and function of tissues and shows how they are organized within the skin, an example of an organ. In a square inch of skin, there are 1,300 receptors that communicate with the brain and spinal cord, enabling us to be sensitive to a lover's touch, among other stimuli.

Since the nervous and endocrine systems coordinate the other systems of the body, they play a pivotal role in maintaining homeostasis, the dynamic equilibrium of the internal environment. All systems of the body contribute to homeostasis, from the digestive system, which provides nutrient molecules, to the muscular system, which enables us to bring food to the mouth. This chapter introduces unique pages which appear throughout the text. These pages tell how the systems of the body help each other maintain homeostasis. ⌇

3.1 Types of Tissues

A **tissue** is composed of similarly specialized cells that perform a common function in the body. The tissues of the human body can be categorized into four major types: *epithelial tissue,* which covers body surfaces and lines body cavities; *connective tissue,* which binds and supports body parts; *muscular tissue,* which moves body parts; and *nervous tissue,* which receives stimuli and conducts impulses from one body part to another.

Cancers are classified according to the type of tissue from which they arise. **Carcinomas,** the most common type, are cancers of epithelial tissues; sarcomas are cancers arising in muscle or connective tissue (especially bone or cartilage); leukemias are cancers of the blood; and lymphomas are cancers of lymphoid tissue. The chance of developing cancer in a particular tissue shows a positive correlation to the rate of cell division; new blood cells arise at a rate of 2,500,000 cells per second, and epithelial cells also reproduce at a high rate.

Epithelial Tissue

Epithelial tissue, also called epithelium, consists of tightly packed cells that form a continuous layer or sheet lining the entire body surface and most of the body's inner cavities. On the external surface, it protects the body from injury, drying out, and possible **pathogen** (virus and bacterium) invasion. On internal surfaces, epithelial tissue may be specialized for other functions in addition to protection. For example, epithelial tissue secretes mucus along the digestive tract and sweeps up impurities from the lungs by means of cilia (sing., **cilium**). It efficiently absorbs molecules from kidney tubules and from the intestine because of minute cellular extensions called **microvilli.**

There are three types of epithelial tissue (Fig. 3.2). **Squamous epithelium** is *composed of flattened cells* and is found lining the lungs and blood vessels. **Cuboidal epithelium** contains *cube-shaped cells* and is found lining the kidney tubules. **Columnar epithelium** has cells *resembling rectangular pillars* or *columns,* and nuclei are usually located near the bottom of each cell. This epithelium is found lining the digestive tract. Ciliated columnar epithelium is found lining the oviducts, where it propels the egg toward the uterus or womb.

An epithelium can be simple or stratified. Simple means the tissue has a single layer of cells, and stratified means that the tissue has layers of cells piled one on top of the other. The walls of the smallest blood vessels, called capillaries, are composed of a single layer of epithelial cells. The permeability of capillaries allows exchange of substances between the blood and tissue cells. The nose, mouth, esophagus, anal canal, and vagina are all lined by stratified squamous epithelium. As we shall see, the outer layer of skin is also stratified squamous epithelium, but the cells have been reinforced by keratin, a protein that provides strength.

Pseudostratified epithelium appears to be layered; however, true layers do not exist because each cell touches the base line. The lining of the windpipe, or trachea, is called *pseudostratified ciliated columnar epithelium.* A secreted covering of mucus traps foreign particles, and the upward motion of the cilia carries the mucus to the back of the throat, where it may either be swallowed or expectorated. Smoking can cause a change in mucus secretion and inhibit ciliary action, and the result is a chronic inflammatory condition called bronchitis.

A so-called **basement membrane** often joins an epithelium to underlying connective tissue. We now know that the basement membrane is glycoprotein, reinforced by fibers that are supplied by connective tissue.

An epithelium sometimes secretes a product, in which case it is described as glandular. A **gland** can be a single epithelial cell, as in the case of mucus-secreting goblet cells found within the columnar epithelium lining the digestive tract, or a gland can contain many cells. Glands that secrete their product into ducts are called *exocrine glands,* and those that secrete their product directly into the bloodstream are called *endocrine glands,* and those that secrete their product directly into the bloodstream are called *endocrine glands.* The pancreas is both an exocrine gland, because it secretes digestive juices into the small intestine via ducts, and an endocrine gland, because it secretes insulin into the bloodstream.

Epithelial tissue is named according to the shape of the cell. These tightly packed protective cells can occur in more than one layer, and the cells lining a cavity can be ciliated and/or glandular.

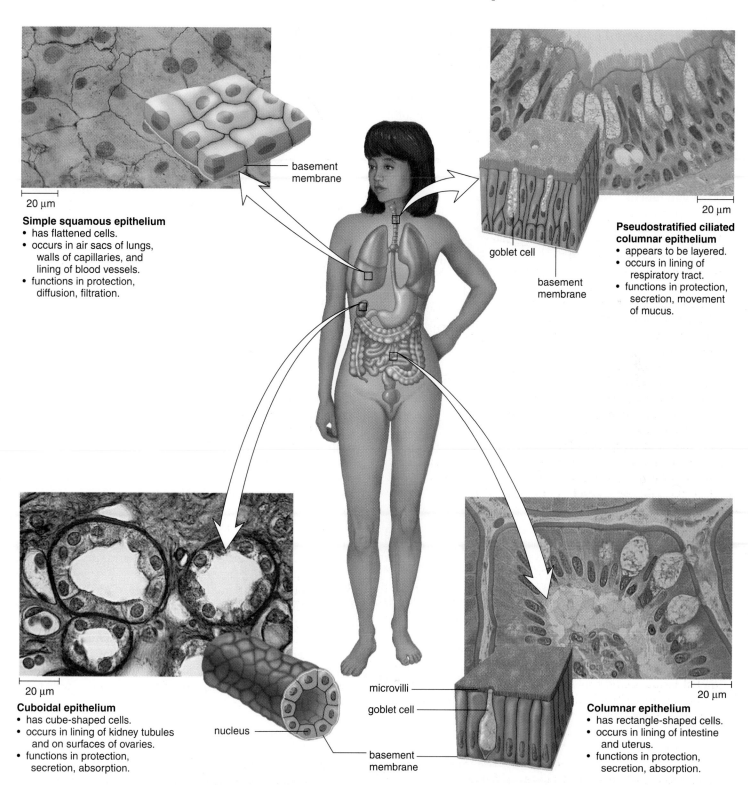

20 µm

Simple squamous epithelium
- has flattened cells.
- occurs in air sacs of lungs, walls of capillaries, and lining of blood vessels.
- functions in protection, diffusion, filtration.

basement membrane

goblet cell

basement membrane

20 µm

Pseudostratified ciliated columnar epithelium
- appears to be layered.
- occurs in lining of respiratory tract.
- functions in protection, secretion, movement of mucus.

20 µm

Cuboidal epithelium
- has cube-shaped cells.
- occurs in lining of kidney tubules and on surfaces of ovaries.
- functions in protection, secretion, absorption.

nucleus

microvilli

goblet cell

basement membrane

20 µm

Columnar epithelium
- has rectangle-shaped cells.
- occurs in lining of intestine and uterus.
- functions in protection, secretion, absorption.

Figure 3.2 Epithelial tissue.
The three types of epithelial tissue—squamous, cuboidal, and columnar—are named for the shape of their cells. They all have a protective function, as well as the other functions noted.

Junctions Between Cells

The cells of a tissue can function in a coordinated manner when the plasma membranes of adjoining cells interact. The junctions that occur between cells help cells function as a tissue (Fig. 3.3). A **tight junction** forms an impermeable barrier because adjacent plasma membrane proteins actually join, producing a zipperlike fastening. In the intestine, the gastric juices stay out of the body, and in the kidneys, the urine stays within kidney tubules because epithelial cells are joined by tight junctions.

A **gap junction** forms when two adjacent plasma membrane channels join. This lends strength, but it also allows ions, sugars, and small molecules to pass between the two cells. Gap junctions in heart and smooth muscle ensure synchronized contraction. In an **adhesion junction** (desmosome), the adjacent plasma membranes do not touch but are held together by intercellular filaments firmly attached to buttonlike thickenings. In some organs—like the heart, stomach, and bladder, where tissues get stretched—adhesion junctions hold the cells together.

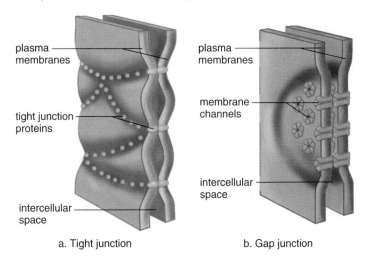

a. Tight junction   b. Gap junction

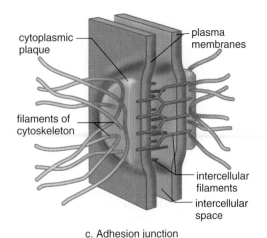

c. Adhesion junction

Figure 3.3   **Junctions between epithelial cells.**
Epithelial tissue cells are held tightly together by **a.** tight junctions; **b.** gap junctions that allow materials to pass from cell to cell; and **c.** adhesion junctions that allow tissues to stretch.

Connective Tissue

Connective tissue binds organs together, provides support and protection, fills spaces, produces blood cells, and stores fat. As a rule, connective tissue cells are widely separated by a **matrix,** consisting of a noncellular material that varies in consistency from solid to semifluid to fluid. The matrix may have fibers of three possible types: **Collagen** (white) **fibers** contain collagen, a protein that gives them flexibility and strength. **Reticular fibers** are very thin collagen fibers that are highly branched and form delicate supporting networks. **Elastic** (yellow) **fibers** contain elastin, a protein that is not as strong as collagen but is more elastic.

Loose Fibrous and Dense Fibrous Tissues

Both loose fibrous and dense fibrous connective tissues have cells called **fibroblasts** that are located some distance from one another and are separated by a jellylike matrix containing white collagen fibers and yellow elastic fibers.

Loose fibrous connective tissue supports epithelium and also many internal organs (Fig. 3.4*a*). Its presence in lungs, arteries, and the urinary bladder allows these organs to expand. It forms a protective covering enclosing many internal organs, such as muscles, blood vessels, and nerves.

Dense fibrous connective tissue contains many collagen fibers that are packed together. This type of tissue has more specific functions than does loose connective tissue. For example, dense fibrous connective tissue is found in **tendons,** which connect muscles to bones, and in **ligaments,** which connect bones to other bones at joints.

Adipose Tissue and Reticular Connective Tissue

In **adipose tissue** (Fig. 3.4*b*), the fibroblasts enlarge and store fat. The body uses this stored fat for energy, insulation, and organ protection. Adipose tissue is found beneath the skin, around the kidneys, and on the surface of the heart. Reticular connective tissue, also called lymphoid tissue, is present in lymph nodes, the spleen, and the bone marrow. These organs are a part of the immune system because they store and/or produce white blood cells, particularly lymphocytes. All types of blood cells are produced in red bone marrow.

Cartilage

In **cartilage,** the cells lie in small chambers called lacunae (sing., **lacuna**), separated by a matrix that is solid yet flexible. Unfortunately, because this tissue lacks a direct blood supply, it heals very slowly. There are three types of cartilage, distinguished by the type of fiber in the matrix.

Hyaline cartilage (Fig. 3.4*c*), the most common type of cartilage, contains only very fine collagen fibers. The matrix has a white, translucent appearance. Hyaline cartilage is found in the nose and at the ends of the long bones and the ribs, and it forms rings in the walls of respiratory passages. The fetal skeleton also is made of this type of cartilage. Later, the cartilaginous fetal skeleton is replaced by bone.

Elastic cartilage has more elastic fibers than hyaline cartilage. For this reason, it is more flexible and is found, for example, in the framework of the outer ear.

Fibrocartilage has a matrix containing strong collagen fibers. Fibrocartilage is found in structures that withstand tension and pressure, such as the pads between the vertebrae in the backbone and the wedges found in the knee joint.

Bone

Bone is the most rigid connective tissue. It consists of an extremely hard matrix of inorganic salts, chiefly calcium salts, deposited around protein fibers, especially collagen fibers. The inorganic salts give bone rigidity, and the protein fibers provide elasticity and strength, much as steel rods do in reinforced concrete.

Compact bone makes up the shaft of a long bone (Fig. 3.4*d*). It consists of cylindrical structural units called osteons (Haversian systems). The central canal of each osteon

is surrounded by rings of hard matrix. Bone cells, called *osteocytes*, are located in spaces called lacunae between the rings of matrix. Blood vessels in the central canal carry nutrients that allow bone to renew itself. The nutrients can reach all of the cells because *canaliculi* (minute canals) containing thin processes of the osteocytes connect the cells with one another and with the central canals.

The ends of a long bone contain spongy bone, which has an entirely different structure. **Spongy bone** contains numerous bony bars and plates, separated by irregular spaces. Although lighter than compact bone, spongy bone still is designed for strength. Just as braces are used for support in buildings, the solid portions of spongy bone follow lines of stress.

Connective tissues, which bind and support body parts, differ according to the type of matrix and the abundance of fibers in the matrix.

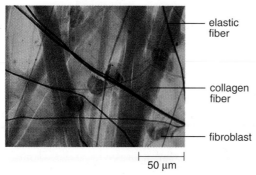

a. Loose fibrous connective tissue
• has space between components.
• occurs beneath skin and most epithelial layers.
• functions in support and binds organs.

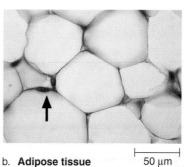

b. Adipose tissue
• cells are filled with fat.
• occurs beneath skin, around organs and heart.
• functions in insulation, stores fat.

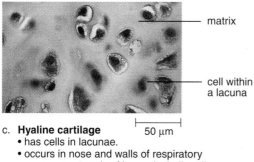

c. Hyaline cartilage
• has cells in lacunae.
• occurs in nose and walls of respiratory passages; at ends of bones including ribs.
• functions in support and protection.

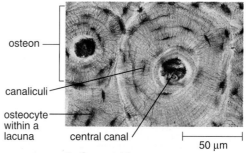

d. Compact bone
• has cells in concentric rings.
• occurs in bones of skeleton.
• functions in support and protection.

Figure 3.4   Connective tissue examples.
a. In loose connective tissue, cells called fibroblasts are separated by a jellylike matrix, which contains both collagen and elastic fibers. **b.** Adipose tissue cells have nuclei (arrow) pushed to one side because the cells are filled with fat. **c.** In hyaline cartilage, the flexible matrix has a white, translucent appearance. **d.** In compact bone, the hard matrix contains calcium salts. Concentric rings of osteocytes in lacunae form an elongated cylinder called an osteon (Haversian system). An osteon has a central canal that contains blood vessels and nerve fibers.

Blood ⚚

The internal environment of the body consists of blood and tissue fluid. The systems of the body help keep blood composition and chemistry within normal limits and blood in turn creates tissue fluid. Blood transports nutrients and oxygen to tissue fluid and removes carbon dioxide and other wastes. It helps distribute heat and also plays a role in fluid, ion, and pH balance. Various components of blood, as discussed below, help protect us from disease, and its ability to clot prevents fluid loss.

If blood is transferred from a person's vein to a test tube and prevented from clotting, it separates into two layers (Fig. 3.5). The upper liquid layer, called **plasma,** represents about 55% of the volume of whole blood and contains a variety of inorganic and organic substances dissolved or suspended in water (Table 3.1). The lower layer consists of red blood cells (erythrocytes), white blood cells (leukocytes), and blood platelets (thrombocytes). Collectively, these are called the formed elements and represent about 45% of the volume of whole blood. Formed elements are manufactured in the red bone marrow of the skull, ribs, vertebrae, and ends of long bones.

The **red blood cells** are small, biconcave, disk-shaped cells without nuclei. The presence of the red pigment hemoglobin makes the cells red, and in turn, makes the blood red. Hemoglobin is composed of four units; each is composed of the protein globin and a complex iron-containing structure called heme. The iron forms a loose association with oxygen, and in this way red blood cells transport oxygen.

White blood cells may be distinguished from red blood cells by the fact that they are usually larger, have a nucleus, and without staining would appear to be translucent. White blood cells characteristically appear bluish because they have been stained that color. White blood cells, which fight infection, function primarily in two ways. Some white blood cells are phagocytic and engulf infectious pathogens, while other white blood cells produce antibodies, molecules that combine with foreign substances to inactivate them.

Platelets are not complete cells; rather, they are fragments of giant cells present only in bone marrow. When a blood vessel is damaged, platelets form a plug that seals the vessel and along with injured tissues release molecules that help the clotting process.

Blood is unlike other types of connective tissue in that the matrix (i.e., plasma) is not made by the cells. Some people do not classify blood as connective tissue; instead, they suggest a separate tissue category for blood called vascular tissue.

Blood is a connective tissue in which the matrix is plasma.

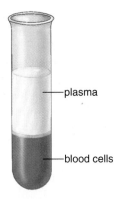

a. Blood sample

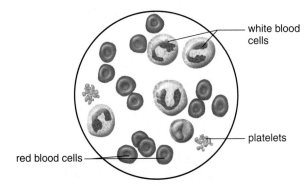

b. Blood smear

Figure 3.5   **Blood, a fluid tissue.**
a. In a test tube, a blood sample separates into its two components: blood cells and plasma. **b.** Microscopic examination of a blood smear shows that there are red blood cells, white blood cells, and platelets. Platelets are fragments of a cell. Red blood cells transport oxygen, white blood cells fight infections, and platelets are involved in initiating blood clotting.

Table 3.1	Blood Plasma
Water (92% of Total)	
Solutes (8% of Total)	
Inorganic ions (salts)	Na^+, Ca^{2+}, K^+, Mg^{2+}, Cl^-, HCO_3^-, HPO_4^{2+}, SO_4^{2+}
Gases	O_2, CO_2
Plasma proteins	Albumin, globulins, fibrinogen
Organic nutrients	Glucose, fats, phospholipids, amino acids, etc.
Nitrogenous waste products	Urea, ammonia, uric acid
Regulatory substances	Hormones, enzymes

Muscular Tissue

Muscular (contractile) tissue is composed of cells that are called muscle fibers. Muscle fibers contain actin filaments and myosin filaments, whose interaction accounts for movement. There are three types of vertebrate muscles: skeletal, smooth, and cardiac.

Skeletal muscle, also called voluntary muscle (Fig. 3.6a), is attached by tendons to the bones of the skeleton, and when it contracts, body parts move. Contraction of skeletal muscle is under voluntary control and occurs faster than in the other muscle types. Skeletal muscle fibers are cylindrical and quite long—sometimes they run the length of the muscle. They arise during development when several cells fuse, resulting in one fiber with multiple nuclei. The nuclei are located at the periphery of the cell, just inside the plasma membrane. The fibers have alternating light and dark bands that give them a **striated** appearance. These bands are due to the placement of actin filaments and myosin filaments in the cell.

Smooth (visceral) muscle is so named because the cells lack striations. The spindle-shaped cells form layers in which the thick middle portion of one cell is opposite the thin ends of adjacent cells. Consequently, the nuclei form an irregular pattern in the tissue (Fig. 3.6b). Smooth muscle is not under voluntary control and therefore is said to be involuntary. Smooth muscle, found in the walls of viscera (intestine, stomach, and other internal organs) and blood vessels, contracts more slowly than skeletal muscle but can remain contracted for a longer time. When the smooth muscle of the intestine contracts, food moves along its lumen (central cavity). When the smooth muscle of the blood vessels contracts, blood vessels constrict, helping to raise blood pressure.

Cardiac muscle (Fig. 3.6c) is found only in the walls of the heart. Its contraction pumps blood and accounts for the heartbeat. Cardiac muscle combines features of both smooth muscle and skeletal muscle. It has striations like skeletal muscle, but the contraction of the heart is involuntary for the most part. Cardiac muscle cells also differ from skeletal muscle cells in that they have a single, centrally placed nucleus. The cells are branched and seemingly fused one with the other, and the heart appears to be composed of one large interconnecting mass of muscle cells. Actually, cardiac muscle cells are separate and individual, but they are bound end to end at **intercalated disks,** areas where folded plasma membranes between two cells contain desmosomes and gap junctions.

All muscular tissue contains actin filaments and myosin filaments; these form a striated pattern in skeletal and cardiac muscle, but not in smooth muscle.

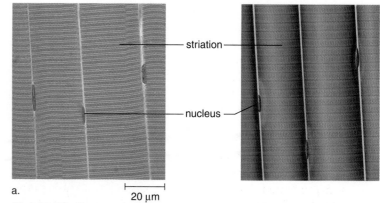

a.   20 μm

Skeletal muscle
- has striated cells with multiple nuclei.
- usually attached to skeleton.
- functions in voluntary movement.

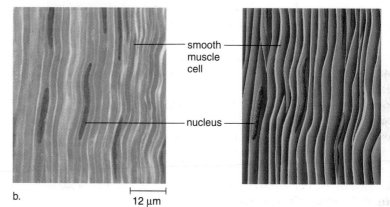

b.   12 μm

Smooth muscle
- has spindle-shaped cells, each with a single nucleus.
- occurs in walls of hollow internal organs.
- functions in movement of substances in lumens of body.
- no cross striations, involuntary.

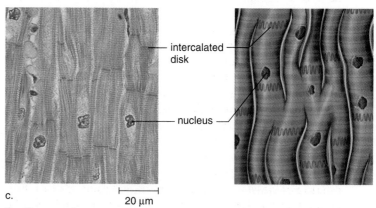

c.   20 μm

Cardiac muscle
- has branching striated cells, each with a single nucleus.
- occurs in the wall of the heart.
- functions in the pumping of blood.
- involuntary.

Figure 3.6   Muscular tissue.
a. Skeletal muscle is voluntary and striated. **b.** Smooth muscle is involuntary and nonstriated. **c.** Cardiac muscle is involuntary and striated. Cardiac muscle cells branch and fit together at intercalated disks.

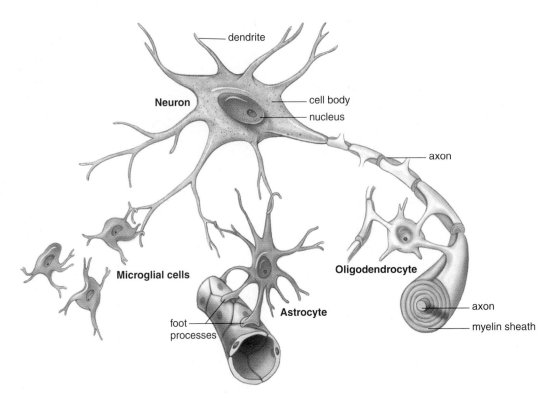

dendrite

Neuron

cell body

nucleus

axon

Microglial cells

Oligodendrocyte

Astrocyte

foot
processes

axon

myelin sheath

Figure 3.7   Neuron and neuroglial cells.
Neurons conduct nerve impulses. Neuroglial cells, which support and service neurons, have various functions: microglial cells are phagocytes that clean up debris. Astrocytes lie between neurons and a capillary; therefore, substances entering neurons from the blood must first pass through astrocytes. Oligodendrocytes form the myelin sheaths around fibers in the brain and spinal cord.

Nervous Tissue

Nervous tissue, which contains nerve cells called neurons, is present in the brain and spinal cord. A **neuron** is a specialized cell that has three parts: dendrites, cell body, and an axon (Fig. 3.7). A dendrite is a process that conducts signals toward the cell body. The cell body contains the major concentration of the cytoplasm and the nucleus of the neuron. An axon is a process that typically conducts nerve impulses away from the cell body. Axons can be quite long, and outside the brain and the spinal cord, long fibers, bound by connective tissue, form **nerves.**

The nervous system has just three functions: sensory input, integration of data, and motor output. Nerves conduct impulses from sensory receptors to the spinal cord and the brain where integration occurs. The phenomenon called sensation occurs only in the brain, however. Nerves also conduct nerve impulses away from the spinal cord and brain to the muscles and glands, causing them to contract and secrete, respectively. In this way, a coordinated response to the stimulus is achieved.

In addition to neurons, nervous tissue contains neuroglial cells.

Neuroglial Cells

There are several different types of neuroglial cells in the brain (Fig. 3.7), and much research is currently being conducted to determine how much "glial" cells contribute to the functioning of the brain. **Neuroglial cells** outnumber neurons nine to one and take up more than half the volume of the brain, but until recently, they were thought to merely support and nourish neurons. Three types of neuroglial cells are oligodendrocytes, microglial cells, and astrocytes. Oligodendrocytes form myelin; and microglial cells, in addition to supporting neurons, phagocytize bacterial and cellular debris. Astrocytes provide nutrients to neurons and produce a hormone known as glial-derived growth factor, which someday might be used as a cure for Parkinson disease and other diseases caused by neuron degeneration. Neuroglial cells don't have a long process, but even so, researchers are now beginning to gather evidence that they do communicate among themselves and with neurons!

Nerve cells, called neurons, have fibers (processes) called axons and dendrites. Axons are found in nerves. Neuroglial cells support and service neurons.

3.2 Body Cavities and Body Membranes

The internal organs are located within specific body cavities (Fig. 3.8). During human development, there is a large ventral cavity called a **coelom,** which becomes divided into the thoracic (chest) and abdominal cavities. Membranes divide the thoracic cavity into the pleural cavities, containing the right and left lungs, and the pericardial cavity, containing the heart. The thoracic cavity is separated from the abdominal cavity by a horizontal muscle called the diaphragm. The stomach, liver, spleen, gallbladder, and most of the small and large intestines are in the upper portion of the abdominal cavity. The lower portion contains the rectum, the urinary bladder, the internal reproductive organs, and the rest of the large intestine. Males have an external extension of the abdominal wall, called the scrotum, containing the testes.

The dorsal cavity also has two parts: the cranial cavity within the skull contains the brain; and the vertebral column, formed by the vertebrae, contains the spinal cord.

Body Membranes

In this context, we are using the term *membrane* to refer to a thin lining or covering composed of an epithelium overlying a loose connective tissue layer. Body membranes line cavities and internal spaces of organs and tubes that open to the outside.

Mucous membranes line the tubes of the digestive, respiratory, urinary, and reproductive systems. The epithelium of this membrane contains goblet cells that secrete mucus. This mucus ordinarily protects the body from invasion by bacteria and viruses; hence, more mucus is secreted and expelled when a person has a cold and has to blow her/his nose. In addition, mucus usually protects the walls of the stomach and small intestine from digestive juices, but this protection breaks down when a person develops an ulcer.

Serous membranes line the thoracic and abdominal cavities and the organs that they contain. They secrete a watery fluid that keeps the membranes lubricated. Serous membranes support the internal organs and compartmentalize the large thoracic and abdominal cavities. This helps to hinder the spread of any infection.

The **pleural membranes** are serous membranes that line the pleural cavity and lungs. **Pleurisy** is a well-known infection of these membranes. The peritoneum lines the abdominal cavity and its organs. In between the organs, there is a double layer of peritoneum called mesentery. **Peritonitis,** a life-threatening infection of the peritoneum, is likely if an inflamed appendix bursts before it is removed.

Synovial membranes line freely movable joint cavities. They secrete synovial fluid into the joint cavity; this fluid lubricates the ends of the bones so that they can move freely. In

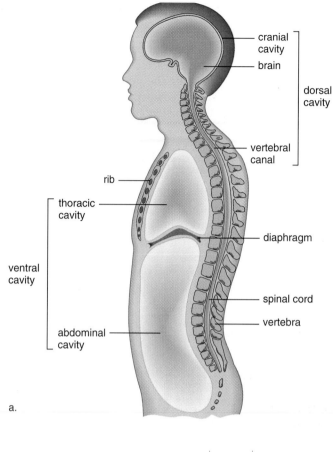

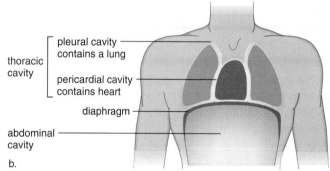

Figure 3.8   Mammalian body cavities.
a. Side view. There is a dorsal (toward the back) cavity, which contains the cranial cavity and the vertebral canal. The brain is in the cranial cavity, and the spinal cord is in the vertebral canal. There is a well-developed ventral (toward the front) cavity, which is divided by the diaphragm into the thoracic cavity and the abdominal cavity. The heart and lungs are in the thoracic cavity, and most other internal organs are in the abdominal cavity. **b.** Frontal view of the thoracic cavity.

rheumatoid arthritis, the synovial membrane becomes inflamed and grows thicker, restricting movement.

The **meninges** are membranes found within the dorsal cavity. They are composed only of connective tissue and serve as a protective covering for the brain and spinal cord. Meningitis is a life-threatening infection of the meninges.

3.3 Organ Systems

The body contains a number of systems that work together to maintain homeostasis. The next page introduces a feature that will be used at the end of each organ system chapter. In this chapter, the box reviews the general functions of the body's organ systems. The corresponding boxes in other chapters will show how a particular organ system interacts with all the other systems.

Maintenance of the Body

The internal environment of the body consists of the blood within the blood vessels and the tissue fluid that surrounds the cells. Five systems add substances to and remove substances from the digestive, cardiovascular, lymphatic, respiratory, and urinary systems.

The **digestive system** consists of the mouth, esophagus, stomach, small intestine, and large intestine (colon) along with the associated organs: teeth, tongue, salivary glands, liver, gallbladder, and pancreas. This system receives food and digests it into nutrient molecules, which can enter the cells of the body.

The **cardiovascular system** consists of the heart and blood vessels that carry blood through the body. Blood transports nutrients and oxygen to the cells, and removes their waste molecules that are to be excreted from the body. Blood also contains cells produced by the lymphatic system.

The **lymphatic system** consists of lymphatic vessels, lymph fluid, lymph nodes, and other lymphoid organs. This system protects the body from disease by purifying lymph and supporting lymphocytes, the white blood cells that produce antibodies. Lymphatic vessels absorb fat from the digestive system and collect excess tissue fluid, which is returned to the blood circulatory system.

The **respiratory system** consists of the lungs and the tubes that take air to and from the lungs. The respiratory system brings oxygen into the lungs and takes carbon dioxide out of the lungs.

The **urinary system** contains the kidneys and the urinary bladder. This system rids the body of nitrogenous wastes and helps regulate the fluid level and chemical content of the blood.

The digestive system, cardiovascular system, lymphatic system, respiratory system, and urinary system all perform specific processing and transporting functions to maintain the normal conditions of the body.

Integumentary System

The skin is sometimes called the **integumentary system** because it contains accessory organs such as hair, nails, sweat glands, and sebaceous glands. The skin provides external support and protects underlying tissues, helps regulate body temperature, contains receptors, and even synthesizes certain chemicals that affect the rest of the body.

Support and Movement

The skeletal system and the muscular system give the body support and are involved in the ability of the body and its parts to move.

The skeletal system, consisting of the bones of the skeleton, protects body parts. For example, the skull forms a protective encasement for the brain, as does the rib cage for the heart and lungs. The skeleton, as a whole, serves as a place of attachment for the skeletal muscles. Contraction of muscles in the muscular system accounts for movement of the body and also body parts.

The skeletal system and the muscular system support the body and permit movement.

Integration and Control of the Body

The **nervous system** consists of the brain, spinal cord, and associated nerves. The nerves conduct nerve impulses from receptors to the brain and spinal cord. They also conduct nerve impulses from the brain and spinal cord to the muscles and glands, allowing us to respond to both external and internal stimuli.

The **endocrine system** consists of the hormonal glands that secrete chemicals that serve as messengers between body parts. **Homeostasis** is the relative constancy of the internal environment. Both the nervous and endocrine systems help maintain homeostasis by coordinating and regulating the functions of the body's other systems. The endocrine system also helps maintain the proper functioning of male and female reproductive organs.

The nervous and endocrine systems coordinate and regulate the activities of the body's other systems.

Continuance of the Species

The **reproductive system** involves different organs in the male and female. The male reproductive system consists of the testes, other glands, and various ducts that conduct semen to and through the penis. The female reproductive system consists of the ovaries, oviducts, uterus, vagina, and external genitals.

The reproductive system in males and in females carries out those functions that give humans the ability to reproduce.

Human Systems Work Together

Integumentary System

External support and protection of body.

Respiratory System

Gaseous exchange between external environment and blood.

Cardiovascular System

Transport of nutrients to body cells and transport of wastes away from cells.

Skeletal System

Internal support and protection; body movement; protection of blood cells.

Lymphatic System/Immunity

Immunity; absorption of fats; drainage of tissue fluid.

Muscular System

Body movement; production of body heat.

Digestive System

Breakdown and absorption of food materials.

Nervous System

Regulation of all body activities; learning and memory.

Urinary System

Maintenance of volume and chemical composition of blood.

Endocrine System

Secretion of hormones for chemical regulation of all tissues.

Reproductive System

Production of sperm and egg; transfer of sperm to female system where development occurs.

3.4 Skin as an Organ System

The outer covering of the body, called **skin,** can be used as an example of an organ system. It is sometimes called the integumentary system because it contains accessory structures such as nails, hair, and glands (Fig. 3.9). Skin covers the body, protecting underlying tissues from physical trauma, pathogen invasion, and water loss. It also helps regulate body temperature in a manner described on the next page. Therefore skin plays a significant role in homeostasis. The skin even synthesizes certain chemicals such as vitamin D that affect the rest of the body. Because skin contains sensory receptors, skin also helps us to be aware of our surroundings and to communicate with others by touch.

Regions of the Skin

The skin has two regions: the epidermis and the dermis. A subcutaneous layer is found between the skin and any underlying structures, such as muscle or bone.

The **epidermis** of skin is made up of stratified squamous epithelium. New cells derived from basal cells become flattened and hardened as they push to the surface.

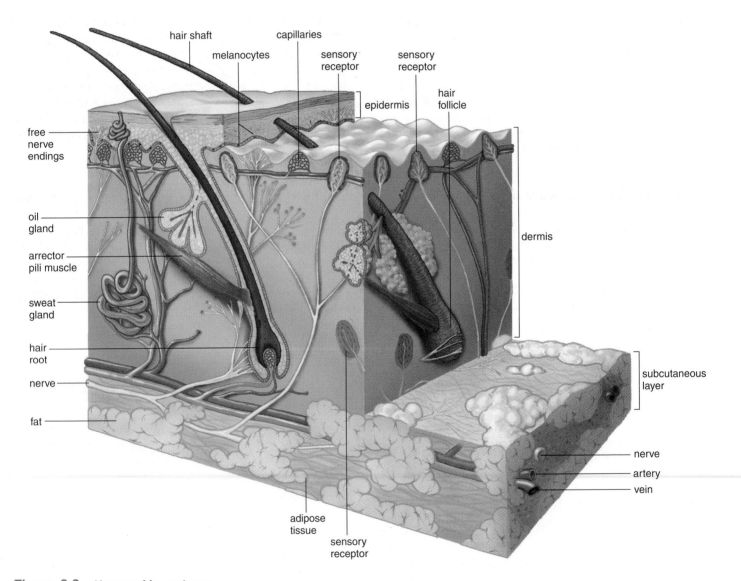

Figure 3.9   Human skin anatomy.
Skin consists of two regions, the epidermis and dermis. A subcutaneous layer lies below the dermis.

Hardening occurs because the cells produce keratin, a waterproof protein. Dandruff occurs when the rate of keratinization is two or three times the normal rate. A thick layer of dead keratinized cells, arranged in spiral and concentric patterns, form fingerprints and footprints. Specialized cells in the epidermis called **melanocytes** produce melanin, the pigment responsible for skin color.

The **dermis** is a region of fibrous connective tissue beneath the epidermis. The dermis contains collagenous and elastic fibers. The collagenous fibers are flexible but offer great resistance to overstretching; they prevent the skin from being torn. The elastic fibers maintain normal skin tension but also stretch to allow movement of underlying muscles and joints. (The number of collagen and elastic fibers decreases with exposure to the sun, and the skin becomes less supple and is prone to wrinkling.) The dermis also contains blood vessels that nourish the skin. When blood rushes into these vessels, a person blushes, and when blood is minimal in them, a person turns "blue."

Sensory receptors are specialized nerve endings in the dermis that respond to external stimuli. There are receptors for touch, pressure, pain, and temperature. The fingertips contain the most touch receptors, and these add to our ability to use our fingers for delicate tasks.

The **subcutaneous layer,** which lies below the dermis, is composed of loose connective tissue, and adipose tissue, which stores fat. Fat is a stored source of energy in the body. Adipose tissue helps to thermally insulate the body from either gaining heat from the outside or losing heat from the inside. A well-developed subcutaneous layer gives the body a rounded appearance and provides protective padding against external assaults. Excessive development of the subcutaneous layer accompanies obesity.

Skin has two regions: the epidermis and the dermis. A subcutaneous layer lies beneath the dermis.

Accessory Structures of the Skin

Nails, hair, and glands are structures of epidermal origin even though some parts of hair and glands are largely found in the dermis.

Nails grow from special epithelial cells at the base of the nail in the portion called the nail root. These cells become keratinized as they grow out over the nail bed. The visible portion of the nail is called the nail body. The cuticle is a fold of skin that hides the nail root. The whitish color of the half-moon-shaped base, or lunula results from the thick layer of cells in this area (Fig. 3.10)

Hair follicles begin in the dermis and continue through the epidermis where the hair shaft extends beyond the skin. Epidermal cells form the root of hair, and their division causes a hair to grow. The cells become keratinized and dead as they are pushed farther from the root. Each hair follicle has one or more **oil (sebaceous) glands,** which secrete sebum, an oily substance that lubricates the hair within the follicle and the skin itself. If the sebaceous glands fail to discharge, the secretions collect and form "whiteheads" or "blackheads." The color of blackheads is due to oxidized sebum. Contraction of the arrector pili muscles attached to hair follicles causes the hairs to "stand on end" and causes goose bumps to develop.

Sweat (sudoriferous) **glands** are quite numerous and are present in all regions of skin. A sweat gland begins as a coiled tubule within the dermis, but then it straightens out near its opening. Some sweat glands open into hair follicles, but most open onto the surface of the skin. Acne is an inflammation of the sebaceous glands that most often occurs during adolescence. Hormonal changes during this time cause the sebaceous glands to become more active.

Regulation of Body Temperature

If the body temperature starts to rise, the blood vessels dilate so that more blood is brought to the surface of the skin and the sweat glands become active. Sweat absorbs body heat as it evaporates. If the outer temperature is cool, the blood vessels constrict so that less blood is brought to the surface of the skin. Whenever the body's temperature falls below normal, the muscles start to contract, causing shivering, which produces heat.

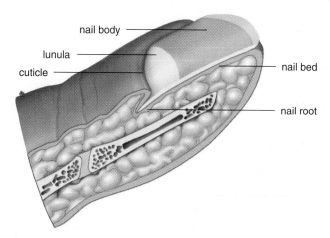

Figure 3.10   **Nail.**
Cells produced by the nail root become keratinized, forming the nail body.

Ecology Focus

Stratospheric Ozone Depletion Threatens the Biosphere

The earth's atmosphere is divided into layers. The troposphere envelops us as we go about our day-to-day lives. When ozone (O_3) is present in the troposphere (called ground-level ozone), it is considered a pollutant because it adversely affects a plant's ability to grow and our ability to breathe oxygen (O_2). In the stratosphere, some 50 kilometers above the earth, ozone forms a shield that absorbs much of the ultraviolet (UV) rays of the sun so that fewer rays strike the earth.

UV radiation causes mutations that can lead to skin cancer and can make the lenses of the eyes develop cataracts. It also is believed to adversely affect the immune system and our ability to resist infectious diseases. Crop and tree growth is impaired, and UV radiation also kills off small plants (phytoplankton) and tiny shrimplike animals (krill) that sustain oceanic life. Without an adequate ozone shield, our health and food sources are threatened.

Depletion of the ozone layer within the stratosphere in recent years is, therefore, of serious concern. It became apparent in the 1980s that some worldwide depletion of ozone had occurred and that there was a severe depletion of some 40–50% above the Antarctic every spring. A vortex of cold wind (a whirlpool in the atmosphere) circles the pole during the winter months, creating ice crystals where chemical reactions occur that break down ozone. Severe depletions of the ozone layer are commonly called "ozone holes." Detection devices now tell us that the ozone hole above the Antarctic is about the size of the United States and growing. Of even greater concern, an ozone hole has now appeared above the Arctic as well, and ozone holes could also occur within northern and southern latitudes, where many people live. Whether or not these holes develop depends on prevailing winds, weather conditions, and the type of particles in the atmosphere. A United Nations Environment Program report predicts a 26% rise in cataracts and nonmelanoma skin cancers for every 10% drop in the ozone level. A 26% increase translates into 1.75 million additional cases of cataracts and 300,000 more skin cancers (see Health reading next page) every year, worldwide.

The cause of ozone depletion can be traced to the release of chlorine atoms (Cl) into the stratosphere (Fig. 3A). Chlorine atoms combine with ozone and strip away the oxygen atoms, one by one. One atom of chlorine can destroy up to 100,000 molecules of ozone before settling to the earth's surface as chloride years later. These chlorine atoms come from the breakdown of chlorofluorocarbons (CFCs), chemicals much in use by humans. The best known CFC is Freon, a heat transfer agent found in refrigerators and air conditioners. CFCs are also used as cleaning agents and foaming agents during the production of Styrofoam found in coffee cups, egg cartons, insulation, and paddings. Formerly, CFCs were used as propellants in spray cans, but this application is now banned in the United States and several European countries.

Most countries of the world have agreed to stop using CFCs by the year 2000. The United States halted production in 1995. Computer projections suggest that an 85% reduction in CFC emissions is needed to stabilize CFC levels in the atmosphere. Otherwise they keep on increasing. Scientists are now searching for CFC substitutes that will not release chlorine atoms (nor bromine atoms) to harm the ozone shield.

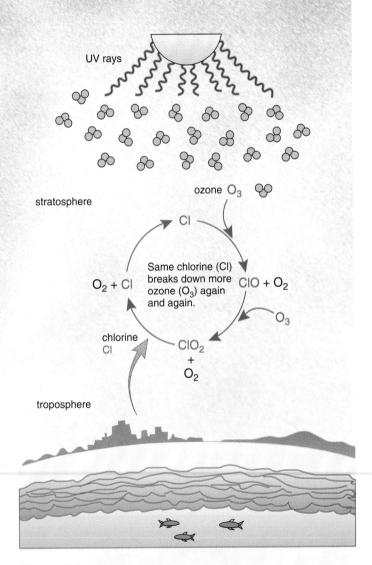

Figure 3A **Ozone depletion.**
CFCs release chlorine atoms that lead to the breakdown of ozone (O_3) and the buildup of oxygen (O_2) in the stratosphere. Oxygen does not absorb UV radiation and does not protect the earth.

Skin Cancer on the Rise

In the nineteenth century, and earlier, it was fashionable for Caucasian women (those who did not labor outdoors) to keep their skin fair by carrying parasols when they went out. But early in this century, some fair-skinned people began to prefer the golden-brown look, and they took up sunbathing as a way to achieve a tan. A few hours after exposure to the sun, pain and redness due to dilation of blood vessels occur. Tanning occurs when melanin granules increase in keratinized cells at the surface of the skin as a way to prevent any further damage by ultraviolet (UV) rays. The sun gives off two types of UV rays: UV-A rays and UV-B rays. UV-A rays penetrate the skin deeply, affect connective tissue, and cause the skin to sag and wrinkle. UV-A rays are also believed to increase the effects of the UV-B rays, which are the cancer-causing rays. UV-B rays are more prevalent at midday.

Skin cancer is categorized as either nonmelanoma or melanoma. Nonmelanoma cancers are of two types. Basal cell carcinoma, the most common type, begins when UV radiation causes epidermal basal cells to form a tumor, while at the same time suppressing the immune system's ability to detect the tumor. The signs of a tumor are varied. They include an open sore that will not heal, a recurring reddish patch, a smooth, circular growth with a raised edge, a shiny bump, or a pale mark (Fig. 3B). In about 95% of patients the tumor can be excised surgically, but recurrence is common.

Squamous cell carcinoma begins in the epidermis proper. Squamous cell carcinoma is five times less common than basal cell carcinoma, but if the tumor is not excised promptly it is more likely to spread to nearby organs. The death rate from squamous cell carcinoma is about 1% of cases. The signs of squamous cell carcinoma are the same as for basal cell carcinoma, except that the former may also show itself as a wart that bleeds and scabs.

Melanoma that starts in pigmented cells often has the appearance of an unusual mole. Unlike a mole that is circular and confined, melanoma moles look like spilled ink spots. A variety of shades can be seen in the same mole, and they can itch, hurt, or feel numb. The skin around the mole turns gray, white, or red. Melanoma is most apt to appear in persons who have fair skin, particularly if they have suffered occasional severe sunburns as children. The chance of melanoma increases with the number of moles a person has. Most moles appear before the age of 14, and their appearance is linked to sun exposure. Melanoma rates have risen since the turn of the century, but the incidence has doubled in the last decade. Most often, malignant moles are removed surgically; if the cancer has spread, chemotherapy and various other treatments are also available.

Since the incidence of skin cancer is related to UV exposure, scientists have developed a UV index to determine how powerful the solar rays are in different U.S. cities. In general, the more southern the city, the higher the UV index, and the greater the risk of skin cancer. Regardless of where you live, for every 10% decrease in the ozone layer, the risk of skin cancer rises 13–20%. To prevent the occurrence of skin cancer, observe the following:

- Use a broad-spectrum sunscreen, which protects you from both UV-A and UV-B radiation, with an SPF (sun protection factor) of at least 15. (This means, for example, that if you usually burn after a 20-minute exposure, it will take 15 times that long before you will burn.)
- Stay out of the sun altogether between the hours of 10 A.M. and 3 P.M. This will reduce your annual exposure by as much as 60%. Wear protective clothing. Choose fabrics with a tight weave and wear a wide-brimmed hat.
- Wear sunglasses that have been treated to absorb both UV-A and UV-B radiation. Otherwise, sunglasses can expose your eyes to more damage than usual because pupils dilate in the shade.
- Avoid tanning machines. Although most tanning devices use high levels of only UV-A, UV-A rays cause the deep layers of the skin to become more vulnerable to UV-B radiation when you are later exposed to the sun.

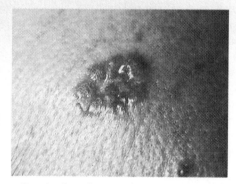
a. Basal cell carcinoma

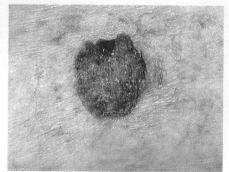
b. Squamous cell carcinoma

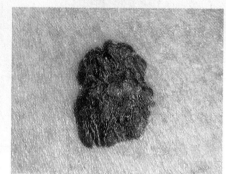
c. Melanoma

Figure 3B **Skin cancer.**
a. Basal cell carcinoma occurs when basal cells proliferate abnormally. **b.** Squamous cell carcinoma arises in epithelial cells derived from basal cells. **c.** Malignant melanoma is due to a proliferation of pigmented cells.

3.5 Homeostasis

Homeostasis means that the internal environment remains within normal limits or values, regardless of the conditions in the external environment. In humans, for example:

1. The blood glucose concentration remains at about 100 mg/100 ml.
2. The pH of blood is always near 7.4.
3. Blood pressure in the brachial artery averages near 120/80 mm Hg.
4. Body temperature averages around 37°C (98.6°F).

Because body conditions do fluctuate somewhat, homeostasis is often called a dynamic equilibrium of normal values. The ability of the body to keep the internal environment within a certain range allows humans to live in a variety of habitats, such as the Arctic regions, the deserts, or the tropics.

This internal environment consists of tissue fluid, which bathes all the cells of the body. Tissue fluid is refreshed when molecules such as oxygen and nutrients exit blood and wastes enter blood (Fig. 3.11). Tissue fluid remains constant only as long as blood composition remains constant. Although we are accustomed to using the word *environment* to mean the external environment of the body, it is important to realize that it is the internal environment of tissues that is ultimately responsible for our health and well-being.

The internal environment of the body consists of tissue fluid, which bathes the cells.

Most systems of the body contribute toward maintaining a relatively constant internal environment. The cardiovascular system conducts blood to and away from capillaries, the smallest of the blood vessels, whose thin walls permit exchanges to occur. Blood pressure aids the movement of water out of capillaries, and osmotic pressure aids the movement of water into capillaries. Blood pressure is created by the pumping of the heart, while osmotic pressure is maintained by the protein content of plasma. The formed elements also contribute to homeostasis. Red blood cells transport oxygen and participate in the transport of carbon dioxide. White blood cells fight infection, and platelets participate in the clotting process. The lymphatic system is accessory to the circulatory system. Lymphatic capillaries collect excess tissue fluid, and this is returned via lymphatic veins to the circulatory veins.

The digestive system takes in and digests food, providing nutrient molecules that enter blood and replace the nutrients that are constantly being used by the body cells. The respiratory system adds oxygen to and removes carbon dioxide from the blood. The chief regulators of blood composition are the liver and the kidneys. They monitor the chemical composition of plasma (see Table 3.1) and alter it as

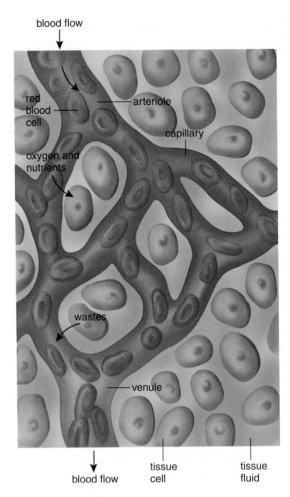

Figure 3.11   Tissue fluid composition.
Cells are surrounded by tissue fluid, which is continually refreshed because oxygen and nutrient molecules constantly exit, and waste molecules continually enter the bloodstream as shown.

required. Immediately after glucose enters the blood, it can be removed by the liver for storage as glycogen. Later, the glycogen can be broken down to replace the glucose used by the body cells; in this way, the glucose composition of blood remains constant. The hormone insulin, secreted by the pancreas, regulates glycogen storage. The liver also removes toxic chemicals, such as ingested alcohol and other drugs. The liver makes urea, a nitrogenous end product of protein metabolism. Urea and other metabolic waste molecules are excreted by the kidneys. Urine formation by the kidneys is extremely critical to the body, not only because it rids the body of unwanted substances, but also because it offers an opportunity to carefully regulate blood volume, salt balance, and the pH of the blood.

Most systems of the body contribute to homeostasis, that is, maintaining the dynamic equilibrium of the internal environment.

Coordination of Organ Systems

The nervous system and endocrine system are ultimately in control of homeostasis. The endocrine system is slower acting than the nervous system, which rapidly brings about a particular response.

Previously, we mentioned that the liver is involved in homeostasis because it stores glucose as glycogen. But actually there is a hormone produced by an endocrine gland that regulates storage of glucose by the liver. When the glucose content of the blood rises after eating, the pancreas secretes insulin, a hormone that causes the liver to store glucose as glycogen. Now the glucose level falls, and the pancreas no longer secretes insulin. This is called control by **negative feedback** because the response (low blood glucose) negates the original stimulus (high blood glucose). In some instances, an endocrine gland is sensitive to the blood level of a hormone whose concentration it regulates. For example, the pituitary gland produces a hormone that stimulates the thyroid gland to secrete its hormone. When the blood level of this hormone rises to a certain level, the pituitary gland no longer stimulates the thyroid gland.

A negative feedback system can regulate itself because it has a sensing device which detects changes in environmental conditions. For example, consider the feedback mechanism that functions to maintain the room temperature of a house. In this feedback system, the thermostat is a device that is sensitive to room temperature. The furnace produces heat, and when the temperature of a room reaches a certain point, the thermostat signals a switching device that turns the furnace off. On the other hand, when the temperature falls below that indicated on the thermostat, it signals the switching device, which turns the furnace on again.

Figure 3.12*a* shows that in the body there are sensory receptors that fulfill the role of sensing devices. When a receptor is stimulated, it signals a regulatory center that then turns on an effector. The effector brings about a response that negates the original conditions that stimulated the receptor. In the absence of suitable stimulation, the receptor no longer signals the regulatory center.

Figure 3.12*b* gives an example involving the nervous system. When blood pressure rises, receptors signal a regulatory center, which then sends out nerve impulses to the arterial walls, causing them to relax, and the blood pressure now falls. Therefore, the sensory receptors are no longer stimulated, and the system shuts down. Notice that negative feedback control results in a fluctuation above and below an average. Thus, there is a dynamic equilibrium of the internal environment.

Positive feedback also occurs on occasion. In these instances, certain events increase the likelihood of a particular

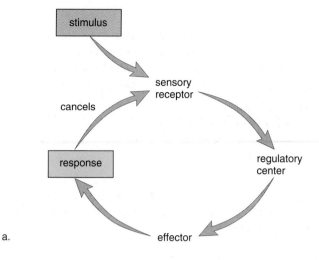

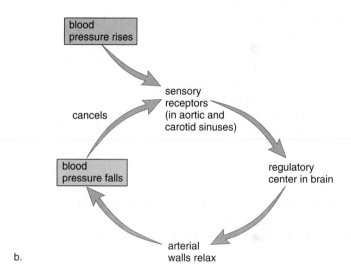

Figure 3.12 **Negative feedback control.**
a. A stimulus causes a receptor to signal a regulatory center in the brain. The regulatory center signals effectors to respond, and the response cancels the stimulus. **b.** For example, when blood pressure rises, special sensory receptors in blood vessels signal a particular center in the brain. The brain signals the arteries to relax, and blood pressure falls.

response. For example, once the childbirth process begins, each succeeding event makes it more likely that the process will continue until completion.

Homeostasis of internal conditions is a self-regulatory mechanism that usually results in slight fluctuations above and below an average.

Bioethical Issue

Transplantation of the kidney, heart, liver, pancreas, lung, and other organs is now possible due to two major breakthroughs. First, solutions have been developed that preserve donor organs for several hours. This made it possible for one young boy to undergo surgery for 16 hours, during which time he received five different organs. Second, rejection of transplanted organs is now prevented by immunosuppressive drugs; therefore, organs can be donated by unrelated individuals, living or dead. After death, it is possible to give the "gift of life" to someone else—over 25 organs and tissues from one cadaver can be used for transplants. Survival rate after a transplant operation is good. So many heart recipients are now alive and healthy they have formed basketball and softball teams, demonstrating the normalcy of their lives after surgery.

One problem persists however, and that is the limited availability of organs for transplantation. At any one time, at least 27,000 Americans are waiting for a donated organ. Keen competition for organs can lead to various bioethical inequities. When the governor of Pennsylvania received a heart and lungs within a relatively short period of time, it appeared that his social status may have played a role. When Mickey Mantle received a liver transplant, people asked if it was right to give an organ to an older man who had a diseased liver due to the consumption of alcohol. If a father gives a kidney to a child, he has to undergo a major surgical operation that leaves him vulnerable to possible serious consequences in the future. If organs are taken from those who have just died, who guarantees that the in-

dividual is indeed dead? And is it right to genetically alter animals to serve as a source of organs for humans? Such organs will most likely be for sale, and does this make the wealthy more likely to receive a transplant than those who cannot pay?

Questions

1. Is it ethical to ask a parent to donate an organ to his or child? Why or why not?
2. Is it ethical to put a famous person at the top of the list for an organ transplant? Why or why not?
3. Is it ethical to remove organs from a newborn who is brain dead but whose organs are still functioning? Why or why not?
4. When xenotransplants (transplants for humans from other animals) are available, should they be for sale? Why or why not?

Summarizing the Concepts

3.1 Types of Tissues

Human tissues are categorized into four groups. Epithelial tissue covers the body and lines its cavities. The different types of epithelial tissue (squamous, cuboidal, and columnar) can be stratified and have cilia or microvilli. Also, columnar cells can be pseudostratified. Epithelial cells sometimes form glands that secrete either into ducts or into blood.

Connective tissues, in which cells are separated by a matrix, often bind body parts together. Loose connective tissue has both white and yellow fibers and may also have fat (adipose) cells. Fibrous connective tissue, such as that of tendons and ligaments, contains closely packed collagen fibers. Both cartilage and bone have cells within lacunae, but the matrix for cartilage is more flexible than that for bone, which contains calcium salts. In bone, the lacunae lie in concentric circles within an osteon (or Haversian system) about a central canal. Blood is a connective tissue in which the matrix is a liquid called plasma.

Muscular tissue is of three types. Both skeletal and cardiac muscle are striated; both cardiac and smooth muscle are involuntary. Skeletal muscle is found in muscles attached to bones, and smooth muscle is found in internal organs. Cardiac muscle makes up the heart.

Nervous tissue has one main type of conducting cell, the neuron, and several types of neuroglial cells. Each neuron has dendrites, a cell body, and an axon. The brain and spinal cord contain complete neurons, while the nerves contain only neuron fibers. Axons are specialized to conduct nerve impulses.

3.2 Body Cavities and Body Membranes

The internal organs occur within cavities; the thoracic cavity contains the heart and lungs; the abdominal cavity contains organs of the digestive, urinary, and reproductive systems, among others. Membranes

line body cavities and internal spaces of organs. As an example, mucous membrane lines the tubes of the digestive system; serous membrane lines the thoracic and abdominal cavities and covers the organs they contain.

3.3 Organ Systems

The skin is sometimes called the integumentary system. The digestive, cardiovascular, lymphatic, respiratory, and urinary systems perform processing and transporting functions that maintain the normal conditions of the body. The nervous system receives sensory input from sensory receptors and directs the musculoskeletal system and glands to respond to outside stimuli. The musculoskeletal system supports the body and permits movement. The endocrine system produces hormones, some of which influence the functioning of the reproductive system, which allows humans to make more of their own kind.

3.4 Skin as an Organ System

The skin can be used as an example of an organ system because it contains accessory structures such as nails, hair, and glands. Skin protects underlying tissues from physical trauma, pathogen invasion, and water loss. Skin helps regulate body temperature, and because it contains sensory receptors, skin also helps us to be aware of our surroundings.

Skin is a two-layered organ that waterproofs and protects the body. The epidermis contains basal cells that produce new epithelial cells that become keratinized as they move toward the surface. The dermis, a largely fibrous connective tissue, contains epidermally derived glands and hair follicles, nerve endings, and blood vessels. Sensory receptors for touch, pressure, temperature, and pain are present. Sweat glands and blood vessels help control body temperature. A subcutaneous layer, which is made up of loose connective tissue containing adipose cells, lies beneath the skin.

3.5 Homeostasis

Homeostasis is the dynamic equilibrium of the internal environment. All organ systems contribute to the constancy of tissue fluid and blood. Special contributions are made by the liver, which keeps blood glucose constant, and the kidneys, which regulate the pH. The nervous and hormonal systems regulate the other body systems. Both of these are controlled by a feedback mechanism, which results in fluctuation above and below the desired levels. Body temperature is regulated by a center in the hypothalamus.

Studying the Concepts

1. Name the four major types of tissues. 62
2. Name the different kinds of epithelial tissue, and give a location and function for each. 62
3. What are the functions of connective tissue? Name the different kinds, and give a location for each. 64–66
4. What are the functions of muscular tissue? Name the different kinds, and give a location for each. 67
5. Nervous tissue contains what type of cell? Which organs in the body are made up of nervous tissue? 68
6. In what cavities are the major organs located? 69
7. Distinguish between plasma membrane and body membrane. 69
8. Describe the structure of skin, and state at least two functions of this organ. 72–73
9. What is homeostasis, and how is it achieved in the human body? 76–77
10. Give an example of a negative feedback system. 77

Testing Your Knowledge of the Concepts

In questions 1–4, match the type of tissue to the functions listed.
a. epithelium
b. connective
c. muscular
d. nervous

_____ 1. Contracts, allowing body parts to move.
_____ 2. Supports and binds body parts to one another.
_____ 3. Lines cavities and protects surfaces.
_____ 4. Conducts messages within and to and from the brain.

In questions 5–7, indicate whether the statement is true (T) or false (F).

_____ 5. The nervous and endocrine systems coordinate the functions of the other systems in the body.
_____ 6. The lymphatic system delivers nutrients to cells and removes their wastes.
_____ 7. Regulation of body temperature involves all systems of the body except the skin.

In questions 8 and 9, fill in the blanks.

8. _____ is the relative constancy of the internal environment.
9. Smooth muscle is _____ and involuntary.
10. Give the name, the location, and the function for each of these tissues.
   a. Type of epithelial tissue
   b. Type of muscular tissue
   c. Type of connective tissue

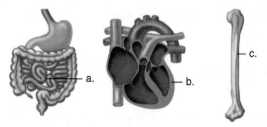

Applying Your Knowledge to the Concepts

These questions pertain to human organization.

1. Materials entering or leaving the human body must always pass through what type of tissue? Give two places in the body where this type of exchange takes place, and indicate the specific type of tissue located there.
2. If you are inactive when you are outside during the winter, you start to shiver, whereas if you are active you do not shiver. Why might that be?
3. The human fetal skeleton is composed of cartilage, much of which is later converted to bone. A fetal skeleton composed of cartilage provides what advantages?
4. What is the advantage of having some structures composed of skeletal muscle and others composed of smooth muscle?

Understanding the Terms

adhesion junction 64
adipose tissue 64
basement membrane 62
blood 66
bone 65
carcinoma 62
cardiac muscle 67
cardiovascular system 70
cartilage 64
cilium 62
coelom 69
collagen fiber 64
columnar epithelium 62
compact bone 65
connective tissue 64
cuboidal epithelium 62
dense fibrous connective
    tissue 64
dermis 73
digestive system 70
elastic cartilage 65
elastic fiber 64
endocrine system 70
epidermis 72
epithelial tissue 62
fibroblast 64
fibrocartilage 65

gap junction 64
gland 62
hair follicle 73
homeostasis 70
hyaline cartilage 64
integumentary system 70
intercalated disk 67
lacuna 64
ligament 64
loose fibrous connective
    tissue 64
lymphatic system 70
matrix 64
melanocyte 73
meninges 69
microvillus 62
mucous membrane 69
muscular (contractile)
    tissue 67
negative feedback 77
nerve 68
nervous system 70
nervous tissue 68
neuroglial cell 68
neuron 68
oil gland 73
pathogen 62

peritonitis 69
plasma 66
platelet 66
pleural membrane 69
pleurisy 69
positive feedback 77
red blood cell 66
reproductive system 70
respiratory system 70
reticular fiber 64
serous membrane 69
skeletal muscle 67
skin 72

smooth (visceral) muscle 67
spongy bone 65
squamous epithelium 62
synovial membrane 69
striated 67
subcutaneous layer 73
sweat gland 73
tendon 64
tight junction 64
tissue 62
urinary system 70
white blood cell 66

Match the terms to these definitions:

a. _____ Fibrous connective tissue that joins bone to bone
at a joint.

b. _____ Outer region of the skin composed of stratified
squamous epithelium.

c. _____ Having bands such as in cardiac and skeletal
muscle.

d. _____ Self-regulatory mechanism that is activated by an
imbalance and results in a fluctuation above and below a mean.

e. _____ Porous bone found at the ends of long bones
where blood cells are formed.

Applying Technology to the Concepts

Your study of homeostasis is supported by these available
technologies:

Essential Study Partner CD-ROM
Animals → Body Organization
Visit the Mader web site for related ESP activities.

Exploring the Internet
The Mader Home Page provides resources and tools as
you study this chapter.

http://www.mhhe.com/biosci/genbio/mader

II

Maintenance of the Human Body

All of the systems of the body help maintain homeostasis, resulting in a dynamic equilibrium of the internal environment. Our internal environment is the blood within blood vessels and the fluid that surrounds the cells of the tissues. The heart pumps the blood and sends it in vessels to the tissues, where exchange of materials occurs with tissue fluid. The composition of blood tends to remain relatively constant as a result of the actions of the digestive, respiratory, and excretory systems. Nutrients enter the blood at the small intestine, external gas exchange occurs in the lungs, and metabolic waste products are excreted at the kidneys. The immune system prevents pathogens from taking over the body and interfering with its proper functioning.

Chapter 4

Digestive System and Nutrition

Chapter Concepts

4.1 The Digestive System
- The human digestive system is an extended tube with specialized parts between two openings, the mouth and the anus. 82
- Food is ingested and then digested to small molecules that are absorbed. Indigestible materials are eliminated. 82

4.2 Three Accessory Organs
- The pancreas, the liver, and the gallbladder are accessory organs of digestion because their activities assist the digestive process. 90

4.3 Digestive Enzymes
- The products of digestion are small molecules, such as amino acids and glucose, that can cross plasma membranes. 92
- The digestive enzymes are specific and have an optimum temperature and pH at which they function. 92

4.4 Homeostasis
- The digestive system works with the other systems of the body to maintain homeostasis. 95

4.5 Nutrition
- Proper nutrition supplies the body with energy and nutrients, including the essential amino acids and fatty acids, and all vitamins and minerals. 95

Has anyone in your family ever had ulcers? Aside from restricting your diet, they can be very painful. Barry Marshall believed that he knew the cause of ulcers and, if he was correct, ulcers would be treatable and curable! One morning in 1984, he walked into his lab, stirred a beaker full of beef soup and *H. pylori,* and gulped the concoction. After five days, he began to vomit. His stomach grew inflamed. With further research, Marshall and others demonstrated that *H. pylori* is responsible for at least 70% of ulcers. Stress and other causes like prescription-drug side effects may also play a role, but these are not usually the direct cause of ulcers.

We have only to consider the frequency of TV commercials concerned with treating gastrointestinal ills in order to conclude that the proper functioning of the digestive system is critical to our everyday lives. This chapter reviews both the anatomy and physiology of our internal tubular digestive tract and its accessory organs. The liver is an accessory organ with a myriad of functions besides its role in digestion, and we will examine many of these. Today, we recognize that in a sense "we are what we eat," and therefore a knowledge of nutrition is essential. This chapter ends with a discussion of the basic principles of nutrition.

4.1 The Digestive System

Digestion takes place within a tube called the digestive tract, which begins with the mouth and ends with the anus (Fig. 4.1). The functions of the digestive system are to ingest food, digest it to nutrients that can cross plasma membranes, absorb nutrients, and eliminate indigestible remains.

The Mouth

The mouth, which receives food, is bounded externally by the lips and cheeks. The lips extend from the base of the nose to the start of the chin. The red portion of the lips is poorly keratinized and this allows blood to show through.

Most people enjoy eating food largely because they like its texture and taste. Sensory receptors called taste buds occur primarily on the tongue, and when these are activated by the presence of food, nerve impulses travel by way of cranial nerves to the brain. The tongue is composed of skeletal muscle whose contraction changes the shape of the tongue. Muscles exterior to the tongue cause it to move about. A fold of mucous membrane on the underside of the tongue attaches it to the floor of the oral cavity.

The roof of the mouth separates the nasal cavities from the oral cavity. The roof has two parts: an anterior (toward the front) **hard palate** and a posterior (toward the back) **soft**

palate (Fig. 4.2*a*). The hard palate contains several bones, but the soft palate is composed entirely of muscle. The soft palate ends in a finger-shaped projection called the *uvula.* The tonsils are in the back of the mouth, on either side of the tongue and in the nasopharynx (called adenoids). The tonsils help protect the body against infections. If the tonsils become inflamed, the person has **tonsillitis.** The infection can spread to the middle ears. If tonsillitis recurs repeatedly, the tonsils may be surgically removed (called a tonsillectomy).

Three pairs of **salivary glands** send juices (saliva) by way of ducts to the mouth. One pair of salivary *glands* lies at the sides of the face immediately below and in front of the ears.

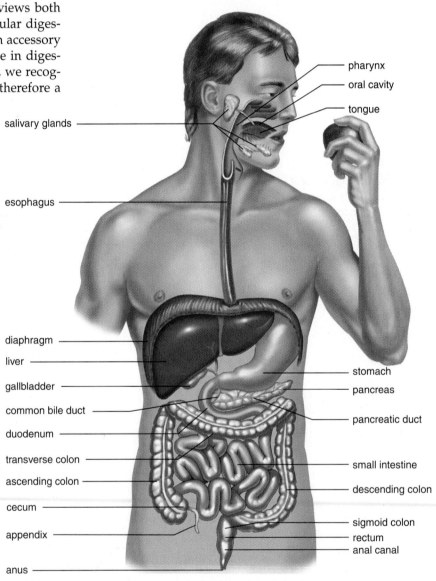

Figure 4.1   Digestive system.

Trace the path of food from the mouth to the anus. The large intestine consists of the cecum; ascending, transverse, descending, and sigmoid colons; plus the rectum and anal canal. Note also the location of the accessory organs of digestion: the pancreas, the liver, and the gallbladder.

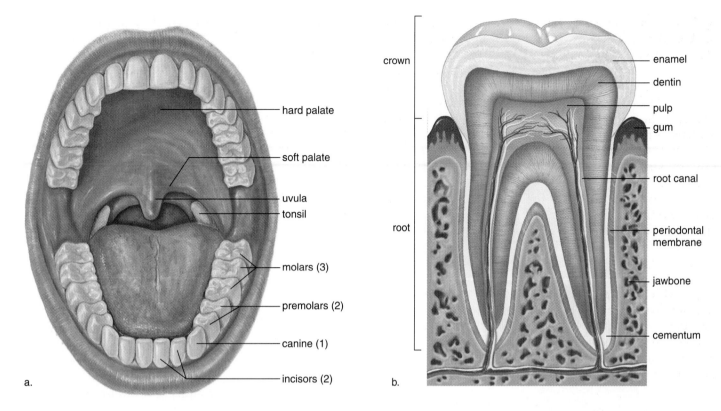

Figure 4.2   Adult mouth and teeth.
a. The chisel-shaped incisors bite; the pointed canines tear; the fairly flat premolars grind; and the flattened molars crush food. The last molar, called a wisdom tooth, may fail to erupt, or if it does, it is sometimes crooked and useless. Often dentists recommend the extraction of the wisdom teeth. **b.** Longitudinal section of a tooth. The crown is the portion that projects above the gum line and is sometimes replaced by a dentist. When a root canal is done, the nerves are removed. When the periodontal membrane is inflamed, the teeth can loosen.

These glands swell when a person has the mumps, a viral infection most often seen in children. Salivary glands have ducts that open on the inner surface of the cheek at the location of the second upper molar. Another pair of salivary glands lies beneath the tongue, and still another pair lies beneath the floor of the oral cavity. The ducts from these salivary glands open under the tongue. You can locate the openings if you use your tongue to feel for small flaps on the inside of your cheek and under your tongue. Saliva contains an enzyme called **salivary amylase** that begins the process of digesting starch.

The Teeth

With our teeth we chew food into pieces convenient for swallowing. During the first two years of life, the smaller 20 deciduous, or baby, teeth appear. These are eventually replaced by 32 adult teeth (Fig. 4.2*a*). The third pair of molars, called the wisdom teeth, sometimes fail to erupt. If they push on the other teeth and/or cause pain, they can be removed by a dentist or oral surgeon.

Each tooth has two main divisions, a crown and a root (Fig. 4.2*b*). The crown has a layer of enamel, an extremely hard outer covering of calcium compounds; dentin, a thick layer of bonelike material; and an inner pulp, which contains the nerves and the blood vessels. Dentin and pulp are also found in the root.

Tooth decay, called **dental caries,** or cavities, occurs when bacteria within the mouth metabolize sugar and give off acids, which erode teeth. Two measures can prevent tooth decay: eating a limited amount of sweets and daily brushing and flossing of teeth. Fluoride treatments, particularly in children, can make the enamel stronger and more resistant to decay. Gum disease is more apt to occur with aging. Inflammation of the gums (*gingivitis*) can spread to the periodontal membrane, which lines the tooth socket. A person then has **periodontitis,** characterized by a loss of bone and loosening of the teeth so that extensive dental work may be required. Stimulation of the gums in a manner advised by your dentist is helpful in controlling this condition.

The tongue, which is composed of striated muscle and an outer layer of mucous membrane, mixes the chewed food with saliva. It then forms this mixture into a mass called a *bolus* in preparation for swallowing.

The salivary glands send saliva into the mouth, where the teeth chew the food and the tongue forms it into a bolus for swallowing.

Table 4.1   Path of Food

Organ Feature(s)	Function of Organ	Special Feature(s)	Function of Special Feature(s)
Oral cavity	Receives food; starts digestion of starch	Teeth Tongue	Chewing of food Formation of bolus
Esophagus	Passageway		
Stomach	Storage of food; acidity kills bacteria; starts digestion of protein	Gastric glands	Release gastric juices
Small intestine	Digestion of all foods; absorption of nutrients	Intestinal glands Villi	Release fluids Absorb nutrients
Large intestine	Absorption of water; storage of indigestible remains		

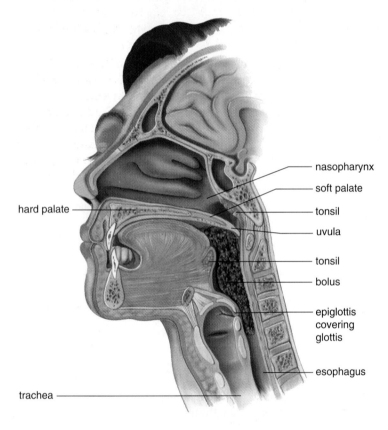

hard palate

trachea

nasopharynx
soft palate
tonsil
uvula
tonsil
bolus
epiglottis covering glottis
esophagus

Figure 4.3   Swallowing.
When food is swallowed, the soft palate closes off the nasopharynx and the epiglottis covers the glottis, forcing the bolus to pass down the esophagus. Therefore, you do not breathe when swallowing.

The Pharynx

The **pharynx** is a region that receives food from the mouth and air from the nasal cavities (Table 4.1). The food passage and air passage cross in the pharynx because the trachea (windpipe) is ventral to (in front of) the esophagus, a long muscular tube that takes food to the stomach.

Swallowing, a process that occurs in the pharynx (Fig. 4.3), is a **reflex action** performed automatically, without conscious thought. During swallowing, food normally enters the esophagus because the air passages are blocked. Unfortunately, we have all had the unpleasant experience of having food "go the wrong way." The wrong way may be either into the nasal cavities or into the trachea. If it is the latter, coughing will most likely force the food up out of the trachea and into the pharynx again. Usually during swallowing, the soft palate moves back to close off the **nasopharynx,** and the trachea moves up under the **epiglottis** to cover the glottis. The **glottis** is the opening to the larynx (voice box). The up and down movement of the Adam's apple, the front part of the larynx, is easy to observe when a person swallows. We do not breathe when we swallow.

The air passage and the food passage cross in the pharynx. When you swallow, the air passage usually is blocked off, and food must enter the esophagus.

The Esophagus

The **esophagus** is a muscular tube that passes from the pharynx through the thoracic cavity and diaphragm into the abdominal cavity where it joins the stomach. The esophagus is ordinarily collapsed, but it opens and receives the bolus when swallowing occurs. A rhythmic contraction

called **peristalsis** pushes the food along the digestive tract. Occasionally, peristalsis begins even though there is no food in the esophagus. This produces the sensation of a lump in the throat.

The esophagus plays no role in the chemical digestion of food. Its sole purpose is to conduct the food bolus from the mouth to the stomach. **Sphincters** are muscles that encircle tubes and act as valves; tubes close when sphincters contract, and they open when sphincters relax. The entrance of the esophagus to the stomach is marked by a constriction, often called a sphincter, although the muscle is not as developed as in a true sphincter. Relaxation of the sphincter allows the bolus to pass into the stomach, while contraction prevents the acidic contents of the stomach from backing up into the esophagus. **Heartburn,** which feels like a burning pain rising up into the throat, occurs when some of the stomach contents escape into the esophagus. When vomiting occurs, a contraction of the abdominal muscles and diaphragm propels the contents of the stomach upward through the esophagus.

The esophagus conducts the bolus of food from the pharynx to the stomach. Peristalsis begins in the esophagus and occurs along the entire length of the digestive tract.

The Wall of the Digestive Tract

The wall of the esophagus in the abdominal cavity is comparable to that of the digestive tract, which has these layers (Fig. 4.4):

Mucosa (mucous membrane layer) A layer of epithelium supported by connective tissue and smooth muscle lines the **lumen** (central cavity) and contains glandular epithelial cells that secrete digestive enzymes and goblet cells that secrete mucus.

Submucosa (submucosal layer) A broad band of loose connective tissue that contains blood vessels. Lymph nodules, called Peyer's patches, are in the submucosa. Like the tonsils, they help protect us from disease.

Muscularis (smooth muscle layer) Two layers of smooth muscle make up this section. The inner, circular layer encircles the gut; the outer, longitudinal layer lies in the same direction as the gut.

Serosa (serous membrane layer) Most of the digestive tract has a serosa, a very thin, outermost layer of squamous epithelium supported by connective tissue. The serosa secretes a serous fluid that keeps the outer surface of the intestines moist so that the organs of the abdominal cavity slide against one another. The esophagus has an outer layer composed only of loose connective tissue called the adventitia.

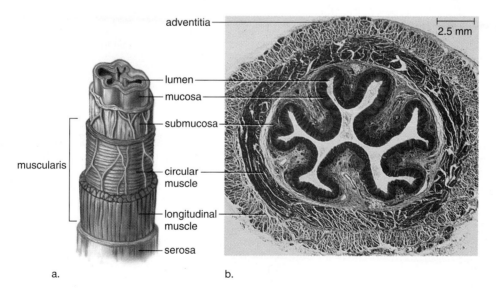

Figure 4.4   Wall of the digestive tract.
a. Several different types of tissues are found in the wall of the digestive tract. Note the placement of circular muscle inside longitudinal muscle.
b. Micrograph of the wall of the esophagus.

The Stomach

The **stomach** (Fig. 4.5) is a thick-walled, J-shaped organ that lies on the left side of the body beneath the diaphragm. The stomach is continuous with the esophagus above and the duodenum of the small intestine below. The stomach stores food and aids in digestion. The wall of the stomach has deep folds, which disappear as the stomach fills to an approximate capacity of one liter. Its muscular wall churns, mixing the food with gastric juice. The term *gastric* always refers to the stomach.

The columnar epithelial lining of the stomach has millions of gastric pits, which lead into **gastric glands.** The gastric glands produce gastric juice. Gastric juice contains an enzyme called **pepsin,** which digests protein, plus hydrochloric acid (HCl) and mucus. HCl causes the stomach to have a high acidity with a pH of about 2, and this is beneficial because it kills most bacteria present in food. Although HCl does not digest food, it does break down the connective tissue of meat and activates pepsin. The wall of the stomach is protected by a thick layer of mucus secreted by goblet cells in its lining. If, by chance, HCl penetrates this mucus, the wall can begin to break down, and an ulcer results. An **ulcer** is an open sore in the wall caused by the gradual disintegration of tissue. It now appears that most ulcers are due to a bacterial (*Helicobacter pylori*) infection that impairs the ability of epithelial cells to produce protective mucus.

Alcohol is absorbed in the stomach, but there is no absorption of food substances. Normally, the stomach empties in about 2–6 hours. When food leaves the stomach, it is a thick, soupy liquid called **chyme.** Chyme leaves the stomach and enters the small intestine in squirts by way of a sphincter that repeatedly opens and closes.

The stomach can expand to accommodate large amounts of food. When food is present, the stomach churns, mixing food with acidic gastric juice.

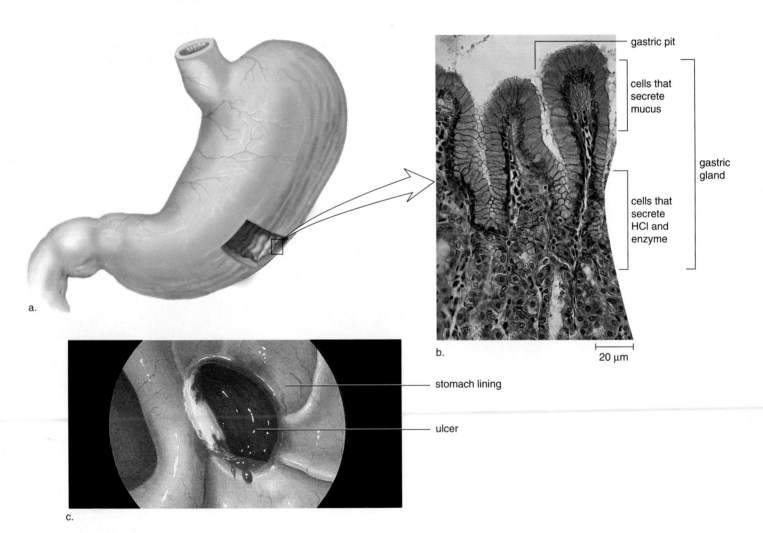

Figure 4.5   Anatomy and histology of the stomach.
a. The stomach has a thick wall with folds that allow it to expand and fill with food. **b.** The mucosa contains gastric glands, which secrete mucus and a gastric juice active in protein digestion. **c.** View of a bleeding ulcer by using an endoscope (a tubular instrument bearing a tiny lens and a light source) that can be inserted into the abdominal cavity.

The Small Intestine

The **small intestine** is named for its small diameter (compared to that of the large intestine); but perhaps it should be called the long intestine. In life, the small intestine averages about 3 meters (9 feet) in length, compared to the large intestine which is about 1.5 meters (4 1/2 ft) in length. (After death, the small intestine becomes as long as 6 meters due to relaxation of muscles.)

The first 25 cm of the small intestine is called the **duodenum**. Ducts from the liver and pancreas join to form one duct that enters the duodenum (see Fig. 4.1). The small intestine receives bile from the liver and pancreatic juice from the pancreas via this duct. **Bile** emulsifies fat—emulsification causes fat droplets to disperse in water. The intestine has a slightly basic pH because pancreatic juice contains sodium bicarbonate (NaHCO3), which neutralizes chyme. The enzymes in pancreatic juice and enzymes produced by the intestinal wall complete the process of digestion.

It's been suggested that the surface area of the small intestine is approximately that of a tennis court. What factors contribute to increasing its surface area? The wall of the small intestine contains fingerlike projections called **villi,** which give the intestinal wall a soft, velvety appearance (Fig. 4.6). Each villus has an outer layer of columnar epithelium and contains blood vessels and a small lymphatic vessel called a **lacteal.** The lymphatic system is an adjunct to the cardiovascular system—its vessels carry a fluid called lymph to the cardiovascular veins.

Each villus has thousands of microscopic extensions called microvilli. Collectively in electron micrographs, microvilli give the villi a fuzzy border known as a "brush border." Since the microvilli bear the intestinal enzymes, these enzymes are called brush-border enzymes. The microvilli greatly increase the surface area of the villus for the absorption of nutrients. Sugars and amino acids pass through the mucosa and enter a blood vessel. The components of fats (glycerol and fatty acids) rejoin in smooth endoplasmic reticulum and are combined with proteins in the Golgi apparatus before they enter a lacteal. After nutrients are absorbed, they are eventually carried to all the cells of the body by the bloodstream. The digestive system contributes to homeostasis by providing the nutrients that cells need.

The large surface area of the small intestine facilitates absorption of nutrients into the cardiovascular (glucose and amino acids) and lymphatic (fats) systems.

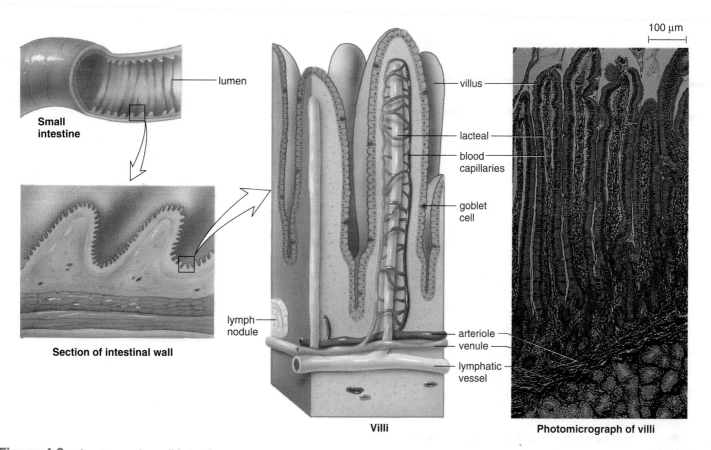

lumen

Small intestine

Section of intestinal wall

lymph nodule

villus

lacteal

blood capillaries

goblet cell

arteriole
venule

lymphatic vessel

100 μm

Villi

Photomicrograph of villi

Figure 4.6   Anatomy of small intestine.
The wall of the small intestine has folds that bear fingerlike projections called villi. The products of digestion are absorbed by villi, which contain blood vessels and a lacteal. Each villus has many microscopic extensions called microvilli.

Regulation of Digestive Secretions

The nervous system promotes the secretion of digestive juices, but so do hormones (Fig. 4.7). A **hormone** is a substance produced by one set of cells that affects a different set of cells, the so-called target cells. Hormones are usually transported by the bloodstream. When a person has eaten a meal particularly rich in protein, the stomach produces the hormone gastrin. Gastrin enters the bloodstream, and soon the stomach is churning, and the secretory activity of gastric glands is increasing. A hormone produced by the duodenal wall, GIP (gastric inhibitory peptide), works opposite from gastrin: it inhibits gastric gland secretion.

Cells of the duodenal wall produce two other hormones that are of particular interest—secretin and CCK (cholecystokinin). Acid, especially hydrochloric acid (HCl) present in chyme, stimulates the release of secretin, while partially digested protein and fat stimulate the release of CCK. Soon after these hormones enter the bloodstream, the pancreas increases its output of pancreatic juice, which helps digest food, and the liver increases its output of bile. The gallbladder contracts to release bile.

Regulation of internal organs is one of the ways hormones contribute to homeostasis. ✗

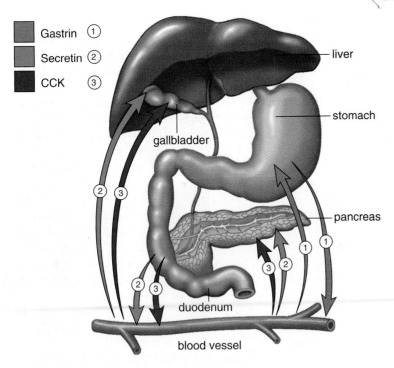

Gastrin ①

Secretin ②

CCK ③

liver

stomach

gallbladder

pancreas

duodenum

blood vessel

Figure 4.7   Hormonal control of digestive gland secretions.
Gastrin ① , produced by the lower part of the stomach, enters the bloodstream and thereafter stimulates the upper part of the stomach to produce more digestive juice. Secretin ② and CCK ③ , produced by the duodenal wall, stimulate the pancreas to secrete its digestive juice and the gallbladder to release bile.

The Large Intestine

The **large intestine,** which includes the cecum, the colon, the rectum, and the anal canal, is larger in diameter than the small intestine (6.5 cm compared to 2.5 cm), but it is shorter in length (1.5 meters compared to 3 meters) (see Fig. 4.1). The large intestine absorbs water, salts, and some vitamins. It also stores indigestible material until it is eliminated at the anus.

The **cecum,** which lies below the junction with the small intestine, is the blind end of the large intestine. The cecum has a small projection called the vermiform **appendix** (*vermiform* means wormlike) (Fig. 4.8). In humans, the appendix also may play a role in fighting infections. This organ is subject to inflammation, a condition called appendicitis. If inflamed, the appendix should be removed before the fluid content rises to the point that the appendix bursts, a situation that may cause **peritonitis,** a generalized infection of the lining of the abdominal cavity. Peritonitis can lead to death.

The **colon** includes the *ascending colon*, which goes up the right side of the body to the level of the liver; the *transverse colon*, which crosses the abdominal cavity just below the liver and the stomach; the *descending colon*, which passes down the left side of the body; and the *sigmoid colon*, which enters the *rectum*, the last 20 cm of the large intestine. The rectum opens at the **anus,** where **defecation,** the expulsion of *feces,* occurs. When feces are forced into the rectum by peristalsis, a defecation reflex occurs. The stretching of the rectal wall initiates nerve impulses to the spinal cord, and shortly thereafter contraction of the rectal muscles and relaxation of anal sphincters occur (Fig. 4.9). Ridding the body of indigestible remains is another way the digestive system helps maintain homeostasis. Feces are three-quarters water and one-quarter solids. Bacteria, fiber (indigestible remains), and other indigestible materials are in the solid portion. The brown color of feces is due to bilirubin (see page 90), and the odor is due to breakdown products as bacteria work on the nondigested remains. This bacterial action also produces gases. ✗

For many years, it was believed that facultative bacteria (bacteria that can live with or without oxygen), such as *Escherichia coli,* were the major inhabitants of the colon, but new culture methods show that over 99% of the colon bacteria are obligate anaerobes (bacteria that die in the presence of oxygen). Not only do the bacteria break down indigestible material, they also produce some vitamins and other molecules that can be absorbed and used by us. In this way, they perform a service for us.

Water is considered unsafe for swimming when the coliform (nonpathogenic intestinal) bacterial count reaches a certain number. A high count is an indication that a significant amount of feces has entered the water. The more feces present, the greater the possibility that disease-causing bacteria are also present.

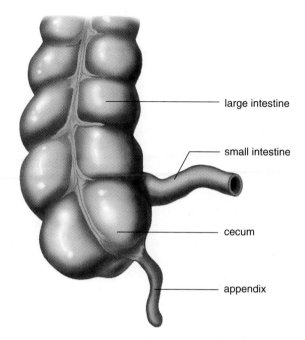

Figure 4.8   Junction of the small intestine and the large intestine.
The cecum is the blind end of the ascending colon. The appendix is attached to the cecum.

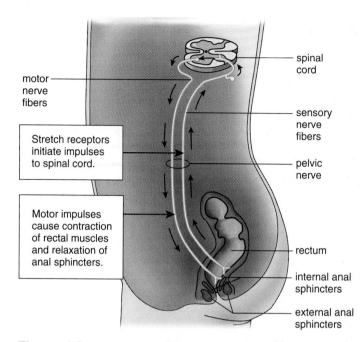

Figure 4.9   Defecation reflex.
The accumulation of feces in the rectum causes it to stretch, which initiates a reflex action resulting in rectal contraction and expulsion of the fecal material.

Polyps

The colon is subject to the development of **polyps**, small growths arising from the epithelial lining. Polyps, whether benign or cancerous, can be removed surgically. If colon cancer is detected while still confined to a polyp, the expected outcome is a complete cure. Some investigators believe that dietary fat increases the likelihood of colon cancer because dietary fat causes an increase in bile secretion. It could be that intestinal bacteria convert bile salts to substances that promote the development of cancer. On the other hand, fiber in the diet seems to inhibit the development of colon cancer. Dietary fiber absorbs water and adds bulk, thereby diluting the concentration of bile salts and facilitating the movement of substances through the intestine. Regular elimination reduces the time that the colon wall is exposed to any cancer-promoting agents in feces.

Diarrhea and Constipation

Two common everyday complaints associated with the large intestine are **diarrhea** and **constipation.** The major causes of diarrhea are infection of the lower tract and nervous stimulation. In the case of infection, such as food poisoning caused by eating contaminated food, the intestinal wall becomes irritated, and peristalsis increases. Water is not absorbed, and the diarrhea that results rids the body of the infectious organisms. In nervous diarrhea, the nervous system stimulates the intestinal wall, and diarrhea results. Prolonged diarrhea can lead to dehydration because of water loss and to disturbances in the heart's contraction due to an imbalance of salts in the blood.

When a person is constipated, the feces are dry and hard. One reason for this condition is that socialized persons have learned to inhibit defecation to the point that the desire to defecate is ignored. Two components of the diet that can help prevent constipation are water and fiber. Water intake prevents drying out of the feces, and fiber provides the bulk needed for elimination. The frequent use of laxatives is discouraged. If, however, it is necessary to take a laxative, a bulk laxative is the most natural because, like fiber, it produces a soft mass of cellulose in the colon. Lubricants, like mineral oil, make the colon slippery, and saline laxatives, like milk of magnesia, act osmotically—they prevent water from being absorbed and, depending on the dosage, may even cause water to enter the colon. Some laxatives are irritants; they increase peristalsis to the degree that the contents of the colon are expelled.

Chronic constipation is associated with the development of hemorrhoids, enlarged and inflamed blood vessels at the anus.

The large intestine does not produce digestive enzymes; it does absorb water, salts, and some vitamins.

4.2 Three Accessory Organs

The pancreas, liver, and gallbladder are accessory digestive organs. Figure 4.1 shows how the pancreatic duct from the pancreas and the common bile duct from the liver and gallbladder join before entering the duodenum.

The Pancreas

The **pancreas** lies deep in the abdominal cavity, resting on the posterior abdominal wall. It is an elongated and somewhat flattened organ that has both an endocrine and an exocrine function. As an endocrine gland it secretes insulin and glucagon, hormones that help keep the blood glucose level within normal limits. We are now interested in its exocrine function. Most pancreatic cells produce pancreatic juice, which contains sodium bicarbonate ($NaHCO_3$) and digestive enzymes for all types of food. Sodium bicarbonate neutralizes chyme; whereas pepsin acts best in an acid pH of the stomach, pancreatic enzymes require a slightly basic pH. **Pancreatic amylase** digests starch, **trypsin** digests protein, and **lipase** digests fat. In cystic fibrosis, a thick mucus blocks the pancreatic duct, and the patient must take supplemental pancreatic enzymes by mouth for proper digestion to occur.

The Liver

The **liver,** which is the largest organ in the body, lies mainly in the upper right section of the abdominal cavity, under the diaphragm (see Fig. 4.1). The liver has two main lobes, the right lobe and the smaller left lobe, which crosses the midline and lies above the stomach. The liver contains approximately 100,000 lobules that serve as the structural functional units of the liver (Fig. 4.10). Triads consisting of these three structures are located between the lobules: (1) a branch of the hepatic artery that brings oxygenated blood to the liver; (2) a branch of the hepatic portal vein that transports nutrients from the intestines; and (3) a bile duct that takes bile away from the liver. The central veins of lobules enter the hepatic vein. Note in Figure 4.11 that the liver lies between the hepatic portal vein (number 2 in the figure), and the hepatic vein (number 4 in the figure) which enters the vena cava.

In some ways, the liver acts as the gatekeeper to the blood. As the blood from the intestines passes through the liver, it removes poisonous substances and works to keep the contents of the blood constant. It also removes and stores iron and the fat-soluble vitamins A, D, E, and K. The liver makes the plasma proteins from amino acids, and lipids from fatty acids. It also produces cholesterol and helps regulate the quantity of this substance in the blood.

The liver maintains the blood glucose level at about 100 mg/100 ml (0.1%), even though a person eats intermittently. Any excess glucose that is present in the hepatic por-

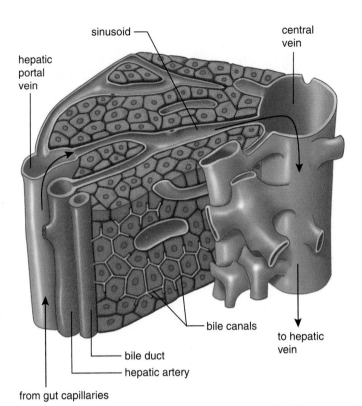

Figure 4.10   Hepatic lobules.
The liver contains over 100,000 lobules. Each lobule contains many cells that perform the various functions of the liver. They remove from and/or add materials to blood and deposit bile in bile ducts.

tal vein is removed and stored by the liver as glycogen. Between eating, glycogen is broken down to glucose, which enters the hepatic vein, and in this way, the blood glucose level remains constant.

If the supply of glycogen is depleted, the liver will convert glycerol (from fats) and amino acids to glucose molecules. The conversion of amino acids to glucose necessitates deamination, the removal of amino acids. By a complex metabolic pathway, the liver then combines ammonia with carbon dioxide to form urea:

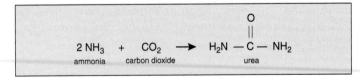

$$2\,NH_3 \;+\; CO_2 \;\longrightarrow\; H_2N - \overset{\displaystyle O}{\overset{\displaystyle \|}{C}} - NH_2$$

ammonia   carbon dioxide   urea

Urea is the usual nitrogenous waste product from amino acid breakdown in humans. After its formation in the liver, urea is excreted by the kidneys.

The liver produces bile, which is stored in the gallbladder. Bile has a yellowish green color because it contains the bile pigment bilirubin, derived from the breakdown of hemoglobin, the red pigment of red blood cells. Bile also contains bile salts, which are derived from cholesterol and emulsify fat in the small intestine. When fat is emulsified, it breaks up into

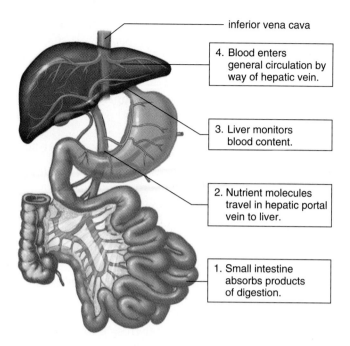

inferior vena cava

4. Blood enters general circulation by way of hepatic vein.

3. Liver monitors blood content.

2. Nutrient molecules travel in hepatic portal vein to liver.

1. Small intestine absorbs products of digestion.

Figure 4.11 Hepatic portal system.
The hepatic portal vein takes the products of digestion from the digestive system to the liver, where they are processed before entering the cardiovascular system proper.

droplets, providing a much larger surface area, which can be acted upon by a digestive enzyme from the pancreas.

Altogether, the following are significant ways in which the liver helps maintain homeostasis.

1. Detoxifies blood by removing and metabolizing poisonous substances.
2. Stores iron (Fe^{2+}) and the fat-soluble vitamins A, D, E, and K.
3. Makes plasma proteins, such as albumins and fibrinogen, from amino acids.
4. Stores glucose as glycogen after eating, and breaks down glycogen to glucose to maintain the glucose concentration of blood between eating periods.
5. Produces urea from the breakdown of amino acids.
6. Removes bilirubin, a breakdown product of hemoglobin from the blood, and excretes it in bile, a liver product.
7. Produces lipids from fatty acids; produces and helps regulate blood cholesterol level, converting some to bile salts.

Liver Disorders

Jaundice, hepatitis, and cirrhosis are three serious diseases that affect the entire liver and hinder its ability to repair itself. Therefore, they are life-threatening diseases. When a person has **jaundice,** there is a yellowish tint to the whites of the eyes and also to the skin of light-pigmented persons. Bilirubin is deposited in the skin due to an abnormally large amount in the blood. In *hemolytic jaundice,* red blood cells have been broken down in abnormally large amounts; in *obstructive jaundice,* bile ducts are blocked or liver cells are damaged.

Jaundice can also result from **hepatitis,** inflammation of the liver. Viral hepatitis occurs in several forms. Hepatitis A is usually acquired from sewage-contaminated drinking water. Hepatitis B, which is usually spread by sexual contact, can also be spread by blood transfusions or contaminated needles. The hepatitis B virus is more contagious than the AIDS virus, which is spread in the same way. Thankfully, however, there is now a vaccine available for hepatitis B. Hepatitis C, which is usually acquired by contact with infected blood and for which there is no vaccine, can lead to chronic hepatitis, liver cancer, and death.

Cirrhosis is another chronic disease of the liver. First the organ becomes fatty, and liver tissue is then replaced by inactive fibrous scar tissue. Cirrhosis of the liver is often seen in alcoholics due to malnutrition and to the excessive amounts of alcohol (a toxin) the liver is forced to break down.

The liver has amazing generative powers and can recover if the rate of regeneration exceeds the rate of damage. During liver failure, however, there may not be enough time to let the liver heal itself. Liver transplantation is usually the preferred treatment for liver failure, but artificial livers have been developed and tried in a few cases. One type is a cartridge that contains liver cells. The patient's blood passes through cellulose acetate tubing of the cartridge and is serviced in the same manner as with a normal liver. In the meantime, the patient's liver has a chance to recover.

The Gallbladder

The **gallbladder** is a pear-shaped, muscular sac attached to the surface of the liver (see Fig. 4.1). About 1,000 ml of bile are produced by the liver each day, and any excess is stored in the gallbladder. Water is reabsorbed by the gallbladder so that bile becomes a thick, mucus-like material. When needed, bile leaves the gallbladder and proceeds to the duodenum via the common bile duct.

The cholesterol content of bile can come out of solution and form crystals. If the crystals grow in size, they form gallstones. The passage of the stones from the gallbladder may block the common bile duct and cause obstructive jaundice. Then the gallbladder must be removed.

The pancreas produces pancreatic juice, which contains enzymes for the digestion of food. Among its many functions, the liver produces bile, which is stored in the gallbladder.

4.3 Digestive Enzymes

The digestive enzymes are **hydrolytic enzymes,** which break down substances by the introduction of water at specific bonds. Digestive enzymes, like other enzymes, are proteins with a particular shape that fits their substrate. They also have an optimum pH, which maintains their shape, thereby enabling them to speed up their specific reaction.

The various digestive enzymes present in the digestive juices, mentioned previously, help break down carbohydrates, proteins, nucleic acids, and fats, the major components of food. Starch is a carbohydrate, and its digestion begins in the mouth. Saliva from the salivary glands has a neutral pH and contains **salivary amylase,** the first enzyme to act on starch:

$$\text{starch} + H_2O \xrightarrow{\text{salivary amylase}} \text{maltose}$$

In this equation, salivary amylase is written above the arrow to indicate that it is neither a reactant nor a product in the reaction. It merely speeds the reaction in which its substrate, starch, is digested to many molecules of maltose, a disaccharide. Maltose molecules cannot be absorbed by the intestine; additional digestive action in the small intestine converts maltose to glucose, which can be absorbed.

Protein digestion begins in the stomach. Gastric juice secreted by gastric glands has a very low pH—about 2—because it contains hydrochloric acid (HCl). Pepsinogen, a precursor that is converted to the enzyme **pepsin** when exposed to HCl, is also present in gastric juice. Pepsin acts on protein to produce peptides:

$$\text{protein} + H_2O \xrightarrow{\text{pepsin}} \text{peptides}$$

Peptides vary in length, but they always consist of a number of linked amino acids. Peptides are usually too large to be absorbed by the intestinal lining, but later they are broken down to amino acids in the small intestine.

Starch, proteins, nucleic acids, and fats are all enzymatically broken down in the small intestine. Pancreatic juice, which enters the duodenum, has a basic pH because it contains sodium bicarbonate ($NaHCO_3$). Sodium bicarbonate neutralizes chyme, producing the slightly basic pH that is optimum for pancreatic enzymes. One pancreatic enzyme, **pancreatic amylase,** digests starch:

$$\text{starch} + H_2O \xrightarrow{\text{pancreatic amylase}} \text{maltose}$$

Another pancreatic enzyme, **trypsin,** digests protein:

$$\text{protein} + H_2O \xrightarrow{\text{trypsin}} \text{peptides}$$

Trypsin is secreted as trypsinogen, which is converted to trypsin in the duodenum.

Lipase, a third pancreatic enzyme, digests fat molecules in the fat droplets after they have been emulsified by bile salts:

$$\text{fat} \xrightarrow{\text{bile salts}} \text{fat droplets}$$

$$\text{fat droplets} + H_2O \xrightarrow{\text{lipase}} \text{glycerol} + \text{fatty acids}$$

The end products of lipase digestion, glycerol and fatty acid molecules, are small enough to cross the cells of the intestinal villi, where absorption takes place. As mentioned previously, glycerol and fatty acids enter the cells of the villi, and within these cells, they are rejoined and packaged as lipoprotein droplets before entering the lacteals (see Fig. 4.6).

Peptidases and **maltase,** two enzymes secreted by the small intestine, complete the digestion of protein to amino acids and starch to glucose, respectively. Amino acids and glucose are small molecules that cross into the cells of the villi. Peptides, which result from the first step in protein digestion, are digested to amino acids by peptidases:

$$\text{peptides} + H_2O \xrightarrow{\text{peptidases}} \text{amino acids}$$

Maltose, a disaccharide that results from the first step in starch digestion, is digested to glucose by maltase:

$$\text{maltose} + H_2O \xrightarrow{\text{maltase}} \text{glucose} + \text{glucose}$$

Other disaccharides, each of which has its own enzyme, are digested in the small intestine. The absence of any one of these enzymes can cause illness. For example, many people, including as many as 75% of African Americans, cannot digest lactose, the sugar found in milk, because they do not produce lactase, the enzyme that converts lactose to its components, glucose and galactose. Drinking untreated milk often gives these individuals the symptoms of **lactose intolerance** (diarrhea, gas, cramps), caused by a large quantity of nondigested lactose in the intestine. In most areas, it is possible to purchase milk made lactose-free by the addition of synthetic lactase or *Lactobacillus acidophilus* bacteria, which break down lactose.

Table 4.2 lists some of the major digestive enzymes produced by the digestive tract, salivary glands, or the pancreas. Each type of food is broken down by specific enzymes.

Digestive enzymes present in digestive juices help break down food to the nutrient molecules: glucose, amino acids, fatty acids, and glycerol. The first two are absorbed into the blood capillaries of the villi, and the last two re-form within epithelial cells before entering the lacteals as lipoprotein droplets.

Table 4.2   Major Digestive Enzymes

Food	Digestion	Enzyme	Optimum pH	Produced by	Site of Action
Starch	Starch + H_2O → maltose	Salivary amylase	Neutral	Salivary glands	Mouth
		Pancreatic amylase	Basic	Pancreas	Small intestine
	Maltose + H_2O → glucose	Maltase	Basic	Small intestine	Small intestine
Protein	Protein + H_2O → peptides	Pepsin	Acidic	Gastric glands	Stomach
		Trypsin	Basic	Pancreas	Small intestine
	Peptides + H_2O → amino acids	Peptidases	Basic	Small intestine	Small intestine
Nucleic Acid	RNA and DNA + H_2O → nucleotides	Nuclease	Basic	Pancreas	Small intestine
	Nucleotides → bases, sugars, phosphate ions	Nucleosidases	Basic	Small intestine	Small intestine
Fat	Fat droplets + H_2O → glycerol + fatty acids	Lipase	Basic	Pancreas	Small intestine

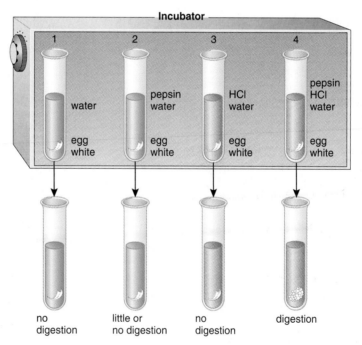

Figure 4.12   Digestion experiment.
This experiment is based on the optimum conditions for digestion by pepsin in the stomach. Knowing that the correct enzyme, optimum pH, optimum temperature, and the correct substrate must be present for digestion to occur, explain the results of this experiment.

Conditions for Digestion

Laboratory experiments can define the necessary conditions for digestion. For example, the four test tubes described in Figure 4.12 can be prepared and observed for the digestion of egg white, a protein digested in the stomach by the enzyme pepsin.

After all tubes are placed in an incubator at body temperature for at least one hour, the results depicted are observed. Tube 1 is a control tube; no digestion has occurred in this tube because the enzyme and HCl are missing. (If a con-

trol gives a positive result, then the experiment is invalidated.) Tube 2 shows limited or no digestion because HCl is missing, and therefore the pH is too high for pepsin to be effective. Tube 3 shows no digestion because although HCl is present, the enzyme is missing. Tube 4 shows the best digestive action because the enzyme is present and the presence of HCl has resulted in an optimum pH. This experiment supports the hypothesis that for digestion to occur, the substrate and enzyme must be present and the environmental conditions must be optimum. The optimal environmental conditions include a warm temperature and the correct pH.

Human Systems Work Together

Integumentary System

Digestive tract provides nutrients needed by skin.

Skin helps to protect digestive organs; helps to provide vitamin D for Ca^{2+} absorption.

Skeletal System

Digestive tract provides Ca^{2+} and other nutrients for bone growth and repair.

Bones provide support and protection; hyoid bone assists swallowing.

Muscular System

Digestive tract provides glucose for muscle activity; liver metabolizes lactic acid following anaerobic muscle activity.

Smooth muscle contraction accounts for peristalsis; skeletal muscles support and help protect abdominal organs.

Nervous System

Digestive tract provides nutrients for growth, maintenance, and repair of neurons and neuroglial cells.

Brain controls nerves, which innervate smooth muscle and permit tract movements.

Endocrine System

Stomach and small intestine produce hormones.

Hormones help control secretion of digestive glands and accessory organs; insulin and glucagon regulate glucose storage in liver.

How the Digestive System works with other body systems

Cardiovascular System

Digestive tract provides nutrients for plasma protein formation and blood cell formation; liver detoxifies blood, makes plasma proteins, destroys old red blood cells.

Blood vessels transport nutrients from digestive tract to body; blood services digestive organs.

Lymphatic System/Immunity

Digestive tract provides nutrients for lymphoid organs; stomach acidity prevents pathogen invasion of body.

Lacteals absorb fats; Peyer's patches prevent invasion of pathogens; appendix contains lymphoid tissue.

Respiratory System

Breathing is possible through the mouth because digestive tract and respiratory tract share the pharynx.

Gas exchange in lungs provides oxygen to digestive tract and excretes carbon dioxide from digestive tract.

Urinary System

Liver synthesizes urea; digestive tract excretes bile pigments from liver and provides nutrients.

Kidneys convert vitamin D to active form needed for Ca^{2+} absorption; compensate for any water loss by digestive tract.

Reproductive System

Digestive tract provides nutrients for growth and repair of organs and for development of fetus.

Pregnancy crowds digestive organs and promotes heartburn and constipation.

4.4 Homeostasis

The previous page tells how the digestive system works with other systems in the body to maintain homeostasis.

Within the digestive tract the food we eat is broken down to nutrients small enough to be absorbed by the villi of the small intestine. Digestive enzymes are produced by the salivary glands, gastric glands, and intestinal glands. Three accessory organs of digestion (the pancreas, the liver, and the gallbladder) also contribute secretions that help break down food. The liver produces bile (stored by the gallbladder), which emulsifies fat. The pancreas produces enzymes for the digestion of carbohydrates, proteins, and fat. Secretions from these glands, which are sent by ducts into the small intestine, are regulated by hormones such as secretin produced by the digestive tract. Therefore, the digestive tract is also a part of the endocrine system.

Blood laden with nutrients passes from the region of the small intestine to the liver by way of the hepatic portal vein. The liver is the most important of the metabolic organs. Aside from making bile, the liver regulates the cholesterol content of the blood, makes plasma proteins, stores glucose as glycogen, produces urea, and metabolizes poisons. Because the liver is such an important organ, diseases affecting the liver such as hepatitis and cirrhosis are extremely dangerous.

4.5 Nutrition

The body requires three major classes of *macronutrients* in the diet: carbohydrate, protein, and fat. These supply the energy and the building blocks that are needed to synthesize cellular contents. *Micronutrients*—especially vitamins and minerals—are also required because they are necessary for optimum cellular metabolism.

Several modern nutritional studies suggest that certain nutrients can protect against heart disease, cancer, and other serious illnesses. These studies include an analysis of the eating habits of healthy people in the United States and from around the world, especially those with lower rates of heart disease and cancer. The result has been the dietary recommendations illustrated by a food pyramid (Fig. 4.13).

The bulk of the diet should consist of bread, cereal, rice, and pasta as energy sources. Whole grains are preferred over those that have been milled because they contain fiber and vitamins and minerals. Vegetables and fruits are another rich source of fiber, vitamins, and minerals. Notice, then, that a largely vegetarian diet is recommended.

Animal products, especially meat, need only be minimally included in the diet; fats and sweets should be used sparingly. Dairy products and meats tend to be high in saturated fats, and an intake of saturated fats increases the risk of cardiovascular disease (see Lipids, p. 98). Low-fat dairy products are available, but there is no way to take much of the fat out of meat. Beef meat, in particular, contains a relatively high fat content. Ironically, the affluence of people in the United States contributes to a poor diet and, therefore, possible illness. Only comparatively rich people can afford fatty meats from grain-fed cattle and carbohydrates that have been highly processed to remove fiber and to add sugar and salt.

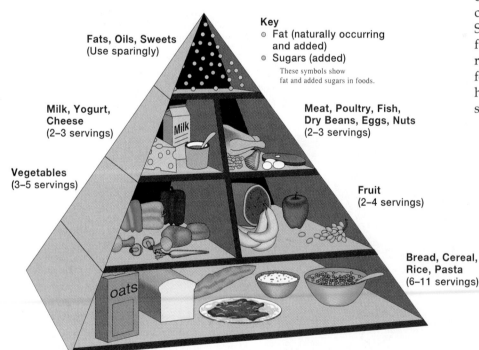

Key
○ Fat (naturally occurring and added)
○ Sugars (added)
These symbols show fat and added sugars in foods.

Fats, Oils, Sweets (Use sparingly)

Milk, Yogurt, Cheese (2–3 servings)

Vegetables (3–5 servings)

Meat, Poultry, Fish, Dry Beans, Eggs, Nuts (2–3 servings)

Fruit (2–4 servings)

Bread, Cereal, Rice, Pasta (6–11 servings)

Figure 4.13   Food guide pyramid: A guide to daily food choices.
The U.S. Department of Agriculture uses a pyramid to show the ideal diet because it emphasizes the importance of including grains, fruits, and vegetables in the diet. Meats and dairy products are needed in limited amounts; fats, oils, and sweets should be used sparingly.
Source: Data from the U.S. Department of Agriculture.

Carbohydrates

The quickest, most readily available source of energy for the body is glucose. Carbohydrates are digested to simple sugars, which are or can be converted to glucose. Glucose is stored by the liver in the form of glycogen. Between eating periods, the blood glucose level is maintained at about 100 mg/100 ml of blood by the breakdown of glycogen or by the conversion of glycerol (from fats) or amino acids to glucose. If necessary, amino acids are taken from the muscles—even from the heart muscle. While body cells can utilize fatty acids as an energy source, brain cells require glucose. For this reason alone, it is necessary to include carbohydrates in the diet. According to Fig. 4.13, carbohydrates should make up the bulk of the diet. Further, these carbohydrates should be complex and not simple carbohydrates. Complex sources of carbohydrates include preferably whole-grain pasta, rice, bread, and cereal (Fig. 4.14). Potatoes and corn, although considered vegetables, are also sources of carbohydrates.

Simple carbohydrates (e.g., sugars) are labeled "empty calories" by some dieticians because they contribute to energy needs and weight gain without supplying any other nutritional requirements. Table 4.3 gives suggestions on how to reduce dietary sugars (simple carbohydrates). In contrast to simple sugars, complex carbohydrates are likely to be accompanied by a wide range of other nutrients and by **fiber,** which is indigestible plant material.

The intake of fiber is recommended because it decreases the risk of colon cancer, a major type of cancer, and cardiovascular disease, the number one killer in the United States. Insoluble fiber, such as that found in wheat bran, has a laxative effect and may guard against colon cancer, because any cancer-causing substances are in contact with the intestinal wall for a limited amount of time. Soluble fiber, such as that found in oat bran, combines with bile acids and cholesterol in the intestine and prevents them from being absorbed. The liver now removes cholesterol from the blood and changes it to bile acids, replacing those that were lost. While the diet should have an adequate amount of fiber, a high-fiber diet can be detrimental. Some evidence suggests that the absorption of iron, zinc, and calcium is impaired by a diet too high in fiber.

Complex carbohydrates, which contain fiber, should form the bulk of the diet.

Proteins

Foods rich in protein include red meat, fish, poultry, dairy products, legumes (i.e., peas and beans), nuts, and cereals. Following digestion of protein, amino acids enter the bloodstream and are transported to the tissues. Ordinarily, amino

Table 4.3	Reducing Dietary Sugar
To reduce dietary sugar:	

1. Eat fewer sweets, such as candy, soft drinks, ice cream, and pastry.
2. Eat fresh fruits or fruits canned without heavy syrup.
3. Use less sugar—white, brown, or raw—and less honey and syrups.
4. Avoid sweetened breakfast cereals.
5. Eat less jelly, jam, preserves
6. Drink pure fruit juices, not imitations.
7. When cooking, use spices like cinnamon instead of sugar to flavor foods.
8. Do not put sugar in tea or coffee.

Figure 4.14 Complex carbohydrates. To meet our energy needs, dieticians recommend consuming foods rich in complex carbohydrates, like those shown here, rather than foods consisting of simple carbohydrates, like candy and ice cream. Simple carbohydrates provide monosaccharides but few other types of nutrients.

acids are not used as an energy source. Most are incorporated into structural proteins found in muscles, skin, hair, and nails. Others are used to synthesize such proteins as hemoglobin, plasma proteins, enzymes, and hormones.

Adequate protein formation requires 20 different types of amino acids. Of these, eight are required from the diet in adults (nine in children) because the body is unable to produce them. These are termed the **essential amino acids.** The body produces the other 11 amino acids by simply transforming one type into another type. Some protein sources, such as meat, are complete; they provide all 20 types of amino acids. Vegetables and grains supply us with amino acids, but each vegetable or grain alone is an *incomplete* protein source because of a deficiency in at least one of the essential amino acids. Absence of one essential amino acid prevents utilization of the other 19 amino acids. Soybeans and tofu, made from soybeans, are rich in amino acids, but it is wise to combine foods to acquire all the essential amino acids. For example, the combinations of cereal with milk, or beans, a legume, with rice, a grain, will provide all the essential amino acids (Table 4.4).

Amino acids are not stored in the body, and a daily supply is needed. However, it does not take very much protein to meet the daily requirement. Two servings of meat a day (equal in total quantity to a deck of cards) is usually enough. Some meats (e.g., hamburger) are high in protein but also high in fat. Everything considered, it is probably a good idea to depend on protein from plant origins (e.g., whole-grain cereals, dark breads, and legumes) to a greater extent than is often the custom in the United States. This can be illustrated by the health statistics of native Hawaiians who no longer eat as their ancestors did (Fig. 4.15). The modern diet depends on animal rather than plant protein and is 42% fat. A statistical study showed that the island's native peoples now have a higher than average death rate from cardiovascular disease and cancer. Diabetes is also common in persons who follow the modern diet. But the health of those who have switched back to the ancient diet has improved immensely!

Nutritionists do not recommend the use of protein and/or amino acid supplements. Protein supplements that people take to build muscle or as part of a diet are not digested as well as protein-rich foods and they cost more than food. Amino acid supplements can be dangerous to your health. An excess of any particular amino acid can lead to a deficiency of absorption of other amino acids present in lesser amounts. Contrary to popular reports, the taking of lysine does not relieve or cure herpes sores. It is also unwise to take tryptophan to

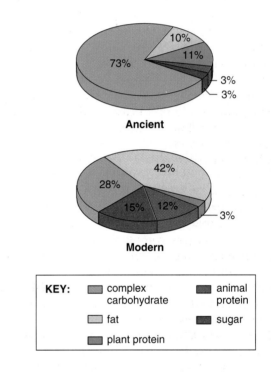

Figure 4.15   **Ancient versus modern diet of native Hawaiians.**
Among those native Hawaiians who have switched back to the native diet, the incidence of cardiovascular disease, cancer, and diabetes has dropped.

Table 4.4	**Complementary Protein Combinations**		
Combine foods from two or more of these columns to obtain complete protein.			
Grains	*Legumes*	*Seeds and Nuts*	*Vegetables*
Barley	Dried beans	Sesame seeds	Leafy greens
Bulgur	Dried lentils	Sunflower seeds	Broccoli
Cornmeal	Dried peas	Walnuts	Others (see exchange list)
Oats	Peanuts	Cashews	
Rice	Soy products	Other nuts	
Whole-grain breads		Nut butters	
Pasta			

From Sizer, F. S. and Whitney, E. N. 1994. Hamilton/Whitney's Nutrition: Concepts and Controversies. 6th ed. Minneapolis: West Publishing Company, page 205.

relieve pain or cure insomnia. Some who have taken supplements of tryptophan have come down with a blood disorder (eosinophilia-myalgia syndrome) characterized by severe muscle and joint pain and swelling of the limbs.

Lipids

Fat and cholesterol are both lipids. Fat is present not only in butter, margarine, and oils, but also in many foods high in animal protein. The body can alter ingested fat to suit the body's needs, except it is unable to produce linoleic acid, which is a polyunsaturated fatty acid. Saturated fatty acids have no double bonds; polyunsaturated fatty acids have many double bonds.

The current guidelines suggest that fat should account for no more than 30% of our daily calories. The chief reason is that an intake of fat not only causes weight gain, it also increases the risk of cancer and cardiovascular disease. Dietary fat apparently increases the risk of colon, hepatic, and pancreatic cancers. Although recent studies suggest no link between dietary fat and breast cancer, other researchers still believe that the matter deserves further investigation.

Cardiovascular disease is often due to arteries blocked by fatty deposits, called **plaque,** that contain saturated fats and cholesterol. Cholesterol is carried in the blood by two types of lipoproteins: low-density lipoprotein (LDL) and high-density lipoprotein (HDL). LDL is thought of as being "bad" because it carries cholesterol from the liver to the cells, while HDL is thought of as being "good" because it carries cholesterol to the liver, which takes it up and converts it to bile salts. Saturated fats, whether in butter or margarine, can raise LDL cholesterol levels, while monounsaturated (one double bond) fats and polyunsaturated (many double bonds) fats lower LDL cholesterol levels. Olive oil and canola oil contain mostly monounsaturated fats; corn oil and safflower oil contain mostly polyunsaturated fats. These oils have a liquid consistency and come from plants. Saturated fats, which are solids at room temperature, usually have an animal origin; two well-known exceptions are palm oil and coconut oil, which contain mostly saturated fats and come from the plants mentioned.

Nutritionists stress that it is more important for the diet to be low in fat rather than be overly concerned about which type fat is in the diet. Still, polyunsaturated fats are nutritionally essential because they are the only type of fat that contains linoleic acid, a fatty acid the body cannot make. Table 4.5 gives suggestions on how to reduce dietary fat.

Fake Fat

Olestra is a substance made to look, taste, and act like real fat but the digestive system is unable to digest it. It travels down the length of the digestive system without being

Table 4.5	Reducing Lipids

To reduce dietary fat:

1. Choose poultry, fish, or dry beans and peas as a protein source.
2. Remove skin from poultry before cooking, and place on a rack so that fat drains off.
3. Broil, boil, or bake rather than fry.
4. Limit your intake of butter, cream, hydrogenated oils, shortenings, and tropical oils (coconut and palm oils).*
5. Use herbs and spices to season vegetables instead of butter, margarine, or sauces. Use lemon juice instead of salad dressing.
6. Drink skim milk instead of whole milk, and use skim milk in cooking and baking.
7. Eat nonfat or low-fat foods.

To reduce dietary cholesterol:

1. Avoid cheese, egg yolks, liver, and certain shellfish (shrimp and lobster). Preferably, eat white fish and poultry.
2. Substitute egg whites for egg yolks in both cooking and eating.
3. Include soluble fiber in the diet. Oat bran, oatmeal, beans, corn, and fruits such as apples, citrus fruits, and cranberries are high in soluble fiber.

*Although coconut and palm oils are from plant sources, they are mostly saturated fats.

absorbed or contributing any calories to the day's total. Therefore it is commonly known as "fake fat." Unfortunately, the fat-soluble vitamins A, D, E, and K tend to be taken up by olestra and thereafter they are not absorbed by the body. Similarly, people using olestra have reduced amounts of carotenoids in the blood. In one study, just a handful of olestra-soaked potato chips caused a 20% decline in blood beta-carotene levels. Manufacturers fortify olestra-containing foods with the vitamins mentioned but not carotenoids.

More apparent, some people who consume olestra have developed anal leakage or underwear staining. Others experience diarrhea, intestinal cramping, and gas. Presently the FDA has limited the use of olestra to potato chips and other salty snacks but the manufacturer wants approval to add it to ice cream, salad dressings and cheese.

Dietary protein supplies the essential amino acids; proteins from plant origins generally have less accompanying fat. A diet no more than 30% fat is recommended because fat intake, particularly saturated fats, is known to be associated with various health problems.

Weight Loss the Healthy Way

For a woman 19 to 22 years, 5 feet 4 inches tall, who exercises lightly, 2,100 Cal[1] per day are normally recommended. For a man the same age, 5 feet 10 inches tall, who exercises lightly, the recommendation is 2,900 Cal. Those who wish to lose weight need to reduce their caloric intake and/or increase their level of exercise. Exercising is a good idea, because to maintain good nutrition, the intake of calories per day should probably not go below 1,200 Cal. Also, for the reasons discussed in this chapter, carbohydrates should still make up at least 58% of these calories, proteins should be no more than 25%, and the rest can be fats. A deficit of 500 Cal a day (through intake reduction or increased exercise) is sufficient to lose a pound of body fat in a week. Once you realize that a diet needs to be judged according to the principles of adequacy of nutrients, of balance in regard to carbohydrates, proteins, and fats, of moderation in number of calories, and of variety of food sources, it is easy to see that many of the diets and gimmicks people use to lose weight are bad for their health. Unhealthy approaches include the following:

Pills

The most familiar pills, and the only ones approved by the FDA, are those that claim to suppress the appetite. They may work at first, but the appetite soon returns to normal and weight lost is regained. Then the user has the problem of trying to get off the drug without gaining more weight. Other types of pills are under investigation and sometimes can be obtained illegally. But, as yet, there is no known drug that is both safe and effective for weight loss.

Low-Carbohydrate Diets

The dramatic weight loss that occurs with a low-carbohydrate diet is not due to a loss of fat; it is due to a loss of muscle mass and water. Glycogen and important minerals are also lost. When a normal diet is resumed, so is the normal weight.

Liquid Diets

Despite the fact that liquid diets provide proteins and vitamins, the number of Cal is so restricted that the body cannot burn fat quickly enough to compensate, and muscle is still broken down to provide energy. A few people on this regime have died, probably because even the heart muscle was not spared by the body.

Single-Category Diets

These diets rely on the intake of only one kind of food, either a fruit or vegetable or rice alone. However, no single type of food provides the balance of nutrients needed to maintain health. Some dieters on strange diets suffer the consequences—in one instance an individual lost hair and fingernails.

[1]Cal = 1,000 calories

Questions to Ask About a Weight-Loss Diet

1. Does the diet have a reasonable number of Cal?
   10 Cal per pound of current weight is suggested. In any case, no fewer than 1,000–1,200 Cal for a normal-sized person.
2. Does the diet provide enough protein?
   For a woman 120 lb, 44 grams protein each day is recommended. For a man 154 lb, 56 grams is recommended. More than twice this amount is too much.
   For reference, 1 c milk and 1 oz meat each has 8 grams protein.
3. Does the diet provide too much fat?
   No more than 20–30% of total Cal is recommended.
   For reference, a pat of butter has 45 Cal. 1 gram fat = 9 Cal.
4. Does the diet provide enough carbohydrates?
   100 grams = 400 Cal is the very least recommended per day; 50% of total Cal should be carbohydrates.
   For reference, a slice of bread contains 14 grams of carbohydrates.
5. Does the diet provide a balanced assortment of foods?
   The diet should include breads, cereals, legumes; vegetables (especially dark-green and yellow ones); low-fat milk products; and meats or a meat substitute.
6. Does the diet make use of ordinary foods that are available locally?
   Diets should not require the purchase of unusual or expensive foods.

Figure 4A Fruit is a healthy and low calorie snack.

Vitamins

Vitamins are organic compounds (other than carbohydrate, fat, and protein) that the body is unable to produce but uses for metabolic purposes. Many vitamins are portions of coenzymes, which are enzyme helpers. For example, niacin is part of the coenzyme NAD, and riboflavin is part of another dehydrogenase, FAD. Coenzymes are needed in only small amounts because each can be used over and over again. Not all vitamins are coenzymes; vitamin A, for example, is a precursor for the visual pigment that prevents night blindness. If vitamins are lacking in the diet, various symptoms develop (Fig. 4.16). Altogether there are 13 vitamins, which are divided into those that are fat soluble (Table 4.6) and those that are water soluble (Table 4.7).

Antioxidants

Over the past 20 years, numerous statistical studies have been done to determine whether a diet rich in fruits and vegetables is protective against cancer. Cellular metabolism generates free radicals, unstable molecules that carry an extra electron. The most common free radicals in cells are the superoxide (O_2^-) and hydroxide (OH^-). In order to stabilize themselves, free radicals donate an electron to DNA, or proteins, including enzymes, or lipids, which are found in plasma membranes. Such donations most likely damage these cellular molecules and thereby may lead to disorders, perhaps even cancer.

Vitamins C, E, and A are believed to defend the body against free radicals, and therefore they are termed antioxidants. These vitamins are especially abundant in fruits and vegetables. The dietary guidelines shown in Figure 4.13 suggest that we eat a minimum of five servings of fruits and vegetables a day. To achieve this goal, include salad greens, raw or cooked vegetables, dried fruit, and fruit juice, in addition to traditional apples and oranges and such.

Dietary supplements may provide a potential safeguard against cancer and cardiovascular disease, but nutritionists do not think it is appropriate to take supplements instead of improving intake of fruits and vegetables. There are many beneficial compounds in fruits that cannot be obtained from a vitamin pill. These compounds enhance each other's absorption or action and also perform independent biological functions.

Vitamin D

Skin cells contain a precursor cholesterol molecule that is converted to vitamin D after UV exposure. Vitamin D leaves the skin and is modified first in the kidneys and then in the liver until finally it becomes calcitriol. Calcitriol promotes the absorption of calcium by the intestines. The lack of vitamin D leads to rickets in children (see Fig. 4.16a). Rickets, characterized by a bowing of the legs, is caused by defective mineralization of the skeleton. Most milk today is fortified with vitamin D, which helps prevent the occurrence of rickets.

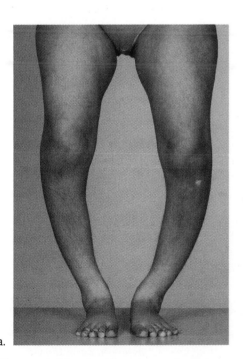

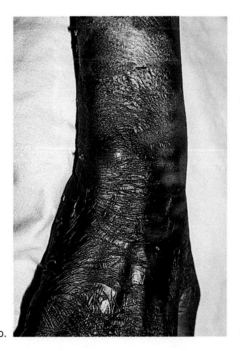

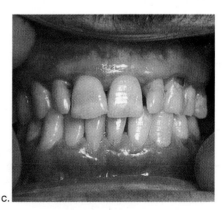

Vitamins are essential to cellular metabolism; many are protective against dentifiable illnesses and conditions.

a.   b.   c.

Figure 4.16   **Illnesses due to vitamin deficiency.**
a. Bowing of bones (rickets) due to vitamin D deficiency. **b.** Dermatitis (pellagra) of areas exposed to light due to niacin (vitamin B₃) deficiency.
c. Bleeding of gums (scurvy) due to vitamin C deficiency.

Table 4.6   Fat-Soluble Vitamins

Vitamin	Functions	Food Sources	Conditions with	
			Too Little	**Too Much**
Vitamin A	Antioxidant synthesized from beta-carotene; needed for healthy eyes, skin, hair, and mucous membranes, and for proper bone growth	Deep yellow/orange and leafy, dark green vegetables, fruits, cheese, whole milk, butter, and eggs	Night blindness, impaired growth of bones and teeth	Headache, dizziness, nausea, hair loss
Vitamin D	A group of steroids needed for development and maintenance of bones and teeth	Milk fortified with vitamin D, fish liver oil; also made in the skin when exposed to sunlight	Rickets, bone decalcification and weakening	Calcification of soft tissues, diarrhea, and possible renal damage
Vitamin E	Antioxidant that prevents oxidation of vitamin A and polyunsaturated fatty acids	Leafy green vegetables, fruits, vegetable oils, nuts, whole-grain breads and cereals	Unknown	Diarrhea, nausea, headaches, fatigue, muscle weakness
Vitamin K	Needed for synthesis of substances active in clotting of blood	Leafy green vegetables, cabbage and cauliflower	Easy bruising and bleeding	Can interfere with anticoagulant medication

Table 4.7   Water-Soluble Vitamins

Vitamin	Functions	Food Sources	Conditions with	
			Too Little	**Too Much**
Vitamin C	Antioxidant; needed for forming collagen; helps maintain capillaries, bones, and teeth	Citrus fruits, leafy green vegetables, tomatoes, potatoes, cabbage	Scurvy, wounds heal slowly, infections	Gout, kidney stones, diarrhea, decreased copper
Thiamine (Vitamin B_1)	Part of coenzyme needed for cellular respiration; also promotes activity of the nervous system	Whole-grain cereals, dried beans and peas, sunflower seeds, and nuts	Beriberi, muscular weakness, heart enlarges	Can interfere with absorption of other vitamins
Riboflavin (Vitamin B_2)	Part of coenzymes, such as FAD; aids cellular respiration, including oxidation of protein and fat	Nuts, dairy products, whole-grain cereals, poultry, and leafy green vegetables	Dermatitis, blurred vision, growth failure	Unknown
Niacin (Nicotinic acid)	Part of coenzymes NAD and NADP; needed for cellular respiration, including oxidation of protein and fat	Peanuts, poultry, whole-grain cereals, leafy green vegetables, and beans	Pellagra, diarrhea, and mental disorders	High blood sugar and uric acid, vasodilation, etc.
Folacin (Folic acid)	Coenzyme needed for production of hemoglobin and the formation of DNA	Dark leafy green vegetables, nuts, beans, whole-grain cereals	Megaloblastic anemia, spina bifida	May mask B_{12} deficiency
Vitamin B_6	Coenzyme needed for the synthesis of hormones and hemoglobin; CNS control	Whole-grain cereals, bananas, beans, poultry, nuts, leafy green vegetables	Rarely, convulsions, vomiting, seborrhea, muscular weakness	Insomnia
Pantothenic acid	Part of coenzyme A needed for oxidation of carbohydrates and fats; aids in the formation of hormones and certain neurotransmitters	Nuts, beans, dark green vegetables, poultry, fruits, and milk	Rarely, loss of appetite, mental depression, numbness	Unknown
Vitamin B_{12}	Complex, cobalt-containing compound; part of the coenzyme needed for synthesis of nucleic acids and myelin	Dairy products, fish, poultry, eggs, fortified cereals	Pernicious anemia	Unknown
Biotin	Coenzyme needed for metabolism of amino acids and fatty acids	Generally in foods, especially eggs	Skin rash, nausea, fatigue	Unknown

Minerals

In addition to vitamins, various **minerals** are required by the body. Minerals are divided into macrominerals and microminerals. The body contains more than 5 grams of each macromineral and less than 5 grams of each micromineral (Fig. 4.17). The macrominerals are constituents of cells and body fluids and are structural components of tissues. For example, calcium (present as Ca^{2+}) is needed for the construction of bones and teeth and for nerve conduction and muscle contraction. Phosphorus (present as PO_4^{3-}) is stored in the bones and teeth and is a part of phospholipids, ATP, and the nucleic acids. Potassium (K^+) is the major positive ion inside cells and is important in nerve conduction and muscle contraction, as is sodium (Na^+). Sodium also plays a major role in regulating the body's water balance, as does chloride (Cl^-). Magnesium (Mg^{2+}) is critical to the functioning of hundreds of enzymes.

The microminerals are parts of larger molecules. For example, iron is present in hemoglobin, and iodine is a part of thyroxin, a hormone produced by the thyroid gland. Zinc, copper, and selenium are present in enzymes that catalyze a variety of reactions. Proteins, called zinc-finger proteins because of their characteristic shapes, bind to DNA when a particular gene is to be activated. As research continues, more and more elements are added to the list of microminerals considered to be essential. During the past three decades, for example, very small amounts of selenium, molybdenum, chromium, nickel, vanadium, silicon, and even arsenic have been found to be essential to good health. Table 4.8 lists the functions of various minerals and gives their food sources, and signs of deficiency and toxicity.

Occasionally, individuals do not receive enough iron (especially women), calcium, magnesium, or zinc in their diet. Adult females need more iron in the diet than males (18 mg compared to 10 mg) because they lose hemoglobin each month during menstruation. Stress can bring on a magnesium deficiency, and due to its high-fiber content, a vegetarian diet may make zinc less available to the body. However, a varied and complete diet usually supplies enough of each type of mineral.

Calcium

Many people take calcium supplements (Fig. 4.17) to counteract **osteoporosis,** a degenerative bone disease that afflicts an estimated one-fourth of older men and one-half of older women in the United States. Osteoporosis develops because bone-eating cells called osteoclasts are more active than bone-forming cells called osteoblasts. Therefore, the bones are porous, and they break easily because they lack sufficient calcium. Due to recent studies that show consuming more calcium does slow bone loss in elderly people, the guidelines have been revised. A calcium intake of 1,000 mg a day is recommended for men and for women who are premenopausal or who use estrogen replacement therapy, and 1,300 mg a day is recommended for postmenopausal women who do not use estrogen replacement therapy. To achieve this amount, supplemental calcium is most likely necessary.

Estrogen replacement therapy and exercise in addition to calcium supplements are effective means to prevent osteoporosis. Medications are also available that slow bone loss while increasing skeletal mass. Etidronate disodium needs to be taken cyclically or else weak, abnormal bone instead of normal bone develops. Fosamax is a new drug that can be taken continually but may accumulate in the skeleton. Therefore, studies are needed to determine if the drug can safely be given for a long time.

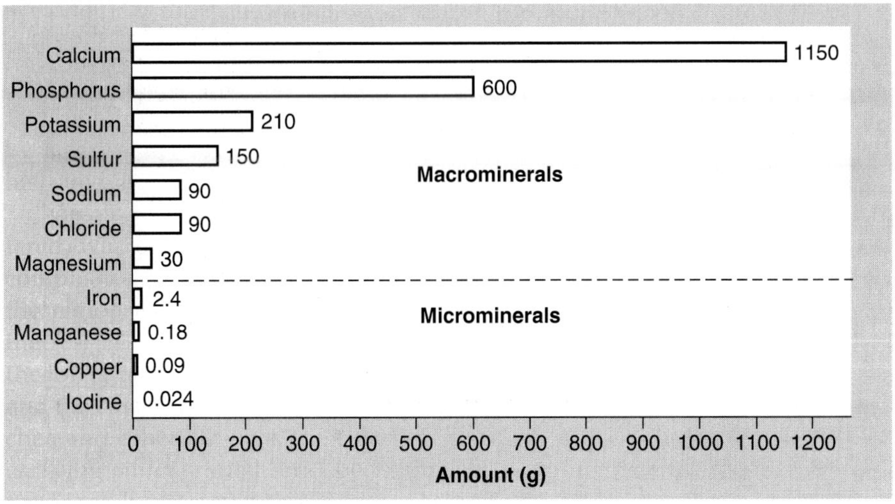

Figure 4.17   Minerals in the body.
This chart shows the usual amount of certain minerals in a 60-kilogram (135 lb) person. The macrominerals are present in amounts larger than 5 grams (about a teaspoon) and the microminerals are present in lesser amounts. The function of these minerals is given in Table 4.8.

Presently, calcium supplements, estrogen therapy for women, and exercise are thought to be the best ways to prevent osteoporosis.

Table 4.8   Minerals

Mineral	Functions	Food Sources	Conditions with	
			Too Little	**Too Much**
Macrominerals (more than 100 mg/day needed)				
Calcium (Ca^{2+})	Strong bones and teeth, nerve conduction, muscle contraction	Dairy products, leafy green vegetables	Stunted growth in children, low bone density in adults	Kidney stones, interferes with iron and zinc absorption
Phosphorus (PO$_4^{3-}$)	Bone and soft tissue growth, part of phospholipids, ATP, and nucleic acids	Meat, dairy products, sunflower seeds, food additives	Weakness, confusion, pain in bones and joints	Low blood and bone calcium levels
Potassium (K$^+$)	Nerve conduction, muscle contraction	Many fruits and vegetables, bran	Paralysis, irregular heartbeat, eventual death	Vomiting, heart attack, death
Sodium (Na$^+$)	Nerve conduction, pH and water balance	Table salt	Lethargy, muscle cramps, loss of appetite	Edema, high blood pressure
Chloride (Cl$^-$)	Water balance	Table salt	Not likely	Vomiting, dehydration
Magnesium (Mg^{2+})	Part of various enzymes for nerve and muscle contraction, protein synthesis	Whole grains, leafy green vegetables	Muscle spasm, irregular heartbeat, convulsions, confusion, personality changes	Diarrhea
Microminerals (less than 20 mg/day needed)				
Zinc (Zn^{2+})	Protein synthesis, wound healing, fetal development and growth, immune function	Meats, legumes, whole grains	Delayed wound healing, night blindness, diarrhea, mental lethargy	Anemia, diarrhea, vomiting, renal failure, abnormal cholesterol levels
Iron (Fe^{2+})	Hemoglobin synthesis	Whole grains, meats, prune juice	Anemia, physical, and mental sluggishness	Iron toxicity disease, organ failure, eventual death
Copper (Cu^{2+})	Hemoglobin synthesis	Meat, nuts, legumes	Anemia, stunted growth in children	Damage to internal organs if not excreted
Iodine (I$^-$)	Thyroid hormone synthesis	Iodized table salt, seafood	Thyroid deficiency	Depressed thyroid, function, anxiety
Selenium (SeO$_4^{2-}$)	Part of antioxidant enzyme	Seafood, meats, eggs	Possible cancer development	Hair and fingernail loss, discolored skin

Source: David C. Nieman, et al., *Nutrition*. Copyright © 1990 Wm. C. Brown.

Sodium

The recommended amount of sodium intake per day is 500 mg, although the average American takes in 4,000–4,700 mg every day. In recent years, this imbalance has caused concern because high sodium intake has been linked to hypertension (high blood pressure) in some people. About one-third of the sodium we consume occurs naturally in foods; another one-third is added during commercial processing; and we add the last one-third either during home cooking or at the table in the form of table salt.

Clearly, it is possible for us to cut down on the amount of sodium in the diet. Table 4.9 gives recommendations for doing so.

Excess sodium in the diet can lead to hypertension; therefore, excess sodium intake should be avoided.

Table 4.9   Reducing Dietary Sodium

To reduce dietary sodium:

1. Use spices instead of salt to flavor foods.
2. Add little or no salt to foods at the table, and add only small amounts of salt when you cook.
3. Eat unsalted crackers, pretzels, potato chips, nuts, and popcorn.
4. Avoid hot dogs, ham, bacon, luncheon meats, smoked salmon, sardines, and anchovies.
5. Avoid processed cheese and canned or dehydrated soups.
6. Avoid brine-soaked foods, such as pickles or olives.
7. Read labels to avoid high salt products.

Persons with obesity have

- weight 20% or more above appropriate weight for height.
- body fat content in excess of that consistent with optimal health; probably due to a diet rich in fats.
- low levels of exercise.

Figure 4.18   Recognizing obesity.

Eating Disorders

Authorities recognize three primary eating disorders: obesity, bulimia nervosa, and anorexia nervosa. Although they exist in a continuum as far as body weight is concerned, they all represent an inability to maintain normal body weight because of eating habits.

Obesity

As indicated in Figure 4.18, **obesity** is most often defined as a body weight 20% or more above the ideal weight for a person's height. By this standard, 28% of women and 10% of men in the United States are obese. Moderate obesity is 41–100% above ideal weight and severe obesity is 100% or more above ideal weight.

Obesity is most likely caused by a combination of factors, including genetic, hormonal, metabolic, and social factors. It's known that obese individuals have more fat cells than normal, and when they lose weight the fat cells simply get smaller; they don't disappear. The social factors that cause obesity include the eating habits of other family members. Consistently eating fatty foods, for example, will make you gain weight. Sedentary activities, like watching television instead of exercising, also determine how much body fat you have. The risk of heart disease is higher in obese individuals, and this alone tells us that excess body fat is not consistent with optimal health.

The treatment depends on the degree of obesity. Surgery to remove body fat may be required for those who are moderately or greatly overweight. For most, a knowledge of good eating habits along with behavior modification may suffice, particularly if a balanced diet is accompanied by a sensible exercise program. A lifelong commitment to a properly planned program is the best way to prevent a cycle of weight gain followed by weight loss. Such a cycle is not conducive to good health. The Health reading on page 99 discusses the proper way to lose weight.

Bulimia Nervosa

Bulimia nervosa can coexist with either obesity or anorexia nervosa, which is discussed next. People with this condition have the habit of eating to excess (called binge eating) and then purging themselves by some artificial means, such as self-induced vomiting or use of a laxative. Bulimic individuals are overconcerned about their body shape and weight, and therefore they may be on a very restrictive diet. A restrictive diet may bring on the desire to binge, and typically the person chooses to consume sweets, like cakes, cookies, and ice cream (Fig. 4.19). The amount of food consumed is far beyond the normal number of calories for one meal, and the person keeps on eating until every bit is gone. Then, a feeling of guilt most likely brings on the next phase, which is a purging of all the calories that have been taken in.

Bulimia can be dangerous to your health. Blood composition is altered, leading to an abnormal heart rhythm, and damage to the kidneys can even result in death. At the very least, vomiting may result in inflammation of the pharynx and esophagus, and stomach acids can cause teeth to erode. The esophagus and stomach may even rupture and tear due to strong contractions during vomiting.

The most important aspect of treatment is to get the patient on a sensible and consistent diet. Again, behavioral modification is helpful and so perhaps is psychotherapy to help the patient understand the emotional causes of the behavior. Medications, including antidepressant medications, have sometimes been helpful to reduce the bulimic cycle and restore normal appetite.

Obesity and bulimia nervosa have complex causes and may be damaging to health. Therefore, they may require competent medical attention.

Anorexia Nervosa

In **anorexia nervosa,** a morbid fear of gaining weight causes the person to be on a very restrictive diet. Athletes such as distance runners, wrestlers, and dancers are at risk of anorexia nervosa because they believe that being thin gives them a competitive edge. In addition to eating only low-calorie foods, the person may induce vomiting and use laxatives to bring about further loss of weight. No matter how thin they have become, people with anorexia nervosa think they are overweight (Fig. 4.20). Such a distorted self-image may prevent recognition of the need for medical help.

Actually, the person is starving and has all the symptoms of starvation, such as low blood pressure, irregular heartbeat, constipation, and constant chilliness. Bone density decreases and stress fractures occur. The body begins to shut down; menstruation ceases in females; the internal organs including the brain don't function well, and the skin dries up. Impairment of the pancreas and digestive tract means that any food consumed does not provide nourishment. Death may be imminent. If so, the only recourse may be hospitalization and force-feeding. Eventually, it is necessary to use behavior therapy and psychotherapy to enlist the cooperation of the person to eat properly. Family therapy may be necessary, because anorexia nervosa in children and teens is believed to be a way for them to gain some control over their lives.

> In anorexia nervosa, the individual has a distorted body image and always feels fat. Competent medical help is often a necessity.

Persons with bulimia nervosa

- often accompanied by lack of exercise

Persons with bulimia nervosa have

- recurrent episodes of binge eating characterized by consuming an amount of food much higher than normal for one sitting and a sense of lack of control over eating during the episode.
- an obsession about their body shape and weight.

Body weight is regulated by

- a restrictive diet, excessive exercise.
- purging (self-induced vomiting or misuse of laxatives).

Figure 4.19   **Recognizing bulimia nervosa.**

Persons with anorexia nervosa have

- a morbid fear of gaining weight; body weight no more than 85% normal.
- a distorted body image so that person feels fat even when emaciated.
- in females, an absence of a menstrual cycle for at least three months.

Body weight is kept too low by either/or

- a restrictive diet, often with excessive exercise.
- binge eating/purging (person engages in binge eating and then self-induces vomiting or misuses laxatives).

Figure 4.20   **Recognizing anorexia nervosa.**

Bioethical Issue

A fat-free fat, called olestra, is now being used to produce foods that are free of calories from fat. A slice of pie ordinarily contains 405 calories, but when the crust is made with olestra, it has only 252 calories. Just three chocolate chip cookies ordinarily adds 63 calories to your daily calorie count, but with olestra it's only 38 calories. And one brownie suddenly goes down from 85 to 49 calories. Olestra molecules made of six or eight fatty acids attached to a sugar molecule are much bigger than a triglyceride, and lipase can't break the molecule down. Olestra goes through the intestines without being absorbed. Proctor and Gamble, who developed olestra, expects to make one billion dollars within ten years on its product.

It is generally recognized that olestra can cause intestinal cramping, flatulence (gas), and diarrhea in a small segment of the population, and can prevent the absorption of some carotenoids, a nutrient that has health benefits in all of us. Even so, the Federal Drug Administration (FDA) approved it for use on the basis that olestra was reasonably harmless. Henry Blackburn, professor of public health at the University of Minnesota, thinks this standard is too low. Instead, the standard for approval ought to be, "Does this product contribute to the nutritional health of the nation?"

David Kessler, FDA commissioner, says, "Ask the American people if government should be off the backs of business,

and you will get a resounding Yes!" However, what is the role of government in protecting public health? Do Americans want more protection than they already have from food additives? Or would they rather be free to make their own choices?

Questions

1. Should the government only approve food additives that will contribute to our health? Why or why not?
2. Are people responsible, themselves, for what they eat? Why or why not?
3. Is there too much emphasis in our culture on staying slim, despite what it may do to our health? Why or why not?

Summarizing the Concepts

4.1 The Digestive System

The salivary glands send saliva into the mouth, where the teeth chew the food and the tongue forms a bolus for swallowing.

The air passage and food passage cross in the pharynx. When one swallows, the air passage is usually blocked off and food must enter the esophagus where peristalsis begins.

The stomach expands and stores food. While food is in the stomach, it churns, mixing food with the acidic gastric juices.

The walls of the small intestine have fingerlike projections called villi where nutrient molecules are absorbed into the cardiovascular and lymphatic systems.

The large intestine consists of the cecum, colons (ascending, transverse, descending, and sigmoid), and the rectum, which ends at the anus.

The large intestine does not produce digestive enzymes; it does absorb water, salts, and some vitamins.

4.2 Three Accessory Organs

The three accessory organs of digestion—the pancreas, liver, and gallbladder—send secretions to the duodenum via ducts. The pancreas produces pancreatic juice, which contains digestive enzymes for carbohydrate, protein, and fat.

The liver produces bile, which is stored in the gallbladder. The liver receives blood from the small intestine by way of the hepatic portal vein. It has numerous important functions, and any malfunction of the liver is a matter of considerable concern.

4.3 Digestive Enzymes

Digestive enzymes are present in digestive juices and break down food into the nutrient molecules glucose, amino acids, fatty acids, and glycerol (see Table 4.2). Glucose and amino acids are absorbed into the blood capillaries of the villi. Fatty acids and glycerol rejoin to produce fat, which enters the lacteals.

Digestive enzymes have the usual enzymatic properties. They are specific to their substrate and speed up specific reactions at body temperature and optimum pH.

4.4 Homeostasis

The digestive system works with the other systems of the body in the ways described in the box on page 94.

4.5 Nutrition

The nutrients released by the digestive process should provide us with an adequate amount of energy, essential amino acids and fatty acids, and all necessary vitamins and minerals.

The bulk of the diet should be carbohydrates (like bread, pasta, and rice) and fruits and vegetables. These are low in saturated fatty acids and cholesterol molecules, whose intake is linked to cardiovascular disease. The vitamins A, E, and C are antioxidants that protect cell contents from damage due to free radicals.

Studying the Concepts

1. List the organs of the digestive tracts, and state the contribution of each to the digestive process. 82–88
2. Discuss the absorption of the products of digestion into the lymphatic and cardiovascular systems. 87
3. Name and state the functions of the hormones that assist the nervous system in regulating digestive secretions. 88
4. Name the accessory organs, and describe the part they play in the digestion of food. 90–91
5. Choose and discuss any three functions of the liver. 90–91
6. Name and discuss three serious illnesses of the liver. 91
7. Discuss the digestion of starch, protein, and fat, listing all the steps that occur to bring about digestion of each of these. 92–93
8. What is the chief contribution of each of these constituents of the diet: a. carbohydrates; b. proteins; c. fats; d. fruits and vegetables? 95–98, 100
9. Why should the amount of saturated fat be curtailed in the diet? 98
10. Name and discuss three eating disorders. 104–105

Testing Your Knowledge of the Concepts

In questions 1–4, match the organ to the functions below.
  a. stomach
  b. small intestine
  c. large intestine
  d. liver

_____ 1. makes bile
_____ 2. digestion of all types of food
_____ 3. absorbs water
_____ 4. produces gastric juice

In questions 5–7, indicate whether the statement is true (T) or false (F).

_____ 5. The pancreas secretes both hormones and digestive enzymes
_____ 6. Lipase is an enzyme that digests carbohydrates.
_____ 7. Only if the diet contains meat, can we be assured that the diet includes all the essential amino acids.

In questions 8 and 9, fill in the blanks.
8. The gallbladder stores _____, a substance that _____ fat.
9. Whereas a(n) _____ pH is optimum for pepsin, a(n) _____ pH is optimum for the enzymes found in pancreatic juice.

10. Label each organ indicated in the diagram. For the arrows, use either glucose, amino acids, lipids, or water.

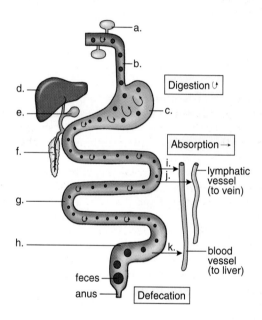

Applying Your Knowledge to the Concepts

These questions pertain to homeostasis.
1. The absorptive surface area of the small intestine is about 600 times more than it would be if the tube were smooth and cylindrical. What accounts for this increased surface area, and what purpose does it serve?
2. How does the liver help maintain the blood glucose level?
3. Some individuals have their gallbladder removed. What effect does this have on their dietary habits?
4. A friend of yours is overweight and wants to lose about 50 pounds. Therefore, she has decided to eliminate fats from her food intake. What advice do you have for her?

Understanding the Terms

anorexia nervosa 105 *morbid fear of getting weight*
anus 88 *the rectum opens at the anus*
appendix 88 *Infectious*
bile 87 *disperse of water (emulsifies)*
bulimia nervosa 104 *Anorexia to obesity*
cecum 88 *lies below junction*
chyme 86 *when the food leaves stomach*
cirrhosis 91 *liver — irreversible injury*
colon 88 *the large portion of the in-*
constipation 89 *less water*
defecation 88 *discharge of feces*
dental caries 83 *tooth decay*
diarrhea 89 *too much water*
duodenum 87 *the liver & pancreas*
epiglottis 84 *cover the glottis closes off the air tract*

esophagus 84 *tube transports food*
essential amino acids 97 *human diet*
fiber 96 *nondigestable plant*
gallbladder 91 *saclike organ*
gastric gland 86 *gastric juices*
glottis 84 *airflow in the pharynx*
hard palate 82 *bony anterior*
heartburn 85 *chest pain*
hepatitis 91 *Inflammation in the liver*
hormone 88 *physiological & development water*
hydrolytic enzyme 92 *substrate breakup*
jaundice 91 *yellowish tint*
lacteal 87 *lymphatic vessel in villus*
lactose intolerance 92 *inability to digest lactose*
large intestine 88 *major portion of the digest tract*

lipase 90, 92 *fat digesting enzyme?*
liver 90 *production of proteins*
lumen 85 *cavity in tubular structure*
maltase 92 *enzyme maltose 2 glucose*
mineral 102 *homogeneous substance*
nasopharynx 84 *Region w/nasal cavity*
obesity 104 *excess adipose tissue*
osteoporosis 102 *bones break easily*
pancreas 90 *elongated (flattened organ)*
pancreatic amylase 90, 92 *starch*
pepsin 86, 92 *protein digesting 6.6*
peptidase 92 *short chain of amino acid*
periodontitis 83 *Inflammation of gums*
peristalsis 85 *Infection of the rhythmic tubular organ*
peritonitis 88

pharynx 84 *Food & Air*
plaque 98 *soft masses of fatty material*
polyp 89 *abnormal growth (epithelial)*
reflex action 84 *automatic inv. respon*
salivary amylase 83, 92 *gland or starch*
salivary gland 82 *oral cavity (saliva)*
small intestine 87 *tubelike chamber*
soft palate 82 *posterior portion*
sphincter 85 *muscle surrounds tube*
stomach 86 *large organ muscular sac*
tonsillitis 82 *Infection of tonsils*
trypsin 90, 92 *protein digests, pancreas*
ulcer 86 *open sore on stomach*
villus 87 *finger like projection*
vitamin 100 *essential requirements on diet*

Match the terms to these definitions:

a. *duodenum* First portion of the small intestine into which secretions from the liver and pancreas enter.

b. *Sphincter* Muscle that surrounds a tube and closes or opens the tube by contracting and relaxing.

c. *lipase* Fat-digesting enzyme secreted by the pancreas.

d. *Defecation* Discharge of feces from the rectum through the anus.

e. *gall bladder* Saclike organ associated with the liver that stores and concentrates bile.

Applying Technology to the Concepts

Your study of the digestive system and nutrition is supported by these available technologies:

Essential Study Partner CD-ROM

Animals → Digestion

Visit the Mader web site for related ESP activities.

Exploring the Internet

The Mader Home Page provides resources and tools as you study this chapter.

http://www.mhhe.com/biosci/genbio/mader

Dynamic Human 2.0 CD-ROM

Digestive System

Virtual Physiology Laboratory CD-ROM

Digestion of Fat
Enzyme Characteristics

HealthQuest CD-ROM

3 Nutrition and Weight Control (all) → Pregnancy and Nutrition

Life Science Animations 3D Video

36 Digestion Overview

Chapter 5

Composition and Function of the Blood

Chapter Concepts

Figure 5.1 **Childhood leukemia.**
Children who are recovering from leukemia are enjoying going to camp despite temporarily losing their hair due to chemotherapy.

Johnny didn't feel well. His mother let him stay home because he had a fever and just wanted to lie on the couch. When he didn't seem better the next day, she took him to the doctor. The doctor was concerned because Johnny was having frequent infections and he was losing weight. He suggested a complete physical examination along with blood work.

In a few days, the response came back that Johnny had acute lymphocyte leukemia (ALL), a type of cancer that occurs in children. *Acute* means that the condition progresses rapidly; a lymphocyte is one of the types of white blood cells, and *leukemia* means that an abnormal increase in immature lymphocytes has occurred. In Johnny's case, as in so many others, the cause of ALL was unknown.

Johnny's parents were really worried, but the doctor was very reassuring. A combination of many different types of cancer drugs works as a cure in 75% of cases at the medical center recommended for Johnny (Fig. 5.1). One year later, Johnny was so well it wasn't possible to tell that he had been through a rough time. He was playing ball and getting into mischief as usual.

Why did Johnny have so many infections? Our white blood cells fight infections, and although in ALL the lymphocyte count is high, the cells are immature and don't fight infections effectively. The other functions of blood were not affected by Johnny's illness. Blood transports various substances to and from the body's cells. Red blood cells are red because they contain hemoglobin, the respiratory pigment that combines with oxygen.

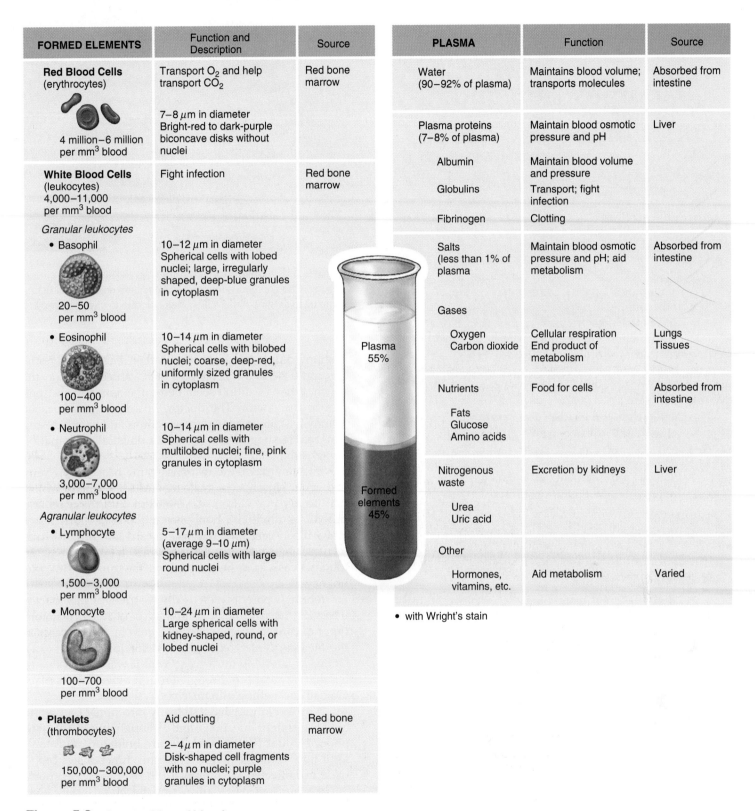

FORMED ELEMENTS	Function and Description	Source
Red Blood Cells (erythrocytes) 4 million–6 million per mm³ blood	Transport O_2 and help transport CO_2 7–8 μm in diameter Bright-red to dark-purple biconcave disks without nuclei	Red bone marrow
White Blood Cells (leukocytes) 4,000–11,000 per mm³ blood	Fight infection	Red bone marrow
Granular leukocytes		
• Basophil 20–50 per mm³ blood	10–12 μm in diameter Spherical cells with lobed nuclei; large, irregularly shaped, deep-blue granules in cytoplasm	
• Eosinophil 100–400 per mm³ blood	10–14 μm in diameter Spherical cells with bilobed nuclei; coarse, deep-red, uniformly sized granules in cytoplasm	
• Neutrophil 3,000–7,000 per mm³ blood	10–14 μm in diameter Spherical cells with multilobed nuclei; fine, pink granules in cytoplasm	
Agranular leukocytes		
• Lymphocyte 1,500–3,000 per mm³ blood	5–17 μm in diameter (average 9–10 μm) Spherical cells with large round nuclei	
• Monocyte 100–700 per mm³ blood	10–24 μm in diameter Large spherical cells with kidney-shaped, round, or lobed nuclei	
• **Platelets** (thrombocytes) 150,000–300,000 per mm³ blood	Aid clotting 2–4 μm in diameter Disk-shaped cell fragments with no nuclei; purple granules in cytoplasm	Red bone marrow

PLASMA	Function	Source
Water (90–92% of plasma)	Maintains blood volume; transports molecules	Absorbed from intestine
Plasma proteins (7–8% of plasma)	Maintain blood osmotic pressure and pH	Liver
Albumin	Maintain blood volume and pressure	
Globulins	Transport; fight infection	
Fibrinogen	Clotting	
Salts (less than 1% of plasma	Maintain blood osmotic pressure and pH; aid metabolism	Absorbed from intestine
Gases		
Oxygen Carbon dioxide	Cellular respiration End product of metabolism	Lungs Tissues
Nutrients	Food for cells	Absorbed from intestine
Fats Glucose Amino acids		
Nitrogenous waste	Excretion by kidneys	Liver
Urea Uric acid		
Other		
Hormones, vitamins, etc.	Aid metabolism	Varied

• with Wright's stain

Figure 5.2   Composition of blood.
When blood is transferred to a test tube and is prevented from clotting, it forms two layers. The transparent, yellow, top layer is plasma, the liquid portion of blood. The formed elements are in the bottom layer. The tables describe these components in detail.

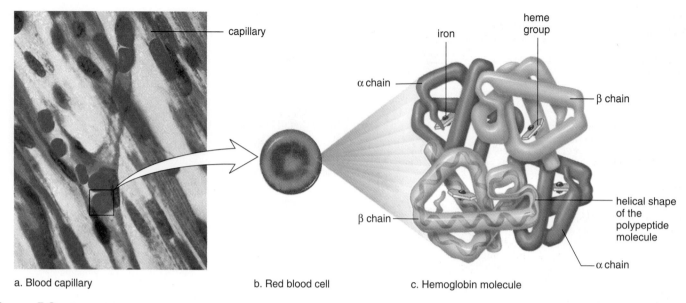

capillary

iron

heme group

α chain

β chain

β chain

helical shape of the polypeptide molecule

α chain

a. Blood capillary

b. Red blood cell

c. Hemoglobin molecule

Figure 5.3   Physiology of red blood cells.
a. Red blood cells move single file through the capillaries. **b.** Each red blood cell is a biconcave disk containing many molecules of hemoglobin, the respiratory pigment. **c.** Hemoglobin contains four polypeptide chains, two of which are alpha (α) chains and two of which are beta (β) chains. There is an iron-containing heme group in the center of each chain. Oxygen combines loosely with iron when hemoglobin is oxygenated. Oxygenated hemoglobin is bright red, and deoxygenated hemoglobin is a darker red color.

This chapter covers the makeup and functions of blood. If blood is transferred from a person's vein to a test tube and is prevented from clotting, it separates into two layers (Fig. 5.2). The lower layer consists of red blood cells (erythrocytes), white blood cells (leukocytes), and blood platelets (thrombocytes). Collectively, these are called the **formed elements**. Formed elements make up about 45% of the total volume of whole blood. The upper layer is plasma, which contains a variety of inorganic and organic molecules dissolved or suspended in water. Plasma accounts for about 55% of the total volume of whole blood.

The functions of blood contribute to homeostasis, a dynamic equilibrium of the internal environment. Cells are surrounded by tissue fluid whose composition must be kept within relatively narrow limits or the cells cease to function in an effective manner. Only if the composition of blood is within the normal range can tissue fluid also have the correct composition. The functions of blood fall into three categories: transport, defense, and regulation. All three are necessary to keep tissue fluid relatively stable.

Blood transports oxygen and nutrients to the tissues and removes wastes for excretion from the body. It also transports hormones, which help control the function of the body's organs. Blood defends the body against invasion by **pathogens** (microscopic infectious agents, such as bacteria and viruses), and it clots, which prevents the loss of blood. Among its various regulatory functions, blood helps maintain normal body temperature and the pH of body fluids.

5.1 The Red Blood Cells

Red blood cells (erythrocytes) are small, biconcave disks that lack a nucleus when mature. They occur in great quantity; there are 4 to 6 million red blood cells per mm^3 of whole blood. The absence of a nucleus provides more space for hemoglobin. **Hemoglobin** is called a respiratory pigment because it carries oxygen, and is red. A red blood cell contains about 200 million hemoglobin molecules. If this much hemoglobin were suspended within the plasma rather than enclosed within the cells, blood would be so viscous the heart would have difficulty pumping it. In a hemoglobin molecule the iron portion of hemoglobin carries oxygen, a molecule that cells require for cellular respiration (Fig. 5.3). The equation for oxygenation of hemoglobin is usually written as

$$Hb + O_2 \underset{\text{tissues}}{\overset{\text{lungs}}{\rightleftharpoons}} HbO_2$$

The hemoglobin on the right, which is combined with oxygen, is called oxyhemoglobin. Oxyhemoglobin, which forms in the lungs, has a bright red color. The hemoglobin on the left, which has given up oxygen to tissue fluid, is called deoxyhemoglobin. Deoxyhemoglobin is a dark purplish color. Unfortunately, as discussed in the Ecology reading on page 114, carbon monoxide combines with hemoglobin more readily than does oxygen, and it stays combined for several hours, making hemoglobin unavailable for oxygen transport.

Blood functions to maintain homeostasis so that the environment of cells (tissue fluid) remains relatively stable.

Because red blood cells contain hemoglobin, they transport oxygen, a molecule that cells use in cellular respiration.

Visual Focus

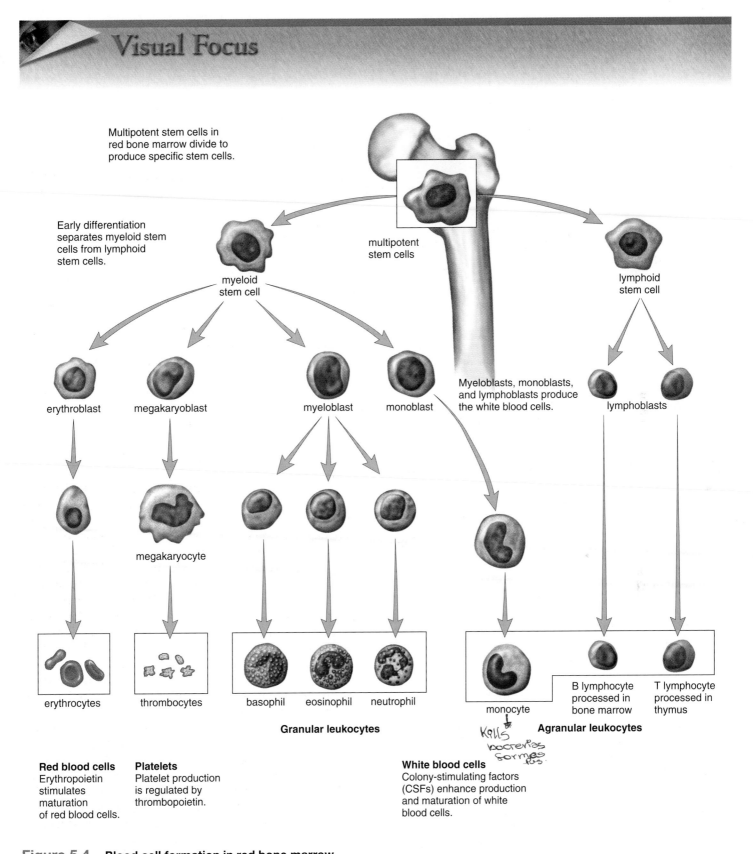

Multipotent stem cells in red bone marrow divide to produce specific stem cells.

multipotent stem cells

Early differentiation separates myeloid stem cells from lymphoid stem cells.

myeloid stem cell

lymphoid stem cell

erythroblast

megakaryoblast

myeloblast

monoblast

Myeloblasts, monoblasts, and lymphoblasts produce the white blood cells.

lymphoblasts

megakaryocyte

erythrocytes

thrombocytes

basophil   eosinophil   neutrophil

monocyte

B lymphocyte processed in bone marrow

T lymphocyte processed in thymus

Granular leukocytes

Agranular leukocytes

Red blood cells
Erythropoietin stimulates maturation of red blood cells.

Platelets
Platelet production is regulated by thrombopoietin.

White blood cells
Colony-stimulating factors (CSFs) enhance production and maturation of white blood cells.

Figure 5.4   Blood cell formation in red bone marrow.
Multipotent stem cells give rise to two specialized stem cells. The myeloid stem cell gives rise to still other cells, which become red blood cells, platelets, and all the white blood cells except lymphocytes. The lymphoid stem cell gives rise to lymphoblasts, which become lymphocytes.

The Life Cycle of Red Blood Cells

In infants, red blood cells are produced in the red bone marrow of all bones, but in adults, production primarily occurs in the red bone marrow of the skull bones, ribs, sternum, vertebrae, and pelvic bones.

All blood cells, including erythrocytes, are formed from special red bone marrow cells called stem cells (Fig. 5.4). A **stem cell** is ever capable of dividing and producing new cells that differentiate into specific type cells. As red blood cells mature, they lose their nucleus and acquire hemoglobin. Possibly because they lack a nucleus, red blood cells live only about 120 days. As they age, they are destroyed in the liver and spleen, where they are engulfed by macrophages, which are large phagocytic cells. It is estimated that about 2 million red blood cells are destroyed per second, and therefore an equal number must be produced to keep the red blood cell count in balance.

When red blood cells are broken down, the hemoglobin is released. The globin portion of the hemoglobin is broken down into its component amino acids, which are recycled by the body. The iron is recovered and is returned to the bone marrow for reuse. The heme portion of the molecule undergoes chemical degradation and is excreted as bile pigments by the liver into the bile. These are the bile pigments bilirubin and biliverdin, which contribute to the color of feces. Chemical breakdown of heme is also what causes a bruise of the skin to change color from red/purple to blue to green to yellow.

The number of red blood cells produced increases whenever arterial blood carries a reduced amount of oxygen, as happens when an individual first takes up residence at a high altitude or loses red blood cells or full use of their lungs. Under these circumstances, the kidneys accelerate their release of **erythropoietin,** a hormone that is carried in blood to red bone marrow (Fig. 5.5). Once there, it speeds up the maturation of cells that are in the process of becoming red blood cells. The liver and other tissues also produce erythropoietin. Erythropoietin, now mass-produced through biotechnology, is sometimes abused by athletes in order to raise their red blood cell counts and thereby increase the oxygen-carrying capacity of their blood.

Red blood cells are produced in red bone marrow after stem cells divide to form cells that undergo maturation in stages. Red blood cells live only 120 days and are destroyed by phagocytic cells in the liver and spleen.

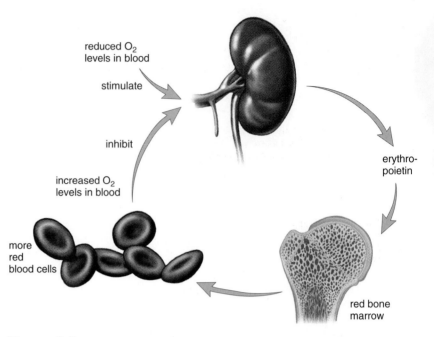

Figure 5.5   Action of erythropoietin.
The kidneys release increased amounts of erythropoietin whenever the oxygen capacity of the blood is reduced. Erythropoietin stimulates the red bone marrow to speed up its production of red blood cells, which carry oxygen. Once the oxygen-carrying capacity of the blood is sufficient to support normal cellular activity, the kidneys cut back on their production of erythropoietin.

Anemia

When there is an insufficient number of red blood cells or the cells do not have enough hemoglobin, the individual suffers from **anemia** and has a tired, run-down feeling. In some types of anemia, the hemoglobin blood level is low. It may be that the diet does not contain enough iron or folic acid. Certain foods, such as whole-grain cereals, are rich in iron and folic acid, and the inclusion of these in the diet can help to prevent anemia.

In another type of anemia, called pernicious anemia, the digestive tract is unable to absorb enough vitamin B_{12}, found in dairy products, fish, eggs, and poultry. This vitamin is essential to the proper formation of red blood cells; without it, immature red blood cells tend to accumulate in the bone marrow in large quantities. A special diet and administration of vitamin B_{12} by injection is an effective treatment for pernicious anemia.

Hemolysis is the rupturing of red blood cells. In hemolytic anemia, there is an increased rate of red blood cell destruction. **Sickle-cell disease** is a hereditary condition in which the individual has sickle-shaped red blood cells that tend to rupture as they pass through the narrow capillaries. Hemolytic disease of the newborn, which is discussed at the end of this chapter (p. 121), is also a type of hemolytic anemia.

Illness (anemia) results when the blood has too few red blood cells and/or not enough hemoglobin.

Ecology Focus

Carbon Monoxide, A Deadly Poison

Carbon monoxide (CO) is an air pollutant that comes primarily from the incomplete combustion of natural gas and gasoline. Figure 5A shows that transportation contributes most of the carbon monoxide to our cities' air. But power plants, factories, waste incineration, and home heating also contribute to the carbon monoxide level. Cigarette smoke contains carbon monoxide and is delivered directly to the smoker's blood and also to nonsmokers nearby.

Because carbon monoxide is a colorless, odorless gas, people can be unaware that it is affecting their systems. But it binds to iron 200 times more tightly than oxygen. Hemoglobin contains iron and so does cytochrome oxidase, the carrier in the electron transport system that passes electrons on to oxygen. When these molecules bind to carbon monoxide preferentially, they cannot perform their usual functions. The end result is that delivery of oxygen to mitochondria is impaired and so is the functioning of mitochondria.

Flushed red skin, especially on facial cheeks, is a first sign of carbon monoxide poisoning, because hemoglobin bound to carbon monoxide is brighter red than oxygenated hemoglobin. Marked euphoria, then sleepiness, coma, and death follow. Removing a person from the carbon monoxide source is not sufficient treatment because carbon monoxide, unlike oxygen, remains tightly bound to iron for many hours. A transfusion of red blood cells will help to increase the carrying capacity of the blood, and pure oxygen given under pressure will displace some carbon monoxide. Despite good medical care, some people still die each year from CO poisoning.

The level of carbon monoxide in polluted air may not be sufficient to kill people, but it does interfere with the ability of the body to function properly. The elderly and those with cardiovascular disease are especially at risk. One study found that even levels once thought to be safe can cause angina patients to experience chest pains, because oxygen delivery to the heart is reduced.

We can all lessen air pollution and reduce the amount of carbon monoxide in the air by doing the following:

- Don't smoke (especially indoors).

- Walk or bicycle instead of driving.

- Use public transportation instead of driving.

- Heat your home with solar energy—not furnaces.

- Support the development of more efficient automobiles, factories, power plants, and home furnaces.

- Support the development of alternative fuels. When hydrogen gas is burned, for example, the result is water, not carbon monoxide and carbon dioxide.

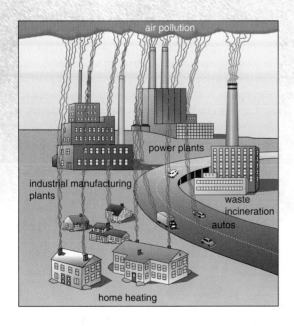

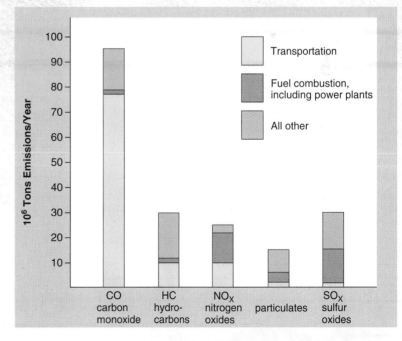

Figure 5A   Air pollution.
Air pollutants (carbon monoxide, hydrocarbons, nitrogen oxides, particulates, and sulfur oxides) enter the atmosphere from the sources noted.

5.2 The White Blood Cells

White blood cells (leukocytes) differ from red blood cells in that they are usually larger, have a nucleus, lack hemoglobin, and without staining are translucent. White blood cells are not as numerous as red blood cells. There are only 5,000–11,000 per mm³ of blood. White blood cells fight infection and in this way are important contributors to homeostasis. This function of white blood cells is discussed at greater length in chapter 7, which concerns immunity.

White blood cells are derived from stem cells in the red bone marrow, and they, too, undergo several maturation stages (see Fig. 5.4). The production of white blood cells increases whenever the body is invaded by pathogens. Hormones called **colony-stimulating factors** are released by white blood cells, and circulate back to the bone marrow, stimulating an increased production.

Red blood cells are confined to the blood, but white blood cells are able to squeeze through pores in the capillary wall, and therefore they are found in tissue fluid and lymph (Fig. 5.6). When there is an infection, white blood cells greatly increase in number. Many white blood cells live only a few days—they probably die while engaging pathogens. Others live months or even years.

White blood cells fight infection. They defend us against pathogens that have invaded the body.

Types of White Blood Cells

White blood cells are classified into the **granular leukocytes** and the **agranular leukocytes.** Both types of cells have granules in the cytoplasm surrounding the nucleus, but the granules are more visible upon staining in granular leukocytes. The granules contain various enzymes and proteins, which help white blood cells defend the body. There are three types of granular leukocytes and two types of agranular leukocytes. They differ somewhat by the size of the cell and the shape of the nucleus (see Fig. 5.2), and they also differ in their functions.

Granular Leukocytes

Neutrophils are the most abundant of the white blood cells. They have a multilobed nucleus joined by nuclear threads; therefore, they are also called *polymorphonuclear.* They have granules that do not significantly take up the stain eosin, a pink to red stain, or a basic stain that is blue to purple. (This accounts for their name, neutrophil.) Neutrophils are the first type of white blood cells to respond to an infection, and they engulf pathogens during **phagocytosis.**

Eosinophils have a bilobed nucleus, and their large, abundant granules take up eosin and become a red color. (This accounts for their name, eosinophil.) Not much is known specifically about the function of eosinophils, but they are known to increase in number when there is a parasitic worm infection or in the case of allergic reactions.

Basophils have a U-shaped or lobed nucleus. Their

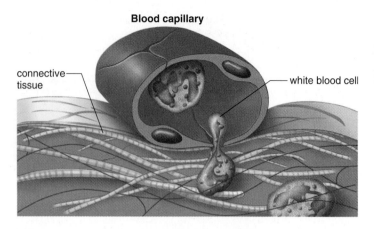

Blood capillary

connective tissue — white blood cell

Figure 5.6   Mobility of white blood cells.
White blood cells can squeeze between the cells of a capillary wall and enter the tissues of the body.

granules take up the basic stain and become a dark blue color. (This accounts for their name, basophil.) Basophils enter the tissues and are believed to become mast cells, which release the histamine associated with allergic reactions. Histamine dilates blood vessels and causes contraction of smooth muscle.

Agranular Leukocytes

The agranular leukocytes include monocytes, which have a kidney-shaped nucleus, and lymphocytes, which have a spherical-shaped nucleus. These cells are responsible for specific defense to particular pathogens and their toxins (poisonous substances). Pathogens have molecules called **antigens** that allow the immune system to recognize them as foreign.

Monocytes are the largest of the white blood cells, and after taking up residence in the tissues, they differentiate into even larger macrophages. Macrophages phagocytize pathogens, old cells, and cellular debris. They also stimulate other white blood cells, including lymphocytes, to defend the body.

The **lymphocytes** are of two types, B lymphocytes and T lymphocytes. B lymphocytes protect us by producing **antibodies** that combine with antigens and thereby target pathogens for destruction. T lymphocytes, on the other hand, directly destroy any cell that has antigens. B lymphocytes and T lymphocytes are discussed more fully in chapter 7.

White blood cells are divided into the granular leukocytes and the agranular leukocytes. Each type of white blood cell has a specific role to play in defending the body against disease.

Leukemia

Leukemia is characterized by an abnormally large number of immature white blood cells that fill the red bone marrow and prevent red blood cell development. Anemia results, and the immature white cells offer little protection from disease. The cause of leukemia, a type of cancer, is unknown, but combined chemotherapy has been most successful, particularly in acute childhood leukemia.

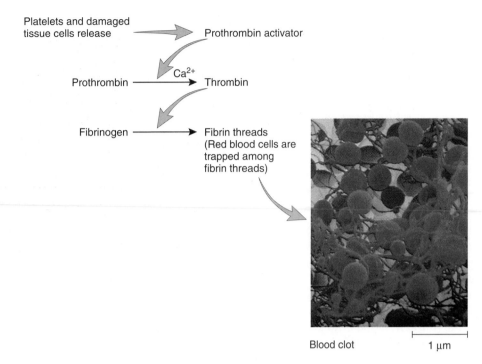

Platelets and damaged tissue cells release → **Prothrombin activator**

Prothrombin —→ (Ca^{2+}) **Thrombin**

Fibrinogen —→ **Fibrin threads** (Red blood cells are trapped among fibrin threads)

Blood clot   1 μm

Figure 5.7   Blood clotting.
Platelets and damaged tissue cells release prothrombin activator, which acts on prothrombin in the presence of calcium ions (Ca^{2+}) to produce thrombin. Thrombin acts on fibrinogen in the presence of Ca^{2+} to form fibrin threads. The scanning electron micrograph of a blood clot shows red blood cells caught in the fibrin threads.

5.3 Blood Clotting

Platelets (thrombocytes) result from fragmentation of certain large cells, called **megakaryocytes,** in the red bone marrow. Platelets are produced at a rate of 200 billion a day, and the blood contains 150,000–300,000 per mm^3. These formed elements are involved in the process of blood **clotting,** or coagulation.

There are at least 12 clotting factors in the blood that participate in the formation of a blood clot. We will discuss the roles played by platelets, prothrombin, and fibrinogen. **Fibrinogen** and **prothrombin** are proteins manufactured and deposited in blood by the liver. Vitamin K, found in green vegetables and also formed by intestinal bacteria, is necessary for the production of prothrombin, and if by chance this vitamin is missing from the diet, hemorrhagic disorders develop.

Stages of Blood Clotting

When a blood vessel in the body is damaged, platelets clump at the site of the puncture and partially seal the leak. They and the injured tissues release a clotting factor called **prothrombin activator** that converts prothrombin to thrombin. This reaction requires calcium ions (Ca^{2+}). **Thrombin,** in turn, acts as an enzyme that severs two short amino acid chains from each fibrinogen molecule. These activated fragments then join end to end, forming long threads of **fibrin.** Fibrin threads wind around the platelet plug in the damaged area of the blood vessel and provide the framework for the clot. Red blood cells also are trapped within the fibrin

Table 5.1	Blood Plasma
Name	**Composition**
Blood	Formed elements and plasma
Plasma	Liquid portion of blood
Serum	Plasma minus fibrinogen
Tissue fluid	Plasma minus most proteins
Lymph	Tissue fluid within lymphatic vessels

threads; these cells make a clot appear red (Fig. 5.7). A fibrin clot is present only temporarily. As soon as blood vessel repair is initiated, an enzyme called plasmin destroys the fibrin network and restores the fluidity of plasma.

If blood is allowed to clot in a test tube, a yellowish fluid develops above the clotted material. This fluid is called **serum,** and it contains all the components of plasma except fibrinogen. Table 5.1 reviews the many different terms we have used to refer to various body fluids related to blood.

Hemophilia

Hemophilia is an inherited clotting disorder due to a deficiency in a clotting factor. The slightest bump can cause the affected person to bleed into the joints. Cartilage degeneration in the joints and resorption of underlying bone can follow. Bleeding into muscles can lead to nerve damage and muscular atrophy. The most frequent cause of death is bleeding into the brain with accompanying neurological damage.

Health Focus

What to Know When Giving Blood

The Procedure

After you register to give blood, you are asked private and confidential questions about your health history and your life-style, and any questions you may have are answered.

Your temperature, blood pressure, and pulse will be checked, and a drop of your blood is tested to ensure that you are not anemic.

You will have several opportunities prior to giving blood and even afterwards to let Red Cross officials know whether your blood is safe to give to another person.

All of the supplies, including the needle, are sterile and are used only once—for YOU. You cannot get infected with HIV (the virus that causes AIDS) or any other disease from donating blood.

When the actual donation is started, you may feel a brief "sting." The procedure takes about 10 minutes, and you will have given about a pint of blood. Your body replaces the liquid part (plasma) in hours and the cells in a few weeks.

After you donate, you are given a card with a number to call if you decide after you leave that your blood may not be safe to give to another person.

An area is provided in which to relax after donating blood. Most people feel fine while they give blood and afterward. A few may have an upset stomach, a faint or dizzy feeling, or a bruise, redness, and pain where the needle was. Very rarely, a person may faint, have muscle spasms, and/or suffer nerve damage.

Your blood is tested for syphilis, AIDS antibodies, hepatitis, and other viruses. You are notified if tests give a positive result. Your blood won't be used if it could make someone ill.

The Cautions

\ \ \ DO NOT GIVE BLOOD / / /

if you have

- ever had hepatitis;
- had malaria or have taken drugs to prevent malaria in the last 3 years; or have traveled to a country where malaria is common.
- been treated for syphilis or gonorrhea in the last 12 months.

if you have AIDS or one of its symptoms:

- unexplained weight loss (4.5 kilograms or more in less than 2 months);
- night sweats;
- blue or purple spots on or under skin;
- long-lasting white spots or unusual sores in mouth;
- lumps in neck, armpits, or groin for over a month;
- diarrhea lasting over a month;
- persistent cough and shortness of breath;
- fever higher than 37°C for more than 10 days.

if you are at risk for AIDS; that is, if you have

- taken illegal drugs by needle, even once;
- taken clotting factor concentrates for a bleeding disorder such as hemophilia;
- tested positive for any AIDS virus or antibody;
- been given money or drugs for sex, since 1977;
- had a sexual partner within the last 12 months who did any of the above things;
- (for men): had sex *even once* with another man since 1977; within the last 12 months had sex with a female prostitute;
- (for women): had sex with a male or female prostitute within the last 12 months; *or* had a male sexual partner who had sex with another man *even once* since 1977.

DO NOT GIVE BLOOD to find out whether you test positive for antibodies to the viruses (HIV) that cause AIDS. Although the tests for HIV are very good, they aren't perfect. HIV antibodies may take weeks to develop after infection with the virus. If you were infected recently, you may have a negative test result yet be able to infect someone. **It is for this reason that you must not give blood if you are at risk of getting AIDS or other infectious diseases.**

Courtesy of the American Red Cross.

5.4 Plasma

Plasma is the liquid portion of blood, and about 92% of plasma is water. The remaining 8% of plasma consists of various salts (ions) and organic molecules. The salts, which are simply dissolved in plasma, help maintain the pH and osmotic pressure of the blood. Small organic molecules like glucose, amino acids, and urea can also dissolve in plasma. Glucose and amino acids are nutrients for cells; urea is a nitrogenous waste product on its way to the kidneys for excretion. The large organic molecules in plasma include hormones and the plasma proteins (Table 5.2).

The Plasma Proteins

The plasma proteins are so named because they usually remain in the plasma. The three major types of plasma proteins are the albumins, globulins, and fibrinogen. Most plasma proteins are made in the liver. An exception is the antibodies produced by B lymphocytes, which function in immunity.

The plasma proteins have many functions that help maintain homeostasis. They are able to take up and release hydrogen ions; therefore, the plasma proteins help buffer the blood and keep the pH of blood around 7.40. They also contribute to the osmotic pressure, which helps keep water in the blood. Certain plasma proteins combine with and transport large organic molecules. For example, **albumin** transports the molecule bilirubin, a breakdown product of hemoglobin. Lipoproteins, whose protein portion is a globulin, transport cholesterol. There are three types of globulins designated alpha, beta, and gamma globulins. Antibodies, which help fight infections by combining with antigens, are gamma globulins. Other plasma proteins also have specific functions. Fibrinogen is necessary to blood clotting, for example. ⌐

The numerous functions of plasma proteins contribute to maintaining homeostasis.

Table 5.2	Blood Plasma Solutes
Plasma proteins	Albumin, globulins, fibrinogen
Inorganic ions (salts)	Na^+, Ca^{2+}, K^+, Mg^{2+}, Cl^-, HCO_3^-, HPO_4^{2-}, SO_4^{2-}
Gases	O_2, CO_2
Organic nutrients	Glucose, fats, phospholipids, amino acids, etc.
Nitrogenous waste products	Urea, ammonia, uric acid
Regulatory substances	Hormones, enzymes

5.5 Capillary Exchange

The internal environment consists of blood and **tissue fluid**. The composition of tissue fluid stays relatively constant because of exchanges with blood in the region of capillaries (Fig. 5.8). Water makes up a large part of tissue fluid, and any excess is collected by lymphatic capillaries, which are always found near blood capillaries. ⌐

Blood Capillaries

A capillary has an arterial end, a midsection, and a venous end.

Arterial End of Capillary

When arterial blood enters the tissue capillaries, it is bright red because red blood cells are carrying oxygen. It is also rich in nutrients, which are dissolved in the plasma. At the arterial end of the capillary, blood pressure (30 mm Hg) is higher than the osmotic pressure of the blood (21 mm Hg). Blood pressure, you recall, is created by the pumping of the heart; the osmotic pressure is caused by the presence of salts and, in particular, by the plasma proteins that are too large to pass through the wall of the capillary. Since the blood pressure is higher than the osmotic pressure, fluid together with nutrients (glucose and amino acids) exit the capillary. Red blood cells and most all plasma proteins generally remain in the capillaries, but small substances leave the capillaries. Therefore, tissue fluid, created by this process, consists of all the components of plasma except the proteins.

Midsection of Capillary

Along the length of the capillary, molecules follow their concentration gradient as diffusion occurs. Diffusion, you recall, is the movement of molecules from an area of greater concentration to an area of lesser concentration. In the tissues, the area of greater concentration for nutrients and oxygen is always blood, because after these molecules have passed into tissue fluid, they are taken up and metabolized by the tissue cells. The cells use glucose ($C_6H_{12}O_6$) and oxygen (O_2) in the process of aerobic cellular respiration, and they use amino acids for protein synthesis. Following aerobic cellular respiration, the cells give off carbon dioxide (CO_2) and water (H_2O). Carbon dioxide and other waste products of metabolism leave the cell by diffusion. Since tissue fluid is always the area of greater concentration for these waste materials, they diffuse into the capillary.

Oxygen and nutrient molecules (e.g., glucose and amino acids) exit a capillary near the arterial end; waste molecules (e.g., carbon dioxide) enter a capillary near the venous end.

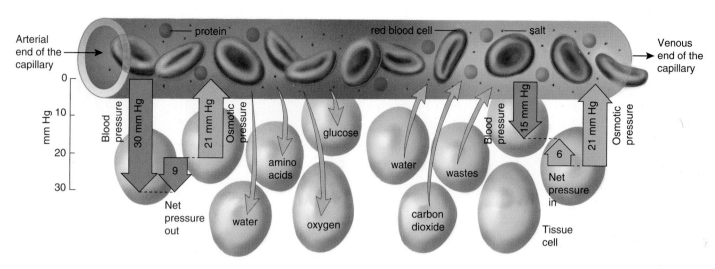

Figure 5.8   **Exchanges between blood and tissue fluid across a capillary wall.**

Venous End of Capillary

At the venous end of the capillary, blood pressure is much reduced (15 mm Hg), as can be verified by reviewing Figure 5.8. However, there is no reduction in osmotic pressure (21 mm Hg), which tends to pull fluid back into the capillary. As water enters a capillary, it brings with it additional waste molecules. Blood that leaves the capillaries is deep purple in color because red blood cells contain reduced hemoglobin.

Retrieving fluid by means of osmotic pressure is not completely effective. There is always some fluid that is not picked up at the venous end. This excess tissue fluid enters the lymphatic capillaries.

Lymphatic Capillaries

Lymphatic capillaries are the smallest of the lymphatic vessels which are a one-way system. The structure of lymphatic vessels is similar to cardiovascular veins, except that their walls are thinner and they have more valves. The valves prevent the backward flow of lymph as lymph flows toward the thoracic cavity. Lymphatic capillaries join to form larger vessels that merge into the lymphatic ducts. The right lymphatic duct empties into a cardiovascular vein within the thoracic cavity (Figure 5.9).

Lymphatic vessels carry **lymph,** which has the same composition as tissue fluid. Why? Because lymphatic capillaries absorb excess tissue fluid at the blood capillaries. The lymphatic system contributes to homeostasis in several ways. One way is to maintain normal blood volume and pressure by returning excess tissue fluid to the blood. Edema is a swelling that occurs when tissue fluid is not collected by the lymphatic capillaries. Edema can be due to

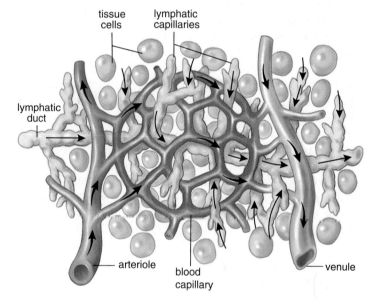

Figure 5.9   **Lymphatic capillaries.**
Arrows indicate that lymph is formed when lymphatic capillaries take up excess tissue fluid. Lymphatic capillaries lie near blood capillaries.

many causes. One dramatic cause is a parasitic infection of lymphatic vessels by a small worm. An effected leg can become so large that the disease is called elephantiasis.

Exchange of nutrients for wastes occurs at the capillaries. Here lymphatic capillaries also collect excess tissue fluid.

5.6 Blood Typing

Blood typing involves the two types of molecules called antigens and antibodies. As mentioned previously, when an *antigen* (foreign substance) is present in the body, an *antibody* reacts with it.

ABO System

The most common system for typing blood is the ABO system. In the ABO system, the presence or absence of type A and type B antigens on red blood cells determines a person's blood type. For example, if a person has type A blood, the A antigen is on his or her red blood cells. This molecule is not an antigen to this individual, although it can be an antigen to a recipient who does not have type A blood.

In the simplified ABO system, there are four types of blood: A, B, AB, and O. Within the plasma, there are antibodies to the antigens that are *not* present on the person's red blood cells. These antibodies are called anti-A and anti-B.

Blood Type	Antigen on Red Blood Cells	Antibody in Plasma
A	A	Anti-B
B	B	Anti-A
AB	A, B	None
O	None	Anti-A and anti-B

It is reasonable that type A blood would have anti-B and not anti-A antibodies in the plasma. If anti-A antibodies were present in plasma, **agglutination,** or clumping of red blood cells, would occur. Agglutination of red blood cells can cause blood to stop circulating in small blood vessels, and this leads to organ damage. It also is followed by hemolysis, which may cause the death of the individual.

For a recipient to receive blood from a donor, the recipient's plasma must not have an antibody that causes the donor's cells to agglutinate. For this reason, it is important to determine each person's blood type. Figure 5.10 demonstrates a way to use the antibodies derived from plasma to determine the blood type. If clumping occurs after a sample of blood is exposed to a particular antibody, the person has that type of blood.

Today, blood transfusions are a matter of concern not only because blood types should match, but also because each person wants to receive blood that is of good quality and free of infectious agents. Blood is tested for the more serious agents such as those that cause AIDS, hepatitis, and syphilis. Donors can help protect the nation's blood supply by knowing when not to give blood. This is the topic of the Health reading on page 117.

For the purpose of blood transfusions, the donor's blood must be compatible with the recipient's blood. The ABO system is used to determine compatibility of donor's and recipient's blood.

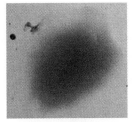
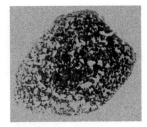

a.   No agglutination   Agglutination

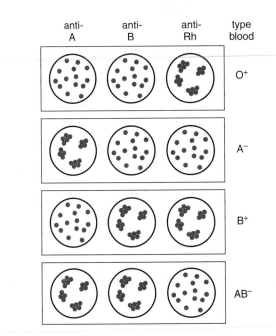

b.

Figure 5.10   Blood typing.

The standard test to determine ABO and Rh blood type consists of putting a drop of anti-A antibodies, anti-B antibodies, and anti-Rh antibodies separately on a slide. To each of these three antibody solutions, a drop of the person's blood is added. **a.** If agglutination occurs, as seen in the *top right* photo, the person has this antigen on red blood cells. **b.** Several possible results. The ABO blood type is indicated by using letters, and the Rh blood type is indicated by using the symbols (+) and (–).

Rh System

Another important antigen in matching blood types is the Rh factor. Eighty-five percent of the U.S. population has this particular antigen on the red blood cells and are Rh^+ (Rh positive). Fifteen percent do not have this antigen and are Rh^- (Rh negative). Rh^- individuals normally do not have antibodies to the Rh factor, but they may make them when exposed to the Rh factor. It is possible to use anti-Rh antibodies for blood testing. When Rh^+ blood is mixed with anti-Rh antibodies, agglutination occurs (Fig. 5.10).

During pregnancy, if the mother is Rh^- and the father is Rh^+, the child may be Rh^+ (Fig. 5.11). The Rh^+ red blood cells may begin leaking across the placenta into the mother's cardiovascular system, as placental tissues normally break down before and at birth. The presence of these Rh antigens

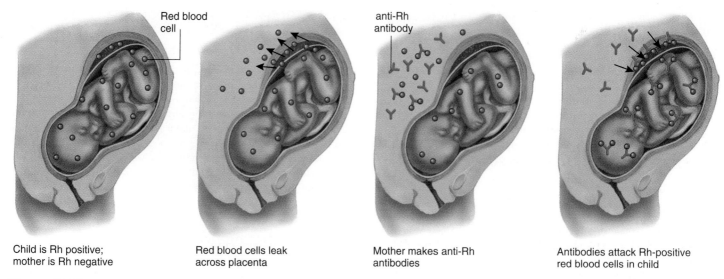

Red blood cell

anti-Rh antibody

| Child is Rh positive; mother is Rh negative | Red blood cells leak across placenta | Mother makes anti-Rh antibodies | Antibodies attack Rh-positive red blood cells in child |

Figure 5.11 **Hemolytic disease of the newborn.**
Due to a pregnancy in which the child is Rh$^+$, an Rh$^-$ mother can begin to produce antibodies against Rh$^+$ red blood cells. In another pregnancy, these antibodies can cross the placenta and cause hemolysis of an Rh$^+$ child's red blood cells.

causes the mother to produce anti-Rh antibodies. In this or a subsequent pregnancy with another Rh$^+$ baby, anti-Rh antibodies produced by the mother may cross the placenta and destroy the child's red blood cells. This is called hemolytic disease of the newborn (HDN) because hemolysis continues after the baby is born. Due to red blood cell destruction, excess bilirubin in the blood can lead to brain damage and mental retardation or even death.

The Rh problem is prevented by giving Rh$^-$ women an Rh immunoglobulin injection either midway through the first pregnancy or no later than 72 hours after giving birth to any Rh$^+$ child. This injection contains anti-Rh antibodies that attack any of the baby's red blood cells in the mother's blood before these cells can stimulate her immune system to produce her own antibodies.

The possibility of hemolytic disease of the newborn exists when the mother is Rh$^-$ and the father is Rh$^+$.

Bioethical Issue

Sickle-cell disease is a prevalent type of debilitating anemia in the African-American community because African Americans are more likely to be carriers of a sickle-cell gene allele. Carriers are not sick, but if two carriers reproduce with one another, each child has a 25% chance of having sickle-cell disease, which can lead to an early death.

In the 1970s, the federal government funded a screening program to detect carriers of the sickle-cell allele gene because they thought that parents might want to know if they were carriers. The procedure is inexpensive and requires the testing of just one drop of blood. At first, members of the black community were in favor of testing; ministers conducted tests on their congregations and the black Panthers offered the test door-to-door in black communities.

Due to a lack of understanding, however, the results of testing were not as expected. The government, employers, and carriers themselves thought that carriers were sick. The Massachusetts legislature required that the schools, for health reasons, identify all carriers as well as students with sickle-cell disease. Some insurance companies began to deny coverage to black carriers on the grounds that they had a preexisting medical condition. The U.S. Air Force Academy rejected black applicants who were carriers. Some commercial airlines refused to hire carriers, thinking that they might faint at high altitudes. Worse yet, prominent scientists suggested that those who were carriers should forego having children.

Those who are not in favor of genetic testing can use this example to suggest that genetic testing should not be done. Those who are in favor of genetic testing can use this example to discover what safeguards are needed to protect the individual from harm due to genetic testing.

Questions

1. If you are in favor of genetic testing, who should know the results beside yourself?
2. Suppose you are a carrier of sickle-cell disease and reproduce with another carrier, would you be in favor of only having offspring free of the disorder, if that were possible? Why or why not?
3. As a society, should we allow people to make up their minds about passing on genetic flaws, or should we try to influence their decision?

Summarizing the Concepts

Blood, which is composed of formed elements and plasma, has several functions. It transports hormones, oxygen, and nutrients to the cells and carbon dioxide and other wastes away from cells. It fights infections and has various regulatory functions. It maintains blood pressure, regulates body temperature, and keeps the pH within normal limits. All of these functions help maintain homeostasis.

All blood cells are produced within red bone marrow from stem cells, which are ever capable of dividing and producing new cells.

5.1 The Red Blood Cells

Red blood cells are small, biconcave disks that lack a nucleus. They live about 120 days and are destroyed in the liver and spleen when they are old or abnormal. The production of red blood cells is controlled by oxygen concentration of the blood. When the oxygen concentration decreases, the kidneys increase their production of erythropoietin and more red blood cells are produced. Red blood cells contain hemoglobin, the respiratory pigment, which combines with oxygen and transports it to the tissues.

5.2 The White Blood Cells

White blood cells are larger than red blood cells, they have a nucleus, and they are translucent unless stained. Like red blood cells, they are produced in the red bone marrow. White blood cells are divided into the granular leukocytes and the agranular leukocytes. The granular leukocytes have conspicuous granules; in eosinophils, granules are red when stained with eosin, and in basophils, granules are blue when stained with a basic dye. The granules in neutrophils don't take up either dye significantly. Neutrophils are the most plentiful of the white blood cells, and they are able to phagocytize pathogens. Many neutrophils die within a few days when they are fighting an infection. The agranulocytes include the lymphocytes and the monocytes. The lymphocytes and monocytes function in specific immunity. On occasion, the monocytes become large phagocytic cells of great significance. They engulf worn-out red blood cells and pathogens at a ferocious rate.

5.3 Blood Clotting

When there is a break in a blood vessel, the platelets clump to form a plug. Blood clotting itself requires a series of enzymatic reactions involving blood platelets, prothrombin, and fibrinogen. In the final reaction, fibrinogen becomes fibrin threads, entrapping cells. The fluid that escapes from a clot is called serum and consists of plasma minus fibrinogen.

5.4 Plasma

Plasma is mostly water (92%) and the plasma proteins (7%). The plasma proteins, most of which are produced by the liver, occur in three categories: albumins, globulins, and fibrinogen. The plasma proteins maintain osmotic pressure, help regulate pH, and transport molecules. Some plasma proteins have specific functions: the gamma globulins, which are antibodies produced by B lymphocytes, function in immunity, and fibrinogen is necessary to blood clotting.

Small organic molecules like glucose and amino acids are dissolved in plasma and serve as nutrients for cells; the gas oxygen is needed for cellular respiration and carbon dioxide is a waste production of this process.

5.5 Capillary Exchange

At the arterial end of a cardiovascular capillary, blood pressure is greater than osmotic pressure; therefore, water leaves the capillary. In the midsection, oxygen and nutrients diffuse out of the capillary; carbon dioxide and other wastes diffuse into the capillary. At the venous end, osmotic pressure created by the presence of proteins exceeds blood pressure, causing water to enter the capillary.

Retrieving fluid by means of osmotic pressure is not completely effective. There is always some fluid that is not picked up at the venous end of the cardiovascular capillary. This excess tissue fluid enters the lymphatic capillaries. Lymph is tissue fluid contained within lymphatic vessels. The lymphatic system is a one-way system and lymph is returned to blood by way of a cardiovascular vein.

5.6 Blood Typing

Red blood cells of an individual are not necessarily received without difficulty by another individual. For example, the membranes of red blood cells may contain type A, B, AB, or no antigens. In the plasma there are two possible antibodies: anti-A or anti-B. If the corresponding antigen and antibody are put together, clumping, or agglutination, occurs; in this way, the blood type of an individual may be determined in the laboratory. After determination of the blood type, it is theoretically possible to decide who can give blood to whom. For this, it is necessary to consider the donor's antigens and the recipient's antibodies.

Another important antigen is the Rh antigen. This particular antigen must also be considered in the transfusing of blood, and it is important during pregnancy because an Rh$^-$ mother may form antibodies to the Rh antigen while carrying or after the birth of the child who is Rh$^+$. These antibodies can cross the placenta to destroy the red blood cells of any subsequent Rh$^+$ child.

Studying the Concepts

1. State the two main components of blood, and give the functions of blood.   111   *Plasma | Funcrfon element.*

2. What is hemoglobin, and how does it function?   111   *Red blood | Carbon oxygen.*

3. Describe the life cycle of red blood cells, and tell how the production of red blood cells is regulated.   113   *4 month 120 day. iĺver & spleer.*

4. Name the five types of white blood cells; describe the structure and give a function for each type.   115

5. Name the steps that take place when blood clots. Which substances are present in blood at all times, and which appear during the clotting process?   116

6. Define blood, plasma, tissue fluid, lymph, and serum.   111, 116, 118, 119

7. List and discuss the major components of plasma. Name several plasma proteins, and give a function for each.   118

8. What forces operate to facilitate exchange of molecules across the capillary wall?   118–19

9. What are the four ABO blood types? For each, state the antigen(s) on the red blood cells and the antibody(ies) in the plasma.   120

10. Explain why a person with type O blood cannot receive a transfusion of type A blood.   120

11. Problems can arise during childbearing if the mother is which Rh type and the father is which Rh type? Explain why this is so.   121

Testing Your Knowledge of the Concepts

In questions 1–4, match the components of blood to the descriptions, and fill in the blanks.

a. red blood cell
b. white blood cells
c. both red and white blood cells
d. plasma

_____ 1. Includes monocytes that _____ debris

_a_ 2. Contains _____ that transports oxygen

_____ 3. Contains fibrinogen that is needed for _____

_____ 4. Antigens in plasma membrane that determine the blood type

In questions 5–7, indicate whether the statement is true (T) or false (F).

_____ 5. Carbon dioxide exits the arteriole end of the capillary and oxygen enters the venule end of the capillary.

_____ 6. Without the work of the lymphatic system, the tissues would fill with fluid.

_____ 7. If a woman is Rh⁺, she doesn't have to worry about hemolytic disease of the newborn.

In questions 8 and 9, fill in the blanks.

8. B lymphocytes produce *plasma cells* that react with antigens.

9. A person with type AB blood has _____ antibodies in the plasma.

10. Label arrows as either blood pressure or osmotic pressure.

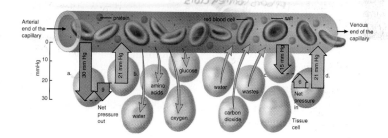

Applying Your Knowledge to the Concepts

These questions pertain to homeostasis.

1. Persons living in Los Angeles have a much higher incidence of respiratory diseases than people living in the rest of California. How would you explain this to people who think of California as the land of suntans and movie stars?

2. Humans and birds, both "warm-blooded" organisms, are the only animals with non-nucleated red blood cells. What could be an advantage to this type of red blood cell over the nucleated red blood cell?

3. Recall that when red blood cells are broken down, the hemoglobin is released and the iron from the heme portion of the molecule is recovered and returned to the bone marrow for use. Why, then, do women need more absorbable iron in their diet than men? What would be another situation in which more iron should be consumed?

4. Tissue fluid (fluid that surrounds cells) is straw-colored, indicating that it does not contain red blood cells. Yet, examination shows that it does contain white blood cells. How can this be explained?

Understanding the Terms

agglutination 120 *Sticking together stop circulating*
agranular leukocyte 115 *not granule*
albumin 118 *plasma protein of blood*
anemia 113 – *low iron white blood*
antibodies 115 – *what fights germs*
antigen 115 *any foreign substance*
basophil 115 *white type cell.*
clotting 116 *stop bleeding*
colony-stimulating factors 115 *cellular or cellular in origin*
eosinophil 115 *white type cell.*
erythropoietin 113 *red blood cell.*
fibrin 116 *fiber*
fibrinogen 116 *fibers formed clots*
formed element 111 *cellular or cellular or germ*
granular leukocyte 115 *granules cytoplasmic*
hemoglobin 111 *red iron.*
hemolysis 113 – *rupture of red blood*
hemophilia 116 *non stopping bleeding*
leukemia 115

lymph 119 *fluid lymphatic vessels*
lymphocyte 115 *specialized w cell*
megakaryocyte 116 *large bone marrow*
monocyte 115 *respond an infection*
neutrophil 115 *granular leukocyte*
pathogen 111 *causes agent*
phagocytosis 115 *cell eating*
plasma 118 *liquid portion of blood*
platelet (thrombocyte) 116 *cell fragment*
prothrombin 116 *activator enzyme plasma protein*
prothrombin activator 116 *activator enzyme.*
red blood cell (erythrocyte) 111 *hemoglobin carries oxygen lungs to tissues.*
serum 116 *light yellow liquid*
sickle-cell disease 113 *genetic disorder hemolysis*
stem cell 113 *undifferentiated cell divides differentiated blood cells germ cells*
thrombin 116 *enzyme of fibrinogen to fibrin*
tissue fluid 118 *solution that bathes every cell in the body.*
white blood cell (leukocyte) 115 *protects body from foreign substances.*

Match the terms to these definitions:

a. *Hemoglobin* Iron-containing protein in red blood cells that combines with and transports oxygen.

b. *Agglutination* Clumping of cells, particularly in reference to red blood cells involved in an antigen-antibody reaction.

c. *Plasma* Liquid portion of blood.

d. *Lymph* Fluid derived from tissue fluid that is carried in lymphatic vessels.

e. *Prothrombin* Plasma protein that is converted to thrombin during the steps of blood clotting.

Applying Technology to the Concepts

Your study of the composition and function of blood is supported by these available technologies:

Essential Study Partner CD-ROM

Animals → Circulatory System

Visit the Mader web site for related ESP activities.

Exploring the Internet

The Mader Home Page provides resources and tools as you study this chapter.

http://www.mhhe.com/biosci/genbio/mader

Dynamic Human 2.0 CD-ROM

Lymphatic System → Clinical Concepts → Blood Type

Chapter 6

Cardiovascular System

Chapter Concepts

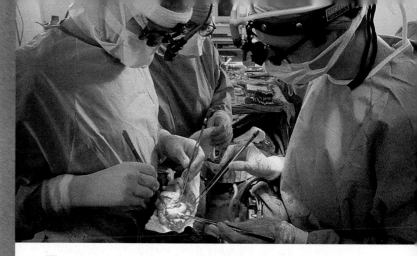

Figure 6.1 Heart transplant operation.
Will it be safe one day to use pigs' hearts for human transplant operations? Pigs have been genetically engineered to be immunocompatible with humans, but there is the possibility of acquiring a virus unique to pigs.

Dolores Manning, a 50-year-old-mother of six, had congestive heart failure. Her heart was unable to beat effectively; blood was backing up in blood vessels, and fluids were collecting in her lungs. She might suffocate. A diuretic helped rid her body of excess fluid but she was still sick. Dolores needed a heart transplant.

Doctors didn't know when a heart might become available, so in the meantime they gave Dolores a LVAD (left ventricular assist device) (Fig. 6.1). This device takes over some of the pumping functions of the heart. Now, the heart doesn't have to work so hard.

Surgeons implanted the device in an abdominal "pocket" of skin they created. It was run by a small electric generator. The device worked so well Dolores was able to exercise and increase her fitness. But she still felt constrained by not having a fully functioning heart. Everyone said the improvement in her general health was bound to increase the chance of a successful heart transplant.

Still there was a problem. No one knew when a heart might become available. The transplanted heart had to be immunologically compatible or it couldn't even be considered. While Dolores was waiting she came across an article about pigs whose tissues had been genetically altered to be compatible with those of any human being. The idea is that some day pigs will be a source of organs for transplantation into humans. Dolores was desperate to lead a normal life with her children, so she asked the

doctors if she could have a pig's heart. "It's experimental right now," the doctors said. "Some investigators are afraid that the tissue of a pig might give the human recipient a virus unique to pigs." "I'm willing to take a chance," Dolores said.

As Dolores knew, the heart is a vital organ because it ordinarily keeps blood moving in the cardiovascular system. Circulation of the blood is so important that if the heart stops beating for only a few minutes, death results. This chapter reviews the cardiovascular system and how it operates to keep blood circulating about the body. In humans, the right side of the heart pumps blood to the lungs and the left side pumps blood to the tissues. The blood never runs free, and is conducted to and from the tissues by blood vessels in these two separate vascular circuits. Among the types of blood vessels, only the capillaries have walls thin enough to allow exchange of molecules with the tissues. This exchange refreshes tissue fluid so that homeostasis is maintained. Otherwise, cells would die from the lack of nutrients and the buildup of waste products.

6.1   The Blood Vessels

The cardiovascular system has three types of blood vessels: the **arteries** (and arterioles), which carry blood away from the heart to the capillaries; the **capillaries,** which permit exchange of material with the tissues; and the **veins** (and venules), which return blood from the capillaries to the heart.

The Arteries

The arterial wall has three layers (Fig. 6.2*a*). The inner layer is a simple squamous epithelium called endothelium with a connective tissue basement membrane that contains elastic fibers. The middle layer is the thickest layer and consists of smooth muscle that can contract to regulate blood flow and blood pressure. The outer layer is fibrous connective tissue near the middle layer, but it becomes loose connective tissue at its periphery. Some arteries are so large that they require their own blood vessels.

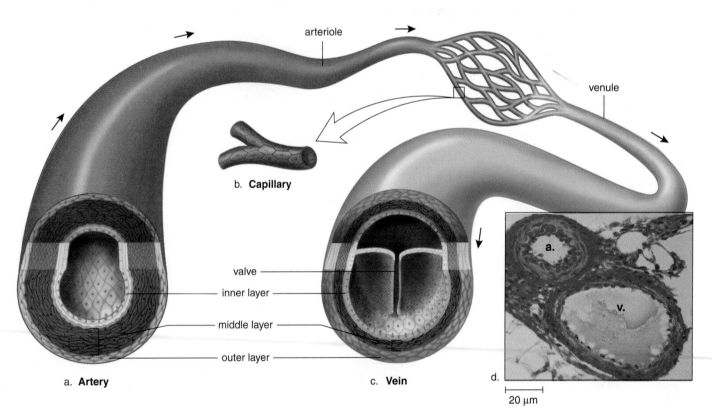

Figure 6.2   Blood vessels.
The walls of arteries and veins have three layers. The inner layer is composed largely of endothelium with a basement membrane that has elastic fibers; the middle layer is smooth muscle tissue; the inner layer is connective tissue (largely collagen fibers). **a.** Arteries have a thicker wall than veins because they have a larger middle layer than veins. **b.** Capillary walls are one-cell-thick endothelium. **c.** Veins are larger in diameter than arteries, so that collectively veins have a larger holding capacity than arteries. **d.** Scanning electron micrograph of an artery and vein.

Arterioles are small arteries just visible to the naked eye. The middle layer of arterioles has some elastic tissue but is composed mostly of smooth muscle whose fibers encircle the arteriole. When these muscle fibers are contracted, the vessel has a smaller diameter (is constricted); and when these muscle fibers are relaxed, the vessel has a larger diameter (is dilated). Whether arterioles are constricted or dilated affects blood pressure. The greater the number of vessels dilated, the lower the blood pressure.

The Capillaries

Arterioles branch into capillaries (Fig. 6.2*b*). Each capillary is an extremely narrow, microscopic tube with one-cell-thick walls composed only of endothelium with a basement membrane. *Capillary beds* (networks of many capillaries) are present in all regions of the body; consequently, a cut to any body tissue draws blood. Capillaries are a very important part of the human cardiovascular system because an exchange of substances takes place across their thin walls. Oxygen and nutrients, such as glucose, diffuse out of a capillary into the tissue fluid that surrounds cells. Wastes, such as carbon dioxide, diffuse into the capillary. The relative constancy of tissue fluid is absolutely dependent upon capillary exchange.

Only certain capillaries are open at any given time. For example, after eating, the capillaries that serve the digestive system are open and those that serve the muscles are closed. When a capillary bed is closed, the precapillary sphincters contract, and the blood moves from arteriole to venule by way of an arteriovenous shunt (Fig. 6.3).

The Veins

Venules are small veins that drain blood from the capillaries and then join to form a vein. The walls of venules (and veins) have the same three layers as arteries, but there is less smooth muscle and connective tissue (Fig. 6.2*c*). Veins often have **valves,** which allow blood to flow only toward the heart when open and prevent the backward flow of blood when closed.

Since walls of veins are thinner, they can expand to a greater extent (Fig. 6.2*d*). At any one time about 70% of the blood is in the veins. In this way, the veins act as a blood reservoir.

Arteries and arterioles carry blood away from the heart toward the capillaries; capillaries join arterioles to venules; veins and venules return blood from the capillaries to the heart.

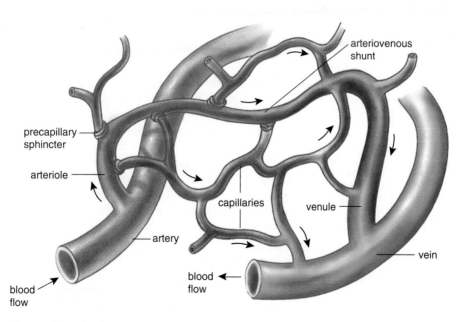

Figure 6.3   Anatomy of a capillary bed.
A capillary bed forms a maze of capillary vessels that lies between an arteriole and a venule. When sphincter muscles are relaxed, the capillary bed is open, and blood flows through the capillaries. When sphincter muscles are contracted, blood flows through a shunt that carries blood directly from an arteriole to a venule. As blood passes through a capillary in the tissues, it gives up its oxygen (O_2). Therefore, blood goes from carrying more oxygen in the arteriole (red color) to carrying less oxygen (blue color) in the vein.

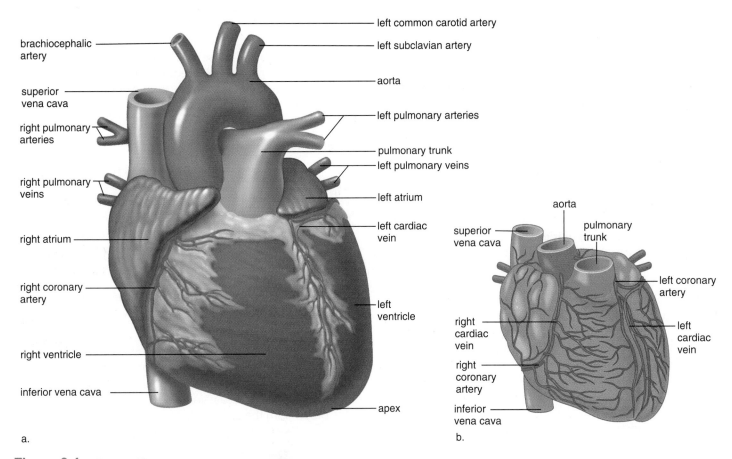

Figure 6.4   **External heart anatomy.**
a. The superior vena cava and the pulmonary arteries are attached to the right side of the heart. The aorta and left pulmonary veins are attached to the left side of the heart. The right ventricle forms most of the anterior surface of the heart, and the left ventricle forms most of the posterior.
b. The coronary arteries and cardiac veins pervade cardiac muscle. They bring oxygen and nutrients to cardiac cells, and return blood to the right atrium.

6.2 The Heart

The **heart** is a cone-shaped, muscular organ about the size of a fist (Fig. 6.4). It is located between the lungs directly behind the sternum (breastbone) and is tilted so that the apex (the pointed end) is oriented to the left. The major portion of the heart, called the **myocardium,** consists largely of cardiac muscle tissue. The muscle fibers of the myocardium are branched and tightly joined to one another. The heart lies within the **pericardium,** a thick, membranous sac that secretes a small quantity of lubricating liquid. The inner surface of the heart is lined with endocardium, which consists of connective tissue and endothelial tissue.

Internally, a wall called the septum separates the heart into a right side and a left side (Fig. 6.5a). The heart has four chambers. The two upper, thin-walled atria (sing., **atrium**) have wrinkled protruding appendages called auricles. The two lower chambers are the thick-walled **ventricles,** which pump the blood.

The heart also has four valves, which direct the flow of blood and prevent its backward movement. The two valves that lie between the atria and the ventricles are called the **atrioventricular valves.** These valves are supported by strong fibrous strings called **chordae tendineae.** The chordae, which are attached to muscular projections of the ventricular walls, support the valves and prevent them from inverting when the heart contracts. The atrioventricular valve on the right side is called the tricuspid valve because it has three flaps, or cusps. The valve on the left side is called the bicuspid (or the mitral) because it has two flaps. The remaining two valves are the **semilunar valves,** whose flaps resemble half-moons, between the ventricles and their attached vessels. The pulmonary semilunar valve lies between the right ventricle and the pulmonary trunk. The aortic semilunar valve lies between the left ventricle and the aorta.

Humans have a four-chambered heart (two atria and two ventricles). A septum separates the right side from the left side.

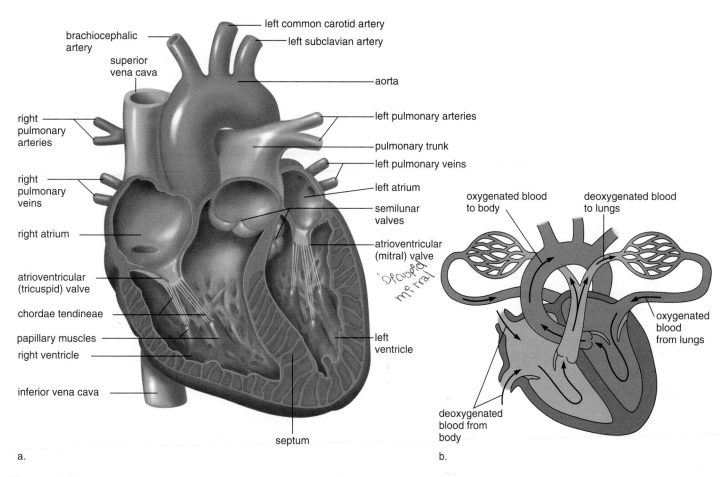

Figure 6.5   Internal view of the heart.
a. The heart has four valves. The atrioventricular valves allow blood to pass from the atria to the ventricles, and the semilunar valves allow blood to pass out of the heart. **b.** This diagrammatic representation of the heart allows you to trace the path of the blood. On the right side of the heart: venae cavae, right atrium, right ventricle, pulmonary arteries to lungs. On the left side of the heart: pulmonary veins, left atrium, left ventricle, aorta to body.

Passage of Blood Through the Heart

We can trace the path of blood through the heart (Fig. 6.5*b*) in the following manner:

The superior vena cava and the inferior **vena cava,** which carry blood that is relatively low in oxygen and relatively high in carbon dioxide, enter the right atrium.

The right atrium sends blood through an atrioventricular valve (the tricuspid valve) to the right ventricle.

The right ventricle sends blood through the pulmonary semilunar valve into the pulmonary trunk and the two **pulmonary arteries** to the lungs.

Four **pulmonary veins,** which carry blood that is relatively high in oxygen and relatively low in carbon dioxide, enter the left atrium.

The left atrium sends blood through an atrioventricular valve (the bicuspid or mitral valve) to the left ventricle.

The left ventricle sends blood through the aortic semilunar valve into the **aorta** to the body proper.

From this description, you can see that deoxygenated blood never mixes with oxygenated blood and that blood must go through the lungs in order to pass from the right side to the left side of the heart. In fact, the heart is a double pump because the right ventricle of the heart sends blood through the lungs, and the left ventricle sends blood throughout the body. Since the left ventricle has the harder job of pumping blood to the entire body, its walls are thicker than those of the right ventricle, which pumps blood a relatively short distance to the lungs.

The right side of the heart pumps blood to the lungs, and the left side of the heart pumps blood throughout the body.

The Heartbeat

Each heartbeat is called a **cardiac cycle** (Fig. 6.6). When the heart beats, first, the two atria contract at the same time; then the two ventricles contract at the same time. Then all chambers relax. The word **systole** refers to contraction of heart muscle, and the word **diastole** refers to relaxation of heart muscle. The heart contracts, or beats, about 70 times a minute, and each heartbeat lasts about 0.85 seconds.

Time	Atria	Ventricles
0.15 sec	Systole	Diastole
0.30 sec	Diastole	Systole
0.40 sec	Diastole	Diastole

A normal adult rate at rest can vary from 60 to 80 beats per minute.

When the heart beats, the familiar lub-dup sound occurs. The longer and lower-pitched lub is caused by vibrations occurring when the atrioventricular valves close due to ventricular contraction. The shorter and sharper dup is heard when the semilunar valves close due to back pressure of blood in the arteries. A heart murmur, or a slight slush sound after the lub, is often due to ineffective valves, which allow blood to pass back into the atria after the atrioventricular valves have closed. Rheumatic fever resulting from a bacterial infection is one possible cause of a faulty valve, particularly the bicuspid valve. Faulty valves can be surgically corrected.

Intrinsic Control of Heartbeat

The rhythmical contraction of the atria and ventricles is due to the instrinsic conduction system of the heart. Nodal tissue, which has both muscular and nervous characteristics, is a unique type of cardiac muscle located in two regions of the heart. The **SA (sinoatrial) node** is located in the upper wall of the right atrium; and the **AV (atrioventricular) node,** is located in the base of the right atrium very near the septum (Fig. 6.7a). The SA node initiates the heartbeat and automatically sends out an excitation impulse every 0.85 seconds; this causes the atria to contract. When impulses reach the AV node there is a slight delay that allows the atria to finish their contraction before the ventricles begin their contraction. The signal for the ventricles to contract travels from the AV node through the two branches of the **atrioventricular bundle** (AV bundle) before reaching the numerous and smaller **Purkinje fibers.** The AV bundle, its branches, and the Purkinje fibers consist of specialized cardiac muscle fibers that efficiently cause the ventricles to contract.

The SA node is called the **pacemaker** because it usually keeps the heartbeat regular. If the SA node fails to work properly, the heart still beats due to impulses generated by the AV node. But the beat is slower (40 to 60 beats per minute). To correct this condition, it is possible to implant an artificial pacemaker, which automatically gives an electric stimulus to the heart every 0.85 seconds.

The intrinsic conduction system of the heart consists of the SA node, the AV node, the atrioventricular bundle, and the Purkinje fibers.

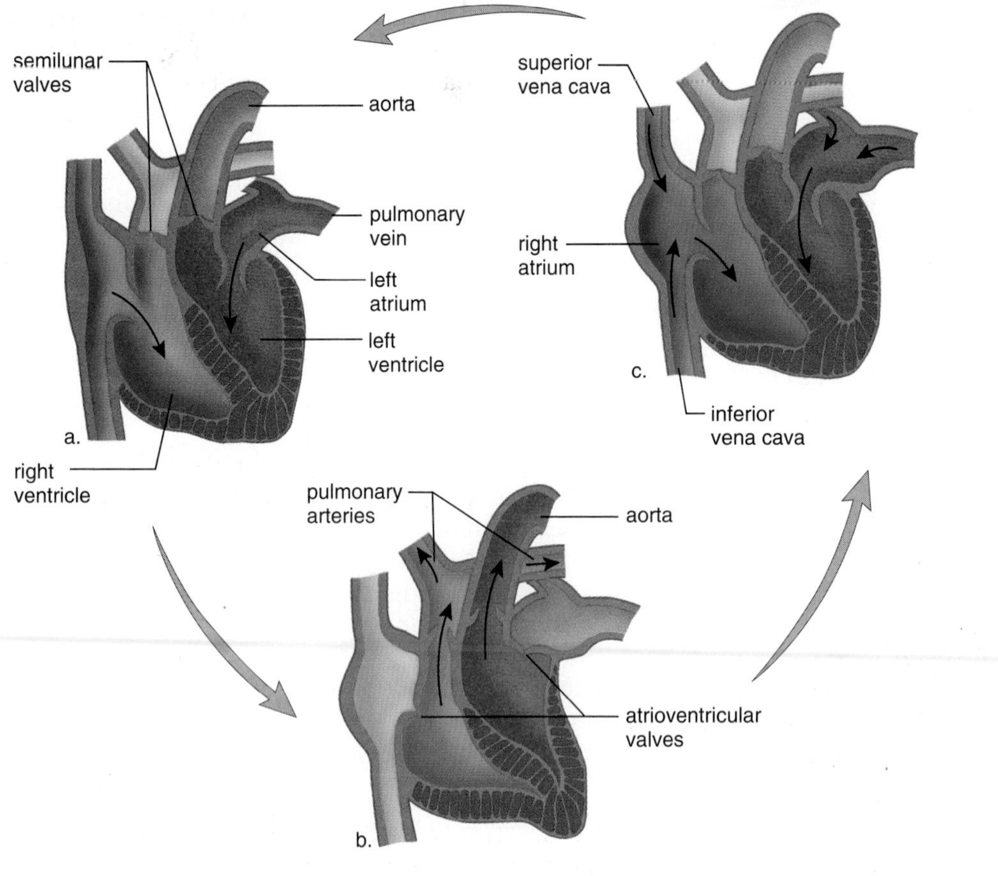

Figure 6.6   Stages in the cardiac cycle.
a. When the atria contract, the ventricles are relaxed and filling with blood. **b.** When the ventricles contract, the atrioventricular valves are closed, the semilunar valves are open, and the blood is pumped into the pulmonary trunk and aorta. **c.** When the heart is relaxed, both atria and ventricles are filling with blood.

Extrinsic Control of Heartbeat

The body has an extrinsic way to regulate the heartbeat. A cardiac control center in the medulla oblongata, a portion of the brain that controls internal organs, can alter the beat of the heart by way of the autonomic system, a division of the nervous system. This system has two divisions: the parasympathetic system, which promotes those functions we tend to associate with a restful state, and the sympathetic system, which brings about those responses we associate with increased activity and/or stress. The parasympathetic system decreases SA and AV nodal activity when we are inactive, and the sympathetic system increases SA and AV nodal activity when we are active or excited.

The hormones epinephrine and norepinephrine, which are released by the adrenal medulla, also stimulate the heart. During exercise, for example, the heart pumps faster and stronger due to sympathetic stimulation and due to the release of epinephrine and norepinephrine.

The body has an extrinsic way to regulate the heartbeat. The autonomic system and hormones can modify the heartbeat rate.

The Electrocardiogram

An **electrocardiogram (ECG)** is a recording of the electrical changes that occur in myocardium during a cardiac cycle. Body fluids contain ions that conduct electrical currents, and therefore the electrical changes in myocardium can be detected on the skin's surface. When an electrocardiogram is being taken, electrodes placed on the skin are connected by wires to an instrument that detects the myocardium's electrical changes. Thereafter a pen rises or falls on a moving strip of paper. Figure 6.7*b* depicts the pen's movements during a normal cardiac cycle.

When the SA node triggers an impulse, the atrial fibers produce an electrical change that is called the P wave. The P wave indicates that the atria are about to contract. After that, the QRS complex signals that the ventricles are about to contract. The electrical changes that occur as the ventricular muscle fibers recover produces the T wave.

Various types of abnormalities can be detected by an electrocardiogram. One of these, called ventricular fibrillation, is caused by uncoordinated contraction of the ventricles (Fig. 6.7*c*). Ventricular fibrillation is of special interest because it can be caused by an injury or drug overdose. It is the most common cause of sudden cardiac death in a seemingly healthy person. Once the ventricles are fibrillating, they have to be defibrillated by applying a strong electric current for a short period of time. Then the SA node may be able to reestablish a coordinated beat.

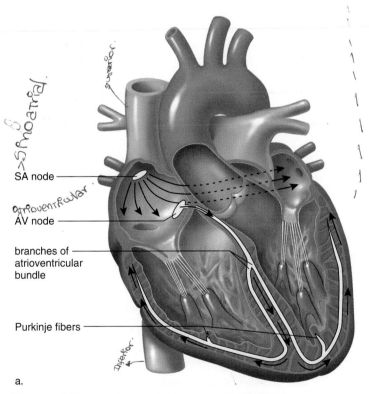

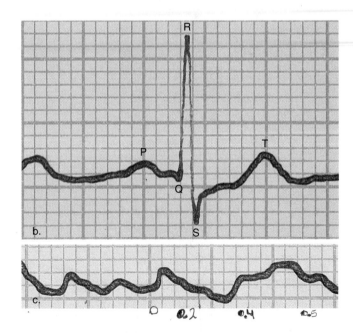

Figure 6.7   Conduction system of the heart.
a. The SA node sends out a stimulus, which causes the atria to contract. When this stimulus reaches the AV node, it signals the ventricles to contract. When this stimulus reaches the AV node, it signals the ventricles to contract. Impulses pass down the two branches of the atrioventricular bundle to the Purkinje fibers and thereafter the ventricles contract. **b.** A normal ECG indicates that the heart is functioning properly. The P wave occurs just prior to atrial contraction; the QRS complex occurs just prior to ventricular contraction; and the T wave occurs when the ventricles are recovering from contraction. **c.** Ventricular fibrillation produces an irregular electrocardiogram due to irregular stimulation of the ventricles.

6.3 Features of the Cardiovascular System

When the left ventricle contracts, blood is sent out into the aorta under pressure.

Pulse

The surge of blood entering the arteries causes their elastic walls to stretch, but then they almost immediately recoil. This alternating expansion and recoil of an arterial wall can be felt as a **pulse** in any artery that runs close to the body's surface. It is customary to feel the pulse by placing several fingers on a radial artery, which lies near the outer border of the palm side of the wrist (Fig. 6.8) A carotid artery, on either side of the trachea in the neck, is another accessible location to feel the pulse. Normally, the pulse rate indicates the rate of the heartbeat because the arterial walls pulse whenever the left ventricle contracts.

Blood Flow

The beating of the heart is necessary to homeostasis because it creates the pressure that propels blood in the arteries and the arterioles, which leads to the capillaries where exchange with tissue fluid takes place.

Blood Flow in Arteries

Blood pressure is the pressure of blood against the wall of a blood vessel. A sphygmomanometer is used to measure blood pressure, as described in Figure 6.9. The highest arterial pressure, called the **systolic pressure,** is reached during ejection of blood from the heart. The lowest arterial pressure is called the **diastolic pressure.** Diastolic pressure occurs while the heart ventricles are relaxing. Normal resting blood pressure for a young adult is said to be 120 mm mercury (Hg) over 80 mm Hg, or simply 120/80. The higher number is the systolic pressure, and the lower number is the diastolic pressure. Actually, 120/80 is the expected blood pressure in the brachial artery of the arm.

Both systolic and diastolic blood pressure decrease with distance from the left ventricle because the total cross-sectional

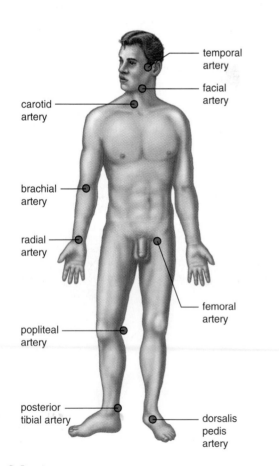

Figure 6.8   Pulse points.
The pulse can be taken at these arterial sites.

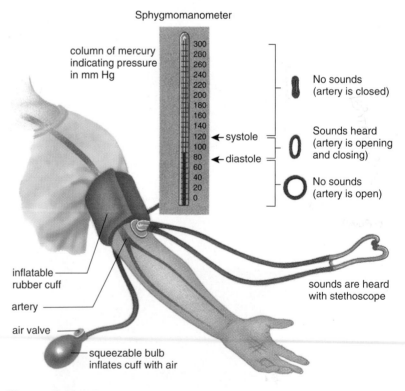

Figure 6.9   Use of a sphygmomanometer.
The technician inflates the cuff with air, gradually reduces the pressure and listens with a stethoscope for the sounds that indicate blood is moving past the cuff in an artery. This is systolic blood pressure. The pressure in the cuff is further reduced until no sound is heard, indicating that blood is flowing freely through the artery. This is diastolic pressure.

area of the blood vessels increase—there are more arterioles than arteries. The decrease in blood pressure causes the blood velocity to gradually decrease as it flows toward the capillaries.

Blood Flow in Capillaries

There are many more capillaries than arterioles, and blood moves slowly through the capillaries (Fig. 6.10). This is important because the slow progress allows time for the exchange of substances between blood in the capillaries and the surrounding tissues.

Blood Flow in Veins

Blood pressure is minimal in venules and veins (20–0 mm Hg). Instead of blood pressure, venous return is dependent upon three factors: skeletal muscle contraction, presence of valves in veins, and respiratory movements. When the skeletal muscles contract, they compress the weak walls of the veins. This causes blood to move past the next valve. Once past the valve, blood cannot flow backward (Fig. 6.11). The importance of muscle contraction in moving blood in

the venous vessels can be demonstrated by forcing a person to stand rigidly still for an hour or so. Frequently, fainting occurs because blood collects in the limbs, depriving the brain of needed blood flow and oxygen. In this case, fainting is beneficial because the resulting horizontal position aids in getting blood to the head.

When inspiration occurs, the thoracic pressure falls and abdominal pressure rises as the chest expands. This also aids the flow of venous blood back to the heart because blood flows in the direction of reduced pressure. Blood velocity increases slightly in the venous vessels due to a progressive reduction in the cross-sectional area as small venules join to form veins.

Blood pressure accounts for the flow of blood in the arteries and the arterioles. Skeletal muscle contraction, valves in veins, and respiratory movements account for the flow of blood in the venules and the veins.

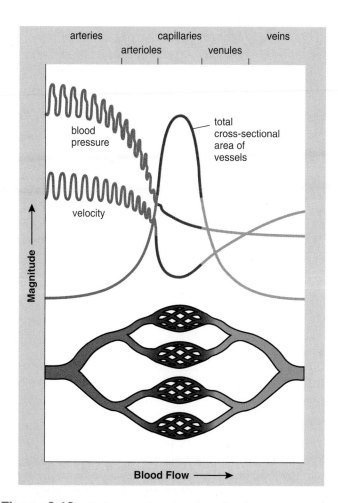

Figure 6.10   Cross-sectional area as it relates to blood pressure and blood velocity.
Blood pressure and blood velocity drop off in capillaries because capillaries have a greater cross-sectional area than arterioles.

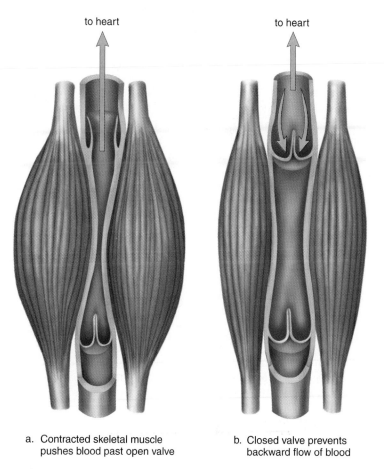

a. Contracted skeletal muscle pushes blood past open valve

b. Closed valve prevents backward flow of blood

Figure 6.11   Skeletal muscle contraction moves blood in veins.
a. Muscle contraction exerts pressure against the vein, and blood moves past the valve. **b.** Blood cannot flow back once it has moved past the valve.

6.4   The Vascular Pathways

The cardiovascular system, which is represented in Figure 6.12, includes two circuits: the **pulmonary circuit,** which circulates blood through the lungs, and the **systemic circuit,** which serves the needs of body tissues. Both circuits, as we shall see, are necessary to homeostasis.

The Pulmonary Circuit

The path of blood through the lungs can be traced as follows. Blood from all regions of the body first collects in the right atrium and then passes into the right ventricle, which pumps it into the pulmonary trunk. The pulmonary trunk divides into the right and left pulmonary arteries, which branch as they approach the lungs. The arterioles take blood to the pulmonary capillaries, where carbon dioxide is given off and oxygen is picked up. Blood then passes through the pulmonary venules, which lead to the four pulmonary veins that enter the left atrium. Since blood in the pulmonary arteries is deoxygenated but blood in the pulmonary veins is oxygenated, it is not correct to say that all arteries carry blood that is high in oxygen and all veins carry blood that is low in oxygen. It is just the reverse in the pulmonary circuit.

The pulmonary arteries take blood that is low in oxygen to the lungs, and the pulmonary veins return blood that is high in oxygen to the heart.

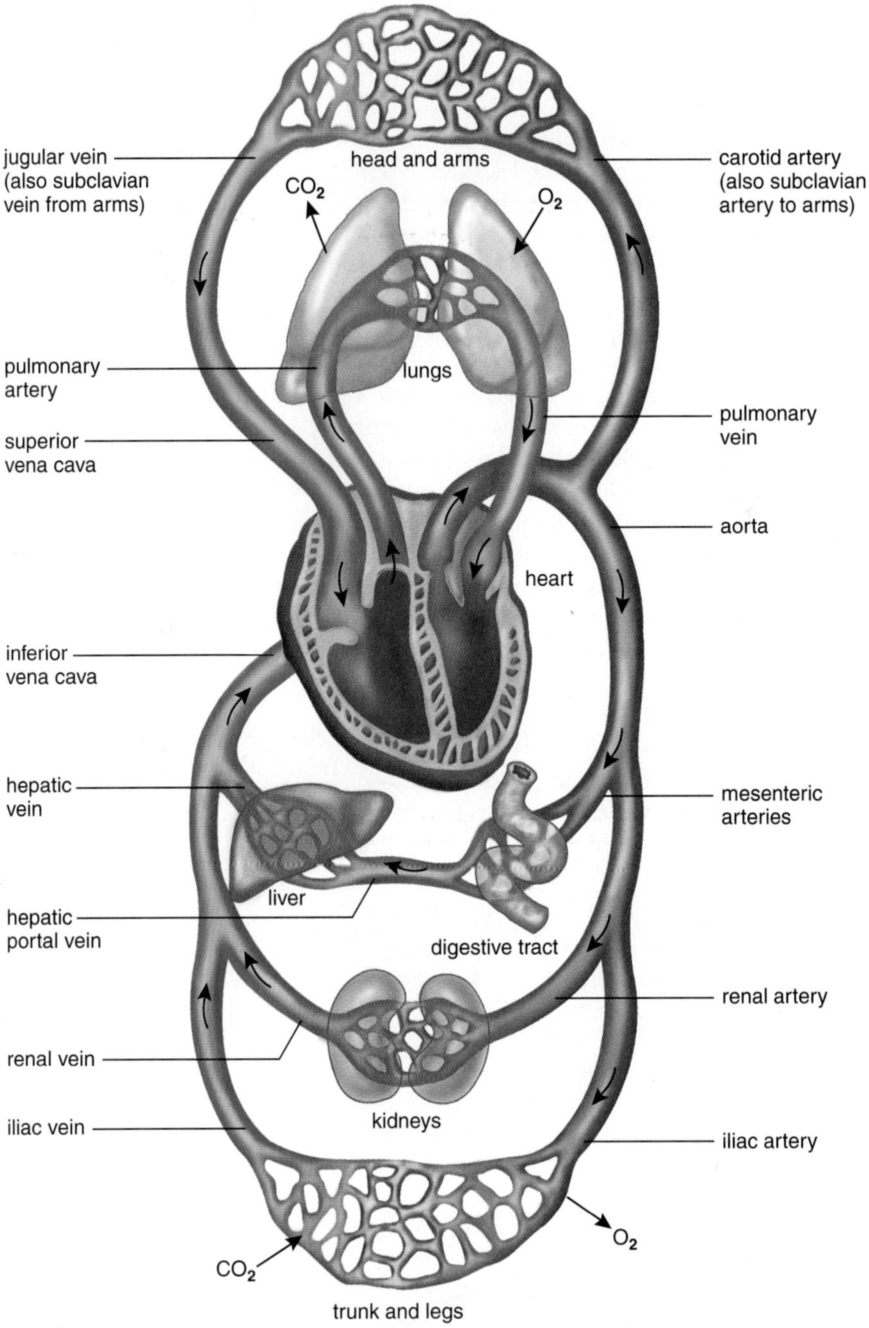

The Systemic Circuit

The systemic circuit includes all of the arteries and veins shown in Figure 6.13. The largest artery in the systemic circuit is the **aorta,** and the largest veins are the **superior** and **inferior venae cavae.** The superior vena cava collects blood from the head, the chest, and the arms, and the inferior vena cava collects blood from the lower body regions. Both enter the right atrium. The aorta and the venae cavae serve as the major pathways for blood in the systemic circuit.

Figure 6.12   Cardiovascular system diagram.
The blue-colored vessels carry deoxygenated blood, and the red-colored vessels carry oxygenated blood; the arrows indicate the flow of blood. Compare this diagram, useful for learning to trace the path of blood, to Figure 6.13 to realize that both arteries and veins go to all parts of the body. Also, there are capillaries in all parts of the body. No cell is located far from a capillary.

The path of systemic blood to any organ in the body begins in the left ventricle, which pumps blood into the aorta. Branches from the aorta go to the organs and major body regions. For example, this is the path of blood to and from the legs:

left ventricle—aorta—common iliac artery—legs—common iliac vein—inferior vena cava—right atrium

Notice when tracing blood, you need only mention the aorta, the proper branch of the aorta, the region, and the vein returning blood to the vena cava. In most instances, the artery and the vein that serve the same region are given the same name (Fig. 6.13). What happens when the blood reaches a particular region? Capillary exchange takes place, refreshing tissue fluid so that its composition stays relatively constant.

The **coronary arteries** (see Fig. 6.4) serve the heart muscle itself. (The heart is not nourished by the blood in its chambers.) The coronary arteries are the first branches off the aorta. They originate just above the aortic semilunar valve, and they lie on the exterior surface of the heart, where they divide into diverse arterioles. Because they have a very small diameter, the coronary arteries may become clogged, as discussed on page 138. The coronary capillary beds join to form venules. The venules converge to form the cardiac veins, which empty into the right atrium.

The body has a portal system called the **hepatic portal system,** which is associated with the liver. A portal system begins and ends in capillaries; in this instance, the first set of capillaries occurs at the villi of the small intestine and the second occurs in the liver. Blood passes from the capillaries of the intestinal villi into venules that join to form the **hepatic portal vein,** a vessel that connects the villi of the intestine with the liver, an organ that monitors the makeup of the blood. The **hepatic vein** leaves the liver and enters the inferior vena cava. While Figure 6.12 is helpful in tracing the path of blood, remember that all parts of the body receive both arteries and veins, as illustrated in Figure 6.13.

The systemic circuit takes blood from the left ventricle of the heart to the right atrium of the heart. It serves the body proper.

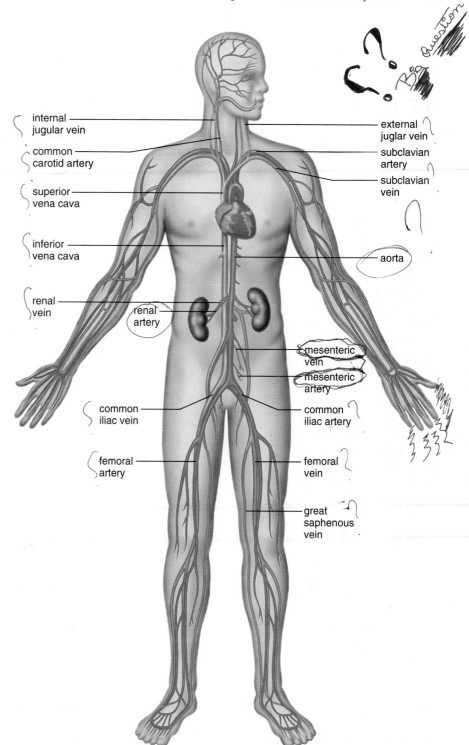

Figure 6.13   Major arteries and veins of the systemic circuit.
A more realistic representation of major blood vessels of the systemic circuit shows how the systemic arteries and veins are actually arranged in the body. The superior and inferior venae cavae take their names from their relationship to which organ?

Prevention of Cardiovascular Disease

All of us can take steps to prevent the occurrence of cardiovascular disease, the most frequent cause of death in the United States. There are genetic factors that predispose an individual to cardiovascular disease, such as family history of heart attack under age 55, male gender, and ethnicity (African Americans are at greater risk). Those with one or more of these risk factors need not despair, however. It only means that they need to pay particular attention to these guidelines for a heart-healthy life-style.

The Don'ts

Smoking

Hypertension is well recognized as a major contributor to cardiovascular disease. When a person smokes, the drug nicotine, present in cigarette smoke, enters the bloodstream. Nicotine causes the arterioles to constrict and the blood pressure to rise. Restricted blood flow and cold hands are associated with smoking by most people. Now, the heart must pump harder to propel the blood through the lungs at a time when the oxygen-carrying capacity of the blood is reduced. Smoking also damages the arterial wall and accelerates the formation of atherosclerosis and plaque (Fig. 6A).

Weight Gain

Hypertension also occurs more often in persons who are more than 20% above the recommended weight for their height. Because more tissue requires servicing, the heart must send extra blood out under greater pressure in those who are overweight. It may be very difficult to lose weight once it is gained, and therefore it is recommended that weight control be a lifelong endeavor. Even a slight decrease in weight can bring with it a reduction in hypertension. A 4.5-kilogram weight loss doubles the chance that blood pressure can be normalized without drugs.

The Do's

Healthy Diet

It was once thought that a low-salt diet was protective against cardiovascular disease, and it still may be in certain persons. Theoretically, hypertension occurs because the more salty the blood, the greater the osmotic pressure and the higher the water content. In recent years, the emphasis has switched to a diet low in saturated fats and cholesterol as protective against cardiovascular disease. Cholesterol is ferried in the blood by two types of plasma lipoproteins called LDL (low-density lipoprotein) and HDL (high-density lipoprotein). LDL (called "bad" lipoprotein) takes cholesterol from the liver to the tissues, and HDL (called "good" lipoprotein) transports cholesterol out of the tissues to the liver. When the LDL level in blood is abnormally high or the HDL level is abnormally low, cholesterol accumulates in the cells. When cholesterol-laden cells line the arteries, plaque develops, which interferes with circulation (Fig. 6A).

It is recommended that everyone know his or her blood cholesterol level. Individuals with a high blood cholesterol level (240 mg/ 100 ml) should be further tested to determine their LDL cholesterol level. The LDL cholesterol level together with other risk factors such as age, family history, general health, and whether the patient smokes will determine who needs dietary therapy to lower their LDL. Drugs are to be reserved for high-risk patients.

Evidence is mounting to suggest a role for antioxidant vitamins (A, E, and C) in the prevention of cardiovascular disease.

6.5 Cardiovascular Disorders

Cardiovascular disease (CVD) is the leading cause of untimely death in the Western countries. Modern research efforts have resulted in improved diagnosis, treatment, and prevention. This section discusses the range of advances that have been made in these areas. The Health reading for this chapter emphasizes how to prevent CVD from developing in the first place.

Hypertension

It is estimated that about 20% of all Americans suffer from **hypertension,** which is high blood pressure. Hypertension is present when the systolic blood pressure is 140 or greater or the diastolic blood pressure is 90 or greater. While both systolic and diastolic pressures are considered important, it is the diastolic pressure that is emphasized when medical treatment is being considered.

Hypertension is sometimes called a silent killer because it may not be detected until a stroke or heart attack occurs. It has long been thought that a certain genetic makeup might account for the development of hypertension. Now researchers have discovered two genes that may be involved in some individuals. One gene codes for angiotensinogen, a plasma protein that is converted to a powerful vasoconstrictor in part by the product of the second gene. Persons with hypertension due to overactivity of these genes might one day be cured by gene therapy.

Antioxidants protect the body from free radicals that may damage HDL cholesterol through oxidation or damage the lining of an artery, leading to a blood clot that can block the vessel. Nutritionists believe that the consumption of at least five servings of fruit and vegetables a day may be protective against cardiovascular disease.

Exercise

Those who exercise are less apt to have cardiovascular disease. One study found that moderately active men who spent an average of 48 minutes a day on a leisure-time activity such as gardening, bowling, or dancing had one-third fewer heart attacks than peers who spent an average of only 16 minutes each day. Exercise helps to keep weight under control, may help minimize stress, and reduces hypertension. The heart beats faster when exercising, but exercise slowly increases its capacity. This means that the heart can beat slower when we are at rest and still do the same amount of work. One physician recommends that his cardiovascular patients walk for one hour, three times a week, and in addition, they are to practice meditation and yoga-like stretching and breathing exercises to reduce stress.

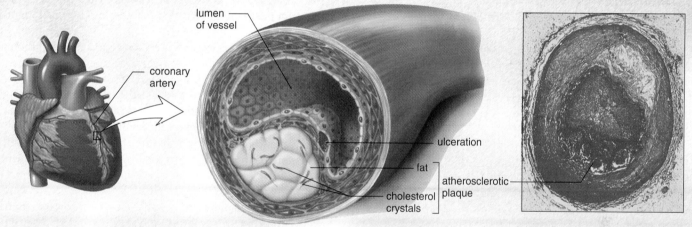

Figure 6A **Coronary arteries and plaque.**
Plaque (in yellow) is an irregular accumulation of cholesterol and other substances. When plaque is present in a coronary artery, a heart attack is more apt to occur because of restricted blood flow.

At present, however, the best safeguard against the development of hypertension is to have regular blood pressure checks and to adopt a life-style that lowers the risk of hypertension. The Health reading for this chapter discusses such health habits.

Atherosclerosis

Hypertension also is seen in individuals who have **atherosclerosis,** an accumulation of soft masses of fatty materials, particularly cholesterol, beneath the inner linings of arteries. Such deposits are called plaque. As it develops, plaque tends to protrude into the lumen of the vessel and interfere with the flow of blood. In certain families, atherosclerosis is due to an inherited condition such as familial hypercholesterolemia.

The presence of the associated mutation can be detected, and this information is helpful if measures are taken to prevent the occurrence of the disease. In most instances, atherosclerosis begins in early adulthood and develops progressively through middle age, but symptoms may not appear until an individual is 50 or older. To prevent the onset and development of plaque, the American Heart Association and other organizations recommend a diet low in saturated fat and cholesterol and rich in fruits and vegetables.

Plaque can cause a clot to form on the irregular arterial wall. As long as the clot remains stationary, it is called a **thrombus,** but when and if it dislodges and moves along with the blood, it is called an **embolus.** If **thromboembolism** is not treated, complications can arise, as mentioned in the following section.

Stroke, Heart Attack, and Aneurysm

Stroke, heart attack, and aneurysm are associated with hypertension and atherosclerosis. A cerebrovascular accident (CVA), also called a **stroke**, often results when a small cranial arteriole bursts or is blocked by an embolus. A lack of oxygen causes a portion of the brain to die, and paralysis or death can result. A person sometimes is forewarned of a stroke by a feeling of numbness in the hands or the face, difficulty in speaking, or temporary blindness in one eye.

A myocardial infarction (MI), also called a **heart attack,** occurs when a portion of the heart muscle dies due to a lack of oxygen. If a coronary artery becomes partially blocked, the individual may then suffer from **angina pectoris,** characterized by a radiating pain in the left arm. Nitroglycerin or related drugs dilate blood vessels and help relieve the pain. When a coronary artery is completely blocked, perhaps because of thromboembolism, a heart attack occurs.

An **aneurysm** is a ballooning of a blood vessel, most often the abdominal artery or the arteries leading to the brain. Atherosclerosis and hypertension can weaken the wall of an artery to the point that an aneurysm develops. If a major vessel like the aorta should burst, death is likely. Since capillaries are the region of exchange with tissues, it is possible to replace a damaged or diseased portion of a vessel, such as an artery with a plastic tube.

Dissolving Blood Clots

Medical treatment for thromboembolism includes the use of t-PA, a biotechnology drug. This drug converts plasminogen, a molecule found in blood, into plasmin, an enzyme that dissolves blood clots. In fact, t-PA, which stands for tissue plasminogen activator, is the body's own way of converting plasminogen to plasmin. t-PA is also being used for thrombolytic stroke patients but with limited success because some patients experience life-threatening bleeding in the brain. A better treatment might be new biotechnology drugs that act on the plasma membrane to prevent brain cells from releasing and/or receiving toxic chemicals caused by the stroke.

If a person has symptoms of angina or a stroke, then aspirin may be prescribed. Aspirin reduces the stickiness of platelets and therefore lowers the probability that a clot will form. There is evidence that aspirin protects against first heart attacks, but there is no clear support for taking aspirin every day to prevent strokes in symptom-free people. Physicians warn that long-term use of aspirin might have harmful effects, including bleeding in the brain.

Coronary Bypass Operations

Each year thousands of persons have coronary bypass surgery. During this operation, a surgeon takes a segment from another blood vessel and stitches one end to the aorta and the other end to a coronary artery past the point of obstruction. Figure 6.14 shows a triple bypass because three blood vessels have been used to allow blood to flow freely from the aorta to cardiac muscle by way of the coronary artery.

Gene therapy is now used to grow new blood vessels which will carry blood to cardiac muscle. The surgeon need only make a small incision and inject many copies of gene that codes for VEGF (vascular endothelial growth factor) between the ribs directly into the area of the heart that most needs improved blood flow. VEGF encourages new blood vessels to sprout out of an artery. If collateral blood vessels do form, they transport blood past clogged arteries, making bypass surgery unnecessary. About 60% of all patients who undergo the procedure do show signs of vessel growth within two to four weeks.

Another way to achieve the growth of new blood vessels is to drill tiny holes directly into the beating heart muscle with a laser. The holes close up immediately with the help of

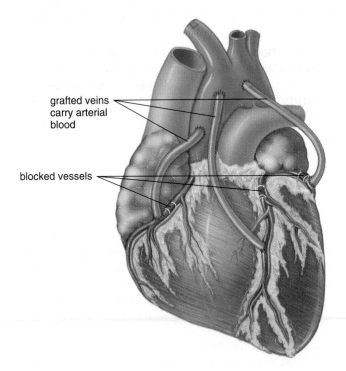

grafted veins carry arterial blood

blocked vessels

Figure 6.14   Coronary bypass operation.
During this operation, the surgeon grafts segments of another vessel, usually a small vein from the leg, between the aorta and the coronary vessels, bypassing areas of blockage. Patients who require surgery often receive two to five bypasses in a single operation.

pressure from the surgeon's fingers. But the channels created inside the muscle do remain open for a time and apparently new blood vessels are formed.

Clearing Clogged Arteries

In **angioplasty,** a cardiologist threads a plastic tube into an artery of an arm or a leg and guides it through a major blood vessel toward the heart. When the tube reaches the region of plaque in a coronary artery, a balloon attached to the end of the tube is inflated, forcing the vessel open (Fig. 6.15). However, the artery may not remain open because the trauma causes smooth muscle cells in the wall of the artery to proliferate and close it.

Two lines of attack are being explored. Small metal devices—either metal coils or slotted tubes called stents—are expanded inside the artery to keep the artery open. When the stents are coated with heparin to prevent blood clotting and with chemicals to prevent arterial closing, results have been promising.

Heart Transplants and Other Treatments

Persons with weakened hearts eventually may suffer from **congestive heart failure,** meaning the heart no longer is able to pump blood adequately, and blood backs up in the heart and lungs. Sometimes it is possible to repair a weak heart. For example, a back muscle can be wrapped around a heart to strengthen it. The muscle's nerve is stimulated with a kind of pacemaker that gives a burst of stimulation every 0.85 second. One day it may be possible to use cardiac cell transplants, because researchers who have injected live cardiac muscle cells into animal hearts find that they will contribute to the pumping of the heart.

On December 2, 1982, Barney Clark became the first person to receive an artificial heart that was driven by bursts of air received from a large external machine. Some clinically used artificial hearts today are driven by small battery-powered systems that can be carried about by a shoulder strap. The National Institutes of Health support the development of a hot-air engine that is fully implantable and is driven by an atomic heat source. The artificial heart is presently used only while the patient is waiting for a heart transplant. The difficulties with a heart transplant are, first, availability, and second, the tendency of the body to reject foreign organs. Recently researchers through genetic engineering altered the immune system of a strain of pigs with the hope that they will soon be a source of donor hearts.

Stroke, heart attack, and an aneurysm are associated with both hypertension and atherosclerosis. Various treatments are available, including both medical and surgical procedures. In the end, a heart transplant may be the only recourse for an ailing heart.

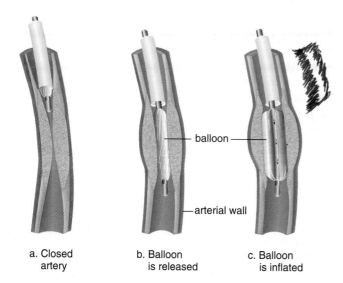

a. Closed
artery

b. Balloon
is released

c. Balloon
is inflated

— balloon —

— arterial wall

Figure 6.15 Angioplasty.
During this procedure **(a)** a plastic tube is inserted into the coronary artery until it reaches the clogged area. **b.** A metal tip with balloon attached is pushed out the end of the plastic tube into the clogged area. **c.** When the balloon is inflated the vessel opens. Sometimes metal coils or slotted tubes, called stents, are inserted to keep the vessel open.

Dilated and Inflamed Veins

Varicose veins develop when the valves of veins become weak and ineffective due to the backward pressure of blood. Abnormal and irregular dilations are particularly apparent in the superficial (near the surface) veins of the lower legs. Crossing the legs or sitting in a chair so that its edge presses against the back of the knees can contribute to the development of varicose veins. Varicose veins also occur in the rectum, where they are called piles, or more properly **hemorrhoids.**

Phlebitis, or inflammation of a vein, is a more serious condition, particularly when a deep vein is involved. Blood in an unbroken but inflamed vessel may clot, and the clot may be carried in the bloodstream until it lodges in a small vessel. If a blood clot blocks a pulmonary vessel, death can result.

6.6   Homeostasis

Homeostasis, the dynamic equilibrium of the internal environment, is possible only if the cardiovascular system delivers oxygen and nutrients and takes away metabolic wastes from tissue fluid that surrounds cells. The next page tells how the cardiovascular system works with the other systems of the body to maintain homeostasis.

The composition of blood is maintained by the other systems of the body. The digestive system absorbs nutrients into blood, and the lungs and kidneys dispose of metabolic wastes. The liver, of course, is a key regulator of blood components by producing plasma proteins, storing glucose until it is needed, transforming ammonia into urea, and changing other poisons into molecules that are also excreted.

The pumping of the heart is critical to creating the blood pressure that moves blood to the lungs and to the tissues. However, the cardiac control center in the medulla oblongata receives sensory input from receptors within the aorta. If blood pressure falls, the cardiac control center signals the heart to beat faster. Constriction or dilation of the arterioles is also directed by the nervous system, helping to regulate blood pressure and determining which capillary beds are open at any point in time. The lymphatic system collects excess tissue fluid at blood capillaries and returns it to cardiovascular veins in the thoracic cavity. In this way the lymphatic system makes an important contribution to regulating blood volume.

The endocrine system assists the nervous system in maintaining homeostasis, so it is not surprising that hormones are also involved in regulating blood pressure. Epinephrine and norephinephrine bring about the constriction of arterioles. Other hormones not yet discussed regulate urine excretion. After all, if water is retained blood volume and pressure will rise, and if water is excreted blood volume and pressure will drop. In fact, some drugs prescribed for hypertension increase the amount of urine excreted. There are also hormones that regulate the manufacture of formed elements in the red bone marrow, which is a lymphoid organ. The previous chapter mentioned that erythropoietin produced by the kidneys controls the speed of red blood cell maturation, and colony-stimulating factors released by whole blood cells control their own production. White blood cells, which fight infection, are intimately associated with all lymphoid organs and lymphatic vessels. Homeostasis would not be possible without our ability to kill off pathogens.

While blood pressure accounts for the arterial flow of blood from the heart to the capillaries, venous return from the capillaries to the heart is dependent on two other systems of the body. Skeletal muscle contraction pushes blood past the valves in the veins, and breathing movements encourage the flow of blood toward the heart in the thoracic cavity.

Bioethical Issue

According to a 1993 study, about one million deaths a year in the United States could be prevented if people adopted the healthy life-style described in the reading on page 136. Tobacco, lack of exercise, and a high-fat diet probably cost the nation about $200 billion per year in health-care costs. To what lengths should we go to prevent these deaths and reduce health-care costs?

E. A. Miller, a meat-packing entity of ConAgra in Hyrum, Utah, charges extra for medical coverage of employees who smoke. Eric Falk, Miller's director of human resources says, "We want to teach employees to be responsible for their behavior." Anthem Blue Cross-Blue Shield of Cincinnati, Ohio, takes a more positive approach. They give insurance plan participants $240 a year in extra benefits, like additional vacation days, if they get good scores in five out of seven health-related categories. The University of Alabama, Birmingham, School of Nursing has a health-and-wellness program that councils employees about how to get into shape in order to keep their insurance coverage. Audrey Brantley is in the program and says she has mixed feelings. She says, "It seems like they are trying to control us, but then, on the other hand, I know of folks who found out they had high blood pressure or were borderline diabetics and didn't know it."

Does it really work is another question. Turner Broadcasting System in Atlanta has a policy that affects all employees hired after 1986. They will be fired if caught smoking—whether at work or at home—but some admit they still manage to sneak a smoke.

Questions

1. Do you think employers who pay for their employees' health insurance have the right to demand, or encourage, or support a healthy life-style?

2. Do you think all participants in a health insurance program should qualify for the same benefits, regardless of their life-style?

3. What steps are ethical to encourage people to adopt a healthy life-style?

Human Systems Work Together

Integumentary System

Blood vessels deliver nutrients and oxygen to skin, carry away wastes; blood clots if skin is broken.

Skin prevents water loss; helps regulate body temperature; protects blood vessels.

How the Cardiovascular System works with other body systems

Lymphatic System/Immunity

Blood vessels transport leukocytes and antibodies; blood services lymphoid organs and is source of tissue fluid that becomes lymph.

Lymphoid organs produce and store formed elements; lymphatic vessels transport leukocytes and return tissue fluid to blood vessels; spleen serves as blood reservoir, filters blood.

Skeletal System

Blood vessels deliver nutrients and oxygen to bones; carry away wastes.

Rib cage protects heart; red bone marrow produces blood cells; bones store Ca^{2+} for blood clotting.

Respiratory System

Blood vessels transport gases to and from lungs; blood services respiratory organs.

Gas exchange in lungs rids body of carbon dioxide, helping to regulate the pH of blood; breathing aids venous return.

Muscular System

Blood vessels deliver nutrients and oxygen to muscles; carry away wastes.

Muscle contraction keeps blood moving in heart and blood vessels.

Digestive System

Blood vessels transport nutrients from digestive tract to body; blood services digestive organs.

Digestive tract provides nutrients for plasma protein formation and blood cell formation; liver detoxifies blood, makes plasma proteins, destroys old red blood cells.

Nervous System

Blood vessels deliver nutrients and oxygen to neurons; carry away wastes.

Brain controls nerves that regulate the heart and dilation of blood vessels.

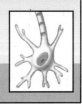

Urinary System

Blood vessels deliver wastes to be excreted; blood pressure aids kidney function; blood services urinary organs.

Kidneys filter blood and excrete wastes; maintain blood volume, pressure, and pH; produce renin and erythropoietin.

Endocrine System

Blood vessels transport hormones from glands; blood services glands; heart produces atrial natriuretic hormone.

Epinephrine increases blood pressure; ADH, aldosterone, and atrial natriuretic hormone factors help regulate blood volume; growth factors control blood cell formation.

Reproductive System

Blood vessels transport sex hormones; vasodilation causes genitals to become erect; blood services reproductive organs.

Sex hormones influence cardiovascular health; sexual activities stimulate cardiovascular system.

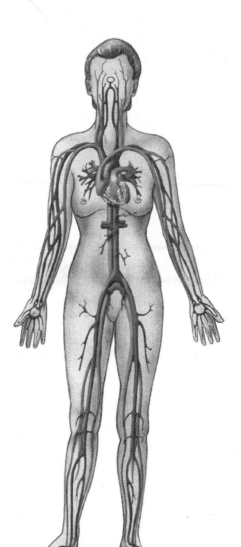

Summarizing the Concepts

6.1 The Blood Vessels
Blood vessels include arteries (and arterioles) that take blood away from the heart; capillaries, where exchange of substances with the tissues occurs; and veins (and venules) that take blood to the heart.

6.2 The Heart
The movement of blood in the cardiovascular system is dependent on the beat of the heart. During the cardiac cycle, the SA node (pacemaker) initiates the beat and causes the atria to contract.

The AV node conveys the stimulus and initiates contraction of the ventricles. The heart sounds, lub-dup, are due to the closing of the atrioventricular valves, followed by the closing of the semilunar valves.

6.3 Features of the Cardiovascular System
The pulse rate indicates the heartbeat rate. Blood pressure accounts for the flow of blood in the arteries, but because blood pressure drops off after the capillaries, it cannot cause blood flow in the veins. Skeletal muscle contraction, presence of valves, and respiratory movements account for blood flow in veins. The reduced velocity of blood flow in capillaries facilitates exchange of nutrients and wastes.

6.4 The Vascular Pathways
The cardiovascular system is divided into the pulmonary circuit and the systemic circuit. In the pulmonary circuit, two pulmonary arteries take blood from the right ventricle to the lungs, and four pulmonary veins return it to the left atrium. To trace the path of blood in the systemic circuit, start with the aorta from the left ventricle. Follow its path until it branches to an artery going to a specific organ. It can be assumed that the artery divides into arterioles and capillaries, and that the capillaries lead to venules. The vein that takes blood to the vena cava most likely has the same name as the artery that delivered blood to the organ. In the adult systemic circuit, unlike the pulmonary circuit, the arteries carry blood that is relatively high in oxygen and relatively low in carbon dioxide, and the veins carry blood that is relatively low in oxygen and relatively high in carbon dioxide.

6.5 Cardiovascular Disorders
Hypertension and atherosclerosis are two cardiovascular disorders that lead to stroke, heart attack, and aneurysm. Medical and surgical procedures are available to control cardiovascular disease, but the best policy is prevention by following a heart-healthy diet, getting regular exercise, maintaining a proper weight, and not smoking cigarettes.

6.6 Homeostasis
Homeostasis is absolutely dependent upon the cardiovascular system because it serves the needs of the cells. However, several other body systems are critical to the functioning of the cardiovascular system. The digestive system supplies nutrients, and the respiratory system supplies oxygen and excretes carbon dioxide from the blood. Like the heart, the nervous and endocrine systems are involved in maintaining the blood pressure that moves blood in the arteries and arterioles. The lymphatic system returns tissue fluid to the veins where blood is propelled by skeletal muscle contraction and breathing movements.

Formed elements are produced in the red bone marrow, which is a lymphoid organ. The lymphatic system also stores white blood cells, which help fight infection. Without the ability to ward off infections, the body would succumb to invasion by pathogens.

Studying the Concepts

1. What types of blood vessels are there? Discuss their structure and function. 126
2. Trace the path of blood in the heart, mentioning the vessels attached to, and the valves within, the heart. 129
3. Describe the cardiac cycle (using the terms systole and diastole), and explain the heart sounds. 130
4. Describe the cardiac conduction system and an ECG. Tell how an ECG is related to the cardiac cycle. 131
5. In what type of vessel is blood pressure highest? Lowest? Why is the slow movement of blood in capillaries beneficial? 132–33
6. What factors assist venous return of blood? 133
7. Trace the path of blood in the pulmonary circuit as it travels from and returns to the heart. 134
8. Trace the path of blood to and from the kidneys in the systemic circuit. 134–35
9. What is atherosclerosis? Name two illnesses associated with hypertension and thromboembolism. 137–38
10. Discuss the medical and surgical treatment of cardiovascular disease. 138–39

Testing Your Knowledge of the Concepts

In questions 1–4, match the circuit to the descriptions below.
a. pulmonary circuit
b. systemic circuit
c. both

_____ 1. Arteries carry blood rich in oxygen.

_____ 2. Carbon dioxide leaves the capillaries and oxygen enters the capillaries.

_____ 3. Arteries carry blood away from the heart, and veins carry blood toward the heart.

_____ 4. Contains the renal arteries and veins.

In questions 5–7, indicate whether the statement is true (T) or false (F).

_____ 5. SA node impulses and the atrioventricular bundle cause the atria to contract.

_____ 6. Venous return is dependent in part on skeletal muscle contraction.

_____ 7. The total cross-sectional area of arteries, arterioles, and capillaries affects the pressure and velocity of blood in these vessels.

In questions 8 and 9, fill in the blanks.

8. Reducing the amount of _____ and _____ in the diet reduces the chance of a heart attack.

9. A _____ occurs when a small cranial arteriole bursts or is blocked by an embolus.

For question 10, label this diagram of the heart.

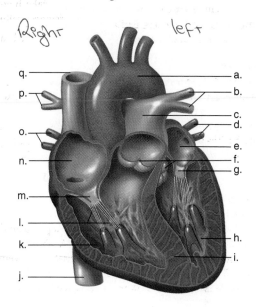

Applying Your Knowledge to the Concepts

These questions pertain to homeostasis.

1. Obesity increases our chance of hypertension and/or a heart attack. What changes in the cardiovascular system occur due to the extra body weight that can produce these results?

2. Patients with an organ transplant are usually given immunosuppressive drugs to help prevent rejection of the foreign cells in the transplanted organ. The level of drugs given is very critical. Explain the reason for this.

3. Research has shown that the atria of the heart can be rendered incapable of contracting and still the heart works quite well in pumping blood to all parts of the body. Explain the reason for this.

4. If an EKG has two P waves for every QRS wave, what does this suggest is wrong with the intrinsic control of the heartbeat?

Understanding the Terms

aneurysm 138 *ballooning of blood vessel*
angina pectoris 138 *by thoracic*
angioplasty 138 *surgical clog arteries*
ventricle aorta 129, 134 *serves blood from left.*
arteriole 127 *vessel takes f. arteries*
artery 126 *vessel take blood f. arterioles*
atherosclerosis 137 *fatty substances*
atrioventricular bundle 130 *node to vent. Heart*
atrioventricular valve 128 *Atrium & venricola*
atrium 128 *chamber above ventricles*
AV (atrioventricular) node 130 *Small region*
    130 *A node to ventricular wall.*
blood pressure 132 *Force of blood*
capillary 126 *Microscopic vessel*
cardiac cycle 130 *myocardial heartbeat*
valves chordae tindineae 128 *tough bands*
capillary *venous* congestive heart failure 139 *no circulation*
coronary artery 135 *supplies blood*
diastole 130 *Relaxation of heart chamber*
diastolic pressure 132 *Arterial blood*
electrocardiogram (ECG or
    EKG) 131 *Recording of electrical*
    activity heartbeat

embolus 138 *moving blood clot (bloodstream)*
heart 128 *muscular organ (blood circulation)*
heart attack 138 *damaged to My Ocardium*
hemorrhoids 139 *blood vessel of rectum*
hepatic portal system 135 *begins at liver*
hepatic portal vein 135 *leading to liver*
hepatic vein 135 *merging blood vessels*
hypertension 136 *evaluated blood pressure (diastolic pressure)*
inferior vena cava 134 *largest vein, systematic circuit*
myocardium 128 *Cardiac muscle in wall*
pacemaker 130 *SA (sinoatrial) node*
pericardium 128 *Protective serous membrane*
phlebitis 139 *Inflammation of vein*
pulmonary artery 129 *blood vessel heart to lungs*
pulmonary circuit 134 *deoxygenated*
pulmonary vein 129 *blood vessel lungs to heart*
pulse 132 *vibration felt in arterial wall*
Purkinje fibers 130 *cardiac impulse AV to wall ventricular*
SA (sinoatrial) node 130 *small region of neromuscular tissue*
semilunar valve 128 *1/2 moon ventricle to vessel*
stroke 138 *Arteriole becomes blocked by embolism*
superior vena cava 134

systemic circuit 134 *circulatory system serves body surfaces in the lungs* ?
systole 130 *contraction of a heart chamber*
systolic pressure 132 *Arterial blood pressure cardiac cycle*
thromboembolism 138 *obstruction*
thrombus 137 *blood vessel by thrombus*
valve 127 *vessel of the heart that opens (one way flow)*

varicose veins 139 *Irregular dilatation*
vein 126 *vessel takes blood to heart, f. venules*
vena cava 129 *systematic vein*
ventricle 128 *cavity in an organ*
venule 127 *takes blood from capillaries to vein*

blood clot remains in blood vessel

Match the terms to these definitions:

a. *diastole* Relaxation of a heart chamber.
b. *vena cava* Large systemic vein that returns blood to the right atrium of the heart.
c. *Fibrinogen* Plasma protein that is converted into fibrin threads during blood clotting.
d. *Hemoglobin* Iron-containing protein in red blood cells that combines with and transports oxygen.
e. *Systematic circuit* That part of the cardiovascular system that serves body parts and does not include the gas-exchanging surfaces in the lungs.

vein that returns blood.

Applying Technology to the Concepts

Your study of the cardiovascular system is supported by these available technologies:

Essential Study Partner CD-ROM
Animals → Circulatory System

Exploring the Internet
The Mader Home Page provides resources and tools as you study this chapter.

http://www.mhhe.com/biosci/genbio/mader

Dynamic Human 2.0 CD-ROM
Cardiovascular System

Virtual Physiology Laboratory CD-ROM
Electrocardiogram
Effects of Drugs on the Frog Heart

HealthQuest CD-ROM
3 Nutrition → Gallery → Occluded Artery
5 Cardiovascular Health

Chapter 7

Lymphatic System and Immunity

Chapter Concepts

Figure 7.1   Allergies.
The immune system protects us from disease but has unwanted side effects such as reactions to environmental substances that will do no harm to the body.

Every spring, Paula C. knows what's coming. The flowers bloom. The wind blows. And she sniffles. Itching, sneezing, rubbing watery eyes with a tissue, Paula endures the onslaught of allergy season.

Ironically, Paula's agony stems from her immune system's admirable capability to fight off pathogens. Allergies are an immune response to substances that most people are insensitive to. But Paula's body senses these substances as foreign, and antibodies kick into gear, replicating like crazy. The well-meaning molecules bind to certain white blood cells in the tissues and in doing so cause the release of a flood of chemicals, such as histamines, that lead to allergy symptoms (Fig. 7.1).

For relief, allergy sufferers turn to antihistamines, or nonprescription pills that block histamine production. And so Paula swallows the medicine, grabs a tissue and waits—fitfully—for cold weather.

Immunity involves some defenses that are termed nonspecific because they immediately respond to any type of pathogen (bacteria and viruses) that enters the body. Cut yourself and the skin becomes inflamed as white blood cells rush to the scene and begin their attack. Then, too, there are plasma proteins that latch onto bacterial surfaces and form pores so that water and salts enter until the bacteria burst. These proteins belong to the "complement system," so called because it complements certain immune responses.

Specific defenses might be slower to start, but they get the job done if you are invaded by a large number of the same type antigen. (An antigen is any foreign substance, usually protein or carbohydrate, that is capable of activating the immune system.) Get the flu and there is nothing to do but wait for your immune system to win

the battle! Unseen in your body, lymphocytes are gearing up to produce vast quantities of just the right type of antibody that will make you well. The antibody combines with the antigen and in the process brings about its own destruction, since the complex is then phagocytized by a macrophage.

So despite the fact that Paula's immune system causes her to have allergies, it otherwise does a magnificent job of keeping her well. Unfortunately, pesticides have been found to suppress the immune system, as discussed in the Ecology reading on page 153.

7.1   Lymphatic System

The **lymphatic system** consists of lymphatic vessels and the lymphoid organs. This system, which is closely associated with the cardiovascular system, has three main functions that contribute to homeostasis: (1) lymphatic capillaries take up excess tissue fluid and return it to the bloodstream; (2) lymphatic capillaries absorb fats at the intestinal villi and transport them to the bloodstream; and (3) the lymphatic system helps to defend the body against disease. ⚡

Lymphatic Vessels

Lymphatic vessels are quite extensive; most regions of the body are richly supplied with lymphatic capillaries (Fig. 7.2). The construction of the larger lymphatic vessels is similar to that of cardiovascular veins, including the presence of valves. Also, the movement of lymph within these vessels is dependent upon skeletal muscle contraction. When the muscles contract, the lymph is squeezed past a valve that closes, preventing the lymph from flowing backwards.

The lymphatic system is a one-way system that begins with lymphatic capillaries. These capillaries take up fluid that has diffused from and has not been reabsorbed by the blood capillaries. **Edema** is localized swelling caused by the accumulation of tissue fluid. This can happen if too much tissue fluid is made and/or not enough of it is drained away. Once tissue fluid enters the lymphatic vessels, it is called **lymph.** The lymphatic capillaries join to form lymphatic vessels that merge before entering one of two ducts: the thoracic duct or the right lymphatic duct. The *thoracic duct* is much larger than the right lymphatic duct. It serves the lower extremities, the abdomen, the left arm, and the left side of both the head and the neck. The *right lymphatic duct* serves the right arm, the right side of both the head and the neck, and the right thoracic area. The lymphatic ducts enter the subclavian veins, which are cardiovascular veins in the thoracic region.

Lymph flows one way from a capillary to ever-larger lymphatic vessels and finally to a lymphatic duct, which enters a subclavian vein.

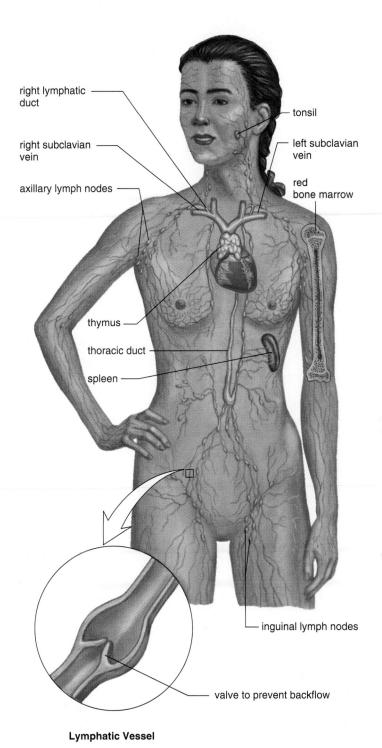

right lymphatic duct

tonsil

right subclavian vein

left subclavian vein

axillary lymph nodes

red bone marrow

thymus

thoracic duct

spleen

inguinal lymph nodes

valve to prevent backflow

Lymphatic Vessel

Figure 7.2   Lymphatic system.
The lymphatic vessels drain excess fluid from the tissues and return it to the cardiovascular system. The enlargement shows that lymphatic vessels have valves to prevent backward flow.

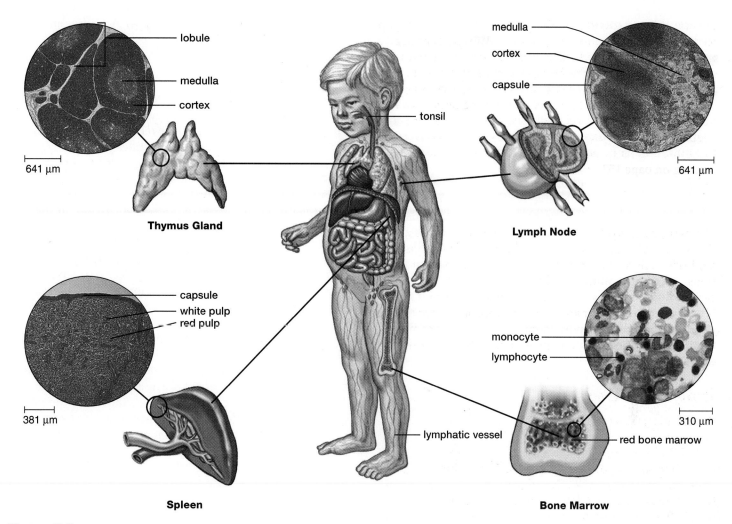

Figure 7.3 The lymphoid organs.
The lymphoid organs include the lymph nodes, the spleen, the thymus gland, and the red bone marrow, which all contain lymphocytes.

Lymphoid Organs

The lymphoid organs of special interest are the lymph nodes, the tonsils, the spleen, the thymus gland, and the bone marrow (Fig. 7.3).

Lymph nodes, which are small (about 1–25 mm) ovoid or round structures, are found at certain points along lymphatic vessels. A capsule surrounds two distinct regions known as the cortex and medulla which contain many lymphocytes. Macrophages, which occur along lymph capillaries called lymph sinuses, purify lymph of infectious organisms and any debris. Antigens which leak into the cortex and medulla activate the lymphocytes to mount an immune response to them. Lymph nodes are named for their location. Inguinal nodes are in the groin and axillary nodes are in the armpits. Physicians often feel for the presence of swollen, tender lymph nodes in the neck as evidence that the body is fighting an infection. This is a noninvasive preliminary way to help make such a diagnosis.

The **tonsils** are partially encapsulated lymphatic tissue located in a ring about the pharynx. The well-known pharyngeal tonsils are also called *adenoids*, while the larger palatine tonsils located on either side of the posterior oral cavity are most apt to be infected. The tonsils perform the same functions as lymph nodes inside the body, but because of their location they are the first to encounter pathogens and antigens that enter the body by way of the nose and mouth.

The **spleen** is located in the upper left region of the abdominal cavity just beneath the diaphragm. It is much larger than a lymph node, about the size of a fist. Whereas the lymph nodes cleanse lymph, the spleen cleanses blood. A capsule surrounds tissue known as white pulp and red pulp. White pulp contains lymphocytes and performs the immune functions of the spleen. The red pulp contains red blood cells and plentiful macrophages. The red pulp helps to purify blood that passes through the spleen by removing bacteria and worn-out or damaged red blood cells.

The spleen's outer capsule is relatively thin, and an infection and/or a blow can cause the spleen to burst. Although its functions are replaced by other organs, a person without a spleen is often slightly more susceptible to infections and may have to receive antibiotic therapy indefinitely.

The **thymus gland** is located along the trachea behind the sternum in the upper thoracic cavity. This gland varies in size, but it is larger in children than in adults and may disappear completely in old age. The thymus is divided into lobules by connective tissue. The T lymphocytes mature in these lobules. The interior (medulla) of the lobule, which consists mostly of epithelial cells, stains lighter. It produces thymic hormones, such as thymosin, that are thought to aid in maturation of T lymphocytes. Thymosin may also have other functions in immunity.

Red bone marrow is the site of origination for all types of blood cells, including the five types of white blood cells pictured in Figure 5.2. The marrow contains stem cells that are ever capable of dividing and producing cells that go on to differentiate into the various types of blood cells (see Figure 5.4). In a child, most bones have red bone marrow, but in an adult it is present only in the bones of the skull, the sternum (breastbone), the ribs, the clavicle, the pelvic bones, and the vertebral column. The red bone marrow consists of a network of connective tissue fibers, called reticular fibers, which are produced by cells called reticular cells. These and the stem cells and their progeny are packed around thin-walled sinuses filled with venous blood. Differentiated blood cells enter the bloodstream at these sinuses.

The lymphoid organs have specific functions that assist immunity. Lymph is cleansed in lymph nodes; blood is cleansed in the spleen; T lymphocytes mature in the thymus; and white blood cells are made in the bone marrow.

7.2   Nonspecific Defenses

Immunity is the ability of the body to defend itself against infectious agents, foreign cells, and even abnormal body cells, such as cancer cells. Thereby the internal environment has a better chance of remaining stable. Immunity includes nonspecific and specific defenses. The four types of nonspecific defenses—barriers to entry, the inflammatory reaction, natural killer cells, and protective proteins—are effective against many types of infectious agents.

Barring Entry

Skin and the mucous membranes lining the respiratory, digestive, and urinary tracts serve as mechanical barriers to entry by pathogens. Oil gland secretions contain chemicals that weaken or kill certain bacteria on skin. The upper respiratory tract is lined by ciliated cells that sweep mucus and trapped particles up into the throat, where they can be swallowed, or expectorated (coughed out). The stomach has an acidic pH, which inhibits the growth of or kills many types of bacteria. The various bacteria that normally reside in the intestine and other areas, such as the vagina, prevent

pathogens from taking up residence. A **pathogen** is any disease-causing agent such as viruses and some bacteria.

Inflammatory Reaction

Whenever the skin is broken due to a minor injury, a series of events occurs that is known as the **inflammatory reaction.** The inflamed area has four outward signs: redness, heat, swelling, and pain. Figure 7.4 illustrates the participants in the inflammatory reaction. **Mast cells,** which occur in tissues, resemble **basophils,** one of the white cells found in the blood.

When an injury occurs, damaged tissue cells and mast cells release chemical mediators, such as **histamine** and **kinins,** which cause the capillaries to dilate and become more permeable. The enlarged capillaries cause the skin to redden, and the increased permeability allows proteins and fluids to escape and swelling results. A rise in temperature increases phagocytosis by white blood cells. The swollen area as well as kinins stimulate free nerve endings, causing the sensation of pain.

Neutrophils and monocytes migrate to the site of injury. They are amoeboid and can change shape to squeeze through capillary walls to enter tissue fluid. Neutrophils, and also mast cells, phagocytize bacteria. The engulfed bacteria are destroyed by hydrolytic enzymes when the endocytic vesicle combines with a lysosome, one of the cellular organelles.

Monocytes differentiate into **macrophages,** large phagocytic cells that are able to devour a hundred bacteria or viruses and still survive. Some tissues, particularly connective tissue, have resident macrophages, which routinely act as scavengers, devouring old blood cells, bits of dead tissue, and other debris. Macrophages can also bring about an explosive increase in the number of leukocytes by liberating colony-stimulating hormones, which pass by way of blood to the red bone marrow, where they stimulate the production and the release of white blood cells, primarily neutrophils.

When a blood vessel ruptures, the blood clots to seal the break. The chemical mediators, mentioned earlier, and antigens move through the tissue fluid and lymph to the lymph nodes. Now lymphocytes can also be activated to react to the threat of an infection. As the infection is being overcome, some neutrophils may die. These—along with dead tissue, cells, bacteria, and living white blood cells—form pus, a whitish material. Pus indicates that the body is trying to overcome the infection.

Sometimes inflammation persists and the result is chronic inflammation that is often treated by the administration of anti-inflammatory agents such as aspirin, ibuprofen, or cortisone. They act against the chemical mediators released by the white blood cells in the area.

The inflammatory reaction is a "call to arms"—it marshals phagocytic white blood cells to the site of bacterial invasion and stimulates the immune system to react against a possible infection.

Visual Focus

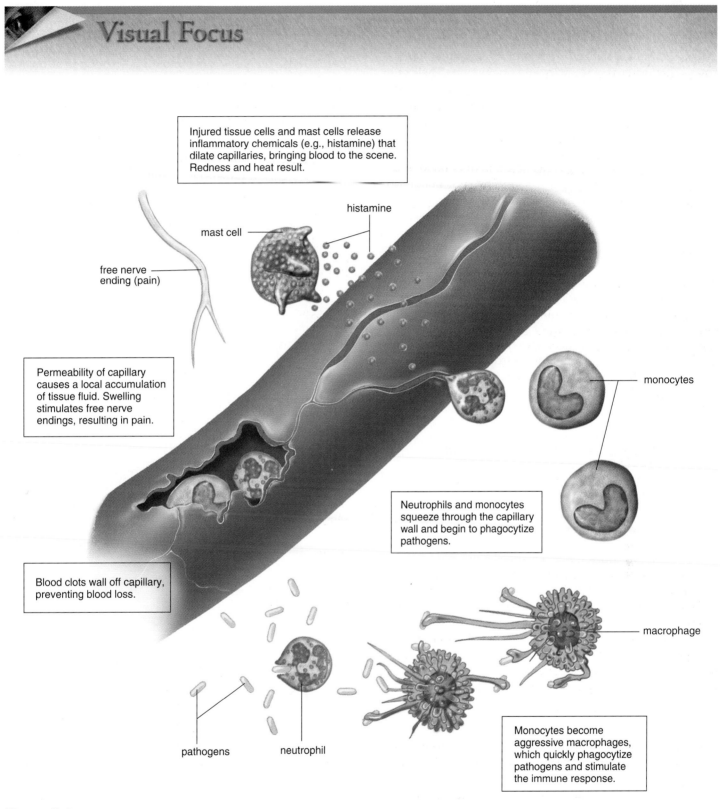

Injured tissue cells and mast cells release inflammatory chemicals (e.g., histamine) that dilate capillaries, bringing blood to the scene. Redness and heat result.

histamine

mast cell

free nerve ending (pain)

Permeability of capillary causes a local accumulation of tissue fluid. Swelling stimulates free nerve endings, resulting in pain.

monocytes

Neutrophils and monocytes squeeze through the capillary wall and begin to phagocytize pathogens.

Blood clots wall off capillary, preventing blood loss.

macrophage

pathogens

neutrophil

Monocytes become aggressive macrophages, which quickly phagocytize pathogens and stimulate the immune response.

Figure 7.4   **Inflammatory reaction.**

Natural Killer Cells

Natural killer (NK) cells kill virus-infected cells and tumor cells by cell-to-cell contact. They are large granular lymphocytes. They have no specificity and no memory. Their number is not increased by immunization.

Protective Proteins

The **complement system,** often simply called complement, is a number of plasma proteins designated by the letter C and a subscript. A limited amount of activated complement protein is needed because a domino effect occurs: each activated protein in a series is capable of activating many other proteins.

Complement is activated when pathogens enter the body. It "complements" certain immune responses, which accounts for its name. For example, it is involved in and amplifies the inflammatory response because complement proteins attract phagocytes to the scene. Some complement proteins bind to the surface of pathogens already coated with antibodies, which ensures that the pathogens will be phagocytized by a neutrophil or macrophage.

Certain other complement proteins join to form a membrane attack complex that produces holes in bacterial cell walls and plasma membranes of bacteria. Fluids and salts then enter the bacterial cell to the point that it bursts (Fig. 7.5).

Interferon is a protein produced by virus-infected cells. Interferon binds to receptors of noninfected cells, causing them to prepare for possible attack by producing substances that interfere with viral replication. Interferon is specific to the species; therefore, only human interferon can be used in humans.

Immunity includes these nonspecific defenses; barriers to entry, the inflammatory reaction, natural killer cells, and protective proteins.

7.3 Specific Defenses

When nonspecific defenses have failed to prevent an infection, specific defenses come into play. An **antigen** is any foreign substance (often a protein or polysaccharide) that stimulates the **immune system** to react to it. Pathogens have antigens, but antigens can also be part of a foreign cell or a cancer cell. Because we do not ordinarily become immune to our own cells, it is said that the immune system is able to distinguish self from nonself. Only in this way can the immune system aid rather than counter homeostasis.

Immunity usually lasts for some time. For example, once we recover from the measles, we usually do not get the illness a second time. Immunity is primarily the result of the action of the **B lymphocytes** and the **T lymphocytes.** B lymphocytes[1] mature in the *bone* marrow, and T lymphocytes mature in the *thymus* gland. B lymphocytes, also called B cells, give rise to plasma cells, which produce **antibodies,** proteins that are capable of combining with and neutralizing antigens. These antibodies are secreted into the blood, lymph, and other body fluids. In contrast, T lymphocytes, also called T cells, do not produce antibodies. Instead, certain T cells directly attack cells that bear antigens. Other T cells regulate the immune response.

Lymphocytes are capable of recognizing an antigen because they have receptor molecules on their surface. The shape of the receptors on any particular lymphocyte is complementary to a specific antigen. It is often said that the receptor and the antigen fit together like *a lock and a key.* It is estimated that during our lifetime, we encounter a million different antigens, so we need a great diversity of lymphocytes to protect us against antigens. It is remarkable that diversification occurs to such an extent during the maturation process that there is a lymphocyte type for any possible antigen.

[1]Historically, the B stands for *bursa of Fabricius,* an organ in the chicken where these cells were first identified. As it turns out, however, the B can conveniently be thought of as referring to bone marrow.

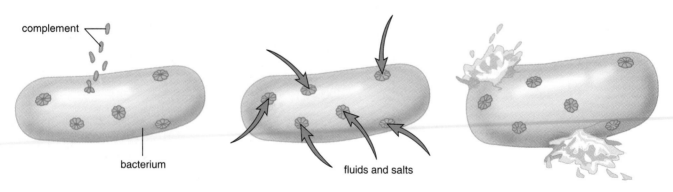

complement

bacterium

fluids and salts

Complement proteins form holes in the bacterial cell wall and membrane.

Holes allow fluids and salts to enter the bacterium.

Bacterium expands until it bursts.

Figure 7.5 **Action of the complement system against a bacterium.**
When complement proteins in the plasma are activated by an immune reaction, they form holes in bacterial cell walls and plasma membranes, allowing fluids and salts to enter until the cell eventually bursts.

B Cells and Antibody-Mediated Immunity

Each type of B cell carries its specific antibody, as a membrane-bound receptor, on its surface. When a B cell in a lymph node or the spleen encounters a bacterial cell or a toxin bearing an appropriate antigen, it becomes activated to divide many times. Most of the resulting cells are plasma cells, which secrete antibodies against this antigen. A **plasma cell** is a mature B cell that mass-produces antibodies in the lymph nodes and in the spleen.

The **clonal selection theory** states that the antigen selects which lymphocyte will undergo clonal expansion and produce more lymphocytes bearing the same type of receptor (Fig. 7.6). Notice that a B cell does not divide until its antigen is present and binds to its receptors. B cells are also stimulated to divide and become plasma cells by helper T cell secretions, as is discussed in the next section. Some members of the clone become *memory cells* which are the means by which long-term immunity is possible. If the same antigen enters the system again, memory cells quickly divide and give rise to more lymphocytes capable of quickly producing antibodies.

Once the threat of an infection has passed, the development of new plasma cells ceases and those present undergo apoptosis. **Apoptosis** is a process of programmed cell death (PCD) involving a cascade of specific cellular events leading to the death and destruction of the cell. The methodology of PCD is still being worked out, but we know it is an essential physiological mechanism regulating the cell population within an organ system. PCD normally plays a central role in maintaining tissue homeostasis.

Defense by B cells is called **antibody-mediated immunity** because the various types of B cells produce antibodies. It is also called *humoral immunity* because these antibodies are present in blood and lymph. A *humor* is any fluid normally occurring in the body.

Characteristics of B cells:

- Antibody-mediated immunity
- Produced and mature in bone marrow
- Reside in spleen and lymph nodes, circulate in blood and lymph
- Directly recognize antigen and then undergo clonal selection
- Clonal expansion produces antibody-secreting plasma cells as well as memory B cells

---

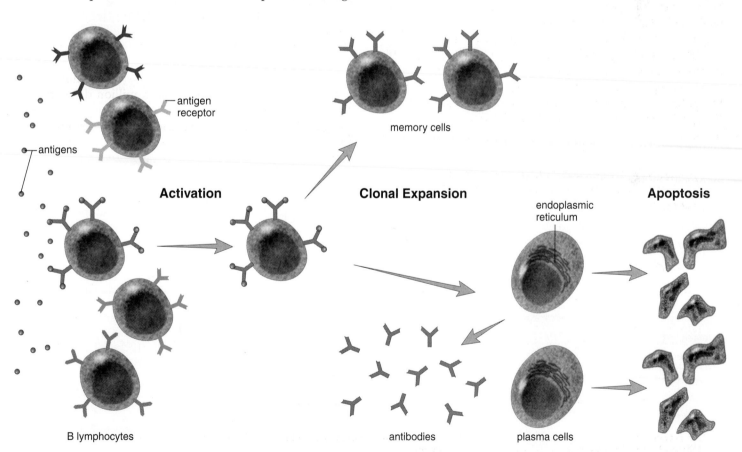

Figure 7.6   Clonal selection theory as it applies to B cells.
An antigen activates only the B cell whose receptors can combine with the antigen. This B cell then undergoes clonal expansion. During the process many plasma cells, which produce specific antibodies against this antigen, are produced. After the infection passes, they undergo apoptosis. Memory cells, which retain the ability to recognize this antigen, are retained in the body.

Structure of IgG

The most common type of antibody (IgG) is a Y-shaped protein molecule with two arms. Each arm has a "heavy" (long) polypeptide chain and a "light" (short) polypeptide chain. These chains have *constant regions,* where the sequence of amino acids is set, and *variable regions,* where the sequence of amino acids varies between antibodies (Fig. 7.7). The constant regions are not identical among all the antibodies. Instead, they are almost the same within different classes of antibodies. The variable regions form an antigen-binding site, and their shape is specific to a particular antigen. The antigen combines with the antibody at the antigen-binding site in a lock-and-key manner.

The antigen-antibody reaction can take several forms, but quite often the reaction produces complexes of antigens combined with antibodies. Such antigen-antibody complexes, sometimes called immune complexes, mark the antigens for destruction. For example, an antigen-antibody complex may be engulfed by neutrophils or macrophages, or it may activate complement. Complement makes pathogens more susceptible to phagocytosis, as discussed previously.

Other Types of Antibodies

There are five different classes of circulating antibody proteins or **immunoglobulins (Igs)** (Table 7.1). IgG antibodies are the major type in blood, and lesser amounts are also found in lymph and tissue fluid. IgG antibodies bind to pathogens and their toxins. A *toxin* is a specific chemical (produced by bacteria, for example) that is poisonous to other living things. IgM antibodies are pentamers, meaning that they contain five of the Y-shaped structures shown in Figure 7.7a. These antibodies appear in blood soon after an infection begins and disappear before it is over. They are good activators of the complement system. IgA antibodies are monomers, dimers, or larger molecules containing two Y-shaped structures. They are the main type of antibody found in bodily secretions. They bind to pathogens before they reach the bloodstream. The main function of IgD antibodies seems to be to serve as receptors for antigens on mature B cells. IgE antibodies, which are responsible for immediate allergic responses, are discussed on page 158.

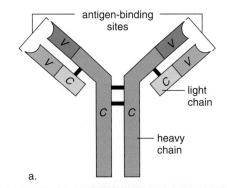

a.

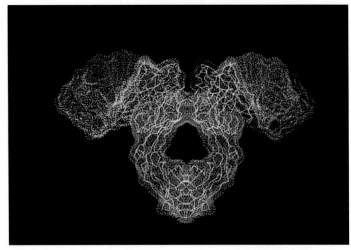

b.

Figure 7.7   **Structure of the most common antibody (IgG).** **a.** An IgG antibody contains two heavy (long) polypeptide chains and two light (short) chains arranged so there are two variable regions, where a particular antigen is capable of binding with an antibody. **b.** Computer model of an antibody molecule. In this model, the antigen combines with the two side branches.

An antigen combines with an antibody at the antigen-binding site in a lock-and-key manner. The reaction can produce antigen-antibody complexes, which contain several molecules of antibody and antigen.

Table 7.1	Antibodies	
Classes	**Presence**	**Function**
IgG	Main antibody type in circulation	Binds to pathogens, activates complement, and enhances phagocytosis
IgM	Antibody type found in circulation; largest antibody	Activates complement; clumps cells
IgA	Main antibody type in secretions such as saliva and milk	Prevents pathogens from attaching to epithelial cells in digestive and respiratory tract
IgD	Antibody type found in circulation in extremely low quantity	Presence signifies maturity of B cell
IgE	Antibody type found as membrane-bound receptor on basophils in blood and on mast cells in tissues	Responsible for immediate allergic response and protection against certain parasitic worms

Ecology Focus

Pesticide: An Asset and a Liability

A pesticide is any one of 55,000 chemical products used to kill insects, plants, fungi, or rodents that interfere with human activities. Increasingly, we have discovered that pesticides are harmful to the environment and humans (Fig 7A). The effect of pesticides on the immune system is a new area of concern; no testing except for skin sensitization has thus far been required. Contact or respiratory allergic responses are among the most immediately obvious toxic effects of pesticides. But pesticides can cause suppression of the immune system leading to increased susceptibility to infection or tumor development. Lymphocyte impairment was found following the worst industrial accident in the world. Nerve gas used to produce an insecticide was released into the air in India; 3,700 people were killed, and 30,000 were injured. Vietnam veterans claim a variety of health effects due to contact with a herbicide called Agent Orange used as a defoliant during the Vietnam War. The EPA says that pesticide residues on food possibly cause suppression of the immune system, disorders of the nervous system, birth defects, and cancer. The immune, nervous, and endocrine systems communicate with one another by way of hormones, and what affects one system can affect the other. Testicular atrophy, low sperm counts, and abnormal sperm in men may be due to the ability of pesticides to mimic the effects of estrogen, a female sex hormone.

The argument for pesticides at first seems attractive. Pesticides are meant to kill off disease-causing agents, increase yield, and work quickly with minimal risk. Over time it has been found that pesticides do not meet these claims. Instead, pests become resistant to pesticides, which kill off natural enemies in addition to the pest. Then the pest population explodes. At first DDT did a marvelous job killing off the mosquitos that carry malaria; now malaria is as big a problem as ever. In the meantime, DDT has accumulated in the tissues of wildlife and humans, causing harmful effects. The problem was made obvious when birds of prey became unable to reproduce due to weak eggshells. The use of DDT is now banned in the United States.

There are alternatives to the addictive use of pesticides. Integrated pest management uses a diversified environment, mechanical and physical means, natural enemies, disruption of reproduction, and resistant plants to control rather than eradicate pest populations. Chemicals are used only as a last resort. Maintaining hedgerows, weedy patches, and certain trees can provide diverse habitats and food for predators and parasites that help control pests. The use of strip farming and crop rotation by farmers denies pests a continuous food source.

Figure 7A **Pesticide warning signs indicate that pesticides are harmful to our health.**

Natural enemies abound in the environment. When lacewings were released in cotton fields, they reduced the boll weevil population by 96% and increased cotton yield threefold. A predatory moth was used to reclaim 60 million acres in Australia overrun by the prickly pear cactus, and another type of moth is now used to control alligator weed in the southern United States. Also in the United States, the Chrysolina beetle controls Klamath weed, except in shady places where a root boring beetle is more effective.

The use of sex attractants and sterile insects are other types of biological control. In Sweden and Norway, scientists synthesized the sex pheromone of the Ips bark beetle, which attacks spruce trees. Almost 100,000 baited traps collected females normally attracted to males emitting this pheromone. Sterile males have also been used to reduce pest populations. The screwworm fly parasitizes cattle in the United States. Flies raised in a laboratory were made sterile by exposure to radiation. The entire Southeast was freed from this parasite when female flies mated with sterile males and then laid eggs that did not hatch.

In the past, only crossbreeding could produce resistant plants. Now genetic engineering, a new technique by which certain genes are introduced into organisms, including plants, may eventually make large-scale use of pesticides unnecessary. Already, 50 types of genetically engineered plants that resist insects, or viruses, have entered small-scale field trials.

In general, biological control is a more sophisticated method of controlling pests than the use of pesticides. It requires an in-depth knowledge of pests and/or their life cycles. Because it does not have an immediate effect on the pest population, the effects of biological control may not be apparent to the farmer or gardener who benefits from it.

It is important for citizens to do all they can to promote biological control of pests instead of relying on pesticides, which are unhealthy for the environment and humans.

- Urge elected officials to support legislation to protect humans and the environment against the use of pesticides.
- Allow native plants to grow on all or most of the land to give natural predators a place to live.
- Cut down on the use of pesticides, herbicides, and fertilizers for the lawn, garden, and house.
- Use alternative methods such as cleanliness and good sanitation to keep household pests under control.
- Dispose of pesticides in a safe manner.

T Cells And Cell-Mediated Immunity

The two main types of T cells are cytotoxic T cells and helper T cells. **Cytotoxic T cells** are the type of T cell responsible for **cell-mediated immunity,** so called because T cells bring about the destruction of specific antigen-bearing cells, such as virus-infected or cancer cells. Cytotoxic T cells have storage vacuoles containing perforin molecules. **Perforin** molecules perforate a plasma membrane, forming a pore that allows water and salts to enter. The cell then swells and eventually bursts (Fig. 7.8).

Helper T cells regulate immunity by enhancing the response of other immune cells. When exposed to an antigen, they enlarge and secrete cytokines, stimulatory molecules that cause helper T cells to divide and other immune cells to perform their functions. For example, cytokines stimulate macrophages to phagocytize and stimulate B cells to become antibody-producing plasma cells. Because HIV, which causes AIDS, infects helper T cells and certain other cells of the immune system, it inactivates the immune response.

As we shall see, there are also memory T cells that remain in the body and can jump-start an immune reaction to an antigen previously present in the body.

Activation of T Cells

When T cells leave the thymus they have unique receptors just as B cells do. Unlike B cells, however, cytotoxic T cells and helper T cells are unable to recognize an antigen present in lymph, blood, or the tissues without help. The antigen must be presented to them by an **antigen-presenting cell (APC).** When an APC, usually a macrophage, engulfs a pathogen, the pathogen is broken down to fragments within an endocytic vesicle. These fragments are antigenic; that is, they have the properties of an antigen. The fragments are linked to a major histocompatibility complex (MHC) protein in the plasma membrane and then they can be presented to a T cell.

Human MHC proteins are called **HLA (human leukocyte-associated) antigens.** Because they mark the cell as belong-

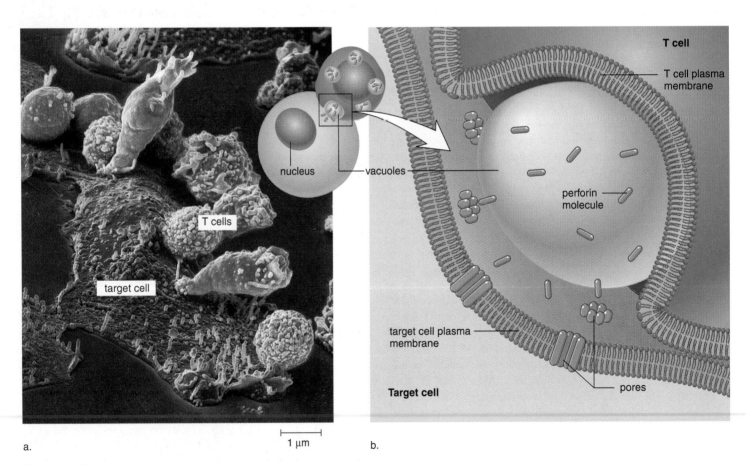

a.

1 μm

b.

Figure 7.8   **Cell-mediated immunity.**

a. The scanning electron microscope shows cytotoxic T cells attacking and destroying a cancer cell. **b.** During the killing process, the vacuoles in a cytotoxic T cell release perforin molecules. These molecules combine to form pores in the target cell plasma membrane. Thereafter, fluid and salts enter so that the target cell eventually bursts.

ing to a particular individual, HLA antigens are self-antigens. The importance of self-antigens in plasma membranes was first recognized when it was discovered that they contribute to the specificity of tissues and make it difficult to transplant tissue from one human (or animal) to another. In other words, when the donor and the recipient are histo (tissue)- compatible (the same or nearly so), a transplant is more likely to be successful.

Figure 7.9 shows a macrophage presenting an antigen to a T cell. Once a helper T cell recognizes an antigen, and is stimulated to do so, it undergoes clonal expansion and produces cytokines that stimulate immune cells to remain active. Once a cytotoxic T cell is activated in this manner, it undergoes clonal expansion and destroys any cell that is infected with the same virus, if the cell bears the correct HLA. As the infection disappears, the immune reaction wanes and fewer cytokines are produced. Now, the activated T cells become susceptible to apoptosis. As mentioned previously, apoptosis is a programmed cell death that contributes to homeostasis by regulating the number of cells that are pres-

ent in an organ, or in this case the immune system. A few of the clonally expanded T cells do not undergo apoptosis. The survivors are memory cells—T cells that can rapidly respond should the same antigen be present at a later time.

Apoptosis also occurs in the thymus as T cells are maturing. A T cell that bears a receptor with the potential to recognize a self-antigen undergoes suicide. When apoptosis does not occur as it should, T-cell cancers (i.e., lymphomas and leukemias) can result.

Characteristics of T cells:

- Cell-mediated immunity
- Produced in bone marrow, mature in thymus
- Antigen must be presented in groove of an HLA molecule
- Cytotoxic T cells destroy antigen-bearing cells
- Helper T cells secrete cytokines that control the immune response

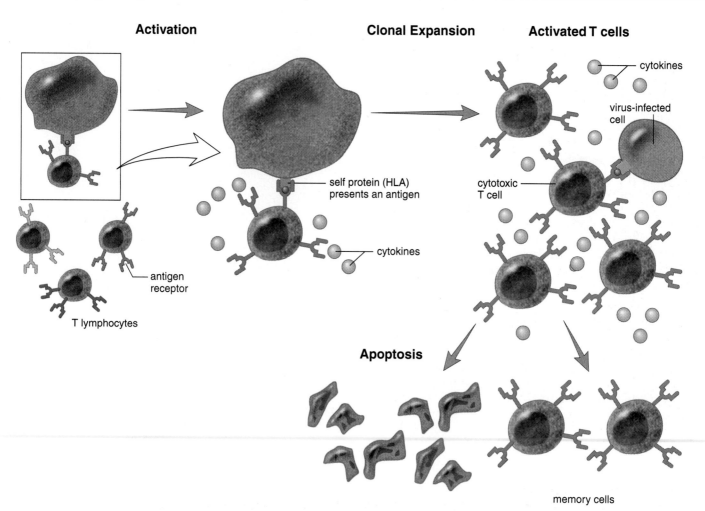

Figure 7.9   Clonal selection theory as it applies to T cells.
Each type of T cell bears a specific antigen receptor. When this receptor binds to an antigen in the groove of an HLA molecule and is stimulated to do so, it undergoes clonal expansion. After the immune response has been successful, the majority of T cells undergo apoptosis while a small number become memory cells. Memory cells provide protection should the same antigen enter the body again at a future time.

7.4   Induced Immunity

Immunity occurs naturally through infection or is brought about artificially by medical intervention. There are two types of induced immunity: active and passive. In active immunity, the individual alone produces antibodies against an antigen; in passive immunity, the individual is given prepared antibodies.

Active Immunity

Active immunity sometimes develops naturally after a person is infected with a pathogen. However, active immunity is often induced when a person is well so that possible future infection will not take place. To prevent infections, people can be artificially immunized against them. The United States is committed to the goal of immunizing all children against the common types of childhood diseases listed in the immunization schedule given in Figure 7.10*a*.

Immunization involves the use of **vaccines,** substances that contain an antigen to which the immune system responds. Traditionally, vaccines are the pathogens themselves, or their products, that have been treated so they are no longer virulent (able to cause disease). Today, it is possible to genetically engineer bacteria to mass-produce a protein from pathogens, and this protein can be used as vaccine. This method now has been used to produce a vaccine against hepatitis B, a viral disease, and is being used to prepare a vaccine against malaria, a protozoan disease.

After a vaccine is given, it is possible to follow an immune response by determining the amount of antibody present in a sample of serum—this is called the *antibody titer.* After the first exposure to a vaccine, a primary response occurs. For a period of several days, no antibodies are present; then, there is a slow rise in the titer, followed by first a plateau and then a gradual decline as the antibodies bind to the antigen or simply break down (Fig. 7.10*b*). After a second exposure, a secondary response is expected. The titer rises rapidly to a plateau level much greater than before. The second exposure is called a "booster" because it boosts the antibody titer to a high level. The high antibody titer now is expected to help prevent disease symptoms even if the individual is exposed to the disease-causing antigen.

Active immunity is dependent upon the presence of memory B cells and memory T cells which are capable of responding to lower doses of antigen. Active immunity is usually long-lived, although a booster may be required every so many years.

Active (long-lived) immunity can be induced by the use of vaccines. Active immunity is dependent upon the presence of memory B cells and memory T cells in the body.

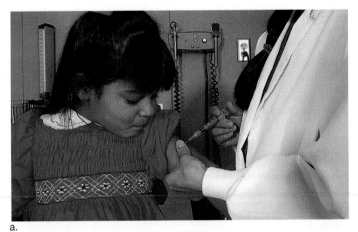

a.

Vaccine	Age (Months)	Age (Years)
HepB* (hepatitis B)	Birth, 2, 4, 6, 12–15	11–12
DTP† (diphtheria, tetanus, whooping cough)	2, 4, 6, 15–18	4–6
Td‡ (adult tetanus)		11–12, 14–16
OPV§ (oral polio vaccine)	2, 4, 6, 12–15	4–6
Hib§ (Haemophilus influenza, type b)	2, 4, 6, 12–15	
MMR¶ (measles, mumps, rubella)	12–15 and 1 month later	4–6, 11–12

* Three doses will be required for kindergarten entry.
† Five doses recommended for school entry.
‡ First Td needed 10 years after last DTP.
§ Doses 3 and 4 should be given according to manufacturer's guidelines.
¶ A second dose given at least 1 month after first dose required for kindergarten entry.
Source: Iowa Department of Public Health, July 1998.

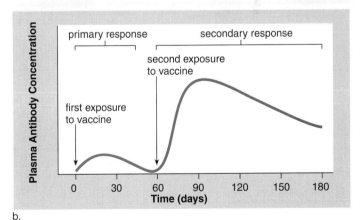

b.

Figure 7.10   Active immunity due to immunizations.
a. Suggested immunization schedule for infants and young children.
b. During immunization, the primary response, after the first exposure to a vaccine, is minimal, but the secondary response, which may occur after the second exposure, shows a dramatic rise in the amount of antibody present in serum.

Passive Immunity

Passive immunity occurs when an individual is given pre-pared antibodies (immunoglobulins) to combat a disease. Since these antibodies are not produced by the individual's B cells, passive immunity is short-lived. For example, new-born infants are passively immune to some diseases because antibodies have crossed the placenta from the mother's blood. These antibodies soon disappear, however, so that within a few months, infants become more susceptible to in-fections. Breast-feeding prolongs the natural passive immu-nity an infant receives from the mother because antibodies are present in the mother's milk (Fig. 7.11).

Even though passive immunity does not last, it some-times is used to prevent illness in a patient who has been un-expectedly exposed to an infectious disease. Usually, the patient receives a gamma globulin injection (serum that con-tains antibodies), perhaps taken from individuals who have recovered from the illness. In the past, horses were immu-nized, and serum was taken from them to provide the needed antibodies against such diseases as diphtheria, botu-lism, and tetanus. In the past, a patient who received these antibodies became ill about 50% of the time, because the serum contained proteins that the individual's immune sys-tem recognized as foreign. This was called serum sickness. But problems can still occur with products produced in other ways. An immunoglobulin intravenous product called Gammagard was withdrawn from the market because of possible implication in the transmission of hepatitis.

Passive immunity provides immediate protection when an individual is in immediate danger of succumbing to an infectious disease. Passive immunity is short-lived because there are no memory cells.

Cytokines and Immunity

Cytokines are messenger molecules produced by lympho-cytes, monocytes, and other cells. Because cytokines regu-late white blood cell formation and/or function, they are being investigated as possible adjunct therapy for cancer and AIDS. Both interferon and **interleukins,** which are cy-tokines produced by various white blood cells, have been used as immunotherapeutic drugs, particularly to enhance the ability of the individual's own T cells (and possibly B cells) to fight cancer.

Interferon, discussed previously on page 150, is a sub-stance produced by leukocytes, fibroblasts, and probably most cells in response to a viral infection. Interferon still is being investigated as a possible cancer drug, but so far it has proven to be effective only in certain patients, and the exact reasons for this as yet cannot be discerned.

When and if cancer cells carry an altered protein on their

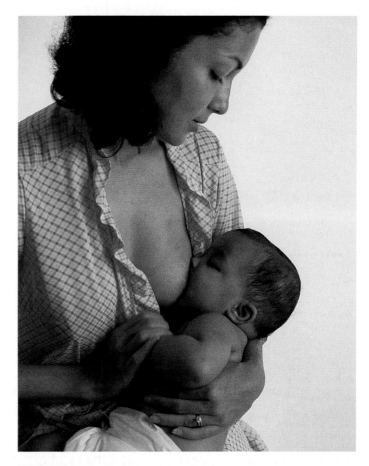

Figure 7.11 **Passive immunity.**
Breast-feeding is believed to prolong the passive immunity an infant receives from the mother because antibodies are present in the mother's milk.

cell surface, they should be attacked and destroyed by cyto-toxic T cells. Whenever cancer does develop, it is possible that the cytotoxic T cells have not been activated. In that case, cytokines might awaken the immune system and lead to the destruction of the cancer. In one technique being in-vestigated, researchers first withdraw T cells from the pa-tient and activate the cells by culturing them in the presence of an interleukin. The cells then are reinjected into the pa-tient, who is given doses of interleukin to maintain the killer activity of the T cells.

Those who are actively engaged in interleukin research believe that interleukins soon will be used as adjuncts for vaccines, for the treatment of chronic infectious diseases, and perhaps for the treatment of cancer. Interleukin antago-nists also may prove helpful in preventing skin and organ rejection, autoimmune diseases, and allergies.

The interleukins and other cytokines show some promise of potentiating the individual's own immune system.

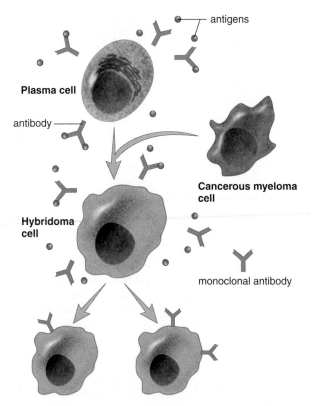

Figure 7.12   **Production of monoclonal antibodies.**
Plasma cells (derived from immunized mice) are fused with myeloma (cancerous) cells, producing hybridoma cells that are "immortal." Hybridoma cells divide and continue to produce the same type of antibody, called monoclonal antibodies.

Monoclonal Antibodies

Every plasma cell derived from the same B cell secretes antibodies against a specific antigen. These are **monoclonal antibodies** because all of them are the same type and because they are produced by plasma cells derived from the same B cell. One method of producing monoclonal antibodies in vitro (outside the body in glassware) is depicted in Figure 7.12. B lymphocytes are removed from an animal (today, usually mice are used) and are exposed to a particular antigen. The activated B lymphocytes are fused with myeloma cells (malignant plasma cells that live and divide indefinitely). The fused cells are called hybridomas; *hybrid* because they result from the fusion of two different cells, and *oma* because one of the cells is a cancer cell.

At present, monoclonal antibodies are being used for quick and certain diagnosis of various conditions. For example, a particular hormone is present in the urine of a pregnant woman. A monoclonal antibody can be used to detect this hormone; if it is present, the woman knows she is pregnant. Monoclonal antibodies also are used to identify infections. And because they can distinguish between cancer and normal tissue cells, they are used to carry radioactive isotopes or toxic drugs to tumors so that they can be selectively destroyed.

7.5   Immunity Side Effects

The immune system usually protects us from disease because it can distinguish self from nonself. Sometimes, however, it responds in a manner that does harm to the body, as when individuals develop allergies, receive the wrong blood type, suffer tissue rejection, or have an autoimmune response.

Allergies

Allergies are hypersensitivities to substances such as pollen or animal hair that ordinarily would do no harm to the body. The response to these antigens, called **allergens,** usually includes some degree of tissue damage. There are four types of allergic responses, but we will consider only two of these: immediate allergic responses and delayed allergic responses.

Immediate Allergic Response
An **immediate allergic response** can occur within seconds of contact with the antigen. As discussed in the reading on page 159, coldlike symptoms are common. Anaphylactic shock is a severe reaction characterized by a sudden and life-threatening drop in blood pressure.

Immediate allergic responses are caused by antibodies known as IgE (see Table 7.1). IgE antibodies are attached to the plasma membrane of mast cells in the tissues and basophils in the blood. When an allergen attaches to the IgE antibodies on these cells, they release histamine and other substances that bring about the coldlike symptoms or, rarely, anaphylactic shock.

Allergy shots sometimes prevent the onset of an allergic response. It's been suggested that injections of the allergen may cause the body to build up high quantities of IgG antibodies, and these combine with allergens received from the environment before they have a chance to reach the IgE antibodies located in the membrane of mast cells and basophils.

Delayed Allergic Response
Delayed allergic responses are initiated by sensitized T cells at the site of allergen in the body. A sensitized T cell is one that is ready to respond to the antigen because it has been present in the body before. T cells initiate the response by recruiting the help of macrophages, which are able to phagocytize offending viral particles or infectious cells. The overall response is regulated by the cytokines secreted by both the T cells and macrophages.

A classic example of a delayed allergic response is the tuberculin skin test. When the result of the test is positive, there is a reddening and hardening of tissue where the antigen was injected. This shows that there was prior exposure to tubercle bacilli which cause TB. Contact dermatitis, such as occurs when one is allergic to poison ivy, jewelry, cosmetics, and so forth, is also an example of a delayed allergic response.

Immediate Allergic Responses

The runny nose and watery eyes of hay fever are often caused by an allergic reaction to the pollen of trees, grasses, and ragweed. Worse, the airways leading to the lungs constrict if one has asthma, resulting in difficult breathing characterized by wheezing. Windblown pollen, particularly in the spring and fall, brings on the symptoms of hay fever. Most people can inhale pollen with no ill effects. But others have developed a hypersensitivity, meaning that their immune system responds in a deleterious manner. The problem stems from a type of antibody called immunoglobulin E (IgE) that causes the release of histamine from mast cells and basophils whenever they are exposed to an allergen. Histamine is a chemical that causes mucosal membranes of the nose and eyes to release fluid as a defense against pathogen invasion. But in the case of allergy, copious fluid is released although no real danger is present.

Most food allergies are also due to the presence of IgE antibodies that bind usually to a protein in the food. The symptoms, such as nausea, vomiting, and diarrhea, are due to the mode of entry of the allergen. Skin symptoms may also occur, however. Adults are often allergic to shellfish, nuts, eggs, cows' milk, fish, and soybeans. Peanut allergy is a common food allergy in the United States possibly because peanut better is a staple in the United States. People seem to outgrow allergies to cows' milk and eggs more often than allergies to peanuts and soybeans.

Celiac disease occurs in people who are allergic to wheat, rye, barley, and sometimes oats—in short, any grain that contains gluten proteins. It is thought that the gluten proteins elicit a delayed cell-mediated immune response by T cells with the resultant production of cytokines. The symptoms of celiac disease can include diarrhea, bloating, weight loss, anemia, bone pain, chronic fatigue, and weakness.

People can reduce the chances of a reaction to airborne and food allergens by avoiding the offending substances. The reaction to peanuts can be so severe that airlines are now required to have a peanut-free zone for those allergic. The people in Figure 7B are trying to avoid windblown allergens. The taking of antihistamines can also be helpful. If these procedures are inadequate, patients can be tested to measure their susceptibility to any number of possible allergens. A small quantity of a suspected allergen is inserted just beneath the skin, and the strength of the subsequent reaction is noted. A wheal-and-flare response at the skin prick site demonstrates that IgE antibodies attached to mast cells have reacted to an allergen. In an immunotherapy called hyposensitization, ever-increasing doses of the allergen are periodically injected subcutaneously with the hope that the body will build up a supply of IgG. IgG, in contrast to IgE, does not cause the release of histamine after it combines with the allergen. If IgG combines first upon exposure to the allergen, the allergic response does not occur. Patients know they are cured when the allergic symptoms no longer occur. Therapy may have to continue for as long as two to three years.

Allergic-type reactions can occur without involving the immune system. Wasp and bee stings contain substances that cause swellings, even in those whose immune system is not sensitized to substances in the sting. Also, jellyfish tentacles and foods such as fish that is not fresh and strawberries contain histamine or closely related substances that can cause a reaction. Immunotherapy is also not possible in those who are allergic to penicillin and bee stings. High sensitivity has built up upon the first exposure, and when reexposed, anaphylactic shock can occur. Among its many effects, histamine causes increased permeability of the smallest blood vessels, called capillaries. In these individuals, there is a drastic decrease in blood pressure that can be fatal within a few minutes. People who know they are allergic to bee stings can obtain a syringe of epinephrine to carry with them. This medication can delay the onset of anaphylactic shock until medical help is available.

Figure 7B   Protection against allergies.
The allergic reactions known as hay fever and asthma attacks, can have many triggers, one of which is the pollen of a variety of plants. A dramatic solution to the problem has been found by these people.

Autoimmune Diseases

When T cells or antibodies mistakenly attack the body's own cells as if they bore foreign antigens, the resulting condition is known as an **autoimmune disease.** Exactly what causes autoimmune diseases is not known. However, sometimes they occur after an individual has recovered from an infection.

In the autoimmune disease myasthenia gravis, neuromuscular junctions do not work properly and muscular weakness results. In multiple sclerosis, the myelin sheath of nerve fibers breaks down, and this causes various neuromuscular disorders. A person with systemic lupus erythematosus has various symptoms prior to death due to kidney damage. In rheumatoid arthritis, the joints are affected. Researchers suggest that heart damage following rheumatic fever and type I diabetes are also autoimmune illnesses. As yet there are no cures for autoimmune diseases, but they can be controlled with drugs.

Autoimmune diseases occur when antibodies and cytotoxic T cells recognize and destroy the body's own cells.

Tissue Rejection

Certain organs, such as skin, the heart, and the kidneys, could be transplanted easily from one person to another if the body did not attempt to *reject* them. Rejection occurs because antibodies and cytotoxic T cells bring about destruction of foreign tissues in the body. When rejection occurs, the immune system is correctly distinguishing between self and nonself.

Organ rejection can be controlled by careful selection of the organ to be transplanted and the administration of immunosuppressive drugs. It is best if the transplanted organ has the same type of HLA antigens as those of the recipient, because cytotoxic T cells recognize foreign HLA antigens. The immunosuppressive drug cyclosporine has been used for many years. A new drug, tacrolimus (formerly known as FK-506), shows some promise, especially in liver transplant patients. However both drugs, which act by inhibiting the response of T cells to cytokines, are known to adversely affect the kidneys.

The hope is that tissue engineering, the production of organs that lack antigens or that can be protected in some way from the immune system, will one day do away with the problem of rejection. For example, pancreatic cells have been placed in protective capsules that are implanted in the abdominal cavity.

When an organ is rejected, the immune system has recognized and destroyed cells that bear HLA antigens different from those of the individual.

7.6   Homeostasis

Blood and tissue fluid provide an internal environment for the body's cells. The lymphatic vessels collect excess tissue fluid and return it as lymph to cardiovascular veins in the thorax. The lymphoid organs, along with the immune system, protect us from infectious diseases.

Nonspecific ways of protecting the body from disease precede specific immunity. The skin, and the mucous membranes of the respiratory tract, the digestive tract, and the urinary system all resist invasion by viruses and bacteria. If a pathogen should enter the body, the infection is localized as much as possible. During the inflammatory reaction, the phagocytic white blood cells immediately rush to the scene and engulf as many pathogens as possible. Macrophages are especially good at devouring viruses and bacteria by phagocytosis. If the infection cannot be confined and pathogens do enter the blood, complement is a series of proteins that work in diverse ways to keep the blood free of disease-causing organisms and their toxins.

Not surprising, specific defenses are dependent upon blood cells; the lymphocytes and macrophages play central roles. B and T cells have receptors that bind to antigens, and in this way these cells distinguish self from nonself. The binding of the antigen selects which specific B or T cells will undergo clonal expansion. B cells are capable of recognizing an antigen directly, but T cells must have the antigen displayed by an APC in the groove of an HLA antigen. Plasma cells (mature B cells) produce antibodies, but T cells kill infected cells outright.

The lymphoid organs play a central role in immunity. White blood cells are made in the red bone marrow where B cells also mature. T cells mature in the thymus. The spleen filters the blood directly. Clonal expansion of lymphocytes occurs in the lymph nodes, which also filter the lymph.

A strong connection exists between the immune, nervous, and endocrine systems. Lymphocytes have receptors for a wide variety of hormones, and the thymus gland produces hormones which influence the immune response. Cytokines help the body recover from disease by affecting the brain's temperature control center. A high body temperature of a fever is thought to create an unfavorable environment for the foreign invaders. Also, cytokines bring about a feeling of sluggishness, sleepiness, and loss of appetite. These behaviors tend to make us take care of ourselves until we feel better. A close connection between the immune and hormonal systems is illustrated by the ability of cortisone to mollify the inflammatory reaction in the joints.

The lymphatic vessels collect excess tissue fluid and return it as lymph to the cardiovascular veins. The immune system normally keeps the body free of infectious diseases.

Human Systems Work Together

Integumentary System

Lymphatic vessels pick up excess tissue fluid; immune system protects against skin infections.

Skin serves as a barrier to pathogen invasion; Langerhans' cells phagocytize pathogens; protects lymphatic vessels.

Skeletal System

Lymphatic vessels pick up excess tissue fluid; immune system protects against infections.

Red bone marrow produces leukocytes involved in immunity.

Muscular System

Lymphatic vessels pick up excess tissue fluid; immune system protects against infections.

Skeletal muscle contraction moves lymph; physical exercise enhances immunity.

Nervous System

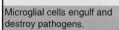

Lymphatic vessels pick up excess tissue fluid; immune system protects against infections of nerves.

Microglial cells engulf and destroy pathogens.

Endocrine System

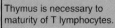

Lymphatic vessels pick up excess tissue fluid; immune system protects against infections.

Thymus is necessary to maturity of T lymphocytes.

How the Lymphatic System works with other body systems

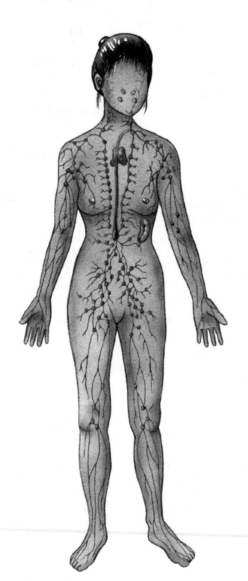

Cardiovascular System

Lymphoid organs produce and store formed elements; lymphatic vessels transport leukocytes and return tissue fluid to blood vessels; spleen serves as blood reservoir, filters blood. Blood vessels transport leukocytes and antibodies; blood services lymphoid organs and is source of tissue fluid that becomes lymph.

Respiratory System

Lymphatic vessels pick up excess tissue fluid; immune system protects against respiratory tract and lung infections.

Tonsils and adenoids occur along respiratory tract; breathing aids lymph flow; lungs carry out gas exchange.

Digestive System

Lacteals absorb fats; Peyer's patches prevent invasion of pathogens; appendix contains lymphoid tissue.

Digestive tract provides nutrients for lymphoid organs; stomach acidity prevents pathogen invasion of body.

Urinary System

Lymphatic system picks up excess tissue fluid, helping to maintain blood pressure for kidneys to function; immune system protects against infections.

Kidneys control volume of body fluids, including lymph.

Reproductive System

Immune system does not attack sperm or fetus, even though they are foreign to the body.

Sex hormones influence immune functioning; acidity of vagina helps prevent pathogen invasion of body; milk passes antibodies to newborn.

Bioethical Issue

The United Nations estimates that 16,000 people become newly infected with the human immunodeficiency virus (HIV) each day, or 5.8 million per year. Ninety percent of these infections occur in the less-developed countries[1] where infected persons do not have access to antiviral therapy. In Uganda, for example, there is only one physician per 100,000 people, and only $6.00 is spent annually on health care, per person. In contrast, in the United States $12,000–$15,000 is sometimes spent on treating an HIV infected person per year.

The only methodology to prevent the spread of HIV in a developing country is counseling against behaviors that increase the risk of infection. Clearly an effective vaccine would be most beneficial to these countries. Several HIV vaccines are in various stages of development, and all need to be clinically tested in order to see if they are effective. It seems reasonable to carry out such trials in developing countries, but there are many ethical questions.

A possible way to carry out the trial is this: vaccinate the uninfected sexual partners of HIV-infected individuals. After all, if the uninfected partner remains free of the disease, then the vaccine is effective. But is it ethical to allow a partner identified as having an HIV infection to remain untreated for the sake of the trial?

And should there be a placebo group—a group that does not get the vaccine? After all, if a greater number of persons in the placebo group become infected than those in the vaccine group, then the vaccine is effective. But if members of the placebo group become infected, shouldn't they be given effective treatment? For that matter, even participants in the vaccine group might become infected. Shouldn't any participant of the trial be given proper treatment if they become infected? Who would pay for such treatment when the trial could involve thousands of persons?

Questions

1. Should HIV vaccine trials be done in developing countries, which stand to gain the most from an effective vaccine? Why or why not?
2. Should the trial be carried out using the same standards as in developed countries? Why or why not?
3. Who should pay for the trial—the drug company, the participants, or the country of the participants?

[1]Country that has only low to moderate industrialization; usually located in the southern hemisphere.

Summarizing the Concepts

7.1 Lymphatic System

The lymphatic system consists of lymphatic vessels and lymphoid organs. The lymphatic vessels collect fat molecules at intestinal villi and excess tissue fluid at blood capillaries, and carry these to the bloodstream.

Lymphocytes are produced and accumulate in the lymphoid organs (red bone marrow, lymph nodes, spleen, and thymus gland). Lymph is cleansed of pathogens and/or their toxins in lymph nodes, and blood is cleansed of pathogens and/or their toxins in the spleen. T lymphocytes mature in the thymus, while B lymphocytes mature in the red bone marrow where all blood cells are produced. White blood cells are necessary for nonspecific and specific defenses.

7.2 Nonspecific Defenses

Immunity involves nonspecific and specific defenses. Nonspecific defenses include barriers to entry, the inflammatory reaction, natural killer cells, and protective proteins.

7.3 Specific Defenses

Specific defenses require lymphocytes, which are produced in the bone marrow. B cells mature in the bone marrow. They undergo clonal selection with production of plasma cells and memory B cells after their specific plasma membrane receptors directly combine with a particular antigen. Plasma cells secrete antibodies and eventually undergo apoptosis. B cells are responsible for antibody-mediated immunity. IgG antibody is a Y-shaped molecule that has two binding sites for a specific antigen. Memory B cells remain in the body and produce antibodies if the same antigen enters the body at a later date.

T cells, which are responsible for cell-mediated immunity, mature in the thymus. The two main types of T cells are cytotoxic T cells and helper T cells. Cytotoxic T cells kill infected cells that bear a foreign antigen on contact; helper T cells stimulate other immune cells and produce cytokines. Like B cells, each T cell bears a specific receptor. However, for a T cell to recognize an antigen, the antigen must be presented by an antigen-presenting cell (APC), usually a macrophage, along with an HLA (human leukocyte-associated) antigen. Thereafter the activated T cell undergoes clonal expansion until the infection has been stemmed. Then most of the activated T cells undergo apoptosis. A few cells remain, however, as memory T cells.

7.4 Induced Immunity

Immunity can be induced in various ways. Vaccines are available to induce long-lived active immunity, and antibodies sometimes are available to provide an individual with short-lived passive immunity.

Cytokines, including interferon, are used in an attempt to promote the body's ability to recover from cancer and to treat AIDS.

7.5 Immunity Side Effects

Allergic responses occur when the immune system reacts vigorously to substances not normally recognized as foreign. Immediate allergic responses, usually consisting of coldlike symptoms, are due to the activity of antibodies. Delayed allergic responses, such as contact dermatitis, are due to the activity of T cells.

7.6 Homeostasis

The lymphatic system works with the other systems of the body in the ways described in the box on page 161.

Studying the Concepts

1. What is the lymphatic system, and what are its three functions? 146
2. Describe the structure and the function of lymph nodes, the spleen, the thymus, and red bone marrow. 147–48
3. What are the body's nonspecific defense mechanisms? 148–49
4. Describe the inflammatory reaction, and give a role for each type of cell and molecule that participates in the reaction. 148
5. What is the clonal selection theory? B cells are responsible for which type of immunity? 151
6. Describe the structure of an antibody, and define the terms variable regions and constant regions. 152
7. Name the two main types of T cells, and state their functions. 154

8. Explain the process by which a T cell is able to recognize an antigen. 154–55
9. How is active immunity achieved? How is passive immunity achieved? 156–57
10. What are cytokines, and how are they used in immunotherapy? 157
11. How are monoclonal antibodies produced, and what are their applications? 158
12. Discuss allergies, tissue rejection, and autoimmune diseases as they relate to the immune system. 158–60

Testing Your Knowledge of the Concepts

In questions 1–4, match the cells to their functions. The answer can require more than one cell.
   a. T cells
   b. B cells
   c. macrophage
   d. neutrophils
      _____ 1. antibody-mediated immunity
      _____ 2. presents antigen to T cells
      _____ 3. phagocytic
      _____ 4. nonspecific defense only

In questions 5–7, indicate whether the statement is true (T) or false (F).
   _____ 5. During the inflammatory reaction, monocytes differentiate into macrophages, cells that release histamine and cause capillary permeability.
   _____ 6. The presentation of the antigen to helper T cells augments the production of antibodies by B cells.
   _____ 7. Mast cells are involved in the inflammatory reaction and aggravate the symptoms of allergies.

In questions 8–18, fill in the blanks.
8. Lymphatic vessels take up excess _____ and return it to the _____ veins.
9. The function of lymph nodes is to _____ lymph.
10. T cells mature in the_____.
11. _____ is a group of proteins present in plasma that plays a role in destroying bacteria.

12. A stimuated B cell becomes antibody-secreting _____ cells and _____ cells, which are ready to produce the same type of antibody at a later time.
13. B cells are responsible for _____-mediated immunity.
14. T cells produce _____, which are stimulatory molecules for all types of immune cells.
15. In order for a T cell to recognize an antigen, it must be presented by a(n) _____ along with an MHC protein.
16. Immunization with _____ brings about active immunity.
17. Antibodies produced by a single clone of cells are called _____ antibodies.
18. Good active immunity lasts as long as clones of _____ B and _____ T cells are present in the body.
19. Use these terms to label this IgG molecule: antigen-binding sites, light chain, heavy chain. d. What does V stand for in the diagram? e. What does C stand for in the diagram? f. What shape antigen would bind to this particular antigen-binding site?

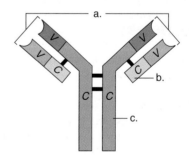

Applying Your Knowledge to the Concepts

These questions pertain to homeostasis.

1. An infection of the throat often results in the lymph nodes in the neck area becoming swollen and sore. Explain what is happening to produce these symptoms.
2. The inflammatory reaction can lead to the clonal expansion of specific B cells and T cells. Explain how this can happen.
3. Immunosuppressive drugs are often used in tissue transplant patients. This puts the patient at risk for what specific types of illnesses and why?
4. Cold medications may contain antihistamines. What effect does this medication have on cold symptoms?

Understanding the Terms

allergen 158 *foreign substance (allergy response)*
allergy 158 *immune response to substances* *Not foreign*
antibody 150 *protein produce in response*
antibody-mediated immunity 151 *specific mechanism of defense*
antigen 150 *Foreign substance*
antigen-presenting cell (APC) 154 *display antigen to defend system*
apoptosis 151 *programmed cell death*
autoimmune disease 160 *attacks body immune system*
basophil 148 *leukocyte granular cytoplasm*
B lymphocyte 150 *matures in bone marrow T cells destroy antigen*
cell-mediated immunity 154
clonal selection theory 151 *produce more lymphocytes*
complement system 150 *plasma proteins*
cytokine 157 *i. attack viroses*
cytotoxic T cell 154 *attacks & kills antigen bearing cells*
delayed allergic response 158 *at the side of allergen sensitized T cells*
edema 146 *swelling tissue fluid*
helper T cell 154 *releases cytokyne*
histamine 148 *substance produce by basophils*
HLA (human leukocyte-associated) antigen 154 *protein in plasma membrane self antigen*
immediate allergic response 158 *occurs in seconds after an allergen contact*

immune system 150 *Population of cells against foreign substances*
immunity 148 *ability of the body to protect itself*
immunization 156 *strategy causing agents*
immunoglobulin (Ig) 152 *globular plasma antibody*
inflammatory reaction 148 *tissue response redness*
interferon 150 *protein infected virus*
interleukin 157 *chemical substance macrophages*
kinin 148 *chemical mediator (capillaries to cticle)*
lymph 146 *Fluid*
lymphatic system 146 *Mammalian organ*
lymph node 147 *Mass lymphoid tissue*
macrophage 148 *phagocytic by monocyte*
mast cell 148 *antibody, allergens*
monoclonal antibody 158 *one type (single plasma cell)*
natural killer (NK) cell 150 *cancerous cell to burst*
pathogen 148 *diseased causing agent*
perforin 154 *Molecule cytotoxic T cells*
plasma cell 151 *cell from b cell (mass produce antibodies)*
red bone marrow 148 *Blood cell forming tissue (spongy bone)*
spleen 147 *Grandular organ (stores & purifies blood)*
thymus gland 148 *Organ lies in the neck & chest (immunity necessary)*
T lymphocyte 150 *matures in the thymus*
tonsils 147 *lymph nodules*
vaccine 156 *antigen w/o causing diseases*

Match the terms to these definitions:

a. *vaccine* Antigens prepared in such a way that they can promote active immunity without causing disease.

b. *lymph* Fluid, derived from tissue fluid, that is carried in lymphatic vessels.

c. *Antigen* Foreign substance, usually a protein or a polysaccharide, that stimulates the immune system to react, such as to produce antibodies.

d. *Apoptosis* Process of programmed cell death involving a cascade of specific cellular events leading to the death and destruction of the cell.

e. *T lymphocyte* Lymphocyte that matures in the thymus and exists in three varieties, one of which kills antigen-bearing cells outright.

Applying Technology to the Concepts

Your study of lymph system and immunity is supported by these available technologies:

Essential Study Partner CD-ROM
Animals → Lymph and Immunity
Visit the Mader web site for related ESP activities.

Exploring the Internet
The Mader Home Page provides resources and tools as you study this chapter.

http://www.mhhe.com/biosci/genbio/mader

Dynamic Human 2.0 CD-ROM
Lymphatic System

HealthQuest CD-ROM
4 Communicable Diseases
→ Gallery
→ Immune Response
→ T-Cells
→ Humoral Immune Response
→ Autoimmune Disorders

Life Science Animations 3D Video
33 Complement System
34 How T Lymphocytes Work
35 Clonal Selection

Chapter 8

Respiratory System

Chapter Concepts

Figure 8.1 The Heimlich maneuver.
This young man shows how he saved his "sweetheart" from choking on a piece of candy by using the Heimlich maneuver.

At one time, Dr. Henry Heimlich thought, like many other people, that choking to death on a piece of food or some other object occurred rarely. But then he learned that choking was the sixth leading cause of accidental death in the United States—more than 20 persons a day die from choking. Many children choke on toys or fragments of balloons that explode when they were trying to blow them up.

Dr. Heimlich started doing his research. Quick action is needed because cessation of breathing causes damage and death to occur within just four minutes. Back slaps don't work because they drive the offending object downward to totally block small air passages instead of sending the object upward, out of the air passages. Heimlich discovered what to do: make a fist; grab it with your other hand and press forcefully upward on the diaphragm. Standing behind someone helps, but you can even do it on yourself. The rush of air from the lungs expels the object every time. The procedure is easy to do—the man in the photograph saved his "sweetheart" using what is now called the Heimlich maneuver (Fig. 8.1).

The air passages and the food passages cross in the pharynx, and that's why you shouldn't eat and talk at the same time. Food can go the wrong way and lodge in the windpipe. The air passages repeatedly divide, getting

smaller and smaller until they reach small air sacs called alveoli in the lungs. Here gas exchange occurs. Food is stored in the body but not oxygen, so you have to keep on breathing to get oxygen into your system. Oxygen is transported in the blood to all the cells where it is used in cellular respiration, the process that produces energy in the form of ATP, the common energy carrier in cells. Carbon dioxide, an end product of cellular respiration, moves in the opposite direction. Blood transports carbon dioxide from the tissues to the lungs where it is expired. The manner in which oxygen and carbon dioxide is carried in the blood is of interest. The red blood cells and respiratory pigment hemoglobin play a role in both of these.

As you would expect, breathing is controlled by the brain, which automatically keeps the rib cage moving up and down. But the breathing rate can be modified, particularly by the amount of carbon dioxide in the blood, and we can voluntarily alter the quantity of air we take in and release. Homeostasis is usually maintained by self-regulatory mechanisms, but it also is possible for us to consciously appraise the need for air and increase the depth of breathing. ✗

8.1 Respiratory Tract

During **inspiration** or inhalation (breathing in) and **expiration** or exhalation (breathing out), air is conducted toward or away from the lungs by a series of cavities, tubes, and openings, illustrated in Figure 8.2.

As air moves in along the airways, it is filtered, warmed, and moistened. Filtering is accomplished by coarse hairs, cilia, and mucus in the region of the nostrils and by cilia alone in the rest of the nasal cavity and the airways of the lower respiratory tract. In the nose, the hairs and the cilia act as a screening device. In the trachea and other airways, the cilia beat upward, carrying mucus, dust, and occasional bits of food that "went down the wrong way" into the pharynx, where the accumulation can be swallowed or expectorated. The air is warmed by heat given off by the blood vessels lying close to the surface of the lining of the airways, and it is moistened by the wet surface of these passages.

Conversely, as air moves out during expiration, it cools and loses its moisture. As the air cools, it deposits its moisture on the lining of the windpipe and the nose, and the nose may even drip as a result of this condensation. The

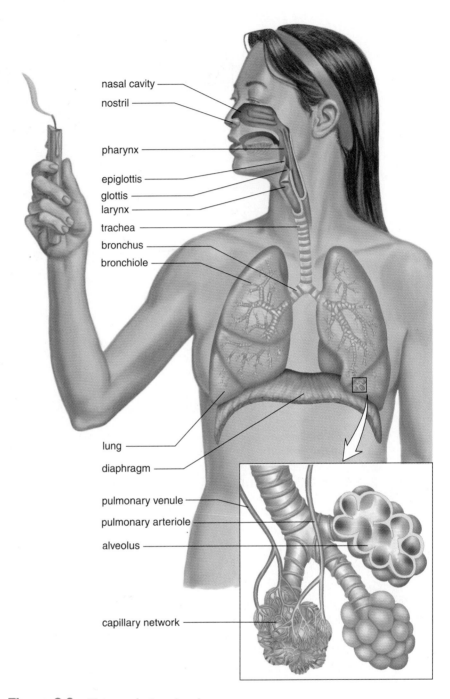

Figure 8.2 **The respiratory tract.**
The respiratory tract extends from the nose to the lungs, which are composed of air sacs called alveoli. Gas exchange occurs between air in the alveoli and blood within a capillary network that surrounds the alveoli. Notice that the pulmonary arteriole is colored blue—it carries blood low in oxygen away from the heart to alveoli. The pulmonary venule is colored red—it carries blood high in oxygen from alveoli toward the heart.

air still retains so much moisture, however, that upon expiration on a cold day, it condenses and forms a small cloud.

Air is filtered, warmed, and moistened as it moves from the nose toward the lungs.

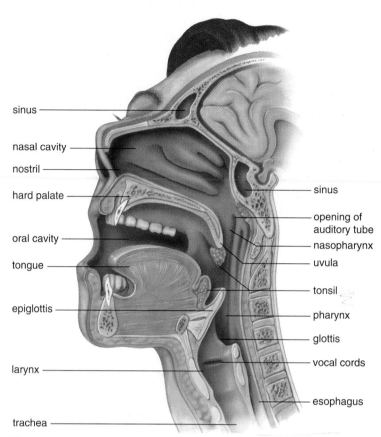

sinus

nasal cavity

nostril

hard palate

oral cavity

tongue

epiglottis

larynx

trachea

sinus

opening of
auditory tube

nasopharynx

uvula

tonsil

pharynx

glottis

vocal cords

esophagus

Figure 8.3 **The upper respiratory tract.**
The upper respiratory tract contains the nasal cavities, pharynx, and larynx.

Table 8.1	Path of Air	
Structure	**Description**	**Function**
Nasal cavities	Hollow spaces in nose	Filter, warm, and moisten air
Pharynx	Chamber behind oral cavity and between nasal cavity and larynx	Connection to surrounding regions
Glottis	Opening into larynx	Passage of air into larynx
Larynx	Cartilaginous organ that contains vocal cords (voice box)	Sound production
Trachea	Flexible tube that connects larynx with bronchi (windpipe)	Passage of air to bronchi
Bronchi	Divisions of the trachea that enter lungs C rings hold it open.	Passage of air to lungs
Bronchioles	Branched tubes that lead from bronchi to the alveoli	Passage of air to each alveolus
Lungs	Soft, cone-shaped organs that occupy a large portion of the thoracic cavity	Gas exchange

C rings hold it open (handwritten annotation)

The Nose

The nose contains two **nasal cavities** (Table 8.1), which are narrow canals separated from one another by a septum composed of bone and cartilage. Special ciliated cells in the narrow upper recesses of the nasal cavities (Fig. 8.3) act as odor receptors. Nerves lead from these cells to the brain, where the impulses generated by the odor receptors are interpreted as smell.

The tear (lacrimal) glands drain into the nasal cavities by way of tear ducts. For this reason, crying produces a runny nose. The nasal cavities also communicate with the cranial sinuses, air-filled mucosa-lined spaces in the skull. If inflammation due to a cold or an allergic reaction blocks the ducts leading from the sinuses, mucus may accumulate, causing a sinus headache.

The nasal cavities empty into the nasopharynx, the upper portion of the pharynx. The auditory tubes lead from the nasopharynx to the middle ears.

The nasal cavities, which receive air, open into the nasopharynx.

The Pharynx

The **pharynx** is a funnel-shaped passageway that connects the nasal and oral cavities to the larynx. Therefore, the pharynx, which is commonly referred to as the "throat," has three parts: the nasopharynx, where the nasal cavities open above the soft palate; the oropharynx, where the oral cavity opens; and the laryngopharynx, which opens into the larynx. The tonsils form a protective ring at the junction of the oral cavity and the pharynx. Being lymphoid tissue, the tonsils contain lymphocytes that protect against invasion of foreign substances that are inhaled. They also prepare B cells and T cells to be able to respond to antigens that may subsequently invade internal tissues and fluids. In this way, the respiratory tract is involved in the role the immune system plays in homeostasis.

In the pharynx, the air passage and the food passage cross because the larynx, which receives air, is ventral to the esophagus, which receives food. The larynx lies at the top of the trachea. The larynx and trachea are normally open, allowing the passage of air, but the esophagus is normally closed and opens only when swallowing occurs.

Air from either the nose or the mouth enters the pharynx, as does food. The passage of air continues in the larynx and then the trachea.

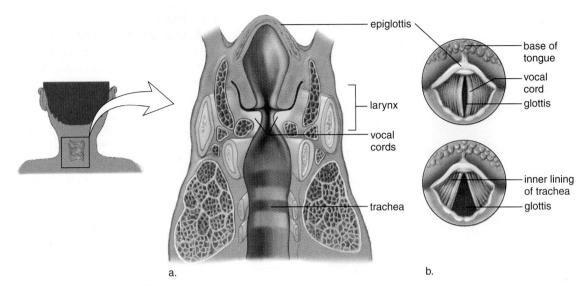

Figure 8.4 Placement of the vocal cords.
a. Frontal section of the larynx shows the location of the vocal cords inside the larynx. The vocal cords viewed from above are stretched across the glottis. When air passes through the glottis, the vocal cords vibrate, producing sound. **b.** The glottis is narrow when we produce a high-pitched sound *(top)* and widens as the pitch deepens *(bottom)*.

The Larynx

The **larynx** can be pictured as a triangular box whose apex, the Adam's apple, is located at the front of the neck. The Adam's apple is more prominent in men than women. At the top of the larynx is a variable-sized opening called the **glottis.** When food is swallowed, the larynx moves upward against the **epiglottis,** a flap of tissue that prevents food from passing into the larynx. You can detect this movement by placing your hand gently on your larynx and swallowing.

The larynx is called the voice box because the vocal cords are inside the larynx. The **vocal cords** are mucosal folds supported by elastic ligaments, which are stretched across the glottis (Fig. 8.4). When air passes through the glottis, the vocal cords vibrate, producing sound. At the time of puberty, the growth of the larynx and the vocal cords is much more rapid and accentuated in the male than in the female, causing the male to have a more prominent Adam's apple and a deeper voice. The voice "breaks" in the young male due to his inability to control the longer vocal cords. These changes cause the lower pitch of the voice in males.

The high or low pitch of the voice is regulated when speaking and singing by changing the tension on the vocal cords. The greater the tension, as when the glottis becomes more narrow, the higher the pitch. When the glottis is wider, the pitch is lower (Fig. 8.4*b*). The loudness, or intensity, of the voice depends upon the amplitude of the vibrations, that is, the degree to which vocal cords vibrate.

The Trachea

The **trachea,** commonly called the windpipe, is a tube connecting the larynx to the primary bronchi. The trachea lies ventral to the esophagus and is held open by C-shaped cartilaginous rings. The open part of the C-shaped rings faces the esophagus and this allows the esophagus to expand when swallowing. The mucosa that lines the trachea has a layer of pseudostratified ciliated columnar epithelium. (Pseudostratified means that while the epithelium appears to be layered, actually each cell touches the basement membrane.) The cilia that project from the epithelium keep the lungs clean by sweeping mucus and debris toward the pharynx. Smoking is known to destroy the cilia, and consequently the soot in cigarette smoke collects in the lungs. Smoking is discussed more fully at the end of this chapter.

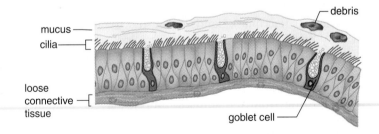

If the trachea is blocked because of illness or the accidental swallowing of a foreign object, it is possible to insert a tube by way of an incision made in the trachea. This tube acts as an artificial air intake and exhaust duct. The operation is called a **tracheostomy.**

The Bronchial Tree

The trachea divides into right and left primary bronchi (sing., **bronchus**), which lead into the right and left lungs (see Fig. 8.2). The bronchi branch into a great number of secondary bronchi that eventually lead to **bronchioles.** The bronchi resemble the trachea in structure, but as the bronchial tubes divide and subdivide, their walls become thinner, and the small rings of cartilage are no longer present. During an asthma attack, the smooth muscle of the bronchioles contracts, causing bronchiolar constriction and characteristic wheezing. Each bronchiole terminates in an elongated space enclosed by a multitude of air pockets, or sacs, called *alveoli* (sing., **alveolus**). The alveoli make up the lungs.

The Lungs

The **lungs** are paired cone-shaped organs within the thoracic cavity. The right lung has three lobes, and the left lung has two lobes, allowing room for the heart, which is on the left side of the body. A lobe is further divided into lobules, and each lobule has a bronchiole serving many alveoli. The lungs lie on either side of the heart in the thoracic cavity. The base of each lung is broad and concave so that it fits the convex surface of the diaphragm. The other surfaces of the lungs follow the contours of the ribs and the diaphragm in the thoracic cavity.

The Alveoli exchange the oxygen

Each alveolar sac is made up of simple squamous epithelium surrounded by blood capillaries. Gas exchange occurs between air in the alveoli and blood in the capillaries (Fig. 8.5). Oxygen diffuses across the alveolar wall and enters the bloodstream, while carbon dioxide diffuses from the blood across the alveolar wall to enter the alveoli.

The alveoli of human lungs are lined with a surfactant, a film of lipoprotein that lowers the surface tension and prevents them from closing. The lungs collapse in some newborn babies, especially premature infants, who lack this film. The condition, called **infant respiratory distress syndrome,** is now treatable by surfactant replacement therapy.

There are approximately 300 million alveoli, with a total cross-sectional area of 50–70 m^2. This is the surface area of a typical classroom and at least 40 times the surface area of the skin. Because of their many air spaces, the lungs are very light; normally, a piece of lung tissue dropped in a glass of water floats.

The trachea divides into the primary bronchi, which divide repeatedly to give rise to the bronchioles. The bronchioles have many branches and terminate at the alveoli, which make up the lungs.

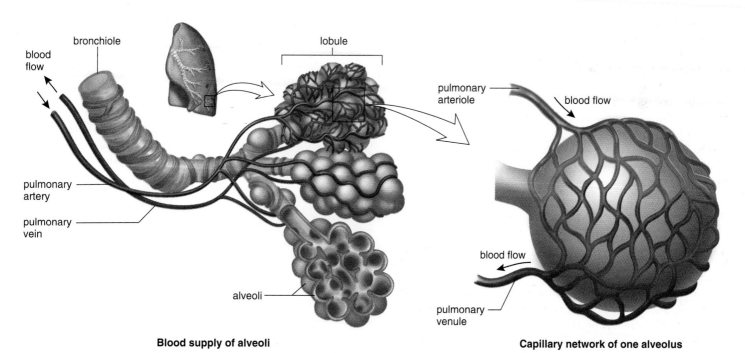

Blood supply of alveoli

Capillary network of one alveolus

Figure 8.5   Gas exchange in the lungs.
The lungs consist of alveoli, surrounded by an extensive capillary network. Notice that the pulmonary arteriole carries blood low in oxygen (colored blue) and the pulmonary venule carries blood high in oxygen (colored red).

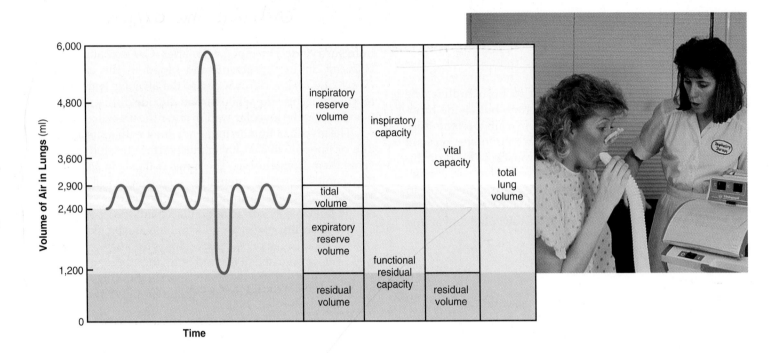

Figure 8.6 **Vital capacity.**
A spirometer measures the maximum amount of air that can be inhaled and exhaled when breathing by way of a tube connected to the instrument. During inspiration, the pen moves up, and during expiration, the pen moves down. The resulting pattern, such as the one shown here, is called a spirograph.

8.2 Mechanism of Breathing

The term respiration refers to the complete process of supplying oxygen to body cells for aerobic cellular respiration and the reverse process of ridding the body of carbon dioxide given off by cells. Respiration includes the following components:

1. Breathing: *inspiration* (entrance of air into the lungs) and *expiration* (exit of air from the lungs).
2. *External respiration:* exchange of the gases oxygen (O_2) and carbon dioxide (CO_2) between air and blood in the lungs.
3. *Internal respiration:* exchange of the gases O_2 and CO_2 between blood and tissue fluid.
4. *Cellular respiration:* production of ATP in cells. to release energy

Respiratory Volumes

When we breathe, the amount of air moved in and out with each breath is called the **tidal volume.** Normally, the tidal volume is about 500 ml, but we can increase the amount inhaled and exhaled by deep breathing. The maximum volume of air that can be moved in and out during a single breath is called the **vital capacity** (Fig. 8.6). First, we can increase inspiration by as much as 3,100 ml of air by

forced inspiration. This is called the **inspiratory reserve volume.**

Even so, some of the inspired air never reaches the lungs; instead it fills the nose, trachea, bronchi, and bronchioles (see Fig. 8.2). These passages are not used for gas exchange, and therefore, they are said to contain dead space air. To ensure that inspired air reaches the lungs, it is better to breathe slowly and deeply. Similarly, we can increase expiration by contracting the abdominal and thoracic muscles. This is called the **expiratory reserve volume,** and it measures approximately 1,400 ml of air. Vital capacity is the sum of tidal, inspiratory reserve, and expiratory reserve volumes.

Note in Figure 8.6 that even after very deep breathing, some air (about 1,000 ml) remains in the lungs; this is called the **residual volume.** This air is no longer useful for gas exchange purposes. In some lung diseases, such as emphysema (p. 179), the residual volume builds up because the individual has difficulty emptying the lungs. This means that the vital capacity is reduced and the lungs tend to be filled with useless air.

The air used for gas exchange excludes both the air in the dead space of the respiratory tract and the residual volume in the lungs.

Ecology Focus

Photochemical Smog Can Kill

Pollution.

Most industrialized cities have photochemical smog at least occasionally. Photochemical smog arises when primary pollutants react with one another under the influence of sunlight to form a more deadly combination of chemicals. For example, the primary pollutants nitrogen oxides (NO_x) and hydrocarbons (HC) react with one another in the presence of sunlight to produce nitrogen dioxide (NO_2), ozone (O_3), and PAN (peroxyacetyl nitrate). Ozone and PAN are commonly referred to as oxidants. Breathing oxidants affects the respiratory and nervous systems, resulting in respiratory distress, headache, and exhaustion.

Cities with warm, sunny climates that are large and industrialized, such as Los Angeles, Denver, and Salt Lake City in the United States, Sydney in Australia, Mexico City in Mexico, and Buenos Aires in Argentina, are particularly susceptible to photochemical smog. If the city is surrounded by hills, a thermal inversion may aggravate the situation. Normally, warm air near the ground rises, so that pollutants are dispersed and carried away by air currents. But sometimes during a thermal inversion, smog gets trapped near the earth by a blanket of warm air (Fig. 8A). This may occur when a cold front brings in cold air, which settles beneath a warm layer. The trapped pollutants cannot disperse, and the results can be disastrous. In 1963, about 300 people died, and in 1966, about 168 people died in New York City when air pollutants accumulated over the city. Even worse were the events in London in 1957, when 700 to 800 people died, and in 1962, when 700 people died, due to the effects of air pollution.

Even though we have federal legislation to bring air pollution under control, more than half the people in the United States live in cities polluted by too much smog. We should place our emphasis on pollution prevention because, in the long run, prevention is usually easier and cheaper than pollution cleanup methods. Some prevention suggestions are as follows:

- Build more efficient automobiles or burn fuels that do not produce pollutants.
- Reduce the amount of waste to be incinerated by recycling materials.
- Reduce our energy use so that power plants need to provide less, and/or use renewable energy sources such as solar, wind, or water power.
- Require industries to meet clean air standards.

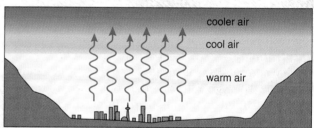

a. Normal pattern

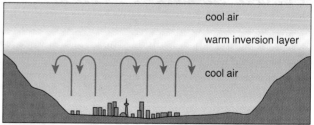

b. Thermal inversion

Figure 8A Thermal inversion.
a. Normally, pollutants escape into the atmosphere when warm air rises. **b.** During a thermal inversion, a layer of warm air (warm inversion layer) overlies and traps pollutants in cool air below. **c.** Los Angeles, a city of 8.5 million cars and thousands of factories, is particularly susceptible to thermal inversions, and this accounts for why this city is the "air pollution capital" of the United States.

Inspiration and Expiration

To understand **ventilation,** the manner in which air enters and exits the lungs, it is necessary to remember first that normally there is a continuous column of air from the pharynx to the alveoli of the lungs.

Secondly, the lungs lie within the sealed-off thoracic cavity. The **rib cage** forms the top and sides of the thoracic cavity. It contains the ribs, hinged to the vertebral column at the back and to the sternum (breastbone) at the front, and the intercostal muscles that lie between the ribs. The **diaphragm,** a dome-shaped horizontal sheet of muscle and connective tissue, forms the floor of the thoracic cavity.

The lungs are enclosed by two membranes called **pleural membranes.** An infection of the pleural membranes is called pleurisy. The parietal pleura adheres to the rib cage and the diaphragm, and the visceral pleura is fused to the lungs. The two pleural layers lie very close to one another, separated only by a small amount of fluid. Normally, the intrapleural pressure (pressure between the pleural membranes) is lower than atmospheric pressure by 4 mm Hg.

The importance of the reduced intrapleural pressure is demonstrated when, by design or accident, air enters the intrapleural space. The affected lobules collapse.

The pleural membranes enclose the lungs and line the thoracic cavity. Intrapleural pressure is lower than atmospheric pressure.

Inspiration

A **respiratory center is** located in the medulla oblongata of the brain. The respiratory center consists of a group of neurons that exhibit an automatic rhythmic discharge that triggers inspiration. Carbon dioxide (CO_2) and hydrogen ions (H^+) are the primary stimuli that directly cause changes in the activity of this center. This center is not affected by low oxygen (O_2) levels. Chemoreceptors in the **carotid bodies,** located in the carotid arteries, and in the **aortic bodies,** located in the aorta, are sensitive to the level of hydrogen ions and also to the levels of carbon dioxide and oxygen in blood. When the concentrations of hydrogen ions and carbon dioxide rise (and oxygen decreases), these bodies communicate with the respiratory center, and the rate and depth of breathing increase.

The respiratory center sends out impulses by way of nerves to the diaphragm and the muscles of the rib cage (Fig. 8.7). In its relaxed state, the diaphragm is dome-shaped, but upon stimulation, it contracts and lowers. Also, the external intercostal muscles contract, causing the rib

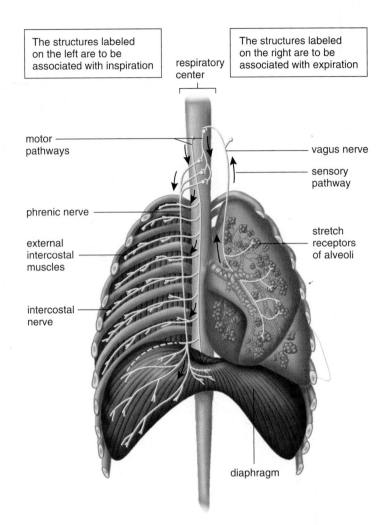

The structures labeled on the left are to be associated with inspiration

respiratory center

The structures labeled on the right are to be associated with expiration

motor pathways

vagus nerve

sensory pathway

phrenic nerve

external intercostal muscles

stretch receptors of alveoli

intercostal nerve

diaphragm

Figure 8.7   Nervous control of breathing.
During inspiration, the respiratory center stimulates the external intercostal (rib) muscles to contract via the intercostal nerves and the diaphragm to contract via the phrenic nerve. Should the tidal volume increase above 1.5 liters, stretch receptors send inhibitory nerve impulses to the respiratory center via the vagus nerve. In any case, expiration occurs due to a lack of stimulation from the respiratory center to the diaphragm and intercostal muscles.

cage to move upward and outward. Now the thoracic cavity increases in size, and the lungs expand. As the lungs expand, air pressure within the enlarged alveoli lowers and air enters through the nose or the mouth.

Inspiration is the active phase of breathing (Fig. 8.8a). During this time, the diaphragm and the rib muscles contract, intrapleural pressure decreases, the lungs expand, and air comes rushing in. Note that air comes in because the lungs already have opened up; air does not force the lungs open. This is why it is sometimes said that *humans breathe by negative pressure.* The creation of a partial vacuum in the alveoli causes air to enter the lungs.

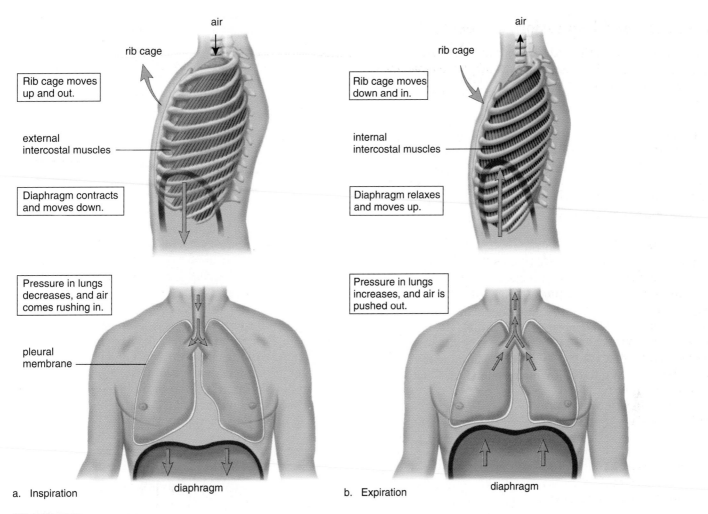

Rib cage moves up and out.

external intercostal muscles

Diaphragm contracts and moves down.

Pressure in lungs decreases, and air comes rushing in.

pleural membrane

a. Inspiration

diaphragm

Rib cage moves down and in.

internal intercostal muscles

Diaphragm relaxes and moves up.

Pressure in lungs increases, and air is pushed out.

b. Expiration

diaphragm

Figure 8.8   Inspiration versus expiration.
a. During inspiration, the thoracic cavity and lungs expand so that air is drawn in. **b.** During expiration, the thoracic cavity and lungs resume their original positions and pressures. Now, air is forced out.

Expiration

When the respiratory center stops sending neuronal signals to the diaphragm and the rib cage, the diaphragm relaxes and it resumes its dome shape. The abdominal organs press up against the diaphragm, and the rib cage moves down and inward (Fig. 8.8*b*). Now, the elastic lungs recoil, and air is pushed out. The respiratory center acts rhythmically to bring about breathing at a normal rate and volume. If by chance we inhale more deeply, the lungs are expanded and the alveoli stretch. This stimulates stretch receptors in the alveolar walls, and they initiate inhibitory nerve impulses that travel from the inflated lungs to the respiratory center. This causes the respiratory center to stop sending out nerve impulses.

While inspiration is the active phase of breathing, expiration is usually passive—the diaphragm and external intercostal muscles are relaxed when expiration occurs. When breathing is deeper and/or more rapid, expiration can also be active. Contraction of internal intercostal muscles can force the rib cage to move downward and inward. Also, when the abdominal wall muscles are contracted, they push on the viscera, which push against the diaphragm, and the increased pressure in the thoracic cavity helps to expel air.

During inspiration, due to nervous stimulation, the diaphragm lowers and the rib cage lifts up and out. During expiration, due to a lack of nervous stimulation, the diaphragm rises and the rib cage lowers.

8.3 Gas Exchanges in the Body

Figure 8.9 shows both external and internal respiration. The act of breathing brings oxygen in air to the lungs and carbon dioxide from the lungs to outside the body. Homeostasis is the constancy of the internal environment, and respiration includes not only the exchange of gases in the lungs, but also the exchange of gases in the tissues. The principles of diffusion alone govern whether O_2 or CO_2 enters or leaves blood in the lungs and in the tissues.

External Respiration

External respiration refers to the exchange of gases between air in the alveoli and blood in the pulmonary capillaries. Gases exert pressure, and the amount of pressure each gas exerts is its partial pressure, symbolized as P_{O_2} and P_{CO_2}. Blood flowing in the pulmonary capillaries has a higher P_{CO_2} than atmospheric air. Therefore, *CO_2 diffuses out of blood into the lungs.* Most of the CO_2 is being carried as bicarbonate ions (HCO_3^-). As the little remaining free CO_2 begins to diffuse out, the following reaction is driven to the right:

$$H^+ \;+\; HCO_3^- \longrightarrow H_2CO_3 \longrightarrow H_2O + CO_2 \uparrow$$

hydrogen bicarbonate   water carbon
ion   ion   dioxide

"Up" arrow indicates carbon
dioxide is leaving the body.

The enzyme **carbonic anhydrase,** present in red blood cells, speeds up the reaction. As the reaction proceeds, the respiratory pigment **hemoglobin,** also present in red blood cells, gives up the hydrogen ions (H^+) it has been carrying; HHb becomes Hb. Hb is called deoxyhemoglobin.

The pressure pattern is the reverse for O_2. Blood flowing in the pulmonary capillaries is low in oxygen, and alveolar air contains a much higher partial pressure of oxygen. Therefore, *O_2 diffuses into blood and red blood cells in the lungs.* Hemoglobin takes up this oxygen and becomes **oxyhemoglobin.**

$$Hb \;+\; \downarrow O_2 \longrightarrow HbO_2$$

deoxyhemoglobin oxygen oxyhemoglobin

"Down" arrow indicates that
oxygen is entering the body.

Internal Respiration

Internal respiration refers to the exchange of gases between blood in systemic capillaries and tissue fluid. Blood that enters the systemic capillaries is bright red in color because red blood cells contain oxyhemoglobin. Oxyhemoglobin gives up O_2, which diffuses out of blood into red blood cells and the tissues.

$$HbO_2 \longrightarrow Hb \;+\; O_2$$

oxyhemoglobin   deoxyhemoglobin   oxygen

Oxygen diffuses out of blood into the tissues because the P_{O_2} of tissue fluid is lower than that of blood. The lower P_{O_2} is due to cells continuously using up oxygen in aerobic cellular respiration. *Carbon dioxide diffuses into blood from the tissues* because the P_{CO_2} of tissue fluid is higher than that of blood. Carbon dioxide, produced continuously by cells, collects in tissue fluid.

After CO_2 diffuses into blood, it enters the red blood cells, where a small amount is taken up by hemoglobin, forming **carbaminohemoglobin.** Most of the CO_2 combines with water, forming carbonic acid (H_2CO_3), which dissociates to hydrogen ions (H^+) and bicarbonate ions (HCO_3^-). The increased concentration of CO_2 in the blood causes the reaction to proceed to the right.

$$CO_2 \;+\; H_2O \xrightleftharpoons[]{\substack{\text{carbonic}\\ \text{anhydrase}}} H_2CO_3 \rightleftharpoons H^+ \;+\; HCO_3^-$$

carbon   water   carbonic hydrogen bicarbonate
dioxide   acid   ion   ion

The enzyme carbonic anhydrase, present in red blood cells, speeds up the first portion of the overall reaction. Bicarbonate ions diffuse out of red blood cells and are carried in the plasma. Blood that leaves the capillaries is deep purple in color because red blood cells contain reduced hemoglobin. The globin portion of hemoglobin combines with excess hydrogen ions produced by the overall reaction, and Hb becomes HHb, called **reduced hemoglobin.** In this way, the pH of blood remains fairly constant.

External and internal respiration are the movement of gases between blood and the alveoli and between blood and the systemic capillaries, respectively. Both processes are dependent on the process of diffusion.

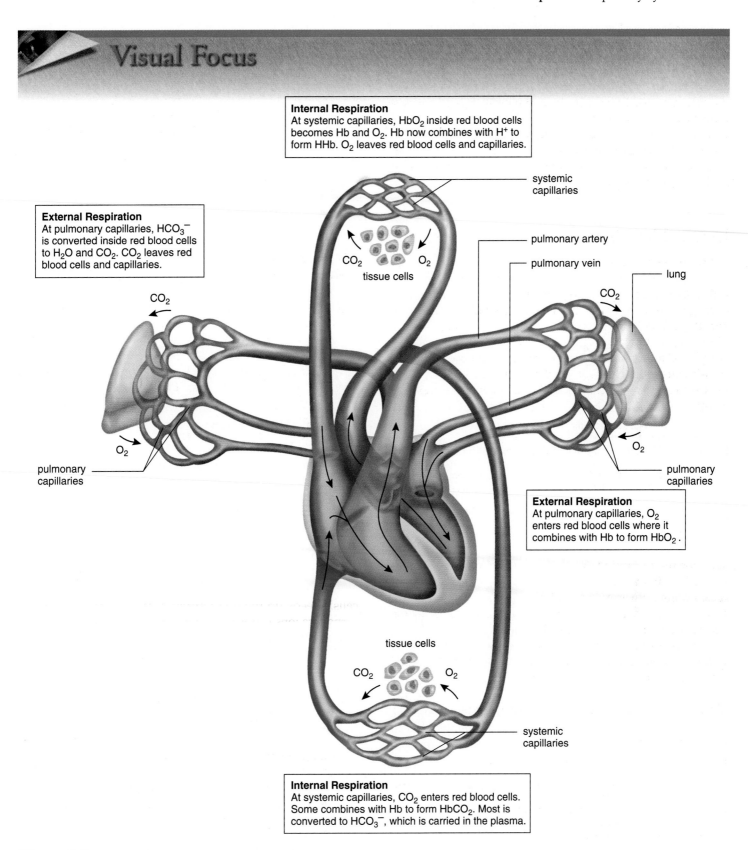

Internal Respiration
At systemic capillaries, HbO_2 inside red blood cells becomes Hb and O_2. Hb now combines with H^+ to form HHb. O_2 leaves red blood cells and capillaries.

systemic capillaries

pulmonary artery

pulmonary vein

lung

External Respiration
At pulmonary capillaries, HCO_3^- is converted inside red blood cells to H_2O and CO_2. CO_2 leaves red blood cells and capillaries.

CO_2

CO_2

CO_2

tissue cells

O_2

O_2

O_2

O_2

pulmonary capillaries

pulmonary capillaries

External Respiration
At pulmonary capillaries, O_2 enters red blood cells where it combines with Hb to form HbO_2.

tissue cells

CO_2

O_2

systemic capillaries

Internal Respiration
At systemic capillaries, CO_2 enters red blood cells. Some combines with Hb to form $HbCO_2$. Most is converted to HCO_3^-, which is carried in the plasma.

Figure 8.9   External and internal respiration.
During external respiration in the lungs, CO_2 leaves blood and O_2 enters blood. During internal respiration in the tissues, O_2 leaves blood and CO_2 enters blood.

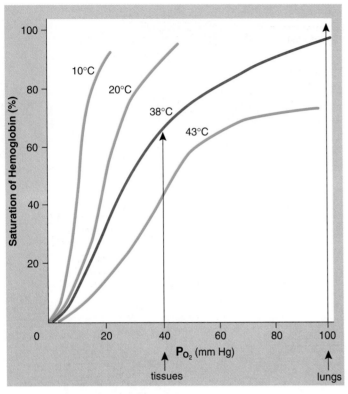

a. Saturation of Hb relative to temperature

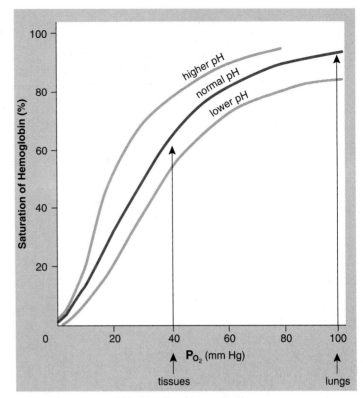

b. Saturation of Hb relative to pH

Figure 8.10   Effect of environmental conditions on hemoglobin saturation.
The partial pressure of oxygen (P_{O_2}) in pulmonary capillaries is about 98–100 mm Hg, but only about 40 mm Hg in tissue capillaries. Hemoglobin is about 98% saturated in the lungs because of P_{O_2}, and also because **(a)** the temperature is cooler and **(b)** the pH is higher in the lungs. On the other hand, hemoglobin is only about 60% saturated in the tissues because of the P_{O_2} and also because **(a)** the temperature is warmer and **(b)** the pH is lower in the tissues.

Binding Capacity of Hemoglobin

The binding capacity of hemoglobin is also affected by partial pressures. The P_{O_2} of air entering the alveoli is about 100 mm Hg, and at this pressure the hemoglobin in the blood becomes saturated with O_2. This means that iron in hemoglobin molecules has combined with O_2. On the other hand, the P_{O_2} in the tissues is about 40 mm Hg, causing hemoglobin molecules to release O_2, and O_2 to diffuse into the tissues.

In addition to the partial pressure of O_2, temperature and pH also affect the amount of oxygen hemoglobin can carry. The lungs have a lower temperature and a higher pH than the tissues:

	pH	Temperature
Lungs	7.40	37°C
Tissues	7.38	38°C

Both Figure 8.10*a* and *b* show that, as expected, hemoglobin is more saturated with O_2 in the lungs than in the tissues. This effect, which can be attributed to the difference in P_{O_2}

between the lungs and tissues, is potentiated by the difference in temperature and pH between the lungs and tissues. Notice in Figure 8.10*a* that the saturation curve for hemoglobin is steeper at 10°C compared to 20°C, and so forth. Also, Figure 8.10*b* shows that the saturation curve for hemoglobin is steeper at higher pH than at lower pH.

This means that the environmental conditions in the lungs are favorable for the uptake of O_2 by hemoglobin, and the environmental conditions in the tissues are favorable for the release of O_2 by hemoglobin. Hemoglobin is about 98–100% saturated in the capillaries of the lungs and about 60–70% saturated in the tissues. During exercise, hemoglobin is even less saturated in the tissues because muscle contraction leads to higher body temperature (up to 103°F in marathoners!) and lowers the pH (due to the production of lactic acid).

The difference in P_{O_2}, temperature, and pH between the lungs and tissues causes hemoglobin to take up oxygen in the lungs and release oxygen in the tissues.

8.4 Respiration and Health

The respiratory tract is constantly exposed to environmental air—the quality of this air, as discussed in the Ecology reading on page 171, can affect our health. The presence of a disease means that homeostasis is threatened, and if the condition is not brought under control, death is a possibility.

Upper Respiratory Tract Infections

The upper respiratory tract consists of the nose, the pharynx, and the larynx. Upper respiratory infections (URI) can spread from the nasal cavities to the sinuses, to the middle ears, and to the larynx (Fig. 8.11). Viral infections sometimes lead to secondary bacterial infections. What we call "strep throat" is a primary bacterial infection caused by *Streptococcus pyogenes* that can lead to a generalized upper respiratory infection and even a systemic (affecting the body as a whole) infection. While antibiotics have no effect on viral infections, they are successfully used for most bacterial infections, including strep throat.

Sinusitis

Sinusitis is an infection of the sinuses, cavities within the facial skeleton that drain into the nasal cavities. Only about 1–3% of upper respiratory infections are accompanied by sinusitis. Sinusitis develops when nasal congestion blocks the tiny openings leading to the sinuses. Symptoms include postnasal discharge as well as facial pain that worsens when the patient bends forward. Pain and tenderness usually occur over the lower forehead or over the cheeks. If the latter, toothache is also a complaint. Successful treatment depends on restoring proper drainage of the sinuses. Even a hot shower and sleeping upright can be helpful. Otherwise, spray decongestants are preferred over oral antihistamines, which thicken rather than liquefy the material trapped in the sinuses.

Otitis Media

Otitis media is a bacterial infection of the middle ear. The middle ear is not a part of the respiratory tract, but this infection is considered here because it is a complication often seen in children who have a nasal infection. Infection can spread by way of the **auditory tube** that leads from the nasopharynx to the middle ear. Pain is the primary symptom of a middle ear infection. A sense of fullness, hearing loss, vertigo (dizziness), and fever may also be present. Antibiotics almost always bring about a full recovery, and a recurrence is most likely due to a new infection. Drainage tubes (called tympanostomy tubes) are sometimes placed in the eardrum of children with multiple recurrences to help prevent the buildup of fluid in the middle ear and the possibility of hearing loss. Normally, the tubes slough out with time.

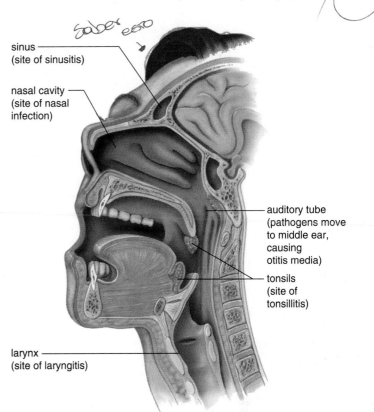

sinus
(site of sinusitis)

nasal cavity
(site of nasal
infection)

auditory tube
(pathogens move
to middle ear,
causing
otitis media)

tonsils
(site of
tonsillitis)

larynx
(site of laryngitis)

Figure 8.11 **Upper respiratory infections.**

Tonsillitis

Tonsillitis occurs when tonsils become inflamed and enlarged. **Tonsils** are masses of lymphatic tissue that occur in the pharynx. The tonsils in the dorsal wall of the nasopharynx are often called adenoids. The tonsils remove many of the pathogens that enter the pharynx; therefore, they are a first line of defense against invasion of the body. If tonsillitis occurs frequently and enlargement makes breathing difficult, the tonsils can be removed surgically in a **tonsillectomy.** Fewer tonsillectomies are performed today than in the past because it is now known that tonsils serve an important function in defending the body against infection.

Laryngitis

Laryngitis is an infection of the larynx with an accompanying hoarseness leading to the inability to talk in an audible voice. Usually laryngitis disappears with treatment of the upper respiratory infection. Persistent hoarseness without the presence of an upper respiratory infection is one of the warning signs of cancer and therefore should be looked into by a physician.

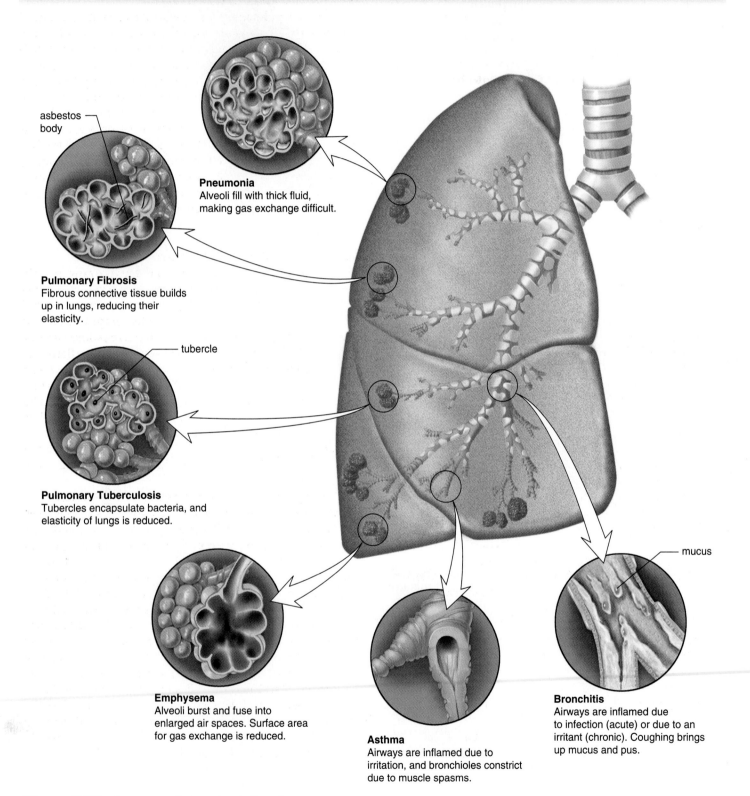

asbestos
body

Pneumonia
Alveoli fill with thick fluid,
making gas exchange difficult.

Pulmonary Fibrosis
Fibrous connective tissue builds
up in lungs, reducing their
elasticity.

tubercle

Pulmonary Tuberculosis
Tubercles encapsulate bacteria, and
elasticity of lungs is reduced.

mucus

Emphysema
Alveoli burst and fuse into
enlarged air spaces. Surface area
for gas exchange is reduced.

Asthma
Airways are inflamed due to
irritation, and bronchioles constrict
due to muscle spasms.

Bronchitis
Airways are inflamed due
to infection (acute) or due to an
irritant (chronic). Coughing brings
up mucus and pus.

Figure 8.12 **Lower respiratory tract disorders.**
Exposure to infectious pathogens and/or air pollutants, including cigarette and cigar smoke, causes the diseases and disorders shown here.

Lower Respiratory Tract Disorders

Lower respiratory tract disorders, which are illustrated in Figure 8.12, include infections, restrictive pulmonary disorders, obstructive pulmonary disorders, and lung cancer.

Lower Respiratory Infections

Acute bronchitis, pneumonia, and tuberculosis are infections of the lower respiratory tract. **Acute bronchitis** is an infection of the primary and secondary bronchi. Usually it is preceded by a viral URI that has led to a secondary bacterial infection. Most likely, a nonproductive cough has become a deep cough that expectorates mucus and perhaps pus.

Pneumonia is a viral or bacterial infection of the lungs in which bronchi and alveoli fill with thick fluid. Most often it is preceded by influenza. Rather than being a generalized lung infection, pneumonia may be localized in specific lobules of the lungs. Obviously the more lobules involved, the more serious the infection. Pneumonia can be caused by a bacterium that is usually held in check, but that has gained the upper hand due to stress and/or reduced immunity. AIDS patients are subject to a particularly rare form of pneumonia caused by the protozoan *Pneumocystis carinii.* Pneumonia of this type is almost never seen in individuals with a healthy immune system. High fever and chills with headache and chest pain are symptoms of pneumonia.

Pulmonary tuberculosis is caused by the tubercle bacillus, a type of bacterium. It is possible to tell if a person has ever been exposed to tuberculosis with a skin test in which a highly diluted extract of the bacillus is injected into the skin of the patient. A person who has never been in contact with tubercle bacillus shows no reaction, but one who has developed immunity to the organism shows an area of inflammation that peaks in about 48 hours. When tubercle bacilli invade the lung tissue, the cells build a protective capsule about the foreigners, isolating them from the rest of the body. This tiny capsule is called a tubercle. If the resistance of the body is high, the imprisoned organisms die, but if the resistance is low, the organisms eventually can be liberated. If a chest X ray detects active tubercles, the individual is put on appropriate drug therapy to ensure the localization of the disease and the eventual destruction of any live bacterial organisms.

Tuberculosis was a major killer in the United States before the middle of this century, after which antibiotic therapy brought it largely under control. In recent years, however, the incidence of tuberculosis is on the rise, particularly among AIDS patients, the homeless, and the rural poor. Worse, the new strains are resistant to the usual antibiotic therapy. Therefore, some physicians would like to again quarantine patients in sanitariums.

Restrictive Pulmonary Disorders

In restrictive pulmonary disorders, vital capacity is reduced not because air does not move freely into and out of the lungs but because the lungs have lost their elasticity. Inhaling particles such as silica (sand), coal dust, asbestos, and, now it seems, fiberglass can lead to **pulmonary fibrosis,** a condition in which fibrous connective tissue builds up in the lungs. The lungs cannot inflate properly and are always tending toward deflation. Breathing asbestos is also associated with the development of cancer. Since asbestos has been used so widely as a fireproofing and insulating agent, unwarranted exposure has occurred. It is projected that two million deaths could be caused by asbestos exposure—mostly in the workplace—between 1990 and 2020.

Obstructive Pulmonary Disorders

In obstructive pulmonary disorders, air does not flow freely in the airways and the time it takes to inhale or exhale maximally is greatly increased. Vital capacity, however, is normal. Several disorders, including chronic bronchitis, emphysema, and asthma, are referred to as chronic obstructive pulmonary disorders (COPD) because they tend to recur and have flare-ups.

In **chronic bronchitis,** the airways are inflamed and filled with mucus. A cough that brings up mucus is common. The bronchi have undergone degenerative changes including the loss of cilia and their normal cleansing action. Under these conditions an infection is more likely to occur. Smoking cigarettes and cigars is the most frequent cause of chronic bronchitis. Exposure to other pollutants can also cause chronic bronchitis.

Emphysema is a chronic and incurable disorder in which the alveoli are distended and their walls damaged so that the surface area available for gas exchange is reduced. Emphysema is often preceded by chronic bronchitis. Air trapped in the lungs leads to alveolar damage and a noticeable ballooning of the chest. The elastic recoil of the lungs is reduced, so not only are the airways narrowed but the driving force behind expiration is also reduced. The victim is breathless and may have a cough. Because the surface area for gas exchange is reduced, oxygen reaching the heart and the brain is reduced. Even so, the heart works furiously to force more blood through the lungs, and an increased workload on the heart can result. Lack of oxygen to the brain can make the person feel depressed, sluggish, and irritable. Exercise, drug therapy, and supplemental oxygen, along with giving up smoking, may relieve the symptoms and possibly slow the progression of emphysema.

Asthma is a disease of the bronchi and bronchioles that is marked by wheezing, breathlessness, and sometimes cough and expectoration of mucus. The airways are unusually sensitive to specific irritants, which can include a wide range of allergens such as pollen, animal dander, dust, cigarette smoke, and industrial fumes. Even cold air, however, can be an irritant. When exposed to the irritant, the smooth muscle in the bronchioles undergoes spasms. It now appears that chemical mediators given off by immune cells in the bronchioles result in the spasms. Most asthma patients have some degree of bronchial inflammation that reduces the diameter of the airways and contributes to the seriousness of an attack. Asthma is not curable but is treatable. There are inhalers that control the inflammation and hopefully prevent an attack, but there are also inhalers that stop the muscle spasms should an attack occur.

Lung Cancer

Lung cancer used to be more prevalent in men than in women, but recently it has surpassed breast cancer as a cause of death in women. This can be linked to an increase in the number of women who smoke today. Autopsies on smokers have revealed the progressive steps by which the most common form of lung cancer develops. The first event appears to be thickening and callusing of the cells lining the airways. (Callusing occurs whenever cells are exposed to irritants.) Then there is a loss of cilia so that it is impossible to prevent dust and dirt from settling in the lungs. Following this, cells with atypical nuclei appear in the callused lining. A tumor (Fig. 8.13*b*) consisting of disordered cells with atypical nuclei is considered to be cancer in situ (at one location). A final step occurs when some of these cells break loose and penetrate other tissues, a process called metastasis. Now the cancer has spread. The original tumor may grow until a bronchus is blocked, cutting off the supply of air to that lung. The entire lung then collapses, the secretions trapped in the lung spaces become infected, and pneumonia or a lung abscess (localized area of pus) results. The only treatment that offers a possibility of cure is to remove a lobe or the lung completely before metastasis has had time to occur. This operation is called **pneumonectomy.**

Current research indicates that *involuntary smoking,* simply breathing in air filled with cigarette smoke, can also cause lung cancer and other illnesses associated with smoking. The Health reading on the next page lists the various illnesses that are apt to occur when a person smokes. If a person stops both voluntary and involuntary smoking, and if the body tissues are not already cancerous, they may return to normal over time.

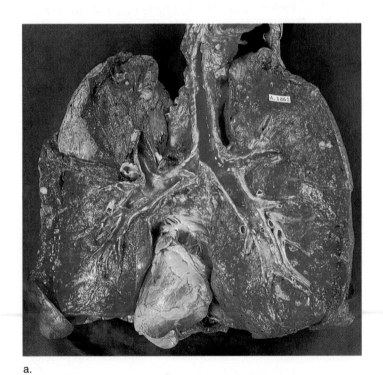

a.

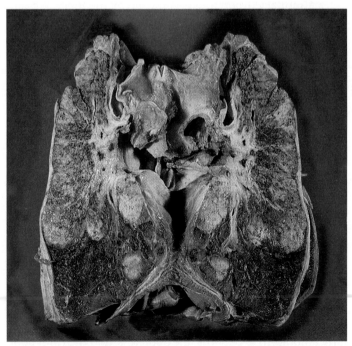

b.

Figure 8.13   Normal lung versus cancerous lung.
a. Normal lung with heart in place. Note the healthy red color. **b.** Lungs of a heavy smoker. Notice how black the lungs are except where cancerous tumors have formed.

The Most Often Asked Questions About Tobacco and Health

Is there a safe way to smoke?

No. All forms of tobacco can cause damage and smoking even a small amount is dangerous. Tobacco is perhaps the only legal product whose advertised and intended use—that is, smoking it—will hurt the body. . . .

Does smoking cause cancer?

Yes, and not only lung cancer. . . . Besides causing lung cancer, smoking a pipe, cigarettes, or cigars is also a major cause of cancers of the mouth, larynx (voice box), and esophagus. In addition, smoking increases the risk of cancer of the bladder, kidney, pancreas, stomach, and the uterine cervix.

What are the chances of being cured of lung cancer?

Very low; the five-year survival rate is only 13%. . . . Fortunately, lung cancer is a largely preventable disease. That is, by not smoking it can probably be prevented. . . .

Does smoking cause other lung diseases?

Yes. . . . It leads to chronic bronchitis—a disease where the airways produce excess mucus, which forces the smoker to cough frequently. Smoking is also the major cause of emphysema—a disease that slowly destroys a person's ability to breathe. . . .

Why do smokers have "smoker's cough"?

. . . Normally, cilia (tiny hairlike formations that line the airways) beat outwards and "sweep" harmful material out of the lungs. Smoke, however, decreases this sweeping action, so some of the poisons in the smoke remain in the lungs. . . .

If you smoke but don't inhale, is there any danger?

Yes. Wherever smoke touches living cells, it does harm. So, even if smokers of pipes, cigarettes, and cigars don't inhale, they are at an increased risk for lip, mouth, and tongue cancer. . . .

Does smoking affect the heart?

Yes. Smoking increases the risk of heart disease, which is America's number one killer. . . . Smoking, high blood pressure, high cholesterol, and lack of exercise are all risk factors for heart disease. Smoking alone doubles the risk of heart disease. . . .

Is there any risk for pregnant women and their babies?

Pregnant women who smoke endanger the health and lives of their unborn babies. When a pregnant woman smokes, she really is smoking for two because the nicotine, carbon monoxide, and other dangerous chemicals in smoke enter the mother's bloodstream and then pass into the baby's body. . . .

Does smoking cause any special health problems for women?

Yes. . . . Women who smoke and use the birth control pill have an increased risk of stroke and blood clots in the legs as well. . . . In addition, women who smoke increase their chances of getting cancer of the uterine cervix.

What are some of the short-term effects of smoking cigarettes?

Almost immediately, smoking can make it hard to breathe. Within a short time, it can also worsen asthma and allergies. Nicotine reaches the brain only seven seconds after a smoker takes a puff where it produces a morphinelike effect.

Are there any other risks to the smoker?

Yes. There are many other risks. As we already mentioned briefly, smoking causes stroke, which is the third leading cause of death in America. Smoking causes lung cancer, but if a person smokes and is exposed to radon or asbestos, the risk increases even more. Smokers are also more likely to have and die from stomach ulcers than nonsmokers. . . .

What are the dangers of passive smoking?

. . . Passive smoking causes lung cancer in healthy nonsmokers. Children whose parents smoke are more likely to suffer from pneumonia or bronchitis in the first two years of life than children who come from smoke-free households. Passive smokers have a 30% greater risk for developing lung cancer than nonsmokers who live in a smoke-free house.

Are chewing tobacco and snuff safe alternatives to cigarette smoking?

No, they are not. Many people who use chewing tobacco or snuff believe it can't harm them because there is no smoke. Wrong. Smokeless tobacco contains nicotine, the same addicting drug found in cigarettes and cigars. Snuff dippers also take in an average of over ten times more cancer-causing substances than cigarette smokers. . . . While not inhaled through the lungs, the juice from smokeless tobacco is absorbed through the lining of the mouth. There it can cause sores and white patches, which often lead to cancer of the mouth. . . . .

Human Systems Work Together

Integumentary System

Gas exchange in lungs provides oxygen to skin and rids body of carbon dioxide from skin.

Skin helps protect respiratory organs and helps regulate body temperature.

How the Respiratory System works with other body systems

Cardiovascular System

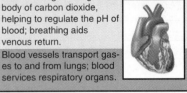

Gas exchange in lungs rids body of carbon dioxide, helping to regulate the pH of blood; breathing aids venous return.

Blood vessels transport gases to and from lungs; blood services respiratory organs.

Skeletal System

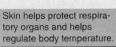

Gas exchange in lungs provides oxygen and rids body of carbon dioxide.

Rib cage protects lungs and assists breathing; bones provide attachment sites for muscles involved in breathing.

Lymphatic System/Immunity

Tonsils and adenoids occur along respiratory tract; breathing aids lymph flow; lungs carry out gas exchange.

Lymphatic vessels pick up excess tissue fluid; immune system protects against respiratory tract and lung infections.

Muscular System

Lungs provide oxygen for contracting muscles and rid the body of carbon dioxide from contracting muscles.

Muscle contraction assists breathing; physical exercise increases respiratory capacity.

Digestive System

Gas exchange in lungs provides oxygen to the digestive tract and excretes carbon dioxide from the digestive tract.

Breathing is possible through the mouth because digestive tract and respiratory tract share the pharynx.

Nervous System

Lungs provide oxygen for neurons and rid the body of carbon dioxide produced by neurons.

Respiratory centers in brain regulate breathing rate.

Urinary System

Lungs excrete carbon dioxide, provide oxygen, and convert angiotensin I to angiotensin II, leading to kidney regulation.

Kidneys compensate for water lost through respiratory tract; work with lungs to maintain blood pH.

Endocrine System

Gas exchange in lungs provides oxygen and rids body of carbon dioxide.

Epinephrine promotes ventilation by dilating bronchioles; growth factors control production of red blood cells that carry oxygen.

Reproductive System

Gas exchange increases during sexual activity.

Sexual activity increases breathing; pregnancy causes breathing rate and vital capacity to increase.

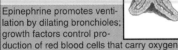

8.5 Homeostasis

The respiratory system contributes to homeostasis in two primary ways. First, the lungs perform gas exchange. Oxygen, a molecule needed for cellular respiration, enters the body, and carbon dioxide, a waste molecule given off by cellular respiration, exits the body at the lungs. Cellular respiration produces ATP, a molecule that allows the body to perform all sorts of work including muscle contraction and nerve conduction. It is estimated that the brain uses 15–20% of the oxygen consumed by the body. Not surprisingly, a lack of oxygen affects the brain, including judgment, first.

The cardiovascular system transports oxygen from the lungs to the tissues and carbon dioxide from the tissues to the lungs. The previous page tells how the act of breathing assists the return of blood to the heart and, therefore, the transport of carbon dioxide to the lungs. (This was discussed more fully in chapter 6).

Second, the respiratory system is involved in regulating the pH of the blood. In the tissues, this reaction in red blood cells produces the bicarbonate ion, which is carried in the plasma and in the lungs.

$$CO_2 + H_2O \underset{\text{lungs}}{\overset{\text{tissues}}{\rightleftharpoons}} H_2CO_3 \underset{\text{lungs}}{\overset{\text{tissues}}{\rightleftharpoons}} H^+ + HCO_3^-$$

In the lungs, the reverse reaction produces carbon dioxide, which exits the body. As the reverse occurs, the $[H^+]$ lowers and the pH rises. If, by chance, the hydrogen ion concentration $[H^+]$ rises above normal, the respiratory center is stimulated by way of chemoreceptors and the breathing rate increases.

The previous page tells how the respiratory system works with the other systems of the body, including the immune system. Now we know that the tonsils serve as a location where T cells are presented with antigens before they enter the body as a whole. This action helps the body prepare to respond to an antigen before it enters the bloodstream! The contribution of the respiratory system to homeostasis cannot be overemphasized.

Bioethical Issue

Since the introduction of the first antibiotics in the 1940s, there has been a dramatic decline in deaths due to respiratory illnesses like pneumonia and tuberculosis. Strep throat and ear infections have also been brought under control with antibiotics, which are chemicals that selectively kill bacteria and not host cells.

There are problems associated with antibiotic therapy, however. Aside from a possible allergic reaction, antibiotics not only kill off disease-causing bacteria, they also reduce the number of beneficial bacteria in the intestinal tract and other locations. These beneficial bacteria hold in check the growth of other microbes that now begin to flourish. Diarrhea can result, as can a vaginal yeast infection. The use of antibiotics can also prevent natural immunity from occurring, leading to the need for recurring antibiotic therapy. Especially alarming at this time is the occurrence of resistance. Resistance takes place when vulnerable bacteria are killed off by an antibiotic, and this allows resistant bacteria to become prevalent. The bacteria that cause ear, nose, and throat infections, and scarlet fever and pneumonia are becoming widely resistant because we have not been using antibiotics properly. Tuberculosis is on the rise, and the new strains are resistant to the usual combined antibiotic therapy. When a disease is caused by a resistant bacterium, it cannot be cured by the administration of any presently available antibiotic.

Although drug companies now recognize the problem and have begun to develop new antibiotics that hopefully will kill bacteria resistant to today's antibiotics, every citizen needs to be aware of our present crisis situation. Stuart Levy, a Tufts University School of Medicine microbiologist says that we should do what is ethical for society and ourselves. What is needed? Antibiotics kill bacteria, not viruses—therefore, we shouldn't take antibiotics unless we know for sure we have a bacterial infection. And we shouldn't take them prophylactically—that is, just in case we might need one. If antibiotics are taken in low dosages and intermittently, resistant strains are bound to take over. Animal and agricultural use should be pared down, and household disinfectants should no longer be spiked with antibacterial agents. Perhaps then, Levy says, vulnerable bacteria will begin to supplant the resistant ones in the population.

Questions

1. With regard to antibiotics, should each person think about the needs of society as well as themselves? Why or why not?
2. Should each person do what they can to help prevent the growing resistance of bacteria to disease? Why or why not?
3. Should you gracefully accept a physician's decision that an antibiotic will not help any illness you may have? Why or why not?

Summarizing the Concepts

8.1 Respiratory Tract
The respiratory tract consists of the nose (nasal cavities), the nasopharynx, the pharynx, the larynx (which contains the vocal cords), the trachea, the bronchi, and the bronchioles. The bronchi, along with the pulmonary arteries and veins, enter the lungs, which consist of the alveoli, air sacs surrounded by a capillary network.

8.2 Mechanism of Breathing
Inspiration begins when the respiratory center in the medulla oblongata sends excitatory nerve impulses to the diaphragm and the muscles of the rib cage. As they contract, the diaphragm lowers and the rib cage moves upward and outward; the lungs expand, creating a partial vacuum, which causes air to rush in. The respiratory center now stops sending impulses to the diaphragm and muscles of the rib cage. As the diaphragm relaxes, it resumes its dome shape, and as the rib cage retracts, air is pushed out of the lungs during expiration.

8.3 Gas Exchanges in the Body
External respiration occurs when CO_2 leaves blood via the alveoli and O_2 enters blood from the alveoli. Oxygen is transported to the tissues in combination with hemoglobin as oxyhemoglobin (HbO_2). Internal respiration occurs when O_2 leaves blood and CO_2 enters blood at the tissues. Carbon dioxide is mainly carried to the lungs within the plasma as the bicarbonate ion (HCO_3^-). Hemoglobin combines with hydrogen ions and becomes reduced (HHb).

8.4 Respiration and Health
A number of illnesses are associated with the respiratory tract. The disorders of the respiratory tract are divided into those that affect the upper respiratory tract and those that affect the lower respiratory tract. Infections of the nasal cavities, sinuses, throat, tonsils, and larynx are all well known. In addition, infections can spread from the nasopharynx to the ears.

The lower respiratory tract is also subject to infections such as acute bronchitis, pneumonia, and pulmonary tuberculosis. In restrictive pulmonary disorders, exemplified by pulmonary fibrosis, the lungs lose their elasticity. In obstructive pulmonary disorders, exemplified by chronic bronchitis, emphysema, and asthma, the bronchi (and bronchioles) do not effectively conduct air to and from the lungs. Smoking, which is associated with chronic bronchitis and emphysema, can eventually lead to lung cancer.

8.5 Homeostasis
The respiratory system works with the other systems of the body in the ways described in the box on page 182.

Studying the Concepts

1. List the parts of the respiratory tract. What are the special functions of the nasal cavity, the larynx, and the alveoli? 167–69

2. Name and explain the four parts of respiration. 170

3. What is the difference between tidal volume and vital capacity? Of the air we breathe, what part is not used for gas exchange? 170

4. What are the steps in inspiration and expiration? How is breathing controlled? 172–73

5. Discuss the events of external respiration, and include two pertinent equations in your discussion. 174

6. What two equations pertain to the exchange of gases during internal respiration? 174

7. State three factors that influence hemoglobin's O_2 binding capacity, and relate them to the environmental conditions in the lungs and tissues. 176

8. Name four respiratory tract infections other than cancer, and explain why breathing is difficult with these conditions. 177–80

9. What are emphysema and pulmonary fibrosis, and how do they affect a person's health? 179

10. List the steps by which lung cancer develops. 180

Testing Your Knowledge of the Concepts

In questions 1–4, match the organ to the description and function.
 a. larynx
 b. bronchi
 c. tonsils
 d. lungs

————— 1. airways to the lungs supported by C-shaped cartilaginous rings

————— 2. contains vocal cords and produces the voice

————— 3. lymphoid tissue that helps fight infections

————— 4. perform gas exchange in alveoli

In questions 5–7, indicate whether the statement is true (T) or false (F).

————— 5. In tracing the path of air, the bronchioles immediately follow the trachea.

————— 6. Reduced hemoglobin becomes oxyhemoglobin in the lungs.

————— 7. The most frequent cause of chronic bronchitis is a strep infection.

In questions 8 and 9, fill in the blanks.

8. Air enters the lungs after they have ——————.

9. The hydrogen ions (H^+) given off when carbonic acid (H_2CO_3) dissociates are carried by ——————.

10. Label this diagram of the human respiratory tract.

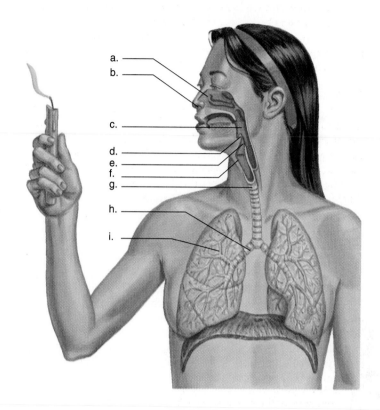

Applying Your Knowledge to the Concepts

These questions pertain to homeostasis.

1. What is the reason one is told to be a nose-breather rather than a mouth-breather?

2. Sometimes, young children hold their breath in an attempt to get their way. Is it possible for a child to hold his or her breath long enough to do bodily damage? Explain.

3. With regard to the effect of temperature and pH on hemoglobin's affinity for oxygen, explain the body's ability to provide extra oxygen to the most active parts of the body.

4. Although we often think of CO_2 as a waste product of respiration, it actually plays a critical role in respiration. Explain.

Understanding the Terms

acute bronchitis 179
alveolus 169
aortic bodies 172
asthma 180
auditory tube 177
bronchiole 169
bronchus 169
carbaminohemoglobin 174
carbonic anhydrase 174
carotid bodies 172
chronic bronchitis 179
diaphragm 172
emphysema 179
epiglottis 168
expiration 166
expiratory reserve
    volume 170
external respiration 174
glottis 168
hemoglobin 174
infant respiratory distress
    syndrome 169
inspiration 166
inspiratory reserve
    volume 170
internal respiration 174
laryngitis 177

larynx 168
lung cancer 180
lungs 169
nasal cavity 167
otitis media 177
oxyhemoglobin 174
pharynx 167
pleural membrane 172
pneumonectomy 180
pneumonia 179
pulmonary fibrosis 179
pulmonary tuberculosis 179
reduced hemoglobin 174
residual volume 170
respiratory center 172
rib cage 172
sinusitis 177
tidal volume 170
tonsillectomy 177
tonsillitis 177
tonsils 177
trachea 168
tracheostomy 169
ventilation 172
vital capacity 170
vocal cords 168

Match the terms to these definitions:

a. _____ Common passageway for both food intake and air movement, located between the mouth and the esophagus.

b. _____ Dome-shaped muscularized sheet separating the thoracic cavity from the abdominal cavity.

c. _____ Form in which most of the carbon dioxide is transported in the bloodstream.

d. _____ Stage during breathing when air is pushed out of the lungs.

e. _____ Terminal, microscopic, grapelike air sac found in lungs.

Applying Technology to the Concepts

Your study of the respiratory system is supported by these available technologies:

Essential Study Partner CD-ROM

Animals → Respiration

Visit the Mader web site for related ESP activities

Exploring the Internet

The Mader Home Page provides resources and tools as you study this chapter.

http://www.mhhe.com/biosci/genbio/mader

Dynamic Human 2.0 CD-ROM

Respiratory System

Virtual Physiology Laboratory CD-ROM

Pulmonary Function
Respiration and Exercise

HealthQuest CD-ROM

2 Fitness
6 Cancer → Gallery → Bronchitis

Life Science Animations 3D Video

37 Gas Exchange

Chapter 9

Urinary System and Excretion

Chapter Concepts

Figure 9.1 **Taking a drink of water.**
Drinking water lowers the osmolarity of the blood. But when a drink is dehydrating, as is seawater, the osmolarity of blood and then urine is increased.

On a mild Saturday morning, Alex heads off in his dad's boat. Once out in the peaceful deep, he cuts the engine, turns on the little portable TV, and falls asleep. Three hours later, he wakes up. Cursing under his breath, Alex turns the key in the boat's ignition. It stalls. He tries again. It stalls again.

Alex decides to stay put and wait for help. Meanwhile, he searches the boat for something to drink. Staring at the ocean, Alex crouches down in the boat and scoops up seawater. He gulps it. That's better. He gets more. Fortunately for Alex, his dad came looking for him, or else he could have died from dehydration.

Ironically, the culprit would probably have been the water he kept drinking. Seawater is far too salty for the human body to process. In trying to flush away the excess salt, Alex's kidneys would have to excrete more liquid than consumed. If you're ever in Alex's position—or even if you just go on a long bike ride or other sports event—make sure you bring water. Your kidneys can't function without it.

When we are able to drink water (Fig. 9.1) as we should, the kidneys maintain the water-salt balance and the acid-base balance of blood within normal limits. The kidneys also excrete the nitrogenous end products urea, uric acid, and creatinine. **Excretion** rids the body of

metabolic wastes, which come from the breakdown of substances that have been metabolized in the body's cells. The kidneys are not the only organs that excrete substances. The skin contains sweat glands that excrete water, salt, and urea. The lungs excrete carbon dioxide and the liver excretes bile. But none of these organs can make up for the work of the kidneys. Excretion should not be confused with secretion, which is the release of a substance useful to the body. Also, defecation is not a form of excretion. Defecation refers only to the elimination of feces from the digestive tract.

The correct functioning of the kidneys is essential to our good health, yet it is something that most of us take for granted until an illness strikes. Urinary tract infections and kidney stones cause much pain and may even result in permanent damage to the kidneys. Nationwide, there is a program of hemodialysis and transplantation for such patients.

9.1 Urinary System

The kidneys are a part of the urinary system because they produce urine, which is sent by way of the ureters to the bladder where it is stored before exiting the body through the urethra (Fig. 9.2).

Urinary Organs

The kidneys are found on either side of the vertebral column, just below the diaphragm. They lie in depressions against the deep muscles of the back beneath the peritoneum, the lining of the abdominal cavity, where they also receive some protection from the lower rib cage. But the kidneys can be damaged by blows on the back; kidney punches are not allowed in boxing.

The **kidneys** are bean-shaped, reddish brown organs, each about the size of a fist, which produce urine. They are covered by a tough capsule of fibrous connective tissue overlaid by adipose tissue. A depression (the hilum) on the concave side is where the **renal artery** enters and the **renal vein** and ureters exit.

The **ureters** are muscular tubes about 25 cm long that convey the urine from the kidneys toward the bladder by peristalsis. Urine enters the bladder by peristaltic contractions, in jets that occur at the rate of five per minute.

The **urinary bladder,** which can hold up to 600 ml of urine, is a hollow, muscular organ that gradually expands as urine enters. There are two sphincters in close proximity where the urethra exits the bladder.

The **urethra,** which extends from the urinary bladder to an external opening, differs in length in the female and the male. In the female, the urethra is only about 4 cm long. The short length of the female urethra makes bacterial invasion easier, and explains why females are more prone to urinary tract infections than males. In the male, the urethra averages 20 cm when the penis is flaccid (limp, nonerect). As the urethra leaves the male urinary bladder, it is encircled by the

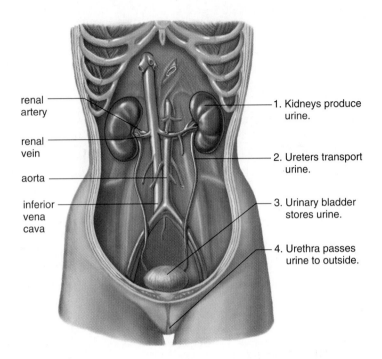

renal artery

renal vein

aorta

inferior vena cava

1. Kidneys produce urine.

2. Ureters transport urine.

3. Urinary bladder stores urine.

4. Urethra passes urine to outside.

Figure 9.2   The urinary system.
Urine is found only within the kidneys, the ureters, the urinary bladder, and the urethra.

prostate gland. In older men, enlargement of the prostate gland can restrict urination, a condition that usually can be corrected surgically.

There is no connection between the genital (reproductive) and urinary systems in females; there is a connection in males because the urethra also carries sperm during ejaculation. This double function does not alter the path of urine, and it is important to realize that urine is found only in those structures noted in Figure 9.3.

Urination and the Nervous System

When the urinary bladder fills with urine to about 250 ml, stretch receptors send sensory nerve impulses to the spinal cord. Subsequently, motor nerve impulses from the spinal cord cause the urinary bladder to contract and the sphincters to relax so that urination is possible. In older children and adults, the brain controls this reflex, delaying urination until a suitable time (Fig. 9.3).

Only the urinary system, consisting of the kidneys, the urinary bladder, the ureters, and the urethra, holds urine.

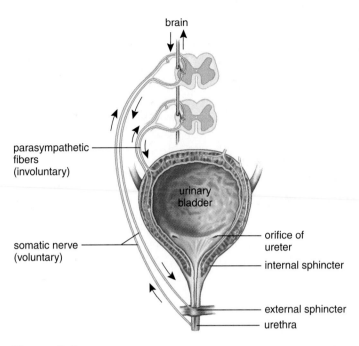

Figure 9.3   Urination.
As the bladder fills with urine, sensory impulses go to the spinal cord and then the brain. The brain can override the urge to urinate. When urination occurs, motor nerve impulses cause the bladder to contract and an internal sphincter to open. Nerve impulses also cause an external sphincter to open.

Functions of the Urinary System

The primary functions of the urinary system are carried out by the kidneys. The kidneys are organs of excretion. **Excretion** is the removal of metabolic wastes. The kidneys also maintain the water-salt balance and the acid-base balance of the body. In addition, they have a hormonal function.

Metabolic Wastes

The kidneys are the primary organs for the excretion of nitrogenous wastes including urea, creatinine, and uric acid.

Urea is the nitrogenous end product of amino acid metabolism. The breakdown of amino acids in the liver releases ammonia, which the liver combines with carbon dioxide to produce urea. Ammonia is very toxic to cells and urea is much less toxic. Urea is the primary nitrogenous end product of human beings.

Two other nitrogenous end products are excreted by the kidneys. **Creatinine** is the end product of creatine phosphate metabolism. Creatine phosphate is a high-energy phosphate reserve molecule in muscles. The breakdown of nucleotides produces **uric acid,** which is rather insoluble. If too much uric acid is present in blood, it precipitates out. Crystals of uric acid sometimes collect in the joints, producing a painful ailment called gout.

Water-Salt Balance

A principal function of the kidneys is to maintain the appropriate water-salt balance of the body. As we shall see, blood volume is intimately associated with the salt balance of the body. As you know, salts, such as NaCl, have the ability to cause osmosis, the diffusion of water—in this case into the blood. The more salts there are in the blood, the greater the blood volume and the greater the blood pressure. Therefore, the kidneys are also involved in regulating blood pressure.

The kidneys maintain the appropriate level of other ions such as potassium ions (K^+), bicarbonate ions (HCO_3^-), and calcium ions (Ca^{2+}) in the blood.

Acid-Base Balance

The kidneys also regulate the blood acid-base balance. In order for us to remain healthy, the blood pH should be just about 7.4. The kidneys monitor and control blood pH, mainly by excreting hydrogen ions (H^+) and reabsorbing the bicarbonate ion (HCO_3^-) as needed. Urine usually has a pH of 6 or lower.

Hormonal Function

The kidneys assist the endocrine system. They secrete the hormone **erythropoietin,** which stimulates red blood cell production, and they help activate the vitamin D precursor from the skin. Vitamin D promotes calcium (Ca^{2+}) reabsorption from the digestive tract.

The kidneys also secrete renin, a substance involved in the secretion of aldosterone from the adrenal cortex. Aldosterone causes the reabsorption of sodium ions (Na^+).

The kidneys are major organs of homeostasis because they excrete nitrogenous wastes. They also regulate the water-salt balance and the acid-base balance of the blood.

Urinary Tract Infections Require Attention

Although males can get a urinary tract infection, the condition is 50 times more common in women. The explanation lies in a comparison of male and female anatomy (Fig. 9A). The female urethral and anal openings are closer together, and the shorter urethra makes it easier for bacteria from the bowels to enter and start an infection. Although it is possible to have no outward signs of an infection, usually urination is painful, and patients often describe a burning sensation. The urge to pass urine is frequent, but it may be difficult to start the stream. Chills with fever, nausea, and vomiting may be present.

Urinary tract infections can be confined to the urethra, in which case urethritis is present. If the bladder is involved, it is called cystitis. Should the infection reach the kidneys, the person has pyelonephritis. *Escherichia coli* (*E. coli*), a normal bacterial resident of the large intestine, is usually the cause of infection. Since the infection is caused by a bacterium, it is curable by antibiotic therapy. The problem is, however, that reinfection is possible as soon as antibiotic therapy is finished.

It makes sense to try to prevent infection in the first place. These tips might help.

Men and women should drink lots of water. Try to drink from 2–2.5 liters of liquid a day. Try to avoid caffeinated drinks, which may be irritating. Cranberry juice is recommended because it contains a substance that stops bacteria from sticking to the bladder wall once an infection has set in. If an attack occurs, testing and antibiotic therapy may be in order. Keep in mind that sexually transmitted diseases such as gonorrhea, chlamydia, or herpes can cause urinary tract infections. All personal behaviors should be examined carefully, and suitable adjustments should be made to avoid urinary tract infections.

Most women have a urinary tract infection for the first time shortly after they become sexually active. Honeymoon cystitis was coined because of the common association of urinary tract infections with sexual intercourse. Washing the genitals before having sex and being careful not to introduce bacteria from the anus into the urethra is recommended. Also, urinating immediately before and after sex will help to flush out any bacteria that are present. A diaphragm may press on the urethra and prevent adequate emptying of the bladder, and estrogen, such as in birth-control pills, can increase the risk of cystitis. A sex partner may have an asymptomatic (no symptoms) urinary infection that causes a woman to become infected repeatedly.

Women should wipe from the front to the back after using the toilet. Perfumed toilet paper and any other perfumed products that come in contact with the genitals may be irritating. Wearing loose clothing and cotton underwear discourages the growth of bacteria, while tight clothing, such as jeans and panty hose, provides an environment for the growth of bacteria.

Personal hygiene is especially important too at the time of menstruation. Hands should be washed before and after changing napkins and/or tampons. Superabsorbent tampons are not best if they are changed infrequently, as this may encourage the growth of bacteria. Also, sexual intercourse may cause menstrual flow to enter the urethra.

In males, the prostate is a gland that surrounds the urethra just below the bladder (Fig. 9A). The prostate contributes secretions to semen whenever semen enters the urethra prior to ejaculation. An infection of the prostate, called prostatitis, is often accompanied by a urinary tract infection. Fever is present and the prostate is tender and inflamed. The patient may have to be hospitalized and treated with a broad spectrum antibiotic. Prostatitis, which in a young person is often preceded by a sexually transmitted disease, can lead to a chronic condition. Chronic prostatitis may be asymptomatic or, as is more typical, there is irritation upon voiding and/or difficulty in voiding. The latter can lead to the need for surgery to remove the obstruction to urine flow.

In both males and females, it is wise to take all necessary steps to avoid urinary tract infections.

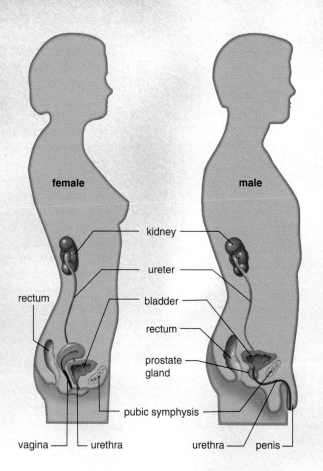

Figure 9A **Female versus male urinary tract.**
Females have a short urinary tract compared to that of males. This means that it is easier for bacteria to invade the urethra and helps explain why females are 50 times more likely than males to get a

9.2 Kidneys

When a kidney is sliced lengthwise, it is possible to see the many branches of the renal artery and vein that reach inside the kidney (Fig. 9.4*a*). If the blood vessels are removed, it is easier to identify three regions of a kidney. The **renal cortex** is an outer granulated layer that dips down in between a radially striated, or lined, inner layer called the renal medulla. The **renal medulla** consists of cone-shaped tissue masses called renal pyramids. The *renal pelvis* is a central space, or cavity, that is continuous with the ureter (Fig. 9.4*b*).

Microscopically, the kidney is composed of over one million **nephrons,** sometimes called renal or kidney tubules (Fig. 9.4*c*). The nephrons produce urine and are positioned so that the urine flows into a collecting duct. Several nephrons enter the same collecting duct; the collecting ducts enter the renal pelvis.

Macroscopically, a kidney has three regions: renal cortex, renal medulla, and a renal pelvis that is continuous with the ureter. Microscopically, a kidney contains over one million nephrons.

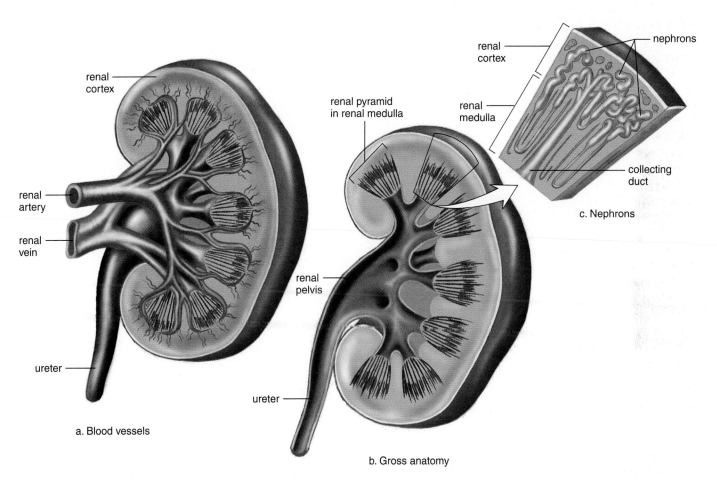

Figure 9.4   Gross anatomy of the kidney.
a. A longitudinal section of the kidney showing the blood supply. Note that the renal artery divides into smaller arteries, and these divide into arterioles. Venules join to form small veins, which join to form the renal vein. **b.** The same section without the blood supply. Now it is easier to distinguish the renal cortex, the renal medulla, and the renal pelvis, which connects with a ureter. The renal medulla consists of the renal pyramids. **c.** An enlargement showing the placement of nephrons.

Anatomy of a Nephron

Each nephron has its own blood supply, including two capillary regions (Fig 9.5). From the renal artery, an afferent arteriole leads to the **glomerulus,** a knot of capillaries inside the glomerular capsule. Blood leaving the glomerulus enters the efferent arteriole and then the **peritubular capillary network,** which surrounds the rest of the nephron. From there the blood goes into a venule that joins the renal vein.

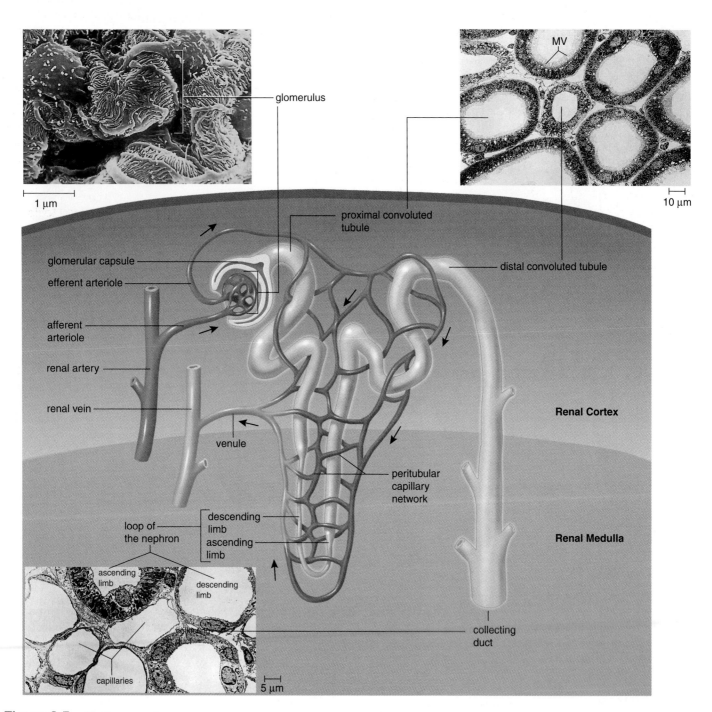

Figure 9.5   Nephron anatomy.

A nephron is made up of a glomerular capsule, the proximal convoluted tubule, the loop of the nephron, the distal convoluted tubule, and the collecting duct. The micrographs show these structures in cross section; MV = microvilli. You can trace the path of blood about the nephron by following the arrows.

Parts of a Nephron

Each nephron is made up of several parts (Fig. 9.5). The structure of each part suits its function.

First, the closed end of the nephron is pushed in on itself to form a cuplike structure called the **glomerular capsule** (Bowman's capsule). The outer layer of the glomerular capsule is composed of squamous epithelial cells; the inner layer is made up of **podocytes** that have long cytoplasmic processes. The podocytes cling to the capillary walls of the glomerulus and leave pores that allow easy passage of small molecules from the glomerulus to the inside of the glomerular capsule. This process, called *glomerular filtration*, produces a filtrate of blood.

Next, there is a **proximal** (meaning near the glomerular capsule) **convoluted tubule.** The cuboidal epithelial cells lining this part of the nephron have numerous microvilli, about 1 μm in length, that are tightly packed and form a brush border (Fig. 9.6). A brush border greatly increases the surface area for the *tubular reabsorption* of filtrate components. Each cell also has many mitochondria, which can supply energy for active transport of molecules from the lumen to the peritubular capillary network.

Simple squamous epithelium appears as the tube narrows and makes a U-turn called the **loop of the nephron** (loop of Henle). Each loop consists of a descending limb that allows water to leave and an ascending limb that extrudes salt (NaCl). Indeed, as we shall see, this activity facilitates the reabsorption of water by the nephron and collecting duct.

The cells of the **distal convoluted tubule** have numerous mitochondria, but they lack microvilli. This is consistent with the active role they play in moving molecules from the blood into the tubule, a process called *tubular secretion*. The distal convoluted tubules of several nephrons enter one collecting duct. A kidney contains many collecting ducts, which carry urine to the renal pelvis.

As shown in Figure 9.5, the glomerular capsule and the convoluted tubules always lie within the renal cortex. The loop of the nephron dips down into the renal medulla; a few nephrons have a very long loop of the nephron, which penetrates deep into the renal medulla. **Collecting ducts** are also located in the renal medulla, and they give the renal pyramids their lined appearance.

Each part of a nephron is anatomically suited to its specific function in urine formation.

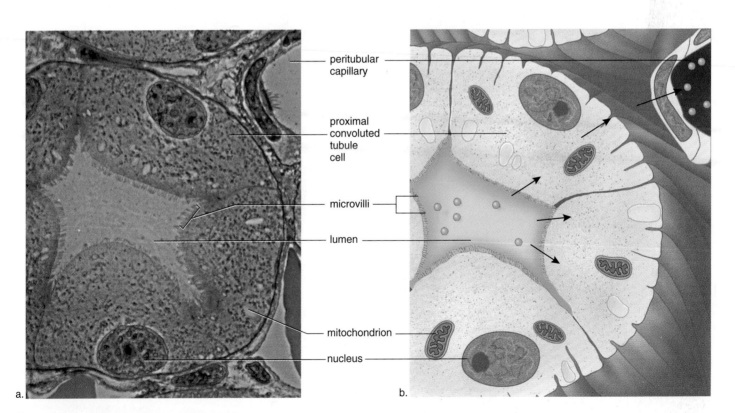

Figure 9.6 Proximal convoluted tubule.
a. This photomicrograph shows that the cells lining the proximal convoluted tubule have a brushlike border composed of microvilli, which greatly increases the surface area exposed to the lumen. The peritubular capillary network surrounds the cells. **b.** Diagrammatic representation of **(a)** shows that each cell has many mitochondria, which supply the energy needed for active transport, the process that moves molecules (green) from the lumen of the tubule to the capillary, as indicated by the arrows.

Visual Focus

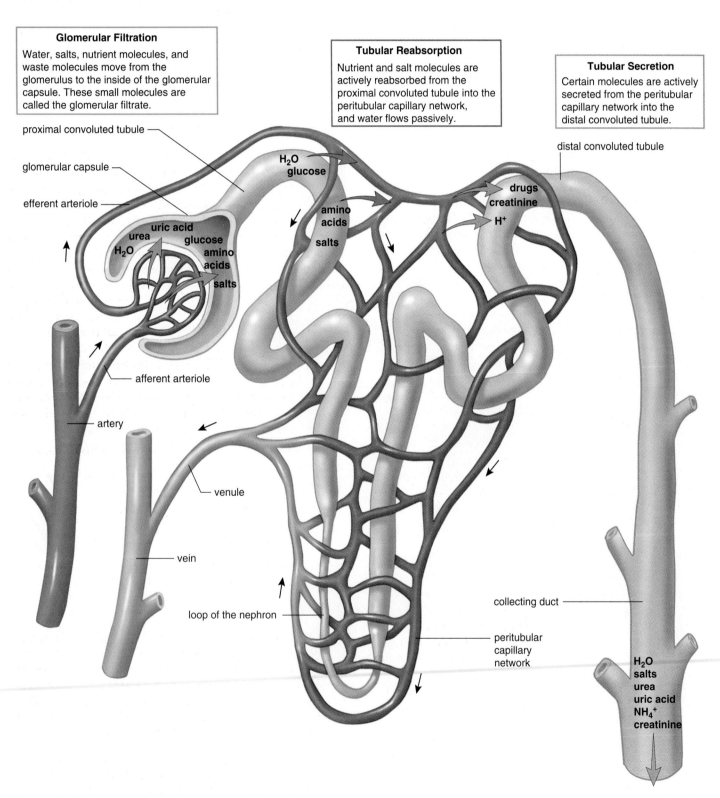

Glomerular Filtration

Water, salts, nutrient molecules, and waste molecules move from the glomerulus to the inside of the glomerular capsule. These small molecules are called the glomerular filtrate.

Tubular Reabsorption

Nutrient and salt molecules are actively reabsorbed from the proximal convoluted tubule into the peritubular capillary network, and water flows passively.

Tubular Secretion

Certain molecules are actively secreted from the peritubular capillary network into the distal convoluted tubule.

proximal convoluted tubule

glomerular capsule

efferent arteriole

H_2O
glucose

amino
acids

salts

uric acid
urea
glucose
H_2O amino
acids

salts

drugs
creatinine

H^+

distal convoluted tubule

afferent arteriole

artery

venule

vein

loop of the nephron

collecting duct

peritubular
capillary
network

H_2O
salts
urea
uric acid
NH_4^+
creatinine

Figure 9.7   Urine formation.

The three steps in urine formation are numbered. Reabsorption of water is not an individual step because it occurs along the length of the nephron and also at the loop of the nephron and collecting duct. Excretion is not a step because it is the end result.

9.3 Urine Formation

Figure 9.7 gives an overview of urine formation, which is divided into these steps: glomerular filtration, tubular reabsorption, and tubular secretion.

Glomerular Filtration

Glomerular filtration occurs when whole blood enters the afferent arteriole and the glomerulus. Due to glomerular blood pressure, which is usually about 60 mm Hg, water and small molecules move from the glomerulus to the inside of the glomerular capsule. This is a filtration process because large molecules and formed elements are unable to pass through the capillary wall. In effect, then, blood in the glomerulus has two portions: the filterable components and the nonfilterable components.

Filterable Blood Components	Nonfilterable Blood Components
Water	Formed elements (blood cells and platelets)
Nitrogenous wastes	Proteins
Nutrients	
Salts (ions)	

The **glomerular filtrate** contains small dissolved molecules in approximately the same concentration as plasma. Small molecules that escape being filtered and the nonfilterable components leave the glomerulus by way of the efferent arteriole.

As indicated in Table 9.1, 180 liters of water are filtered per day along with a considerable amount of small molecules, such as glucose and amino acids. If the composition of urine were the same as that of the glomerular filtrate, the body would continually lose water, salts, and nutrients. Death from dehydration, starvation, and low blood pressure would quickly follow. Therefore, we can conclude that the composition of the filtrate must be altered as this fluid passes through the remainder of the tubule.

Tubular Reabsorption

Tubular reabsorption occurs as molecules and ions are both passively and actively reabsorbed from the nephron into the blood of the peritubular capillary network. The osmolarity of the blood is maintained by the presence of plasma proteins and also by salt. When sodium ions (Na^+) are actively reabsorbed, chloride ions (Cl^-) follow passively. The reabsorption of salt (NaCl) increases the osmolarity of the blood compared to the filtrate, and therefore water moves passively from the tubule into the blood. About 67% of Na^+ is reabsorbed at the proximal convoluted tubule.

Nutrients such as glucose and amino acids also return to the blood at the proximal convoluted tubule. This is a selective process because only molecules recognized by carrier molecules are actively reabsorbed. Glucose is an example of a molecule that ordinarily is completely reabsorbed because there is a plentiful supply of carrier molecules for it. However, every substance has a maximum rate of transport, and

Table 9.1	Reabsorption from Nephron		
Substance	Amount Filtered (Per Day)	Amount Excreted (Per Day)	Reabsorption (%)
Water, L	180	1.8	99.0
Sodium, g	630	3.2	99.5
Glucose, g	180	0.0	100.0
Urea, g	54	30.0	44.0

L = liters, g = grams

From A. J. Vander, et al. *Human Physiology*, 4th ed. © 1985. The McGraw-Hill Publishing Companies, Inc. All Rights Reserved. Reprinted by permission.

after all its carriers are in use, any excess in the filtrate will appear in the urine. For example, as reabsorbed levels of glucose approach 180–200 mg/100 ml plasma, the rest will appear in the urine. In diabetes mellitus, excess glucose occurs in the blood, and then in the filtrate, and then in the urine, because the liver and muscles fail to store glucose as glycogen and the kidneys cannot reabsorb all of it. The presence of glucose in the filtrate increases its osmolarity compared to blood, and therefore less water is reabsorbed into the peritubular capillary network. The frequent urination and increased thirst experienced by patients with uncontrolled diabetes mellitus is due to the fact that water is remaining in the filtrate and is not being reabsorbed.

We have seen that the filtrate that enters the proximal convoluted tubule is divided into two portions: components that are reabsorbed from the tubule into blood, and components that are nonreabsorbed and continue to pass through the nephron to be further processed into urine.

Reabsorbed Filtrate Components	Nonreabsorbed Filtrate Components
Most water	Some water
Nutrients	Much nitrogenous waste
Required salts (ions)	Excess salts (ions)

The substances that are not reabsorbed become the tubular fluid, which enters the loop of the nephron.

Tubular Secretion

Tubular secretion is a second way by which substances are removed from blood and added to the tubular fluid. Hydrogen ions, creatinine, and drugs such as penicillin are some of the substances that are moved by active transport from blood into the distal convoluted tubule. In the end, urine contains substances that underwent glomerular filtration but were not reabsorbed, and substances that underwent tubular secretion.

Urine formation requires glomerular filtration (small molecules enter tubule), tubular reabsorption (many molecules are reabsorbed), and tubular secretion (substances are actively added to tubule).

9.4 Maintaining Water-Salt Balance

The kidneys maintain the water-salt balance of the blood within normal limits. In this way, they also maintain the blood volume and blood pressure. Most of the water and salt (NaCl) present in the filtrate is reabsorbed across the wall of the proximal convoluted tubule. Reabsorption also occurs along the remainder of the nephron.

Reabsorption of Water

The excretion of a hypertonic urine (one that is more concentrated than blood) is dependent upon the reabsorption of water from the loop of the nephron (loop of Henle) and the collecting duct.

A long loop of the nephron, which typically penetrates deep into the renal medulla, is made up of a *descending* (going down) *limb* and an *ascending* (going up) *limb*. Salt (NaCl) passively diffuses out of the lower portion of the ascending limb, but the upper, thick portion of the limb actively extrudes salt out into the tissue of the outer renal medulla (Fig. 9.8). Less and less salt is available for transport as fluid moves up the thick portion of the ascending limb. Because of these circumstances, the loop of the nephron establishes an *osmotic gradient* within the tissues of the renal medulla: the concentration of salt is greater in the direction of the inner medulla. (Note that water cannot leave the ascending limb because the limb is impermeable to water.)

Also, if you examine Figure 9.8 carefully, you can see that the innermost portion of the inner medulla has the highest concentration of solutes. This cannot be due to salt because active transport of salt does not start until the thick portion of the ascending limb. Urea is believed to leak from the lower portion of the collecting duct, and it is this molecule that contributes to the high solute concentration of the inner medulla.

Because of the osmotic gradient within the renal medulla, water leaves the descending limb of the loop of the nephron along its length. This is a countercurrent mechanism: as water diffuses out of the descending limb, the remaining solution within the limb encounters an even greater osmotic concentration of solute; therefore, water will continue to leave the descending limb from the top to the bottom.

Fluid entering a collecting duct comes from the distal convoluted tubule. This fluid is now isotonic to the cells of the cortex. This means that to this point, the net effect of reabsorption of water and salt is the production of a fluid that has the same tonicity as blood. However, the filtrate within the collecting duct also encounters the same osmotic gradient mentioned earlier (Fig. 9.8). Therefore, water diffuses out of the collecting duct into the renal medulla, and the urine within the collecting duct becomes hypertonic to blood plasma.

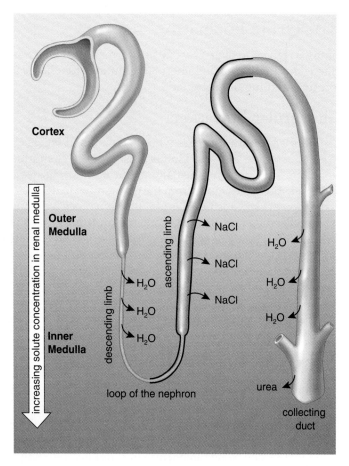

Figure 9.8   Reabsorption of water at the loop of the nephron and the collecting duct.
Salt (NaCl) diffuses and is actively transported out of the ascending limb of the loop of the nephron into the renal medulla; also, urea is believed to leak from the collecting duct and to enter the tissues of the renal medulla. This creates a hypertonic environment, which draws water out of the descending limb and the collecting duct. This water is returned to the cardiovascular system. (The thick line means the ascending limb is impermeable to water.)

Antidiuretic hormone (ADH) released by the posterior lobe of the pituitary plays a role in water reabsorption at the collecting duct. In order to understand the action of this hormone, consider its name. Diuresis means increased amount of urine, and antidiuresis means decreased amount of urine. When ADH is present, more water is reabsorbed (blood volume and pressure rise), and a decreased amount of urine results. In practical terms, if an individual does not drink much water on a certain day, the posterior lobe of the pituitary releases ADH, causing more water to be reabsorbed and less urine to form. On the other hand, if an individual drinks a large amount of water and does not perspire much, ADH is not released. Now more water is excreted, and more urine forms.

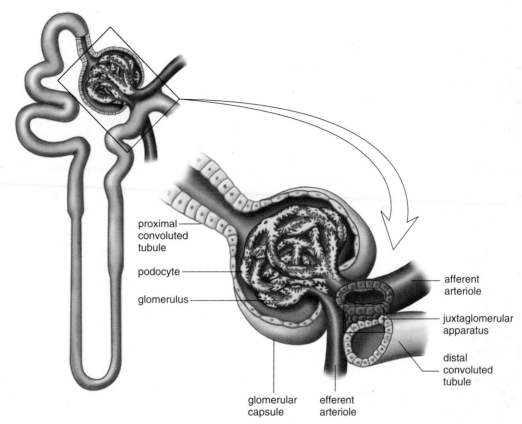

proximal convoluted tubule

podocyte

glomerulus

afferent arteriole

juxtaglomerular apparatus

distal convoluted tubule

glomerular capsule

efferent arteriole

Figure 9.9 Juxtaglomerular apparatus.
This drawing shows that the afferent arteriole and the distal convoluted tubule usually lie next to each other. The juxtaglomerular apparatus occurs where they touch.

Reabsorption of Salt

Usually, more than 99% of sodium (Na^+) filtered at the glomerulus is returned to the blood. Most sodium (67%) is reabsorbed at the proximal tubule, and a sizable amount (25%) is extruded by the ascending limb of the loop of the nephron. The rest is reabsorbed from the distal convoluted tubule and collecting duct.

Hormones regulate the reabsorption of sodium at the distal convoluted tubule. **Aldosterone** is a hormone secreted by the adrenal cortex, the outer portion of the adrenal glands, which lie atop the kidneys. Aldosterone promotes the excretion of potassium ions (K^+) and the reabsorption of sodium ions (Na^+). The release of aldosterone is set in motion by the kidneys themselves. The **juxtaglomerular apparatus** is a region of contact between the afferent arteriole and the distal convoluted tubule (Fig. 9.9). When blood volume, and therefore blood pressure, is not sufficient to promote glomerular filtration, the juxtaglomerular apparatus secretes renin. **Renin** is an enzyme that changes angiotensinogen (a large plasma protein produced by the liver) into angiotensin I. Later, angiotensin I is converted to angiotensin II, a powerful vasoconstrictor that also stimulates the adrenal cortex to release aldosterone. The reabsorption of sodium ions is followed by the reabsorption of water. Therefore, blood volume and blood pressure increase.

Atrial natriuretic hormone (ANH) is a hormone secreted by the atria of the heart when cardiac cells are stretched due to increased blood volume. ANH inhibits the secretion of renin by the juxtaglomerular apparatus and the secretion of aldosterone by the adrenal cortex. Its effect, therefore, is to promote the excretion of Na^+, that is, natriuresis. When Na^+ is excreted, so is water, and therefore blood volume and blood pressure decrease.

These examples show that the kidneys regulate the salt balance in blood by controlling the excretion and the reabsorption of various ions. Sodium (Na^+) is an important ion in plasma that must be regulated, but the kidneys also excrete or reabsorb other ions, such as potassium ions (K^+), bicarbonate ions (HCO_3^-), and magnesium ions (Mg^{2+}), as needed.

Diuretics

Diuretics are agents that increase the flow of urine. Drinking alcohol causes diuresis because it inhibits the secretion of ADH. The dehydration that follows is believed to contribute to the symptoms of a hangover. Caffeine is a diuretic because it increases the glomerular filtration rate and decreases the tubular reabsorption of Na^+. Diuretic drugs developed to counteract high blood pressure in patients inhibit active transport of Na^+ at the loop of the nephron or at the distal convoluted tubule. A decrease in water reabsorption and a decrease in blood volume follow.

Human Systems Work Together

Integumentary System

Kidneys compensate for water loss due to sweating; activate vitamin D precursor made by skin.

Skin helps regulate water loss; sweat glands carry on some excretion.

Skeletal System

Kidneys provide active vitamin D for Ca^{2+} absorption and help maintain blood level of Ca^{2+}, needed for bone growth and repair.

Bones provide support and protection.

Muscular System

Kidneys maintain blood levels of Na^+, K^+, and Ca^{2+}, which are needed for muscle innervation, and eliminate creatinine, a muscle waste.

Smooth muscular contraction assists voiding of urine; skeletal muscles support and help protect urinary organs.

Nervous System

Kidneys maintain blood levels of Na^+, K^+, and Ca^{2+}, which are needed for nerve conduction.

Brain controls nerves, which innervate muscles that permit urination.

Endocrine System

Kidneys keep blood values within normal limits so that transport of hormones continues.

ADH and aldosterone, and atrial natriuretic hormone regulate reabsorption of Na^+ by kidneys.

How the Urinary System works with other body systems

Cardiovascular System

Kidneys filter blood and excrete wastes; maintain blood volume, pressure, and pH; produce renin and erythropoietin.

Blood vessels deliver waste to be excreted; blood pressure aids kidney function; heart produces atrial natriuretic hormone.

Lymphatic System/Immunity

Kidneys control volume of body fluids, including lymph.

Lymphatic system picks up excess tissue fluid, helping to maintain blood pressure for kidneys to function; immune system protects against infections.

Respiratory System

Kidneys compensate for water lost through respiratory tract; work with lungs to maintain blood pH.

Lungs excrete carbon dioxide, provide oxygen, and convert angiotensin I to angiotensin II, leading to kidney regulation.

Digestive System

Kidneys convert vitamin D to active form needed for Ca^{2+} absorption; compensate for any water loss by digestive tract.

Liver synthesizes urea; digestive tract excretes bile pigments from liver and provides nutrients.

Reproductive System

Semen is discharged through the urethra in males; kidneys excrete wastes and maintain electrolyte levels for mother and child.

Penis in males contains the urethra and performs urination; prostate enlargement hinders urination.

9.5 Maintaining Acid-Base Balance

The bicarbonate (HCO_3^-) buffer system and breathing work together to maintain the pH of the blood. Central to the mechanism is this reaction, which you have seen before:

$$H^+ + HCO_3^- \rightleftharpoons H_2CO_3 \rightleftharpoons H_2O + CO_2$$

The excretion of carbon dioxide (CO_2) by the lungs helps keep the pH within normal limits, because when carbon dioxide is exhaled this reaction is pushed to the right and hydrogen ions are tied up in water. Indeed, when blood pH decreases, chemoreceptors in the carotid bodies (located in the carotid arteries) and in aortic bodies (located in the aorta) stimulate the respiratory center, and the rate and depth of breathing increases. On the other hand, when blood pH begins to rise, the respiratory center is depressed and the bicarbonate ion increases in the blood.

As powerful as this system is, only the kidneys can rid the body of a wide range of acidic and basic substances. The kidneys are slower acting than the buffer/breathing mechanism, but they have a more powerful effect on pH. For the sake of simplicity, we can think of the kidneys as reabsorbing bicarbonate ions and excreting hydrogen ions as needed to maintain the normal pH of the blood. If the blood is acidic, hydrogen ions are excreted and bicarbonate ions are reabsorbed. If the blood is basic, hydrogen ions are not excreted and bicarbonate ions are not reabsorbed. Since the urine is usually acidic, it shows that usually an excess of hydrogen ions are excreted. Ammonia (NH_3) provides a means for buffering these hydrogen ions in urine: ($NH_3 + H^+ \rightarrow NH_4^+$). Ammonia (whose presence is quite obvious in the diaper pail or kitty litter box) is produced in tubule cells by the deamination of amino acids. Phosphate provides another means of buffering hydrogen ions in urine.

The acid-base balance of the blood is adjusted by the reabsorption of the bicarbonate ions (HCO_3^-) and the secretion of hydrogen ions (H^+) as appropriate.

9.6 Homeostasis

The kidneys are primary organs of homeostasis because they maintain the water-salt balance and the acid-base balance of the blood. The previous page tells how the urinary

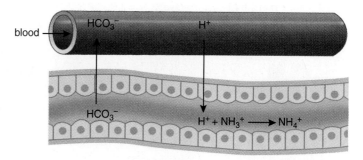

Figure 9.10 Acid-base balance.
In the kidneys, bicarbonate ions are reabsorbed and the hydrogen ions are excreted as needed to maintain the pH of the blood. Excess hydrogen ions are buffered, for example, by ammonia (NH_3), which is produced in tubule cells by the deamination of amino acids.

system works with the other systems of the body, including the musculoskeletal system and the nervous system.

If blood does not have the usual osmolarity and blood pressure, exchange across capillary walls cannot take place nor is glomerular filtration possible in the kidneys themselves. The production of renin by the kidneys and subsequently the renin-angiotensin-aldosterone sequence helps ensure that the sodium concentration of the blood stays normal. The kidneys maintain the normal concentration of sodium ions (Na^+) and also the concentration of potassium ions (K^+) and calcium ions (Ca^{2+}). All these ions are necessary to the contraction of the heart and other muscles in the body, and are also needed for nerve conduction.

The bicarbonate ion is a part of a buffer system that helps maintain the pH of the blood. The kidneys can reabsorb the bicarbonate ion and excrete hydrogen ions (H^+) as needed to maintain the pH (Fig. 9.10). The kidneys have ultimate control over the pH of the blood, and the importance of this function cannot be overemphasized. The enzymes of cells cannot continue to function if the internal environment does not have near normal pH.

The kidneys also excrete nitrogenous wastes, which are end products of metabolism. The excretion of nitrogenous wastes may not be as critical as maintaining the water-salt and the acid-base balance, but it still is a necessity. If the urea produced day by day were not removed, in about four years we would be nothing but urea.

Finally, the kidneys assist the endocrine system. They produce erythropoietin, which stimulates red bone marrow to produce red blood cells, and they help convert vitamin D to an active hormone that stimulates the reabsorption of calcium from the digestive tract.

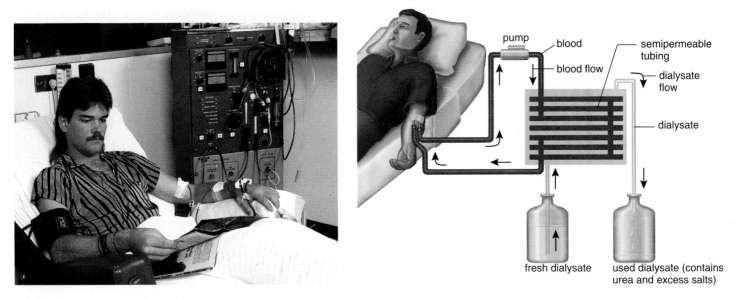

Figure 9.11   An artificial kidney machine.
As the patient's blood is pumped through dialysis tubing, it is exposed to a dialysate (dialysis solution). Wastes exit from blood into the solution because of a preestablished concentration gradient. In this way, blood is not only cleansed, but its water-salt and acid-base balance can also be adjusted.

9.7 Problems with Kidney Function

Many types of illnesses cause progressive renal disease and renal failure. Urinary tract infections include urethritis, infection of the urethra; cystitis, infection of the bladder; and pyelonephritis, infection of the kidneys. The Health reading on page 190 suggests ways to prevent urinary tract infections.

Glomerular damage sometimes leads to blockage of the glomeruli so that glomerular filtration does not occur or allows large substances to pass through. This is detected when a urinalysis is done. If the glomeruli are too permeable, albumin, white blood cells, or even red blood cells appear in the urine. A trace amount of protein in the urine is not a matter of concern, however.

When glomerular damage is so extensive that more than two-thirds of the nephrons are inoperative, waste substances accumulate in blood. This condition is called uremia because urea is one of the substances that accumulates. Imbalance in the ionic composition of body fluids is even more serious because it can lead to loss of consciousness and to heart failure.

Patients in renal failure sometimes undergo a kidney transplant operation. Rejection is less likely if the donor is a close relative.

Hemodialysis

While waiting for a kidney transplant, it is probably necessary for the patient to undergo **hemodialysis,** utilizing either an artificial kidney machine or continuous ambulatory peritoneal (abdominal) dialysis (CAPD). Dialysis is defined as the diffusion of dissolved molecules through a semipermeable membrane (an artificial membrane with pore sizes that allow only small molecules to pass through). In an artificial kidney machine (Fig. 9.11), the patient's blood is passed through a membranous tube, which is in contact with a dialysis solution, or dialysate. Substances more concentrated in blood diffuse into the dialysate, and substances more concentrated in the dialysate diffuse into blood. Accordingly, the artificial kidney can be utilized either to extract substances from blood, including waste products or toxic chemicals and drugs, or to add substances to blood—for example, bicarbonate ions (HCO_3^-) if blood is acidic. In the course of a three- to six-hour hemodialysis, from 50–250 grams of urea can be removed from a patient, which greatly exceeds the amount excreted by normal kidneys. Therefore, a patient needs to undergo treatment only about twice a week.

In the case of CAPD, a fresh amount of dialysate is introduced directly into the abdominal cavity from a bag attached to a permanently implanted plastic tube. Waste and salt molecules pass from the blood vessels in the abdominal wall into the dialysate before the fluid is collected four or eight hours later. The individual can go about his or her normal activities during CAPD, unlike during hemodialysis.

Bioethical Issue

As a society we are accustomed to thinking that as we grow older, diseases like urinary disorders will begin to occur. Almost everyone is aware that most males are subject to enlargement of the prostate as they age, and that cancer of the prostate is not uncommon among elderly men. However, like many illnesses associated with aging, medical science now knows how to treat or even cure prostate problems. Because of these successes, medical science has lengthened our life span. A child born in the United States in 1900 lived to, say, 47 years. If that same child were born today, it would probably live to at least 76. Even more exciting is the probability that scientists will improve the life span. People could live beyond 100 years and have the same vigor and vitality they had when they were young.

Most people are appreciative of living longer, especially if they can expect to be free of the illnesses and inconveniences associated with aging. But have we examined how we feel about longevity as a society? Whereas we are accustomed to considering that if the birthrate increases so does the size of a population, what about the death rate? If the birthrate stays constant and the death rate decreases, obviously population size also increases. Most experts agree that population growth depletes resources and increases environmental degradation. An older population can also put a strain on the economy if they are unable to meet their financial, including medical, needs without governmental assistance.

What is the ethical solution to this problem? Should we just allow the population to increase due to older people living longer? Should we decrease the birthrate? Should we reduce governmental assistance to older people so they realize that they must be able to take care of themselves? Should we call a halt to increasing the life span through advancements in medical science?

Questions

1. Do you feel that older people make a significant contribution to or a drain on society? Explain.
2. Would you be willing to have fewer children in order to hold the population in check if more people live longer? Why or why not?
3. Should the elderly expect governmental or family assistance as they age? Why or why not?

Summarizing the Concepts

9.1 Urinary System
The kidneys produce urine, which is conducted by the ureters to the bladder where it is stored before being released by way of the urethra.

The kidneys maintain the normal water-salt balance and the acid-base balance of the blood. They also excrete nitrogenous wastes, including urea, uric acid, and creatinine.

9.2 Kidneys
Macroscopically, the kidneys are divided into the renal cortex, renal medulla, and renal pelvis. Microscopically, they contain the nephrons.

Each nephron has its own blood supply; the afferent arteriole approaches the glomerular capsule and divides to become the glomerulus, a capillary tuft. The spaces between the podocytes of the glomerular capsule allow small molecules to enter the capsule from the glomerulus. The efferent arteriole leaves the capsule and immediately branches into the peritubular capillary network .

Each region of the nephron is anatomically suited to its task in urine formation. The spaces between the podocytes of the glomerular capsule allow small molecules to enter the capsule from the glomerulus, a capillary knot. The cuboidal epithelial cells of the proximal convoluted tubule have many mitochondria and microvilli to carry out active transport (following passive transport) from the tubule to blood. In contrast, the cuboidal epithelial cells of the distal convoluted tubule have numerous mitochondria but lack microvilli. They carry out active transport from the blood to the tubule.

9.3 Urine Formation
Urine is composed primarily of nitrogenous waste products and salts in water.

The steps in urine formation are glomerular filtration, tubular reabsorption, and tubular secretion, as explained in Figure 9.7.

9.4 Maintaining Water-Salt Balance
The kidneys regulate the water-salt balance of the body. Water is reabsorbed from all parts of the tubule, and the loop of the nephron establishes an osmotic gradient that draws water from the descending loop of the nephron and also the collecting duct. The permeability of the collecting duct is under the control of the hormone ADH.

The reabsorption of salt increases blood volume and pressure because more water is also reabsorbed. Two other hormones, aldosterone and ANH, control the kidneys' reabsorption of sodium (Na^+).

9.5 Maintaining Acid-Base Balance
The kidneys keep blood pH within normal limits. They reabsorb HCO_3^- and excrete H^+ as needed to maintain the pH at about 7.4.

9.6 Homeostasis
The urinary system works with the other systems of the body to maintain homeostasis in the ways described in the box on page 198.

9.7 Problems with Kidney Function
Various types of problems, including repeated urinary infections, can lead to kidney failure, which necessitates receiving a kidney from a donor or undergoing hemodialysis by utilizing a kidney machine or CAPD.

Studying the Concepts

1. State the path of urine and the function of each organ mentioned. 188
2. Explain how urination is controlled. 189
3. List and explain four functions of the urinary system. 189
4. Describe the macroscopic anatomy of a kidney. 191
5. Trace the path of blood about a nephron. 192
6. Name the parts of a nephron, and tell how the structure of the convoluted tubules suits their function. 193
7. State and describe the three steps of urine formation. 194–95

8. Where in particular is water and salt reabsorbed along the length of the nephron? Describe the contribution of the loop of the nephron. 196–97
9. Name and describe the action of antidiuretic hormone (ADH), the renin-aldosterone connection, and the atrial natriuretic hormone (ANH). 196–97
10. How do the kidneys maintain the pH of the blood within normal limits? 199
11. Explain how the artificial kidney machine works. 200

Testing Your Knowledge of the Concepts

In questions 1-4, match the structure to the functions below.
a. glomerulus
b. proximal convoluted tubule
c. distal convoluted tubule
d. collecting duct
    _____ 1. Regulation of water reabsorption
    _____ 2. Reabsorption of vital molecules
    _____ 3. Formation of filtrate
    _____ 4. Secretion of toxins and poisons

In questions 5–7, indicate whether the statement is true (T) or False (F).
_____ 5. The ureters conduct urine from the bladder to outside the body.
_____ 6. Amino acids are filtered, reabsorbed, and not in urine.
_____ 7. When antidiuretic hormone (ADH) is present, water is maximally reabsorbed.

In question 8 and 9, fill in the blanks.
8. The ascending limb of the loop of the nephron extrudes _____ into the renal medulla, making the renal medulla hypertonic to fluid in the collecting duct.
9. _____ is a molecule that is found in the filtrate, is reabsorbed, and is concentrated in urine.

10. Label this diagram of a nephron.

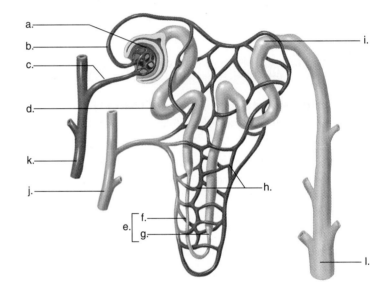

Applying Your Knowledge to the Concepts

These questions pertain to homeostasis.
1. Explain why pregnant women have to urinate more often than they did before becoming pregnant.
2. A common symptom experienced by heavy drinkers is a dry mouth and a feeling of thirst when they awaken the next morning. Is this just a psychological phenomenon? Explain.
3. During World War II, recruits who tested positive for diabetes were held overnight and retested the next day. The repeat test

often showed no diabetes. How would it be possible to fake the condition of diabetes?

4. Materials can be extracted from a person's blood or can be added to the blood during dialysis. The functioning kidney can also extract or add substances to the blood. What is the major physiological difference between the artificial kidney and the human kidney?

Understanding the Terms

aldosterone 197
antidiuretic hormone
  (ADH) 196
atrial natriuretic
  hormone (ANH) 197
collecting duct 193
creatinine 189
distal convoluted
  tubule 193
diuretics 197
erythropoietin 189
excretion 189
glomerular capsule 193
glomerular filtrate 195
glomerular filtration 195
glomerulus 192

hemodialysis 200
juxtaglomerular apparatus 197
kidney 188
loop of the nephron 193
nephron 191
peritubular capillary
  network 192
podocyte 193
proximal convoluted
  tubule 193
renal artery 188
renal cortex 191
renal medulla 191
renal vein 188
renin 197
tubular reabsorption 195

tubular secretion 195
urea 189
ureter 188

urethra 188
uric acid 189
urinary bladder 188

Match the terms to these definitions:

a. _____ Portion of the nephron lying between the proximal convoluted tubule and the distal convoluted tubule that functions in water reabsorption.

b. _____ Movement of molecules from the contents of the nephron into blood at the proximal convoluted tubule.

c. _____ Hormone secreted by the adrenal cortex that regulates the sodium and potassium balance of the blood.

d. _____ Outer portion of the kidney that appears granular.

e. _____ Network capillary that surrounds a nephron and functions in reabsorption during urine formation.

Applying Technology to the Concepts

Your study of the urinary system and excretion is supported by these available technologies:

Essential Study Partner CD-ROM

Animals → Osmoregulation

Visit the Mader web site for related ESP activities.

Exploring the Internet

The Mader Home Page provides resources and tools as you study this chapter.

http://www.mhhe.com/biosci/genbio/mader

Dynamic Human 2.0 CD-ROM

Urinary System

Life Science Animations 3D Video

38 Kidney Function

Further Readings for Part 2

Abraham, S. N. September/October 1997. Discovering the benign traits of the mast cell. *Science & Medicine* 4(5):46. Recent investigations suggest mast cells have roles in immune surveillance and control of the immune response.

Arakawa, T. and Langridge, W. R. H. May 1998. Plants are not just passive creatures! *Nature Medicine* 4(5):550. Plants are being used as bioreactors to produce foreign proteins useful for human immunity.

Beck, G., and Habicht, G. S. November 1996. Immunity and the invertebrates. *Scientific American* 275(5):60. Nearly all aspects of the human immune system appear to have a cellular or chemical parallel among the invertebrates.

Becker, R. C. July/August 1996. Antiplatelet therapy. *Science & Medicine* 3(4):12. This article discusses the control of platelet aggregation.

Benjamini, E., and Leskowitz, S. 1996. *Immunology: A short course.* 3d ed. New York: John Wiley & Sons. Presents the essential principles of immunology.

Bikle, D. D. March/April 1995. A bright future for the sunshine hormone. *Science & Medicine* 2(2):58. Vitamin D receptors exist in many types of cells, suggesting therapeutic uses for the hormonally active metabolite of vitamin D.

Blaser, M. J. February 1996. The bacteria behind ulcers. *Scientific American* 274(2):104. Acid-loving microbes are linked to stomach ulcers and stomach cancer.

Brown, J. L., and Pollitt, E. February 1996. Malnutrition, poverty and intellectual development. *Scientific American* 274(2):38. Article discusses the complex role of essential nutrients in a child's mental development.

Further Readings for Part 2 (continued)

Dickman, S. July 1997. Mysteries of the heart. *Discover* 18(7):117. Article discusses why coronary arteries may still become blocked after treatment for atherosclerosis.

Gibbs, W. W. August 1996. Gaining on fat. *Scientific American* 275(2):88. Some weight problems are genetic or physiological in origin. New treatments might help.

Glausiusz, J. September 1998. Infected hearts. *Discover* 19(9):30. Infectious bacteria may play a role in heart disease; antibiotics could prevent the need for heart surgery.

Glausiusz, J. October 1997. The good bugs on our tongues. *Discover* 18(10):32. Without the friendly bacteria that live on our tongues, we would be vulnerable to bacteria such as Salmonella.

Guyton, A. C., and Hall, J. E. 1996. *Textbook of medical physiology.* Philadelphia: W. B. Saunders Co. Presents physiological principles for those in the medical fields.

Haen, P. J. 1995. *Principles of hematology.* Dubuque, Iowa: Wm. C. Brown Publishers. An introductory text for students planning a career in the medical sciences.

Hanson, L. A. November/December 1997. Breast feeding stimulates the infant immune system. *Science & Medicine* 4(6):12. Long-lasting protection against some infectious diseases has been reported in breast-fed infants.

Hotez, P. J., and Pritchard, D. I. June 1995. Hookworm infection. *Scientific American* 272(6):68. Discusses how the biology of parasites offers clues to possible vaccines and also for new treatments for heart disease and immune disorders.

Klatsky, A. L. March/April 1995. Cardiovascular effects of alcohol. *Science & Medicine* 2(2):28. The effects of moderate and heavy alcohol use on the cardiovascular system is discussed.

Kooyman, G. L., and Ponganis, P. J. November/December 1997. The challenges of diving to depth. *American Scientist* 85(6):530. Marine animals have ways to control their oxygen supply, and do not experience problems associated with pressure at depth.

Little, R. C., and Little, W. C. 1989. *Physiology of the heart and circulation.* 4th ed. Chicago: Year Book Medical Publishers, Inc. A good reference that gives an in-depth look at cardiovascular physiology.

Mader, S. S. 1997. *Understanding anatomy and physiology.* 3d ed. Dubuque, Iowa: Wm. C. Brown Publishers. A text that emphasizes the basics for beginning allied health students.

Mader, S. S. 1998. *Human biology.* 5th ed. Dubuque, Iowa: WCB/McGraw-Hill, Inc. A student-friendly text that covers the principles of biology with emphasis on human anatomy and physiology.

Nature Medicine Vaccine Supplement, May 1998, Vol. 4 No. 5. Entire issue is devoted to the topic of vaccines, including history, recent developments and research in malaria, cancer, and HIV vaccines.

Newman, J. December 1995. How breast milk protects newborns. *Scientific American* 273(6):76. Human milk contains special antibodies that boost the newborn's immune system.

Nucci, M. L., and Abuchowski, A. February 1998. The search for blood substitutes. *Scientific American* 278(2):72. Artificial blood substitutes are being developed from synthetic chemicals, and some are based on hemoglobin.

Roitt, I., et al. 1998. *Immunology.* 5th ed. London, Mosby International, Ltd.: For the advanced student, this text features a clear description of the scientific principles involved in immunology, combined with clinical examples.

Superko, H. R. September/October 1997. The atherogenic lipid profile. *Science & Medicine* 4(5):36. Gel electrophoresis is used to relate the risk of heart disease to cholesterol lowering therapy.

Sussman, N. L., and Kelly, J. H. May/June 1995. The artificial liver. *Science & Medicine* 2(3):68. An artificial liver assists temporarily while the natural liver regenerates, restoring normal function.

Valtin, H. 1995. *Renal function.* 3d ed. Boston: Little, Brown and Co. A good reference resource that discusses renal mechanisms for preserving fluid and solute balance.

Wardlaw, G., et al. 1994. *Contemporary nutrition.* 2d ed. St. Louis: Mosby-Year Book, Inc. This text gives a clear understanding of nutritional information found on product labels.

Weindruch, R. January 1996. Caloric restriction and aging. *Scientific American* 274(1):64. Consuming fewer calories may increase longevity.

West, J. B. 1994. *Respiratory physiology—The essentials.* 5th ed. Baltimore: Williams & Wilkins. A good reference resource that discusses all aspects of respiratory physiology including breathing, and external and internal respiration.

White, R. J. September 1998. Weightlessness and the human body. *Scientific American* 279(3):58. Space medicine is providing new ideas about treatment of anemia and osteoporosis.

Yock, P., et al. September/October 1995. Intravascular ultrasound. *Science & Medicine* 2(3):68. Ultrasound images of coronary arteries helps diagnose atherosclerosis.

The skeletal and muscular systems give the body its shape and allow it to move. The skeletal system protects and supports other organs—the skull protects the brain, and the rib cage protects the lungs and heart; the pelvis supports the organs of the abdominal cavity. This system also contributes to homeostasis because it stores minerals and produces the blood cells.

Contraction of the skeletal muscles allows the body to move but also accounts for facial expressions and our ability to speak. Homeostasis would be impossible without movement of the rib cage up and down and contraction of the heart to keep the blood moving. Smooth muscle contraction allows the other internal organs to function; it helps move food along the digestive tract and urine along the urinary tract. Indeed, life as we know it is dependent on the skeletal and muscular systems.

Chapter 10

Skeletal System

Chapter Concepts

10.1 Tissues of the Skeletal System
- Various connective tissues are necessary to the anatomy of bones. 206

10.2 Bone Growth and Repair
- Bone is a living tissue; therefore, it develops and undergoes repair. 208
- The fetal skeleton is cartilaginous, and then it is replaced by bone. 208
- The adult bones undergo remodeling—they are constantly being broken down and rebuilt. 209
- The mending of a fracture requires certain identifiable steps. 209

10.3 Bones of the Skeleton
- The bones of the skeleton are divided into those of the axial skeleton and those of the appendicular skeleton. 211

10.4 Articulations
- Joints are classified according to their anatomy, and only one type is freely movable. 218

10.5 Homeostasis
- The skeletal system works with the other systems of the body to maintain homeostasis. 221

Ellen B. knew there was something different about the way she stood or walked. She leaned to one side, and her knees didn't line up properly because she had scoliosis, a sideways curvature of the spine that throws the body out of line. So, at 17, while most kids dwelled on high-school graduation, Ellen checked into the hospital. During surgery, doctors carefully hooked steel rods to vertebrae at the top and bottom of Ellen's spine, fusing the bones together with fragments taken from her hip. The fusions healed in a straightened position.

With her spine corrected, Ellen found herself two inches taller, virtually free of back pain, and physically similar to the other students in her class. In addition to being able to proceed with graduation plans, she now could go shopping for clothes without the concern that they wouldn't "hang" correctly. After the operation, Ellen knew that the skeleton ordinarily permits flexible body movement without any pain. It serves as an attachment for muscles, whose contraction makes the bones move so that we can walk, play tennis, type papers, and do all manner of activities. The bones also support and protect. The large heavy bones of the legs support the entire body against the pull of gravity. The skull protects our brain, and the rib cage protects the heart and lungs. Bones have other functions too: they are the site of blood cell formation, and they store mineral salts.

This chapter reviews the structure of bones and the manner in which they perform these numerous functions.

10.1   Tissues of the Skeletal System

The tissues of the skeletal system are largely composed of connective tissues. Connective tissue contains cells separated by a matrix that contains fibers.

Bone

Bones are strong because their matrix contains mineral salts, notably calcium phosphate. **Compact bone** is highly organized and composed of tubular units called osteons. In cross section of an osteon, bone cells called osteocytes lie in lacunae, which are tiny chambers arranged in concentric circles around a central canal (Fig. 10.1). The mineralized matrix, which also contains protein fibers, fills the spaces between the lacunae. Tiny canals called canaliculi run through the matrix, connecting the lacunae with each other and with the central canal. Central canals contain blood vessels, lymphatic vessels, and nerves. Canaliculi bring nutrients from the blood vessel in the central canal to the cells in the lacunae.

Compared to compact bone, **spongy bone** has an unorganized appearance (Fig. 10.1). It contains numerous thin plates called *trabeculae* separated by unequal spaces. Although this makes spongy bone lighter than compact bone,

spongy bone is still designed for strength. Just as braces are used for support in buildings, the trabeculae follow lines of stress. The spaces of spongy bone are often filled with **red bone marrow,** a specialized tissue that produces all types of blood cells that have homeostatic functions. The osteocytes of spongy bone are irregularly placed within the trabeculae, and canaliculi bring them nutrients from the red bone marrow.

Cartilage

Cartilage is not as strong as bone, but it is more flexible because the matrix is gel-like and contains many collagenous and elastic fibers. The cells, called chondrocytes, lie within lacunae that are irregularly grouped. Cartilage has no blood vessels, and therefore injured cartilage is slow to heal.

The three types of cartilage differ according to the type and arrangement of fibers in the matrix. *Hyaline cartilage* is firm and somewhat flexible. The matrix appears uniform and glassy, but actually it contains a generous supply of collagenous fibers. Hyaline cartilage is found at the ends of long bones and in the nose, at the ends of the ribs, and in the larynx and trachea.

Fibrocartilage is stronger than hyaline cartilage because the matrix contains wide rows of thick collagenous fibers. Fibrocartilage is able to withstand both tension and pressure, and this type of cartilage is found where support is of prime importance—in the disks located between the vertebrae and also in the cartilage of the knee.

Elastic cartilage is more flexible than hyaline cartilage because the matrix contains mostly elastin fibers. This type of cartilage is found in the ear flaps and epiglottis.

Fibrous Connective Tissue

Fibrous connective tissue contains rows of cells called fibroblasts separated by bundles of collagenous fibers. This tissue makes up the **ligaments** that connect bone to bone and the **tendons** that connect muscles to a bone at **joints,** which are also called articulations.

Structure of a Long Bone

Figure 10.1 shows how the tissues we have been discussing are arranged in a long bone. The expanded region at the end of a long bone is called an epiphysis (pl., epiphyses). The epiphyses are composed largely of spongy bone that contains red bone marrow where blood cells are made. The epiphyses are coated with a thin layer of hyaline cartilage, which is called **articular cartilage** because it occurs at a joint.

The shaft, or main portion of the bone, is called the diaphysis. The diaphysis has a large medullary that is filled with fatty **yellow bone marrow** in adults. A thin, vascular

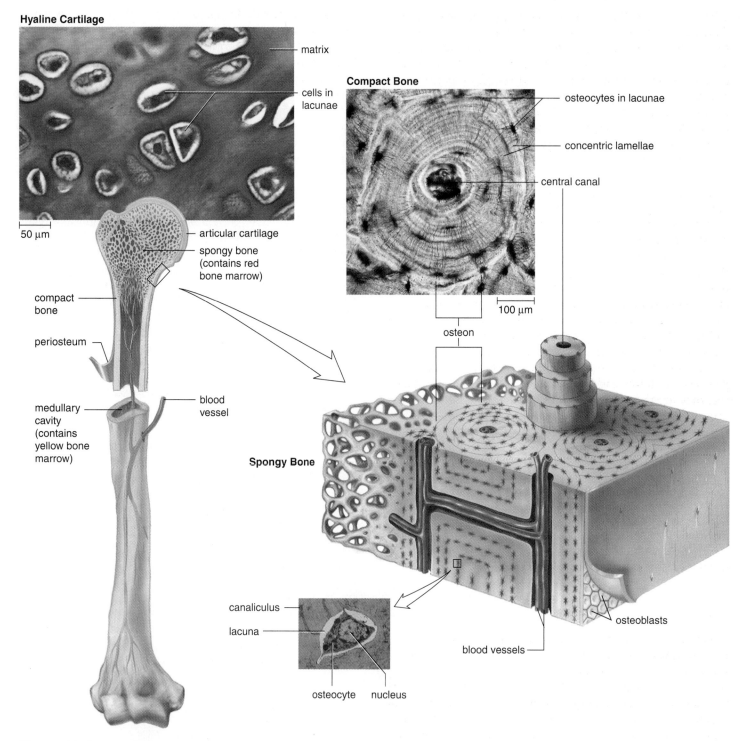

Hyaline Cartilage

matrix

cells in
lacunae

50 µm

Compact Bone

osteocytes in lacunae

concentric lamellae

central canal

100 µm

osteon

articular cartilage

spongy bone
(contains red
bone marrow)

compact
bone

periosteum

medullary
cavity
(contains
yellow bone
marrow)

blood
vessel

Spongy Bone

canaliculus

lacuna

osteocyte   nucleus

osteoblasts

blood vessels

Figure 10.1   **Anatomy of a long bone.**

A long bone is encased by the periosteum except where it is covered by hyaline (articular) cartilage (see micrograph). Spongy bone located in each epiphysis may contain red bone marrow. The central shaft contains yellow bone marrow and is bordered by compact bone, which is shown in the enlargement and micrograph.

membrane called the *endosteum* lines the **medullary cavity,** whose walls are composed of compact bone.

Except for the articular cartilage on its ends, a long bone is completely covered by a layer of fibrous connective tissue called the **periosteum.** This covering contains blood vessels, lymphatic vessels, and nerves. Note in Figure 10.1 how a blood vessel penetrates the periosteum and enters the bone where it gives off branches within the central canals. The periosteum is continuous with ligaments and tendons that are connected to this bone.

10.2 Bone Growth and Repair

Bones are composed of living tissues, as exemplified by their ability to grow and undergo repair. Several different types of cells are involved in bone growth and repair.

Osteogenitor cells are unspecialized cells present in the inner portion of the periosteum, in the endosteum, and in the central canal of compact bone.

Osteoblasts are bone-forming cells derived from osteogenitor cells. They are responsible for secreting the matrix, characteristic of bone.

Osteocytes are mature bone cells derived from osteoblasts. Once the osteoblasts are surrounded by matrix, they become the osteocytes found in bone.

Osteoclasts are thought to be derived from monocytes, a type of white blood cell present in red bone marrow. Osteoclasts perform bone resorption; that is, they break down bone and deposit calcium and phosphate in the blood. The work of osteoclasts is important to the growth and repair of bone.

Bone Development and Growth

The bones of the skeleton form during embryonic development in two distinctive ways. The term **ossification** refers to the formation of bone.

Some bones develop between sheets of fibrous connective tissue during a process called intramembranous ossification. The bones of the skull are examples of intramembranous bones. Osteogenitor cells derived from connective tissue cells become osteoblasts that begin to lay down matrix in

various directions, forming the trabeculae of spongy bone. Other osteoblasts associated with the periosteum formed from the fibrous membrane lay down compact bone over the surface of the spongy bone. When the osteoblasts are surrounded by matrix, they become osteocytes.

Most of the bones of the human skeleton first appear as hyaline cartilage during prenatal development. Since these cartilaginous structures are shaped like the future bones, they provide "models" of these bones. The replacement of cartilaginous models by bone is called endochondral ossification (Fig. 10.2).

During endochondral ossification of a long bone, the cartilage begins to break down in the center of the diaphysis, which is now covered by a periosteum. Osteoblasts invade the region and begin to lay down spongy bone in what is called a primary ossification center. Other osteoblasts lay down compact bone beneath the periosteum. As the compact bone thickens, the spongy bone of the diaphysis is broken down by osteoclasts, and the cavity created becomes the medullary cavity.

The epiphyses (ends) of developing bone continue to grow, but soon secondary ossification centers appear in these regions. Here spongy bone forms and does not break down. Also, a band of cartilage called the **epiphyseal (growth) plate** remains between the primary ossification center and each secondary center. The limbs keep increasing in length as long as the epiphyseal plates are still present. The rate of growth is controlled by hormones, such as growth hormones and the sex hormones. Eventually the growth plates become ossified, and the bone stops growing.

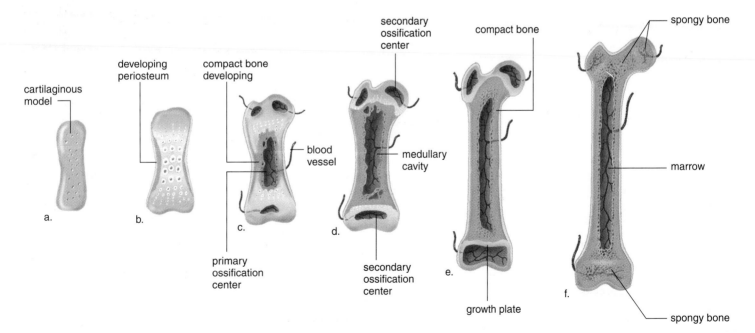

Figure 10.2   Endochondral bone formation.
a. A cartilaginous model develops during fetal development. **b.** A periosteum develops. **c.** A primary ossification center contains spongy bone surrounded by compact bone. **d.** Medullary cavity forms in diaphysis, and secondary ossification centers develop in epiphyses. **e.** Growth is still possible as long as cartilage remains at epiphyseal (growth) plate. **f.** When the bone is fully formed, the epiphyseal plate becomes an epiphyseal line.

Remodeling of Bones

In the adult, bone is continually being broken down and built up again. Osteoclasts derived from monocytes in red bone marrow break down bone, remove worn cells, and deposit calcium in the blood. After a period of about three weeks, the osteoclasts disappear, and the bone is repaired by the work of osteoblasts. As they form new bone, osteoblasts take calcium from the blood. Eventually some of these cells get caught in the matrix they secrete and are converted to osteocytes, the cells found within the lacunae of osteons.

Thus, through this process of *remodeling,* old bone tissue is replaced by new bone tissue. Because of continual remodeling, the thickness of bones can change. Physical use and hormone balance affect the thickness of bones. Strange as it may seem, adults apparently require more calcium in the diet (about 1,000 to 1,500 mg daily) than do children in order to promote the work of osteoblasts. Otherwise, *osteoporosis,* a condition in which weak and thin bones easily fracture, may develop. Osteoporosis is discussed in the Health reading on page 210.

Bone Repair

Repair of a bone is required after it breaks or fractures. In some ways, bone repair parallels the development of a bone (Fig. 10.3) except that the first step mentioned here indicates that injury has occurred and fibrocartilage instead of hyaline cartilage precedes the production of compact bone.

1. *Hematoma forms* six to eight hours after fracture. Blood escapes from ruptured blood vessels and forms a hematoma (mass of clotted blood) in the space between the broken bones. The area is inflamed and swollen.
2. *Fibrocartilaginous callus* lasts about three weeks. Tissue repair begins, and fibrocartilage now fills the space between the ends of the broken bone.
3. *Bony callus* lasts about three to four months. Osteoblasts produce trabeculae of spongy bone and convert the fibrocartilage callus to a bony callus that joins the broken bones together.
4. *Remodeling.* Osteoblasts build new compact bone at the periphery, and osteoclasts reabsorb the spongy bone, creating a new medullary cavity.

The naming of fractures tells you what kind of break occurred. A fracture is complete if the bone is broken clear through and incomplete if the bone is not separated into two parts. A fracture is simple if it does not pierce the skin and compound if it does pierce the skin. Impacted means that the broken ends are wedged into each other, and a spiral fracture occurs when there is a ragged break due to twisting of a bone.

Bone is living tissue. It develops, grows, remodels, and repairs itself. In all these processes, osteoclasts break down bone, and osteoblasts build bone.

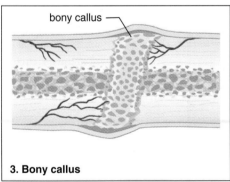

periosteum — hematoma
compact bone — medullary cavity
1. Hematoma

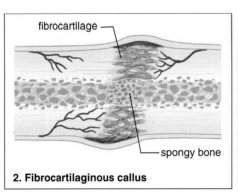

fibrocartilage —
— spongy bone
2. Fibrocartilaginous callus

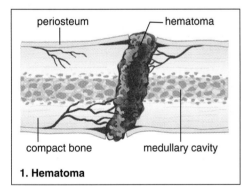

bony callus —
3. Bony callus

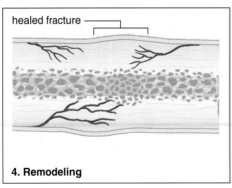

healed fracture —
4. Remodeling

a.

b.

Figure 10.3   Bone fracture and repair.
a. Steps in the repair of a fracture. **b.** A plaster of Paris cast helps stabilize the bones while repair takes place.

You Can Avoid Osteoporosis

Osteoporosis is a condition in which the bones are weakened due to a decrease in the bone mass that makes up the skeleton. Throughout life, bones are continuously remodeled. While a child is growing, the rate of bone formation is greater than the rate of bone breakdown. The skeletal mass continues to increase until ages 20 to 30. After that, there is an equal rate of formation and breakdown of bone mass until ages 40 to 50. Then, reabsorption begins to exceed formation, and the total bone mass slowly decreases.

Over time, men are apt to lose 25% and women lose 35% of their bone mass. But we have to consider that men tend to have denser bones than women anyway, and their testosterone (male sex hormone) level generally does not begin to decline significantly until after age 65. In contrast, the estrogen (female sex hormone) level in women begins to decline at about age 45. Since sex hormones play an important role in maintaining bone strength, this difference means that women are more likely than men to suffer a higher incidence of fractures, involving especially the hip, vertebrae, long bones, and pelvis. Although osteoporosis may at times be the result of various disease processes, it is essentially a disease of aging.

There are measures that everyone can take to avoid osteoporosis when they get older. Adequate dietary calcium throughout life is an important protection against osteoporosis. The U.S. National Institutes of Health recommend a calcium intake of 1,200–1,500 mg per day during puberty. Males and females require 1,000 mg per day until the age 65 and 1,500 mg per day after age 65. In postmenopausal women not receiving estrogen replacement therapy, 1,500 mg per day is desirable.

A small daily amount of vitamin D is also necessary in order to use calcium correctly. Exposure to sunlight is required to allow skin to synthesize a precursor to vitamin D. If you reside on or north of a "line" drawn from Boston to Milwaukee, to Minneapolis, to Boise, chances are you're not getting enough vitamin D during the winter months. Therefore, you should avail yourself of vitamin D present in fortified foods such as low-fat milk and cereal.

Very inactive people, such as those confined to bed, lose bone mass 25 times faster than people who are moderately active. On the other hand, a regular, moderate, weight-bearing exercise like walking or jogging is another good way to maintain bone strength (Fig. 10A).

Postmenopausal women with any of these risk factors should have an evaluation of their bone density:

- white or Asian race
- thin body type
- family history of osteoporosis
- early menopause (before age 45)
- smoking
- a diet low in calcium, or excessive alcohol consumption and caffeine intake
- sedentary life-style

Presently bone density is measured by a method called dual energy X-ray absorptiometry (DEXA). This test measures bone density based on the absorption of photons generated by an X-ray tube. Soon there may be a blood and urine test to detect the biochemical markers of bone loss. Then it may be possible for physicians to screen all older women and at-risk men for osteoporosis.

If the bones are thin, it is worthwhile to take other measures to gain bone density because even a slight increase can significantly reduce fracture risk. Usually, experts believe that estrogen therapy in women is the treatment of choice, but there are other options. Evidence shows that estrogen replacement therapy will delay accelerated bone loss in postmenopausal women. Hormone replacement is most effective when begun at the start of menopause and continued over the long term. Some benefit, however, may still be obtained when treatment is begun later. A combination of hormone replacement and exercise apparently yields the best results. But there are other options, also. Calcitonin is available as a nasal spray. Raloxifene is a drug with a selective estrogen-like effect on bone that has now been approved for use in postmenopausal women, and alendonate is a drug that has increased bone density in women with osteoporosis.

Figure 10A **Preventing osteoporosis.**
A dietary intake of calcium and vitamin D can help prevent osteoporosis. Exercise is also important. When playing golf, you should carry your own clubs and walk instead of using a golf cart.

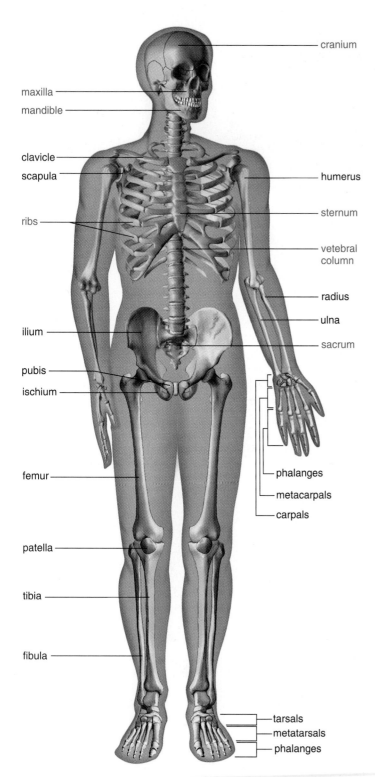

cranium

maxilla

mandible

clavicle

scapula

ribs

ilium

pubis

ischium

femur

patella

tibia

fibula

humerus

sternum

vetebral column

radius

ulna

sacrum

phalanges

metacarpals

carpals

tarsals

metatarsals

phalanges

Figure 10.4 **The skeleton.**
The skeleton of a human adult contains bones that belong to the axial skeleton (red labels) and those that belong to the appendicular skeleton (black labels).

10.3 Bones of the Skeleton

Let's discuss the functions of the skeleton in relation to particular bones.

The skeleton supports the body. The bones of the legs (the femur in particular and also the tibia) support the entire body when we are standing, and the coxal bones of the pelvic girdle support the abdominal cavity.

The skeleton protects soft body parts. The bones of the skull protect the brain; and the rib cage, composed of the ribs, thoracic vertebrae, and sternum, protects the heart and lungs.

The skeleton produces blood cells. All bones in the fetus have spongy bone with red bone marrow that produces blood cells. In the adult, the flat bones of the skull, ribs, sternum, clavicles, and also the vertebrae and pelvis produce blood cells. Fat is stored in yellow bone marrow.

The skeleton stores minerals and fat. All bones have a matrix that contains calcium phosphate. When bones are remodeled, osteoclasts break down bone and return calcium ions and phosphorus ions to the bloodstream.

The skeleton, along with the muscles, permits flexible body movement. While articulations (joints) occur between all the bones, we can associate body movement in particular with the bones of the legs (especially the femur and tibia) and the feet (tarsals, metatarsals, and phalanges) because we use them when walking.

Classification of the Bones

The bones are classified according to their shape. Long bones, exemplified by the humerus and femur, are longer than they are wide. Short bones, such as the carpals and tarsals, are cube shaped—their lengths and widths are about equal. Flat bones, like those of the skull, are platelike with broad surfaces. Round bones, exemplified by the patella, are circular in shape. Irregular bones, such as the vertebrae and facial bones, have varied shapes that permit connections with other bones.

The 206 bones of the skeleton are also classified according to whether they occur in the axial skeleton or the appendicular skeleton. The axial skeleton is in the midline of the body, and the appendicular skeleton is the limbs along with their girdles (Fig. 10.4).

The bones of the skeleton are not smooth; they have articulating depressions and protuberances at various joints. And they have projections, often called processes, where the muscles attach. Also, there are openings for nerves and/or blood vessels.

The skeleton is divided into the axial and appendicular skeleton. Each has different types of bones with protuberances at joints and processes where the muscles attach.

The Axial Skeleton

The **axial skeleton** lies in the midline of the body and consists of the skull, hyoid bone, vertebral column, and rib cage.

The Skull

The **skull** is formed by the cranium (braincase) and the facial bones. It should be noted, however, that some cranial bones contribute to the face.

The **cranium** protects the brain and is composed of eight flat bones fitted tightly together in adults. In newborns, certain bones are not completely formed and instead are joined by membranous regions called **fontanels.** The fontanels usually close by the age of 16 months by the process of intramembranous ossification.

Some of the bones of the cranium contain the **sinuses,** air spaces lined by mucous membrane, which reduce the weight of the skull and give a resonant sound to the voice. Two sinuses called the mastoid sinuses drain into the middle ear. **Mastoiditis,** a condition that can lead to deafness, is an inflammation of these sinuses.

The major bones of the cranium have the same names as the lobes of the brain: frontal, parietal, occipital, and temporal. On the top of the cranium (Fig. 10.5*a*), the **frontal bone** forms the forehead, the parietal bones extend to the sides, and the **occipital bone** curves to form the base of the skull. Here there is a large opening, the **foramen magnum** (Fig. 10.5*b*), through which the spinal cord passes and becomes the brain stem. Below the much larger parietal bones, each temporal bone has an opening (external auditory canal) that leads to the middle ear.

The **sphenoid bone,** which is shaped like a bat with wings outstretched, extends across the floor of the cranium from one side to the other. The sphenoid is considered to be the keystone bone of the cranium because all the other bones articulate with it. The sphenoid completes the sides of the skull and also contributes to forming the *orbits* (eye sockets). The **ethmoid bone,** which lies in front of the sphenoid, also

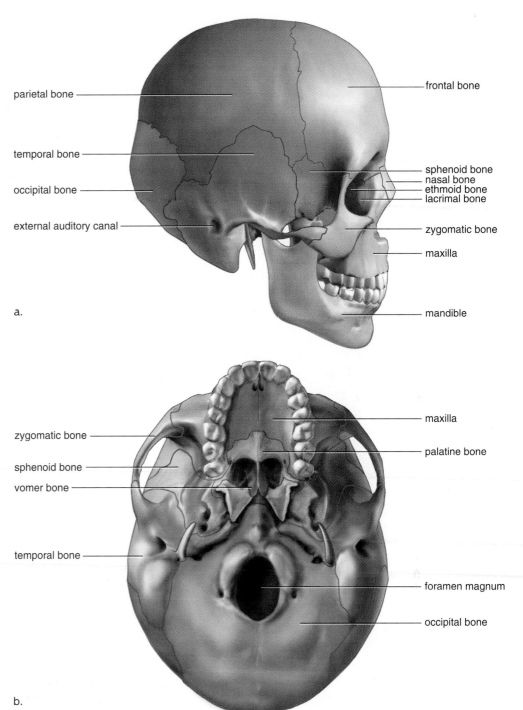

Figure 10.5   Bones of the skull.
a. Lateral view. **b.** Inferior view.

helps form the orbits and the nasal septum. The orbits are completed by various facial bones. Eye sockets are called orbits because of our ability to rotate the eyes.

The cranium contains eight bones: the frontal, two parietal, the occipital, two temporal, the sphenoid, and the ethmoid.

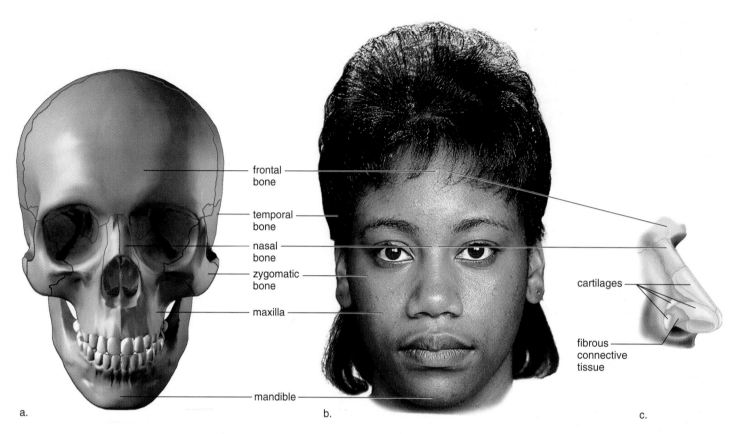

Figure 10.6 **Bones of the face, including the nose.**
a. The frontal bone forms the forehead and eyebrow ridges; the zygomatic bones form the cheekbones, and maxillae form the upper jaw. The maxillae are the most expansive facial bones, extending from the forehead to the lower jaw. The mandible has a projection we call the chin. **b.** The maxillae, frontal, and nasal bones help form the external nose. **c.** The rest of the nose is formed by cartilages and fibrous connective tissue.

The Facial Bones

The most prominent of the facial bones are the mandible, maxillae (maxillary bones), the zygomatic bones, and the nasal bones.

The **mandible,** or lower jaw, is the only movable portion of the skull (Fig. 10.6*a*), and its action permits us to chew our food. It also forms the "chin." Tooth sockets are located on the mandible and on the **maxillae,** the upper jaw that also forms the anterior portion of the hard palate. The palatine bones make up the posterior portion of the hard palate and the floor of the nasal cavity (see Fig. 10.5*b*).

The lips and cheeks have a core of skeletal muscle. The **zygomatic bones** are the cheekbone prominences, and the **nasal bones** form the bridge of the nose. Other bones (e.g., lacrimal bone and vomer) are a part of the nasal septum which divides the nose cavity into two regions. The lacrimal bone (see Fig. 10.5*a*) contains the opening for the nasolacrimal canal, which brings tears from the eyes to the nose.

The temporal and frontal bones are cranial bones that contribute to the face. The temporal bones account for the flattened areas we call the temples. The frontal bone forms the forehead and has supraorbital ridges where the eyebrows are located. Glasses sit where the frontal bone joins the nasal bones.

While the ears are formed only by elastic cartilage and not by bone, the nose (Fig. 10.6*c*) is a mixture of bones and cartilages and fibrous connective tissue. The cartilages complete the tip of the nose, and fibrous connective tissue forms the flared sides of the nose.

Among the facial bones, the mandible is the lower jaw where the chin is located, the two maxillae form the upper jaw, the two zygomatic bones are the cheekbones, and the two nasal bones form the bridge of the nose.

The Hyoid Bone

Although the **hyoid bone** is not part of the skull, it will be mentioned here because it is a part of the axial skeleton. The larynx is the voice box at the top of the trachea in the neck region. The hyoid bone which is located superior to the larynx is the only bone in the body that does not articulate with another bone. It is attached to the temporal bones by muscles and ligaments. The hyoid bone anchors the tongue and serves as the site for the attachment of muscles associated with swallowing.

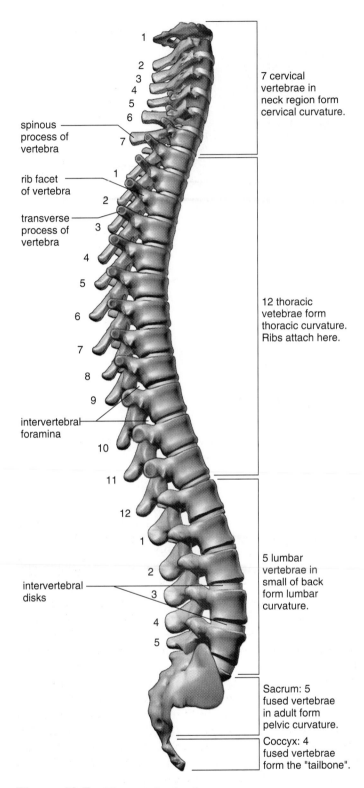

Labels on figure:
- 7 cervical vertebrae in neck region form cervical curvature.
- spinous process of vertebra
- rib facet of vertebra
- transverse process of vertebra
- 12 thoracic vetebrae form thoracic curvature. Ribs attach here.
- intervertebral foramina
- 5 lumbar vertebrae in small of back form lumbar curvature.
- intervertebral disks
- Sacrum: 5 fused vertebrae in adult form pelvic curvature.
- Coccyx: 4 fused vertebrae form the "tailbone".

Figure 10.7   **The vertebral column.**
The vertebrae are named according to their location in the vertebral column, which is flexible due to the intervertebral disks. Note the presence of the coccyx, also called the tailbone.

The Vertebral Column

The **vertebral column** consists of 33 vertebrae (Fig. 10.7). The vertebral column has many functions, including:

- supports the head and trunk, allowing movement.
- protects the spinal cord and roots of spinal nerves.
- serves as a site for muscle attachment.

Normally, the vertebral column has four curvatures that provide more resiliency and strength in an upright posture than a straight column could. As discussed in the introduction to this chapter, *scoliosis* is an abnormal lateral (sideways) curvature of the spine. There are two other well-known abnormal curvatures: *kyphosis* is an abnormal posterior curvature that often results in a hunchback, and *lordosis* is an abnormal anterior curvature resulting in a swayback.

The vertebrae join at articular processes and form the vertebral canal through which the spinal cord passes. The spinal nerves which pass out of the canal at the intervertebral foramina function to control skeletal muscle contraction and the internal organs. The spinous processes of the vertebrae can be felt as bony projections along the midline of the back. The spinous processes and also transverse processes, which extend laterally, serve as attachment sites for the muscles that move the vertebral column.

The various vertebrae are named according to their location in the vertebral column. The cervical vertebrae are located in the neck. The first *cervical vertebra*, called the **atlas,** holds up the head. It is so-named because Atlas, of Greek mythology, held up the world. Movement of the atlas permits the "yes" motion of the head. It also allows the head to tilt from side to side. The second cervical vertebra is called the **axis** because it allows a degree of rotation as when we shake the head "no." The *thoracic vertebrae* have long, thin spinous processes, and they have extra articular facets for the attachment of the ribs (Fig. 10.8). *Lumbar vertebrae* have a large body and thick processes. The five *sacral vertebrae* are fused together in the sacrum, which is a part of the pelvic girdle. The *coccyx,* or tailbone, is also composed of fused vertebrae.

There are **intervertebral disks** composed of fibrocartilage between the vertebrae that act as a kind of padding. They prevent the vertebrae from grinding against one another and absorb shock caused by movements such as running, jumping, and even walking. The presence of the disks allows motion between vertebrae so that we can bend forward, backward, and from side to side. Unfortunately, these disks become weakened with age and can even slip and rupture. Pain will result if a slipped disk presses against the spinal cord and/or spinal nerves. If so, surgical removal of the disk may relieve the pain.

The vertebral column consists of the vertebrae and serves as the backbone for the body. Disks between the vertebrae provide padding and account for flexibility of the column.

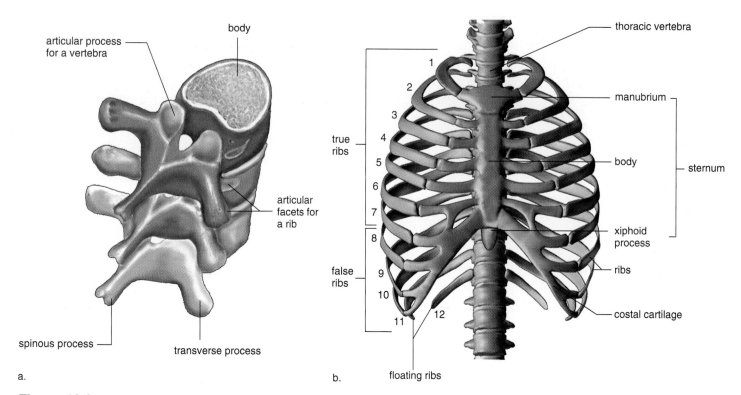

Figure 10.8   **Thoracic vertebrae and the rib cage.**
a. The thoracic vertebrae articulate with each other and with the ribs. A thoracic vertebra has two facets for articulation with a rib; one is on the body, and the other is on the transverse process. **b.** The rib cage consists of the thoracic vertebrae, the ribs, the costal cartilages, and the sternum.

The Rib Cage

The rib cage is composed of the thoracic vertebrae, the ribs and their associated cartilages, and the sternum (Fig. 10.8b).

The rib cage demonstrates how the skeleton is protective but also flexible. The rib cage protects the heart and lungs; yet it swings outward and upward upon inspiration and then downward and inward upon expiration.

The Ribs

There are twelve pairs of ribs. All twelve pairs connect directly to the thoracic vertebrae in the back. A rib articulates with the body and transverse process of its corresponding thoracic vertebra. Each rib curves outward and then forward and downward.

The upper seven pairs of ribs connect directly to the sternum by means of costal cartilages. These are called the "true ribs." The lower five pairs of ribs do not connect directly to the sternum and they are called the "false ribs." Three pairs of false ribs attach to the sternum by means of a common cartilage. The other two pairs are called "floating ribs" because they do not attach to the sternum at all.

The Sternum

The **sternum,** or breastbone, is a flat bone that has the shape of a blade. The sternum, along with the ribs, helps protect the heart and lungs.

The sternum is composed of three bones that fuse during development. These bones are the manubrium, the body, and the xiphoid process. An elevation called the *sternal angle* occurs where the *manubrium* joins with the body of the sternum. This is an important anatomical landmark because it occurs at the level of the second rib and therefore allows the ribs to be counted. Counting the ribs is sometimes used to determine where the apex of the heart is located. The apex of the heart is usually between the fifth and sixth ribs.

The *xiphoid process* is the third part of the sternum. Composed of hyaline cartilage in the child, it becomes ossified in the adult. The variably shaped xiphoid process serves as an attachment site for the diaphragm, the structure that divides the thoracic cavity from the abdominal cavity.

The rib cage, consisting of the thoracic vertebrae, the ribs, and the sternum, protects the heart and lungs in the thoracic cavity.

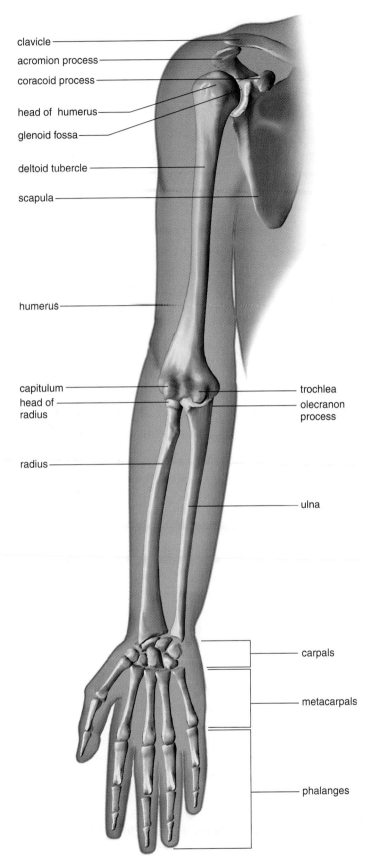

clavicle
acromion process
coracoid process

head of humerus
glenoid fossa

deltoid tubercle

scapula

humerus

capitulum
head of radius

radius

trochlea
olecranon process

ulna

carpals

metacarpals

phalanges

Figure 10.9   **Bones of the pectoral girdle and arm.**

The Appendicular Skeleton

The **appendicular skeleton** consists of the bones within the pectoral and pelvic girdles and their attached limbs. The pectoral (shoulder) girdle and arm are specialized for flexibility; the pelvic (hip) girdle and legs are specialized for strength.

The Pectoral Girdle and Arm

A **pectoral girdle** consists of a scapula (shoulder blade) and a clavicle (collarbone) (Fig. 10.9). The **clavicle** extends across the top of the thorax; it articulates with the *acromion process* of the scapula and also with the sternum. The **scapulae,** which are quite visible in the back, are held in place only by muscles. Muscles of the arm and chest attach to the *coracoid process* of the scapula. The **glenoid fossa** articulates with the head of the humerus. The glenoid fossa, a very shallow cavity, is much smaller than the head of the humerus. Although this means that the arm can move in almost any direction, there is little stability. Therefore, this is the joint that is most apt to dislocate. The components of the pectoral girdle are loosely linked together by ligaments, and this allows the girdle to follow freely the movements of the arm.

The **humerus,** the single long bone in the upper arm, has a smoothly rounded head that fits into the glenoid fossa of the scapula as mentioned. The shaft of the humerus has a *tubercle* (protuberance) where the deltoid, the prominent muscle of the chest, attaches. Upon death, enlargement of this tubercle can be used as evidence that the person did a lot of heavy lifting.

The far end of the humerus has two protuberances called the *capitulum* and the *trochlea,* which articulate respectively with the **ulna** and the radius, at the elbow. The "funny bone" of the elbow is the *olecranon process* of the ulna.

When the arm is held so that the palm is turned upward, the radius and ulna are about parallel to one another. When the arm is turned so that the palm is next to the body, the radius crosses in front of the ulna, a feature that contributes to the easy twisting motion of the forearm.

The hand has many bones, and this increases its flexibility. The wrist has eight **carpal** bones, which look like small pebbles. From these, five **metacarpal** bones fan out to form a framework for the palm. The metacarpal bone that leads to the thumb is placed in such a way that the thumb can reach out and touch the other digits. (**Digits** is a term that refers to either fingers or toes.) Your knuckles are the enlarged distal ends of the metacarpals. Beyond the metacarpals are the **phalanges,** the bones of the fingers and the thumb. The phalanges of the hand are long, slender, and lightweight.

The pectoral girdle and arm are specialized for flexibility of movement.

The Pelvic Girdle and Leg

Figure 10.10 shows how the leg is attached to the pelvic girdle. The **pelvic (hip) girdle** consists of two heavy, large coxal bones (hipbones). The pelvic cavity is bordered by the bones of the pelvis: the sacrum and the two coxal bones. The pelvis bears the weight of the body, protects the organs within the pelvic cavity, and serves as the place of attachment for the legs.

Each **coxal bone** has three parts: the ilium, the ischium, and the pubis, which are fused in the adult (Fig. 10.10). The hip socket, called the *acetabulum,* occurs where these three bones meet. The *ilium* is the largest part of the coxal bones, and our hips occur where it flares out. We sit on the *ischium,* which has a posterior spine called the ischial spine that projects into the pelvic cavity. The *pubis* (referring to pubic hair) is the anterior part of a coxal bone. The two pubic bones are joined together by a fibrocartilage disk at the pubic symphysis.

The male and female pelvis differ from one another. In the female, the iliac bones are more flared; the pelvic cavity is more shallow, but the outlet is wider. These adaptations facilitate giving birth.

The **femur** (thighbone) is the longest and strongest bone in the body. The head of the femur articulates with the coxal bones at the acetabulum, and the short *neck* better positions the legs for walking. The femur has two large processes, the *greater* and *lesser trochanter,* which are places of attachment for the muscles of the legs and buttocks. At its distal end, the femur has a *medial* and *lateral condyle* that articulate with the **tibia** of the lower leg. This is the region of the knee, and the **patella,** or kneecap, is held in place here by tendons that attach to the *tibial tuberosity.* At the distal end, the *medial malleolus* of the tibia causes the inner bulge of the ankle. The **fibula** is the more slender bone in the lower leg. The fibula has a head that articulates with the tibia and a distal *lateral malleolus* that forms the outer bulge of the ankle.

Each foot has an ankle, an instep, and five toes. The many bones of the foot give it considerably more flexibility, especially on rough surfaces. The ankle contains seven **tarsal** bones, one of which (the talus) can move freely where it joins the tibia and fibula. Strange to say, the *calcaneus,* or heel bone, is also considered to be part of the ankle. The talus and calcaneus support the weight of the body.

The instep has five elongated **metatarsal** bones. The distal end of the metatarsals forms the ball of the foot. If the ligaments that bind the metatarsals together become weakened, flat feet are apt to result. The bones of the toes are called **phalanges,** just like those of the fingers, but in the foot, the phalanges are stout and extremely sturdy.

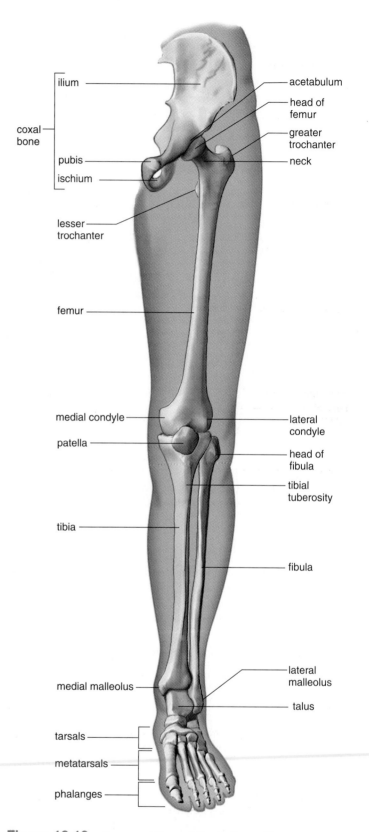

Figure 10.10   **Bones of the pelvic girdle and leg.**

The pelvic girdle and leg are adapted to supporting the weight of the body. The femur is the longest and strongest bone in the body.

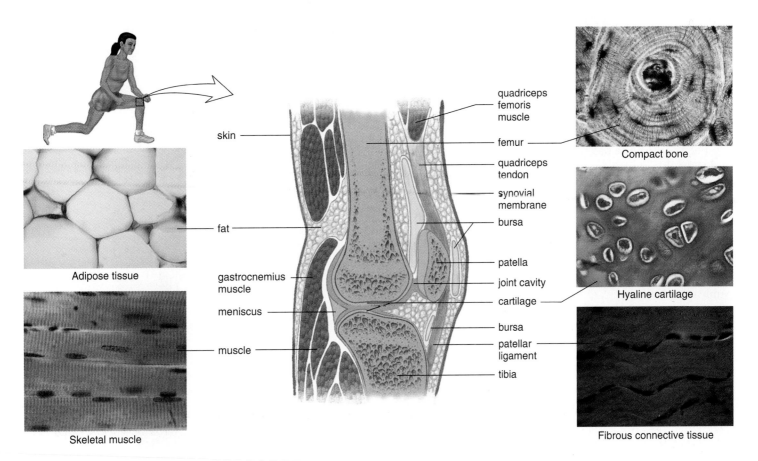

Figure 10.11   Knee joint.
The knee joint is a synovial joint. Notice the cavity between the bones, which is encased by ligaments and lined by synovial membrane. The patella (kneecap) serves to guide the quadriceps tendon over the joint when flexion or extension occurs

10.4 Articulations

Bones are joined at the joints, which are classified as fibrous, cartilaginous, and synovial. Fibrous joints, such as the **sutures** between the cranial bones, are *immovable*. Cartilaginous joints are connected by hyaline cartilage, as in the costal cartilages that join the ribs to the sternum, or by fibrocartilage, as seen in the intervertebral disks. Cartilaginous joints are *slightly movable*.

In *freely movable* **synovial joints,** the two bones are separated by a cavity. Ligaments hold the two bones in place as they form a capsule. Tendons also help stabilize joints. The joint capsule is lined by a *synovial membrane,* which produces *synovial fluid,* a lubricant for the joint. The knee is an example of a synovial joint (Fig. 10.11). Aside from articular cartilage, the knee contains **menisci** (sing., meniscus), crescent-shaped pieces of hyaline cartilage between the bones. These give added stability and act as shock absorbers. Unfortunately, athletes often suffer injury of the menisci, known as torn cartilage. The knee joint also contains 13 fluid-filled sacs called bursae (sing., **bursa**), which ease friction between tendons and ligaments. Inflammation of the bursae is called **bursitis.** Tennis elbow is a form of bursitis.

There are different types of movable joints. The knee and elbow joints are **hinge joints** because, like a hinged door, they largely permit movement in one direction only. The joint between the radius and ulna is a pivot joint in which only rotation is possible. More movable are the **ball-and-socket joints;** for example, the ball of the femur fits into a socket on the hipbone. Ball-and-socket joints allow movement in all planes and even a rotational movement. The various movements of body parts at synovial joints are depicted in Figure 10.12.

Synovial joints are subject to **arthritis.** In *rheumatoid arthritis*, the synovial membrane becomes inflamed and grows thicker. Degenerative changes take place that make the joint almost immovable and painful to use. Evidence indicates that these effects are brought on by an autoimmune reaction. In *osteoarthritis*, the cartilage at the ends of the bones eventually disintegrates so that the two bones become rough and irregular. The pain of arthritis, however, is believed to come from the soft tissues associated with joints and not from bone.

Joints are regions of articulations between bones. Synovial joints are freely movable and allow particular types of movements. Unfortunately, synovial joints are subject to various disorders.

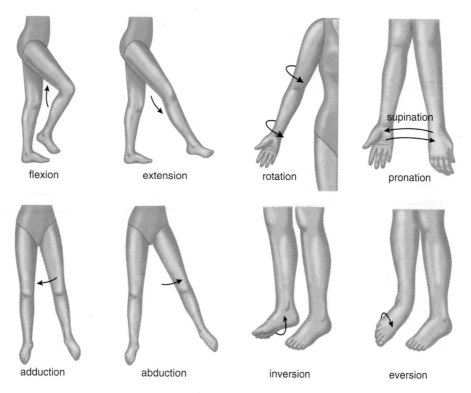

flexion   extension   rotation   supination / pronation

adduction   abduction   inversion   eversion

Figure 10.12   Joint movement.
Flexion decreases and extension increases the joint angle. Rotation is movement of a body part around its own axis. Supination is rotation of the lower arm so that the palm is upward; pronation is the opposite. Abduction is movement away from and adduction is movement toward the midline. Inversion is turning the foot so that the sole is inward, and eversion is the opposite.

Movements Permitted by Synovial Joints

Intact skeletal muscles are attached to bones by tendons that span joints. When a muscle contracts, one bone moves in relation to another bone.

Angular movements increase or decrease the joint angle between the bones of a joint:

Flexion (Fig. 10.12) decreases the joint angle. Flexion of the elbow moves the forearm toward the upper arm; flexion of the knee moves the lower leg toward the upper leg. *Dorsiflexion* is flexion of the foot upward, as when you stand on your heels; *plantar flexion* is flexion of the foot downward, as when you stand on your toes.

Extension increases the joint angle. Extension of the flexed elbow straightens the arm so that there is a 180° angle at the elbow. Hyperextension occurs when a portion of the body part is extended beyond 180°. It is possible to hyperextend the head and the trunk of the body.

Abduction is the movement of a body part laterally away from the midline. Abduction of the arms or legs moves them to the side, away from the body.

Adduction is the opposite of abduction. It is the movement of a body part toward the midline. For example, adduction of the arms or legs moves them back to the sides, toward the body.

Circular movements occur at ball-and-socket joints:

Rotation (Fig. 10.12) is the movement of a body part around its own axis, as when the head is turned to answer "no" or when the arm is twisted one way and then the other.

Supination is the rotation of the lower arm so that the palm is upward; **pronation** is the opposite—the movement of the lower arm so that the palm is downward.

Circumduction is the movement of a body part in a wide circle, as when a person makes arm circles. If the motion is observed carefully, one can see that, because the proximal end of the arm is stationary, the shape outlined by the arm is actually a cone.

Inversion and **eversion** are terms that apply only to the feet. Inversion is turning the foot so that the sole is inward, and eversion is turning the foot so that the sole is outward.

Elevation and **depression** are the lifting up and down, respectively, of a body part, such as when you shrug your shoulders.

Movements at joints are broadly classified as angular and circular.

Human Systems Work Together

Integumentary System

Bones provide support for skin.

Skin protects bones; helps provide vitamin D for Ca^{2+} absorption.

Muscular System

Bones provide attachment sites for muscles; store Ca^{2+} for muscle function.

Muscular contraction causes bones to move joints; muscles help protect bones.

Nervous System

Bones protect sense organs, brain, and spinal cord; store Ca^{2+} for nerve function.

Receptors send sensory input from bones to joints.

Endocrine System

Bones provide protection for glands; store Ca^{2+} used as second messenger.

Growth hormone regulates bone development; parathyroid hormone and calcitonin regulate Ca^{2+} content.

Cardiovascular System

Rib cage protects heart; red bone marrow produces blood cells; bones store Ca^{2+} for blood clotting.

Blood vessels deliver nutrients and oxygen to bones, carry away wastes.

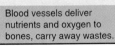

How the Skeletal System works with other body systems

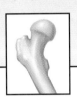

Lymphatic System/Immunity

Red bone marrow produces white blood cells involved in immunity.

Lymphatic vessels pick up excess tissue fluid; immune system protects against infections.

Respiratory System

Rib cage protects lungs and assists breathing; bones provide attachment sites for muscles involved in breathing.

Gas exchange in lungs provides oxygen and rids body of carbon dioxide.

Digestive System

Jaws contain teeth that chew food; hyoid bone assists swallowing.

Digestive tract provides Ca^{2+} and other nutrients for bone growth and repair.

Urinary System

Bones provide support and protection.

Kidneys provide active vitamin D for Ca^{2+} absorption and help maintain blood level of Ca^{2+}, needed for bone growth and repair.

Reproductive System

Bones provide support and protection of reproductive organs.

Sex hormones influence bone growth and density in males and females.

10.5 Homeostasis

The previous page tells how the skeletal system works with the other systems of the body to maintain homeostasis.

The rib cage assists the breathing process, enabling oxygen to enter the blood where it is transported by red blood cells to the tissues. Red bone marrow produces the red blood cells that transport oxygen. Without a supply of oxygen the cells of the body could not efficiently produce ATP. ATP is needed for muscle contraction and for nerve conduction but also for all the synthetic reactions that occur in cells.

Red bone marrow produces not only red blood cells but also produces the white blood cells. The white cells are involved in the defense of the body against pathogens and also cancerous cells. Without the ability to withstand foreign invasion, the body may soon succumb to disease and die.

The jaws contain sockets for the teeth which chew food. Chewing breaks food into pieces small enough to be swallowed and chemically digested. Without digestion, nutrients would not enter the body to serve as building blocks for repair and a source of energy for the production of ATP.

The bones also protect the internal organs. The rib cage protects the heart and lungs; the skull protects the brain; and the vertebrae protect the spinal cord. The endocrine organs such as the thymus and the adrenal glands are also protected by bone. The nervous system and the endocrine system work together to control the other organs and therefore ultimately homeostasis.

The storage of calcium in the bones is under hormonal control. There is a dynamic equilibrium between the concentration of calcium in the bones and in the blood. Calcium ions play a major role in muscle contraction and nerve conduction. Calcium ions also help regulate cellular metabolism. Protein hormones which cannot enter cells are called the first messenger, and a second messenger such as calcium ions jump-starts cellular metabolism, directing it to proceed in a particular way.

Locomotion is efficient in human beings because they have a jointed skeleton for the attachment of muscles which move the bones. Our jointed skeleton allows us to seek out and move to a more suitable external environment in order to maintain the internal environment within reasonable limits.

The skeleton has many and diverse roles in homeostasis.

Bioethical Issue

Aches and pains bothering you? Are your lower back, hips, and knees acting up? If you go to a conventional doctor, he might just put you on pain medication, or give you a cortisone shot, or worse, recommend surgery. If you don't find these remedies appealing, you might turn to an alternative form of medicine. Alternative medicine includes such nonconventional therapies as chiropractic therapy, herbal supplements, acupuncture, homeopathy, osteopathy, and therapeutic touch (e.g., the laying on of hands).

Advocates of alternative medicine have made some headway in having alternative medical practices accepted by most anyone. In 1992, congress established what is now called the National Center for Complementary and Alternative Medicine (NCCAM), whose budget has grown from $2 million to its present $50 million for 1999. Then in 1994, the Dietary Supplement Health and Education Act allowed the marketing of vitamins, minerals, and herbs without the requirement that they be approved by the Food and Drug Administration (FDA).

Many scientists feel that alternative medicine should be subjected to the same rigorous clinical testing as traditional medicine. They approve of a study that is now testing the efficacy of the herb St. John's wort; for example. In the study, 336 patients diagnosed with moderate depression will be divided into three groups. One group will receive the herb St. John's wort; another will be given sertraline, a traditional medicine, and a third group will receive a placebo. The results are expected to tell whether St. John's wort, a relatively inexpensive remedy, works just as well as a relatively expensive drug produced by pharmaceutical firms.

Questions

1. Do you believe that alternative medical practices should be subjected to clinical testing, or do you believe the public should simply rely on "word of mouth" recommendations?

2. What changes do you anticipate if alternative medical practices were to become a regular part of traditional medicine?

3. Would it be acceptable to you if it could be shown that alternative medical practices succeed simply for psychological reasons rather than their physical benefits?

Summarizing the Concepts

10.1 Tissues of the Skeletal System
The tissues of the skeletal system are largely composed of connective tissue, particularly bone, cartilage, and fibrous connective tissue. Hyaline (articular) cartilage covers the ends, while periosteum (fibrous connective tissue) covers the rest of a long bone. Spongy bone, which may contain red bone marrow, is in the epiphyses, and yellow bone marrow is in the medullary cavity of the diaphysis. The wall of the diaphysis is compact bone.

10.2 Bone Growth and Repair
Bone is a living tissue that can grow and repair. The prenatal human skeleton is at first cartilaginous, but it is later replaced by a bony skeleton. During adult life, bone is constantly being broken down by osteoclasts and then rebuilt by osteoblasts that become osteocytes in lacunae. Repair of a fracture requires four steps: (1) hematoma, (2) fibrocartilage callus, (3) bony callus, and (4) remodeling.

10.3 Bones of the Skeleton
The skeleton supports and protects the body, permits flexible movement, produces blood cells, and serves as a storehouse for mineral salts, particularly calcium phosphate.

The axial skeleton lies in the midline of the body and consists of the skull, the hyoid bone, the vertebral column, and the rib cage. The skull contains the cranium, which protects the brain and the facial bones. Considering the face, the frontal bone forms the forehead; the maxillae extend from the frontal bone to the upper row of teeth; the zygomatic bones are the cheekbones; and the nasal bones form the bridge of the nose. The rest of the nose is hyaline cartilage, except the flares are fibrocartilage. The ears are elastic cartilage. The mandible is the lower jaw and accounts for the chin.

The appendicular skeleton consists of the bones of the pectoral girdle, arms, pelvic girdle, and legs. The pectoral girdle and arms are adapted for flexibility. The glenoid fossa of the scapula is barely large enough to receive the head of the humerus, and this means that dislocated shoulders are more likely to happen than dislocated hips. The pelvic girdle and the legs are adapted for strength: the femur is the strongest bone in the body. However, the foot, like the hand, is flexible because it contains so many bones. Both fingers and toes are digits, and the term phalanges refers to the bones of each.

10.4 Articulations
There are three types of joints: fibrous, like the sutures of the cranium, are immovable; cartilaginous, like those between the ribs and sternum and the pubic symphysis, are slightly movable; and the synovial joints, consisting of a membrane-lined (synovial membrane) capsule, are freely movable. There are different kinds of synovial joints, and the movements they permit are varied.

10.5 Homeostasis
The skeletal system works with the other systems of the body in the ways described in the box on page 220.

Studying the Concepts

1. Bones are organs composed of what types of tissues? What are some differences between compact bone and spongy bone? 206

2. Describe the makeup of long bone. 206–7

3. What are the two types of ossification? Describe endochondral ossification. 208

4. How does a long bone grow in children, and how is it remodeled in all age groups? 208–9

5. What are the four steps required for fracture repair? 209

6. What are the bones of the cranium and the face? Associate the parts of a face with particular bones or with cartilages. 212–13

7. What are the parts of the vertebral column, and what are its curvatures? Distinguish between the atlas, axis, sacrum, and coccyx. 214

8. What are the bones of the rib cage, and what are several of its functions? 215

9. What are the bones of the pectoral girdle? Give examples to demonstrate the flexibility of the pectoral girdle. Where does the scapula articulate with the clavicle? 216

10. What are the bones of the arm? Name several processes of the humerus; of the elbow. 216

11. What are the bones of the pelvic girdle, and what are their functions? Give examples to demonstrate the strength and stability of the pelvic girdle. 217

12. What are the bones of the leg? Name several processes of the femur; of the ankle. 217

13. How are joints classified? Give examples of the different types of synovial joints and the movements they permit. 218–19

Testing Your Knowledge of the Concepts

In questions 1–6, match the location to the bones.
  a. forehead
  b. chin
  c. cheekbone
  d. elbow
  e. shoulder blade
  f. hip
  g. ankle
  _____ 1. zygomatic
  _____ 2. tibia and fibula
  _____ 3. frontal bone
  _____ 4. humerus
  _____ 5. coxal bone
  _____ 6. scapula

Match the items in the key to the bones listed in quesitons 7–12.
Key:
  a. glenoid fossa
  b. olecranon process
  c. acetabulum
  d. spinous process
  e. greater and lesser trochanter
  f. xiphoid process
  _____ 7. femur
  _____ 8. scapula
  _____ 9. ulna
  _____ 10. coxal bone
  _____ 11. sternum
12. vertebra

In questions 13 and 14, indicate whether the statement is true (T) or false (F).
  _____ 13. The pectoral girdle is specialized for weight-bearing, while the pelvic girdle is specialized for flexibility of movement.
  _____ 14. The term phalanges refers to the bones in both the fingers and toes.

For questions 15–20, fill in the blanks.
15. Compact bone contains tubular units called _____.
16. The epiphysis of a long bone contains _____ bone, where blood cells are produced.
17. During bone remodeling, cells called _____ break down bone and return calcium to the blood.
18. The _____ are air-filled spaces in the cranium.
19. The sacrum is a part of the _____, and the sternum is a part of the _____.
20. Label this diagram of the pelvis and leg.

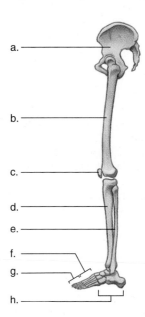

a. ———
b. ———
c. ———
d. ———
e. ———
f. ———
g. ———
h. ———

Applying Your Knowledge to the Concepts

These questions pertain to homeostasis.
1. Osteoporosis is a common disease, affecting 50% of women over the age of 45 and 90% of women over the age of 75. It is due to loss of bone mass. How is bone mass decreased?
2. A friend of mine was having severe neck and back pains. A surgeon at the Mayo Hospital inserted a metal rod alongside the cervical vertebrae and fused three cervical vertebrae. This took away the pain. What was her condition that warranted this operation?

3. The knee is a weak spot in many sports, but particularly in football. Explain the reason for this.
4. Jane's mother measured Jane's height each year on her birthday. Just for fun, they also measured the height of Jane's mother and grandmother. A curious thing happened: Jane's height increased with age, her grandmother's height decreased slightly, and her mother's height remained the same. Explain. What can women do to prevent a decrease in height as they age?

Understanding the Terms

appendicular skeleton 216
arthritis 218
articular cartilage 206
axial skeleton 212
ball-and-socket joint 218
bursa 218
bursitis 218
cartilage 206
compact bone 206
epiphyseal (growth) plate 208
fibrous connective tissue 206
fontanel 212
foramen magnum 212
hinge joint 218
intervertebral disk 214
joint 206
ligament 206
mastoiditis 212

medullary cavity 207
menisci 218
ossification 208
osteoblast 208
osteoclast 208
osteocyte 208
pectoral girdle 216
pelvic girdle 217
periosteum 207
red bone marrow 206
sinus 212
spongy bone 206
suture 218
synovial joint 218
tendon 206
vertebral column 214
yellow bone marrow 206

Match the terms to these definitions:

a. _____ Freely movable joint.

b. _____ Portion of the skeleton that lies in the midline: the skull, the vertebral column, and the rib cage.

c. _____ Crescent-shaped pieces of hyaline cartilage between the bones at a synovial joint.

d. _____ Membranous region located between certain cranial bones in the skull of a fetus or infant.

e. _____ Portion of the skeleton to which the arms are attached.

Applying Technology to the Concepts

Your study of skeletal systems in humans is supported by these available technologies:

Essential Study Partner CD-ROM

Animals → Support / Locomotion

Visit the Mader web site for related ESP activities.

Exploring the Internet

The Mader Home Page provides resources and tools as you study this chapter.

http://www.mhhe.com/biosci/genbio/mader

Dynamic Human 2.0 CD-ROM

Skeletal System

Chapter *11*

Muscular System

Chapter Concepts

Figure 11.1   How important is a smile?
Facial expressions are dependent on muscles that allow us to open our mouths, wink our eyes, and of course, smile.

When most kids were dreaming of new toys, sugar cookies, and other holiday treats, 7-year-old Chelsey Thomas found herself hoping for just one thing—the chance to smile.

Chelsey was born with a rare disorder called Moebius syndrome, which causes paralysis of facial muscles. All through childhood—as her friends giggled and grinned—Chelsey could not force even the smallest of smiles. She stubbornly managed to stay happy. But it was hard. And Chelsey's parents worried that she would eventually become depressed and withdrawn.

So after much preparation and talk with doctors, Chelsey and her parents drove to a medical center in California. There, a team of surgeons began a 10-hour operation designed to give Chelsey her smile.

Moving cautiously, doctors extracted a strip of muscle, vein, and artery from Chelsey's thigh. Making an incision along her cheek, the team carefully sewed the extractions into place along Chelsey's jaw. The tiny sutures, thinner than a human hair, were sewn while viewing them under a microscope.

The painstaking surgery allowed the transplanted muscle in the left side of Chelsey's face to work. She gave a beautiful, if lopsided, smile. Then later, doctors performed the same surgery on the right side of her face. Finally, her long-awaited smile was complete. Today, Chelsey grins nonstop (Fig. 11.1). Her newfound smile dazzles her family and helps her fit in with kids at school.

Researchers are now learning to perform many kinds of muscle and nerve transplants. These operations can save or improve the use of a patient's arms or legs, for example. As knowledge of the muscular system improves, transplants will improve and become much easier and cheaper to perform. And that, say doctors, is something everyone can smile about.

All the activities of the body from smiling to balancing on tiptoe are dependent on muscles whose sole ability is to contract. That's why surgeons could take a muscle from the leg and transplant it to the face to allow Chelsey to smile. The location and function of certain well-known muscles are reviewed in this chapter, which concentrates on the structure of skeletal muscles from the macroscopic to the microscopic level. When muscles contract, two protein filaments slide past one another within individual muscle fibers. Much is known about the interaction of these filaments, and we will examine in detail how muscle contraction is activated.

Muscle contraction requires the presence of ATP, which is produced most efficiently when oxygen is present. Some muscle fibers, however, are adapted to anaerobic conditions, and these fibers may be associated with particular sports that require endurance, like weight lifting. Muscle anatomy and physiology are of extreme interest to sports enthusiasts, and this chapter emphasizes sports-related topics.

11.1 Skeletal Muscles

Muscles have various functions, which they are more apt to perform well if they are exercised regularly.

Skeletal muscles support the body. Skeletal muscle contraction opposes the force of gravity and allows us to remain upright.

Skeletal muscles make bones move. Muscle contraction accounts not only for the movement of arms and legs but also for movements of the eyes, facial expressions, and breathing.

Skeletal muscles help maintain a constant body temperature. Skeletal muscle contraction causes ATP to break down, releasing heat that is distributed about the body. ⚚

Skeletal muscle contraction assists movement in cardiovascular and lymphatic vessels. The pressure of skeletal muscle contraction keeps blood moving in cardiovascular veins and lymph moving in lymphatic vessels. ⚚

Skeletal muscles help protect internal organs. Skeletal muscles cover the bones and underlying organs such as the kidneys and blood vessels.

Cardiac and smooth muscles have specific functions. The contraction of cardiac muscle causes the heart to pump blood. Smooth muscle contraction propels food in the digestive tract and urine in the ureters. ⚚

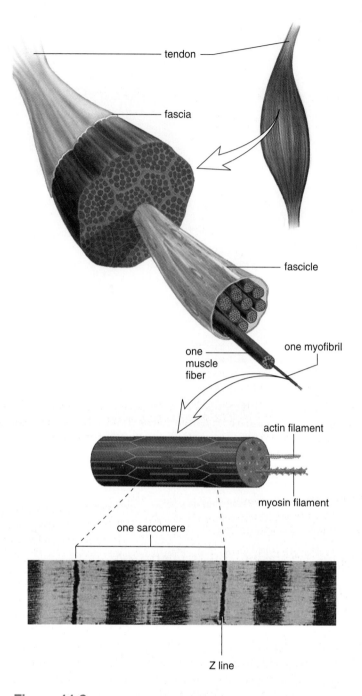

Figure 11.2 Anatomy of a muscle.

A fascia covers the surface of a muscle and contributes to the tendon, which attaches the muscle to a bone. Fascia also separates bundles of muscle fibers that contain contractile elements called myofibrils. The arrangement of myofilaments within a myofibril accounts for skeletal muscle striations. A sarcomere is a contractile unit of a myofibril.

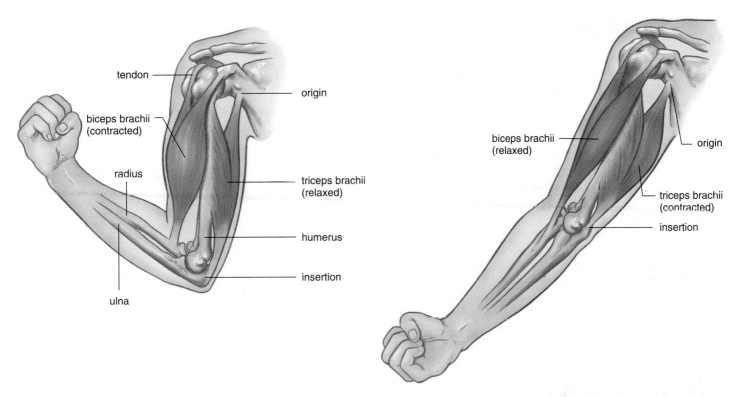

Figure 11.3   **Attachment of the skeletal muscles as exemplified by the biceps brachii and the triceps brachii.**
The origin of a muscle is on a bone that remains stationary, and the insertion of a muscle is on a bone that moves when the muscle contracts.
The muscles in this drawing are antagonistic. When the biceps brachii contracts, the lower arm flexes, and when the triceps brachii contracts, the
lower arm extends.

Muscles Work in Pairs

This chapter is primarily concerned with skeletal muscles—those muscles that make up the bulk of the human body. Muscles are covered by several layers of fibrous connective tissue called fascia, which extends beyond the muscle to become its **tendon** (Fig. 11.2). Fascia also surrounds fascicles, which are bundles of muscle fibers, and separates the muscle fibers within a fascicle. A muscle fiber contains many contractile elements called myofibrils. Myofibrils run the length of a muscle fiber, and the striations seen in observations of muscular tissue are due to the placement of myofilaments within sarcomeres, the units of a myofibril. There are thin (actin) filaments and thick (myosin) filaments in the myofibril.

Skeletal muscles are attached to the skeleton, and their contraction causes the movement of bones at a joint. When muscles contract, one bone remains fairly stationary, and the other one moves. The **origin** of a muscle is on the stationary bone, and the **insertion** of a muscle is on the bone that moves.

Frequently, a body part is moved by a group of muscles working together. Even so, one muscle does most of the work and is called the *prime mover.* The assisting muscles are called the *synergists.* When muscles contract, they shorten; therefore, muscles can only pull; they cannot push.

Muscles have antagonists, and *antagonistic pairs* work opposite to one another to bring about movement in opposite directions. For example, the biceps brachii and the triceps brachii are antagonists; one flexes the forearm, and the other extends the forearm (Fig. 11.3).

When muscles cooperate to achieve movement,
some act as prime movers, others are synergists,
and still others are antagonists.

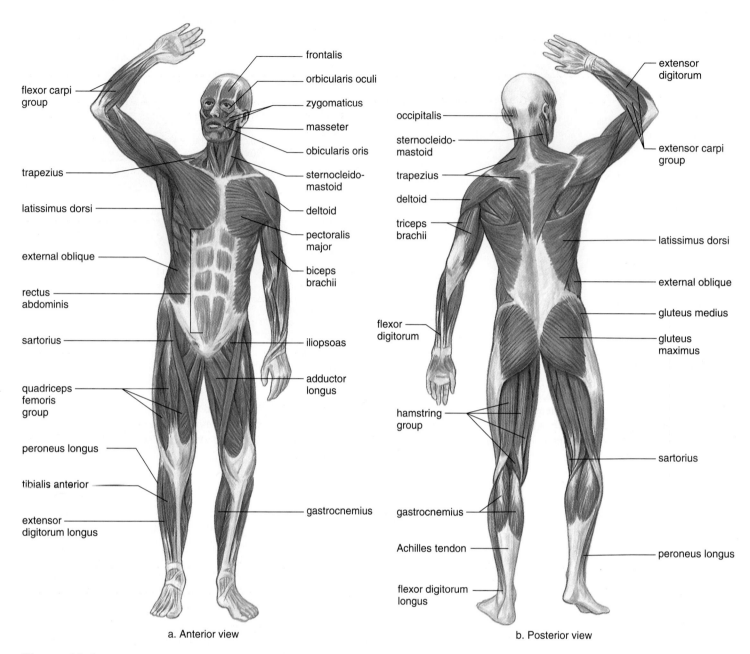

a. Anterior view

b. Posterior view

Figure 11.4   Human musculature.
Superficial skeletal muscles. In **(a)** anterior and **(b)** posterior view.

Skeletal Muscles of the Body

Skeletal muscles (Fig. 11.4 and Tables 11.1 and 11.2) are given names based on the following characteristics and examples:

1. Size. The gluteus maximus is the largest muscle that makes up the buttocks.
2. Shape. The deltoid is shaped like a delta, or triangle.
3. Location. The frontalis overlies the frontal bone.
4. Direction of fibers. The rectus abdominis is a longitudinal muscle of the abdomen (rectus means straight).

5. Number of attachments. The biceps brachii has two attachments, or origins.
6. Action. The extensor digitorum extends the fingers and digits. Extension increases the joint angle; flexion decreases the joint angle; abduction is the movement of a body part sideways away from the midline, and adduction is the movement of a body part toward the midline.

Table 11.1   Muscles (anterior view)

Name	Function
Head and neck	
Frontalis	Wrinkles forehead and lifts eyebrows
Orbicularis oculi	Closes eye (winking)
Zygomaticus	Raises corner of mouth (smiling)
Masseter	Closes jaw
Orbicularis oris	Closes and protrudes lips (kissing)
Arms and trunk	
External oblique	Compresses abdomen; rotates trunk
Rectus abdominis	Flexes spine
Pectoralis major	Flexes and adducts shoulder and arm ventrally (pulls arm across chest)
Deltoid	Abducts and raises arm at shoulder joint
Biceps brachii	Flexes forearm and supinates hand
Legs	
Adductor longus	Adducts thigh
Iliopsoas	Flexes thigh
Sartorius	Rotates thigh (sitting cross-legged)
Quadriceps femoris group	Extends lower leg
Peroneus longus	Everts foot
Tibialis anterior	Dorsiflexes and inverts foot
Flexor and extensor digitorum longus	Flex and extend toes

Table 11.2   Muscles (posterior view)

Name	Function
Head and neck	
Occipitalis	Moves scalp backward
Sternocleidomastoid	Turns head to side; flexes neck and head
Trapezius	Extends head; raises and adducts shoulders dorsally (shrugging shoulders)
Arms and trunk	
Latissimus dorsi	Extends and adducts shoulder and arm dorsally (pulls arm across back)
Deltoid	Abducts and raises arm at shoulder joint
External oblique	Rotates trunk
Triceps brachii	Extends forearm
Flexor and extensor carpi group	Flex and extend hand
Flexor and extensor digitorum	Flex and extend fingers
Buttocks and legs	
Gluteus medius	Abducts thigh
Gluteus maximus	Extends thigh (forms buttocks)
Hamstring group	Flexes lower leg
Gastrocnemius	Extends foot (tiptoeing)

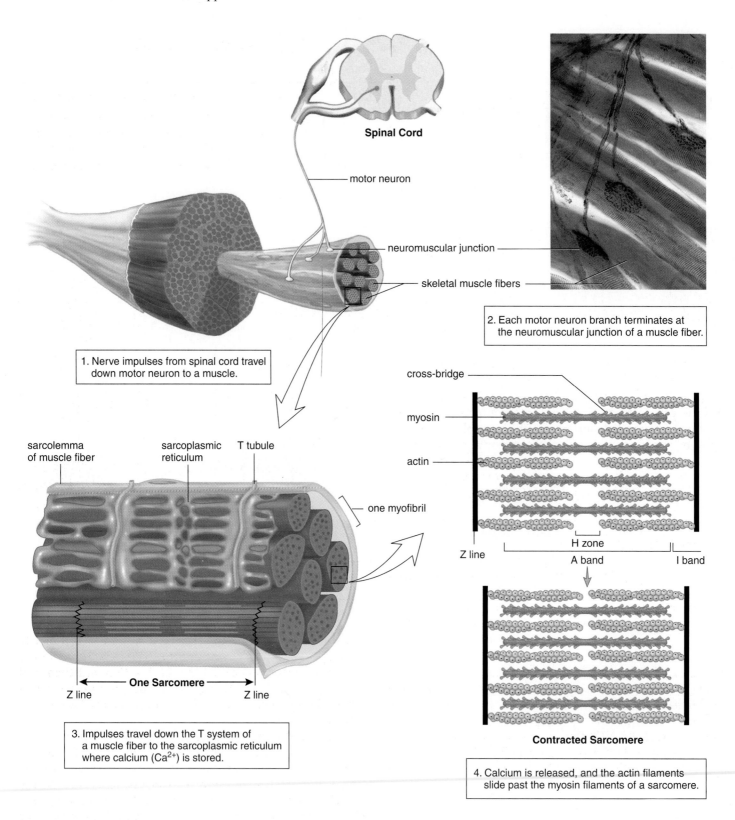

Spinal Cord

motor neuron

neuromuscular junction

skeletal muscle fibers

1. Nerve impulses from spinal cord travel down motor neuron to a muscle.

2. Each motor neuron branch terminates at the neuromuscular junction of a muscle fiber.

sarcolemma of muscle fiber

sarcoplasmic reticulum

T tubule

one myofibril

One Sarcomere

Z line   Z line

3. Impulses travel down the T system of a muscle fiber to the sarcoplasmic reticulum where calcium (Ca^{2+}) is stored.

cross-bridge

myosin

actin

Z line

H zone

A band

I band

Contracted Sarcomere

4. Calcium is released, and the actin filaments slide past the myosin filaments of a sarcomere.

Figure 11.5   Contraction of a muscle.
A whole muscle contains bundles of skeletal muscle fibers. A motor neuron ends in neuromuscular junctions at these fibers. A fiber contains many myofibrils. Each myofibril is divided into sarcomeres, where actin and myosin filaments are precisely arranged. When a motor impulse reaches a neuromuscular junction, the impulse travels down the T system and calcium is released from the sarcoplasmic reticulum. Then the actin filaments slide past myosin filaments and the sarcomere contracts.

11.2 Mechanism of Muscle Fiber Contraction

A series of events leads up to muscle fiber contraction. After discussing these events, we will examine a neuromuscular junction and filament sliding in detail.

Overview of Muscular Contraction

Figure 11.5 shows the steps that lead to muscle contraction. Muscles are stimulated to contract by nerve impulses that begin in the brain or spinal cord. These nerve impulses travel down a motor neuron to a neuromuscular junction, a region where a motor neuron fiber meets a muscle fiber.

Each muscle fiber is a cell containing the usual cellular components, but special names have been assigned to some of these components. The plasma membrane is called the **sarcolemma,** the cytoplasm is the sarcoplasm, and the endoplasmic reticulum is the **sarcoplasmic reticulum.** A muscle fiber also has some unique anatomical characteristics. For one thing, it has a T (for transverse) system; the sarcolemma forms **T (transverse) tubules** that penetrate, or dip down, into the cell so that they come into contact—but do not fuse—with expanded portions of the sarcoplasmic reticulum. The expanded portions of the sarcoplasmic reticulum contain calcium ions (Ca^{2+}), which are essential for muscle contraction. The sarcoplasmic reticulum encases hundreds and sometimes even thousands of **myofibrils,** which are the contractile portions of the fibers.

Myofibrils and Sarcomeres

Myofibrils are cylindrical in shape and run the length of the muscle fiber. The light microscope shows that muscle fibers have light and dark bands called striations (Fig. 11.6). The electron microscope shows that the striations of myofibrils are formed by the placement of myofilaments within contractile units called **sarcomeres.** A sarcomere extends between two dark lines called the Z lines. A sarcomere contains two types of protein myofilaments. The thick filaments are made up of a protein called **myosin,** and the thin filaments are made up of a protein called **actin.** Other proteins in addition to actin are also present. The I band is light colored because it contains only actin filaments attached to a Z line. The dark regions of the A band contain overlapping actin and myosin filaments, and its H zone has only myosin filaments.

Sliding Filaments

Impulses generated at a neuromuscular junction travel down a T tubule, and calcium is released from the sarcoplasmic reticulum. Now the muscle fiber contracts as the sarcomeres within the myofibrils shorten. When a sarcomere shortens, the actin (thin) filaments slide past the myosin (thick) filaments and approach one another. This causes the

Table 11.3	Muscle Contraction
Name	**Function**
Actin filaments	Slide past myosin, causing contraction
Ca^{2+}	Needed for myosin to bind the actin
Myosin filaments	Pull actin filaments by means of cross-bridges; are enzymatic and split ATP
ATP	Supplies energy for muscle contraction

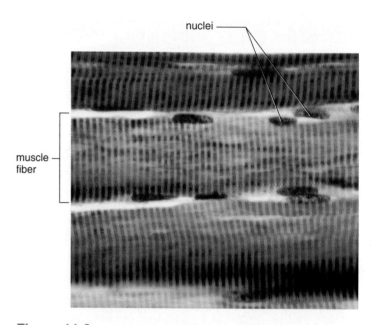

Figure 11.6   Light micrograph of skeletal muscle.
The striations of skeletal muscle tissue are produced by alternating dark A bands and light I bands. See the sarcomere in Figure 11.5.

I band to shorten and the H zone to almost or completely disappear. The movement of actin filaments in relation to myosin filaments is called the **sliding filament theory** of muscle contraction. During the sliding process, the sarcomere shortens even though the filaments themselves remain the same length.

The participants in muscle contraction have the functions listed in Table 11.3. ATP supplies the energy for muscle contraction. Although the actin filaments slide past the myosin filaments, it is the myosin filaments that do the work. Myosin filaments break down ATP and have cross-bridges that pull the actin filaments toward the center of the sarcomere.

Muscle fibers are innervated, and when they are stimulated to contract, myofilaments slide past one another causing sarcomeres to shorten.

Muscle Innervation

Muscles are stimulated to contract by motor nerve fibers. Each nerve fiber has several branches ending in an axon bulb that lies in close proximity to the sarcolemma of a muscle fiber. A small gap, called a synaptic cleft, separates the axon bulb from the sarcolemma. This entire region is called a **neuromuscular junction** (Fig. 11.17).

Axon bulbs contain synaptic vesicles that are filled with neurotransmitter molecules called acetylcholine (ACh). When nerve impulses traveling down a motor neuron arrive at an axon bulb, the synaptic vesicles release ACh into the synaptic cleft. ACh quickly diffuses across the cleft and binds to receptor proteins in the sarcolemma. Now the sarcolemma generates impulses that spread over the sarcolemma and down T tubules to the sarcoplasmic reticulum. The release of calcium from the sarcoplasmic reticulum causes the myofibrils within a muscle fiber to contract as discussed previously.

At a neuromuscular junction, nerve impulses bring about the release of neurotransmitter molecules that signal a muscle fiber to contract.

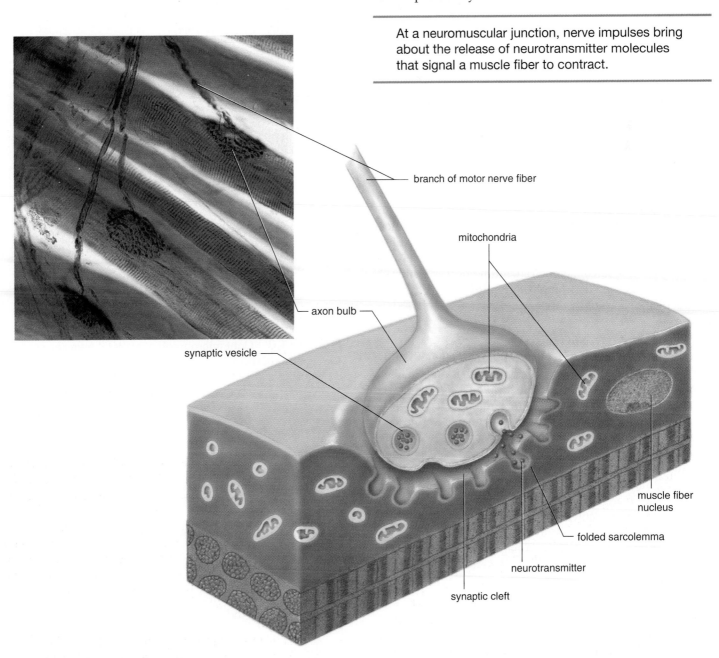

branch of motor nerve fiber

mitochondria

axon bulb

synaptic vesicle

muscle fiber nucleus

folded sarcolemma

neurotransmitter

synaptic cleft

Figure 11.7   Neuromuscular junction.
The branch of a motor nerve fiber terminates in an axon bulb that meets but does not touch a muscle fiber. A synaptic cleft separates the axon bulb from the sarcolemma of the muscle fiber. Nerve impulses traveling down a motor neuron cause synaptic vesicles to discharge acetylcholine, which diffuses across the synaptic cleft. When this neurotransmitter is received by the sarcolemma of a muscle fiber, contraction follows.

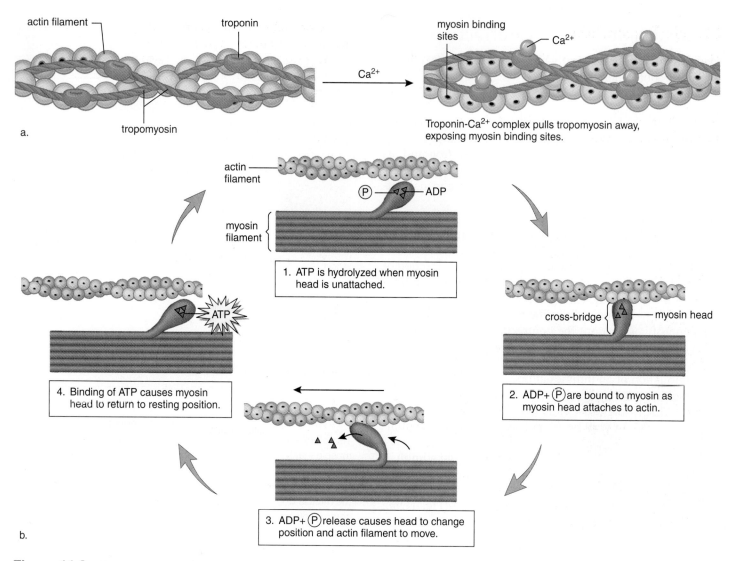

Figure 11.8   The role of calcium and myosin in muscle contraction.
a. Upon release, calcium binds to troponin, exposing myosin binding sites. **b.** After breaking down ATP, myosin heads bind to an actin filament, and later, a power stroke causes the actin filament to move.

Figure 11.8 shows the placement of two other proteins associated with a thin filament, which is composed of a double row of twisted actin molecules. Threads of **tropomyosin** wind about an actin filament, and **troponin** occurs at intervals along the threads. Calcium ions (Ca^{2+}) that have been released from the sarcoplasmic reticulum combine with troponin. After binding occurs, the tropomyosin threads shift their position, and myosin binding sites are exposed.

The thick filament is actually a bundle of myosin molecules, each having a double globular head with an ATP binding site. The heads function as ATPase enzymes, splitting ATP into ADP and ℗. This reaction activates the head so that it will bind to actin. The ADP and ℗ remain on the myosin heads until the heads attach to actin, forming a cross-bridge. Now, ADP and ℗ are released, and this causes the cross-bridges to change their positions. This is the *power stroke* that pulls the thin filaments toward the middle of the sarcomere. When another ATP molecule binds to a myosin head, the cross-bridge is broken as the head detaches from actin. The cycle begins again; the actin filaments move nearer the center of the sarcomere each time the cycle is repeated.

Contraction continues until nerve impulses cease and calcium ions are returned to their storage sacs. The membranes of the sarcoplasmic reticulum contain active transport proteins that pump calcium ions back into the sarcoplasmic reticulum.

Myosin filament heads break down ATP and then attach to an actin filament, forming cross-bridges that pull the actin filament to the center of a sarcomere.

11.3 Whole Muscle Contraction

Observation of muscle contraction in the laboratory applies to muscle contraction in the body.

Basic Laboratory Observations

To study muscle contraction in the laboratory, the gastrocnemius (calf muscle) is usually removed from a frog and attached to a movable lever. The muscle is stimulated, and the mechanical force of contraction is recorded as a visual pattern called a **myogram.**

At first, the stimulus may be too weak to cause a contraction, but as soon as the strength of the stimulus reaches a threshold stimulus, the muscle contracts and then relaxes. This action—a single contraction that lasts only a fraction of a second—is called a **muscle twitch.** Figure 11.9*a* is a myogram of a twitch, which is customarily divided into the latent period, or the period of time between stimulation and initiation of contraction; the contraction period, when the muscle shortens; and the relaxation period, when the muscle returns to its former length.

Stimulation of an individual fiber within a muscle usually results in a maximal, *all-or-none contraction.* But the contraction of a whole muscle, as evidenced by the size of a muscle twitch, can vary in strength depending on the number of fibers contracting.

If a muscle is given a rapid series of threshold stimuli, it can respond to the next stimulus without relaxing completely. In this way, muscle contraction *summates* until maximal sustained contraction, called **tetanus,** is achieved (Fig. 11.9*b*). The myogram no longer shows individual twitches; rather, the twitches are fused and blended completely into a straight line. Tetanus continues until the muscle fatigues due to depletion of energy reserves. *Fatigue* is apparent when a muscle relaxes even though stimulation continues.

Muscle Tone in the Body

Tetanic contractions ordinarily occur in the body's muscles whenever skeletal muscles are actively used. But while some fibers are contracting, others are relaxing. Because of this, intact muscles rarely fatigue completely. Even when muscles appear to be at rest, they exhibit **tone** in which some of their fibers are contracting. Muscle tone is particularly important in maintaining posture. If all the fibers within the muscles of the neck, trunk, and legs were to suddenly relax, the body would collapse.

Maintenance of the right amount of tone requires the use of special receptors called muscle spindles. A muscle spindle consists of a bundle of modified muscle fibers, with sensory nerve fibers wrapped around a short, specialized region. A muscle spindle contracts along with muscle fibers, but thereafter it sends sensory nerve impulses to the central nervous system, which then regulates muscle contraction so that tone is maintained (see Fig. 13.3).

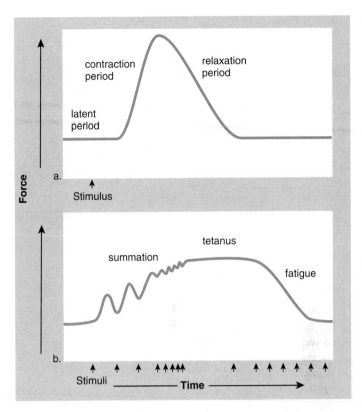

Figure 11.9   Physiology of skeletal muscle contraction.
These are myograms, visual representations of the contraction of a muscle that has been dissected from a frog. **a.** Simple muscle twitch is composed of three periods: latent, contraction, and relaxation. **b.** Summation and tetanus. When stimulation frequency increases, the muscle does not relax completely between stimuli, and the contraction gradually increases in intensity. The muscle becomes maximally contracted until it fatigues.

Recruitment and the Strength of Contraction

Each motor neuron innervates several muscle fibers at a time. A motor neuron together with all of the muscle fibers that it innervates is called a **motor unit.** As the intensity of nervous stimulation increases, more and more motor units are activated. This phenomenon, known as recruitment, results in stronger and stronger contractions.

Another variable of importance is the number of fibers within a motor unit. In the ocular muscles that move the eyes, the innervation ratio is one motor neuron per 23 muscle fibers, while in the gastrocnemius muscle of the buttocks, the ratio is about one neuron per 1,000 muscle fibers. Moving the eyes requires finer control than moving the legs.

Observation of muscle contraction in the laboratory applies to muscle contraction in the body. A muscle at rest exhibits tone dependent on tetanic contractions; a contracting muscle exhibits degrees of contraction dependent on recruitment.

11.4 Energy for Muscle Contraction

ATP produced previous to strenuous exercise lasts a few seconds, and then muscles acquire new ATP in three different ways (Fig. 11.10). There are two anaerobic ways by which muscles can acquire ATP immediately. The first method depends on **creatine phosphate,** a high-energy compound built up when a muscle is resting. Creatine phosphate cannot participate directly in muscle contraction. Instead, it is used to regenerate ATP by the following reaction:

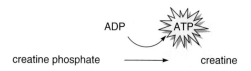

This reaction, which occurs in the midst of sliding filaments, and therefore this method of supplying ATP is the speediest way to make ATP available to muscles. Creatine phosphate provides enough energy for only about eight seconds of intense activity, and then it is spent. Creatine phosphate is rebuilt when a muscle is resting by transferring a phosphate group from ATP to creatinine.

Fermentation also supplies ATP without consuming oxygen. During fermentation, glucose is broken down to lactate (lactic acid):

The accumulation of lactate in a muscle fiber makes the cytoplasm more acidic, and eventually enzymes will cease to function well. If fermentation continues longer than two or three minutes, cramping and fatigue will set in. Cramping seems to be due to lack of ATP needed to pump calcium ions back into the sarcoplasmic reticulum and to break the linkages between the actin and myosin filaments so that muscle fibers can relax.

Fortunately, aerobic respiration occurring in mitochondria usually provides most of a muscle's ATP. A muscle cell can make use of glycogen and fatty acids stored in the muscle as fuel, but still also needs a supply of oxygen in order to produce ATP

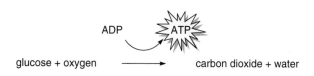

Myoglobin, an oxygen carrier similar to hemoglobin, is synthesized in muscle cells, and its presence accounts for the reddish brown color of skeletal muscle fibers. Myoglobin has a higher affinity of oxygen than does hemoglobin. Therefore, myoglobin can pull oxygen out of blood and make it available to muscle mitochondria which are carrying on aerobic respiration. Then, too, the ability of myoglobin to temporarily store oxygen reduces a muscle's immediate need for oxygen when aerobic respiration begins. The end products (carbon dioxide and water) can be rapidly disposed of, and the by-product heat keeps the entire body warm.

The three pathways for acquiring ATP work together during muscle contraction. But the anaerobic pathways are usually no longer needed as the body achieves an aerobic steady state. At this point, some lactate has accumulated but not enough to bring on exhaustion.

People who train rely even more heavily on aerobic respiration than people who do not train. In people who train, the number of muscle mitochondria increases, and so fermentation is not needed to produce ATP. Their mitochondria can start consuming oxygen as soon as ADP concentration starts rising during muscle contraction. Because mitochondria can break down fatty acid, instead of glucose, blood glucose is spared for the activity of the brain. (The brain, unlike other organs, can only utilize glucose to produce ATP.) Because less lactate is produced in people who train, the pH of the blood remains steady. And, there is less of an oxygen debt.

Oxygen Debt

When a muscle uses the anaerobic means of supplying energy needs, it incurs an **oxygen debt.** Oxygen debt is obvious when a person continues to breathe heavily after exercising. The ability to run up an oxygen debt is one of muscle tissue's greatest assets. Brain tissue cannot last nearly as long as muscles can without oxygen.

Repaying an oxygen debt requires replenishing creatine phosphate supplies and disposing of lactic acid. Lactic acid can be changed back to pyruvic acid and metabolized completely in mitochondria, or it can be sent to the liver to reconstruct glycogen. A marathon runner who has just crossed the finish line is not exhausted due to oxygen debt. Instead, the runner has used up all the muscles', and probably the liver's, glycogen supply. It takes about two days to replace glycogen stores on a high-carbohydrate diet.

Working muscles require a supply of ATP. Anaerobic creatine phosphate breakdown and glycolysis can quickly generate ATP. Aerobic respiration in mitochondria is best for sustained exercise.

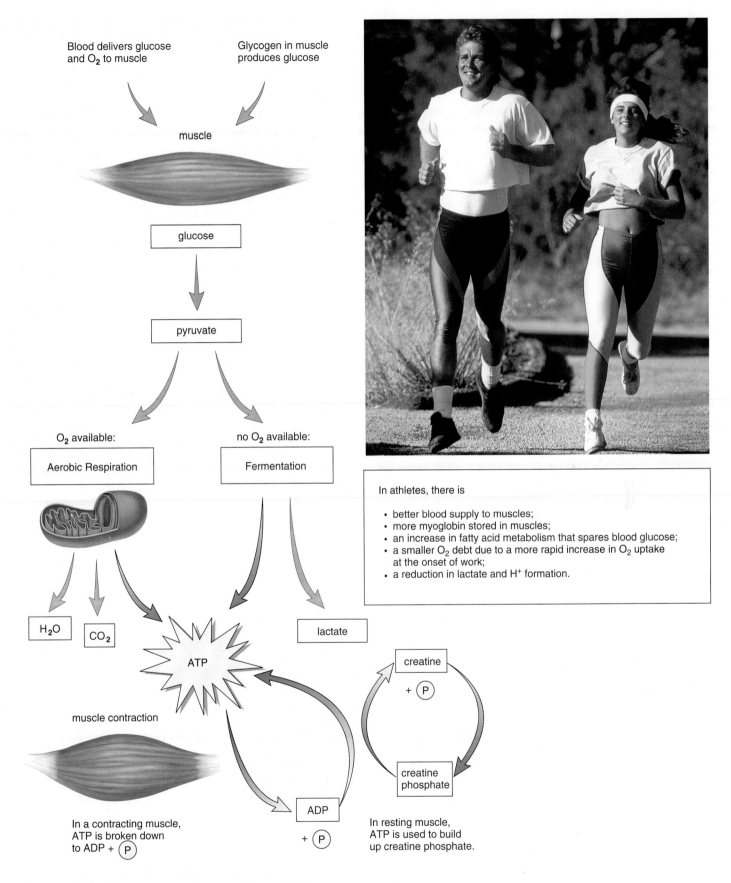

Blood delivers glucose and O_2 to muscle

Glycogen in muscle produces glucose

muscle

glucose

pyruvate

O_2 available:

Aerobic Respiration

no O_2 available:

Fermentation

H_2O

CO_2

lactate

ATP

muscle contraction

In a contracting muscle, ATP is broken down to ADP + (P)

ADP

+ (P)

creatine

+ (P)

creatine phosphate

In resting muscle, ATP is used to build up creatine phosphate.

In athletes, there is

- better blood supply to muscles;
- more myoglobin stored in muscles;
- an increase in fatty acid metabolism that spares blood glucose;
- a smaller O_2 debt due to a more rapid increase in O_2 uptake at the onset of work;
- a reduction in lactate and H^+ formation.

Figure 11.10   **Energy sources for muscle contraction.**

Athletics and Muscle Contraction

Athletes who excel in a particular sport, and much of the general public are interested today in staying fit by exercising. The Health reading on page 240 gives suggestions for exercise programs according to age.

Exercise and Size of Muscles

Muscles that are not used or that are used for only very weak contractions decrease in size, or atrophy. **Atrophy** can occur when a limb is placed in a cast or when the nerve serving a muscle is damaged. If nerve stimulation is not restored, muscle fibers gradually are replaced by fat and fibrous tissue. Unfortunately, atrophy can cause muscle fibers to shorten progressively, leaving body parts contracted in contorted positions.

Forceful muscular activity over a prolonged period causes muscle to increase in size as the number of myofibrils within muscle fibers increases. Increase in muscle size, called **hypertrophy,** occurs only if the muscle contracts to at least 75% of its maximum tension.

Some athletes take anabolic steroids, either testosterone or related chemicals, to promote muscle growth. This practice has many undesirable side effects such as cardiovascular disease, liver and kidney dysfunction, impotency and sterility, and even an increase in rash behavior called "roid mania."

Slow-Twitch and Fast-Twitch Muscle Fibers

We have seen that all muscle fibers metabolize both aerobically and anaerobically. Some fibers, however, utilize one method more than the other to provide myofibrils with ATP. Slow-twitch fibers tend to be aerobic, and fast-twitch fibers tend to be anaerobic (Fig. 11.11).

Slow-Twitch Fibers Slow-twitch fibers have a steadier tug and have more endurance despite motor units with a smaller number of fibers. These fibers are most helpful in sports like long-distance running, biking, jogging, and swimming. Because they produce most of their energy aerobically, they tire only when their fuel supply is gone. Slow-twitch fibers have many mitochondria and are dark in color because they contain **myoglobin,** a respiratory pigment found in muscles. They are also surrounded by dense capillary beds and draw more blood and oxygen than fast-twitch fibers. Slow-twitch fibers have a low maximum tension, which develops slowly, but these fibers are highly resistant to fatigue. Because slow-twitch fibers have a substantial reserve of glycogen and fat, their abundant mitochondria can maintain a steady, prolonged production of ATP when oxygen is available.

Fast-Twitch Fibers Fast-twitch fibers are those that tend to be anaerobic and seem to be designed for strength because their motor units contain many fibers. They provide explosions of energy and are most helpful in sports like sprinting, weight lifting, swinging a golf club, or putting a shot. Fast-twitch fibers are white in color because they have fewer mitochondria, little or no myoglobin, and fewer blood vessels than slow-twitch fibers do. Fast-twitch fibers can develop maximum tension more rapidly than slow-twitch fibers can, and their maximum tension is greater. However, their dependence on anaerobic energy leaves them vulnerable to an accumulation of lactic acid that causes them to fatigue quickly.

Success in sports is in part determined by the proportion of slow-twitch and fast-twitch fibers in a person's muscles.

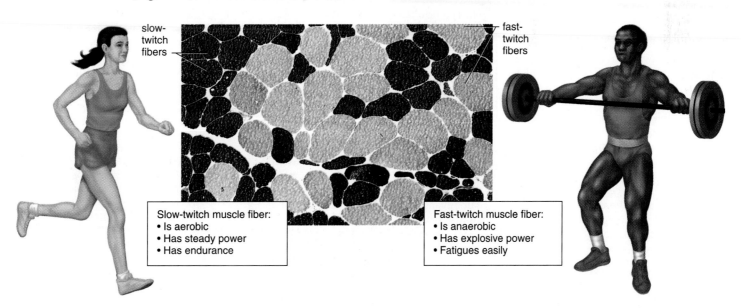

slow-twitch fibers

fast-twitch fibers

Slow-twitch muscle fiber:
• Is aerobic
• Has steady power
• Has endurance

Fast-twitch muscle fiber:
• Is anaerobic
• Has explosive power
• Fatigues easily

Figure 11.11 Slow- and fast-twitch fibers.
If your muscles contain many slow-twitch fibers (dark color), you would probably do better at a sport like cross-country running. But if your muscles contain many fast-twitch fibers (light color), you would probably do better at a sport like weight lifting.

11.5 Homeostasis

The previous page tells how the muscular system works with other systems of the body to maintain homeostasis.

Cardiac muscle contraction accounts for the heartbeat which creates blood pressure, the force that propels blood in the arteries and arterioles. The arterioles branch into the capillaries where exchange takes place that creates and cleanses tissue fluid. Blood and tissue fluid are the internal environment of the body, and without cardiac muscle contraction blood would never reach the capillaries for exchange to take place. Blood is returned to the heart in cardiovascular veins, and excess tissue fluid is returned to the cardiovascular system within lymphatic vessels. Skeletal muscle contraction presses on the cardiovascular veins and lymphatic vessels, and this creates the pressure that moves fluids in both types of vessels.

Constriction of arteriole smooth muscle walls regulates the flow of blood within arterioles, and contraction of sphincters temporarily prevents the flow of blood into a capillary. This is an important homeostatic mechanism because in times of emergencies it is more important, for example, for blood to be directed toward the skeletal muscles than to the tissues of the digestive tract.

Smooth muscle contraction accounts for peristalsis, the process that moves food along the digestive tract. Without this action, food would never reach the various organs of the digestive tract where digestion releases nutrients that enter the bloodstream. Smooth muscle contraction assists the voiding of urine. The formation of urine is necessary for ridding the body of metabolic wastes and for the regulation of blood volume, salt concentration, and pH of internal fluids.

Skeletal muscles protect the bones and other internal organs. Skeletal muscle contraction raises the rib cage and lowers the diaphragm during the active phase of breathing. As we breathe, oxygen enters the blood and is delivered to the tissues, including the muscles where ATP is produced in mitochondria with heat as a by-product. The heat produced by skeletal muscle contraction allows the body temperature to remain within the normal range for human beings.

The overall importance of muscle contraction to the maintenance of good health cannot be overestimated. Exercise improves the functioning of the heart, increases the respiratory volume, enhances immunity, and in general improves the health of the individual.

Muscle contraction is involved in the functioning of the organs of the respiratory system, the digestive system, and the urinary system. Cardiac muscle contraction keeps the heart pumping so that blood is delivered to the tissues. Skeletal muscle contraction presses on the veins so that blood is returned to the heart.

Bioethical Issue

On page 237, we learned that athletes are better at one sport or another, depending on whether their muscles contain fast- or slow-twitch fibers. A natural advantage of this sort does not bar an athlete from participating in and winning a medal in a particular sport at the Olympic games. Nor are athletes restricted to a certain amount of practice or required to eliminate certain foods from their diet.

Athletes are, however, prevented from participating in the Olympic games if they have taken certain performance enhancing drugs. There is no doubt that regular use of drugs like anabolic steroids leads to kidney disease, liver dysfunction, hypertension, and a myriad of other undesirable side effects. Even so, shouldn't the individual be allowed to take these drugs if they want to? Anabolic steroids are synthetic forms of the male sex hormone testosterone. Large doses along with strength training leads to much larger muscles than otherwise. Extra strength and endurance can give an athlete an advantage in certain sports such as racing, swimming, and weight lifting.

Should the Olympic committee outlaw the taking of anabolic steroids and, if so, on what basis? The basis can't be an unfair advantage, because some athletes naturally have an unfair advantage over other athletes. Should these drugs be outlawed on the basis of health reasons? Excessive practice alone and a purposeful decrease or increase in weight to better perform in a sport can also be injurious to one's health. In other words, how can you justify allowing some behaviors that enhance performance and not others?

Questions

1. Do you believe that the manner in which athletes wish to train and increase their performance should be regulated in any way? Why or why not?
2. Is it all right for athletes to endanger their health by excessive practice, gaining or losing pounds, or the taking of drugs? Why or why not?
3. Who should be in charge of regulating the behavior of athletes so that they do not do harm to themselves?

Human Systems Work Together

Integumentary System

Muscle contraction provides heat to warm skin.

Skin protects muscles; rids the body of heat produced by muscle contraction.

Skeletal System

Muscle contraction causes bones to move joints; muscles help protect bones.

Bones provide attachment sites for muscles; store Ca^{2+} for muscle function.

Nervous System

Muscle contraction moves eyes, permits speech, creates facial expressions.

Brain controls nerves that innervate muscles; receptors send sensory input from muscles to brain.

Endocrine System

Muscles help protect glands.

Androgens promote growth of skeletal muscle; epinephrine stimulates heart and constricts blood vessels.

Cardiovascular System

Muscle contraction keeps blood moving in heart and blood vessels.

Blood vessels deliver nutrients and oxygen to muscles, carry away wastes.

How the Muscular System works with other body systems

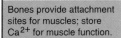

Lymphatic System/Immunity

Skeletal muscle contraction moves lymph; physical exercise enhances immunity.

Lymphatic vessels pick up excess tissue fluid; immune system protects against infections.

Respiratory System

Muscle contraction assists breathing; physical exercise increases respiratory capacity.

Lungs provide oxygen for, and rid the body of, carbon dioxide from contracting muscles.

Digestive System

Smooth muscle contraction accounts for peristalsis; skeletal muscles support and help protect abdominal organs.

Digestive tract provides glucose for muscle activity; liver metabolizes lactic acid following anaerobic muscle activity.

Urinary System

Smooth muscle contraction assists voiding of urine; skeletal muscles support and help protect urinary organs.

Kidneys maintain blood levels of Na^+, K^+, and Ca^{2+}, which are needed for muscle innervation, and eliminate creatinine, a muscle waste.

Reproductive System

Muscle contraction occurs during orgasm and moves gametes; abdominal and uterine muscle contraction occurs during childbirth.

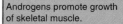

Androgens promote growth of skeletal muscle.

Exercise, Exercise, Exercise

Exercise programs improve muscular strength, muscular endurance, and flexibility. Muscular strength is the force a muscle group (or muscle) can exert against a resistance in one maximal effort. Muscular endurance is judged by the ability of a muscle to contract repeatedly or sustain a contraction for an extended period. Flexibility is tested by observing the range of motion about a joint.

Exercise also improves cardiorespiratory endurance. The heart rate and capacity increase, and the air passages dilate so that the heart and lungs are able to support prolonged muscular activity. The blood level of high-density lipoprotein (HDL), the molecule that prevents the development of plaque in blood vessels, increases. Also, body composition, the proportion of protein to fat, changes favorably when you exercise.

Exercise also seems to help prevent certain kinds of cancer. Cancer prevention involves eating properly, not smoking, avoiding cancer-causing chemicals and radiation, undergoing appropriate medical screening tests, and knowing the early warning signs of cancer. However, studies show that people who exercise are less likely to develop colon, breast, cervical, uterine, and ovarian cancers.

Physical training with weights can improve bone density and strength and muscular strength and endurance in all adults regardless of age. Even men and women in their eighties and nineties make substantial gains in bone and muscle strength, which can help them lead more independent lives. Exercise helps prevent osteoporosis, a condition in which the bones are weak and tend to break. Exercise promotes the activity of osteoblasts in young people as well as older people. The stronger the bones when a person is young, the less chance of osteoporosis as a person ages. Exercise helps prevent weight gain not only because the level of activity increases but also because muscles metabolize faster than other tissues. As a person becomes more muscular, it is less likely that fat will accumulate.

Exercise relieves depression and enhances the mood. Some report exercise actually makes them feel more energetic, and after exercising, particularly in the late afternoon, they sleep better that night. Self-esteem rises not only because of improved appearance, but also due to other factors that are not well understood. It is known that vigorous exercise releases endorphins, hormone-like chemicals that are known to alleviate pain and provide a feeling of tranquility.

A sensible exercise program is one that provides all the benefits without the detriments of a too strenuous program. Overexertion can actually be harmful to the body and might result in sports injuries such as a bad back or bad knees. The programs suggested in Table 11A are tailored according to age and, if followed, are beneficial.

Dr. Arthur Leon at the University of Minnesota performed a study involving 12,000 men, and the results showed that only moderate exercise is needed to lower the risk of a heart attack by one-third. In another study conducted by the Institute for Aerobics Research in Dallas, Texas, which included 10,000 men and more than 3,000 women, even a little exercise was found to lower the risk of death from circulatory diseases and cancer. Increasing daily activity by walking to the corner store instead of driving and by taking the stairs instead of the elevator can improve your health.

Table 11A	A Checklist for Staying Fit		
Children, 7–12	**Teenagers, 13–18**	**Adults, 19–55**	**Seniors, 55 and up**
Vigorous activity 1–2 hours daily	Vigorous activity 1 hour 3–5 days a week, otherwise 1/2 hour daily moderate activity	Vigorous activity 1 hour 3 days a week, otherwise 1/2 hour daily moderate activity	Moderate exercise 1 hour daily 3 days a week, otherwise 1/2 hour daily moderate activity
Free play	Build muscle with calisthenics	Exercise to prevent lower back pain: aerobics, stretching, yoga	Plan a daily walk
Build motor skills through team sports, dance, swimming	Plan aerobic exercise to control buildup of fat cells	Take active vacations: hike, bicycle, cross-country ski	Daily stretching exercise
Encourage more exercise outside of physical education classes	Pursue tennis, swimming, horseback riding—sports that can be enjoyed for a lifetime	Find exercise partners: join a running club, bicycle club, outing group	Learn a new sport or activity: golf, fishing, ballroom dancing
Initiate family outings: bowling, boating, camping, hiking	Continue team sports, dancing, hiking, swimming		Try low-impact aerobics. Before undertaking new exercises, consult your doctor

Summarizing the Concepts

11.1 Skeletal Muscles

Skeletal muscles have levels of organization. A whole muscle contains bundles of muscle fibers; muscle fibers contain myofibrils; and myofibrils contain actin and myosin filaments. Skeletal muscles are usually attached to the skeleton, where some are prime movers, some are synergists, and others are antagonists. Muscles have various functions; they provide movement and heat, help maintain posture, and protect underlying organs.

Muscles are named for their size, shape, direction of fibers, location, number of attachments, and action.

11.2 Mechanism of Muscle Fiber Contraction

Nerve impulses travel down motor neurons and meet muscle fibers at neuromuscular junctions. The sarcolemma of a muscle fiber forms T tubules that extend into the fiber and almost touch the sarcoplasmic reticulum which stores calcium ions. When calcium ions are released into muscle fibers, actin filaments slide past myosin filaments within the sarcomeres of a myofibril.

At a neuromuscular junction, synaptic vesicles release acetylcholine (ACh), which binds to protein receptors on the sarcolemma, causing impulses to travel down the T tubules and calcium to leave the sarcoplasmic reticulum. Myofibril contraction follows.

Calcium ions bind to troponin and cause tropomyosin threads winding around actin filaments to shift their position, revealing myosin binding sites. The myosin filament is composed of many myosin molecules, each containing a head with an ATP binding site. Myosin is an ATPase, and once it breaks down ATP, the myosin head is ready to attach to actin. The release of ADP + ⓟ causes the head to change its position. This is the power stroke that causes the actin filament to slide toward the center of a sarcomere. When myosin catalyzes another ATP, the head detaches from actin, and the cycle begins again.

11.3 Whole Muscle Contraction

In the laboratory, muscle contraction is described in terms of a muscle twitch, summation, and tetanus. In the body, muscles exhibit tone, in which tetanic contraction involving a number of fibers is the rule. The strength of muscle contraction varies according to recruitment of motor units.

11.4 Energy for Muscle Contraction

A muscle fiber has three ways to acquire ATP after muscle contraction begins: (1) creatine phosphate, built up when a muscle is resting, donates phosphates to ADP, forming ATP; (2) fermentation with the concomitant accumulation of lactic acid quickly produces ATP; and (3) oxygen-dependent aerobic respiration occurs within mitochondria. Two of these ways (1 and 2) are anaerobic (they do not require oxygen). Fermentation can result in oxygen debt because oxygen is needed to complete the metabolism of lactic acid.

Exercise results in muscle fiber increase called hypertrophy. Certain sports like running and swimming can be associated with slow-twitch fibers, which rely on aerobic respiration to acquire ATP. They have a plentiful supply of mitochondria and myoglobin, which gives them a dark color. Other sports like weight lifting can be associated with fast-twitch fibers, which rely on an anaerobic means of acquiring ATP. They have few mitochondria and myoglobin, and their motor units contain more muscle fibers. Fast-twitching fibers are known for their explosive power, but they fatigue quickly.

11.5 Homeostasis

The skeletal system works with the other systems of the body in the ways described in the box on page 239.

Studying the Concepts

1. List and discuss the functions of muscles. 226
2. Describe the levels of structure of a muscle from the whole muscle level to the myofilaments within a myofibril. 227
3. Describe the steps resulting in muscle contraction by starting with the motor neuron and ending with the sliding of actin filaments. 231
4. Describe the structure and function of a neuromuscular junction. 232
5. Describe the cyclical events as myosin pulls actin toward the center of a sarcomere. 233
6. Contrast a muscle twitch with summation and tetanus. 234
7. What is tone, and how is it maintained? 234
8. By what mechanism does the strength of muscle contraction vary? 234
9. What are the three ways a muscle fiber can acquire ATP for muscle contraction? How are the three ways interrelated? 235–36
10. What is atrophy? Hypertrophy? 237
11. Contrast slow-twitch and fast-twitch fibers in as many ways as possible. 237

Testing Your Knowledge of the Concepts

In questions 1–4, match the muscles to a region of the body, and fill in the blanks.

a.   head and neck
b.   trunk
c.   arms
d.   legs

_____ 1. hamstring group
_____ 2. trapezius
_____ 3. rectus abdominis
_____ 4. triceps brachii

In questions 5–7, indicate whether the statement is true (T) or false (F).

_____ 5. Myosin breaks down ATP and pulls actin filaments.
_____ 6. Slow-twitch muscle fibers are associated with weight lifting.
_____ 7. Muscle cramping appears to be due to a depletion of oxygen and a buildup of lactic acid.

In questions 8 and 9, fill in the blanks.

8. _____ stores high-energy phosphate bonds in muscle fibers.

9. _____ is the transmitter released by a motor neuron at a neuromuscular junction.

10. Label this diagram of a muscle fiber, using these terms: myofibril, mitochondrion, T tubule, sarcomere, sarcolemma, sarcoplasmic reticulum.

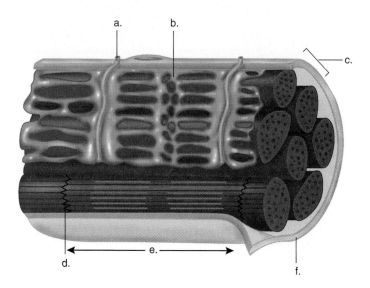

Applying Your Knowledge to the Concepts

These questions pertain to homeostasis.

1. Which type exercise—isotonic or isometric—would you recommend for someone with arthritis? Why?

2. Under laboratory conditions, nervous stimulation no longer causes a muscle to contract, but direct stimulation of the muscle does produce contraction. Explain.

3. Muscles store glucose as glycogen and use it as a source of glucose for muscle contraction. The liver stores glucose as glycogen and then returns glucose to the bloodstream. Explain the advantage of this difference in the two organs.

4. Bodybuilders concentrate on exercises that increase muscle size and strength but do not increase muscle endurance. This type of exercise is often referred to as anaerobic or isometric. Isotonic exercises do not increase the size of the muscle but do increase endurance. This type of exercise is referred to as aerobic. Which type would place stress on the cardiovascular and respiratory systems? Give some examples of this type of exercise.

Understanding the Terms

actin 231
atrophy 237
creatine phosphate 235
hypertrophy 237
insertion 227
motor unit 234
muscle twitch 234
myofibril 231
myoglobin 237
myogram 234
myosin 231
neuromuscular junction 232

origin 227
oxygen debt 235
sarcolemma 231
sarcomere 231
sarcoplasmic reticulum 231
sliding filament theory 231
tendon 227
tetanus 234
tone 234
tropomyosin 233
troponin 233
T (transverse) tubule 231

Match the terms to these definitions:
a. _____ Structural and functional unit of a myofibril; contains actin and myosin filaments.
b. _____ End of a muscle that is attached to a movable bone.
c. _____ Sustained maximal muscle contraction.
d. _____ Oxygen that is needed to metabolize lactate, a compound that accumulates during vigorous exercise.
e. _____ Motor neuron and the muscle fibers associated with it.

Applying Technology to the Concepts

Your study of the muscular system is supported by these available technologies:

Essential Study Partner CD-ROM

Animals → Support / Locomotion

Visit the Mader web site for related ESP activities.

Exploring the Internet

The Mader Home Page provides resources and tools as you study this chapter.

http://www.mhhe.com/biosci/genbio/mader

Dynamic Human 2.0 CD-ROM

Muscular System
Nervous System

Further Readings for Part 3

Applegate, E. J. 1995. *The anatomy and physiology learning system.* Philadelphia: W. B. Saunders Publishing. Designed for a one-semester introductory course, this text provides fundamental information for students with minimal science background.

Armbruster, P., and Hessberger, F. P. September 1998. Making new elements. *Scientific American* 279(3):72. Three new elements have been added to the periodic table; article shows the process of creating artificial elements.

Clemente, C. D. 1996. *Anatomy: A regional atlas of the human body.* 4th ed. Philadelphia: Lea and Febiger. This atlas contains both drawings and photographs of anatomical regions of the human body.

Deyo, R. A. August 1998. Low-back pain. *Scientific American* 279(2):48. Treatment options for low-back pain which don't involve bed rest or surgery are improving.

Fox, S. I. 1996. *Human physiology.* 5th ed. Dubuque, Iowa: Wm. C. Brown Publishers. This is an introductory physiology text.

Frank, J. September/October 1998. How the ribosome works. *American Scientist* 86(5):428. New imaging techniques using cryo-electron microscopy allows researchers to study a three-dimensional map of the ribosome.

Grillner, S. January 1996. Neural networks for vertebrate locomotion. *Scientific American* 274(1):64. Discoveries about how the brain coordinates muscle movement raise hopes for restoration of mobility for some accident victims.

Gunstream, S. E. 1995. *Anatomy and physiology.* 2d ed. Dubuque, Iowa: Wm. C. Brown Publishers. This is an introductory anatomy and physiology text-workbook.

Guyton, A. C., and Hall, J. E. 1996. *Textbook of medical physiology.* Philadelphia: Saunders College Publishing. Presents physiological principles for those in the medical fields.

Halstead, L. S. April 1998. Post-polio syndrome. *Scientific American* 278(4):42. Recovered polio victims are experiencing fatigue, pain, and weakness, resulting from degeneration of motor neurons.

Mader, S. S. 1997. *Understanding anatomy and physiology.* 3d ed. Dubuque, Iowa: Wm. C. Brown Publishers. A text that emphasizes the basics for beginning allied health students.

Marieb, E. N. 1997. *Human anatomy and physiology.* 4th ed. Redwood City, Calif.: Benjamin/Cummings Publishing. A thorough anatomy and physiology text that can safely be used as a complete and accurate reference.

Rischer, C. E., and Easton, T. A. 1995. *Focus on human biology*. 2d ed. New York: HarperCollins College Publishers. This comprehensive introductory textbook stresses basic human anatomy and physiology.

Rome, L. C. July/August 1997. Testing a muscle's design. *American Scientist* 85(4):356. Muscular systems adapt to specific functions, such as the specialized muscles found in frogs.

Scerri, E. R. September 1998. The evolution of the periodic system. *Scientific American* 279(3):78. Article discusses the history and evolution of the periodic table.

Shier, D., et al. 1996. *Hole's human anatomy & physiology*. 7th ed. Dubuque, Iowa: Wm. C. Brown Publishers. An introductory anatomy and physiology text that has proved useful to beginning students.

Thomas, E. D. September/October 1995. Hematopoietic stem cell transplantation. *Scientific American Science & Medicine* 2(5):38. Article discusses reconstituting marrow from cultured stem cells for bone marrow transplants.

Tortora, G. J. 1996. *Atlas of the human skeleton*. New York: HarperCollins Publishers. This supplement contains carefully selected and labeled photographs of the human skeleton.

Van De Graaff, K.M. 1998. *Human anatomy*. 5th ed. Dubuque, Iowa: WCB/McGraw-Hill. This introductory text provides a balanced presentation of anatomy at various levels.

Vander, A., et al. 1998. *Human physiology: The mechanisms of body function*. 7th ed. Dubuque: Iowa: WCB:McGraw-Hill. Presents the principles of human physiology with an emphasis on physiological mechanisms and homeostasis.

White, R. J. September 1998. Weightlessness and the human body. *Scientific American* 279(3):58. Space medicine is providing new ideas about treatment of osteoporosis and anemia.

Wolkomir, R. August 1998. Oh, my aching back. *Smithsonian* 29(5):36. Researchers try to pin-point the source of back pain using a Virtual Corset to monitor the subject's activities.

The nervous system is the ultimate coordinator of homeostasis, that is, an internal environment that is in dynamic equilibrium. Nerves bring information to the brain and spinal cord from sensory receptors that detect changes both inside and outside the body. Then, nerves take the commands given by the brain and spinal cord to effectors, allowing the body to respond to these changes.

The endocrine system, like the nervous system, regulates other organs, but it acts more slowly and brings about a response that lasts longer. The endocrine organs secrete chemical messengers called hormones into the bloodstream. After arriving at their target organs, hormones alter cellular metabolism.

We now know that the nervous system and the endocrine system are joined in numerous ways despite their very different modes of operation.

Chapter 12

Nervous System

Chapter Concepts

12.1 Neurons and How They Work
- The nervous system contains cells called neurons, which are specialized to carry nerve impulses. 246
- A nerve impulse is an electrochemical change that travels along the length of a neuron fiber. 248
- Transmission of signals between neurons is dependent on neurotransmitter molecules. 251

12.2 Peripheral Nervous System
- The peripheral nervous system contains nerves that conduct nerve impulses toward and away from the central nervous system. 252

12.3 Central Nervous System
- The central nervous system is made up of the spinal cord and the brain. 256
- The parts of the brain are specialized for particular functions. 259
- The reticular formation contains fibers that arouse the brain when active and account for sleep when they are inactive. 259

12.4 The Cerebral Hemispheres
- The cerebral cortex contains motor areas, sensory areas, and association areas which are in communication with each other. 260
- The limbic system contains cortical and subcortical areas that are involved in higher mental functions and emotional responses. 261

12.5 Higher Mental Functions
- Long-term memory depends upon association areas that are in contact with the limbic system. 262
- There are particular areas in the left hemisphere that are involved in language and speech. 263

12.6 Homeostasis
- The nervous system works with the other systems of the body to maintain homeostasis. 265

12.7 Drug Abuse
- The use of psychoactive drugs such as alcohol, nicotine, marijuana, cocaine, and heroin is detrimental to the body. 265

It was a warm spring Saturday in 1995 when a crowd gathered in Culpeper, Virginia, to enjoy a horse riding competition. Everything was fine until—suddenly—one horse stopped dead in its tracks. The horse's rider, Superman star Christopher Reeve, went tumbling through the air. Hitting the ground, Reeve crushed the top two vertebrae in his neck, damaging the sensitive spinal cord underneath. In a split second, he was paralyzed. Doctors say he is lucky to be alive.

The spinal cord is a ropelike bundle of long nerve tracts that shuttle messages between the brain and the rest of the body. Together the brain and spinal cord compose the **central nervous system (CNS),** which interprets sensory input before coordinating a response that helps maintain homeostasis. The **peripheral nervous system (PNS)** consists of nerves, which carry sensory information to the CNS and also motor commands from the CNS to the muscles and glands (Fig. 12.1).

When Reeve hurt his spinal cord, the CNS lost its avenue of communication with the portion of his body located below the site of damage. He receives no sensation from most of his body nor can he command his arms and legs to move. But cranial nerves from his eyes and ears still allow him to see and hear and his brain still enables him to have emotions, remember, and reason. Also his internal organs still function normally, a sign that his injury was not as severe as it could have been. In this chapter we will examine the structure of the nervous system and how it carries out its numerous functions.

12.1 Neurons and How They Work

Nervous tissue contains two types of cells: neuroglial cells and neurons. **Neuroglial cells** support and service **neurons**, the cells that actually transmit nerve impulses (see Fig. 3.7).

Neuron Structure

Despite their varied appearance, all neurons have just three parts: dendrites, cell body, and an axon. It becomes apparent from studying the motor neuron in Figure 12.2 that **dendrites** are processes that send signals toward the cell body. The **cell body** is the part of a neuron that contains the nucleus and other organelles. An **axon** conducts nerve impulses along its entire length. Axons are sometimes referred to as long fibers.

There are three classes of neurons: sensory neurons, motor neurons, and interneurons, whose functions are best described in relation to the CNS (Fig. 12.2). A **sensory neuron** takes information from a sensory receptor to the CNS, and a **motor neuron** takes information away from the CNS to an effector (muscle fiber or gland). An **interneuron** conveys information between neurons in the CNS. An interneuron can receive input from sensory neurons and also from other interneurons in the CNS. Thereafter, they sum up all these signals before sending commands out to the muscles and glands by way of motor neurons.

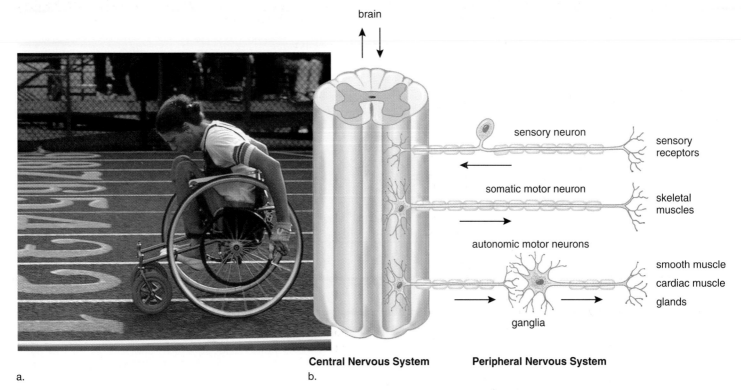

brain

sensory neuron

sensory receptors

somatic motor neuron

skeletal muscles

autonomic motor neurons

smooth muscle
cardiac muscle
glands

ganglia

Central Nervous System   **Peripheral Nervous System**

a.   b.

Figure 12.1   Organization of nervous system.

a. In paraplegics, messages no longer flow between the legs and the central nervous system (the spinal cord and brain). **b.** The sensory neurons of the peripheral nervous system take nerve impulses from sensory receptors to the central nervous system (CNS), and motor neurons (both somatic to skeletal muscles and autonomic to internal organs) take nerve impulses from the CNS to the organs mentioned.

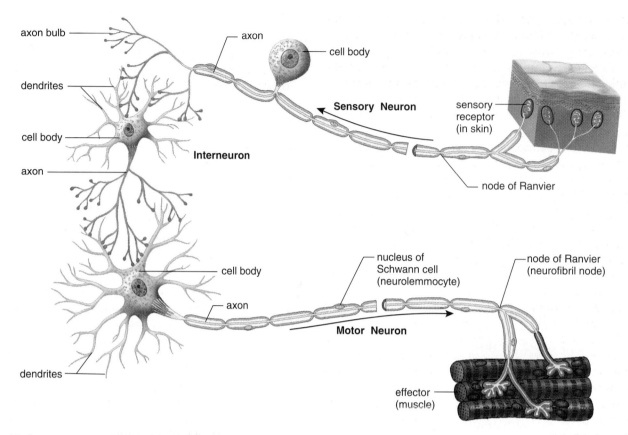

Figure 12.2   **Types of neurons.**

A sensory neuron, an interneuron, and a motor neuron are drawn here to show their arrangement in the body. (The breaks indicate that the fibers are much longer than shown.) How does this arrangement correlate with the function of each neuron?

Myelin Sheath

Long axons are covered by a protective **myelin sheath** formed by neuroglial cells called **Schwann cells** (neurolemmocytes). The myelin sheath develops when Schwann cells wrap themselves around an axon many times and in this way lay down several layers of plasma membrane. Schwann cell plasma membrane contains myelin, a lipid substance that is an excellent insulator. A myelin sheath, which is interrupted by gaps called **nodes of Ranvier**, gives nerve fibers their white, glistening appearance (Fig. 12.3).

Multiple sclerosis (MS) is a disease of the myelin sheath. Lesions develop and become hardened scars that interfere with normal conduction of nerve impulses, and the result is various neuromuscular symptoms. On the other hand, the myelin sheath plays an important role in nerve regeneration. If an axon is accidentally severed, the distal part of the axon degenerates but the myelin sheath remains and serves as a passageway for new fiber growth.

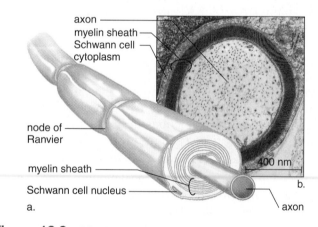

Figure 12.3   **Myelin sheath.**

a. A myelin sheath forms when Schwann cells wrap themselves around a nerve fiber. **b.** Electron micrograph of a cross section of an axon surrounded by a myelin sheath.

All neurons have three parts: dendrites, cell body, and axon. Sensory neurons take information to the CNS, and interneurons sum up sensory input before motor neurons take commands away from the CNS.

Nerve Impulse

A **nerve impulse** is the way a neuron transmits information. The nature of a nerve impulse has been studied by using excised axons and a voltmeter called an **oscilloscope.** Voltage, often measured in millivolts (mV), is a measure of the electrical potential difference between two points, which in this case are the inside and the outside of the axon. Voltage is displayed on the oscilloscope screen as a trace, or pattern over time.

Resting Potential

In the experimental setup shown in Figure 12.4, an oscilloscope is wired to two electrodes: one electrode is placed inside and the other electrode is placed outside an axon. The axon is essentially a membranous tube filled with axoplasm (cytoplasm of the axon). When the axon is not conducting an impulse, the oscilloscope records a potential difference across a membrane equal to about −65 mV. This reading indicates that the inside of the axon is negative compared to the outside. This is called the **resting potential** because the axon is not conducting an impulse.

The existence of this polarity (charge difference) can be correlated with a difference in ion distribution on either side of the axomembrane (plasma membrane of the axon). As Figure 12.4*a* shows, the concentration of sodium ions (Na^+) is greater outside the axon than inside, and the concentration of potassium ions (K^+) is greater inside the axon than outside. The unequal distribution of these ions is due to the action of the **sodium-potassium pump,** a membrane protein in the membrane that actively transports Na^+ out of and K^+ into the axon. The work of the pump maintains the unequal distribution of Na^+ and K^+ across the membrane.

The pump is always working because the membrane is somewhat permeable to these ions, and they tend to diffuse toward their lesser concentration. Since the membrane is more permeable to K^+ than to Na^+, there are always more positive ions outside the membrane than inside. This accounts for the polarity recorded by the oscilloscope. Large, negatively charged organic ions in the axoplasm also contribute to the polarity across a resting axomembrane.

Because of the sodium-potassium pump there is a concentration of Na^+ outside an axon and K^+ inside an axon. An unequal distribution of ions causes the inside of an axon to be negative compared to the outside.

Action Potential

An **action potential** is a rapid change in polarity across an axomembrane as the nerve impulse occurs. An action potential is an all or none phenomenon. If a stimulus causes the axomembrane to depolarize to a certain level, called **threshold,** an action potential occurs. The strength of an action potential does not change, but an intense stimulus can cause an axon to fire (start an axon potential) at a greater frequency than a weak stimulus.

The action potential requires two types of gated channel proteins in the membrane. There is a gated channel protein that opens allowing Na^+ to pass through the membrane, and another that opens allowing K^+ to pass through the membrane. The sodium channel is faster to open than the potassium channel.

Sodium Gates Open   When an action potential occurs, the gates of sodium channels open first and Na^+ flows into the axon. As Na^+ moves to inside the axon, the membrane potential changes from −65 mV to +40 mV. This is a *depolarization* because the charge inside the axon changes from negative to positive as Na^+ enters the axon.

Potassium Gates Open   Second, the gates of potassium channels open and K^+ flows to outside the axon. As K^+ moves to outside the axon, the action potential changes from +40 mV back to −65 mV. This is a *repolarization* because the inside of the axon resumes a negative charge as K^+ exits the axon.

The nerve impulse consists of an electrochemical change that occurs across an axomembrane. During depolarization, Na^+ moves to inside the axon, and during repolarization K^+ moves to outside the axon.

Propagation of an Action Potential

When an action potential travels down an axon, each successive portion of the axon undergoes a depolarization and then a repolarization. Like a domino effect, each preceding portion causes an action potential in the next portion of an axon.

As soon as an action potential has moved on, the previous portion of an axon undergoes a **refractory period** during which the sodium gates are unable to open. This ensures that the action potential cannot move backwards and instead always moves down an axon toward its branches.

In myelinated axons, the gated ion channels that produce an action potential are concentrated at the nodes of Ranvier. Since ion exchange occurs only at the nodes, the action potential travels faster than in nonmyelinated axons. This is called saltatory conduction, meaning that the action potential "jumps" from node to node. Speeds of 200 meters per second (450 miles per hour) have been recorded.

An action potential travels along the length of an axon.

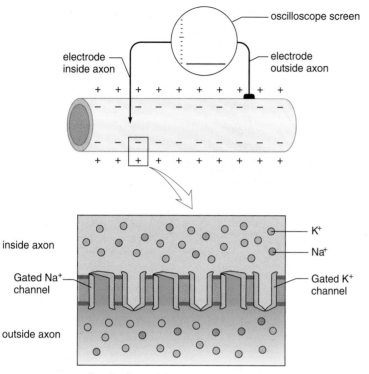

a. Resting Potential

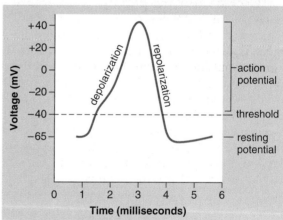

c. Enlargement of action potential

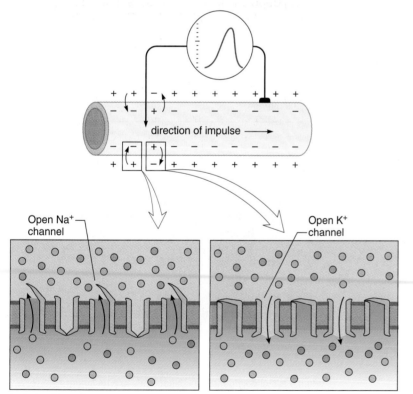

b. Action Potential

Figure 12.4   Resting and action potential.
a. Resting potential. An oscilloscope records a resting
potential of −65 mV. There is a preponderance of Na$^+$
outside the axon and preponderance of K$^+$ inside the
axon. The permeability of the membrane to K$^+$ compared
to Na$^+$ causes the inside to be negative compared to the
outside. **b.** Action potential. A depolarization occurs when
Na$^+$ gates open and Na$^+$ moves to inside the axon, and a
repolarization occurs when K$^+$ gates open and K$^+$ moves
to outside the axon. **c.** Enlargement of the action potential,
which is seen by an experimenter using an oscilloscope,
an instrument that records voltage changes.

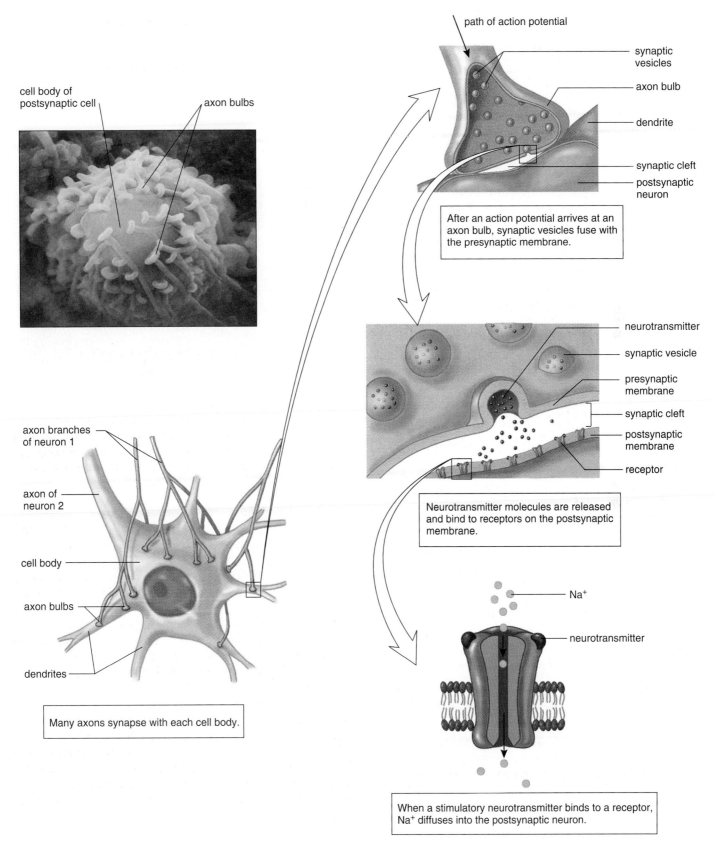

path of action potential

synaptic vesicles

axon bulb

dendrite

synaptic cleft

postsynaptic neuron

cell body of postsynaptic cell

axon bulbs

After an action potential arrives at an axon bulb, synaptic vesicles fuse with the presynaptic membrane.

neurotransmitter

synaptic vesicle

presynaptic membrane

synaptic cleft

postsynaptic membrane

receptor

Neurotransmitter molecules are released and bind to receptors on the postsynaptic membrane.

axon branches of neuron 1

axon of neuron 2

cell body

axon bulbs

dendrites

Many axons synapse with each cell body.

Na⁺

neurotransmitter

When a stimulatory neurotransmitter binds to a receptor, Na⁺ diffuses into the postsynaptic neuron.

Figure 12.5   **Synapse structure and function.**

Transmission across a synapse from one axon to another occurs when a neurotransmitter is released by the presynaptic neuron and diffuses across a synaptic cleft and binds to a receptor in the postsynaptic neuron.

Transmission Across a Synapse

Every axon branches into many fine endings, each tipped by a small swelling, called an **axon bulb** (Fig. 12.5*a, b*). Each bulb lies very close to the dendrite (or the cell body) of another neuron. This region of close proximity is called a **synapse.** At a synapse, the membrane of the first neuron is called the *pre*synaptic membrane, and the membrane of the next neuron is called the *post*synaptic membrane. The small gap between is the **synaptic cleft.**

Transmission across a synapse is carried out by molecules called **neurotransmitters,** which are stored in synaptic vesicles (Fig. 12.5*b, c*). When nerve impulses traveling along an axon reach an axon bulb, gated channels for calcium ions (Ca^{2+}) open and calcium enters the bulb. This sudden rise in Ca^{2+} stimulates synaptic vesicles to merge with the presynaptic membrane, and neurotransmitter molecules are released into the synaptic cleft. They diffuse across the cleft to the postsynaptic membrane, where they bind with specific receptor proteins (Fig. 12.5*c*).

Depending on the type of neurotransmitter and/or the type of receptor, the response of the postsynaptic neuron can be toward excitation or toward inhibition. Excitatory neurotransmitters that utilize gated ion channels are fast acting. Other neurotransmitters affect the metabolism of the postsynaptic cell and therefore they are slower acting.

Neurotransmitter Molecules

At least 25 different neurotransmitters have been identified, but two very well-known neurotransmitters are **acetylcholine (ACh)** and **norepinephrine (NE).**

Once a neurotransmitter has been released into a synaptic cleft and has initiated a response, it is removed from the cleft. In some synapses, the postsynaptic membrane contains enzymes that rapidly inactivate the neurotransmitter. For example, the enzyme **acetylcholinesterase (AChE)** breaks down acetylcholine. In other synapses, the presynaptic membrane rapidly reabsorbs the neurotransmitter, possibly for repackaging in synaptic vesicles or for molecular breakdown. The short existence of neurotransmitters at a synapse prevents continuous stimulation (or inhibition) of postsynaptic membranes.

It is of interest to note here that many drugs that affect the nervous system act by interfering with or potentiating the action of neurotransmitters. As described in Figure 12.18, drugs can enhance or block the release of a neurotransmitter, mimic the action of a neurotransmitter or block the receptor, or interfere with the removal of a neurotransmitter from a synaptic cleft.

Transmission across a synapse is dependent on the release of neurotransmitters, which diffuse across the synaptic cleft from one neuron to the next.

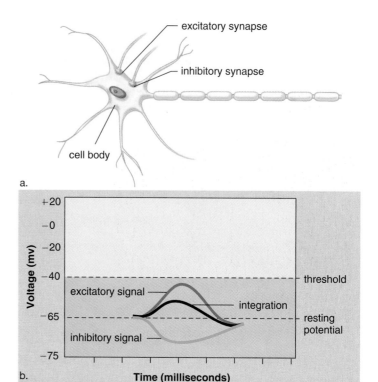

a.

b.

Figure 12.6 Integration.
a. Inhibitory signals and excitatory signals are summed up in the dendrite and cell body of the postsynaptic neuron. Only if the combined signals cause the membrane potential to rise above threshold does an action potential occur. **b.** In this example, threshold was not reached.

Synaptic Integration

A single neuron has many dendrites plus the cell body and both can have synapses with many other neurons. One thousand to ten thousand synapses per a single neuron is not uncommon. Therefore, a neuron is on the receiving end of many excitatory and inhibitory signals. An excitatory neurotransmitter produces a potential change called a signal that drives the neuron closer to an action potential, and an inhibitory neurotransmitter produces a signal that drives the neuron farther from an action potential. Excitatory signals have a depolarizing effect, and inhibitory signals have a hyperpolarizing effect (Fig. 12.6).

Neurons integrate these incoming signals. **Integration** is the summing up of excitatory and inhibitory signals. If a neuron receives many excitatory signals (either from different synapses or at a rapid rate from one synapse), the chances are the axon will transmit a nerve impulse. On the other hand, if a neuron receives both inhibitory and excitatory signals, the summing up of these signals may prohibit the axon from firing.

Integration is the summing up of inhibitory and excitatory signals received by a postsynaptic neuron.

12.2 Peripheral Nervous System

The *peripheral nervous system* (*PNS*) lies outside the central nervous system and contains **nerves** which are bundles of axons. Axons that occur in nerves are also called nerve fibers.

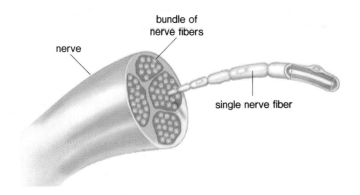

The cell bodies of neurons are found in the CNS—that is, the brain and spinal cord—or in ganglia. Ganglia (sing., **ganglion**) are collections of cell bodies within the PNS.

Humans have 12 pairs of **cranial nerves** attached to the brain (Fig. 12.7*a*). Some of these are sensory nerves; that is, they contain only sensory nerve fibers. Some are motor nerves that contain only motor fibers, and others are mixed nerves that contain both sensory and motor fibers. Cranial nerves are largely concerned with the head, neck, and facial regions of the body. However, the vagus nerve has branches not only to the pharynx and larynx but also to most of the internal organs.

Humans have 31 pairs of **spinal nerves** (Fig. 12.7*b*). The paired spinal nerves emerge from the spinal cord by two short branches, or roots. The *dorsal root* contains the axons of sensory neurons, which are conducting impulses to the spinal cord from sensory receptors. The cell body of a sensory neuron is in the **dorsal-root ganglion.** The *ventral root* contains the axons of motor neurons, which are conducting impulses away from the cord to effectors. These two roots join to form a spinal nerve. All spinal nerves are mixed nerves that contain many sensory and motor fibers. Each spinal nerve serves the particular region of the body in which it is located.

In the PNS, cranial nerves take impulses to and/or from the brain, and spinal nerves take impulses to and from the spinal cord.

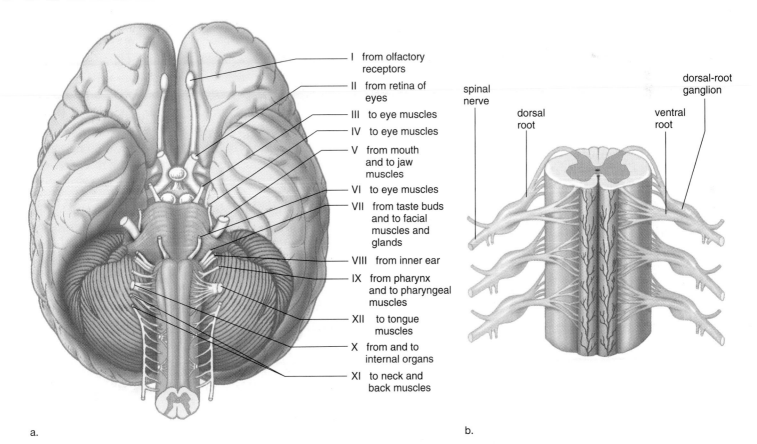

a.

b.

Figure 12.7   Cranial and spinal nerves.
a. Ventral surface of the brain showing the attachment of the 12 pairs of cranial nerves. **b.** Cross section of the spinal cord, showing a few spinal nerves. The human body has 31 pairs of spinal nerves and each spinal nerve has a dorsal root and a ventral root attached to the spinal cord.

Somatic System

The **somatic system** includes the nerves that take sensory information from external sensory receptors to the CNS and motor commands away from the CNS to skeletal muscles. Voluntary control of skeletal muscles always originates in the brain. Involuntary responses to stimuli, called **reflexes,** can involve either the brain or just the spinal cord. Flying objects cause eyes to blink and sharp tacks cause hands to jerk away even without us having to think about it.

The Reflex Arc

Figure 12.8 illustrates the path of a reflex that involves only the spinal cord. If your hand touches a sharp tack, sensory receptors in the skin generate nerve impulses that move along sensory axons toward the spinal cord. Sensory neurons which enter the cord dorsally pass signals on to many interneurons. Some of these interneurons synapse with motor neurons. The short dendrites and the cell bodies of motor neurons are in the spinal cord but their axons leave the cord ventrally. Nerve impulses travel along motor axons to an effector, which brings about a response to the stimulus. In this case a muscle contracts so that you withdraw your hand from the tack. Various other reactions are possible—you will most likely look at the tack, wince, and cry out in pain. This whole series of responses is explained by the fact some of the interneurons involved carry nerve impulses to the brain. The brain makes you aware of the stimulus and directs these other reactions to it.

In the somatic system, nerves take information from external sensory receptors to the CNS and motor commands to skeletal muscles. Involuntary reflexes allow us to respond rapidly to external stimuli.

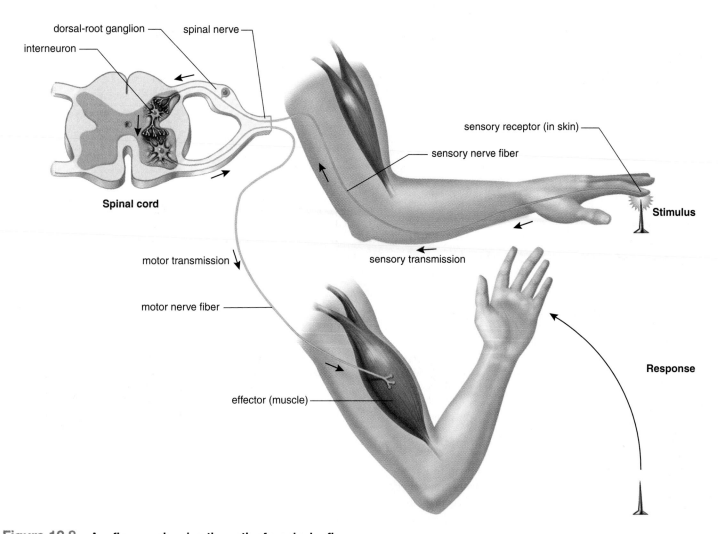

Figure 12.8 **A reflex arc showing the path of a spinal reflex.**
A stimulus (e.g., sharp tack) causes sensory receptors in the skin to generate nerve impulses that travel in sensory nerve fibers to the spinal cord. Interneurons integrate data from sensory neurons and then relay signals to motor neurons. Motor nerve fibers convey nerve impulses from the spinal cord to a skeletal muscle which contracts. Movement of the hand away from the tack is the response to the stimulus.

Sympathetic Division

Parasympathetic Division

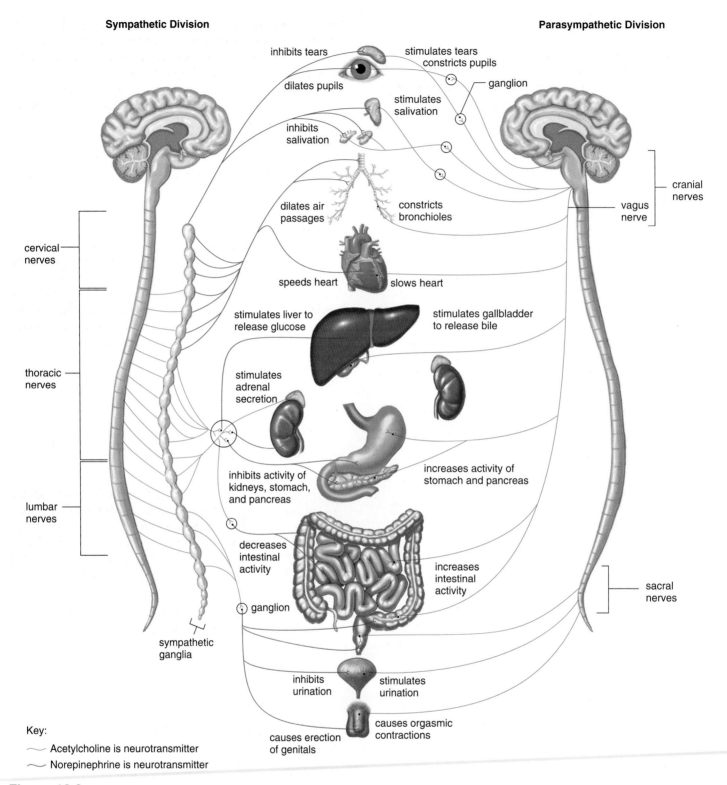

inhibits tears

stimulates tears
constricts pupils

dilates pupils

ganglion

stimulates
salivation

inhibits
salivation

cranial
nerves

cervical
nerves

dilates air
passages

constricts
bronchioles

vagus
nerve

speeds heart

slows heart

thoracic
nerves

stimulates liver to
release glucose

stimulates gallbladder
to release bile

stimulates
adrenal
secretion

lumbar
nerves

inhibits activity of
kidneys, stomach,
and pancreas

increases activity of
stomach and pancreas

decreases
intestinal
activity

increases
intestinal
activity

sacral
nerves

ganglion

sympathetic
ganglia

inhibits
urination

stimulates
urination

causes orgasmic
contractions

causes erection
of genitals

Key:

~ Acetylcholine is neurotransmitter

~ Norepinephrine is neurotransmitter

Figure 12.9   Autonomic system structure and function.
Sympathetic preganglionic fibers arise from the thoracic and lumbar portions of the spinal cord; parasympathetic preganglionic fibers arise from the brain and the sacral portion of the spinal cord. Each system innervates the same organs but has contrary effects as described.

Autonomic System

The **autonomic system** of the PNS regulates the activity of cardiac and smooth muscle and glands. The system is divided into the sympathetic and parasympathetic divisions (Fig. 12.9 and Table 12.1). Both of these divisions (1) function automatically and usually in an involuntary manner; (2) innervate all internal organs; and (3) utilize two neurons and one ganglion for each impulse. The first neuron has a cell body within the CNS and a *preganglionic fiber*. The second neuron has a cell body within the ganglion and a *postganglionic fiber*.

Reflex actions, such as those that regulate the blood pressure and breathing rate, are especially important to the maintenance of homeostasis. These reflexes begin when the sensory neurons in contact with internal organs send information to the CNS. They are completed by motor neurons within the autonomic system.

Sympathetic Division

Most preganglionic fibers of the **sympathetic division** arise from the middle, or *thoracic-lumbar*, portion of the spinal cord and almost immediately terminate in ganglia that lie near the cord. Therefore, in this division, the preganglionic fiber is short, but the postganglionic fiber that makes contact with an organ is long.

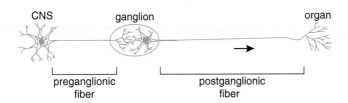

The sympathetic division is especially important during emergency situations and is associated with **"fight or flight."** If you need to fend off a foe or flee from danger, active muscles require a ready supply of glucose and oxygen. The sympathetic division accelerates the heartbeat and di-

lates the bronchi. On the other hand, the sympathetic division inhibits the digestive tract—digestion is not an immediate necessity if you are under attack. The neurotransmitter released by the postganglionic axon is primarily norepinephrine (NE). The structure of NE is like that of epinephrine (adrenaline), an adrenal medulla hormone that usually increases heart rate and contractility.

The sympathetic division brings about those responses we associate with "fight or flight."

Parasympathetic Division

The **parasympathetic division** includes a few cranial nerves (e.g., the vagus nerve) and also fibers that arise from the sacral (bottom) portion of the spinal cord. Therefore, this division often is referred to as the craniosacral portion of the autonomic system. In the parasympathetic division, the preganglionic fiber is long, and the postganglionic fiber is short because the ganglia lie near or within the organ.

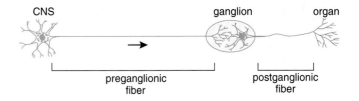

The parasympathetic division, sometimes called the "housekeeper division," promotes all the internal responses we associate with a relaxed state; for example, it causes the pupil of the eye to contract, promotes digestion of food, and retards the heartbeat. The neurotransmitter utilized by the parasympathetic division is acetylcholine (ACh).

The parasympathetic division brings about the responses we associate with a relaxed state.

Table 12.1	Comparison of Somatic Motor and Autonomic Motor Pathways		
Items	**Somatic Motor Pathway**	**Autonomic Motor Pathways**	
		Sympathetic	**Parasympathetic**
Type of control	Voluntary/involuntary	Involuntary	Involuntary
Number of neurons per message	One	Two (preganglionic shorter than postganglionic)	Two (preganglionic longer than postganglionic)
Location of motor fiber	Most cranial nerves and all spinal nerves	Thoracolumbar spinal nerves	Cranial (e.g., vagus) and sacral spinal nerves
Neurotransmitter	Acetylcholine	Norepinephrine	Acetylcholine
Effectors	Skeletal muscles	Smooth and cardiac muscle, glands	Smooth and cardiac muscle, glands

12.3 Central Nervous System

The *central nervous system (CNS)* consists of the spinal cord and the brain where sensory information is received and motor control is initiated. Both the spinal cord and the brain are protected by bone; the brain is enclosed by the skull and the spinal cord is surrounded by vertebrae. Also, both the spinal cord and brain are wrapped in protective membranes known as **meninges** (sing., meninx). Meningitis is an infection of these coverings. The spaces between the meninges are filled with **cerebrospinal fluid,** which cushions and protects the CNS. A small amount of this fluid sometimes is withdrawn from around the cord for laboratory testing when a spinal tap (i.e., lumbar puncture) is performed.

Cerebrospinal fluid is also contained within the ventricles of the brain and in the central canal of the spinal cord. The brain's **ventricles** are interconnecting cavities that produce and serve as a reservoir for cerebrospinal fluid. Normally, any excess cerebrospinal fluid drains away into the circulatory system. However, blockages can occur. In an infant, the brain can enlarge due to cerebrospinal fluid accumulation, and this condition is called "water on the brain." If cerebrospinal fluid collects in an adult, the brain cannot enlarge and instead, the brain is pushed against the skull, possibly causing injury.

The CNS, which lies in the midline of the body and consists of the brain and the spinal cord, receives sensory information and initiates voluntary motor control.

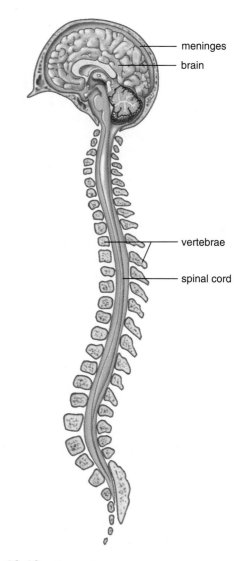

Figure 12.10   Central nervous system.
The central nervous system consists of the brain and spinal cord. The brain is protected by the skull and the spinal cord is protected by the vertebrae.

The Spinal Cord

The **spinal cord** extends from the base of the brain through a large opening in the skull called the foramen magnum and into the vertebral canal formed by the vertebrae (Fig. 12.10).

Structure of the Spinal Cord

Figure 12.11*a* shows how the individual vertebrae join to form the vertebral canal which protects the spinal cord. The spinal nerves which project from the cord pass through openings between the vertebrae of the vertebral canal.

A cross section of the spinal cord shows that the spinal cord has a central canal, gray matter, and white matter (Fig. 12.11*b, c*). The central canal contains cerebrospinal fluid, as do the meninges that protect the spinal cord. The

gray matter is centrally located and shaped like the letter H. It is gray because it contains cell bodies and short, non-myelinated fibers. Portions of sensory neurons and motor neurons are found here, as are interneurons that communicate with these two types of neurons. The dorsal root of a spinal nerve contains sensory nerve fibers entering the gray matter, and the ventral root of a spinal nerve contains motor nerve fibers exiting the gray matter. The dorsal and

ventral roots join before the spinal nerve leaves the vertebral canal. Spinal nerves are a part of the PNS.

The **white matter** of the spinal cord occurs in areas around the gray matter. The white matter is white because it contains myelinated axons of interneurons that run together in bundles called **tracts.** Ascending tracts taking information to the brain are primarily located dorsally, and descending tracts taking information from the brain are primarily located ventrally. Because the tracts cross just after they enter and exit the brain, the left side of the brain controls the right side of the body, and the right side of the brain controls the left side of the body.

The spinal cord extends from the base of the brain into the vertebral canal formed by the vertebrae. A cross section shows that the spinal cord has a central canal, gray matter, and white matter.

Functions of the Spinal Cord

The spinal cord is the center for thousands of reflex arcs. Figure 12.11 indicates the path of a spinal reflex from sensory receptors to muscle effectors. Each interneuron in the spinal cord has synapses with many other neurons, and therefore they carry out integration of incoming information before sending signals to other interneurons and motor neurons (see Fig. 12.6).

The spinal cord provides a means of communication between the brain and the peripheral nerves that leave the cord. When someone touches your hand, sensory information passes from sensory receptors through sensory nerve fibers to the spinal cord and up ascending tracts to the brain. When we voluntarily move our limbs, motor impulses originating in the brain pass down descending tracts to the spinal cord and out to our muscles by way of motor nerve fibers. Therefore, if the spinal cord is severed, we suffer a loss of sensation and a loss of voluntary control— that is, we suffer a paralysis. If the spinal cord is completely cut across in the thoracic region, paralysis of the lower body and legs occurs. This condition is known as paraplegia. If the injury is in the neck region, the four limbs are usually affected. This condition is called quadriplegia.

The spinal cord is a center for reflex action. The spinal cord also serves as a means of communication between the brain and much of the body. Because tracts to and from the brain cross over, the left side of the brain controls the right side of the body and vice versa.

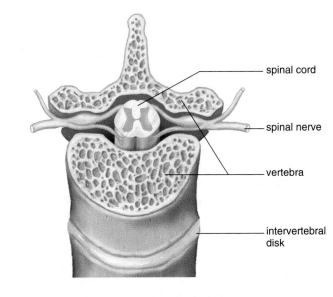

a.

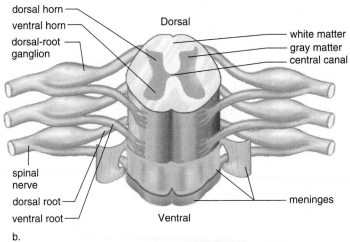

b.

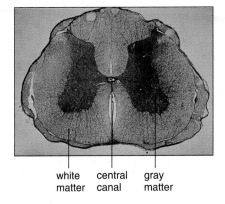

c.

Figure 12.11   Spinal cord.

a. The spinal cord passes through the vertebral canal formed by the vertebrae. **b.** The spinal cord has a central canal filled with cerebrospinal fluid, H-shaped gray matter, and white matter. The white matter contains tracts that take nerve impulses to and from the brain. **c.** Photomicrograph.

The Brain

The human **brain** has been called the last great frontier of biology. The goal of modern neuroscience is to understand the structure and function of the brain's various parts so well that it will be possible to prevent or correct the more than 1,000 mental disorders that rob human beings of a normal life. This section gives only a glimpse of what is known about the brain and the modern avenues of research.

We will discuss the parts of the brain with reference to the brain stem, the diencephalon, and the cerebrum. The brain has four **ventricles** called, in turn, the fourth ventricle, the third ventricle, and the two lateral ventricles. It may be helpful to you to associate the brain stem with the fourth ventricle, the diencephalon with the third ventricle, and the cerebrum with the two lateral ventricles (Fig. 12.12*a*).

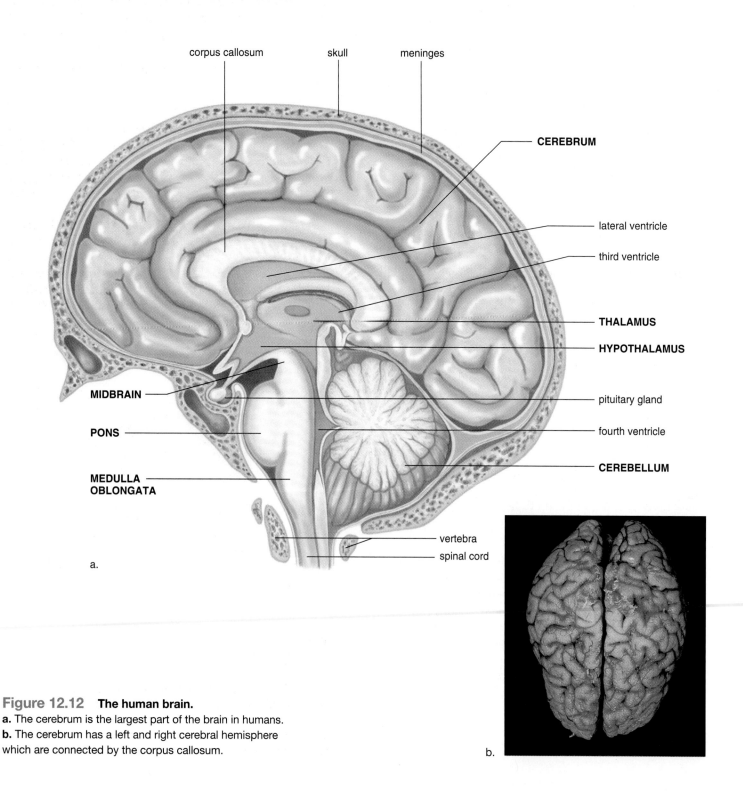

Figure 12.12   The human brain.
a. The cerebrum is the largest part of the brain in humans.
b. The cerebrum has a left and right cerebral hemisphere which are connected by the corpus callosum.

The Brain Stem

The **brain stem,** the initial portion of the brain, contains the medulla oblongata, the cerebellum, the pons, and the midbrain. The **medulla oblongata** lies between the spinal cord and the pons. It contains a number of *vital centers* for regulating heartbeat, breathing, and vasoconstriction (blood pressure). It also contains the reflex centers for vomiting, coughing, sneezing, hiccuping, and swallowing. The medulla contains tracts that ascend or descend between the spinal cord and higher brain centers.

The word **pons** means bridge in Latin, and true to its name, the pons contains bundles of axons traveling between the cerebellum and the rest of the CNS. In addition, the pons functions with the medulla to regulate breathing rate and has reflex centers concerned with head movements in response to visual and auditory stimuli.

The **midbrain** acts as a relay station for tracts passing between the cerebrum and the spinal cord or cerebellum. It also has reflex centers for visual, auditory, and tactile responses.

The Cerebellum

The **cerebellum** is separated from the brain stem by the fourth ventricle. The cerebellum has two portions that are joined by a narrow median portion. The surface of the cerebellum is gray matter, and the interior is largely white matter. The cerebellum integrates and passes on both sensory and motor information. It maintains normal muscle tone, posture, and balance and ensures that all of the skeletal muscles work together to produce smooth and coordinated motions. The cerebellum is necessary for the learning of new motor skills like playing the piano or hitting a baseball.

The Diencephalon

The hypothalamus and the thalamus are in the **diencephalon,** a region that encircles the third ventricle. The **hypothalamus** forms the floor of the third ventricle. The hypothalamus is the integrating center for the autonomic system. It also helps maintain homeostasis by regulating hunger, sleep, thirst, body temperature, and water balance. The hypothalamus controls the pituitary gland and thereby serves as a link between the nervous and endocrine systems.

The **thalamus** consists of two masses of gray matter located in the sides and roof of the third ventricle. The thalamus integrates sensory information and serves as a central relay station for sensory impulses traveling upward from other parts of the brain to the cerebrum. The thalamus is also involved in arousal and higher mental functions such as memory and emotion.

The pineal gland, which secretes the hormone melatonin, is located in the diencephalon. Presently there is much popular interest in the role of melatonin in our daily rhythms and whether it can help meliorate jet lag or insomnia. Scientists are also interested in the possibility that the hormone may regulate the onset of puberty.

The Cerebrum

The **cerebrum,** also called the telencephalon, is the foremost and largest portion of the brain in humans. Just as the human body has two halves, so does the cerebrum. These halves are called the left and right **cerebral hemispheres** (Fig. 12.12*b*). These two cerebral hemispheres are connected by a bridge of tracts within the corpus callosum.

The cerebrum is the highest center to receive sensory input and carry out integration before commanding voluntary motor responses. It is in communication with and coordinates the activities of the other parts of the brain. As we shall see, the cerebrum carries out higher thought processes required for learning and memory and for language and speech.

The Reticular Formation

The **reticular formation** is a complex network of nuclei and fibers that extend the length of the brain stem (Fig. 12.13). In this context, the term **nuclei** means masses of cell bodies in the CNS. The reticular formation receives sensory signals which it sends up to higher centers and motor signals which it sends to the spinal cord.

One portion of the reticular formation called the reticular activating system (RAS) arouses the cerebrum via the thalamus and causes a person to be alert. It is believed to filter out unnecessary sensory stimuli and this may account for why you can study with the TV on. An inactive reticular formation results in sleep; a severe injury to the RAS can cause a person to be comatose.

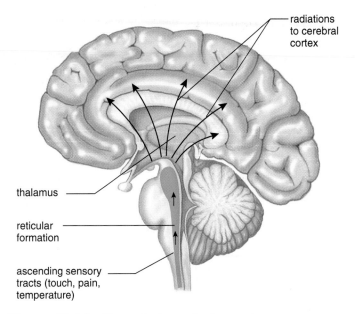

Figure 12.13   The reticular activating system.
The reticular formation receives and sends on motor and sensory information to various parts of the CNS. One portion, the reticular activating system (see arrows), arouses the cerebrum and in this way controls alertness versus sleep.

12.4 The Cerebral Hemispheres

A deep groove called the longitudinal fissure divides the left from the right cerebral hemisphere; shallow grooves called sulci (sing., sulcus) divide each hemisphere into lobes (Fig. 12.14). The frontal lobe is toward the front of a cerebral hemisphere and the parietal lobe is toward the back of a cerebral hemisphere. The occipital lobe is dorsal to (behind) the parietal lobe and the temporal lobe lies below the frontal and parietal lobes.

The Cerebral Cortex

The **cerebral cortex** is a thin but highly convoluted outer layer of gray matter that covers the cerebral hemispheres. The cerebral cortex contains over one billion cell bodies and is the region of the brain that accounts for sensation, voluntary movement, and all the thought processes we associate with consciousness.

The cerebral cortex contains motor areas and sensory areas and also association areas. The **primary motor area** is in the frontal lobe just ventral to (before) the central sulcus. Voluntary commands begin in the primary motor area and each part of the body is controlled by a certain section. Our versatile hand takes up an especially large area of the primary motor area. Ventral to the primary motor area is a premotor area. The *premotor area* organizes motor functions for skilled motor activities before the primary motor area sends signals to the cerebellum which integrates them. The unique ability of humans to speak is partially dependent upon *Broca's area,* a motor speech area located in the left frontal lobe. Signals originating here pass to the premotor area before reaching the primary motor area.

The **primary somatosensory area** is just dorsal to the central sulcus. Sensory information from the skin and skeletal muscles arrives here, where each part of the body is sequentially represented. A primary visual area in the occipital lobe receives information from our eyes, and a primary auditory area in the temporal lobe receives information from our ears. A primary taste area in the parietal lobe accounts for taste sensations.

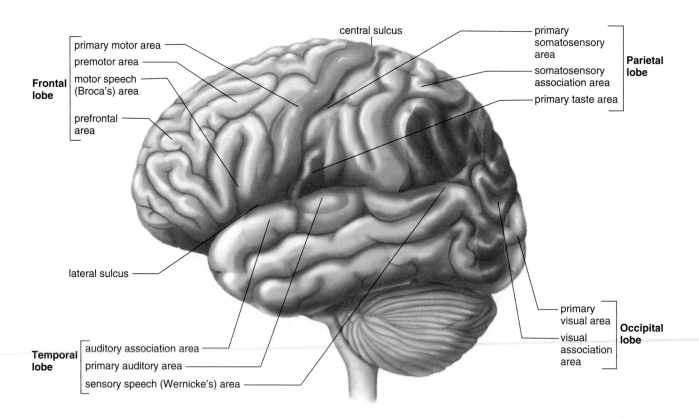

Figure 12.14   **The cerebral cortex.**
The cortex of the cerebrum is divided into four lobes: frontal, parietal, temporal, and occipital. The frontal lobe has motor areas and an association area called the prefrontal area. The other lobes have sensory areas and also association areas.

True to their name, **association areas** are places where cell bodies integrate information. The somatosensory association area is located just dorsal to the primary somatosensory area. This area processes and analyzes sensory information from the skin and muscles. The visual association area in the occipital lobe associates new visual information with previously received visual information. It might "decide," for example, if we have seen this face or tool or whatever before. The auditory association area in the temporal lobe performs the same functions with regard to sounds. The **prefrontal area,** an association area in the frontal lobe, receives information from the other association areas and uses this information to reason and plan our actions. Integration in this area accounts for our most cherished human abilities to think critically and to formulate appropriate behaviors.

The parietal, temporal, and occipital association areas meet near the dorsal end of the lateral sulcus. This region is called the general interpretation area because it receives information from all the sensory association areas and allows us to quickly integrate incoming signals and send them on to the prefrontal area so an immediate response is possible. Heroes are people that can quickly assess a situation and take actions that save others from dangerous situations.

Limbic System

The **limbic system** is a complex network of tracts and nuclei that incorporates medial portions of the cerebral lobes, subcortical nuclei, and the diencephalon (Fig. 12.15). The limbic system blends higher mental functions and primitive emotions into a united whole. It accounts for why activities like sexual behavior and eating seem pleasurable and also why, say, mental stress can cause high blood pressure.

Two significant structures within the limbic system are the hippocampus and the amygdala, which are essential for learning and memory. The hippocampus is well situated in the brain to make the prefrontal area aware of past experiences stored in association areas. The amygdala, in particular, can cause these experiences to have emotional overtones. The inclusion of the frontal lobe in the limbic system means that reason can keep us from acting out strong feelings.

The gray matter of the cerebrum consists of the cerebral cortex and also nuclei. The white matter consists of tracts. The limbic system is a unique combination of various portions of the brain which unify brain functions and sensations.

White Matter

Much of the rest of the cerebrum is composed of white matter. From your understanding of the spinal cord you know that white matter in the CNS consists of long myelinated fibers organized into tracts. Descending tracts from the primary motor area communicate with lower brain centers, and ascending tracts from lower brain centers send sensory information up to the primary somatosensory area. Tracts within the cerebrum take information between the different sensory, motor, and association areas pictured in Figure 12.14. The corpus callosum, you will recall, contains tracts that join the two cerebral hemispheres.

While the bulk of the cerebrum is composed of tracts, there are subcortical (below the cortex) nuclei deep within the white matter. The **basal nuclei** serve as relay stations for motor impulses from the primary motor area. The basal nuclei produce dopamine, an inhibitory neurotransmitter that helps control various skeletal muscle activities. Huntington disease and Parkinson disease, which are both characterized by uncontrollable movements, are believed to be due to malfunctioning of the basal nuclei.

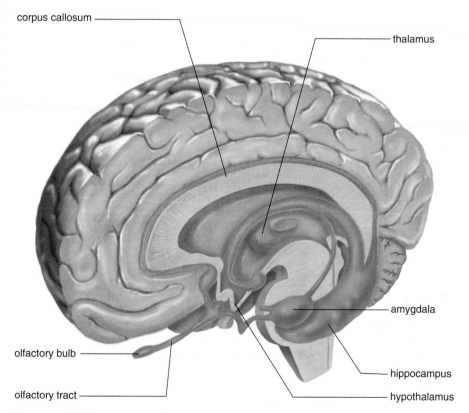

Figure 12.15   The limbic system.
Structures deep within the cerebral hemisphere and surrounding the diencephalon join higher mental functions like reasoning with more primitive feelings like fear and pleasure.

12.5 Higher Mental Functions

As in other areas of biological research, brain research has progressed due to technological breakthroughs. Neuroscientists now have a wide range of techniques at their disposal for studying the human brain, including modern technologies that allow us to record the functioning of the brain.

Learning and Memory

Just as the corpus callosum gives evidence that the two cerebral hemispheres work together, so the limbic system indicates that cortical areas possibly work with lower centers to produce learning and memory (Fig. 12. 16). **Memory** is the ability to hold a thought in mind or recall events from the past, ranging from a word we learned only yesterday to an early emotional experience that has shaped our lives. Learning takes place when we retain and utilize past memories.

Types of Memories

We have all had the experience of trying to remember a seven-digit telephone number for a very short period of time. If we say we are trying to keep it in the forefront of our brain, we are exactly correct because the prefrontal area is active during **short-term memory.** The prefrontal area is the association area that lies just dorsal to our forehead! There are some telephone numbers that you know by heart; in other words, they have gone into **long-term memory.** Think of a telephone number you know by heart and see if you can bring it to mind without also thinking about the place or person associated with that number. Most likely you cannot because typically long-term memory is a mixture of what is called **semantic memory** (numbers, words, etc.) and **episodic memory** (persons, events, etc.). The flow chart for long-term memory in Figure 12.16 has two sets of arrows—one for semantic memory and the other for episodic memory. Due to brain damage, some persons lose one type of memory ability but not the other. Without a working episodic memory they can carry on a conversation but have no recollection of recent events. If you are talking to them and leave the room, they don't remember you when you come back!

Skill memory is another type of memory that can exist independent of episodic memory. Skill memory is being able to perform motor activities like riding a bike or playing ice hockey. When a person first learns a skill, more areas of the cerebral cortex are involved than after the skill is perfected. In other words, you have to think about what you are doing when you learn a skill, but later the actions become automatic. This shows that the preprimary motor area can communicate with the primary motor area below the level of consciousness.

Memory Storage and Retrieval

The first step toward being able to cure memory disorders is to know what parts of the brain are functioning when we remember something. Investigators have been able to work it out pretty well. The **hippocampus,** a seahorse-shaped structure that lies deep in the temporal lobe, is in a unique position to serve as a bridge to the sensory association areas where memories are stored and the prefrontal area where memories are utilized. The prefrontal area communicates with the hippocampus when memories are stored and when these memories are brought to mind. Why are some memories so emotionally charged? The **amygdala** seems to be responsible for fear conditioning and associating danger with sensory stimuli received from both the diencephalon and the cortical sensory areas.

Our long-term memories are stored in bits and pieces throughout the sensory association areas of the cerebral cortex. Visions are stored in the vision association area, sounds are stored in the auditory association area, and so forth. The hippocampus gathers this information together for use by the prefrontal cortex when we remember Uncle Frank or our summer holiday. And the amygdala adds emotional overtones to memories.

Long-Term Potentiation

While it is helpful to know the memory functions of various portions of the brain, an important step toward curing mental disorders is understanding memory on the cellular

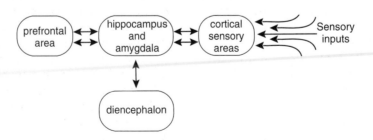

Figure 12.16   Memory circuits.
The hippocampus and amygdala are believed to be involved in the storage and retrieval of memories.

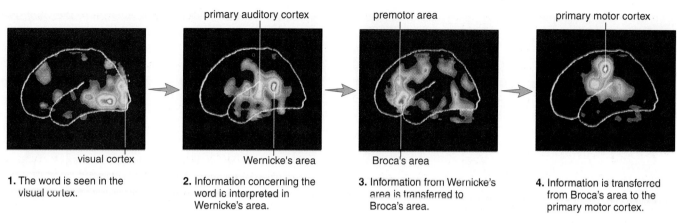

primary auditory cortex | premotor area | primary motor cortex

visual cortex | Wernicke's area | Broca's area

1. The word is seen in the visual cortex.

2. Information concerning the word is interpreted in Wernicke's area.

3. Information from Wernicke's area is transferred to Broca's area.

4. Information is transferred from Broca's area to the primary motor cortex.

Figure 12.17 Language and speech.

These functional images were captured by a high speed computer during PET (position emission tomography) scanning of the brain. Injected radioactively labeled water is preferentially taken up by active brain tissue and not by inactive brain tissue. The computer uses this information to generate cross-sectional images of the brain while subjects are asked to perform the activities noted. Red indicates the most active areas of the brain and blue indicates the least active areas.

level. **Long-term potentiation (LTP)** is an enhanced response at synapses within the hippocampus. LTP is most likely essential to memory storage but, unfortunately, it sometimes causes a postsynaptic neuron to become so excited, it undergoes apoptosis, a form of cell death. This phenomenon called excitotoxicity may develop due to a mutation. (The longer we live the more likely that any particular mutation will occur.) Excitotoxicity is due to the action of the neurotransmitter glutamate which is active in the hippocampus. When glutamate binds with a specific type of receptor in the postsynaptic membrane, calcium (Ca^{2+}) may rush in too fast and the influx is lethal to the cell. A gradual extinction of brain cells in the hippocampus appears to be the underlying cause of Alzheimer disease (AD).

AD is characterized by gradual loss of memory and ends with an inability to perform any type of daily activity. Personality changes signal the onset of AD. A normal 50–60-year-old adult might forget the name of a friend not seen for years. However, people with AD forget the name of a neighbor who visits daily. With time, they have trouble finding their way and cannot perform simple errands. People afflicted with AD become confused and tend to repeat the same questions over and over. Other signs of mental disturbances eventually appear, and patients gradually become bedridden and die of a complication such as pneumonia.

AD neurons have neurofibrillary tangles (bundles of fibrous protein) surrounding the nucleus and protein-rich accumulations called amyloid plaques, enveloping the axon branches. Although it is not yet known how excitotoxicity is related to structural abnormalities of AD neurons, some researchers are trying to develop neuroprotective drugs that can possibly guard brain cells against damage due to glutamate. One type of neuroprotective drug blocks glutamate-receptors and thereby prevents the entry of Ca^{2+} and excitotoxicity.

Language and Speech

Language is obviously dependent upon semantic memory. Therefore, we would expect some of the same areas in the brain to be involved in both memory and language. Seeing and hearing words depends on sensory centers in the occipital and temporal lobes respectively. And generating and speaking words depends on motor centers in the frontal lobe (Fig. 12.17).

From studies of patients with speech disorders, it has been known for some time that damage to a motor speech area called Broca's area results in an inability to speak. Broca's area is located just in front of the primary motor zone for speech musculature (lips, tongue, larynx, and so forth). Damage to a sensory speech area in the temporal lobe called Wernicke's area results in the inability to comprehend speech. Any disruption of pathways between various regions of the brain could very well contribute to an inability to comprehend our environment and use speech correctly. Indeed, damage to lower centers, especially the thalamus, can also cause speech difficulties. Remember that the thalamus passes on sensory information, and if the cerebral cortex does not receive the necessary sensory input, the motor areas cannot formulate the proper output.

One interesting aside pertaining to language and speech is the recognition that the left brain and right brain have different functions. Only the left hemisphere and not the right contains a Broca's area and Wernicke's area! Indeed, the left hemisphere plays a role of great importance in language functions in general and not just in speech. In an attempt to

Human Systems Work Together

Integumentary System

Brain controls nerves that regulate size of cutaneous blood vessels, activate sweat glands and arrector pili muscles.

Skin protects nerves, helps regulate body temperature; skin receptors send sensory input to brain.

How the Nervous System works with other body systems

Lymphatic System/Immunity

Microglial cells engulf and destroy pathogens.

Lymphatic vessels pick up excess tissue fluid; immune system protects against infections of nerves.

Skeletal System

Receptors send sensory input from bones and joints to brain.

Bones protect sense organs, brain, and spinal cord; store Ca^{2+} for nerve function.

Respiratory System

Respiratory centers in brain regulate breathing rate.

Lungs provide oxygen for neurons and rid the body of carbon dioxide produced by neurons.

Muscular System

Brain controls nerves that innervate muscles; receptors send sensory input from muscles to brain.

Muscle contraction moves eyes, permits speech, and creates facial expressions.

Digestive System

Brain controls nerves that innervate smooth muscle and permit digestive tract movements.

Digestive tract provides nutrients for growth, main- tenance, and repair of neurons and neuroglial cells.

Endocrine System

Hypothalamus is part of endocrine system; nerves innervate certain glands of secretion.

Sex hormones affect development of brain.

Urinary System

Brain controls nerves that innervate muscles that permit urination.

Kidneys maintain blood levels of Na^+, K^+, and Ca^{2+}, which are needed for nerve conduction.

Cardiovascular System

Brain controls nerves that regulate the heart and dilation of blood vessels.

Blood vessels deliver nutrients and oxygen to neurons, carry away wastes.

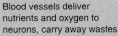

Reproductive System

Brain controls onset of puberty; nerves are involved in erection of penis and clitoris, contraction of ducts that carry gametes, and contraction of uterus.
Sex hormones masculinize or feminize the brain, exert feedback control over the hypothalamus, and influence sexual behavior.

cure epilepsy in the early 1940s, the corpus callosum was surgically severed in some patients. Later studies showed that these split-brain patients could name objects only if they were seen by the left hemisphere. If objects were viewed only by the right hemisphere, a split-brain patient could choose the proper object for a particular use but was unable to name it. Based on these and various types of studies, the popular idea developed that the left brain can be contrasted with the right brain along these lines:

Left Hemisphere	Right Hemisphere
Verbal	Nonverbal, visuo-spatial
Logical, analytical	Intuitive
Rational	Creative

Further, it became generally thought that one hemisphere was dominant in each person, accounting in part for personality traits. However, recent studies suggest that the hemispheres process the same information differently. The right hemisphere is more global, whereas the left hemisphere is more specific in its approach.

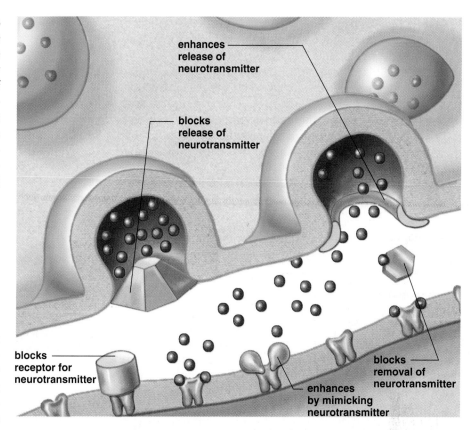

Figure 12.18   **Drug actions at a synapse.**
Each drug typically has one of these specific functions.

Memory has been studied at various levels: behavioral, structural, and cellular. Knowledge of memory on the cellular level may offer treatments and cures for mental disorders. Language is dependent upon memory; special areas in the left hemisphere help account for our ability to comprehend and use speech.

12.6   Homeostasis

The nervous system and the endocrine system coordinate the functioning of the other systems in the body. The governance of internal organs and the regulation of the composition of blood and tissue fluid usually take place below the level of consciousness. Subconscious control is dependent on reflex actions that involve the hypothalamus and the medulla oblongata. The hypothalamus and the medulla oblongata act by way of the autonomic nervous system to control such important parameters as the heart rate, the constriction of the blood vessels, and the breathing rate. The cardiovascular system works harder to carry oxygen to the skeletal muscles when we are in a "fight or flight" mode than when we are in a relaxed state.

Page 264 tells how the nervous system works with other systems in the body to maintain homeostasis. The hypothalamus works closely with the endocrine system and even pro-

duces the hormone ADH which helps regulate the osmolarity of the blood. The kidneys are under hormonal control as they regulate the salt and water balance of blood. If blood osmolarity is too low, tissue fluid will collect in the tissues to the degree that the lymphatic system is unable to transport it away.

You might think that voluntary movements don't play a role in homeostasis, but actually we usually modify our behavior to stay in as moderate an environment as possible. Otherwise we are testing the ability of the nervous system to maintain homeostasis despite extreme conditions.

12.7   Drug Abuse

A wide variety of drugs can be used to alter the mood and/or emotional state (see Appendix D). Drugs that affect the nervous system have two general effects: (1) they impact the limbic system, and (2) they either promote or decrease the action of a particular neurotransmitter (Fig. 12.18). Stimulants increase the likelihood of neuron excitation, and depressants decrease the likelihood of excitation. Increasingly, researchers believe that dopamine, a neurotransmitter in the brain, is primarily responsible for mood. Cocaine is known to potentiate the effects of dopamine by interfering with its uptake from synaptic clefts. Many new medications developed to counter drug addiction and mental illness affect the release, reception, or breakdown of dopamine.

Drug abuse is apparent when a person takes a drug at a dose level and under circumstances that increase the potential for a harmful effect. Drug abusers are apt to display either a psychological and/or a physical dependence on the drug. Dependence has developed when the person spends much time thinking about the drug or arranging to get it and often takes more of the drug than was intended. With physical dependence, formerly called an addiction to the drug, the person is tolerant to the drug—that is, must increase the amount of the drug to get the same effect and has withdrawal symptoms when he or she stops taking the drug.

Drugs that affect the nervous system can cause physical dependence and withdrawal symptoms.

Alcohol

It is possible that *alcohol* influences the action of GABA, an inhibiting transmitter, or glutamate, an excitatory neurotransmitter. Once imbibed, alcohol is primarily metabolized in the liver, where it disrupts the normal workings of this organ so that fats cannot be broken down. Fat accumulation, the first stage of liver deterioration, begins after only a single night of heavy drinking. If heavy drinking continues, fibrous scar tissue appears during a second stage of deterioration. If heavy drinking stops, the liver can still recover and become normal once again. If not, the final and irrevocable stage, cirrhosis of the liver, occurs: liver cells die, harden, and turn orange (cirrhosis means orange).

Alcohol is used by the body as an energy source but it lacks the vitamins, minerals, essential amino acids, and fatty acid the body needs to stay healthy. Many alcoholics are undernourished and prone to illness for this reason alone.

The surgeon general recommends that pregnant women drink no alcohol at all. Alcohol crosses the placenta freely and causes *fetal alcohol syndrome,* which is characterized by mental retardation and various physical defects.

Nicotine

Nicotine, an alkaloid derived from tobacco, is a widely used neurological agent. When a person smokes a cigarette, nicotine is quickly distributed to the central and peripheral nervous systems. In the central nervous system, nicotine causes neurons to release dopamine, a neurotransmitter mentioned earlier. The excess of dopamine has a reinforcing effect that leads to dependence on the drug. In the peripheral nervous system, nicotine stimulates the same postsynaptic receptors as acetylcholine and leads to increased skeletal muscular activity. It also increases the heart rate and blood pressure, and digestive tract mobility.

Many cigarette and cigar smokers find it difficult to give up the habit because nicotine induces both physiological and psychological dependence. Withdrawal symptoms include headache, stomach pain, irritability, and insomnia. Tobacco contains not only nicotine, but many other harmful substances. Cigarette and cigar smoking contributes to early death from cancer, including not only lung cancer but also cancer of the

Figure 12.19   Drug use.
Users often smoke crack in a glass water pipe. The high produced consists of a "rush" lasting a few seconds, followed by a few minutes of euphoria. Continuous use makes the user extremely dependent on the drug.

larynx, mouth, throat, pancreas, and urinary bladder. Now that women are as apt to smoke as men, lung cancer has surpassed breast cancer as a cause of death. Cigarette smoking in young women who are sexually active is most unfortunate because if they become pregnant, nicotine, like other psychoactive drugs, adversely affects a developing embryo and fetus.

Cocaine

Cocaine is an alkaloid derived from the shrub *Erythroxylum cocoa.* Cocaine is sold in powder form and as *crack,* a more potent extract (Fig. 12.19). Cocaine prevents the synaptic uptake of dopamine and this causes the user to experience the sensation of a rush. The epinephrine-like effects of dopamine account for the state of arousal that lasts for some minutes after the rush experience.

A cocaine binge can go on for days, after which the individual suffers a crash. During the binge period, the user is hyperactive and has little desire for food or sleep but has an increased sex drive. During the crash period, the user is fatigued, depressed, and irritable, has memory and concentration problems, and displays no interest in sex. Indeed, men are often impotent.

With continued cocaine use, the body begins to make less dopamine to compensate for a seemingly excess supply. The user, therefore, now experiences *tolerance, withdrawal symptoms,* and an intense *craving* for cocaine. These are indications that the person is highly dependent upon the drug or, in other words, that cocaine is extremely addictive. Overdosing on cocaine can cause seizures and cardiac and respiratory

arrest. It is possible that long-term cocaine abuse causes brain damage; babies born to addicts suffer withdrawal symptoms and may suffer neurological and developmental problems.

Heroin

Heroin is derived from morphine, an alkaloid of *opium.* After an intravenous injection, there is a feeling of euphoria along with relief of pain within 3 to 6 minutes. Side effects can include nausea, vomiting, dysphoria, and respiratory and circulatory depression leading to death.

Heroin binds to receptors meant for the **endorphins,** the special neurotransmitters that kill pain and produce a feeling of tranquility. With time, the body's production of endorphins decreases. *Tolerance* develops so that the user needs to take more of the drug just to prevent *withdrawal* symptoms. The euphoria originally experienced upon injection is no longer felt.

Heroin withdrawal symptoms include perspiration, dilation of pupils, tremors, restlessness, abdominal cramps, gooseflesh, defecation, vomiting, and increase in systolic pressure and respiratory rate. Those who are excessively dependent may experience convulsions, respiratory failure, and death. Infants born to women who are physically dependent also experience these withdrawal symptoms.

Marijuana

The dried flowering tops, leaves, and stems of the Indian hemp plant *Cannabis sativa* contain and are covered by a resin that is rich in THC (tetrahydrocannabinol). The names *cannabis* and *marijuana* apply to either the plant or THC. Usually marijuana is smoked in a cigarette form called a joint.

The occasional user reports experiencing a mild euphoria along with alterations in vision and judgment, which result in distortions of space and time. Motor incoordination, including the inability to speak coherently, takes place. Heavy use can result in hallucinations, anxiety, depression,

rapid flow of ideas, body image distortions, paranoid reactions, and similar psychotic symptoms. The terms *cannabis psychosis* and *cannabis delirium* refer to such reactions.

Recently, researchers have found that marijuana binds to a receptor for anandamide, a normal molecule in the body. Craving and difficulty in stopping usage can occur as a result of regular use. Some researchers believe that long-term marijuana use leads to brain impairment. *Fetal cannabis syndrome,* which resembles fetal alcohol syndrome, has been reported. Some psychologists believe that marijuana use among adolescents prevents them from dealing with the personal problems that often develop during that stage of life.

Methamphetamine (Ice)

Methamphetamine is related to amphetamine, a well-known stimulant. Both methamphetamine and amphetamine have been drugs of abuse for some time, but a new form of methamphetamine known as "ice" is now used as an alternative to cocaine. Ice is a pure, crystalline hydrochloride salt that has the appearance of sheetlike crystals. Unlike cocaine, ice can be illegally produced in this country in laboratories and does not need to be imported.

Ice, like crack, will vaporize in a pipe, so it can be smoked, avoiding the complications of intravenous injections. After rapid absorption into the bloodstream, the drug moves quickly to the brain. It has the same stimulatory effect as cocaine, and subjects report they cannot distinguish between the two drugs after intravenous administration. Methamphetamine effects, however, persist for hours instead of a few seconds. Therefore, it is the preferred drug of abuse by many.

Neurological drugs either potentiate or dampen the effect of the body's neurotransmitters.

Bioethical Issue

To control their weight, many people in the United States have turned to a new diet pill called Redux, which stimulates the production and availability of the neurotransmitter serotonin in the brain. This very same neurotransmitter is released when we eat a high carbohydrate-rich meal. Redux also prevents serotonin from being reabsorbed at presynaptic membranes. The result is spirits are lifted and appetite is squelched. It's not uncommon for patients to lose 20 or more pounds a week, simply because they have lost all interest in eating!

Unfortunately, Redux has side effects. Some, like fatigue, diarrhea, vivid dreams, and a dry mouth can be tolerated. Others, however, are low in incidence but very serious. Studies suggest that the drug increases the incidence of primary pulmonary hypertension from 1 to 2 per one million to as much as 46 per one million. Primary pulmonary hypertension destroys blood vessels in the lungs and heart, and can lead to death. Also, the drug causes significant, and possibly permanent, brain damage in lab animals. An abundance of serotonin makes the neurons that ordinarily produce the neurotransmitter swell, wither, and then die according to Dr. Mark Molliver, a Johns Hopkins neurologist.

Are you acting recklessly, and therefore unethically, if you take this drug? To decide, a risk-benefit analysis may be appropriate. Severe obesity puts people at risk for hypertension, heart attacks, diabetes, and some cancers. And these illnesses contribute to 300,000 deaths a year in the United States. The slim risk of serious side effects may be worth it for the obese, but the same conclusion does not hold for those who are merely overweight. Even so, most doctors will probably prescribe the drug for anyone who asks for it. Just three months after the introduction of Redux, doctors were writing 85,000 prescriptions a week!

Questions

1. Is there any difference in smoking to control weight gain and taking Redux to control weight gain? Explain your answer.
2. Is there any difference in drinking alcohol to feel better and taking Redux to feel better? Explain your answer.
3. Is it ever ethical to take drugs to control our everyday behavior? Why or why not?

Summarizing the Concepts

12.1 Neurons and How They Work

The nervous system contains neurons and neuroglial cells which service neurons. Sensory neurons take information from sensory receptors to the CNS; interneurons occur within the CNS; and motor neurons take information from the CNS to effectors (muscles or glands). A motor neuron can be used to demonstrate that neurons are composed of dendrites, a cell body, and an axon. Long axons are covered by a myelin sheath.

When an axon is not conducting a nerve impulse, the inside of an axon is negative (–65mV) compared to the outside. The sodium-potassium pump actively transports Na$^+$ out of an axon and K$^+$ to inside an axon. The resting potential is due to the leakage of K$^+$ to the outside of the neuron. When an axon is conducting a nerve impulse (action potential), Na$^+$ first moves into the axoplasm and then K$^+$ moves out of the axoplasm.

Transmission of the nerve impulse from one neuron to another takes place when a neurotransmitter molecule is released into a synaptic cleft. The binding of the neurotransmitter to receptors in the postsynaptic membrane causes excitation or inhibition. Integration is the summing of excitatory and inhibitory signals. Neurotransmitter molecules are removed from the cleft by enzymatic breakdown or by reabsorption.

12.2 Peripheral Nervous System

The peripheral nervous system contains only nerves and ganglia. The brain is always involved in voluntary actions but reflexes are automatic, and some do not require involvement of the brain. In the somatic system, for example, a stimulus causes sensory receptors to generate nerve impulses which are conducted by sensory nerve fibers to interneurons in the spinal cord. Interneurons signal motor neurons which conduct nerve impulses to a skeletal muscle that contracts, giving the response to the stimulus.

The autonomic (involuntary) system controls smooth muscle of the internal organs and glands. The sympathetic division is associated responses that occur during times of stress, and the parasympathetic system is associated responses that occur during times of relaxation.

12.3 Central Nervous System

The CNS consists of the spinal cord and brain, which are both protected by bone. The CNS receives and integrates sensory input and formulates motor output. The gray matter of the spinal cord contains neuron cell bodies; the white matter consists of myelinated axons that occur in bundles called tracts. The spinal cord carries out reflex actions and sends sensory information to the brain and receives motor output from the brain. Because tracts cross over, the left side of the brain controls the right side of the body and vice versa.

In the brain, the medulla oblongata and pons have centers for vital functions, like breathing and the heartbeat. The cerebellum primarily coordinates muscle contractions. The hypothalamus controls homeostasis, and the thalamus specializes in sending on sensory input to the cerebrum. The cerebrum has two cerebral hemispheres connected by the corpus callosum. Sensation, reasoning, learning and memory, and also language and speech take place in the cerebrum.

12.4 The Cerebral Hemispheres

Each cerebral hemisphere contains a frontal, parietal, occipital, and temporal lobe. The cerebral cortex is a thin layer of gray matter covering the cerebrum. The primary motor area in the frontal lobe sends out motor commands to lower brain centers which pass them on to motor neurons. The primary somatosensory area in the parietal lobe receives sensory information from lower brain centers in communication with sensory neurons. A visual area occurs in the occipital lobe, an auditory area occurs in the temporal lobe, and so forth for the other senses. Association areas are located in all the lobes; the prefrontal area of the frontal lobe is especially necessary to higher mental functions.

12.5 Higher Mental Functions

Short-term memory and long-term memory are dependent upon the prefrontal area. The hippocampus acts as conduit for sending information to long-term memory and retrieving it once again. The amygdala adds emotional overtones to memories. On the cellular level, long-term potentiation seems to be required for long-term memory. Unfortunately long-term potentiation can go awry when neurons become overexcited and die. Neuroprotective drugs are being developed in the hope they will prevent disorders like Alzheimer disease. Language and speech are dependent upon Broca's area (a motor speech area) and Wernicke's area (a sensory speech area). These two areas communicate with one another. Interestingly enough, these two areas are located only in the left hemisphere.

12.6 Homeostasis

Homeostasis is ultimately under the control of the nervous system and the endocrine system. The hypothalamus and the medulla oblongata have various centers that control the functioning of internal organs and the internal environment. These parts of the brain send commands to internal organs by way of the autonomic system. The hypothalamus works closely with endocrine glands to control blood composition and therefore tissue fluid.

12.7 Drug Abuse

Although neurological drugs are quite varied, each type has been found to either promote or prevent the action of a particular neurotransmitter.

Studying the Concepts

1. With reference to a motor neuron, describe the structure and function of the three parts of a neuron. What are the three types of neurons and what is their relationship to the CNS?   246

2. What is the sodium-potassium pump? What is the resting potential, and how is it brought about?   248

3. Describe the two parts of an action potential and the changes that can be associated with each part.   248

4. What is a neurotransmitter, where is it stored, how does it function, and how is it destroyed? Name two well-known neurotransmitters.   250–51

5. The peripheral nervous system contains what three types of nerves? What is meant by a mixed nerve?   252

6. Trace the path of a somatic reflex.   253

7. What is the autonomic system, and what are its two major divisions? Give several similarities and differences between these divisions.   255

8. Describe the structure and function of the spinal cord.   256–57

9. Name the major parts of the brain, and give a function for each. What is the reticular formation?   258–59

10. Name the lobes of the cerebral hemispheres and describe the function of motor, sensory, and association areas. What is the limbic system?   260–61

11. Name several different types of memory and explain how long-term memory is thought to occur. What is long-term potentiation?   262–63

12. Language and speech require what portions of the brain? What is the left brain/right brain hypothesis?   263

13. Describe the physiological effects and mode of action of alcohol, marijuana, cocaine, and heroin.   265–67

Testing Your Knowledge of the Concepts

In questions 1–4, match the functions to a region of the brain and fill in the blanks.
a. medulla oblongata
b. thalamus
c. hypothalamus
d. cerebrum

_____ 1. Interprets sensory input and initiates voluntary muscular movements.

_____ 2. Contains centers for regulating the heart rate, breathing, and blood pressure.

_____ 3. Receives sensory information and channels it to the cerebrum.

_____ 4. Works with the endocrine system to control the salt and water balance of the blood.

In questions 5–7, indicate whether the statement is true (T) or false (F).

_____ 5. There is a preponderance of K^+ outside a resting axon and a preponderance of Na^+ inside an axon.

_____ 6. Transmission of signals at a synapse is carried out by neurotransmitter molecules.

_____ 7. The spinal cord is a part of the peripheral nervous system.

In questions 8 and 9, fill in the blanks.

8. The space between the axon of one neuron and the dendrite of another is called a _____.

9. In the peripheral nervous system a mixed nerve contains _____ nerve fibers and _____ nerve fibers.

10. Label this diagram.

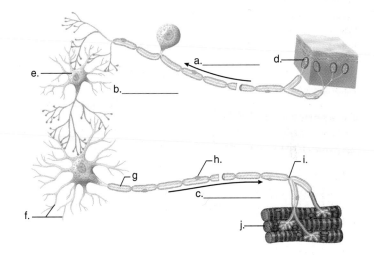

Applying Your Knowledge to the Concepts

These questions pertain to homeostasis.

1. When you are angry or excited during the eating of a meal, you might subsequently have indigestion. Discuss how this is possible by linking together the limbic system, the hypothalamus, and the sympathetic division of the autonomic nervous system.

2. Poliomyelitis, or polio, is a viral disease that selectively destroys the cell bodies of motor neurons in the spinal cord. What effects would you expect from this disease and what effects would you not expect? Would homeostasis still be possible? Why?

3. What is the advantage of a reflex action for homeostasis? Describe a situation in which a reflex action might be disadvantageous.

4. A person who drinks alcohol shows the following symptoms as the blood alcohol level increases: impaired judgment, slurred speech, inability to walk a straight line, and finally, difficulty in breathing. What does this tell you about how alcohol sequentially affects the brain?

Understanding the Terms

acetylcholine (ACh) 251
acetylcholinesterase (AChE) 251
action potential 248
amygdala 262
association area 261
autonomic system 255
axon 246
axon bulb 251
basal nuclei 261
brain 258
brain stem 259
cell body 246
central nervous system (CNS) 246
cerebellum 259
cerebral cortex 260
cerebral hemisphere 259
cerebrospinal fluid 256
cerebrum 259
cranial nerve 252
dendrite 246
diencephalon 259
dorsal-root ganglion 252
drug abuse 265
endorphin 267
episodic memory 262

fight or flight 255
ganglion 252
gray matter 256
hippocampus 262
hypothalamus 259
integration 251
interneuron 246
limbic system 261
long-term potentiation (LTP) 263
medulla oblongata 259
memory 262
meninges (sing., meninx) 256
midbrain 259
motor neuron 246
myelin sheath 247
nerve 252
nerve impulse 248
neuroglial cell 246
neuron 246
neurotransmitter 251
node of Ranvier 247
norepinephrine (NE) 251
nuclei 259
oscilloscope 248
parasympathetic division 255
peripheral nervous system

(PNS) 246
pons 259
prefrontal area 261
primary motor area 260
primary somatosensory area 260
reflex 253
refractory period 248
resting potential 248
reticular formation 259
Schwann cell 247
semantic memory 262
sensory neuron 246
skill memory 262

sodium-potassium pump 248
somatic system 253
spinal cord 256
spinal nerve 252
sympathetic division 255
synapse 251
synaptic cleft 251
thalamus 259
threshold 248
tract 257
ventricle 256
white matter 257

Match the terms to these definitions:

a. _____ Action potential (an electrochemical change) traveling along an axon.

b. _____ One of the large paired structures making up the cerebrum of the brain.

c. _____ The summing of excitatory and inhibitory signals received at synapses by a neuron.

d. _____ Neuron that takes information from the central nervous system to effectors.

e. _____ A network of nuclei and tracts extending from the brain stem to the thalamus that passes on sensory stimuli and arouses the cerebral cortex.

Applying Technology to the Concepts

Your study of the nervous system is supported by these available technologies:

Essential Study Partner CD-ROM
Animals → Nervous System
Visit the Mader web site for related ESP activities.

Exploring the Internet
The Mader Home Page provides resources and tools as you study this chapter.

http://www.mhhe.com/biosci/genbio/mader

Dynamic Human 2.0 CD-ROM
Nervous System

Virtual Physiology Laboratory CD-ROM
Action Potential
Synaptic Transmission

HealthQuest CD-ROM
1 Stress Management & Mental Health → Gallery → Transmitting Neurons

Chapter *13*

Senses

Chapter Concepts

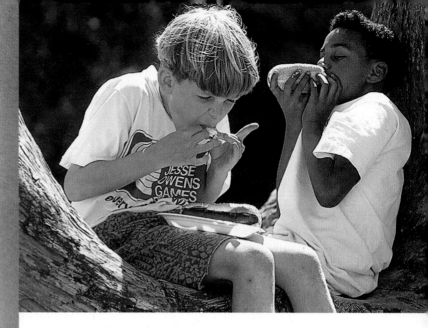

Figure 13.1   **Eating habits.**
Do we like certain foods because we associate their taste and smell with fun?

After a bad day, what's your favorite thing to eat? Chocolate? Mom's homemade pasta? So-called comfort food soothes our spirits and—if only for a minute—makes the world seem okay again.

That's because we learn to link certain tastes and smells with emotion (Fig. 13.1). Sensory cells, like those found in the nose, send messages to the parts of our brain that control emotion and memory. So we remember those freshly baked cookies that once brightened a depressing day. We may then reach for a cookie when we feel down.

Taste and smell are not only essential to experience pleasure, they also help ensure our very survival. In nature, animals learn to avoid tempting substances that lead to illness. Likewise, humans shy away from foods linked to negative experiences. A person who comes down with food poisoning at a Japanese restaurant, for example, is unlikely to crave the taste of sushi in the near future. Smell alone can sometimes be protective. Many lives have been saved by a sensitive nose picking up a whiff of smoke or poisonous vapor or detecting spoilage in food.

The senses work together. Once an animal has experienced a noxious substance, it may recognize it by sight alone in the future. Eyes watch out for imminent dangers

and ears hear warning sounds or we would never learn to drive in traffic. These same organs allow us to communicate with others and enrich our lives as when we see a play or go to the ballet.

Sense organs at the periphery of the body are the "windows of the brain" because they keep the brain aware of what is going on in the external world. When stimulated, sensory receptors generate nerve impulses that travel to the central nervous system (CNS). Nerve impulses arriving at the cerebral cortex of the brain result in sensation. We see, hear, taste, and smell with our brain, not our sense organs.

13.1 Sensory Receptors and Sensations

Sensory receptors are specialized to receive certain types of stimuli (sing., **stimulus**), or particular forms of energy. Eventually these stimuli result in a sensation within the cerebral cortex.

Types of Sensory Receptors

Sensory receptors in humans can be classified into just four types.

Chemoreceptors respond to chemical substances in the immediate vicinity. The senses of taste and smell are well known to have this type of sensory receptor, but there are also chemoreceptors in various other organs that are sensitive to internal conditions. Chemoreceptors in certain blood vessels monitor the hydrogen ion concentration $[H^+]$ in the blood, and if the pH lowers, the breathing rate increases. As more carbon dioxide is expired, the blood pH will rise.

Pain receptors are a type of chemoreceptor. They are naked dendrites that respond to chemicals released by damaged tissues. Pain receptors are protective because they alert us to possible danger. Without the pain of appendicitis, we may never seek the medical help that is needed to avoid a ruptured appendix.

Mechanoreceptors are stimulated by mechanical forces, which are most often pressure of some sort. The sense of touch is dependent on pressure receptors that are sensitive to either strong or slight pressures. Pressure receptors located in certain arteries detect changes in blood pressure, and stretch receptors in the lungs detect the degree of lung inflation. Proprioceptors which respond to the stretching of muscle fibers, tendons, joints, and ligaments, make us aware of the position of our limbs. Even hearing is dependent on mechanoreceptors. In this case, the receptors are sensitive to pressure waves in inner ear fluid. Pressure receptors that provide information regarding equilibrium are also located in the inner ear.

Thermoreceptors are stimulated by changes in temperature. Those that respond when temperatures rise are called warmth receptors, and those that respond when temperatures lower are called cold receptors. There are internal thermoreceptors in the hypothalamus and surface thermoreceptors in the skin.

Photoreceptors respond to light energy. Our eyes contain photoreceptors that are sensitive to light and thereby provide us with a sense of vision. Stimulation of the photoreceptors, known as rod cells, results in black and white vision, while stimulation of the photoreceptors, known as cone cells, results in color vision.

The sensory receptors of humans respond to four types of stimuli: chemical, mechanical, temperature, and light.

How Sensation Occurs

Sensory receptors respond to environmental stimuli by generating nerve impulses. **Sensation** occurs when nerve impulses arrive at the cerebral cortex of the brain. **Perception** occurs when the cerebral cortex interprets the meaning of sensations.

Most sensory receptors are free nerve endings or encapsulated nerve endings, but some are specialized cells closely associated with neurons. The plasma membrane of a sensory receptor contains receptor proteins that react to the stimulus. For example, the receptor proteins in the plasma membrane of chemoreceptors bind to certain molecules. When this happens, ion channels open and ions flow across the plasma membrane. If the stimulus is sufficient, nerve impulses begin and are carried by a sensory nerve fiber to the CNS (Fig. 13.2). The stronger the stimulus, the greater the frequency of nerve impulses.

Although sensory receptors simply initiate nerve impulses we have different senses. The brain, as you know, is responsible for sensation and perception. Nerve impulses that begin in the optic nerve eventually reach the visual areas of the cerebral cortex and, thereafter, we see objects. Nerve impulses that begin in the auditory nerve eventually reach the auditory areas of the cerebral cortex and, thereafter, we hear sounds. If it were possible to switch these nerves, then stimulation of the eyes would result in hearing! On the other hand, when a blow to the eye stimulates photoreceptors, we "see stars" because nerve impulses from the eyes can only result in sight.

Sensory receptors carry out **integration,** the summing of signals. In this instance, the signals are in the form of environmental stimuli. One example of integration is **sensory adaptation,** a decrease in response to a stimulus. We have all had the experience of smelling an odor when we first enter a room and then later not being aware of it at all. Some au-

thorities believe that when sensory adaptation occurs, sensory receptors have stopped sending impulses to the brain. Others believe that interneurons in the CNS have filtered out the ongoing stimuli (see Fig. 12.13).

Sensation occurs when nerve impulses reach the cerebral cortex of the brain. Perception, which also occurs in the cerebral cortex, is an interpretation of the meaning of sensations.

Types of Senses

Senses are divided into the somatic senses and the special senses.

Somatic Senses

Senses associated with the skin, muscles, joints, and internal organs are called **somatic senses.** Mechanoreceptors in the skin give us a sense of touch and throughout the body a sense of pressure. A sense of temperature is due to receptors located in the skin and the brain. A sense of pain is due to pain receptors located in the skin and also among internal organs. Proprioception, which helps maintain posture and balance, is due to proprioceptors like Golgi tendon organs in tendons and muscle spindles in skeletal muscles.

Special Senses

The sense organs for taste, smell, vision, balance, and hearing contain a number of receptors all specialized to give us a particular sense. Table 13.1 lists the **special senses** and their associated sense organs.

The sensory receptors, whether existing independently or in sense organs, account for our various senses.

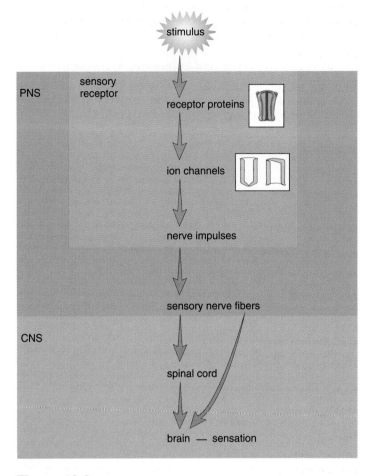

Figure 13.2 Sensation.
The stimulus is received by a receptor protein which causes ion channels to open and nerve impulses (action potentials) to be generated. Conduction of nerve impulses to the CNS by sensory nerve fibers results in sensation.

Table 13.1	Special Sense Organs		
Sense	**Type of Sensory Receptor**	**Specific Sensory Receptors**	**Sense Organ**
Taste	Chemoreceptor	Taste cells	Taste buds
Smell	Chemoreceptor	Olfactory cells	Olfactory epithelium
Vision	Photoreceptor	Rod cells and cone cells in retina	Eye
Hearing	Mechanoreceptor	Hair cells in spiral organ	Ear
Balance	Mechanoreceptor	Hair cells in utricle, saccule, and semicircular canals	Ear

13.2 Senses

Somatic senses are dependent upon receptors in the muscles, joints and tendons, and in the skin. The sensory receptors for somatic senses send nerve impulses to the spinal cord. From there, they travel up the spinal cord in tracts to the somatosensory areas of the cerebral cortex.

Proprioception

Proprioceptors detect the degree of muscle relaxation, the stretch of tendons, and the movement of ligaments. Muscle spindles act to increase and Golgi tendon organs act to decrease the degree of muscle contraction. The result is a muscle that has the proper length and tension, or muscle tone.

Figure 13.3 illustrates the activity of a muscle spindle. In a muscle spindle, sensory nerve endings are wrapped around thin muscle cells within a connective tissue sheath.

When the muscle relaxes and undue stretching of the muscle spindle occurs, nerve impulses are generated. The rapidity of the nerve impulses generated by the muscle spindle is proportional to the stretching of a muscle. A reflex action occurs which results in contraction of muscle fibers adjoining the muscle spindle. The knee-jerk reflex, which involves muscle spindles, offers an opportunity for physicians to test a reflex action.

The information sent by muscle spindles to the CNS is used to maintain the body's balance and posture despite the force of gravity always acting upon the skeleton and muscles. Proprioception also helps us know the position of our limbs in space.

Proprioceptors are involved in reflex actions that maintain muscle tone and thereby the body's balance and posture.

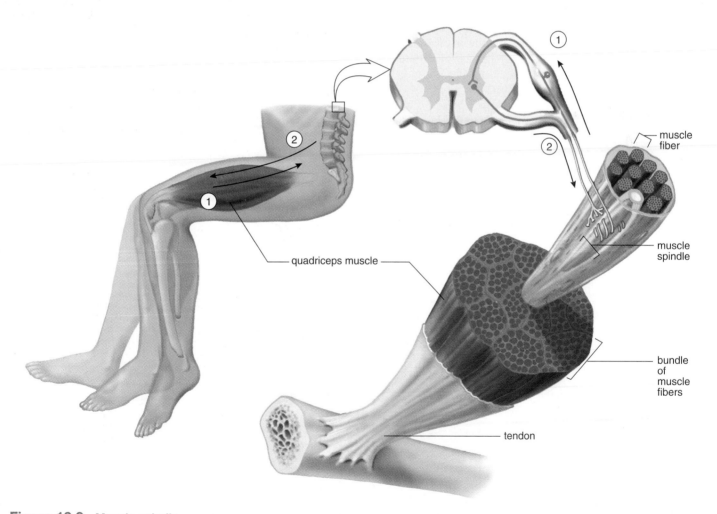

Figure 13.3   Muscle spindle.
① When a muscle is stretched, a muscle spindle sends sensory nerve impulses to the cord. ② Motor nerve impulses from the cord result in muscle fiber contraction so that muscle tone is maintained.

Cutaneous Senses

The skin is composed of two layers: the epidermis and the dermis. In Figure 13.4, the artist has dramatically indicated these two layers by separating the epidermis from the dermis in one location. The epidermis is stratified squamous epithelium in which cells become keratinized as they rise to the surface where they are sloughed off. The dermis of skin is a thick connective tissue layer. The dermis contains receptors for touch, pressure, pain, and temperature. It is a mosaic of these tiny receptors, as you can determine by slowly passing a metal probe over your skin. At certain points, there is a feeling of pressure, and at others, there is a feeling of hot or cold (depending on the probe's temperature).

Two types of receptors are responsive to fine touch: the Meissner's corpuscles and the Merkel's disks. The Meissner's corpuscles are concentrated in the fingertips, the palms, the lips, the tongue, the nipples, the penis and the clitoris. Their prevalence provides these regions with special sensitivity. Pacinian corpuscles, Ruffini's endings, and Krause end bulbs are three different types of pressure receptors. Pacinian corpuscles are onion-shaped sensory receptors that lie deep inside the dermis. Ruffini's endings and Krause end bulbs are encapsulated by sheaths of connective tissue and contain lacy networks of nerve fibers. Temperature and pain receptors are simply free nerve endings in the epidermis. Some free nerve endings are responsive to cold; others are responsive to warmth. Cold receptors are far more numerous than warmth receptors, but there are no known structural differences between the two.

Specialized receptors in the human skin respond to touch, pressure, pain, and temperature (warmth and cold).

Referred Pain

Like the skin, many internal organs have pain receptors. Sometimes, stimulation of internal pain receptors is felt as pain from the skin. This is called **referred pain.** Some internal organs have a referred pain relationship with areas located in the skin of the back, groin, and abdomen; pain from the heart is felt in the left shoulder and arm. This most likely happens when nerve impulses from the pain receptors of internal organs travel to the spinal cord and synapse with neurons also receiving impulses from the skin. The brain perceives this as pain in the skin.

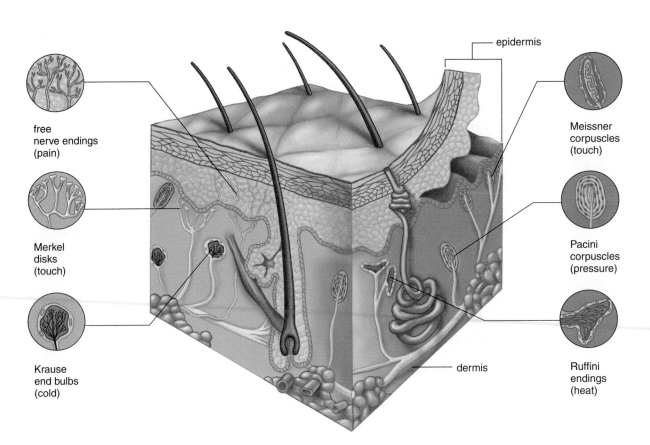

Figure 13.4   Receptors in human skin.
The classical view shown here is that each receptor has the main function indicated. However, investigations in this century indicate that matters are not so clear-cut. For example, microscopic examination of the skin of the ear shows only free nerve endings (pain receptors), and yet the skin of the ear is sensitive to all sensations. Therefore, it appears that the receptors of the skin are somewhat, but not completely, specialized.

13.3 Chemical Senses

There are chemoreceptors located in the carotid arteries and in the aorta that are primarily sensitive to the pH of the blood. These bodies communicate via sensory nerve fibers with the respiratory center located in the medulla oblongata. When the pH drops, they signal this center, and immediately thereafter the breathing rate increases. The expiration of CO_2 raises the pH of the blood.

Taste and smell are called the chemical senses because the receptors for these senses are sensitive to molecules in the food we eat and air we breathe.

Sense of Taste

Taste buds are sense organs located primarily on the tongue (Fig. 13.5). Many lie along the walls of the papillae, the small elevations on the tongue that are visible to the naked eye. Isolated ones are also present on the hard palate, the pharynx, and the epiglottis.

Taste buds are embedded in tongue epithelium and open at a taste pore. They have supporting cells and a number of elongated taste cells that end in microvilli. The microvilli bear receptor proteins for certain molecules. When these molecules bind to receptor proteins, nerve impulses are generated in associated sensory nerve fibers. These nerve impulses go to the brain, including cortical areas which interpret them as tastes.

There are four primary types of tastes (sweet, sour, salty, bitter) and taste buds for each are concentrated on the tongue in particular regions (Fig. 13.5a). Sweet receptors are most plentiful near the tip of the tongue. Sour receptors occur primarily along the margins of the tongue. Salty receptors are most common on the tip and the upper front portion of the tongue. Bitter receptors are located toward the back of the tongue. Actually, the response of taste buds can result in a range of sweet, sour, salty, and bitter tastes. The brain appears to survey the overall pattern of incoming sensory impulses and to take a "weighted average" of their taste messages as the perceived taste.

Taste buds are sense organs that contain taste cells. The microvilli of taste cells have receptor proteins for molecules that cause the brain to distinguish between sweet, sour, salty, and bitter tastes.

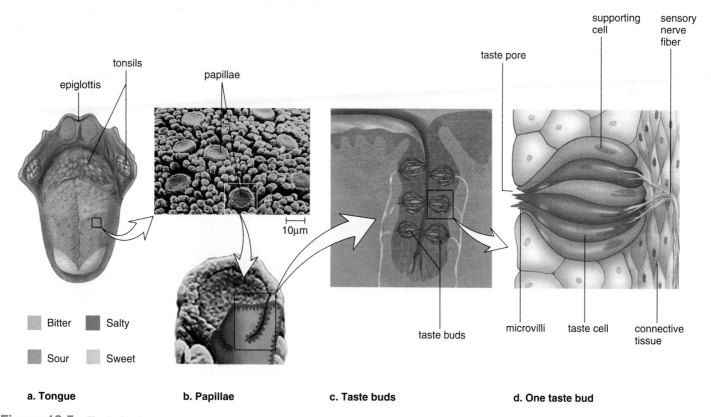

■ Bitter	■ Salty
■ Sour	■ Sweet

a. Tongue   **b. Papillae**   **c. Taste buds**   **d. One taste bud**

Figure 13.5   **Taste buds.**
a. Papillae on the tongue contain taste buds that are sensitive to sweet, sour, salty, and bitter tastes as indicated. **b.** Enlargement of papillae. **c.** Taste buds occur along the walls of the papillae. **d.** Taste cells end in microvilli that bear receptor proteins for certain molecules. When molecules bind to the receptor proteins, nerve impulses are generated that go to the brain where the sensation of taste occurs.

Sense of Smell

Our sense of smell is dependent on **olfactory cells** located within olfactory epithelium high in the roof of the nasal cavity (Fig. 13.6). Olfactory cells are modified neurons. Each cell ends in a tuft of about five olfactory cilia which bear receptor proteins for odor molecules. Each olfactory cell has only one out of 1,000 different types of receptor proteins. Nerve fibers from like olfactory cells lead to the same interneuron in the olfactory bulb, an extension of the brain. An odor contains many odor molecules which activate a characteristic combination of receptor proteins. A rose might stimulate olfactory cell type A, B, C, while a daffodil might stimulate olfactory cell type B, C, E . An odor's signature in the olfactory bulb is determined by which corresponding interneurons are stimulated. When the interneurons communicate this information via the olfactory tract to the olfactory areas of the cerebral cortex, we know we have smelled a rose or a daffodil.

Have you ever noticed that an aroma will bring to mind a vivid memory of a person or place? A person's perfume may remind you of someone else, or the smell of boxwood may re-mind you of your grandfather's farm. Look again at Figure 12.15 and notice that the olfactory bulbs have direct connections with the limbic system and its centers for emotions and memory. One investigator showed that when subjects smelled an orange when viewing a painting, they not only remembered the painting, they had many deep feelings about it.

Actually, the sense of taste and the sense of smell work together to create a combined effect when interpreted by the cerebral cortex. For example, when you have a cold, you think that food has lost its taste, but most likely you have lost the ability to sense its smell. This method works in reverse also. When you smell something, some of the molecules move from the nose down into the mouth region and stimulate the taste buds there. Therefore, part of what we refer to as smell may in fact be taste.

The olfactory epithelium is the sense organ that contains olfactory cells. The cilia of olfactory cells have receptor proteins for odor molecules that cause the brain to distinguish odors.

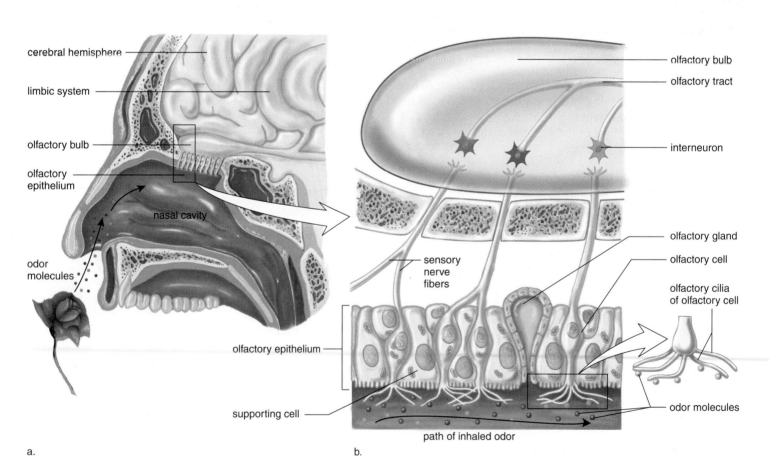

Figure 13.6 **Olfactory cell location and anatomy.**
a. The olfactory epithelium in humans is located high in the nasal cavity. **b.** Olfactory cells end in cilia that bear receptor proteins for specific odor molecules. The cilia of each olfactory cell can bind to only one type of odor molecule signified here by color. If a rose causes A, B, C type olfactory cells to be stimulated, then interneurons A, B, C in the olfactory bulb are activated. The primary olfactory area of the cerebral cortex interprets the pattern of interneurons stimulated as the scent of a rose.

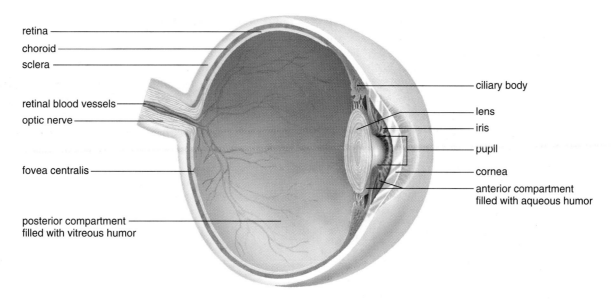

retina
choroid
sclera
retinal blood vessels
optic nerve
fovea centralis
posterior compartment
filled with vitreous humor

ciliary body
lens
iris
pupil
cornea
anterior compartment
filled with aqueous humor

Figure 13.7   Anatomy of the human eye.
Notice that the sclera, the outer layer of the eye, becomes the cornea and that the choroid, the middle layer, is continuous with the ciliary body and the iris. The retina, the inner layer, contains the photoreceptors for vision; the fovea centralis is the region where vision is most acute.

13.4 Sense of Vision

Vision requires the work of the eyes and the brain. As we shall see, much processing of stimuli occurs in the eyes before nerve impulses are sent to the brain. Still, researchers estimate that at least a third of the cerebral cortex takes part in processing visual information.

Anatomy of the Eye

The eyeball, which is an elongated sphere about 2.5 cm in diameter, has three layers, or coats: the sclera, the choroid, and the retina (Fig. 13.7 and Table 13.2). The outer layer, the **sclera,** is white and fibrous except for the transparent **cornea** which is made of transparent collagen fibers. The cornea is the window of the eye.

The middle, thin, darkly pigmented layer, the **choroid,** is vascular and absorbs stray light rays that photoreceptors have not absorbed. Toward the front, the choroid becomes the donut-shaped **iris.** The iris regulates the size of the **pupil,** a hole in the center of the iris through which light enters the eyeball. The color of the iris (and therefore the color of your eyes) correlates with its pigmentation. Heavily pigmented eyes are brown while lightly pigmented eyes are green or blue. Behind the iris, the choroid thickens and forms the circular ciliary body. The **ciliary body** contains the ciliary muscle, which controls the shape of the lens for near and far vision.

The **lens,** attached to the ciliary body by ligaments, di-

vides the eye into two compartments; the one in front of the lens is the anterior compartment and the one behind the lens is the posterior compartment. The anterior compartment is filled with a clear, watery fluid called the **aqueous humor.** A small amount of aqueous humor is continually produced each day. Normally, it leaves the anterior compartment by way of tiny ducts. When a person has **glaucoma,** these drainage ducts are blocked, and aque-

Table 13.2	Function of Parts of the Eye
Part	**Function**
Sclera	Protects and supports eyeball
Cornea	Refracts light rays
Choroid	Absorbs stray light
Retina	Contains receptors for sight
Rods	Make black-and-white vision possible
Cones	Make color vision possible
Fovea centralis	Makes acute vision possible
Lens	Refracts and focuses light rays
Ciliary body	Holds lens in place, accommodation
Iris	Regulates light entrance
Pupil	Admits light
Humors	Transmit light rays and support eyeball
Optic nerve	Transmits impulse to brain

ous humor builds up. If glaucoma is not treated, the resulting pressure compresses the arteries that serve the nerve fibers of the retina, where photoreceptors are located. The nerve fibers begin to die due to lack of nutrients, and the person becomes partially blind. Eventually, total blindness can result.

The third layer of the eye, the **retina,** is located in the posterior compartment which is filled with a clear gelatinous material called the **vitreous humor.** The retina contains photoreceptors called rod cells and cone cells. The rods are very sensitive to light but they do not see color; therefore, at night or in a darkened room we see only shades of gray. The cones, which require bright light, are sensitive to different wavelengths of light and, therefore, we have the ability to distinguish colors. The retina has a very special region called the **fovea centralis** where cone cells are densely packed. Light is normally focused on the fovea when we look directly at an object. This is helpful because vision is most acute in the fovea centralis. Sensory fibers from the retina form the **optic nerve** which takes nerve impulses to the brain.

The eye has three layers: the outer sclera, the middle choroid, and the inner retina. Only the retina contains photoreceptors for light energy.

Focusing

When we look at an object, light rays pass through the pupil and are **focused** on the retina (Fig. 13.8a). The image produced is much smaller than the object because light rays are bent (refracted) when they are brought into focus. Focusing starts with the cornea and continues as the rays pass through the lens and the humors. Notice that the image on the retina is inverted (it is upside down) and reversed from left to right.

The lens provides additional focusing power as **visual accommodation** occurs for close vision. The shape of the lens is controlled by the ciliary muscle within the ciliary body. When we view a distant object, the ciliary muscle is relaxed, causing the suspensory ligaments attached to the ciliary body to be taut; therefore, the lens remains relatively flat (Fig. 13.8b). When we view a near object, the ciliary muscle contracts, releasing the tension on the suspensory ligaments, and the lens rounds up due to its natural elasticity (Fig. 13.8c). Because close work requires contraction of the ciliary muscle, it very often causes muscle fatigue known as eyestrain.

Usually after the age of 40, the lens loses some of its elasticity and is unable to accommodate. Bifocal lenses are then usually necessary for those who already have correc-

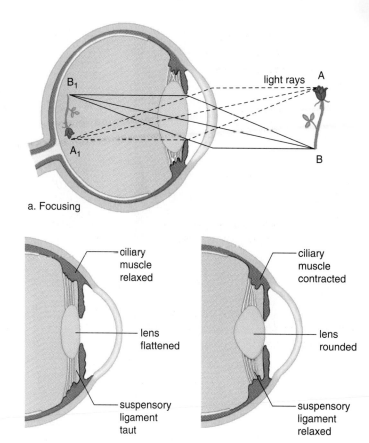

a. Focusing

b. Focusing on distant object   c. Focusing on near object

Figure 13.8   Focusing.
a. Light rays from each point on an object are bent by the cornea and the lens in such a way that an inverted and reversed image of the object forms on the retina. **b.** When focusing on a distant object, the lens is flat because the ciliary muscle is relaxed and the suspensory ligament is taut. **c.** When focusing on a near object, the lens accommodates; it becomes rounded because the ciliary muscle contracts, causing the suspensory ligament to relax.

tive lenses. Also with aging or possibly exposure to the sun (see the Health reading on page 284), the lens is subject to **cataracts.** The lens becomes opaque and therefore incapable of transmitting rays of light. Usually today, the lens is replaced with an artificial lens during surgery. In the future, it may be possible to restore the original configuration of the proteins making up the lens.

The lens, assisted by the cornea and the humors, focuses images on the retina.

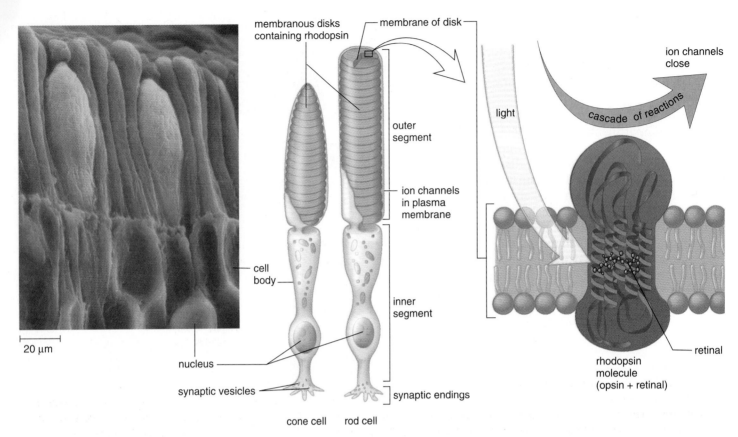

Figure 13.9   Structure and function of rod cells and cone cells.
The outer segment of rods and cones contains stacks of membranous disks, which contain visual pigments. In rods, the membrane of each disk contains rhodopsin, a complex molecule containing the protein opsin and the pigment retinal. When rhodopsin absorbs light energy, it splits, releasing opsin, which sets in motion a cascade of reactions that ends when ion channels in the plasma membrane close.

Photoreceptors

Vision begins once light has been focused on the photoreceptors in the retina. Figure 13.9 illustrates the structure of our photoreceptors called **rod cells** and **cone cells.** Both rods and cones have an outer segment joined to an inner segment by a stalk. The outer segment contains many membranous disks. And many pigment molecules are embedded in the membrane of these disks. Synaptic vesicles are located at the synaptic endings of the inner segment

The visual pigment in rods is a deep purple pigment called rhodopsin. **Rhodopsin** is a complex molecule made up of the protein opsin and a light absorbing molecule called **retinal,** which is a derivative of vitamin A. When a rod absorbs light, rhodopsin splits into opsin and retinal, leading to a cascade of reactions and the closure of ion channels in the rod cell's plasma membrane. This stops the release of inhibitory transmitter molecules from the rod's synaptic vesicles and starts the signals that result in nerve impulses going to the brain. Rods are very sensitive to light and therefore are suited to night vision. (Since carrots are rich in vitamin

A, it is true that eating carrots can improve your night vision.) Rod cells are plentiful throughout the entire retina; therefore, they also provide us with peripheral vision and perception of motion.

The cones, on the other hand, are located primarily in the fovea and are activated by bright light. They allow us to detect the fine detail and the color of an object. **Color vision** depends on three different kinds of cones, which contain pigments called the B (blue), G (green), and R (red) pigments. Each pigment is made up of retinal and opsin, but there is a slight difference in the opsin structure of each, which accounts for their individual absorption patterns. Various combinations of cones are believed to be stimulated by in-between shades of color.

The receptors for sight are the rods and the cones. The rods permit vision in dim light at night, and the cones permit vision in bright light needed for color vision.

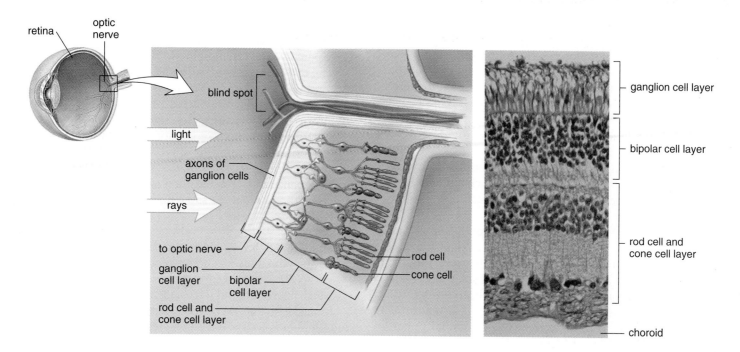

Figure 13.10 **Structure and function of retina.**

The retina is the inner layer of the eye. Rod cells and cone cells located at the back of the retina synapse with bipolar cells which synapse with ganglion cells. Integration of signals occurs at these synapses; therefore much processing occurs in bipolar and ganglion cells. Further, notice that many rod cells share one bipolar cell but cone cells do not. Each cone synapses with only one ganglion cell. Cone cells, therefore, distinguish more detail than do rod cells.

Integration of Visual Signals in the Retina

The retina has three layers of neurons (Fig. 13.10). The layer closest to the choroid contains the rods and cones; the middle layer contains bipolar cells; and the innermost layer contains ganglion cells, whose fibers become the optic nerve. Only the rod cells and the cone cells are sensitive to light, and therefore light must penetrate to the back of the retina before they are stimulated.

The rod cells and the cone cells synapse with the bipolar cells, which in turn synapse with ganglion cells that initiate nerve impulses. Notice in Figure 13.10 that there are many more rod cells and cone cells than ganglion cells. In fact, the retina has as many as 150 million rod cells and 6 billion cone cells but only one million ganglion cells. The sensitivity of cones in the fovea versus rods is mirrored by how directly they connect to ganglion cells. As many as 100 rods may synapse with the same ganglion cell. No wonder that stimulation of rods results in vision that is only blurred and indistinct. In contrast, each cone in the fovea synapses with only one ganglion cell. This accounts for why cones particularly in the fovea provide us with a sharper, more detailed image of an object.

As signals pass to bipolar cells and ganglion cells, integration occurs. Each ganglion cell receives signals from rod cells covering about one square millimeter of retina (about the size of a thumbtack hole). This region is called the ganglion cell's receptive field. Some time ago, scientists discovered that a ganglion cell is stimulated when light hits the center of its receptive field and is inhibited when light hits the area of the receptive field surrounding the center. If all the rod cells in the receptive field are stimulated, the ganglion cell responds in a neutral way—it reacts only weakly or perhaps not at all. This supports the hypothesis that considerable processing occurs in the retina before nerve impulses are sent to the brain.

Synaptic integration and processing begins in the retina before nerve impulses are sent to the brain.

Blind Spot

Figure 13.10 provides an opportunity to point out that there are no rods and cones where the optic nerve exits the retina. Therefore, no vision is possible in this area. You can prove this to yourself by putting a dot to the right of center on a piece of paper. Use the right hand to move the paper slowly toward the right eye while you look straight ahead. The dot will disappear at one point—this is your **"blind spot."**

Integration of Visual Signals in the Brain

Sensory fibers from ganglion cells assemble to form the optic nerves. At the X-shaped optic chiasma, fibers from the right half of each retina converge and continue on together in the right optic tract, and fibers from the left half of each retina converge and continue on together in the left optic tract. This is the first way in which the brain divides up the visual field (Fig. 13.11).

Because the image is reversed, the left optic tract carries information about the right portion of the **visual field,** and the right optic tract carries information about the left portion of the visual field.

The optic tracts sweep around the hypothalamus, and most fibers synapse with interneurons in a nucleus (mass of neuron cell bodies) of the thalamus. Axons from the thalamic nucleus form an optic radiation that takes nerve impulses to the primary visual area of the cerebral cortex. Since each primary visual area receives information regarding only half the visual field, these areas must eventually share information to form a unified image. Also, the inverted and reversed image must be righted in the brain for us to correctly perceive the visual field.

The most surprising finding has been that the brain has a second way of taking the field apart. Each primary visual area of the cerebral cortex acts like a post office, parceling out information regarding color, form, motion, and possibly other attributes to different portions of the adjoining visual association area. Therefore, the brain has taken the field apart even though we see a unified visual field. The cerebral cortex is believed to rebuild the visual field and give us an understanding of it at the same time.

The visual pathway which begins in the retina passes through the thalamus before reaching the cerebral cortex. The pathway and the visual cortex take the visual field apart, but the visual association areas rebuild it so that we correctly perceive the entire field.

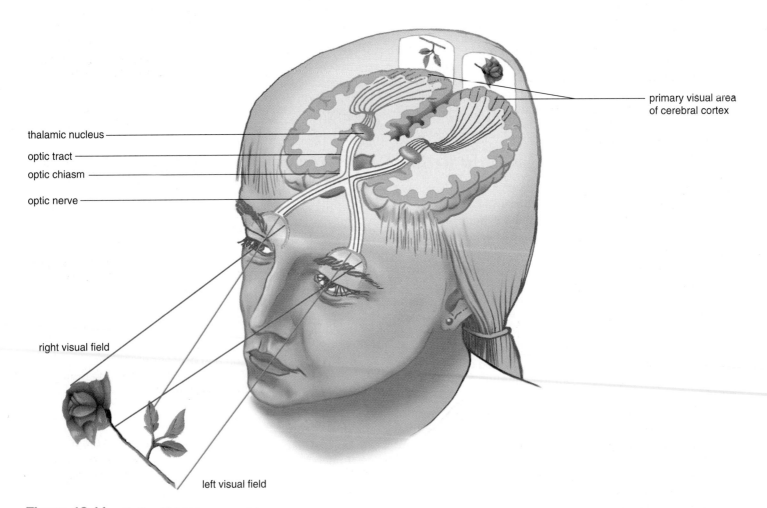

thalamic nucleus

optic tract

optic chiasm

optic nerve

primary visual area of cerebral cortex

right visual field

left visual field

Figure 13.11   Optic chiasma.
Both eyes "see" the entire visual field. Because of the optic chiasma, data from the right half of each retina go to the right visual areas of the cerebral cortex, and data from the left half of the retina go to the left visual areas of the cerebral cortex. This data is combined to allow us to see the entire visual field. Note that the visual pathway to the brain includes the thalamus which has the ability to filter sensory stimuli.

Abnormalities of the Eye

Color blindness and misshaped eyeballs are two common abnormalities of the eyes. More serious abnormalities are discussed in the reading on page 284.

Color Blindness

Complete color blindness is extremely rare. In most instances, a particular type of cone is lacking or deficient in number. The most common mutation is a lack of red or green cones. This abnormality affects 5–8% of the male population. If the eye lacks red cones, the green colors are accentuated, and vice versa.

Distance Vision

The majority of people can see what is designated as a size 20 letter 20 feet away, and so are said to have 20/20 vision. Persons who can see close objects but cannot see the letters from this distance are said to be nearsighted. Nearsighted people can see close objects better than they can see objects at a distance. These individuals have an elongated eyeball, and when they attempt to look at a distant object, the image is brought to focus in front of the retina (Fig. 13.12). They can see close objects because they can adjust the lens to allow the image to focus on the retina, but to see distant objects, these people must wear concave lenses, which diverge the light rays so that the image can be focused on the retina.

There is a treatment for nearsightedness called radial keratotomy, or RK. From four to eight cuts are made in the cornea so that they radiate out from the center like spokes in a wheel. When the cuts heal, the cornea is flattened. Lasers are now also used to flatten or change the shape of the cornea. Although some patients are satisfied with the result, others complain of glare and varying visual acuity.

Persons who can easily see the optometrist's chart but cannot see close objects well are farsighted; these individuals can see distant objects better than they can see close objects. They have a shortened eyeball, and when they try to see close objects, the image is focused behind the retina. When the object is distant, the lens can compensate for the short eyeball, but when the object is close, these persons must wear a convex lens to increase the bending of light rays so that the image can be focused on the retina.

When the cornea or lens is uneven, the image is fuzzy. The light rays cannot be evenly focused on the retina. This condition, called **astigmatism,** can be corrected by an unevenly ground lens to compensate for the uneven cornea.

The shape of the eyeball determines the need for corrective lenses; the inability of the lens to accommodate as we age also requires corrective lenses for close vision.

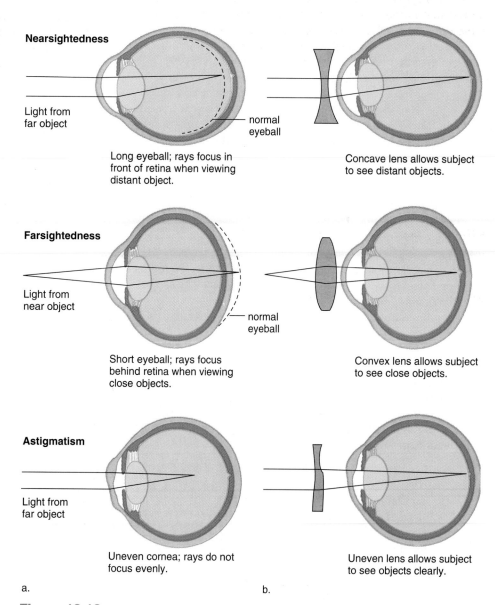

Nearsightedness

Light from far object

normal eyeball

Long eyeball; rays focus in front of retina when viewing distant object.

Concave lens allows subject to see distant objects.

Farsightedness

Light from near object

normal eyeball

Short eyeball; rays focus behind retina when viewing close objects.

Convex lens allows subject to see close objects.

Astigmatism

Light from far object

Uneven cornea; rays do not focus evenly.

Uneven lens allows subject to see objects clearly.

a.   b.

Figure 13.12   Common abnormalities of the eye, with possible corrective lenses.
a. The cornea and the lens function in bringing light rays (lines) to focus, but sometimes they are unable to compensate for the shape of the eyeball or for an irregular curvature of the cornea. **b.** In these instances, corrective lenses can allow the individual to see normally.

Protecting Vision and Hearing

Preventing a Loss of Vision

The eye is subject to both injuries and disorders. Although flying objects sometimes penetrate the cornea and damage the iris, lens, or retina, careless use of contact lenses is the most common cause of injuries to the eye. Injuries cause only 4% of all cases of blindness; the most frequent causes are retinal disorders, glaucoma, and cataracts, in that order. Retinal disorders are varied. In diabetic retinopathy, which blinds many people between the ages of 20 and 74, capillaries to the retina burst and blood spills into the vitreous fluid. Careful regulation of blood glucose levels in these patients may be protective. In macular degeneration, the cones are destroyed because thickened choroid vessels no longer function as they should. Glaucoma occurs when the drainage system of the eyes fails, so that fluid builds up and destroys nerve fibers responsible for peripheral vision. Eye doctors always check for glaucoma, but it is advisable to be aware of the disorder in case it comes on quickly. Those who have experienced acute glaucoma report that the eyeball feels as heavy as a stone. In cataracts, cloudy spots on the lens of the eye eventually pervade the whole lens. The milky yellow-white lens scatters incoming light and blocks vision.

Regular visits to an eye-care specialist, especially by the elderly, are a necessity in order to catch conditions, such as glaucoma, early enough to allow effective treatment. It has not been proven, however, that consuming particular foods or vitamin supplements will reduce the risk of cataracts. Even so, it's possible that estrogen replacement therapy may be somewhat protective in postmenopausal women.

Accumulating evidence suggests that both macular degeneration and cataracts, which tend to occur in the elderly, are caused by long-term exposure to the ultraviolet rays of the sun. It is recommended, therefore, that everyone, especially those who live in sunny climates or work outdoors, wear sunglasses that absorb ultraviolet light. Large lenses worn close to the eyes offer further protection. The Sunglass Association of America has devised the following system for categorizing sunglasses:

- Cosmetic lenses absorb at least 70% of UV-B and 20% of UV-A and 60% of visible light. Such lenses are worn for comfort rather than protection.
- General purpose lenses absorb at least 95% of UV-B and 60% of UV-A and 60–92% of visible light. They are good for outdoor activities in temperate regions.
- Special purpose lenses block at least 99% of UV-B and 60% of UV-A, and 20–97% of visible light. They are good for bright sun combined with sand, snow, or water.

Health care providers have found an increased incidence of cataracts in heavy cigarette smokers. In men, smoking 20 cigarettes or more a day, and in women, smoking more than 35 cigarettes a day doubles the risk of cataracts. It is possible that smoking reduces the delivery of blood and therefore nutrients to the lens.

Preventing a Loss of Hearing

Especially when we are young, the middle ear is subject to infections that can lead to hearing impairments if the infections are not treated promptly by a physician. The mobility of ossicles decreases with age, and in otosclerosis, new filamentous bone grows over the stirrup, impeding its movement. Surgical treatment is the only remedy for this type of conduction deafness. However, age-associated nerve deafness due to stereocilia damage from exposure to loud noises is preventable. Hospitals are now aware that even the ears of the newborn need to be protected from noise and are taking steps to make sure neonatal intensive care units and nurseries are as quiet as possible.

In today's society, exposure to the types of noises listed in Table 13A is a common occurrence. Noise is measured in decibels, and any noise above a level of 80 decibels could result in damage to the hair cells of the organ of Corti. Eventually, the stereocilia and then the hair cells disappear completely (Fig. 13A). If listening to city traffic for extended periods can damage hearing, it stands to reason that frequent attendance at rock concerts, constantly playing a stereo loudly, or using earphones at high volume is also damaging to hearing. The first hint of danger could be temporary hearing loss, a "full" feeling in the ears, muffled hearing, or tinnitus (e.g., ringing in the ears). If you have any of these symptoms, modify your listening habits immediately to prevent further damage. If exposure to noise is unavoidable, specially designed noise-reduction earmuffs are available, and it is also possible to purchase earplugs made from a compressible, spongelike material at the drugstore or sporting-goods store. These earplugs are not the same as those worn for swimming, and they should not be used interchangeably.

Aside from loud music, noisy indoor or outdoor equipment, such as a rug-cleaning machine or a chain saw, is also troublesome. Even motorcycles and recreational vehicles such as snowmobiles and motocross bikes can contribute to a gradual loss of hearing. Exposure to intense sounds of short duration, such as a burst of gunfire, can result in an immediate hearing loss. Hunters may have a significant hearing reduction in the ear opposite to the shoulder where the gun is carried. The butt of the rifle offers some protection to the ear nearest the gun when it is shot.

Finally, people need to be aware that some medicines are ototoxic. Anticancer drugs, most notably cisplatin, and certain antibiotics (e.g., streptomycin, kanamycin, gentamicin) make ears especially susceptible to a hearing loss. Anyone taking such medications needs to be especially careful to protect the ears from any loud noises.

Table 13A Noises That Affect Hearing

Type of Noise	Sound Level (decibels)	Effect
"Boom car," jet engine, shotgun, rock concert	over 125	Beyond threshold of pain; potential for hearing loss high
Discotheque, "boom box," thunderclap	over 120	Hearing loss likely
Chain saw, pneumatic drill, jackhammer, symphony orchestra, snowmobile, garbage truck, cement mixer	100–200	Regular exposure of more than one minute risks permanent hearing loss
Farm tractor, newspaper press, subway, motorcycle	90–100	Fifteen minutes of unprotected exposure potentially harmful
Lawn mower, food blender	85–90	Continuous daily exposure for more than eight hours can cause hearing damage
Diesel truck, average city traffic noise	80–85	Annoying; constant exposure may cause hearing damage

Source: National Institute on Deafness and Other Communication Disorders, January 1990, National Institute of Health.

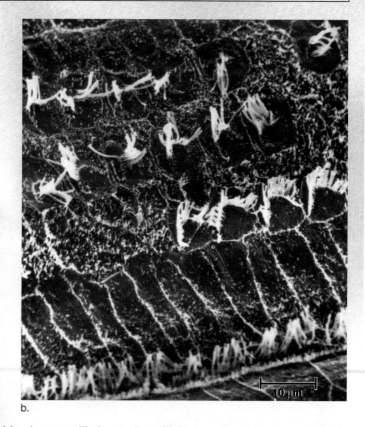

a.   b.

Figure 13A The higher the decibel reading noted in the table, the more likely a noise will damage hearing.
a. Normal hair cells in the spiral organ of a guinea pig. **b.** Damaged cells. This damage occurred after 24-hour exposure to a noise level like that at heavy-metal rock concerts. Hearing is permanently impaired because lost cells will not be replaced, and damaged cells may also die.

13.5 Sense of Hearing

The ear has two sensory functions: hearing and balance (equilibrium). The receptors for both of these are located in the inner ear, and each consists of **hair cells** with stereocilia that respond to mechanical stimulation.

Anatomy of the Ear

Figure 13.13 shows that the ear has three divisions: outer, middle, and inner. The **outer ear** consists of the pinna (external flap) and the auditory canal. The opening of the auditory canal is lined with fine hairs and sweat glands. Modified sweat glands are located in the upper wall of the canal; they secrete earwax, a substance that helps to guard the ear against the entrance of foreign materials, such as air pollutants.

The **middle ear** begins at the **tympanic membrane** (eardrum) and ends at a bony wall containing two small openings covered by membranes. These openings are called the **oval window** and the **round window**. Three small bones are found between the tympanic membrane and the oval window. Collectively called the **ossicles,** individually they are the **malleus** (hammer), the **incus** (anvil), and the **stapes** (stirrup) because their shapes resemble these objects. The

malleus adheres to the tympanic membrane, and the stapes touches the oval window. An **auditory** (eustachian) **tube,** which extends from each middle ear to the nasopharynx, permits equalization of air pressure. Chewing gum, yawning, and swallowing in elevators and airplanes help to move air through the auditory tubes upon ascent and descent. As this occurs we often hear the ears "pop."

Whereas the outer ear and the middle ear contain air, the inner ear is filled with fluid. The **inner ear,** anatomically speaking, has three areas: the semicircular canals and the **vestibule** are concerned with equilibrium; the **cochlea** is concerned with hearing. The cochlea resembles the shell of a snail because it spirals.

Process of Hearing

The process of hearing begins when sound waves enter the auditory canal (Fig. 13.14). Just as ripples travel across the surface of a pond, sound waves travel by the successive vibrations of molecules. Ordinarily, sound waves do not carry much energy, but when a large number of waves strike the tympanic membrane, it moves back and forth (vibrates) ever so slightly. The malleus then takes the pressure from the inner surface of the tympanic membrane and passes it by means of

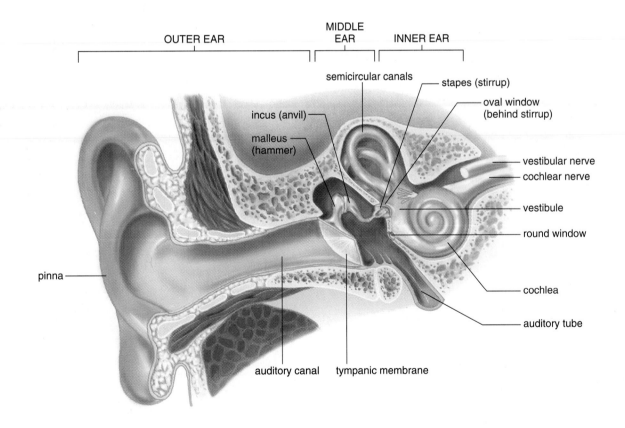

Figure 13.13   **Anatomy of the human ear.**
In the middle ear, the malleus (hammer), the incus (anvil), and the stapes (stirrup) amplify sound waves. The inner ear contains the sensory receptors for balance in the semicircular canals and the vestibule and the sensory receptors for hearing in the cochlea.

the incus to the stapes in such a way that the pressure is multiplied about 20 times as it moves from the tympanic membrane to the stapes. The stapes strikes the membrane of the oval window, causing it to vibrate, and in this way, the pressure is passed to the fluid within the cochlea.

If the cochlea is unwound and examined in cross section (Fig. 13.14) you can see that it has three canals: the vestibular canal, the **cochlear canal,** and the tympanic canal. The vestibular canal connects with the tympanic canal, which leads to the round window membrane. Along the length of the basilar membrane, which forms the lower wall of the cochlear canal, are little hair cells whose stereocilia are embedded within a gelatinous material called the **tectorial membrane.** The hair cells of the cochlear canal, called the **spiral organ** (organ of Corti), synapse with nerve fibers of the **cochlear** (auditory) **nerve.**

When the stapes strikes the membrane of the oval window, pressure waves move from the vestibular canal to the tympanic canal and across the basilar membrane, and the round window bulges. As the basilar membrane moves up and down, the stereocilia of the hair cells embedded in the tectorial membrane bend. Then, nerve impulses begin in the cochlear nerve and travel to the brain stem. When they reach the auditory areas of the cerebral cortex they are interpreted as a sound.

Each part of the spiral organ is sensitive to different wave frequencies, or pitch. Near the tip, the spiral organ responds to low pitches, such as a tuba, and near the base, it responds to higher pitches, such as a bell or a whistle. The nerve fibers from each region along the length of the spiral organ lead to slightly different areas in the brain. The pitch sensation we experience depends upon which region of the basilar membrane vibrates and which area of the brain is stimulated.

Volume is a function of the amplitude of sound waves. Loud noises cause the fluid of the cochlea to vibrate to a greater degree, and this, in turn, causes the basilar membrane to move up and down to a greater extent. The resulting increased stimulation is interpreted by the brain as volume. It is believed that the tone of a sound is an interpretation of the brain based on the distribution of hair cells stimulated.

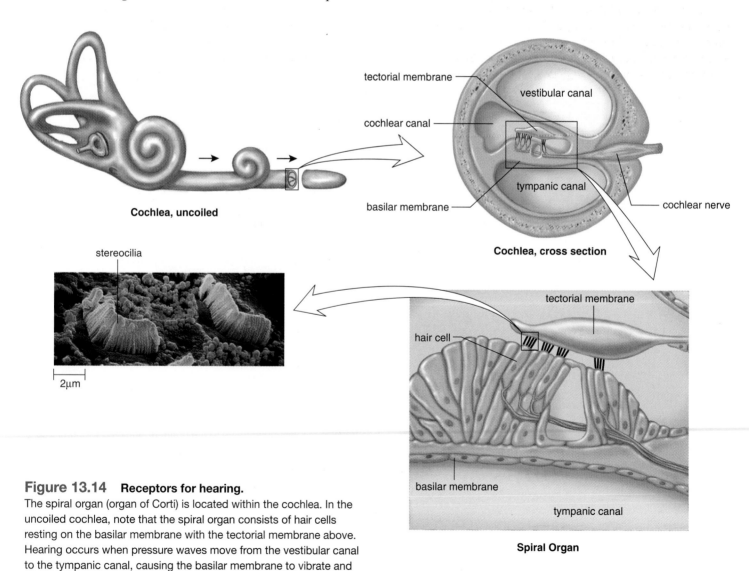

Figure 13.14   Receptors for hearing.
The spiral organ (organ of Corti) is located within the cochlea. In the uncoiled cochlea, note that the spiral organ consists of hair cells resting on the basilar membrane with the tectorial membrane above. Hearing occurs when pressure waves move from the vestibular canal to the tympanic canal, causing the basilar membrane to vibrate and the stereocilia (or at least a portion of the more than 20,000 hair cells) to bend within the tectorial membrane. Nerve impulses traveling in the cochlear nerve result in hearing.

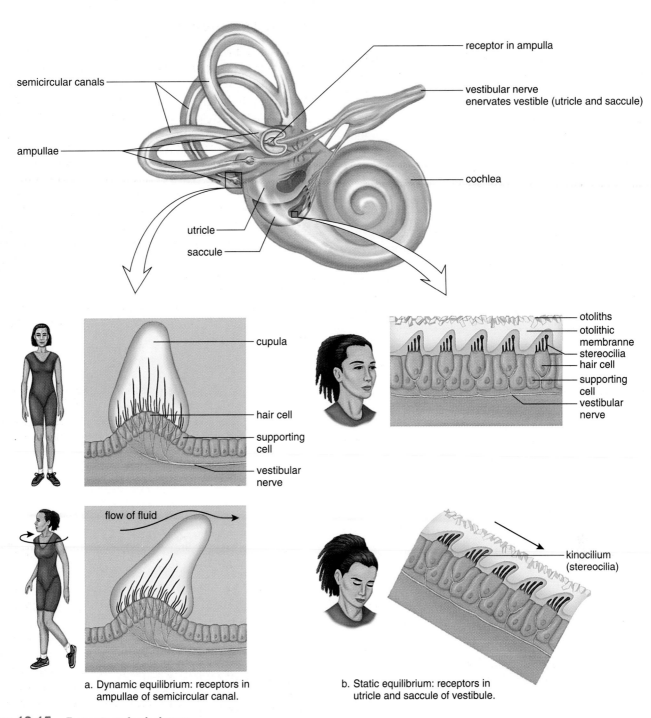

a. Dynamic equilibrium: receptors in
ampullae of semicircular canal.

b. Static equilibrium: receptors in
utricle and saccule of vestibule.

Figure 13.15   Receptors for balance.

a. Dynamic equilibrium. The ampullae of the semicircular canals contain hair cells with stereocilia embedded in a cupula. When the head rotates, the cupula is displaced, bending the stereocilia. Thereafter, nerve impulses travel in the vestibular nerve to the brain. **b.** Static equilibrium. The utricle and the saccule contain hair cells with stereocilia embedded in an otolithic membrane. When the head bends, otoliths are displaced, causing the membrane to sag and the stereocilia to bend. The rapidity of nerve impulses in the vestibular nerve tells the brain how much the head has moved.

13.6 Sense of Balance

The sense of balance has been subdivided into two senses: **dynamic equilibrium,** involving angular and/or rotational movement of the head, and **static equilibrium,** involving movement of the head in one plane, either vertical or horizontal (Fig. 13.15).

Dynamic Equilibrium

Dynamic equilibrium utilizes the **semicircular canals** which are arranged so that there is one in each dimension of space. The base of each of the three canals, called the **ampulla,** is slightly enlarged. Little hair cells whose stereocilia are embedded within a gelatinous material called a cupula are found within the ampullae. Because there are three semicircular canals, each ampulla responds to head rotation in a different plane of space. As fluid within a semicircular canal flows over and displaces a cupula, the stereocilia of the hair cells bend, and the pattern of impulses carried by the vestibular nerve to the brain changes. Continuous movement of fluid in the semicircular canals causes one form of motion sickness.

  Vertigo is dizziness and a sensation of rotation. It is possible to simulate a feeling of vertigo by spinning rapidly and stopping suddenly. When the eyes are rapidly jerked back to a midline position, the person feels like the room is spinning. This shows that the eyes are also involved in our sense of balance.

Static Equilibrium

Static equilibrium depends on the **utricle** and **saccule,** two membranous sacs located in the vestibule. Both of these sacs contain little hair cells, whose stereocilia are embedded within a gelatinous material called an otolithic membrane. Calcium carbonate ($CaCO_3$) granules, or **otoliths,** rest on this membrane. The utricle is especially sensitive to horizontal movements and the bending of the head, while the saccule responds best to vertical (up-down) movements. When the body is still, the otoliths in the utricle and the saccule rest on the otolithic membrane above the hair cells. When the head bends or the body moves in the horizontal and vertical planes, the otoliths are displaced and the otolithic membrane sags, bending the stereocilia of the hair cells beneath. If the stereocilia move toward the kinocilium, the largest stereocilium, nerve impulses in the vestibular nerve increase. If the stereocilia move away from the kinocilium, nerve impulses in the vestibular nerve decrease. These data tell the brain the direction of the movement of the head.

Movement of a cupula within the semicircular canals contributes to the sense of dynamic equilibrium. Movement of the otolithic membrane within the utricle and the saccule accounts for static equilibrium.

Bioethical Issue

A cataract is a cloudiness of the lens that occurs in 50% of people between the ages 65 and 74, and in 70% of those age 75 or older. The extent of visual impairment depends on the size, density of the cataract, and where it is located in the lens. A dense centrally placed cataract causes severe blurring of vision.

  Are cataracts preventable? For most people the answer is yes, they are preventable. These factors have been identified as contributing to the chances of having a cataract:

• Smoking 20 or more cigarettes a day doubles the risk of cataracts in men—women have to smoke more than 30 cigarettes a day to increase their chances of a cataract.

• Exposure to the ultraviolet radiation in sunlight can more than double the risk of a cataract. In addition, this relationship is dose-dependent—the more sunlight, the higher the risk.

• Also, as many as one-third of cataracts may be caused by being overweight. Diet, rather than exercise, seems to reduce cataract formation, perhaps through lower blood sugar levels or improved antioxidant properties of the blood.

  It's clear, then, that our own behavior contributes to the occurrence of cataracts for most of us. Should we be responsible in our actions and take all possible steps to prevent developing a cataract, such as not smoking, wearing sunglasses and a wide-brim hat, and watching our weight? Or should we

simply rely on medical science to restore our eyesight? Most cataract operations today are performed on an outpatient basis with minimal postoperative discomfort and with a high expectation of restoration of sight. Perhaps it's better, though, to take all possible steps to prevent the occurrence of cataracts, just in case our experience is atypical.

Questions

1. To what extent do you feel each person is responsible for his or her own health? Explain your answer.
2. Should Medicare pay for cataract surgery in heavy smokers?
3. At what age should we make people aware that their behavior today can affect their health tomorrow?

Summarizing the Concepts

13.1 Sensory Receptors and Sensations

Each type of sensory receptor responds to a particular kind of stimulus. When stimulation occurs, sensory receptors initiate nerve impulses that are transmitted to the spinal cord and/or brain. Sensation occurs when nerve impulses reach the cerebral cortex. Perception is an interpretation of the meaning of sensations.

The senses are divided into the somatic (general) senses and the special senses (taste, smell, vision, hearing, balance).

13.2 Somatic Senses

Proprioception is illustrated by the action of muscle spindles which are stimulated when muscle fibers stretch. A reflex action, which is illustrated by the knee-reflex, causes the muscle fibers to contract. Proprioception helps maintain balance and posture.

The skin contains sensory receptors for touch, pressure, pain, and temperature (warmth and cold). The pain of internal organs is sometimes felt in the skin and is called referred pain.

13.3 Chemical Senses

Taste and smell are due to chemoreceptors that are stimulated by molecules in the environment. The taste buds contain taste cells that communicate with sensory nerve fibers, while the receptors for smell are neurons.

After molecules bind to plasma membrane receptor proteins on the microvilli of taste cells and the cilia of olfactory cells, nerve impulses eventually reach the cerebral cortex, which determines the taste and odor according to the pattern of receptors stimulated.

13.4 Sense of Vision

Vision is dependent on the eye, the optic nerves, and the visual areas of the cerebral cortex. The eye has three layers. The outer layer, the sclera, can be seen as the white of the eye; it also becomes the transparent bulge in the front of the eye called the cornea. The middle pigmented layer, called the choroid, absorbs stray light rays. The rod cells (sensory receptors for dim light) and the cone cells (sensory receptors for bright light and color) are located in the retina, the inner layer of the eyeball. The cornea, the humors, and especially the lens bring the light rays to focus on the retina. To see a close object, accommodation occurs as the lens rounds up.

When light strikes rhodopsin within the membranous disks of rod cells, rhodopsin splits into opsin and retinal. A cascade of reactions leads to the closing of ion channels in a rod cell's plasma membrane. Inhibitory transmitter molecules are no longer released and nerve impulses are carried in the optic nerve to the brain.

Integration occurs in the retina which is composed of three layers of cells: the rod and cone layer, the bipolar cell layer, and the ganglion cell layer. Integration also occurs in the brain, especially because the visual field is first taken apart by the optic chiasma, and the primary visual area in the cerebral cortex parcels out signals for color, form, and motion to the visual association area.

13.5 Sense of Hearing

Hearing is a specialized sense dependent on the ear, the cochlear nerve, and the auditory areas of the cerebral cortex.

The ear is divided into three parts: outer, middle, and inner. The outer ear consists of the pinna and the auditory canal, which direct sound waves to the middle ear. The middle ear begins with the tympanic membrane and contains the ossicles (malleus, incus, stapes). The malleus is attached to the tympanic membrane, and the stapes is attached to the oval window, which is covered by membrane. The inner ear contains the cochlea and the semicircular canals, plus the utricle and saccule.

Hearing begins when the outer and middle portions of the ear convey and amplify the sound waves that strike the oval window. Its vibrations set up pressure waves within the cochlea, which contains the spiral organ, consisting of hair cells whose stereocilia are embedded within the tectorial membrane. When the stereocilia of the hair cells bend, nerve impulses begin in the cochlear nerve and are carried to the brain.

13.6 Sense of Balance

The ear also contains receptors for our sense of balance. Dynamic equilibrium is dependent on the stimulation of hair cells within the ampullae of the semicircular canals. Static equilibrium relies on the stimulation of hair cells within the utricle and the saccule.

Studying the Concepts

1. What are the four types of receptors in the human body? 272
2. Explain sensation from the reception of stimuli to the passage of nerve impulses to the brain. 272
3. What are the somatic senses? Explain how muscle spindles are involved in a reflex action. What are the cutaneous senses? 273–75
4. Describe the structure of a taste bud, and tell how a taste cell functions. 276
5. Describe the structure and function of the olfactory epithelium. How does the sense of smell come about? 277
6. Describe the anatomy of the eye, and explain focusing and accommodation. 278–79
7. Describe the structure and function of rod cells and cone cells. 280
8. Explain the process of integration in the retina and the brain. 281–82
9. Relate the need for corrective lenses to three possible eye shapes. 283
10. Describe the anatomy of the ear and how we hear. 286–87
11. Describe the role of the semicircular canals, the utricle, and the saccule in balance. 289

Testing Your Knowledge of the Concepts

In questions 1-4, match the sense to the descriptions.
a. vision
b. smell
c. hearing
d. somatic

_____ 1. A chemical sense that utilizes modified neurons that send nerve impulses directly to the brain.

_____ 2.. Much integration occurs in the PNS before nerve impulses are sent to the brain.

_____ 3. Includes the five types of senses associated with the skin.

_____ 4. Sensory receptors are hair cells that communicate with sensory nerve fibers.

In questions 5–7, indicate whether the statement is true (T) or false (F).

_____ 5. The spiral organ is located in the semicircular canals.

_____ 6. The choroid, the middle layer of the eye, becomes the cornea of the eyeball.

_____ 7. The cone cells function in bright light and are sensitive to color.

For questions 8 and 9, fill in the blanks.

8. The _____ and the _____ of the inner ear contain sensory receptors for our sense of balance.

9. Because the image is inverted and because nerve fibers cross at the optic chiasma, the right cerebral hemisphere receives data about the _____ side of the visual field.

10. Label this diagram of an eye.

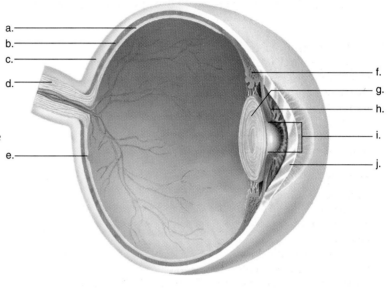

Applying Your Knowledge to the Concepts

These questions pertain to the senses.

1. All sensory receptors are transducers and they convert one form of energy into another form. Explain.

2. A friend of yours is red–green color blind. Will eating carrots help your friend? Why or why not?

3. A blow to the eye produces a sensation of flashing lights ("seeing stars"). What does this tell you about the receptors in an eye?

4. The inner ear is involved in balance; why does it help to open the eyes when standing on one foot?

Applying Technology to the Concepts

Your study of the senses is supported by these available technologies:

Essential Study Partner CD-ROM
Animals → Sense Organs
Visit the Mader web site for related ESP activities.

Dynamic Human 2.0 CD-ROM
Nervous System

Exploring the Internet
The Mader Home Page provides resources and tools as you study this chapter.

http://www.mhhe.com/biosci/genbio/mader

Understanding the Terms

ampulla 289
aqueous humor 278
astigmatism 283
auditory tube 286
blind spot 281
cataract 279
chemoreceptor 272
choroid 278
ciliary body 278
cochlea 286
cochlear canal 287
cochlear nerve 287
color vision 280
cone cell 280
cornea 278
dynamic equilibrium 289
focused 279
fovea centralis 279
glaucoma 278
hair cell 286
incus 286
inner ear 286
integration 272
iris 278

lens 278
malleus 286
mechanoreceptor 272
middle ear 286
olfactory cell 277
optic nerve 279
ossicle 286
otolith 289
outer ear 286
oval window 286
pain receptor 272
perception 272
photoreceptor 272
proprioceptor 274
pupil 278
referred pain 275
retina 279
retinal 280
rhodopsin 280
rod cell 280
round window 286
saccule 289
sclera 278
semicircular canal 289

sensation 272
sensory adaptation 272
sensory receptor 272
somatic sense 273
special sense 273
spiral organ 287
stapes 286
static equilibrium 289
stimulus 272
taste bud 276

tectorial membrane 287
thermoreceptor 272
tympanic membrane 286
utricle 289
vertigo 289
vestibule 286
visual accommodation 279
visual field 282
vitreous humor 279

Match the terms to these definitions:

a. _____ Portion of the inner ear that resembles a snail's shell and contains the spiral organ, the sense organ for hearing.

b. _____ Sensory receptor that responds to chemical stimulation—for example, receptors for taste and smell.

c. _____ Visual pigment found in the rods whose activation by light energy leads to vision.

d. _____ An awareness of a stimulus; occurs only in cerebral cortex.

e. _____ A mechanoreceptor that gives rise to nerve impulses when its stereocilia are bent or tilted.

Chapter 14

Endocrine System

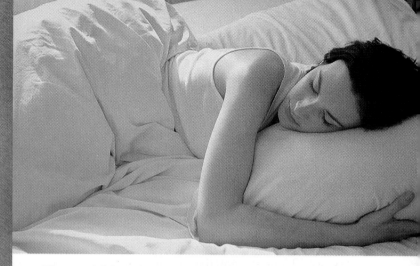

Figure 14.1 **Hormone regulation.**
Coordination by the nervous and endocrine systems never ceases, even when we sleep. Melatonin, the sleep hormone, and many other hormones are now coursing through this person's body even as she sleeps.

It was another sleepless night of TV reruns for Joanne. For the past month, Joanne had tossed and turned, occasionally peering out the window at darkened houses down her street. She was exhausted.

Finally, Joanne decided to call a friend who was a budding neurobiologist. That's when she learned about a natural hormone called melatonin. Produced by the brain's pea-sized pineal gland, melatonin lulls us to sleep. At night, our eyes signal a biological clock located in the brain that darkness is upon us, and the clock directs the pineal gland to produce melatonin. Our metabolism slows. Our body temperature falls. Blissfully, we fall asleep (Fig. 14.1).

Some researchers believe that a melatonin supplement—often taken about an hour before bedtime—can launch such nighttime reactions in restless bodies. Joanne, like many other Americans, began taking over-the-counter melatonin pills. Sure enough, she did fall asleep faster.

A nation of well-rested people sounds ideal, but melatonin is useful only for those with delayed sleep syndrome, and there are many other types of sleep disorders. And the longtime effects of melatonin on other hormonal levels is not yet known. Therefore, it's important to learn about the possible side effects of melatonin before becoming a true believer and taking the medication.

Chapter Concepts

A **hormone** is a substance secreted by an endocrine gland that is transported in blood to a target organ. The study of hormones is one of the most exciting disciplines within modern biology; it has produced many useful advances such as the synthesis of melatonin and other hormones, contraceptive methods, treatment of infertility and disorders such as diabetes, which was once fatal but now is controlled by injections of the hormone insulin. Melatonin is only one of the many hormones produced by the body. The presence or absence of hormones affects our metabolism, our appearance, and our behavior.

The endocrine system, like the nervous system, coordinates the functioning of body parts. Melatonin affects the brain and various other regions of the body. The nervous system is fast acting and utilizes nerve impulses that travel rapidly along nerve fibers. The endocrine system is slower acting because hormones have to be produced and then transported in the blood to target organs where they influence metabolism. Only then is their effect made known.

14.1 Environmental Signals

Even though the endocrine system and the nervous system have distinct differences, they both utilize chemical messengers—hormones versus neurotransmitter molecules. This realization supports a general interest in environmental signals that can be categorized in these three ways (Fig. 14.2).

Environmental signals that act at a distance between organisms. **Pheromones** are chemical messengers that pass between members of a species. As an example, ants lay down a pheromone trail to direct other ants to food, and female silkworm moths release a sex attractant that is received by male moth antennae even several kilometers away. Mammals also release pheromones, as when dogs use their urine to serve as a territorial marker. Some studies are being conducted to determine if humans have pheromones. In one investigation, it was observed that women who live in close quarters tend to have coinciding menstrual cycles. It is possible that this is due to a pheromone.

Environmental signals that act at a distance between body parts. This category includes **hormones** that are produced by endocrine glands or by neurosecretory cells of the hypothalamus. This is an example of how the nervous and endocrine systems work together to maintain homeostasis. An overlap between the nervous and endocrine systems is also exemplified by the secretion of epinephrine and norepinephrine at sympathetic nerve endings and also by the adrenal medulla, an endocrine gland.

Environmental signals that act locally between adjacent cells. Neurotransmitters belong in this category, as do substances that are sometimes called local hormones. For example, when the skin is cut, histamine, released by mast cells, promotes the inflammatory response.

Today, hormones are categorized as one type of environmental signal, a term which includes molecules that work at a distance between individuals or body parts or locally between adjacent cells.

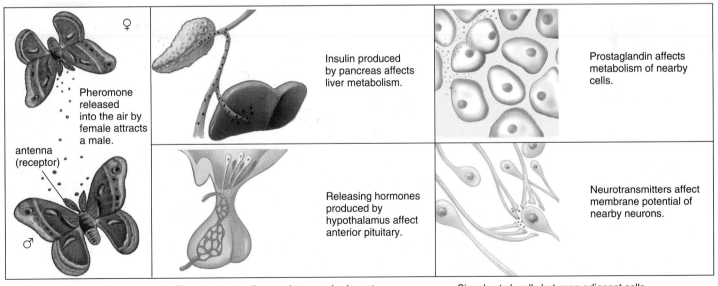

a. Signal acts at a distance between individuals.

b. Signal acts at a distance between body parts.

c. Signal acts locally between adjacent cells.

Pheromone released into the air by female attracts a male.

antenna (receptor)

Insulin produced by pancreas affects liver metabolism.

Releasing hormones produced by hypothalamus affect anterior pituitary.

Prostaglandin affects metabolism of nearby cells.

Neurotransmitters affect membrane potential of nearby neurons.

Figure 14.2   Environmental signals.
There are three categories of environmental signals (red dots). **a.** Pheromones are chemical messengers that act at a distance between individuals. **b.** Endocrine hormones and neurosecretions typically are carried in the bloodstream and act at a distance within the body of a single organism. **c.** Some chemical messengers have local effects only; they pass between cells that are adjacent to one another. This, of course, includes neurotransmitter molecules.

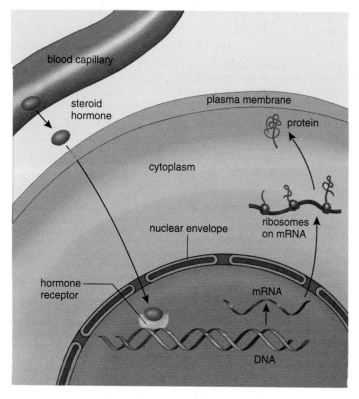

a. Action of steroid hormone

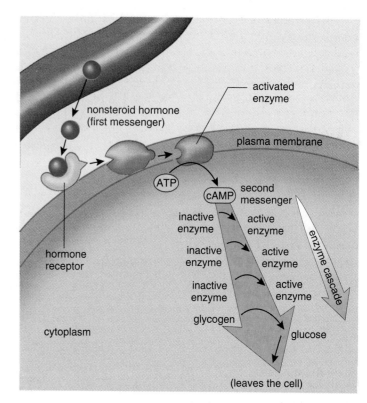

b. Action of nonsteroid hormone

Figure 14.3 **Cellular activity of hormones.**
a. After passing through the plasma membrane and nuclear envelope, a steroid hormone binds to a receptor inside the nucleus. The hormone-receptor complex then binds to DNA, and this leads to activation of certain genes and protein synthesis. **b.** Nonsteroid hormones, called first messengers, bind to a specific receptor protein in the plasma membrane. A protein relay ends when an enzyme converts ATP to cAMP, the second messenger, which activates an enzyme cascade.

The Action of Hormones

Hormones fall into two basic categories: (1) a **nonsteroid hormone** can be an amino acid, a peptide, or a protein composed of one or more polypeptides; a **steroid hormone** is always the same complex of four-carbon rings, but each type has different side chains. Because their effect is amplified through cellular mechanisms, either type of hormone can function at extremely low concentrations.

A hormone does not seek out a particular organ; rather the organ is awaiting the arrival of the hormone. Cells that can react to a hormone have hormone receptor proteins that combine with the hormone in a lock-and-key manner. Steroid hormones are lipids and therefore they cross cell membranes (Fig. 14.3a). Only after they are inside the nucleus do steroid hormones, such as estrogen and progesterone, bind to hormone receptor proteins. The hormone-receptor complex then binds to DNA, activating particular genes. Activation leads to production of a cellular enzyme in multiple quantities.

Most nonsteroid hormones cannot pass through the plasma membrane, and instead they bind to a receptor protein in the membrane (Fig. 14.3b). After epinephrine binds to a receptor protein, a relay system leads to the conversion of ATP to **cyclic AMP** (cyclic adenosine monophosphate). Cyclic AMP (cAMP) is made from ATP, but it contains only one phosphate group, which is attached to the adenosine portion of the molecule at two spots. Thus the molecule is cyclic. The nonsteroid hormone is called the *first messenger* and cAMP—or some other molecule—is called the *second messenger*. Calcium is also a common second messenger, and this helps explain why calcium regulation in the body is so important.

The second messenger sets in motion an *enzyme cascade*. In muscle cells epinephrine leads to the breakdown of glycogen to glucose (Fig. 14.3b). An enzyme cascade is so called because each enzyme in turn activates another. Because enzymes work over and over, every step in an enzyme cascade leads to more reactions—the binding of a single nonsteroid hormone molecule can result even in a thousandfold response.

Hormones are chemical messengers that influence the metabolism of the cell either indirectly by regulating the production of a particular protein (steroid hormone) or directly by activating an enzyme cascade (nonsteroid hormone).

Table 14.1   Principal Endocrine Glands and Hormones

Endocrine Gland	Hormone(s) Released	Target Tissues/Organ	Chief Function(s) of Hormone
Hypothalamus	Hypothalamic-releasing and release-inhibiting hormones	Anterior pituitary	Regulate anterior pituitary hormones
Anterior pituitary	Thyroid-stimulating (TSH, thyrotropic)	Thyroid	Stimulates thyroid
	Adrenocorticotropic (ACTH)	Adrenal cortex	Stimulates adrenal cortex
	Gonadotropic [follicle-stimulating (FSH), luteinizing (LH)]	Gonads	Egg and sperm production, and sex hormone production
	Prolactin (PRL)	Mammary glands	Milk production
	Growth (GH, somatotropic)	Soft tissues, bones	Cell division, protein synthesis, and bone growth
	Melanocyte-stimulating (MSH)	Melanocytes in skin	Unknown function in humans; regulates skin color in lower vertebrates
Posterior pituitary	Antidiuretic (ADH, vasopressin)	Kidneys	Stimulates water reabsorption by kidneys
	Oxytocin	Uterus, mammary glands	Stimulates uterine muscle contraction and release of milk by mammary glands
Pineal gland	Melatonin	Brain	Circadian and circannual rhythms; possibly involved in maturation of sex organs
Thyroid	Thyroxine (T_4) and triiodothyronine (T_3)	All tissues	Increases metabolic rate; regulates growth and development
	Calcitonin	Bones, kidneys, intestine	Lowers blood calcium level
Parathyroids	Parathyroid (PTH)	Bones, kidneys, intestine	Raises blood calcium level
Thymus	Thymosins	T lymphocytes	Production and maturation of T lymphocytes
Adrenal cortex	Glucocorticoids (cortisol)	All tissues	Raise blood glucose level; stimulate breakdown of protein
	Mineralocorticoids (aldosterone)	Kidneys	Reabsorb sodium and excrete potassium
	Sex hormones	Gonads, skin, muscles, bones	Stimulate sex characteristics
Adrenal medulla	Epinephrine and norepinephrine	Cardiac and other muscles	Emergency situations; raise blood glucose level
Pancreas	Insulin	Liver, muscles, adipose tissue	Lowers blood glucose level; promotes formation of glycogen
	Glucagon	Liver, muscles, adipose tissue	Raises blood glucose level
Gonads			
Testes	Androgens (testosterone)	Gonads, skin, muscles, bones	Stimulate secondary male sex characteristics
Ovaries	Estrogens and progesterone	Gonads, skin, muscles, bones	Stimulate female sex characteristics

Endocrine Glands

Endocrine glands can be contrasted with exocrine glands. The latter have ducts and secrete their products into these ducts for transport into body cavities. For example, the salivary glands send saliva into the mouth by way of the salivary ducts. **Endocrine glands** are ductless; they secrete their hormones directly into the bloodstream for distribution throughout the body.

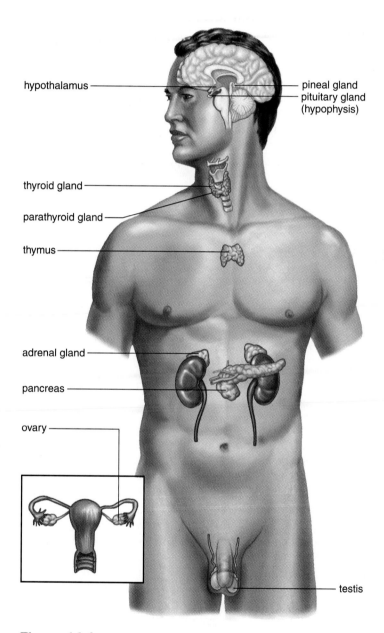

Figure 14.4   **The endocrine system.**
Anatomical location of major endocrine glands in the body.

Table 14.1 lists the hormones released by the principal endocrine glands, which are depicted in Figure 14.4. The hypothalamus, a part of the brain, is in close proximity to the pituitary. The hypothalamus controls the pituitary gland and this, too, exemplifies the close association between the nervous and endocrine systems. The pineal gland is also located in the brain. The thyroid and parathyroids are located in the neck, and the thymus lies just beneath the sternum in the thoracic cavity. The adrenal glands and pancreas are located in the abdominal cavity. The gonads include the ovaries, located in the pelvic cavity, and the testes, located outside this cavity in the scrotum.

Like the nervous system, the endocrine system is especially involved in homeostasis, that is, the dynamic equilibrium of the internal environment. The internal environment is the blood and tissue fluid that surrounds the body's cells. Notice that several hormones directly affect the osmolarity of the blood. Others control the calcium and glucose levels. Several hormones are involved in the maturation and function of the reproductive organs. In fact, many people are most familiar with the effect of hormones on sexual functions.

There are two mechanisms that control the effect of endocrine glands. Quite often a negative feedback mechanism controls the secretion of hormones. An endocrine gland can be sensitive to either the condition it is regulating or to the blood level of the hormone it is producing. For example, when the blood glucose level rises, the pancreas produces insulin, which causes the cells to take up glucose and the liver to store glucose. The stimulus for the production of insulin has thereby been dampened, and therefore the pancreas stops producing insulin. On the other hand, when the blood level of thyroid hormones rises, the anterior pituitary stops producing thyroid-stimulating hormone. We will discuss these examples in more detail later.

The presence of contrary hormonal actions is a way the effect of a hormone is controlled. The action of insulin, for example, is offset by the production of glucagon by the pancreas. Notice there are other examples of contrary hormonal actions in Table 14.1. The thyroid lowers the blood calcium level, but the parathyroids raise the blood calcium level. We will also have the opportunity to point out other instances in which hormones work opposite to one another and thereby bring about the regulation of a substance in the blood.

The secretion of a hormone is often controlled by negative feedback, and the effect of a hormone is often opposed by a contrary hormone. The end result is homeostasis and the normal functioning of body parts.

14.2 Hypothalamus and Pituitary Gland

The **hypothalamus** regulates the internal environment through the autonomic system. For example, it helps control heartbeat, body temperature, and water balance. The hypothalamus also controls the glandular secretions of the **pituitary gland.** The pituitary, a small gland about 1 cm in diameter, is connected to the hypothalamus by a stalklike structure. The pituitary has two portions: the anterior pituitary and the posterior pituitary (Fig. 14.5).

Posterior Pituitary

There are neurons in the hypothalamus called neurosecretory cells which produce the hormones **antidiuretic hormone (ADH)** and oxytocin. These hormones pass through axons into the **posterior pituitary** where they are stored in axon endings. ADH promotes the reabsorption of water from the collecting ducts attached to nephrons within the kidneys. There are neurons in the hypothalamus that act as a sensor because they are sensitive to the osmolarity of the blood. When these cells determine that the blood is too concentrated, ADH is released into the bloodstream from the axon endings in the posterior pituitary, and upon reaching the kidneys, ADH causes water to be reabsorbed. As the blood becomes dilute, ADH is no longer released. This is an example of control by negative feedback because the effect of the hormone (to dilute blood) acts to shut down the release of the hormone.

Inability to produce ADH causes diabetes insipidus (watery urine), in which a person produces copious amounts of urine with a resultant loss of ions from the blood. The condition can be corrected by the administration of ADH.

Oxytocin is the other hormone that is made in the hypothalamus and stored in the posterior pituitary. When labor begins during childbirth, pressure receptors in the uterine wall send nerve impulses to the hypothalamus and thereafter oxytocin is released from the posterior pituitary. Oxytocin then causes the uterus to contract more forcefully. This is an example of control by positive feedback: uterine contractions (the condition) brings about a result that increases uterine contractions. Oxytocin also stimulates the release of milk from the mammary glands when a baby is nursing.

The posterior pituitary stores two hormones, ADH and oxytocin, both of which are produced by and released from neurosecretory cells in the hypothalamus.

Anterior Pituitary

A portal system, consisting of two capillary systems connected by a vein, lies between the hypothalamus and the anterior pituitary (Fig. 14.5). The hypothalamus controls the anterior pituitary by producing **hypothalamic-releasing hormones** and **hypothalamic-inhibiting hormones.** For example, there is a thyroid-releasing hormone (TRH) and a thyroid-inhibiting hormone (TIH). TRH stimulates the anterior pituitary to release thyroid-stimulating hormone, and TIH inhibits the pituitary from releasing thyroid-stimulating hormone.

Three of the six hormones produced by the **anterior pituitary** (hypophysis) have an effect on other glands: (1) **thyroid-stimulating hormone (TSH)** stimulates the thyroid to produce the thyroid hormones; (2) **adrenocorticotropic hormone (ACTH)** stimulates the adrenal cortex to produce cortisol; and (3) **gonadotropic hormones (FSH and LH)** stimulate the gonads—the testes in males and the ovaries in females—to produce gametes and sex hormones. In each instance, the blood level of the last hormone in the sequence exerts negative feedback control over the secretion of the first two hormones:

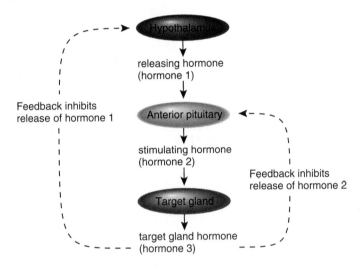

The other three hormones produced by the anterior pituitary do not affect other endocrine glands. **Prolactin (PRL)** is produced in quantity only after childbirth. It causes the mammary glands in the breasts to develop and produce milk. It also plays a role in carbohydrate and fat metabolism.

Melanocyte-stimulating hormone (MSH) causes skin-color changes in many fishes, amphibians, and reptiles that have melanophores, special skin cells that produce color variations. The concentration of this hormone in humans is very low.

Growth hormone (GH), or somatotropic hormone, promotes skeletal and muscular growth. It stimulates the rate at which amino acids enter cells and protein synthesis occurs. It also promotes fat metabolism as opposed to glucose metabolism.

The hypothalamus, the anterior pituitary, and other glands controlled by the anterior pituitary are all involved in a self-regulating negative feedback mechanism.

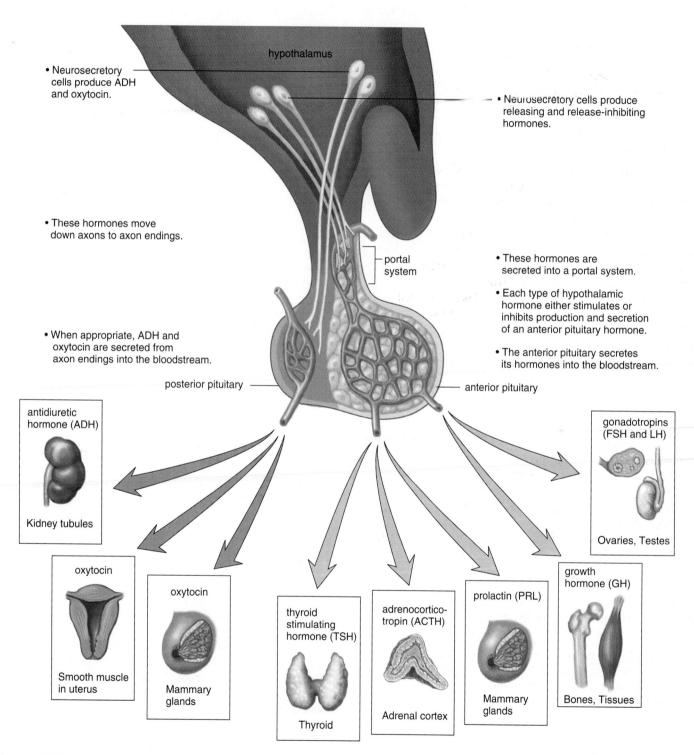

- Neurosecretory cells produce ADH and oxytocin.

hypothalamus

- Neurosecretory cells produce releasing and release-inhibiting hormones.

- These hormones move down axons to axon endings.

portal system

- These hormones are secreted into a portal system.

- Each type of hypothalamic hormone either stimulates or inhibits production and secretion of an anterior pituitary hormone.

- When appropriate, ADH and oxytocin are secreted from axon endings into the bloodstream.

- The anterior pituitary secretes its hormones into the bloodstream.

posterior pituitary

anterior pituitary

antidiuretic hormone (ADH)

Kidney tubules

gonadotropins (FSH and LH)

Ovaries, Testes

oxytocin

Smooth muscle in uterus

oxytocin

Mammary glands

thyroid stimulating hormone (TSH)

Thyroid

adrenocortico-tropin (ACTH)

Adrenal cortex

prolactin (PRL)

Mammary glands

growth hormone (GH)

Bones, Tissues

Figure 14.5   Hypothalamus and the pituitary.
The hypothalamus produces two hormones, ADH and oxytocin, which are stored and secreted by the posterior pituitary. The hypothalamus controls the secretions of the anterior pituitary, and the anterior pituitary controls the secretions of the thyroid, adrenal cortex, and gonads, which are also endocrine glands.

Effects of Growth Hormone

GH is produced in greatest quantities during childhood and adolescence, when most body growth is occurring. If too little GH is produced during childhood, the individual becomes a **pituitary dwarf,** characterized by perfect proportions but small stature. If too much GH is secreted, a person can become a giant (Fig. 14.6). Giants usually have poor health, primarily because GH has a secondary effect on the blood sugar level, promoting an illness called diabetes mellitus.

On occasion, there is overproduction of growth hormone in the adult, and a condition called **acromegaly** results. Long bone growth is no longer possible in adults. Only the feet, hands, and face (particularly the chin, nose, and eyebrow ridges) can respond, and these portions of the body become overly large (Fig. 14.7).

The amount of growth hormone during childhood and adulthood affects the height of an individual.

Figure 14.6 **Effect of growth hormone.**
The amount of growth hormone production during childhood affects the height of an individual. Excessive growth hormone produces very tall basketball players and even giants. Little growth hormone results in limited stature and even pituitary dwarfism.

Age 9   Age 16   Age 33   Age 52

Figure 14.7 **Acromegaly.**
Acromegaly is caused by overproduction of GH in the adult. It is characterized by an enlargement of the bones in the face, the fingers, and the toes of an adult.

14.3 Thyroid and Parathyroid Glands

The **thyroid gland** is a large gland located in the neck, where it is attached to the trachea just below the larynx (see Fig. 14.4). The parathyroid glands are imbedded in the posterior surface of the thyroid gland.

Thyroid Gland

The thyroid gland is composed of a large number of follicles, each a small spherical structure made of thyroid cells filled with triiodothyronine (T$_3$), which contains three iodine atoms and **thyroxine (T$_4$),** which contains four iodine atoms.

Effects of Thyroid Hormones

To produce thyroxine and triiodothyronine, the thyroid gland actively acquires iodine. The concentration of iodine in the thyroid gland can increase to as much as 25 times that of blood. If iodine is lacking in the diet, the thyroid gland is unable to produce the thyroid hormones. In response to constant stimulation by the anterior pituitary, it enlarges, resulting in a **simple goiter** (Fig. 14.8). Some years ago it was discovered that the use of iodized salt allows the thyroid to produce the thyroid hormones, and therefore helps prevent simple goiter.

Thyroid hormones increase the metabolic rate. They do not have one target organ; instead, they stimulate all organs of the body to metabolize at a faster rate. More glucose is broken down and more energy is utilized.

If the thyroid fails to develop properly, a condition called **cretinism** results (Fig. 14.9). Individuals with this condition are short and stocky and have had extreme hypothyroidism since infancy or childhood. Thyroid hormone therapy can initiate growth, but unless treatment is begun within the first two months, mental retardation results. The occurrence of hypothyroidism in adults produces the condition known as **myxedema,** which is characterized by lethargy, weight gain, loss of hair, slower pulse rate, lowered body temperature, and thickness and puffiness of the skin. The administration of adequate doses of thyroid hormones restores normal function and appearance.

In the case of hyperthyroidism, or *Graves' disease,* the thyroid gland is enlarged and overactive, causing a goiter to form. The eyes protrude because of edema in eye socket tissues and swelling of muscles that move the eyes. This type of goiter is called **exophthalmic goiter.** The patient usually becomes hyperactive, nervous, irritable, and suffers from insomnia. Removal or destruction of a portion of the thyroid by means of radioactive iodine is sometimes effective in curing the condition. Hyperthyroidism can also be caused by a thyroid tumor, which is usually detected as a lump during physical examination. Again, the treatment is surgery in combination with administration of radioactive iodine. The prognosis for most patients is excellent.

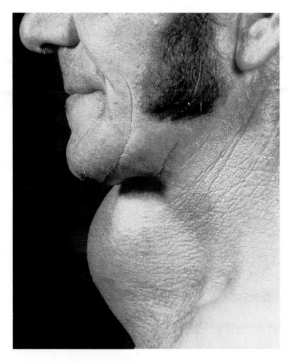

Figure 14.8   Simple goiter.
An enlarged thyroid gland often is caused by a lack of iodine in the diet. Without iodine, the thyroid is unable to produce thyroxine, and continued anterior pituitary stimulation causes the gland to enlarge.

Figure 14.9   Cretinism.
Individuals who have hypothyroidism since infancy or childhood do not grow and develop as others do. Unless medical treatment is begun, the body is short and stocky; mental retardation is also likely.

Calcitonin

Calcium (Ca^{2+}) plays a significant role in both nervous conduction and muscle contraction. It is also necessary to blood clotting. Blood calcium level is regulated in part by **calcitonin,** a hormone secreted by the thyroid gland when the blood calcium level rises (Fig. 14.10). The primary effect of calcitonin is to bring about the deposit of calcium in the bones. It does this by temporarily reducing the activity and number of osteoclasts. When the blood calcium lowers to normal, the release of calcitonin by the thyroid is inhibited, but a low level stimulates the release of **parathyroid hormone (PTH)** by the parathyroid glands.

Parathyroid Glands

Many years ago, the four parathyroid glands were sometimes mistakenly removed during thyroid surgery because they are so small. Parathyroid hormone (PTH), the hormone produced by the **parathyroid glands,** causes the blood phosphate (HPO_4^{2-}) level to decrease and the blood calcium level to increase.

A low blood calcium level stimulates the release of PTH, a hormone that has a powerful effect on the body. PTH promotes the activity of osteoclasts and the release of calcium from the bones. PTH also promotes the reabsorption of calcium by the kidneys where it activates vitamin D. Vitamin D, in turn, stimulates the absorption of calcium from the intestine. These effects bring the blood calcium level back to the normal range so that the parathyroid glands no longer secrete PTH.

When an insufficient parathyroid hormone production leads to a dramatic drop in the blood calcium level, tetany results. In **tetany,** the body shakes from continuous muscle contraction. The effect is brought about by increased excitability of the nerves, which initiate nerve impulses spontaneously and without rest.

The contrary actions of calcitonin, from the thyroid gland, and parathyroid hormone, from the parathyroid glands, maintain the blood calcium level within normal limits.

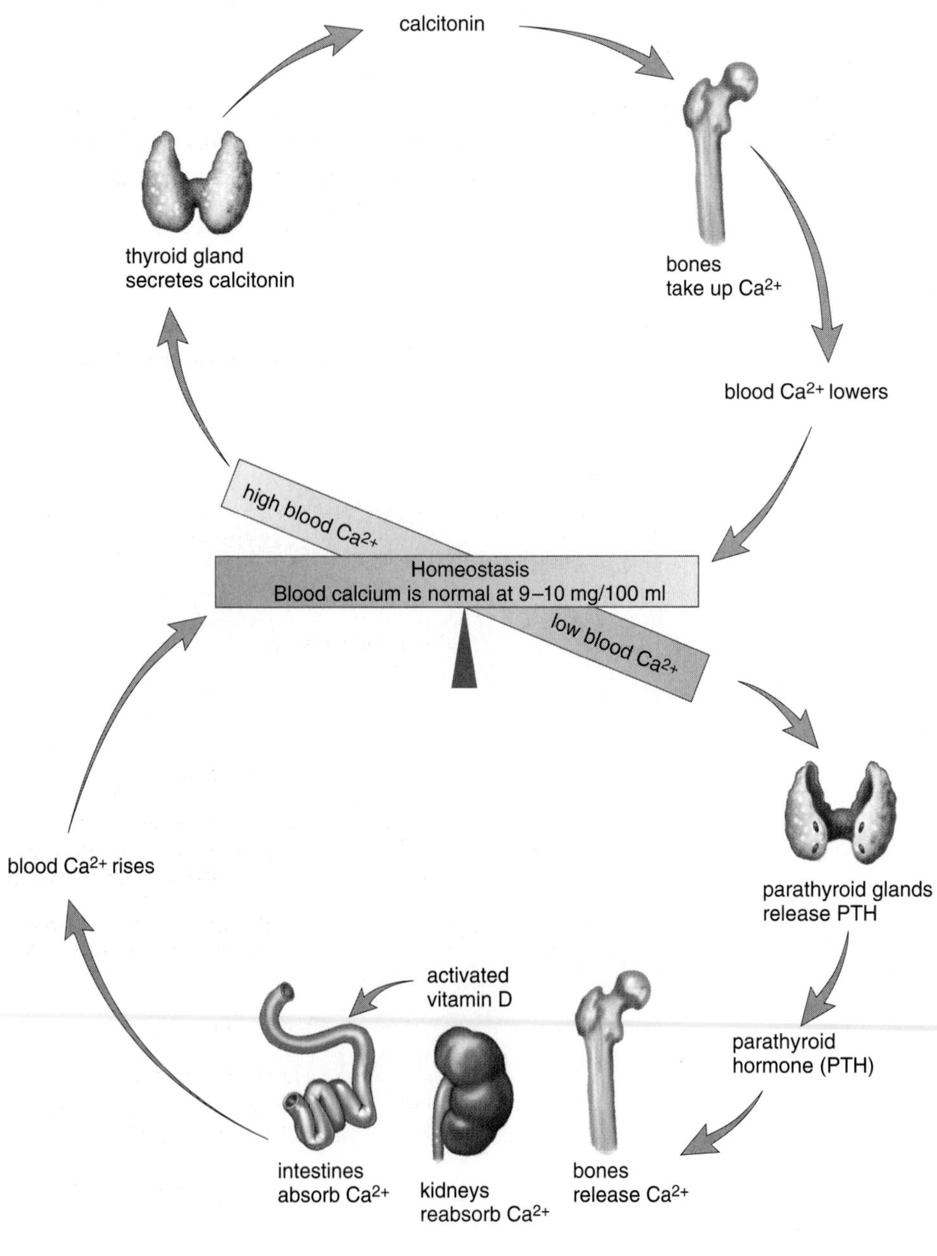

Figure 14.10   Regulation of blood calcium level.
A negative feedback system and the opposing action of calcitonin and the parathyroid hormone regulate the blood calcium (Ca^{2+}) level.

14.4 Adrenal Glands

We have two **adrenal glands** that sit atop the kidneys (see Fig. 14.4). Each adrenal gland consists of an inner portion called the **adrenal medulla** and an outer portion called the **adrenal cortex.** These portions, like the anterior pituitary and the posterior pituitary, have no physiological connection with one another.

The hypothalamus exerts control over the activity of both portions of the adrenal glands. It can initiate nerve impulses that travel by way of the brain stem, spinal cord, and sympathetic nerve fibers to the adrenal medulla, which then secretes its hormones. The hypothalamus, by means of ACTH-releasing hormone, controls the anterior pituitary's secretion of ACTH, which, in turn, stimulates the adrenal cortex. Stress of all types, including both emotional and physical trauma, prompts the hypothalamus to stimulate the adrenal glands. The adrenal hormones increase during times of stress.

Epinephrine (adrenaline) and **norepinephrine** (noradrenaline) produced by the adrenal medulla rapidly bring about all the bodily changes that occur when an individual reacts to an emergency situation as listed in Figure 14.11. In contrast, the hormones produced by the adrenal cortex provide a sustained response to stress. The two major types of hormones produced by the adrenal cortex are the mineralocorticoids and the glucocorticoids. The **mineralocorticoids** regulate salt and water balance leading to increase in blood volume and blood pressure. The **glucocorticoids** regulate carbohydrate, protein, and fat metabolism, leading to an increase in blood glucose level. Cortisone, the medication that is often administered for inflammation of joints, is a glucocorticoid.

The adrenal cortex also secretes a small amount of male sex hormones and a small amount of female sex hormones in both sexes—that is, in the male, both male and female sex hormones are produced by the adrenal cortex, and in the female, both male and female sex hormones are also produced by the adrenal cortex.

The adrenal medulla is under nervous control, and the adrenal cortex is under the control of ACTH, a hormone of the anterior pituitary. Their hormonal secretions help us respond to stress.

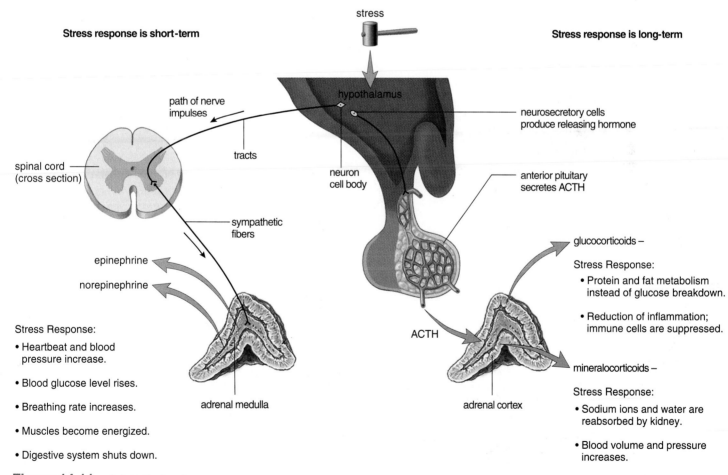

Figure 14.11   Adrenal glands.
Both the adrenal medulla and the adrenal cortex are under the control of the hypothalamus when they help us respond to stress. The adrenal medulla provides a rapid but short-lived emergency response, while the adrenal cortex provides a sustained stress response.

Mineralocorticoids

Aldosterone is the most important of the mineralocorticoids. The primary target organ of aldosterone is the kidney, where it promotes renal absorption of sodium (Na^+) and renal excretion of potassium (K^+).

The secretion of mineralocorticoids is not under the control of the anterior pituitary. When the blood sodium level and therefore blood pressure are low, the kidneys secrete **renin** (Fig. 14.12). Renin is an enzyme that converts the plasma protein angiotensinogen to angiotensin I, which is changed to angiotensin II by a converting enzyme found in the lungs. Angiotensin II stimulates the adrenal cortex to release aldosterone. The effect of this system, called the renin-angiotensin-aldosterone system, is to raise blood pressure in two ways. Angiotensin II constricts the arterioles, and aldosterone causes the kidneys to reabsorb sodium. When the blood sodium level rises, water is reabsorbed in part because the hypothalamus secretes ADH (see page 298). Then blood pressure increases to normal.

There is a contrary hormone to aldosterone, as you might suspect. When the atria of the heart are stretched due to increased blood volume, cardiac cells release a hormone called **atrial natriuretic hormone (ANH),** which inhibits the secretion of aldosterone from the adrenal cortex. The effect of this hormone is, therefore, to cause the excretion of sodium, that is, *natriuresis*. When sodium is excreted, so is water and therefore blood pressure lowers to normal.

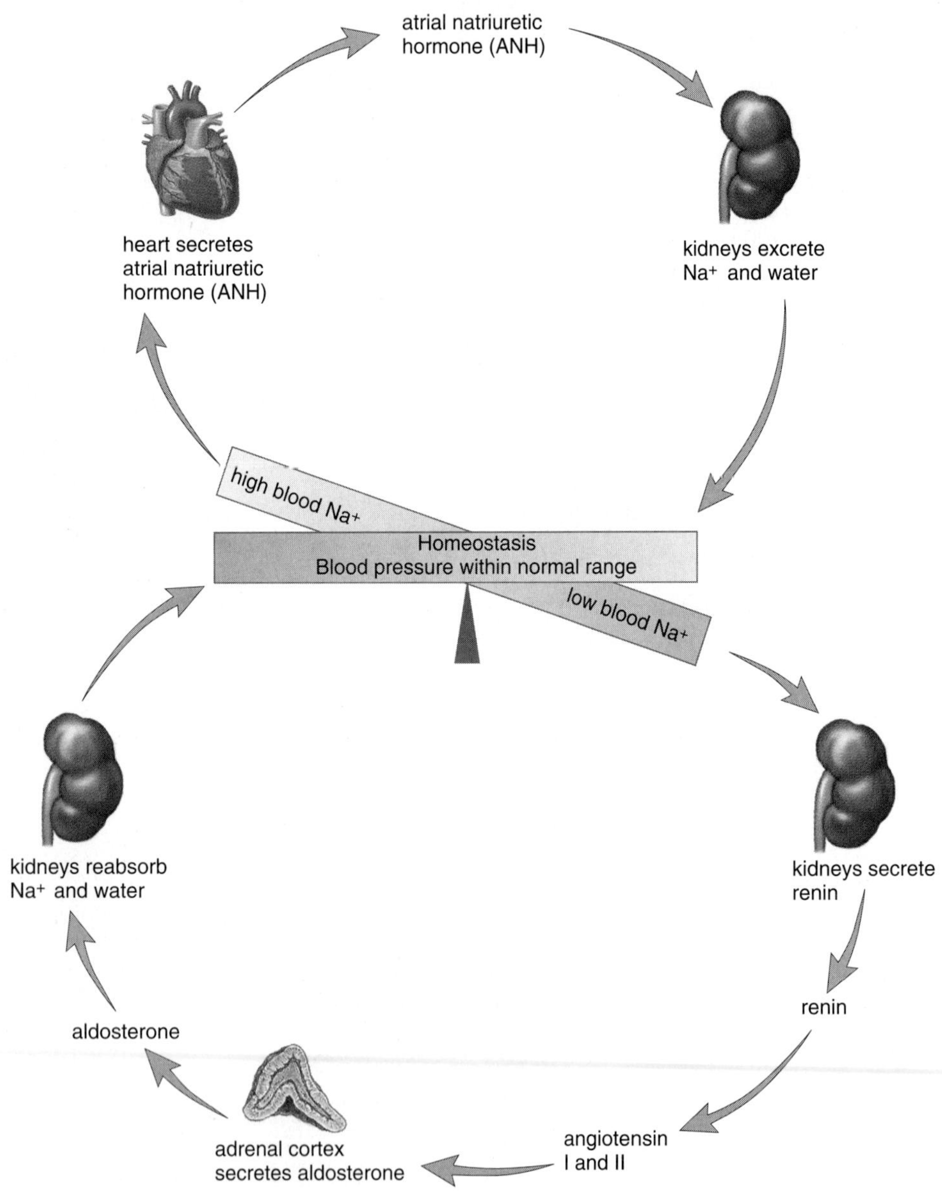

Glucocorticoids

There are several glucocorticoids, one of which, **cortisol,** is most important biologically. Cortisol promotes the hydrolysis of muscle protein to amino acids, which enter the bloodstream. This leads to a higher blood glucose level when the liver converts these amino acids to glucose. Cortisol also favors metabolism of fatty acids rather than carbohydrate. In opposition to insulin, therefore, cortisol raises the blood glucose level. Cortisol also counteracts the inflammatory response that leads to the pain and the swelling of joints in arthritis and bursitis. The administration of cortisol aids these conditions because it reduces inflammation.

Very high levels of glucocorticoids in the blood can suppress the body's defense system, including the inflammatory response that occurs at infection sites. Cortisone and other glucocorticoids can relieve swelling and pain from inflammation, but by suppressing pain and immunity they can also make a person highly susceptible to injury and infection.

Figure 14.12   Regulation of blood pressure and volume.
A low blood sodium (Na^+) level, and therefore low blood volume and pressure, stimulates the kidneys to release renin, which leads to the secretion of aldosterone from the adrenal cortex. Aldosterone causes the kidneys to reabsorb Na^+, and therefore blood volume and pressure rise. A high sodium level stimulates the release of atrial natriuretic hormone from the heart. Thereafter, the kidneys excrete Na^+ and water, and the blood volume and pressure lower.

Malfunction of the Adrenal Cortex

When there is a low level of adrenal cortex hormones due to hyposecretion, a person develops **Addison disease.** The presence of ACTH which is in excess but ineffective causes a bronzing of the skin because ACTH like MSH can lead to a buildup of melanin (Fig. 14.13). The lack of cortisol results in an inability to replenish the blood glucose level when a stressful situation arises. Even a mild infection can lead to death. The lack of aldosterone results in a loss of sodium and water and the development of low blood pressure and possibly severe dehydration. Left untreated, Addison disease can be fatal.

When there is a high level of adrenal cortex hormones due to hypersecretion, a person develops **Cushing syndrome** (Fig. 14.14). The excess of cortisol results in a tendency toward diabetes mellitus as muscle protein is metabolized and subcutaneous fat is deposited in the midsection. The trunk is obese while the arms and legs remain a normal size. An excess of aldosterone and reabsorption of sodium and water by the kidneys leads to a basic blood pH and hypertension. The face is moonshaped due to edema. Masculinization may occur in women because of excess adrenal male sex hormones.

The adrenal cortex hormones are essential to homeostasis. Addison disease is due to adrenal cortex hyposecretion, and Cushing syndrome is due to adrenal cortex hypersecretion.

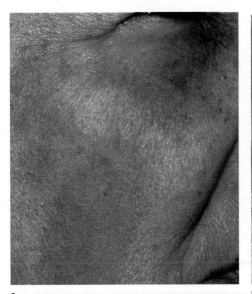
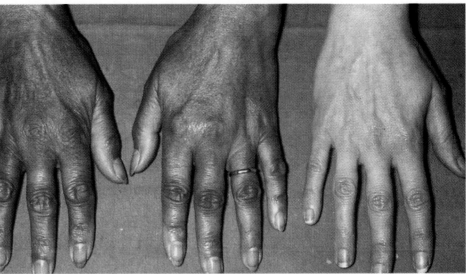

Figure 14.13   Addison disease.
Addison disease is characterized by a peculiar bronzing of the skin, particularly noticeable in these light-skinned individuals. Note the color of **(a)** the face and **(b)** the hands compared to the hand of an individual without the disease.

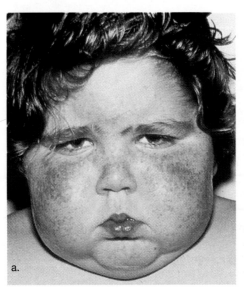
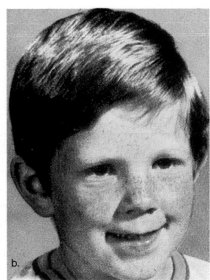

Figure 14.14   Cushing syndrome.
Cushing syndrome results from hypersecretion due to an adrenal cortex tumor. **a.** First diagnosed with Cushing syndrome. **b.** Four months later, after therapy.

14.5 Pancreas

The **pancreas** is a long organ that lies transversely in the abdomen between the kidneys and near the duodenum of the small intestine. It is composed of two types of tissue. Exocrine tissue produces and secretes digestive juices that go by way of ducts to the small intestine. Endocrine tissue, called the **pancreatic islets** (islets of Langerhans), produces and secretes the hormones **insulin** and **glucagon** directly into the blood.

Insulin is secreted when there is a high blood glucose level, which usually occurs just after eating. Insulin stimulates the uptake of glucose by cells, especially liver cells, muscle cells, and adipose tissue cells. In liver and muscle cells, glucose is then stored as glycogen. In muscle cells the breakdown of glucose supplies energy for protein metabolism, and in fat cells the breakdown of glucose supplies glyc-

erol for the formation of fat. In these various ways insulin lowers the blood glucose level.

Glucagon is secreted from the pancreas, usually in between eating, when there is a low blood glucose level. The major target tissues of glucagon are the liver and adipose tissue. Glucagon stimulates the liver to break down glycogen to glucose and to use fat and protein in preference to glucose as energy sources. Adipose tissue cells break down fat to glycerol and fatty acids. The liver takes these up and uses them as substrates for glucose formation. In these various ways glucose raises the blood glucose level.

The two contrary hormones insulin and glucagon, both produced by the pancreas, maintain the normal level of glucose in the blood.

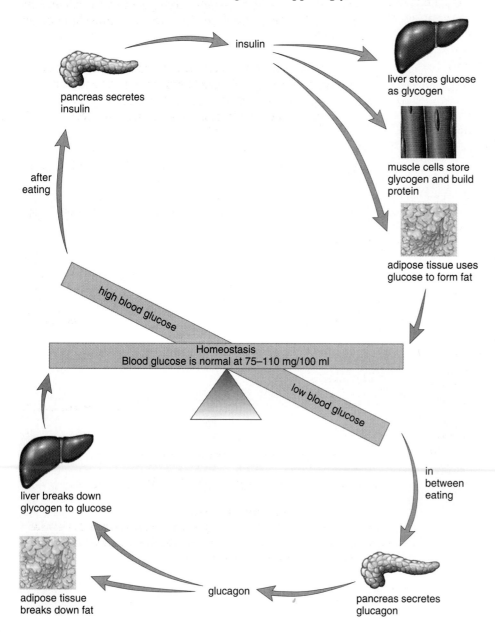

insulin

pancreas secretes insulin

after eating

high blood glucose

Homeostasis
Blood glucose is normal at 75–110 mg/100 ml

low blood glucose

liver stores glucose as glycogen

muscle cells store glycogen and build protein

adipose tissue uses glucose to form fat

liver breaks down glycogen to glucose

adipose tissue breaks down fat

glucagon

in between eating

pancreas secretes glucagon

Figure 14.15   Regulation of blood glucose level.
When the blood glucose level is high, the pancreas secretes insulin. Insulin promotes the storage of glucose as glycogen and the synthesis of proteins and fats (as opposed to their use as energy sources). Therefore, insulin lowers the blood glucose level. When the blood glucose level is low, the pancreas secretes glucagon. Glucagon acts opposite to insulin; therefore, glucagon raises the blood glucose level.

Diabetes Mellitus

Diabetes mellitus is a fairly common hormonal disease in which liver, and indeed all body, cells are unable to take up and/or metabolize glucose. Therefore cellular famine exists in the midst of plenty. As blood glucose level rises, glucose, along with water, is excreted in the urine. The loss of water in this way causes the diabetic to be extremely thirsty. Since glucose is not being metabolized, the body turns to the breakdown of protein and fat for energy. The metabolism of fat leads to the buildup of ketones in the blood and acidosis (acid blood) that can eventually cause coma and death. The symptoms of hyperglycemia (high blood sugar), such as fruity breath odor, lack of appetite, and stupor, develop slowly, so there is time to get adequate medical care.

The glucose-tolerance test is often used to assist the diagnosis of diabetes mellitus. After the patient is given 100 g of glucose, blood glucose concentration is measured at intervals. In a diabetic, the blood glucose level rises greatly and remains elevated for several hours. In a nondiabetic, the blood glucose level rises somewhat and the level returns to normal in about one and one-half hours (Fig. 14.16).

There are two types of diabetes mellitus. In *type I (insulin-dependent) diabetes,* the pancreas is not producing insulin. The condition is believed to be brought on by exposure to an environmental agent, most likely a virus, whose presence causes cytotoxic T cells to destroy the pancreatic islets. As a result, the individual must have daily insulin injections. These injections control the diabetic symptoms but still can cause inconveniences, since either an overdose of insulin or the absence of regular eating can bring on the symptoms of hypoglycemia (low blood sugar). These symptoms include perspiration, pale skin, shallow breathing, and anxiety. Because the brain requires a constant supply of sugar, unconsciousness can result. The cure is quite simple: an immediate ingestion of a sugar cube or fruit juice can very quickly counteract hypoglycemia.

It's possible to transplant a working pancreas into patients with type I diabetes. To do away with the necessity to take immunosuppresssive drugs, fetal pancreatic islet cells have been injected into patients. Another experimental procedure is to place pancreatic islet cells in a capsule that allows insulin to get out but prevents antibodies and

T lymphocytes from getting in. This artificial organ is implanted in the abdominal cavity.

Of the 16 million people who now have diabetes in the United States, most have *type II (noninsulin-dependent) diabetes.* This type of diabetes mellitus usually occurs in people of any age who are obese and inactive. The pancreas produces insulin, but the liver and muscle cells do not respond to it in the usual manner. They may increasingly lack the receptor proteins necessary to detect the presence of insulin. If type II diabetes is untreated, the results can be as serious as type I diabetes. (Diabetics are prone to blindness, kidney disease, and circulatory disorders. Pregnancy carries an increased risk of diabetic coma, and the child of a diabetic is somewhat more likely to be stillborn or to die shortly after birth.) It is possible to prevent or at least control type II diabetes by adhering to a low-fat diet and exercising regularly. If this fails, oral drugs that stimulate the pancreas to secrete more insulin and enhance the metabolism of glucose in the liver and muscle cells are available.

Diabetes mellitus is caused by the lack of insulin or the insensitivity of cells to insulin, a hormone that lowers the blood glucose level, particularly by causing the liver to store glucose as glycogen.

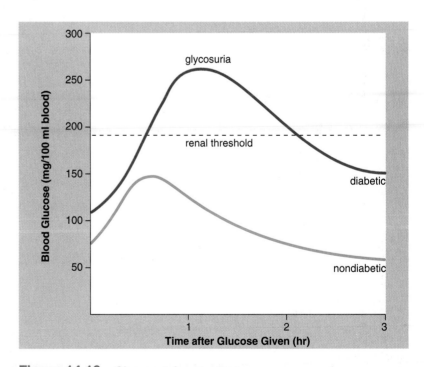

Figure 14.16 Glucose tolerance test.
Following the administration of 100 g of glucose, the glucose blood level rises dramatically in the diabetic but not in the nondiabetic. Glucose will appear in the urine when its level exceeds 190 mg/100 ml.

Health Focus

Dangers of Anabolic Steroids

Anabolic steroids are synthetic forms of the male sex hormone testosterone. Trainers may have been the first to acquire anabolic steroids for weight lifters, bodybuilders, and other athletes such as professional football players. When taken in large doses (10 to 100 times the amount prescribed by doctors for illnesses) and accompanied by exercise, anabolic steroids promote larger muscles. Occasionally, steroid abuse makes the news because an Olympic winner tests positive for the drug and must relinquish a medal. Steroid use has been outlawed by the International Olympic Committee.

The U.S. Food and Drug Administration bans the importation of most steroids, but they are brought into the country illegally and sold through the mail or in gyms and health clubs.

According to federal officials, 1 to 3 million Americans now take anabolic steroids. Their increased use by teenagers wishing to build bulk quickly is of special concern. Some attribute this to society's emphasis on physical appearance and the need of insecure youngsters to feel better about how they look.

Physicians, teachers, and parents are quite alarmed about anabolic steroid abuse. It's even predicted that two or three months of high-dosage use in a youngster can cause death two or three decades later. The many harmful effects of anabolic steroids on the body are listed in Figure 14A. In addition, these drugs increase aggression and make a person feel invincible. One abuser even had his friend videotape him as he drove his car at 40 miles per hour into a tree!

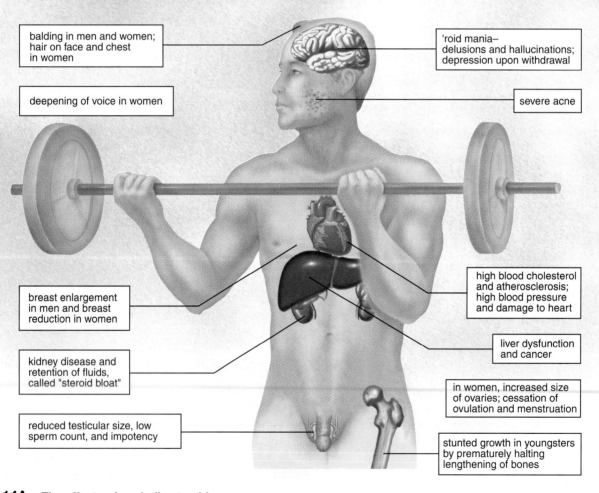

balding in men and women; hair on face and chest in women

deepening of voice in women

breast enlargement in men and breast reduction in women

kidney disease and retention of fluids, called "steroid bloat"

reduced testicular size, low sperm count, and impotency

'roid mania– delusions and hallucinations; depression upon withdrawal

severe acne

high blood cholesterol and atherosclerosis; high blood pressure and damage to heart

liver dysfunction and cancer

in women, increased size of ovaries; cessation of ovulation and menstruation

stunted growth in youngsters by prematurely halting lengthening of bones

Figure 14A   The effects of anabolic steroid use.

14.6 Other Endocrine Glands

The testes and ovaries are endocrine glands. There are also lesser known glands and some tissues that produce hormones.

Testes and Ovaries

The gonads are the testes in males and the ovaries in females. The **testes** are located in the scrotum, and the **ovaries** are located in the pelvic cavity. The testes produce **androgens** (e.g., **testosterone**), which are the male sex hormones, and the ovaries produce estrogens and progesterone, the female sex hormones. The hypothalamus and the pituitary gland control the hormonal secretions of these organs in the same manner that was previously described for the thyroid gland.

The male sex hormone, testosterone, has many functions. It is essential for the normal development and functioning of the sex organs in males. It is also necessary for the maturation of sperm.

Greatly increased testosterone secretion at the time of puberty stimulates the growth of the penis and the testes. Testosterone also brings about and maintains the secondary sex characteristics in males that develop at the time of puberty. Testosterone causes growth of a beard, axillary (underarm) hair, and pubic hair. It prompts the larynx and the vocal cords to enlarge, causing the voice to change. It is partially responsible for the muscular strength of males, and this is the reason some athletes take supplemental amounts of **anabolic steroids,** which are either testosterone or related chemicals. The contraindications of taking anabolic steroids are discussed in the reading on the previous page. Testosterone also stimulates oil and sweat glands in the skin; therefore, it is largely responsible for acne and body odor. Another side effect of testosterone is baldness. Genes for baldness probably are inherited by both sexes, but baldness is seen more often in males because of the presence of testosterone.

Testosterone is believed to be largely responsible for the sex drive. It may even contribute to the suggested aggressiveness of males.

The female sex hormones, **estrogens** and **progesterone,** have many effects on the body. In particular, estrogens secreted at the time of puberty stimulate the growth of the uterus and the vagina. Estrogen is necessary for egg maturation and is largely responsible for the secondary sex characteristics in females. It is responsible for female body hair and fat distribution. In general, females have a more rounded appearance than males because of a greater accumulation of fat beneath the skin. Also, the pelvic girdle is wider in females than in males, resulting in females having a larger pelvic cavity. Both estrogen and progesterone are required for breast development and regulation of the uterine cycle, which includes monthly menstruation (discharge of blood and mucosal tissues from the uterus).

Pineal Gland

The **pineal gland,** which is located in the brain (see Fig. 14.4), produces the hormone called **melatonin,** primarily at night. Melatonin is involved in daily cycles, or **circadian rhythms.** Normally we grow sleepy at night when melatonin levels increase and awaken once daylight returns and melatonin levels are low. Shift work is usually troublesome because it upsets this normal daily rhythm. Similarly, travel to another time zone, as when going to Europe from the United States, results in jet lag because the body is still producing melatonin according to the old schedule. Some people even have **seasonal affective disorder (SAD);** they become depressed and have an uncontrollable desire to sleep with the onset of winter. Receiving melatonin makes their symptoms worse, but exposure to a bright light improves them.

Based on animal research, it appears that melatonin also regulates sexual development. It is of interest that children whose pineal gland has been destroyed due to a brain tumor experience early puberty.

Thymus Gland

The **thymus** is a lobular gland that lies just beneath the sternum (see Fig. 14.4). This organ reaches its largest size and is most active during childhood. With aging, the organ gets smaller and becomes fatty. Lymphocytes that originate in the bone marrow and then pass through the thymus are transformed into T lymphocytes. The lobules of the thymus are lined by epithelial cells that secrete hormones called thymosins. These hormones aid in the differentiation of lymphocytes packed inside the lobules. Although the hormones secreted by the thymus ordinarily work in the thymus, there is hope that these hormones could be injected into AIDS or cancer patients where they would increase T lymphocyte function.

Nontraditional Sources

Some organs that are usually not considered endocrine glands do indeed secrete hormones. We have already mentioned that the heart produces atrial natriuretic hormone. And you will recall that the stomach and the small intestine produce peptide hormones that regulate digestive secretions. There are a number of other types of tissues that produce hormones.

Leptin

Leptin is a protein hormone produced by adipose tissue that acts on the hypothalamus where it signals satiety—that the individual has had enough to eat. Strange to say, the blood of obese individuals may be rich in leptin. The possibility exists that the leptin they produce is ineffective because of a genetic mutation or else their hypothalamic cells lack a suitable number of receptors for leptin.

Growth Factors

A number of different types of organs and cells produce peptide **growth factors,** which stimulate cell division and mitosis. They are like hormones in that they act on cell types with specific receptors to receive them. Some, like lymphokines, are released into the blood; others diffuse to nearby cells. Growth factors of particular interest are:

> *Granulocyte and macrophage colony-stimulating factor (GM-CSF)* is secreted by many different tissues. GM-CSF causes a common stem cell to form either granulocyte or macrophage cells, depending on whether the concentration is low or high.

> *Platelet-derived growth factor* is released from platelets and from many other cell types. It helps in wound healing and causes an increase in the number of fibroblasts, smooth muscle cells, and certain cells of the nervous system.

> *Epidermal growth factor and nerve growth factor* stimulate the cells indicated by their names, as well as many others. These growth factors are also important in wound healing.

> *Tumor angiogenesis factor* stimulates the formation of capillary networks and is released by tumor cells. One treatment for cancer is to prevent the activity of this growth factor.

Prostaglandins

Prostaglandins (PG) are produced in the same locale where they act. PG is produced and released by many different types of tissues. In the uterus, prostaglandins cause muscles to contract; therefore, they are implicated in the pain and discomfort of menstruation in some women. Also, prostaglandins mediate the effects of pyrogens, chemicals that are believed to reset the temperature regulatory center in the brain. Aspirin reduces body temperature and controls pain because of its effect on prostaglandins.

Certain prostaglandins reduce gastric secretion and have been used to treat ulcers; others lower blood pressure and have been used to treat hypertension, and yet others inhibit platelet aggregation and have been used to prevent thrombosis. However, different prostaglandins have contrary effects and it has been very difficult to successfully standardize their use. Therefore, prostaglandin therapy is still considered experimental.

Many tissues aside from the traditional endocrine glands produce hormones. Some of these enter the bloodstream and some act only locally.

14.7 Homeostasis

The hypothalamus, the part of the brain most involved in maintaining homeostasis, is also involved in regulating the endocrine system. This is a tip-off that hormones affect the composition and characteristics of blood and tissue fluid, which, of course, is the internal environment of cells. Consider that if the osmolarity of the blood were to rise above normal, tissue fluid would decrease and dehydration of cells would result. On the other hand, if the osmolarity of the blood were to fall below normal, tissue fluid would collect and edema would result. Several hormones keep the salt/water balance and blood volume within a normal range. Aldosterone promotes the reabsorption of sodium (Na^+), and antidiuretic hormone causes water to be reabsorbed. Atrial natriuretic hormone promotes the excretion of sodium, and water follows passively so that osmolarity remains stable.

Through its effect on the autonomic system, the hypothalamus controls the secretion of epinephrine and norepinephrine by sympathetic nerve endings and by the adrenal medulla. Epinephrine and norepinephrine allow the body to respond to emergency situations in a fight or flight manner. The stress response is sustained by cortisol secreted by the adrenal cortex. Cortisol permits the body to ignore any injuries that may have occurred during a confrontation and keeps the blood glucose level high so that energy is available for a continued struggle.

The concentration of calcium (Ca^{2+}) in the blood is critical because of the importance of this ion to nervous conduction and muscle contraction. As you know, the bones serve as a reservoir for calcium. When the blood concentration lowers, parathyroid hormone promotes the breakdown of bone and the reabsorption of calcium by the kidneys and the intestines. Opposing the action of parathyroid hormone, calcitonin secreted by the thyroid brings about the deposit of calcium in the bones.

Cells usually break down glucose, and indeed the brain can only break down glucose to acquire ATP. Cells cannot function without a continual supply of ATP. Just after eating, insulin encourages the uptake of glucose by cells and the storage of glucose as glycogen in the liver and muscles. In between eating, glucagon stimulates the liver to break down glycogen to glucose so that the blood level stays constant.

Once we realize how many glands, organs, and tissues produce hormones, we begin to realize their importance in keeping the body working in an efficient and productive manner, as discussed on the next page.

Hormones are intimately involved in regulating the internal environment.

Human Systems Work Together

How the Endocrine System works with other body systems

Integumentary System

Androgens activate sebaceous glands and help regulate hair growth.

Skin provides sensory input that results in the activation of certain endocrine glands.

Skeletal System

Growth hormone regulates bone development; parathyroid hormone and calcitonin regulate Ca^{2+} content.

Bones provide protection for glands; store Ca^{2+} used as second messenger.

Muscular System

Androgens promote growth of skeletal muscle; epinephrine stimulates heart and constricts blood vessels.

Muscles help protect glands.

Nervous System

Sex hormones affect development of brain.

Hypothalamus is part of endocrine system; nerves innervate glands of secretion.

Cardiovascular System

Epinephrine increases blood pressure; ADH, aldosterone, and atrial natriuretic hormone help regulate blood volume; growth factors control blood cell formation.

Blood vessels transport hormones from glands; blood services glands; heart produces atrial natriuretic hormones.

Lymphatic System/Immunity

Thymus is necessary for maturity of T lymphocytes.

Lymphatic vessels pick up excess tissue fluid; immune system protects against infections.

Respiratory System

Epinephrine promotes ventilation by dilating bronchioles; growth factors control production of red blood cells that carry oxygen.

Gas exchange in lungs provides oxygen and rids body of carbon dioxide.

Digestive System

Hormones help control secretion of digestive glands and accessory organs; insulin and glucagon regulate glucose storage in liver.

Stomach and small intestine produce hormones.

Urinary System

ADH, aldosterone, and atrial natriuretic hormone regulate reabsorption of water and Na^+ by kidneys.

Kidneys keep blood values within normal limits so that transport of hormones continues.

Reproductive System

Hypothalamic, pituitary, and sex hormones control sex characteristics and regulate reproductive processes.

Gonads produce sex hormones.

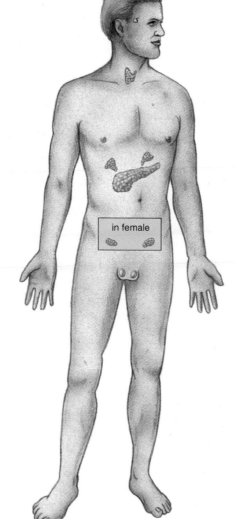

in female

Bioethical Issue

Hormone therapy saves lives. Diabetics and those with Addison disease would soon succumb if hormone therapy was not available for their conditions. Other hormone therapies are a matter of choice and raise ethical questions. If you were the parent of a child who was short for his or her age group, on what basis might you decide to have the child undergo hormone therapy?

Researchers have found that human growth hormone doesn't work. It causes puberty to arrive earlier than usual, and this does away with any gain in height that occurred before puberty. However, studies are being conducted with a synthetic growth hormone called somatotropin that seems to work better. Re-

searchers in England conducted a survey of about 14,000 girls, and decided to treat seven girls who were short for their age. They compared their progress to a group of untreated girls. All girls started puberty at about the same age—13.5 years. The girls in the treated group received a daily injection of somatotropin between the ages of 8 and 14. They grew to 5'1" by age 16, when most girls stop growing. That was 2.4 to 3.0 inches taller than the girls in the untreated group.

The researchers were interested in whether height affected psychological outlook. They found that girls who were 4'10" appeared to be just as happy and well balanced as those who were taller. Is psychological outlook a good criteria to use when

deciding whether to use hormone therapy to increase height? Is this criteria more appropriately applied to boys than to girls? When does short height become a disability in our society? Are tall people likely to have more successful careers than short people?

Questions

1. Do you approve of using hormone therapy to increase the growth of a child? Why or why not?
2. On what basis would you decide to use hormone therapy for your child in order to increase height? Should it matter whether the child is a girl or boy?
3. What assurances would you need before you enter your child in an experimental study?

Summarizing the Concepts

14.1 Environmental Signals

There are three categories of environmental signals: those that act at a distance between individuals (pheromones); those that act at a distance within the individual (traditional endocrine hormones and secretions of neurosecretory cells); and local messengers (such as prostaglandins, growth factors, and neurotransmitters). Since there is great overlap between these categories, perhaps the definition of a hormone should be expanded to include all of them.

Steroid hormones enter the nucleus and combine with a receptor hormone, and the complex attaches to and activates DNA. Transcription and translation lead to protein synthesis. Nonsteroid hormones are usually received by a hormone receptor located in the plasma membrane. Most often their reception leads to activation of an enzyme that changes ATP to cyclic AMP (cAMP). cAMP then activates an enzyme cascade. Hormones work in small quantities because their effect is amplified.

14.2 Hypothalamus and Pituitary Gland

Neurosecretory cells in the hypothalamus produce antidiuretic hormone (ADH) and oxytocin, which are stored in axon endings in the posterior pituitary until they are released.

The hypothalamus produces hypothalamic-releasing and hypothalamic-inhibiting hormones, which pass to the anterior pituitary by way of a portal system. The anterior pituitary produces at least six types of hormones, and some of these stimulate other hormonal glands to secrete hormones.

14.3 Thyroid and Parathyroid Glands

The thyroid gland requires iodine to produce thyroxine and triiodothyronine, which increase the metabolic rate. If iodine is available in limited quantities, a simple goiter develops; if the thyroid is overactive, an exophthalmic goiter develops. The thyroid gland also produces calcitonin, which helps lower the blood calcium level. The parathyroid

glands secrete parathyroid hormone which raises the blood calcium and decreases the blood phosphate level.

14.4 Adrenal Glands

The adrenal glands respond to stress: immediately, the adrenal medulla secretes epinephrine and norepinephrine, which bring about responses we associate with emergency situations. On a long-term basis, the adrenal cortex produces the glucocorticoids (e.g., cortisol) and the mineralocorticoids (e.g., aldosterone). Cortisol stimulates hydrolysis of proteins to amino acids that are converted to glucose; in this way, it raises the blood glucose level. Aldosterone causes the kidneys to reabsorb sodium ions (Na^+) and excrete potassium ions (K^+). Addison disease develops when the adrenal cortex is underactive, and Cushing syndrome develops when the adrenal cortex is overactive.

14.5 Pancreas

The pancreatic islets secrete insulin, which lowers the blood glucose level, and glucagon, which has the opposite effect. The most common illness due to hormonal imbalance is diabetes mellitus, which is due to the failure of the pancreas to produce insulin or the cells to take it up.

14.6 Other Endocrine Glands

The gonads produce the sex hormones; the pineal gland produces melatonin, which may be involved in circadian rhythms and the development of the reproductive organs and the thymus secretes thymosins, which stimulate T lymphocyte production and maturation.

Tissues also produce hormones. Adipose tissue produces leptin which acts on the hypothalamus and various tissues produce growth factors. Prostaglandins are produced and act locally.

14.7 Homeostasis

The endocrine system works with the other systems of the body in the ways described in the box on page 311.

Studying the Concepts

1. Categorize chemical messengers into three groups based upon the distance between site of secretion and receptor site, and give examples of each group. 294
2. Give examples to show that there is an overlap between the mode of operation of the nervous system and that of the endocrine system. Explain why the traditional definition of a hormone may need to be expanded. 294
3. Explain how steroid hormones and nonsteroid hormones affect the metabolism of the cell. 295
4. Describe a mechanism by which the production of a hormone is regulated and another by which the effect of a hormone is controlled. 297
5. Explain the relationship of the hypothalamus to the posterior pituitary gland and to the anterior pituitary gland. List the hormones secreted by the posterior and anterior pituitary 298
6. Give an example of the three-tier relationship among the hypothalamus, the anterior pituitary, and other endocrine glands. 298
7. Discuss the effect of growth hormone on the body and the result of there being too much or too little growth hormone when a young person is growing. What is the result if the anterior pituitary produces growth hormone in an adult? 300
8. What types of goiters are associated with a malfunctioning thyroid? Explain each type. 301
9. How do the thyroid and the parathyroid work together to control the blood calcium level? 302
10. How do the adrenal glands respond to stress? What hormones are secreted by the adrenal medulla, and what effects do these hormones have? 303
11. Name the most significant glucocorticoid and mineralocorticoid, and discuss their functions. Explain the symptoms of Addison disease and Cushing syndrome. 304–5
12. Draw a diagram to explain how insulin and glucagon maintain the blood glucose level. Use your diagram to explain the major symptoms of type I diabetes mellitus. 306–7
13. Name the other endocrine glands discussed in this chapter, and discuss the function of the hormones they secrete. 309
14. What are leptin, growth factors, and prostaglandins? How do these substances act? 309–10

Understanding the Terms

acromegaly 300
Addison disease 305
adrenal cortex 303
adrenal gland 303
adrenal medulla 303
adrenocorticotropic hormone (ACTH) 298
aldosterone 304
anabolic steroid 309
androgens 309
anterior pituitary 298
antidiuretic hormone (ADH) 298
atrial natriuretic hormone (ANH) 304
calcitonin 302
circadian rhythm 309
cortisol 304
cretinism 301
Cushing syndrome 305
cyclic AMP 295
diabetes mellitus 307
endocrine gland 297
epinephrine 303

estrogen 309
exophthalmic goiter 301
glucagon 306
glucocorticoids 303
gonadotropic hormone 298
growth factor 310
growth hormone (GH) 298
hormone 294
hypothalamic-inhibiting hormone 298
hypothalamic-releasing hormone 298
hypothalamus 298
insulin 306
leptin 309
melanocyte-stimulating hormone (MSH) 298
melatonin 309
mineralocorticoids 303
myxedema 301
nonsteroid hormone 295
norepinephrine (NE) 303
ovaries 309
oxytocin 298

pancreas 306
pancreatic islets (of Langerhans) 306
parathyroid gland 302
parathyroid hormone (PTH) 302
pheromone 294
pineal gland 309
pituitary dwarf 300
pituitary gland 298
posterior pituitary 298
progesterone 309
prolactin (PRL) 298
prostaglandin (PG) 310

renin 304
seasonal affective disorder (SAD) 309
simple goiter 301
steroid hormone 295
testes 309
testosterone 309
tetany 302
thymus 309
thyroid gland 301
thyroid-stimulating hormone (TSH) 298
thyroxine (T_4) 301

Match the terms to these definitions:

a. _____ Organ that is in the neck and secretes several important hormones, including thyroxine and calcitonin.
b. _____ Condition characterized by high blood glucose level and the appearance of glucose in the urine.
c. _____ Hormone secreted by the anterior pituitary that stimulates activity in the adrenal cortex.
d. _____ Type of hormone that binds to a plasma membrane receptor and results in activation of an enzyme cascade.
e. _____ Hormone released by the posterior pituitary that causes contraction of uterus and milk letdown.

Testing Your Knowledge of the Concepts

In questions 1–6, match the endocrine gland to its hormone(s).

  a. anterior pituitary
  b. posterior pituitary
  c. thyroid
  d. adrenal medulla
  e. adrenal cortex
  f. pancreas

  _____ 1. insulin
  _____ 2. growth hormone
  _____ 3. cortisol
  _____ 4. oxytocin
  _____ 5. norepinephrine
  _____ 6. calcitonin

In questions 7–10, indicate whether the statement is true (T) or false (F).

  _____ 7. The hypothalamus produces the hormones secreted by the anterior pituitary.
  _____ 8. Growth hormone is sometimes produced in adults.
  _____ 9. Parathyroid hormone acts on the bones, intestines, and the kidneys to cause the secretion of sodium.
  _____ 10. The release of insulin just after eating keeps the blood glucose level within normal range.

In questions 11–13, fill in the blanks.

11. Aldosterone and its opposing hormone _____ help keep the osmolarity of the blood within its normal range.
12. The secretion of many hormones is regulated by a _____ feedback system.
13. The male sex hormone _____ and the female sex hormones _____ and _____ maintain the secondary sex characteristics.

14. Fill in this diagram to explain the three-tier relationship between the hypothalamus, the anterior pituitary, and the target gland.

Applying Technology to the Concepts

Your study of the endocrine system is supported by these available technologies:

Essential Study Partner CD-ROM
Animals → Endocrine System
Visit the Mader web site for related ESP activities.

Exploring the Internet
The Mader Home Page provides resources and tools as you study this chapter.

http://www.mhhe.com/biosci/genbio/mader

Dynamic Human 2.0 CD-ROM
Endocrine System

HealthQuest CD-ROM
1 Stress Management & Mental Health
→ Gallery → Stressors Affect Body
6 Cancer → Gallery → Diabetes Mellitus

Life Science Animations 3D Video
41 Hormone Action

Applying Your Knowledge to the Concepts

These questions pertain to homeostasis.

1. Explain the fact that an injection of epinephrine (adrenaline) causes almost an immediate effect on the body, but an injection of testosterone takes a much longer time to produce an effect.

2. The anterior pituitary is often referred to as the master gland. Offer an argument that this term should be reserved for the hypothalamus.

3. In the early twentieth century, thyroxine was included in diet pills. Since then this practice has been banned. What was the reason for including thyroxine in diet pills?

4. Overseas travelers are encouraged to take melatonin to enable them to get on the sleep schedule of the foreign country. Explain the reason for this treatment and the time of day that one would take the hormone.

Further Readings for Part 4

Axel, R. October 1995. The molecular logic of smell. *Scientific American* 273(4):154. Article discusses how the brain identifies a scent by neuron activation.

Barkley, R. A. September 1998. Attention-deficit hyperactivity disorder. *Scientific American* 279(3):66. ADHD may result from neurological abnormalities with a genetic basis.

Beardsley, T. August 1997. The machinery of thought. *Scientific American* 277(2):78. Researchers have identified the area of the brain responsible for memory.

Debinski, W. May/June 1998. Anti-brain tumor cytotoxins. *Science & Medicine* 5(3):36. Delivery of bacterial toxins specifically to tumor cells is a new therapeutic strategy for the treatment of brain tumors.

Drollette, D. October 1997. The next hop: Can wallabies replace the lab rat? *Scientific American* 277(4):14. Wallaby embryos may be the ideal model in mammalian neurobiology.

Duke, R. C., et al. December 1996. Cell suicide in health and disease. *Scientific American* 275(6):80. Failures in the processes of cellular self-destruction may give rise to cancer, AIDS, Alzheimer disease, and some genetic diseases.

Gazzaniga, M. S. July 1998. The split brain revisited. *Scientific American* 279(1):50. Recent research on split brains has led to new insights into brain organization and consciousness.

Grillner, S. January 1996. Neural networks for vertebrate locomotion. *Scientific American* 274(1):64. Discoveries about how the brain coordinates muscle movement raise hopes for restoration of mobility for some accident victims.

Hadley, M. E. 1996. *Endocrinology*. 4th ed. Upper Saddle River, NJ: Prentice-Hall. This text discusses the role of chemical messengers in the control of neurological processes.

Halstead, L. S. April 1998. Post-polio syndrome. *Scientific American* 278(4):42. Recovered polio victims are experiencing fatigue, pain, and weakness, resulting from degeneration of motor neurons.

Harvard Health Letter. April 1998. A special report: Parkinson's disease. This overview presents the symptoms and diagnosis of this disease, and discusses medications and surgical methods of treatment.

Jordan, V. C. October 1998. Designer estrogens. *Scientific American* 279(4):60. Selective estrogen receptor modulators may protect against breast and endometrial cancers, osteoporosis, and heart disease.

Julien, R. M. 1997. *A primer of drug action.* 8th ed. New York: W. H. Freeman and Company. A concise, nontechnical guide to the actions, uses, and side effects of psychoactive drugs.

Karch, S. B. 1996. *The pathology of drug abuse.* 2d ed. Boca Raton, Fl: CRC Press, Inc. This book, which can be used by students and medical professionals, contains the history, cultivation or manufacture, and effects of psychoactive drugs on the body.

Mader, S. S. 1997. *Understanding anatomy and physiology.* 3d ed. Dubuque, Iowa: Wm. C. Brown Publishers. A text that emphasizes the basics for beginning allied health students.

Marcus, D. M. and Camp, M. W. May/June 1998. Age-related macular degeneration. *Science & Medicine* 5(3):10. New therapies are needed for this common cause of vision loss in the elderly.

Mattson, M. P. March/April 1998. Experimental models of Alzheimer's disease. *Science & Medicine* 5(2):16. In Alzheimer's disease, mutations accelerate changes that occur during normal aging.

Maudgil, D. D. and Shorvon, S. D. September/October 1997. Locating the epileptogenic focus by MRI. *Science & Medicine* 4(5):26. Use of MRI should improve treatment of certain kinds of epilepsy.

McLachlan, J. A. and Arnold, S. F. September/October 1996. *American Scientist* 84(5):452. Environmental estrogens. Research suggests that ecoestrogens can mimic molecules involved in environmental signaling.

Nolte, J. 1993. *The human brain.* 3d ed. St. Louis: Mosby-Year Book, Inc. Beginners are guided through the basic aspects of brain structure and function.

Sack, R. L. September/October 1998. Melatonin. *Science & Medicine* 5(5): 8. Certain mood and sleep disorders can be managed with melatonin treatments.

Schwartz, W. J. May/June 1996. Internal timekeeping. *Science & Medicine* 3(3):44. Article discusses circadian rhythm mechanisms.

Swerdlow, J. L. June 1995. The brain. *National Geographic* 187(6):2. New research leads to treatments for many age-old disorders.

Synder, S. H. 1996. Drugs and the brain. New York: *Scientific American Library.* The effect of drugs on brain function is discussed.

Thomas, E. D. September/October 1995. Hematopoietic stem cell transplantation. *Scientific American Science & Medicine* 2(5):38. Article discusses reconstituting marrow from cultured stem cells for bone marrow transplants.

Youdim, M. B., and Riederer, P. January 1997. Understanding Parkinson's disease. *Scientific American* 276(1):52. The tremors and immobility of Parkinson's disease can be traced to damage in a part of the brain that regulates movement.

PART

V

Reproduction in Humans

Human beings are one of two sexes, male and female. The reproductive organs of each sex function to produce the sex cells that join prior to the development of a new individual. The embryo develops into a fetus within the body of the female, and birth occurs when there is a reasonable chance for independent existence.

We are in the midst of a sexual revolution. We have the freedom to engage in varied sexual practices and to reproduce by alternative methods of conception, such as in vitro fertilization. With freedom comes a responsibility to be familiar with the biology of reproduction and health-related issues, such as sexually transmitted diseases, not only for ourselves but for our potential offspring.

Chapter 15

Reproductive System

Chapter Concepts

It had seemed simple enough. Leigh Anne and Joe graduated from college, launched their careers, and got married. Some years later, they bought a house. Then, they decided to have a baby.

That's when things got complicated.

For some reason, Leigh Anne just didn't get pregnant. After two years of trying to conceive, the couple headed to a well-known fertility specialist. After some tests, the doctor explained a variety of fertility treatments and drugs designed to help couples conceive.

Leigh Anne and Joe weighed their options and decided to try in vitro fertilization. During a series of visits to the clinic, a doctor removed eggs from Leigh Anne and combined them with sperm from Joe. Nurtured in the lab, the combination formed fertilized eggs, which the doctor then placed back into Leigh Anne's uterus.

Fortunately, the procedure worked. Leigh Anne's pregnancy was normal and healthy. Today, the couple's three-year-old races around the house, plays leapfrog over the family dog, and eats everything in sight. At some point, Leigh Anne and Joe might try in vitro fertilization again. For now, though, they'd just like their toddler to try a nap.

15.1 Male Reproductive System

The male reproductive system includes the organs depicted in Figure 15.1 and listed in Table 15.1. The male **gonads** are paired testes (sing., **testis**), which are suspended within the *scrotal sacs* of the **scrotum.**

Sperm produced by the testes mature within the **epididymis** (pl., epididymides), which is a tightly coiled tubule lying just outside each testis. Maturation seems to be required for the sperm to swim to the egg. Each epididymis joins with a **vas deferens** (pl., vasa deferentia), which descends through a canal called the inguinal canal and enters the abdominal cavity where it curves around the bladder and empties into the **urethra.** Sperm are stored in both the epididymides and the vas deferens.

At the time of ejaculation, sperm leave the penis in a fluid called seminal fluid **(semen).** The seminal vesicles, the prostate gland, and the bulbourethral glands (Cowper's glands) add secretions to seminal fluid. The pair of **seminal vesicles** lie at the base of the bladder and each has a duct that joins with a vas deferens. The **prostate gland** is a single doughnut-shaped gland that surrounds the upper

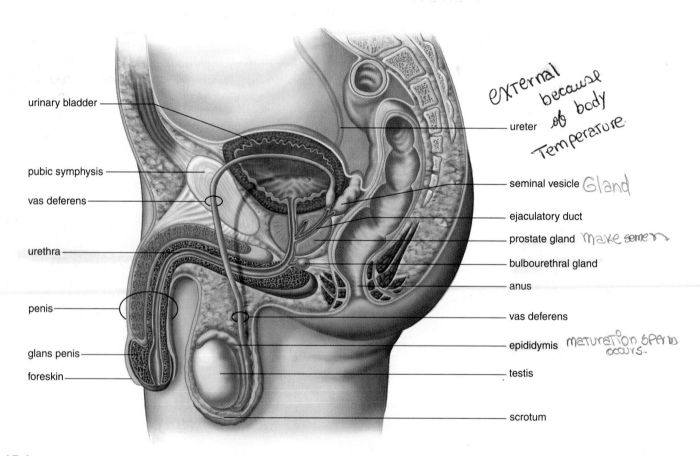

Figure 15.1   The male reproductive system.
The testes produce sperm. The seminal vesicles, the prostate gland, and the bulbourethral gland provide a fluid medium for the sperm. Circumcision is the removal of the foreskin. Notice that the penis in this drawing is not circumcised because the foreskin is present.

portion of the urethra just below the bladder. In older men, the prostate can enlarge and squeeze off the urethra, making urination painful and difficult. The condition can be treated medically. **Bulbourethral glands** are pea-sized organs that lie posterior to the prostate on either side of the urethra.

Each component of seminal fluid seems to have a particular function. Sperm are more viable in a basic solution, and seminal fluid, which is milky in appearance, has a slightly basic pH (about 7.5). Swimming sperm require energy, and seminal fluid contains the sugar fructose, which presumably serves as an energy source. Seminal fluid also contains prostaglandins, chemicals that cause the uterus to contract. Some investigators believe that uterine contractions help propel the sperm toward the egg.

Orgasm in Males

The **penis** (Fig. 15.2) is the male organ of sexual intercourse. The penis has a long shaft and an enlarged tip called the glans penis. The glans penis is normally covered by a layer of skin called the foreskin. Circumcision is the surgical removal of the foreskin, usually soon after birth.

Spongy, erectile tissue containing distensible blood spaces extends through the shaft of the penis. During sexual arousal, nerve impulses stimulate the release of cGMP (cyclic guanosine monophosphate) and the erectile tissue fills with blood. The veins that take blood away from the penis are compressed and the penis becomes erect. Impotency exists when the erectile tissue doesn't expand enough to compress the veins. The new drug Viagra inhibits an enzyme that breaks down cGMP, ensuring that a full erection will take place. Vision problems may occur because the same enzyme occurs in the retina.

As sexual stimulation intensifies, sperm enter the urethra from each vas deferens and the glands contribute secretions to seminal fluid (semen). Once seminal fluid is in the urethra, rhythmic muscle contractions cause it to be expelled from the penis in spurts. During ejaculation, a sphincter closes off the bladder so that no urine enters the urethra. (Notice that the urethra carries either urine or semen at different times.)

The contractions that expel seminal fluid from the penis are a part of male orgasm, the physiological and psychological sensations that occur at the climax of sexual stimulation. The psychological sensation of pleasure is centered in the brain, but the physiological reactions involve the genital (reproductive) organs and associated muscles, as well as the entire body. Marked muscular tension is followed by contraction and relaxation.

Following ejaculation and/or loss of sexual arousal, the penis returns to its normal flaccid state. After ejaculation, a male typically experiences a period of time, called the refractory period, during which stimulation does not bring about an erection. The length of the refractory period increases with age.

There may be in excess of 400 million sperm in the 3.5 ml of semen expelled during ejaculation. The sperm count can be much lower than this, however, and fertilization of the egg by a sperm still can take place.

Sperm are produced in the testes, mature in the epididymis, and pass from the vas deferens to the urethra. After glands add fluid to sperm, semen is ejaculated from the penis at the time of male orgasm.

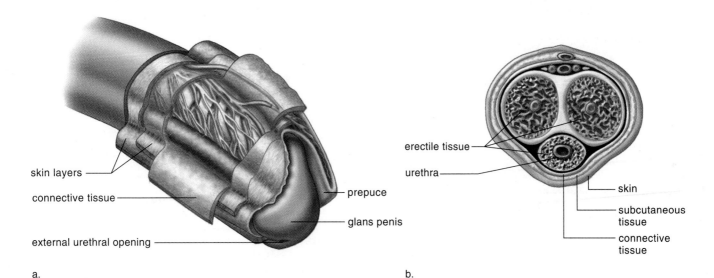

a.

erectile tissue

urethra

skin

subcutaneous tissue

connective tissue

skin layers

connective tissue

external urethral opening

prepuce

glans penis

b.

Figure 15.2   Penis anatomy.
a. Beneath the skin and the connective tissue lies the urethra, surrounded by erectile tissue. This tissue expands to form the glans penis, which in uncircumcised males is partially covered by the foreskin. **b.** Two other columns of erectile tissue in the penis are located dorsally.

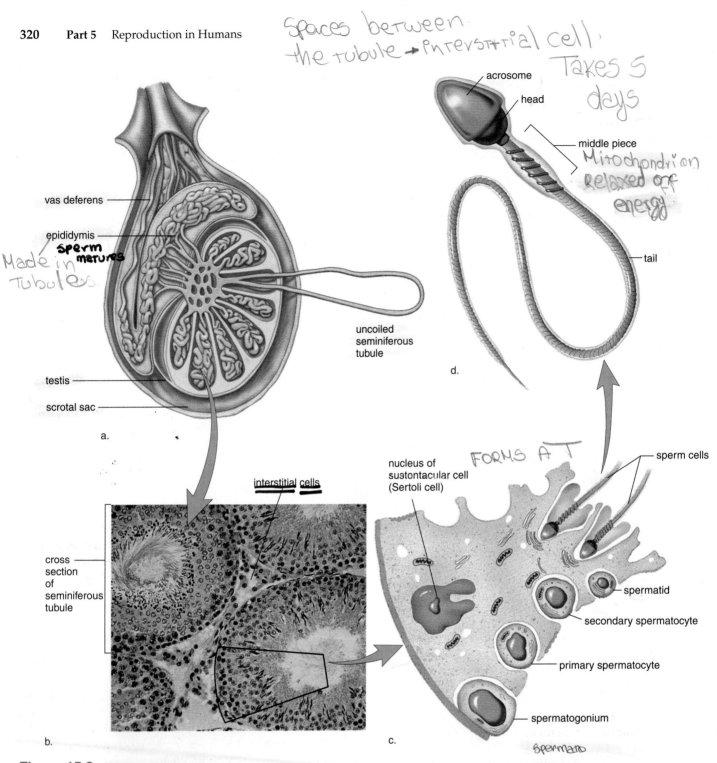

(handwritten annotations)
Spaces between the tubule → interstitial cells
Takes 5 days
Mitochondrion Relaxed off energy
Made in Tubules. sperm matures
interstitial cells
FORMS AT
Spermato

Figure 15.3   Testis and sperm.
a. The lobules of a testis contain seminiferous tubules. **b.** Light micrograph of cross section of seminiferous tubules where spermatogenesis occurs. **c.** Diagrammatic representation of spermatogenesis, which occurs in the wall of the tubules. **d.** A sperm has a head, a middle piece, and a tail. The nucleus is the head, capped by the enzyme-containing acrosome.

Male Gonads, the Testes

The **testes** lie outside the abdominal cavity of the male within the scrotum. The testes begin their development inside the abdominal cavity but descend into the scrotal sacs during the last two months of fetal development. If, by chance, the testes do not descend and the male is not treated or operated on to place the testes in the scrotum, sterility— the inability to produce offspring—usually follows. This is because the internal temperature of the body is too high to produce viable sperm. The scrotum helps regulate the temperature of the testes by holding them closer or further away from the body.

A longitudinal section of a testis shows that it is composed of compartments called lobules, each of which contains one to three tightly coiled **seminiferous tubules**

Table 15.1	Male Reproductive System
Organ	**Function**
Testes	Produce sperm and sex hormones
Epididymides	Maturation and some storage of sperm
Vasa deferentia	Conduct and store sperm
Seminal vesicles	Contribute nutrients and fluid to semen
Prostate gland	Contributes basic fluid to semen
Urethra	Conducts sperm
Bulbourethral glands	Contribute mucoid fluid to semen
Penis	Organ of sexual intercourse

(Fig. 15.3*a*). Altogether, these tubules have a combined length of approximately 250 meters. A microscopic cross section of a seminiferous tubule shows that it is packed with cells undergoing **spermatogenesis** (Fig. 15.3*b*), the production of sperm. Also present are **sustentacular (Sertoli) cells,** which support, nourish, and regulate the spermatogenic cells (Fig. 15.3*c*).

Mature **sperm,** or spermatozoa, have three distinct parts: a head, a middle piece, and a tail (Fig. 15.3*d*). There are mitochondria in the middle piece that provide the energy for the movement of the tail which has the structure of a flagellum. The head contains a nucleus covered by a cap called the **acrosome,** which stores enzymes needed to penetrate the egg. The ejaculated semen of a normal human male contains several hundred million sperm, assuring an adequate number for fertilization to take place. Only one sperm normally enters an egg.

Hormonal Regulation in Males

The hypothalamus has ultimate control of the testes' sexual function because it secretes a hormone called **gonadotropin-releasing hormone,** or **GnRH,** that stimulates the anterior pituitary to secrete the gonadotropic hormones. There are two gonadotropic hormones—**follicle-stimulating hormone (FSH)** and **luteinizing hormone (LH)**—in both males and females. In males, FSH promotes the production of sperm in the seminiferous tubules, which also release the hormone inhibin.

LH in males is sometimes given the name *interstitial cell-stimulating hormone (ICSH)* because it controls the production of testosterone by the **interstitial cells,** which are found in the spaces between the seminiferous tubules. All these hormones are involved in a negative feedback relationship that maintains the fairly constant production of sperm and testosterone (Fig. 15.4).

Testosterone, the main sex hormone in males, is essential for the normal development and functioning of the organs listed in Table 15.1. Testosterone also brings about and maintains the male secondary sex characteristics that develop at the time of puberty. Males are generally taller than females and have broader shoulders and longer legs relative

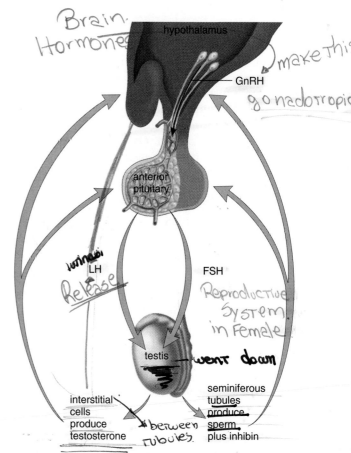

Figure 15.4   Hormonal control of testes.
GnRH (gonadotropin-releasing hormone) stimulates the anterior pituitary to secrete the gonadotropic hormones: FSH stimulates the production of sperm, and LH stimulates the production of testosterone. Testosterone and inhibin exert negative feedback control over the hypothalamus and the anterior pituitary, and this regulates the level of testosterone in blood.

to trunk length. The deeper voice of males compared to females is due to a larger larynx with longer vocal cords. Since the so-called Adam's apple is a part of the larynx, it is usually more prominent in males than in females. Testosterone causes males to develop noticeable hair on the face, chest, and occasionally on other regions of the body such as the back. Testosterone also leads to the receding hairline and pattern baldness that occurs in males.

Testosterone is responsible for the greater muscular development in males. Knowing this, males and females sometimes take anabolic steroids, which are either testosterone or related steroid hormones resembling testosterone. Health problems involving the kidneys, the circulatory system, and hormonal imbalances can arise from such use. The testes shrink in size, and feminization in regard to other male traits occurs.

The gonads in males are the testes, which produce sperm as well as testosterone, the most significant male sex hormone.

15.2 Female Reproductive System

The female reproductive system includes the organs depicted in Figure 15.5 and listed in Table 15.2. The female **gonads** are paired **ovaries** that lie in shallow depressions, one on each side of the upper pelvic cavity. **Oogenesis** is production of an **egg,** the female gonad. The ovaries alternate in producing one egg a month. **Ovulation** is the process by which an egg bursts from an ovary and usually enters an oviduct.

The Genital Tract

The **oviducts,** also called uterine or fallopian tubes, extend from the uterus to the ovaries; however, the oviducts are not attached to the ovaries. Instead, they have fingerlike projections called fimbriae (sing., **fimbria**) that sweep over the ovaries. When an egg bursts from an ovary during ovulation, it usually is swept into an oviduct by the combined action of the fimbriae and the beating of cilia that line the oviducts.

Once in the oviduct, the egg is propelled slowly by cilia movement and tubular muscle contraction toward the uterus. Fertilization and **zygote** formation occurs in an oviduct because the egg only lives approximately 6 to 24 hours. The developing embryo normally arrives at the uterus after several days and then embeds, or implants, itself in the uterine lining, which has been prepared to receive it.

The **uterus** is a thick-walled, muscular organ about the size and shape of an inverted pear. Normally, it lies above and is tipped over the urinary bladder. The oviducts join the uterus at its upper end, while at its lower end the **cervix** enters the vagina nearly at a right angle. A small

Table 15.2	**Female Reproductive Organs**
Organ	**Function**
Ovaries	Produce egg and sex hormones
Oviducts (fallopian tubes)	Conduct egg; location of fertilization
Uterus (womb)	Houses developing fetus
Cervix	Contains opening to uterus
Vagina	Receives penis during sexual intercourse; serves as birth canal, and as an exit for menstrual flow.

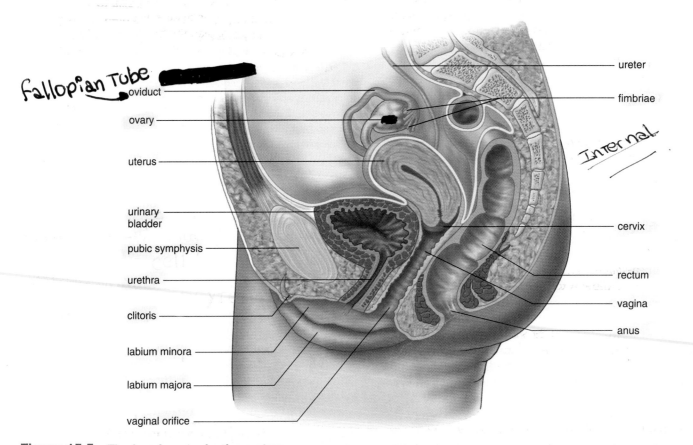

Figure 15.5 **The female reproductive system.**
The ovaries release one egg a month; fertilization occurs in the oviduct, and development occurs in the uterus. The vagina is the birth canal and the organ of sexual intercourse.

opening in the cervix leads to the vaginal canal. Development of the embryo normally takes place in the uterus. This organ, sometimes called the womb, is approximately 5 cm wide in its usual state but is capable of stretching to over 30 cm wide to accommodate the growing baby. The lining of the uterus, called the **endometrium,** participates in the formation of the placenta (p. 327) which supplies nutrients needed for embryonic and fetal development. The endometrium has two layers, a basal layer and an inner functional layer. In the nonpregnant female, the functional layer of the endometrium varies in thickness according to a monthly reproductive cycle, called the uterine cycle.

Cancer of the cervix is a common form of cancer in women. Early detection is possible by means of a **Pap test,** which requires the removal of a few cells from the region of the cervix for microscopic examination. If the cells are cancerous, a hysterectomy may be recommended. A hysterectomy is the removal of the uterus, including the cervix. Removal of the ovaries in addition to the uterus is technically termed an ovariohysterectomy. Because the vagina remains, the woman still can engage in sexual intercourse.

The **vagina** is a tube at a 45° angle with the small of the back. The mucosal lining of the vagina lies in folds and can extend. This is especially important when the vagina serves as the birth canal. It also facilitates sexual intercourse, when the vagina receives the penis, and acts as an exit for menstrual flow.

External Genitals

The external genital organs of the female are known collectively as the **vulva** (Fig. 15.6). The vulva includes two large, hair-covered folds of skin called the labia majora. They extend backward from the mons pubis, a fatty prominence underlying the pubic hair. The labia minora are two small folds lying just inside the labia majora. They extend forward from the vaginal opening to encircle and form a foreskin for the clitoris, an organ that is homologous to the penis. Although quite small, the clitoris has a shaft of erectile tissue and is capped by a pea-shaped glans. The glans clitoris also has sense receptors that allow it to function as a sexually sensitive organ.

The vestibule, a cleft between the labia minor, contains the openings of the urethra and the vagina. The vagina may be partially closed by a ring of tissue called the hymen. The hymen ordinarily is ruptured by initial sexual intercourse; however, it also can be disrupted by other types of physical activities. If the hymen persists after sexual intercourse, it can be surgically ruptured.

Notice that the urinary and reproductive systems in the female are entirely separate. For example, the urethra carries only urine, and the vagina serves only as the birth canal and the organ for sexual intercourse.

Orgasm in Females

Sexual response in the female may be more subtle than in the male, but there are certain corollaries. The clitoris is believed to be an especially sensitive organ for initiating sexual sensations. It is possible for the clitoris to become ever so slightly erect as its erectile tissues become engorged with blood, but vasocongestion is more obvious in the labia minora, which expand and deepen in color. Erectile tissue within the vaginal wall also expands with blood, and the added pressure in these blood vessels causes small droplets of fluid to squeeze through the vessel walls and to lubricate the vagina. Another possible source of lubrication is from mucus-secreting glands beneath the labia minora on either side of the vagina.

Release from muscular tension occurs in females, especially in the region of the vulva and vagina but also throughout the entire body. Increased uterine motility may assist the transport of sperm toward the oviducts. Since female orgasm is not signaled by ejaculation, there is a wide range in normalcy of sexual response.

Once each month, an egg produced by an ovary enters an oviduct. If fertilization occurs, the developing embryo is propelled by cilia to the uterus where it implants itself in the uterine lining. The vagina (which is also the birth canal) and the external genitals play an active role in the sexual response of females.

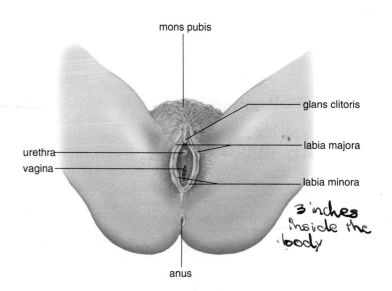

mons pubis · glans clitoris · labia majora · urethra · vagina · labia minora · anus

3 inches inside the body

Figure 15.6 External genitals of the female.
At birth, the opening of the vagina is partially blocked by a membrane called the hymen. Physical activities and sexual intercourse disrupt the hymen.

Visual Focus

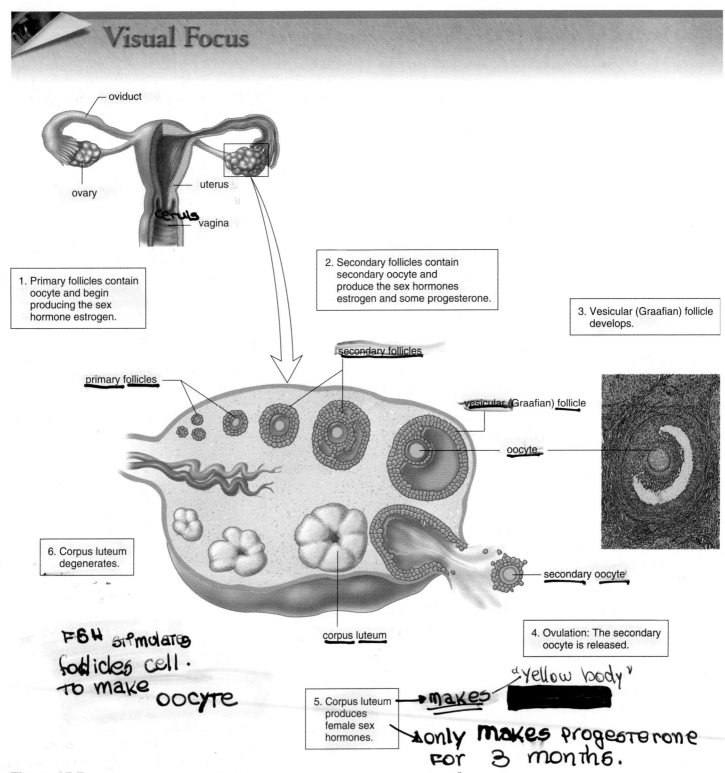

1. Primary follicles contain oocyte and begin producing the sex hormone estrogen.

2. Secondary follicles contain secondary oocyte and produce the sex hormones estrogen and some progesterone.

3. Vesicular (Graafian) follicle develops.

4. Ovulation: The secondary oocyte is released.

5. Corpus luteum produces female sex hormones.

6. Corpus luteum degenerates.

oviduct
ovary
uterus
cervix
vagina

primary follicles
secondary follicles
vesicular (Graafian) follicle
oocyte
secondary oocyte
corpus luteum

(handwritten notes)

FSH stimulates follicles cell. to make oocyte

makes "yellow body"
↓only makes progesterone for 3 months.

If Progesterone does not produce enough abortion would occur

Figure 15.7   Anatomy of ovary and follicle.
As a follicle matures, the oocyte enlarges and is surrounded by layers of follicular cells and fluid. Eventually, ovulation occurs, the mature follicle ruptures, and the secondary oocyte is released. A single follicle actually goes through all stages in one place within the ovary.

Generic eMaterial come from the grandparents. (handwritten)

15.3 Female Hormone Levels

Hormone levels cycle in the female on a monthly basis, and the ovarian cycle drives the uterine cycle as discussed in this section.

The Ovarian Cycle

A longitudinal section through an ovary shows that it is made up of an outer cortex and an inner medulla (Fig. 15.7). There are many **follicles** in the cortex and each one contains an immature egg, called an oocyte. A female is born with as many as 2 million follicles, but the number is reduced to 300,000–400,000 by the time of puberty. Only a small number of follicles (about 400) ever mature because a female usually produces only one egg per month during her reproductive years. Since oocytes are present at birth, they age as the woman ages. This may be one possible reason why older women are more likely to produce children with genetic defects.

As the follicle undergoes maturation, it develops from a primary follicle to a secondary follicle to a vesicular (Graafian) follicle (Fig. 15.7). In a primary follicle, a primary oocyte divides into two cells. One of these cells, termed the secondary oocyte, receives almost all the cytoplasm. A secondary follicle contains the secondary oocyte pushed to one side of a fluid-filled cavity. In a Graafian follicle, the fluid-filled cavity increases to the point that the follicle wall balloons out on the surface of the ovary and bursts, releasing the secondary oocyte (often called an egg for convenience) surrounded by a clear membrane. As mentioned, this is referred to as ovulation. Once a follicle has lost its egg, it develops into a **corpus luteum,** a glandlike structure. If pregnancy does not occur, the corpus luteum begins to degenerate after about 10 days.

These events, called the **ovarian cycle,** are under the control of the gonadotropic hormones, *follicle-stimulating hormone (FSH)* and *luteinizing hormone (LH)* (Fig. 15.8.) The gonadotropic hormones are not present in constant amounts but instead are secreted at different rates during the cycle. For simplicity's sake, it can be emphasized that during the first half, or *follicular phase,* of the ovarian cycle, FSH promotes the development of a follicle in the ovary, which secretes estrogen. As the estrogen level in the blood rises, it exerts feedback control over the anterior pituitary secretion of FSH so that the follicular phase comes to an end.

Presumably, the high level of estrogen in the blood also causes a sudden secretion of a large amount of GnRH from the hypothalamus. This leads to a surge of LH production by the anterior pituitary and to ovulation at about the 14th day of a 28-day cycle.

During the second half, or *luteal phase,* of the ovarian cycle, LH promotes the development of the corpus luteum, which secretes progesterone. Progesterone causes the uterine lining to build up. As the blood level of progesterone rises, it exerts feedback control over the anterior pituitary secretion of LH so that the corpus luteum in the ovary begins to degenerate. As the luteal phase comes to an end, menstruation occurs.

1 - 5. (handwritten)

One ovarian follicle per month produces a secondary oocyte. Following ovulation, the follicle develops into the corpus luteum.

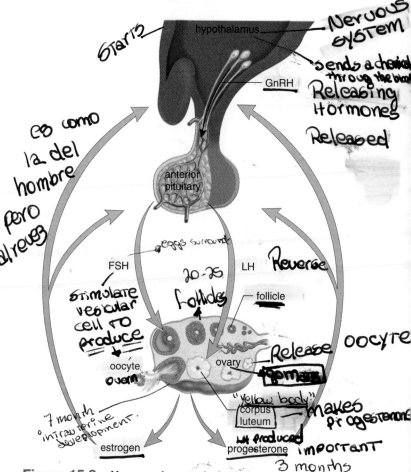

Figure 15.8 Hormonal control of ovaries.
The hypothalamus produces GnRH (gonadotropin-releasing hormone). GnRH stimulates the anterior pituitary to produce FSH (follicle-stimulating hormone) and LH (luteinizing hormone). FHS stimulates the follicle to produce estrogen, and LH stimulates the corpus luteum to produce progesterone. Estrogen and progesterone maintain the sex organs (e.g., uterus) and the secondary sex characteristics and exert feedback control over the hypothalamus and the anterior pituitary.

Follicles Cell is formed by the 7th month before the baby is born. Formed around the oocyte & mature from puberty to menopause. (handwritten)

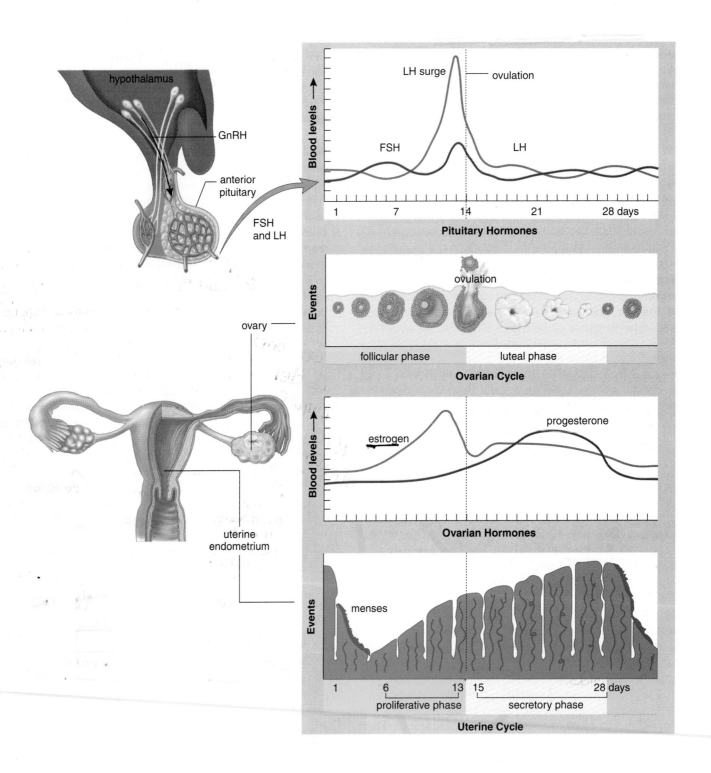

Figure 15.9   Female hormone levels.
During the follicular phase of the ovarian cycle, FSH released by the anterior pituitary promotes the maturation of a follicle in the ovary. The ovarian follicle produces increasing levels of estrogen, which causes the endometrium to thicken during the proliferative phase of the uterine cycle. After ovulation and during the luteal phase of the ovarian cycle, LH promotes the development of the corpus luteum. This structure produces increasing levels of progesterone, which causes the endometrial lining to become secretory. Menstruation begins when progesterone production declines to a low level.

Table 15.3   Ovarian and Uterine Cycles

Ovarian Cycle	Events	Uterine Cycle	Events
Follicular phase—Days 1–13	FSH	Menstruation—Days 1–5	Endometrium breaks down
	Follicle maturation	Proliferative phase—Days 6–13	Endometrium rebuilds
	Estrogen		
Ovulation—Day 14*	LH spike		
Luteal phase—Days 15–28	LH	Secretory phase—Days 15–28	Endometrium thickens and glands are secretory
	Corpus luteum		
	Progesterone		

*Assuming 28-day cycle.

The Uterine Cycle

The female sex hormones, **estrogen** and **progesterone,** have numerous functions. The effects of these hormones on the endometrium of the uterus cause the uterus to undergo a cyclical series of events known as the **uterine cycle** (Table 15.3 and Fig. 15.9). Twenty-eight-day cycles are divided as follows.

During *days 1–5,* a low level of female sex hormones in the body causes the endometrium to disintegrate and its blood vessels to rupture. On day one of the cycle, a flow of blood and tissues, known as the menses, passes out of the vagina during **menstruation,** also called the menstrual period.

During *days 6–13,* increased production of estrogen by a new ovarian follicle in the ovary causes the endometrium to thicken and to become vascular and glandular. This is called the proliferative phase of the uterine cycle.

Ovulation usually occurs on the fourteenth day of the twenty-eight-day cycle.

During *days 15–28,* increased production of progesterone by the corpus luteum in the ovary causes the endometrium of the uterus to double or triple in thickness (from 1 mm to 2–3 mm) and the uterine glands to mature, producing a thick mucoid secretion. This is called the secretory phase of the uterine cycle. The endometrium now is prepared to receive the developing embryo. If pregnancy does not occur, the corpus luteum in the ovary degenerates, and the low level of sex hormones in the female body results in the uterine lining breaking down.

During the uterine cycle, the endometrium of the uterus builds up and then is broken down during menstruation.

Fertilization and Pregnancy

[handwritten: 6–10 days after conception the human Chorionic Hormone come be measure]

If fertilization does occur, an embryo begins development even as it travels down the oviduct to the uterus. The endometrium is now prepared to receive the developing embryo, which becomes embedded in the lining several days following fertilization (Fig. 15.10). The **placenta** originates from both maternal and fetal tissues. It is the region of exchange of molecules between fetal and maternal blood, although there is rarely any mixing of the two. At first, the placenta produces **human chorionic gonadotropin (HCG),** which maintains the corpus luteum in the ovary until the placenta begins its own production of progesterone and estrogen. Progesterone and estrogen have two effects: they shut down the anterior pituitary so that no new follicle in the ovaries matures, and they maintain the lining of the uterus so that the corpus luteum in the ovary is not needed. Usually, there is no menstruation during pregnancy.

[handwritten: 2 days circulatory]

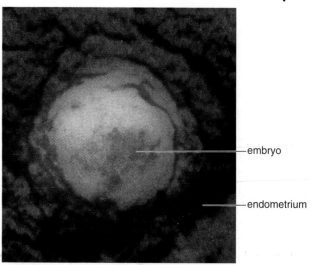

Figure 15.10   Implantation.
A scanning electron micrograph showing embryo implanted on day 12 following fertilization.

Estrogen and Progesterone

Estrogen and progesterone affect not only the uterus but other parts of the body as well. Estrogen is largely responsible for the secondary sex characteristics in females, including body hair and fat distribution. In general, females have a more rounded appearance than males because of a greater accumulation of fat beneath the skin. Like males, females develop axillary and pubic hair during puberty. In females, the upper border of pubic hair is horizontal, but in males, it tapers toward the navel. Both estrogen and progesterone are also required for breast development. Other hormones are involved in milk production and letdown following a pregnancy.

The pelvic girdle is wider and deeper in females, so the pelvic cavity usually has a larger relative size compared to males. This means that females have wider hips than males and that the thighs converge at a greater angle toward the knees. Because the female pelvis tilts forward, females tend to have protruding buttocks, more of a lower back curve than males, an abdominal bulge, and a tendency to be somewhat knock-kneed.

Menopause

Menopause, the period in a woman's life during which the ovarian and uterine cycles cease, is likely to occur between ages 45 and 55. The ovaries are no longer responsive to the gonadotropic hormones produced by the anterior pituitary, and the ovaries no longer secrete estrogen or progesterone. At the onset of menopause, the uterine cycle becomes irregular, but as long as menstruation occurs, it is still possible for a woman to conceive. Therefore, a woman usually is not considered to have completed menopause until there has been no menstruation for a year.

The hormonal changes during menopause often produce physical symptoms, such as "hot flashes," (caused by circulatory irregularities), dizziness, headaches, insomnia, sleepiness, and depression. These symptoms may be mild or even absent. If they are severe, medical attention should be sought. Women sometimes report an increased sex drive following menopause. It has been suggested that this may be due to androgen production by the adrenal cortex.

> Estrogen and progesterone produced by the ovaries are the female sex hormones responsible for the primary sex characteristics, the uterine cycle, and the secondary sex characteristics of females.

15.4 Development of Male and Female Sex Organs

The sex of an individual is determined at the moment of fertilization. Males have the chromosomes X and Y, while girls have two X chromosomes. During the first few months of development, it is impossible to tell whether the unborn child is a boy or girl by external inspection. Gonads don't start developing until the seventh week of development. The tissue that gives rise to the gonads is called indifferent because it can become testes or ovaries depending on the action of hormones. Genes on the Y chromosome cause testes to develop and produce androgenic hormones, and these determine the course of development.

In Figure 15.11*a*, notice that at six weeks both males and females have the same type tissues and ducts. During this indifferent stage an embryo has the potential to develop into a male or female. If a Y chromosome is present, androgenic (male) hormones stimulate the mesonephric ducts to become male genital ducts. The mesonephric ducts enter the urethra, which belongs to both the urinary and reproductive systems in males. Androgenic hormones suppress development of paramesonephric ducts in males.

In the absence of a Y chromosome and in the presence of two X chromosomes, ovaries develop instead of testes from the same indifferent tissue. Now the mesonephric ducts regress, and the paramesonephric ducts develop into the uterus and uterine tubes. A developing vagina also extends from the uterus. There is no connection between the urinary and genital systems in females.

At fourteen weeks, both the primitive testes and ovaries are located deep inside the abdominal cavity. An inspection of the interior of the testes would show that sperm are even now starting to develop, and similarly, the ovaries already contain large numbers of tiny follicles, each containing an ovum. Toward the end of development, the testes descend into the scrotal sac; the ovaries remain in the abdominal cavity.

Figure 15.11*b* shows the development of the external genitals. These tissues are also indifferent at first—they can develop into either male or female genitals. At six weeks, a small bud appears between the legs that can develop into the male penis or the female clitoris, depending on the presence or absence of the Y chromosome and androgenic hormones. At nine weeks, there is a groove called the urogenital groove bordered by two swellings. By fourteen weeks, in males, this groove has disappeared, and the scrotum has formed from the original swellings. In females, the groove persists and becomes the vaginal opening. Labia majora and labia minora are present instead of a scrotum.

Taking the first week implantation occurs

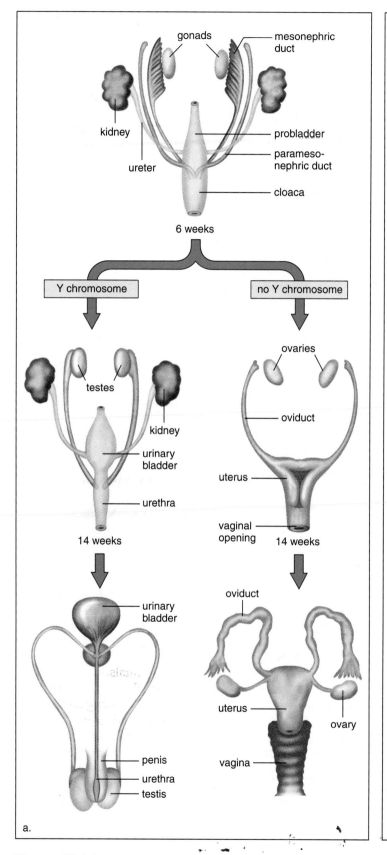

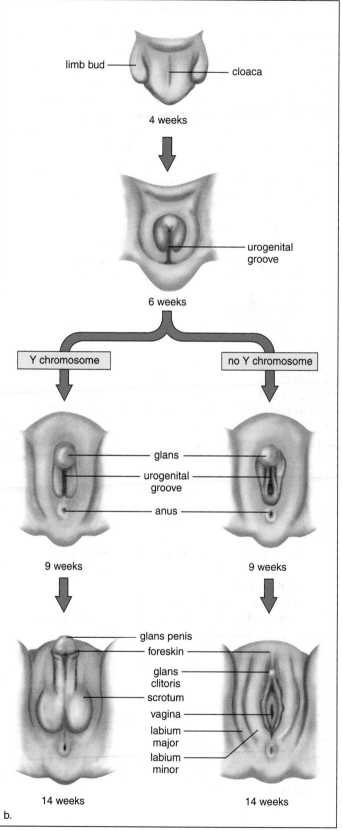

Figure 15.11   Male and female organs.

Development of gonads and ducts in **(a)** female and **(b)** male.

15.5 Control of Reproduction

Several means are available to dampen or enhance our reproductive potential. **Contraceptives** are medications and devices that reduce the chance of pregnancy.

Birth Control Methods

The most reliable method of birth control is abstinence, that is, the absence of sexual intercourse. This form of birth control has the added advantage of preventing transmission of a sexually transmitted disease. Other means of birth control used in this country are given in Table 15.4. The table gives the effectiveness for the various birth-control methods listed. For example, with the least effective method given in the table, we expect that within a year, 70 out of 100, or 70%, of sexually active women will not get pregnant, while 30 women will get pregnant.

Figure 15.12 features some of the most effective and commonly used means of birth control. *Oral contraception*

(birth-control pills) usually involves taking a combination of estrogen and progesterone for 21 days of a 28-day cycle. The estrogen and progesterone in the birth-control pill effectively shut down the pituitary production of both FSH and LH so that no follicle in the ovary begins to develop in the ovary; and since ovulation does not occur, pregnancy cannot take place. Since there are possible side effects, those taking birth-control pills should see a physician regularly.

An **intrauterine device (IUD)** is a small piece of molded plastic that is inserted into the uterus by a physician. IUDs are believed to alter the environment of the uterus and oviducts so that fertilization probably does not occur—but if fertilization should occur, implantation cannot take place. The type of IUD featured in Figure 15.12 has copper wire wrapped around the plastic.

The **diaphragm** is a soft latex cup with a flexible rim that lodges behind the pubic bone and fits over the cervix. Each woman must be properly fitted by a physician, and the diaphragm can be inserted into the vagina two hours at most before sexual relations. It must be used with spermicidal

Table 15.4	Common Birth-Control Methods			
Name	**Procedure**	**Methodology**	**Effectiveness**	**Risk**
Abstinence	Refrain from sexual intercourse	No sperm in vagina	100%	None
Vasectomy	Vasa deferentia cut and tied	No sperm in seminal fluid	Almost 100%	Irreversible sterility
Tubal ligation	Oviducts cut and tied	No eggs in oviduct	Almost 100%	Irreversible sterility
Oral contraception	Hormone medication is taken daily	Anterior pituitary does not release FSH and LH	Almost 100%	Thromboembolism, especially in smokers
Depo-Provera injection	Four injections of progesterone-like steroid given per year	Anterior pituitary does not release FSH and LH	About 99%	Breast cancer? Osteoporosis?
Contraceptive implants	Tubes of progestin (form of progesterone) implanted under skin	Anterior pituitary does not release FSH and LH	More than 90%	Presently none known
Intrauterine device (IUD)	Plastic coil is inserted into uterus by physician	Prevents implantation	More than 90%	Infection (pelvic inflammatory disease, PID)
Diaphragm	Latex cup is inserted into vagina to cover cervix before intercourse	Blocks entrance of sperm to uterus	With jelly, about 90%	Presently none known
Cervical cap	Latex cap is held by suction over cervix	Delivers spermicide near cervix	Almost 85%	Cancer of cervix
Male condom	Latex sheath is fitted over erect penis	Traps sperm and prevents STDs	About 85%	Presently none known
Female condom	Polyurethane liner fitted inside vagina	Blocks entrance of sperm to uterus and prevents STDs	About 85%	Presently none known
Coitus interruptus	Penis withdrawn before ejaculation	Prevents sperm from entering vagina	About 75%	Presently none known
Jellies, creams, foams	These spermicidal products inserted before intercourse	Kills a large number of sperm	About 75%	Presently none known
Natural family planning	Day of ovulation determined by record keeping; various methods of testing	Intercourse avoided on certain days of the month	About 70%	Presently none known
Douche	Vagina is cleansed after intercourse	Washes out sperm	Less than 70%	Presently none known

jelly or cream and should be left in place at least six hours after sexual relations. The cervical cap is a minidiaphragm.

A male **condom** is a thin skin (lambskin) or latex sheath that fits over the erect penis. The ejaculate is trapped inside the sheath and, thus, does not enter the vagina. When used in conjunction with a spermicide, the protection is better than with the condom alone. The condom is generally recognized as giving protection against sexually transmitted diseases.

Contraceptive implants utilize a synthetic progesterone to prevent ovulation by disrupting the ovarian cycle. Six match-sized, time-release capsules are surgically inserted under the skin of a woman's upper arm. The effectiveness of this system may last five years. *Depo-Provera* injections, which change the endometrium, utilize a synthetic progesterone that must be administered every three months. Changes occur in the endometrium that make it less likely that pregnancy will occur.

There has been a revival of interest in *barrier methods* of birth control, including the male condom, because these methods offer some protection against sexually transmitted diseases. A female condom, now available, consists of a large polyurethane tube with a flexible ring that fits onto the cervix. The open end of the tube has a ring that covers the external genitals.

Investigators have long searched for a *"male pill."* Analogues of gonadotropic-releasing hormone have been used to prevent the hypothalamus from stimulating the anterior pituitary. Inhibin has also been used to prevent the anterior pituitary from producing FSH. Testosterone and/or related chemicals have been used to inhibit spermatogenesis in males, but this hormone must be administered by injection.

Contraceptive vaccines are now being developed. For example, a vaccine developed to immunize women against HCG, the hormone so necessary to maintaining the **implantation** of the embryo, was successful in a limited clinical trial. Since HCG is not normally present in the body, no untoward autoimmune reaction is expected, and the immunization does wear off with time. Others believe that it would also be possible to develop a safe antisperm vaccine that would be used in women.

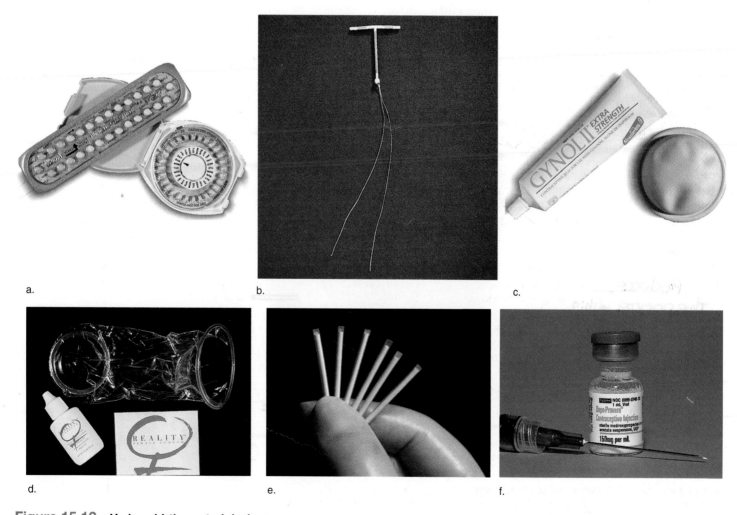

Figure 15.12   **Various birth-control devices.**
a. Oral contraception (birth-control pills). **b.** Intrauterine device. **c.** Spermicidal jelly and diaphragm. **d.** Female condom. **e.** Contraceptive implants. **f.** Depo-Provera injection.

Health Focus

Endometriosis

In the female reproductive tract, there is a small space between an ovary and the oviduct, the tube that leads to the uterus. The lining of the uterus, called the endometrium, loses its outer layer during menstruation, and sometimes a portion of the menstrual discharge is carried backwards up the oviduct and into the abdominal cavity instead of being discharged through the opening in the cervix (Fig. 15A). There, endometrial tissue can become attached to and implanted in various organs such as the ovaries; the wall of the vagina, bowel, or bladder; or even on the nerves that serve the lower back or legs. This painful condition is called endometriosis, and it affects 1–3% of women of reproductive age.

Women with uterine cycles of less than 27 days and with a menstrual flow lasting longer than one week have an increased chance of endometriosis. Women who have taken the birth-control pill for a long time or have had several pregnancies have a decreased chance of endometriosis.

When a woman has endometriosis, the displaced endometrial tissue reacts as if it were still in the uterus—thickening, becoming secretory, and then breaking down. The discomfort of menstruation is then felt in other organs of the abdominal cavity, resulting in pain. An area of endometriosis can degenerate and become a scar. Scars that hold two organs together are called adhesions, which can distort organs and lead to infertility.

Only direct observation of the abdominal organs can confirm endometriosis. First, a half-inch incision is made near the navel, and the gas carbon dioxide is injected into the abdominal cavity to separate the organs. Then, a laparoscope (optical telescope) is inserted into the abdomen, allowing the physician to see the organs. Patches of endometriosis show up as purple, blue, or red spots, and there may be dark brown cysts filled with blood. When these rupture, there is a great deal of pain. In the severest cases, there is scarring and the formation of adhesions and abnormal masses around the pelvic organs. A second incision, usually at the pubic hairline, allows the insertion of other instruments that can be used to remove endometrial implants.

Further treatment can take one of two courses. The drug nafarelin can be administered as a nasal spray, which acts through hormonal controls to stop the production of estrogen and, therefore, to stop the uterine cycle. Or, the ovaries can be removed, stopping the uterine cycle. In either case, the woman will suffer symptoms of menopause that can be relieved in the first instance by withdrawing the drug nafarelin, or in the second instance by giving the woman estrogen in doses that do not reactivate the uterine cycle.

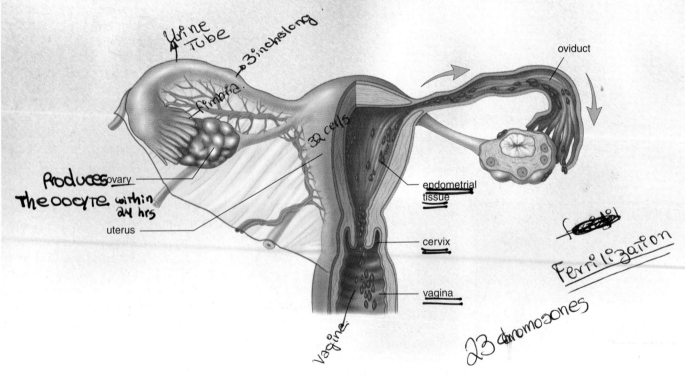

Figure 15A   Endometriosis.
It has been suggested that endometriosis is caused by a backward menstrual flow, as represented by the arrows in this drawing. This backward flow allows endometrial cells to enter the abdominal cavity, where they take up residence and respond to monthly cyclic changes in hormonal levels. The result is extreme discomfort.

Morning-after Pills

The expression "morning-after pill" refers to a medication that will prevent pregnancy after unprotected intercourse. The expression is a misnomer in that medication can begin one to several days after unprotected intercourse.

A kit, now called Preven, is made up of four synthetic progesterone pills; two are taken up to 72 hours after unprotected intercourse, and two more are taken 12 hours later. The medication upsets the normal uterine cycle, making it difficult for the embryo to implant itself in the endometrium. In a recent study, it was estimated that the medication was 85% effective in preventing unintended pregnancies.

Mifepristone, better known as RU-486, is a pill that is presently used to cause the loss of an implanted embryo by blocking the progesterone receptors of endometrial cells. Without functioning receptors for progesterone, the endometrium sloughs off, carrying the embryo with it. When taken in conjunction with a prostaglandin to induce uterine contractions, RU-486 is 95% effective. It is possible that some day this medication will also be a "morning-after pill," taken when menstruation is late without evidence that pregnancy has occurred.

There are numerous well-known birth-control methods and devices available to those who wish to prevent pregnancy. Their effectiveness varies. In addition, new methods are expected to be developed.

Infertility

Sometimes couples do not need to prevent pregnancy; conception or fertilization does not occur despite frequent intercourse. The American Medical Association estimates that 15% of all couples in this country are unable to have any children and therefore are properly termed *sterile*; another 10% have fewer children than they wish and therefore are termed *infertile*. The latter assumes that the couple has been unsuccessfully trying to become pregnant for at least one year.

Causes of Infertility

The most common causes of infertility in females are blocked oviducts and endometriosis. **Endometriosis** is the presence of uterine tissue outside the uterus, particularly in the oviducts and on the abdominal organs. As discussed in the Health reading on page 332, endometriosis can contribute to infertility. Endometriosis occurs when the menstrual discharge flows up into the oviducts and out into the abdominal cavity. This backward flow allows living uterine cells to establish themselves in the abdominal cavity where

they go through the usual uterine cycle, causing pain and structural abnormalities that make it more difficult for a woman to conceive.

Sometimes the causes of infertility can be corrected by medical intervention so that couples can have children (Fig. 15.13). If no obstruction is apparent and body weight is normal, it is possible to give females HCG, extracted from the urine of pregnant women, along with gonadotropins extracted from the urine of postmenopausal women. This treatment may cause multiple ovulations and sometimes multiple pregnancies.

The most frequent cause of infertility in males is low sperm count and/or a large proportion of abnormal sperm. Disease, radiation, chemical mutagens, high testes temperature, and the use of psychoactive drugs can contribute to this condition.

When reproduction does not occur in the usual manner, many couples adopt a child. Others sometimes first try one of the alternative reproductive methods discussed in the following paragraphs. If all the alternative methods discussed are considered, it is possible for a baby to have five parents: (1) sperm donor, (2) egg donor, (3) surrogate mother, and (4) and (5) adoptive mother and father.

Alternative Methods of Reproduction

Artificial Insemination by Donor (AID). During artificial insemination, sperm are placed in the vagina by a physician. Sometimes, a woman is artificially inseminated by her husband's sperm. This is especially helpful if the husband has a low sperm count—the sperm can be collected over a period of time and concentrated so that the sperm count is sufficient to result in fertilization. Often, however, a woman is inseminated by sperm acquired from a donor who is a complete stranger to her. At times, a combination of husband and donor sperm is used.

A variation of AID is intrauterine insemination (IUI). IUI involves hormonal stimulation of the ovaries, followed by placement of the donor's sperm in the uterus rather than in the vagina.

In Vitro Fertilization (IVF). During IVF, conception occurs in laboratory glassware. The newer ultrasound machines can spot follicles in the ovaries that hold immature eggs; therefore, the latest method is to forego the administration of fertility drugs and retrieve immature eggs by using a needle. The immature eggs are then brought to maturity in glassware before concentrated sperm from the male are added. After about two to four days, the embryos are inserted into the uterus of the woman, who is now in the secretory phase of her uterine cycle. If implantation is successful, development is normal and continues to term.

Gamete Intrafallopian Transfer (GIFT). The term **gamete** refers to a sex cell, either a sperm or an egg. Gamete

Figure 15.13 Mother and child.
Sometimes couples rely on medical intervention in order to experience parenthood.

15.6 Homeostasis

Regulation of sex hormone blood level is an example of homeostatic control. Figure 15.4 shows how the blood level of testosterone is maintained, and Figure 15.8 shows how the blood levels of estrogen and progesterone are maintained within normal limits. Negative feedback results in a self-regulatory mechanism that maintains the appropriate level of these hormones in the blood.

The next page shows how the reproductive system works with the other systems of the body to maintain homeostasis. Most of our knowledge about the function of sex hormones concerns their role in the maturation of sexual organs and the maintenance of the secondary sex hormones. It would be an exaggeration to say that these functions have to do with homeostasis. Why? Because homeostasis has to do with the constancy of the internal environment of cells. Nevertheless, certain activities of the sex hormones do affect homeostasis. For example, estrogens promote fat deposition, which ensures that cells will have a source of energy in lean times and helps maintain the normal temperature of the body.

In recent years, it's been discovered that besides the action of estrogens on sexual organs and sexuality of females, estrogens have effects on the liver, bones, and kidneys which are important to homeostasis. Estrogens induce the liver to produce many types of proteins that transport substances in the blood. These include proteins that bind iron and copper and lipoproteins that transport cholesterol. Iron and copper are enzyme cofactors necessary to cellular metabolism. While we associated cholesterol with cardiovascular diseases, in fact, it is a substance that contributes to the functioning of the plasma membrane. Estrogens induce synthesis of bone matrix proteins and counteract the loss of bone mass. At menopause, when the rate of estrogen secretion is drastically reduced, osteoporosis (decrease in bone density) may develop.

Similarly, besides the action of androgens on sexual organs and function of males, androgens play a metabolic role in cells. They stimulate synthesis of structural proteins in skeletal muscles and bone and affect the activity of various enzymes in the liver and kidneys. In the kidney, androgens stimulate synthesis of erythropoietin, the protein that signals the bone marrow to increase production of red blood cells. We are just now beginning to discover the role that estrogens and androgens play in the metabolism of cells and therefore their role in homeostasis in general.

intrafallopian transfer (GIFT) was devised as a means to overcome the low success rate (15–20%) of in vitro fertilization. The method is exactly the same as in vitro fertilization, except the eggs and the sperm are placed in the oviducts immediately after they have been brought together. GIFT has an advantage in that it is a one-step procedure for the woman—the eggs are removed and are reintroduced all in the same time period. For this reason, it is less expensive—approximately $1,500 compared to $3,000 and up for in vitro fertilization.

Surrogate Mothers. In some instances, women are paid to have babies. These women are called surrogate mothers. Other individuals contribute sperm (or eggs) to the fertilization process in such cases.

When corrective procedures fail to reverse infertility, it is possible to consider an alternative method of reproduction.

The sex hormones greatly affect our appearance because they maintain the secondary sex characteristics. Some of these characteristics, such as fat deposition in females, help maintain homeostasis. In addition, it is now known that both estrogens and androgens have general metabolic effects that pertain to homeostasis.

Human Systems Work Together

Integumentary System

Androgens activate oil glands; sex hormones stimulate fat deposition, affect hair distribution in males and females.

Skin receptors respond to touch; modified sweat glands produce milk; skin stretches to accommodate growing fetus.

How the Reproductive System works with other body systems

Cardiovascular System

Sex hormones influence cardiovascular health; sexual activities stimulate **cardio-vascular** system.

Blood vessels transport sex hormones; vasodilation causes genitals to become erect; blood services the reproductive organs.

Skeletal System

Sex hormones influence bone growth and density in males and females.

Bones provide support and protection of reproductive organs.

Lymphatic System/Immunity

Sex hormones influence immune functioning; acidity of vagina helps prevent pathogen invasion of body; milk passes antibodies to newborn.

Immune system does not attack sperm or fetus, even though they are foreign to the body.

Muscular System

Androgens promote growth of skeletal muscle.

Muscle contraction occurs during orgasm and moves gametes; abdominal and uterine muscle contractions occur during childbirth.

Respiratory System

Sexual activity increases breathing; pregnancy causes breathing rate and vital capacity to increase.

Gas exchange increases during sexual activity.

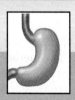

Nervous System

Sex hormones masculinize or feminize the brain, exert feedback control over the hypothalamus, and influence sexual behavior.

Brain controls onset of puberty; nerves are involved in erection of penis and clitoris, movement of gametes along ducts, and contraction of uterus.

Digestive System

Pregnancy crowds digestive organs and promotes heartburn and constipation.

Digestive tract provides nutrients for growth and repair of organs and for development of fetus.

Endocrine System

Gonads produce the sex hormones.

Hypothalamic, pituitary, and sex hormones control sex characteristics and regulate reproductive processes.

Urinary System

Penis in males contains the urethra and performs urination; prostate enlargement hinders urination.

Semen is discharged through the urethra in males; kidneys excrete wastes and maintain electrolyte levels for mother and child.

Bioethical Issue

The dizzying array of reproductive techniques has progressed from simple in vitro fertilization to the ability to freeze eggs or sperm or even embryos for future use. Older women who never had the opportunity to freeze their eggs can still have children if they use donated eggs—perhaps today harvested from a fetus.

Legal complications abound from which mother has first claim to the child—the surrogate mother, the woman who donated the egg, or the primary care giver—to which partner has first claim to frozen embryos following a divorce. Legal decisions about who has the right to use what techniques have rarely been discussed, much less decided upon. Some clinics will help anyone, male or female, no questions asked, as long as they have the ability to pay. And most clinics are heading toward doing any type of procedure, including guaranteeing the sex of the child, and making sure the child will be free from some particular genetic disorder. It would not be surprising if, in the future, zygotes could be engineered to have any particular trait desired by the parents.

Even today eugenic (good gene) goals are evidenced by the fact that reproductive clinics advertise for egg and sperm donors, primarily in elite college newspapers. The question becomes "Is it too late for us as a society to make ethical decisions about reproductive issues?" Should we come to a consensus about what techniques should be allowed and who should be able to use them? We all want to avoid, if possible, what happened to Jonathan Alan Austin. Jonathan, who was born to a surrogate mother, later died from injuries inflicted by his father. Perhaps if a background check were legally required, surrogate mothers would only make themselves available to individuals or couples who are known to have certain psychological characteristics.

Questions

1. As a society we have never been in favor of regulating reproduction. Should we regulate reproduction by an alternative method if we do not regulate unassisted reproduction? Why or why not?
2. Should the state be the guardian of frozen embryos and make sure they all get a chance to life? Why or why not?
3. Is it appropriate for physicians and parents to select which embryos will be implanted in the uterus? On the basis of sex? On the basis of genetic inheritance? Why or why not?

Summarizing the Concepts

15.1 Male Reproductive System

In males, spermatogenesis, occurring in seminiferous tubules of the testes, produces sperm that mature and are stored in the epididymides and may be stored in the vasa deferentia before entering the urethra, along with secretions produced by seminal vesicles, the prostate gland, and bulbourethral glands. Semen contains sperm and these secretions.

The external genitals of males are the penis, the organ of sexual intercourse and the scrotum which contains the testes. Orgasm in males is a physical and emotional climax during sexual intercourse that results in ejaculation of semen from the penis.

Hormonal regulation, involving secretions from the hypothalamus, the anterior pituitary, and the testes, maintains testosterone, produced by the interstitial cells of the testes, at a fairly constant level. FSH from the anterior pituitary promotes spermatogenesis in the seminiferous tubules, and LH promotes testosterone production by the interstitial cells.

15.2 Female Reproductive System

In females, oogenesis occuring within the ovaries typically produces one mature follicle each month. This follicle balloons out of the ovary and bursts, releasing an egg which enters an oviduct. The oviducts lead to the uterus where implantation and development occur. The external genital area includes the vaginal opening, the clitoris, the labia minora, and the labia majora.

The vagina is the organ of sexual intercourse in females. In females orgasm is not signaled by ejaculation, and normalcy of sexual response varies greatly.

15.3 Female Hormone Levels

In the nonpregnant female, the ovarian and uterine cycles are under hormonal control of the hypothalamus, anterior pituitary, and the female sex hormones estrogen and progesterone. During the first half of the ovarian cycle, FSH from the anterior pituitary causes maturation of a follicle which secretes estrogen. After ovulation, and during the second half of the cycle, LH from the anterior pituitary converts the follicle into the corpus luteum which produces progesterone. Estrogen and progesterone regulate the uterine cycle. Estrogen causes the endometrium to rebuild. Ovulation usually occurs on the fourteenth day of a twenty-eight-day cycle. As progesterone is produced by the corpus luteum, the endometrium thickens and becomes secretory. Then, a low level of hormones causes the endometrium to break down, as menstruation occurs.

If fertilization occurs, the corpus luteum in the ovary is maintained because of HCG production by the placenta. Progesterone production does not cease, and the embryo implants itself in the thick uterine lining. Menstruation does not occur during pregnancy because of HCG production.

15.4 Development of Male and Female Sex Organs

Male and female organs develop from the same indifferent tissue depending on whether the chromosomes are XY or XX. External examination does not reveal the sex of the fetus until after the third month.

15.5 Control of Reproduction

Corrective medical and surgical procedures can help people who are infertile but wish to have a child. Alternative methods of reproduction may also be considered. Numerous birth-control methods and devices are available for those who wish to prevent pregnancy. Infertile couples are increasingly resorting to alternative methods of reproduction.

15.6 Homeostasis

The reproductive system works with the other systems of the body in the ways described in the box on page 335.

Studying the Concepts

1. Outline the path of sperm. What glands contribute fluids to semen? 318–19
2. Discuss the anatomy and physiology of the testes. Describe the structure of sperm. 320–21
3. Name the endocrine glands involved in maintaining the sex characteristics of males and the hormones produced by each. 321
4. Describe the organs of the female genital tract. Where do fertilization and implantation occur? Name two functions of the vagina. 322–23
5. Name and describe the external genitals in females. 323
6. Discuss the anatomy and the physiology of the ovaries. Describe the ovarian cycle. 324–25
7. Describe the uterine cycle and relate it to the ovarian cycle. In what way is menstruation prevented if pregnancy occurs? 327
8. Name three functions of the female sex hormones. 328
9. Discuss the various means of birth control and their relative effectiveness in preventing pregnancy 330–33
10. Describe how in vitro fertilization is carried out. 333

Testing Your Knowledge of the Concepts

In questions 1–4, match the organs to the functions listed.
a. testes and ovaries
b. clitoris and penis
c. oviducts and vas deferens
d. scrotum and labium majora
_____ 1. part of the genital tract that conducts gametes
_____ 2. develop from swellings on either side of the urogenital groove
_____ 3. gonads that produce gametes
_____ 4. organs involved in sexual response

In questions 5–7, indicate whether the statement is true (T) or false (F).
_____ 5. In tracing the path of sperm, the structure that follows the epididymis is the bladder.
_____ 6. The testes are an endocrine gland.
_____ 7. In the ovarian cycle, estrogen causes the endometrial lining to become secretory.

In questions 8–16, fill in the blanks.
8. An erection is caused by the entrance of _____ into sinuses within the penis.
9. Estrogen is to females as _____ is to males.
10. The prostate gland, the bulbourethal glands, and the _____ all contribute to seminal fluid.
11. The main male sex hormone is _____.
12. In the female reproductive system, the uterus lies between the oviducts and the _____.

13. In the ovarian cycle, once each month a(n) _____ releases an egg. In the uterine cycle, the _____ lining of the uterus is prepared to receive the embryo.
14. The female sex hormones are _____ and _____.
15. Pregnancy in the female is detected by the presence of _____ in blood or urine.
16. In vitro fertilization occurs in _____.
17. Label this diagram of the male reproductive system, and trace the path of sperm.

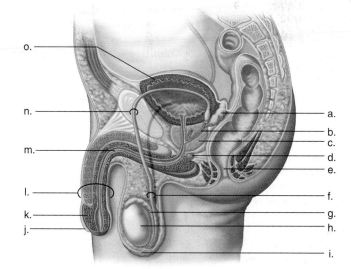

Applying Your Knowledge to the Concepts

These questions pertain to homeostasis.
1. The male reproductive system is often referred to as a closed system, whereas the female reproductive system is referred to as an open system. To what aspect(s) of the system does this refer?
2. Using Figures 15.4 and 15.8, explain what is meant by a negative feedback mechanism.
3. The female vagina produces an acidic secretion. Why is a vaginal acidic environment advantageous and by what mechanism do sperm survive it?
4. Women sometimes fail to ovulate because of low body weight. What does this possibly tell you about body weight and homeostasis?

Understanding the Terms

acrosome 321
birth control pill 330
bulbourethral gland 319
cervix 322 •
condom 331
contraceptive 330
corpus luteum 325
diaphragm 330
egg 322
endometriosis 333
endometrium 323
epididymis 318
estrogen 327
fimbria 322
follicle 325
follicle-stimulating
    hormone (FSH) 321
gamete 333 *sex cell*
gonad 318
gonadotropin-releasing
    hormone (GnRH) 321
human chorionic
    gonadotropin (HCG) 327
implantation 331
interstitial cell 321
intrauterine device (IUD) 330
luteinizing hormone (LH) 321
menopause 328

menstruation 327
oogenesis 322
ovarian cycle 325
ovary 322
oviduct 322
ovulation 322
Pap test 323
penis 319
placenta 327
progesterone 327
prostate gland 318
scrotum 318
semen 318
seminal vesicle 318
seminiferous tubule 320
sperm 321
spermatogenesis 321 *making of sper·*
sustentacular (Sertoli) cell
    321
testis 318
testosterone 321
urethra 318
uterine cycle 327
uterus 322
vagina 323
vas deferens 318
vulva 323
zygote 322

Match the terms to these definitions.

a. _vagina_ Organ that leads from the uterus to the vestibule and serves as the birth canal and organ of sexual intercourse in females.

b. _endometrium_ Lining of the uterus, which becomes thickened and vascular during the uterine cycle.

c. _Prostate_ Gland located around the male urethra below the urinary bladder; adds secretions to semen.

d. _FSH_ Hormone secreted by the anterior pituitary gland that stimulates the development of an ovarian follicle in a female or the production of sperm in a male.

e. _vulva ·_ External genitals of the female that surround the opening of the vagina.

Applying Technology to the Concepts

Your study of the reproductive system is supported by these available technologies:

Essential Study Partner CD-ROM
Animals → Reproduction
Visit the Mader web site for related ESP activities.

Exploring the Internet
The Mader Home Page provides resources and tools as you study this chapter.

http://www.mhhe.com/biosci/genbio/mader

Dynamic Human 2.0 CD-ROM
Reproductive System

Chapter *16*

Sexually Transmitted Diseases

Chapter Concepts

Figure 16.1   Sexual relationships.
Everyone who has sexual relationships should be aware of the possibility of acquiring a sexually transmitted disease and take all necessary precautions.

By the time Jennifer W. realizes what has happened, it's too late. She is sitting in a doctor's office, quietly explaining that every time she goes to the bathroom, she has this terrible burning pain. She doesn't want to tell her friends. She can't tell her parents.

At least she is telling the doctor.

The source of Jennifer's agony is genital herpes, a very common—and very embarrassing—condition. Genital herpes is a sexually transmitted disease (STD) caused by a virus. As in Jennifer's case, most people catch the disease by having sex with someone who's infected but has no symptoms (Fig. 16.1). That's because herpes viruses—as well as AIDS and a host of other STDs—can lie latent in the body, hiding from sight. Even if you can't see the disease, you should try to prevent passage by (1) practicing abstinence; or (2) having a monogamous (always the same partner) sexual relationship with someone who does not have an STD, and in the case of AIDS and hepatitis B, is not an intravenous drug user; or (3) always using a female or male latex condom in the proper manner with a water-based vaginal spermicide containing nonoxynol-9. You should also avoid oral/genital contact—just touching the genitals can transfer an STD in some cases.

With a prescription drug, Jennifer recovers from her present symptoms of a herpes infection. But she'll never be free of the virus, a possible recurrence of symptoms, or the knowledge that her pain could have been avoided. Perhaps if Jennifer had taken Human Biology she would have known about STDs and how to protect herself.

STDs are now a major worldwide health problem, especially because many young people are sexually active and the age of sexual intercourse has been declining. This chapter discusses pathogens such as viruses, bacteria, and fungi and the STDs they cause. In addition, more information about a human immunodeficiency viral (HIV) infection and AIDS can be found in the supplement that follows this chapter.

16.1 Viral in Origin

Sexually transmitted diseases (STDs) are contagious diseases caused by pathogens that are passed from one human to another by sexual contact. **Pathogens** are viruses, bacteria, and other organisms that cause diseases in humans. Viruses cause numerous diseases in humans (Table 16.1), including AIDS, herpes, genital warts, and hepatitis B, four sexually transmitted diseases of great concern today.

Viruses are incapable of independent reproduction and reproduce only inside a living **host** cell. For this reason, viruses are called *obligate intracellular parasites.* To maintain viruses in the laboratory, they are injected into laboratory-bred animals, live chick embryos, or animal cells maintained in tissue culture. Viruses infect all sorts of cells—from bacterial cells to human cells—but they are very specific. For example, viruses called bacteriophages infect only bacteria, the tobacco mosaic virus infects only plants, and the rabies virus infects only mammals. Human viruses even specialize in a particular tissue. Human immunodeficiency virus (HIV) enters certain blood cells, the polio virus reproduces in spinal nerve cells, the hepatitis viruses infect only liver cells.

Viruses are noncellular, and they have a unique construction. These tiny particles always have at least two parts: an outer capsid composed of protein subunits and an inner

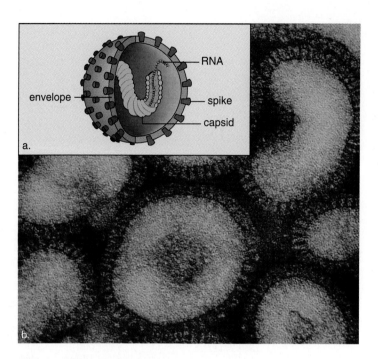

Figure 16.2 **Influenza virus.**
This RNA virus causes influenza in humans. **a.** Diagram of virus. **b.** Electron micrograph of actual virus.

core of nucleic acid—either DNA or RNA (Fig. 16.2). The capsid of a virus that infects animals is often surrounded by an outer envelope, which is derived from the host-cell plasma membrane and contains viral glycoprotein spikes. The glycoprotein spikes within the envelope allow the virus to adhere to plasma membrane receptors. Then the entire virus enters the host cell by endocytosis. When the virus is uncoated, the envelope and capsid are removed. Once the viral genes are free inside the cell, viral components are synthesized and assembled into new viruses (Fig. 16.3). Release of the virus from the host cell occurs by *budding.* During budding, the virus gets its envelope, which consists of host plasma membrane components and glycoproteins that were coded for by viral genes. These glycoproteins are the spikes that allow the virus to attach to the plasma membrane of a host cell. The process of budding does not necessarily kill the host cell.

Some DNA viruses that enter human cells—for example, the papillomaviruses, the herpesviruses, the hepatitis viruses, and the adenoviruses—can undergo a period of latency. While latent, the viral genes are integrated into the host cell chromosome DNA, and they are copied whenever the host cell reproduces. Certain environmental factors, such as ultraviolet radiation, can induce the virus to replicate and bud from the cell. Latent viruses are of special concern because their presence in the host-cell chromosome can alter the cell and make it become cancerous. Some viruses are cancer-producing because they bring with them *oncogenes,* cancer-causing genes.

Table 16.1	Infectious Diseases Caused by Viruses	
Respiratory Tract		**Nervous System**
Common colds		Encephalitis
Flu*		Polio*
Viral pneumonia		Rabies*
Skin Reactions		**Liver**
Measles*		Yellow fever*
German measles*		Hepatitis A, C, and D
Chicken pox*		**Other**
Shingles		Mumps*
Warts		Cancer
Sexually Transmitted		
AIDS		
Genital herpes		
Genital warts		
Hepatitis B*		

*Vaccines available. Yellow fever, rabies, and flu vaccines are given only if the situation requires them. Smallpox vaccinations are no longer required.

Retroviruses are RNA viruses that have a DNA stage in their replication, and these viruses can also become latent. A retrovirus contains a special enzyme called reverse transcriptase that carries out RNA → cDNA transcription. This DNA is called cDNA because it is a DNA copy of the viral RNA genes. During latency, the cDNA is integrated into host DNA and then it leaves when viral reproduction and budding occur. Retroviruses are of interest because human immunodeficiency virus (HIV), which causes AIDS, is a retrovirus. Retroviruses also cause certain forms of cancer.

Viral diseases are controlled by preventing transmission, by administering vaccines, and only recently by administering antiviral drugs as discussed in the Health reading on page 345. Knowing how a particular pathogen is transmitted can help prevent its spread. Covering the mouth and nose when coughing or sneezing helps prevent the spread of a cold, and use of a condom during intercourse helps prevent the transmission of sexually transmitted diseases. Vaccines, which are antibodies administered to stimulate immunity to a pathogen so that it cannot later cause disease, are available for some viral diseases such as polio, measles, mumps, and hepatitis B, a sexually transmitted disease (Table 16.1). Antibiotics, which are designed to interfere with bacterial metabolism, have no effect on viral illnesses. Instead, drugs must be designed that interfere with either entry of the virus into the host cell, replication of its DNA when copies are made, or exit of the virus from the cell.

A number of viruses cause diseases in humans. Some of these are significant sexually transmitted diseases (STDs).

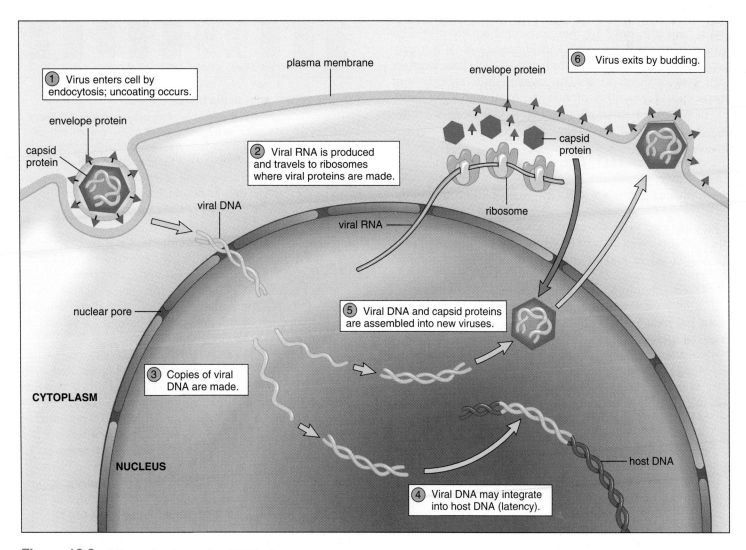

Figure 16.3   Life cycle of an animal DNA virus.
DNA viruses reproduce within a cell and become latent by inserting their DNA into host DNA.

HIV Infections

AIDS (acquired immunodeficiency syndrome) is caused by a human immunodeficiency virus (HIV) that infects specific blood cells. The brief discussion here may be supplemented by the discussion beginning on page 355.

Prevalence of an HIV infection and AIDS

Estimates vary, but as many as 40 million people worldwide may have become infected with HIV, and of these, 12 million have died of AIDS. A new HIV infection is believed to occur every 15 seconds, the majority in heterosexuals. HIV infections are not distributed equally throughout the world. Most infected people live in Africa (66%) where it is believed HIV infections first began, but new infections are now occurring at the fastest rate in Southeast Asia and the Indian subcontinent.

In the United States, HIV infections are concentrated in large cities but are now spreading to small towns and rural areas. New infections are more prevalent among particular groups; African Americans and Hispanics have a higher proportionate number of cases than do Caucasians. Even so, HIV poses a threat to sexually active adults regardless of ethnicity and sexual orientation and to all who inject themselves with drugs intravenously. Figure 16.4 shows the number of AIDS cases in the United States since 1988.

Symptoms of an HIV Infection

The primary hosts for HIV are certain immune cells: macrophages, the white cell type that first encounters the virus; *helper T lymphocytes,* the white cell type that stimulates B lymphocytes to produce antibodies; and cytotoxic T lymphocytes that attack and kill virus-infected cells. HIV progressively alters and often destroys a person's immune system.

During an initial acute phase (called category A), there are usually no symptoms, yet the person is highly infectious. Immediately after infection and before the blood test becomes positive, there are a large number of infectious viruses that could be passed on to another person. After the blood test becomes positive, the person remains well as long as the body stays ahead of the hordes of viruses entering the blood. Millions of helper T lymphocytes are most likely produced each day, and their count is higher than 500 per mm^3.

Several months to several years after infection, the helper T lymphocyte count falls below 500 per mm^3 in untreated persons, and the symptoms of a chronic infection (called category B) begin to appear. Lymph nodes are swollen, severe fatigue is common, and fever with night sweats and diarrhea are present. There may be indications that the virus has entered the brain: loss of memory, inability to think clearly, loss of judgment, and/or depression.

If the individual develops non-life-threatening but recurrent infections, it most likely means that full-blown AIDS

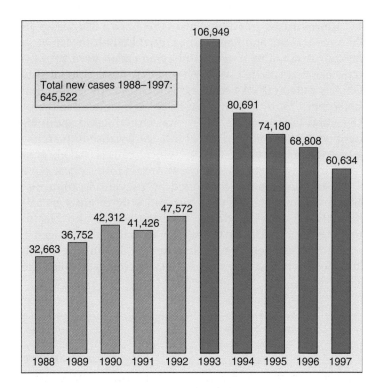

Figure 16.4   AIDS in the United States.
The dramatic increase in the number of reported cases in 1993 was caused by a modification of the definition of AIDS—anyone with an opportunistic disease and less than 200 helper T lymphocytes per mm^3 in their blood is now considered to have AIDS. The recent decline in the number of cases is due to a new multidrug therapy that keeps persons with an HIV infection from developing AIDS.
Source: HIV/AIDS Surveillance Report (7) 2:5, 19. Atlanta: Centers for Disease Control and Prevention.

will occur shortly. One possible infection is thrush, an infection caused by the fungus *Candida albicans,* which is identified by the presence of white spots and ulcers on the tongue and inside the mouth. The *Candida* infection may also spread to the vagina, resulting in a chronic infection there. Another frequent infection is oral and genital herpes, a disease discussed on page 339.

Until recently, the majority of infected persons eventually developed AIDS, which is the final stage (called category C) of an HIV infection. In 1993 the definition of AIDS was broadened to include those persons with a severe depletion of helper T lymphocytes (less than 200 per mm^3 of blood) and/or who have an opportunistic infection. An **opportunistic infection** is one that has the opportunity to occur only because the immune system is severely weakened. Persons with AIDS die from one or more opportunistic diseases, such as *P. carinii* pneumonia and Kaposi's sarcoma, a form of cancer—not from the HIV infection itself.

Treatment for an HIV Infection

There is no cure for AIDS, but a treatment called highly active antiretroviral therapy (HAART) is usually able to stop HIV replication to the extent the virus becomes undetectable

in the blood. Therapy usually consists of two drugs that in-hibit reverse transcriptase and one that inhibits protease, an enzyme needed for viral assembly. This multidrug therapy, when taken according to the manner prescribed, seems to usually prevent mutation of the virus to a resistant strain. Unfortunately, mutation of the virus to an HIV strain resis-tant to all known drugs has been reported—persons who become infected with this strain have no drug therapy avail-able to them.

The sooner drug therapy begins after infection, the bet-ter the chances that the immune system will not be de-stroyed by HIV. And medication must be continued indefinitely. Investigators have found that when HAART is discontinued, the virus rebounds.

Many investigators are working on an AIDS vaccine. Some are trying to develop a vaccine in the traditional way. Traditionally, vaccines are made by weakening or killing a pathogen so that it can be injected into the body without causing disease. Others are working on subunit vaccines that utilize just a single HIV protein as the vaccine. They have found that DNA for this protein can also act as a vac-cine because it enters and causes cells to produce and dis-play the protein at the plasma membrane. The various vaccines are at different stages in their development. Thus far, certain vaccines have increased the immune response in the blood, but none of the vaccines has prevented future infection.

Transmission of HIV

The largest proportion of people with AIDS in the United States are homosexual men, but the proportions attributed to intravenous drug users and heterosexuals are rising. Women now account for 20% of all newly diagnosed cases of AIDS. An infected woman can pass HIV to her unborn children by way of the placenta or to a newborn through breast milk. Transmission at birth can be prevented if the mother takes AZT, and delivers by planned cesarean section.

It is clear that sexual behaviors and drug-related activi-ties are the major means by which AIDS is transmitted in the United States. Essentially, HIV is spread by passing virus-infected T lymphocytes found in body fluids, such as se-men or blood, from one person to another. (Blood and blood products are now tested for the presence of HIV, so the risk of contracting an infection in this manner is now considered very unlikely.) In order to prevent transmission of HIV, you should follow the general advice given in the introduction to this chapter (see p. 339).

More people, many of whom are heterosexuals, are infected with the HIV virus each year. All possible steps should be taken to avoid transmission because treatment is costly and there is no cure.

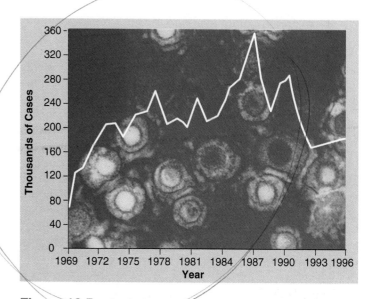

Figure 16.5   Genital warts.
A graph depicting the incidence of new cases of genital warts in the United States from 1969 to 1996 is superimposed on a photomicrograph of human papillomaviruses.

Genital Warts

Human papillomaviruses (HPVs) cause warts, including common warts, plantar warts, and also genital warts, which are sexually transmitted (Fig. 16.5). Over one million per-sons become infected each year with a form of HPV, which causes **genital warts,** but only a portion seek medical help.

Transmission and Symptoms

Quite often, carriers of genital warts do not detect any sign of warts, although flat lesions may be present. When pres-ent, the warts commonly are seen on the penis and foreskin of men and near the vaginal opening in women. A newborn can become infected by passage through the birth canal.

Genital warts are associated with cancer of the cervix, as well as tumors of the vulva, the vagina, the anus, the penis, and the mouth. Some researchers believe that HPVs are in-volved in 90–95% of all cases of cancer of the cervix. Teenagers with multiple sex partners seem to be particularly susceptible to HPV infections. More cases of cancer of the cervix are being seen among this age group.

Treatment

Presently, there is no cure for an HPV infection, but it can be treated effectively by surgery, freezing, application of an acid, or laser burning, depending on severity. If the warts are removed, they may recur. Also, even after treatment, the virus can be transmitted. Therefore, abstinence or use of a condom with vaginal spermicide containing nonoxynol-9 is necessary to prevent the spread of genital warts.

Genital warts is a prevalent STD associated with cervical cancer in women.

Herpes Infections

The herpesviruses cause various illnesses. Chicken pox and shingles are caused by a type of herpesvirus called the varicella-zoster virus, and mononucleosis is caused by another type of herpesvirus called the Epstein-Barr virus. These conditions do not appear to be spread by sexual contact. The herpes simplex viruses that do cause sexually transmitted disease are large DNA viruses that infect the mucosal linings of the body such as the mouth and vagina. There are two types of herpes simplex viruses: type 1 usually causes cold sores and fever blisters, while type 2 more often causes genital herpes. Cross-over infections do occur, however. That is, type 1 has been known to cause a genital infection, while type 2 has been known to cause cold sores and fever blisters.

Cold Sores and Fever Blisters

Cold sores are usually caused by herpes simplex virus type 1. Infection with this virus usually occurs during childhood with symptoms of a mild upper respiratory illness. In some cases, however, painful blisters may appear in the throat, on or below the tongue, and on other mouth parts, including the lips. The sores heal, but the virus remains latent in ganglia outside the central nervous system and spinal cord. Thereafter, a recurrence of symptoms takes the form of a cold sore around the lips. A high fever, stress, colds, menstruation, or exposure to sunlight seems to trigger a reactivation of the virus. It is important to realize that herpes lesions are infectious for at least three to four days until the sores begin to heal. Contact with the sores or any contaminated object can cause the virus to be transmitted. Oral/genital contact can lead to a genital herpes infection.

Genital Herpes

Genital herpes, usually caused by the herpes simplex virus type 2, is as prevalent as genital warts, with over a million or more new cases appearing each year (Fig. 16.6). At any one time, millions of persons could be having recurring symptoms.

Persons usually get infected with herpes simplex virus type 2 when they are adults. In some people there are no symptoms. Or there may be a tingling or itching sensation before blisters appear on the genitals (within 2–20 days). Once the blisters rupture, they leave painful ulcers that may take as long as three weeks or as little as five days to heal. The blisters may be accompanied by fever, pain on urination, swollen lymph nodes in the groin, and in women, a copious vaginal discharge.

After the ulcers heal, the disease is only latent. Blisters can recur, although usually at less frequent intervals and with milder symptoms. Again, fever, stress, sunlight, and menstruation are associated with a recurrence of symptoms. When there are no symptoms, the virus primarily resides in the ganglia of sensory nerves associated with the affected skin. Although herpes simplex type 2 was formerly thought to be a cause of cervical cancer, this is no longer believed to be the case.

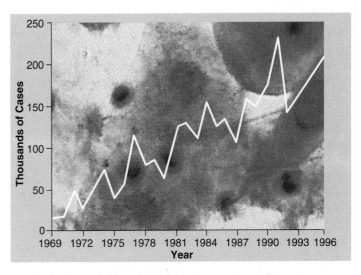

Figure 16.6   Genital herpes.
A graph depicting the incidence of new cases of genital herpes in the United States from 1969 to 1996 is superimposed on a photomicrograph of cells infected with the herpesvirus.

Infection of the newborn can occur if the child comes in contact with a lesion in the birth canal. At worst, the infant can be gravely ill, become blind, have neurological disorders including brain damage, or die. Birth by cesarean section prevents these occurrences; therefore, all pregnant women infected with the virus should be sure to tell their health-care provider.

Transmission

Genital herpes lesions shed infective viruses. Live viruses have occasionally been cultured from the skin of persons with no lesions; therefore, it is believed that the virus can be spread even when there are no visible lesions. Persons who are infected should take extreme care to prevent the possibility of spreading the infection to other parts of their own body, such as the eyes. Certainly, sexual contact should be avoided until all lesions are completely healed, and then the general directions given in the introduction for avoiding STD transfer should be followed (see p. 339).

Treatment

Presently, there is no cure for genital herpes. The drugs acyclovir and vidarabine disrupt viral reproduction. The ointment form of acyclovir relieves initial symptoms, and the oral form prevents the recurrence of symptoms as long as it is being taken. Research is being conducted in an attempt to develop a vaccine.

Herpes simplex viruses cause genital herpes, an extremely infectious disease whose symptoms recur and for which there is no cure.

STDs and Medical Treatment

Treatment of STDs is troublesome at best. Presently, there are no vaccines available except for a hepatitis B infection, and to date, not many persons have been inoculated. Unfortunately, there is no treatment once a hepatitis B infection has occurred. The antiviral drugs acyclovir and vidarabine are helpful against genital herpes, but they do not cure herpes. Once an individual stops the therapy, the lesions are apt to recur. The current recommended therapy of AIDS includes an expensive combination of drugs that prevent HIV reproduction and curtail its ability to infect other cells.

Antibiotics are used to treat sexually transmitted diseases caused by bacteria. Antibiotics such as penicillin, streptomycin, and tetracycline interfere with metabolic pathways unique to bacteria, and therefore, they are not expected to harm host cells. Still, there are problems associated with antibiotic therapy. Some individuals are allergic to antibiotics, and the reaction may even be fatal. Antibiotics not only kill off disease-causing bacteria, they also reduce the number of beneficial bacteria in the intestinal tract. The use of antibiotics sometimes prevents natural immunity from occurring, leading to the necessity for recurring antibiotic therapy.

Most important, perhaps, is the growing resistance of certain strains of bacteria to a particular antibiotic. Antibiotics were introduced in the 1940s, and for several decades they worked so well it appeared that infectious diseases had been brought under control. However, we now know that bacterial strains can mutate and become resistant to a particular antibiotic. Worse yet, bacteria can swap bits of DNA, and in this way, resistance can pass to other strains of bacteria. Penicillin and tetracycline, long used to cure gonorrhea, now have a failure rate of more than 20% against certain strains of gonococcus. To help prevent resistant bacteria, antibiotics should be administered only when absolutely necessary, and the prescribed therapy should be finished. Also, as a society we have to continue to develop new antibiotics to which resistance has not yet occurred.

An alarming new concern has been the upsurge of tuberculosis in persons with sexually transmitted diseases. Tuberculosis is a chronic lung infection caused by the bacterium *Mycobacterium tuberculosis*. This bacteria is often called the tubercle bacillus (TB) because usually the bacilli are walled off within tubercles that can sometimes be seen in X rays of the lungs. TB can pass from an HIV-infected person to a healthy person more easily than HIV. Sexual contact is not needed—only close personal contact, and therefore, it is possible that TB will now spread from those with STDs to the general populace. Some of the modern strains of TB are resistant to antibiotic therapy; therefore, just as with HIV, research is needed for the development of an effective vaccine.

Prevention is still the best way to manage STDs. Several STDs are primarily transmitted by infected white blood cells that penetrate the single layer of cells which lines the inside of the uterine cervix and uterus. The use of a condom along with a spermicide that contains nonoxynol-9 and the avoidance of oral/genital contact is essential unless you have a monogamous relationship (always the same partner) with someone who is free of STDs and is not an intravenous drug user.

Hepatitis Infections

There are several types of **hepatitis.** Hepatitis A, caused by HAV (hepatitis A virus), is usually acquired from sewage-contaminated drinking water. Hepatitis A can also be sexually transmitted through oral/anal contact.

Hepatitis C, caused by HCV, is called the post-transfusion form of hepatitis. This type of hepatitis, which is usually acquired by contact with infected blood, is also of great concern. Infection can lead to chronic hepatitis, liver cancer, and death.

Hepatitis E, caused by HEV, is usually seen in developing countries. Only imported cases, in travelers to the country, or visitors to endemic regions, have been reported in the United States.

Hepatitis B

HBV is a DNA virus that is spread in the same way as HIV, the cause of AIDS—sharing needles by drug abusers and sexual contact between either heterosexuals or homosexual men transmits the disease. Therefore, it is common for an AIDS patient to also have an HBV infection. Also, like HIV, HBV can be passed from mother to child by way of the placenta.

Only about 50% of infected persons have flulike symptoms, including fatigue, fever, headache, nausea, vomiting, muscle aches, and dull pain in the upper right of the abdomen. Jaundice, a yellowish cast to the skin, can also be present. Some persons have an acute infection that lasts only three to four weeks. Others have a chronic form of the disease that leads to liver failure and a need for a liver transplant.

Since there is no treatment for an HBV infection, prevention is imperative. The general directions given on page 339 should be followed, but inoculation with the HBV vaccine is the best protection. The vaccine, which is safe and does not have any major side effects, is now on the list of recommended immunizations for children.

Hepatitis B is an infection that can lead to liver failure. Because it is spread in the same way as AIDS, many persons are infected with both viruses at the same time.

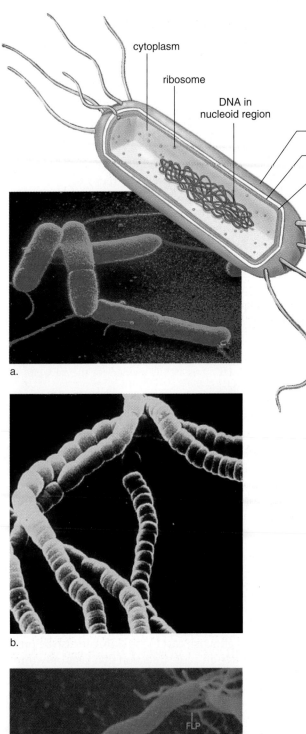

a.

b.

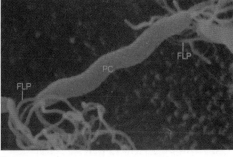

c.

Figure 16.7   Bacteria.

Bacteria occur as **(a)** a bacillus (rod shape), **(b)** a coccus (round shape), and **(c)** a spirillum. The structure of a bacterium is also shown in **(a)**.

16.2 Bacterial in Origin

Although **bacteria** are generally larger than viruses, they are still microscopic. As a result, it is not always obvious that they are abundant in the air, water, and soil and on most objects. It has even been suggested that the combined weight of all bacteria would exceed that of any other type of organism on earth. Bacteria have long been used by humans to produce various products commercially. Chemicals, such as ethyl alcohol, acetic acid, butyl alcohol, and acetone, are produced by bacteria. Bacterial action is involved in the production of butter, cheese, sauerkraut, rubber, cotton, silk, coffee, wine, and cocoa. By means of genetic engineering, bacteria are now used to produce human insulin and interferon, as well as other types of proteins. Even certain antibiotics are produced by bacteria.

Bacteria occur in three basic shapes: rod (bacillus); round, or spherical (coccus); and spiral (a curved shape called a spirillum) (Fig. 16.7). Human cells contain a membrane-bound nucleus and several kinds of membranous organelles; therefore, they are called eukaryotic (true nucleus) cells. Bacterial cells lack a membrane-bound nucleus and the other membranous organelles typical of human cells; therefore, they are called prokaryotic (before nucleus) cells. A nucleoid region contains their DNA, and a cytoplasm contains the enzymes of their many metabolic pathways. Bacteria are enclosed in a cell wall, and some are also surrounded by a polysaccharide or polypeptide capsule that inhibits destruction by the host. Motile bacteria often have flagella (sing., flagellum).

Bacteria usually reproduce asexually by **binary fission.** First, the single chromosome duplicates, and then the two chromosomes move apart into separate areas. Next, the plasma membrane and cell wall grow inward and partition the cell into two daughter cells, each of which has its own chromosome (Fig. 16.8). Under favorable conditions, growth may be very rapid, with cell division occurring as fast as every 20 minutes in *E. coli.* When faced with unfavorable environmental conditions, some bacteria can form **endospores.** During spore formation, the cell shrinks, rounds up within the former plasma membrane, and secretes a new, thicker wall inside the old one. Endospores are amazingly resistant to extreme temperatures, drying out, and harsh chemicals, including acids and bases. When conditions are again suitable for growth, the spore absorbs water, breaks out of the inner shell, and becomes a typical bacterial cell capable of reproducing.

Most bacteria are free-living organisms that perform many useful services in the environment. They are able to live in just about any habitat and use just about any substance as a food source because of their wide metabolic ability. Most bacteria are decomposers—they break down dead organic matter in the environment by secreting digestive enzymes, and then they absorb the nutrient molecules.

Many bacteria are *symbiotic,* that is, they live in close as-

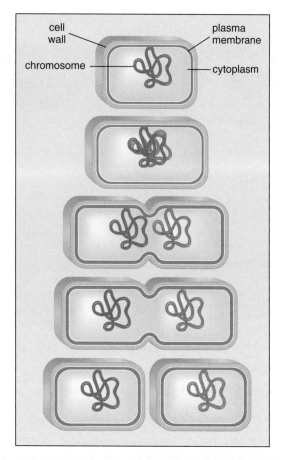

Figure 16.8 **Bacterial reproduction.**
When bacteria reproduce, the single chromosome is attached to the plasma membrane, where it is replicating. As the plasma membrane and the cell wall lengthen, the two chromosomes separate. Once fission has taken place, each bacterium has its own chromosome.

Table 16.2	Some Infectious Diseases Caused by Bacteria	
Respiratory Tract		**Nervous System**
Strep throat		Tetanus*
Pneumonia*		Botulism
Whooping cough*		Meningitis
Tuberculosis*		**Digestive Tract**
Skin Reactions		Food poisoning
Staph (pimples and boils)		Dysentery
Other		Cholera*
Gas gangrene* (wound infections)		**Sexually Transmitted**
Diphtheria*		Gonorrhea
Typhoid fever*		Chlamydia
		Syphilis

*Vaccines are available. Tuberculosis vaccine is not used in this country. Typhoid fever, cholera, and gas gangrene vaccines are given if the situation requires it. Others are routinely given.

sociation with another species. There are a number of bacteria that are normal inhabitants of our bodies. These bacteria, called the *normal microbial flora,* cause disease only when the opportunity arises. The bacterium *Escherichia coli,* which lives in our intestines, breaks down remains that we have not digested. The microflora of our intestines, including *E. coli,* provide us with the vitamins thiamine, riboflavin, pyridoxine, B_{12}, and K. The last two vitamins are necessary for the production of blood components. *E. coli* causes cystitis if it gets into the urinary tract. And *Staphylococcus aureus,* which lives on our skin, causes an infection if it gets into a wound.

The diseases listed in Table 16.2 are caused by bacteria that are ordinarily pathogenic; that is, they usually do cause an illness. Most bacteria are aerobic, and they require a supply of oxygen just as we do, but several human diseases are caused by anaerobic bacteria. Among them are botulism, gas gangrene, and tetanus, which are listed in Table 16.2.

Diseases caused by pathogens are termed communicable because the pathogen must be transmitted to the new host. Pathogens are transferred by a variety of mechanisms,

including food and water, human or animal bites, contact, and aerosols (droplets in the air). Sexually transmitted diseases require intimate contact between persons for their successful transmission.

To be a pathogen, a bacterium must

- have an ability to pass from one host to the next
- penetrate into the host's tissues
- withstand the host's defense mechanisms
- induce illness in the host

Bacterial diseases are controlled by preventing transmission, by administering vaccines, and by antibiotic therapy. **Antibiotics** are medications that kill bacteria by interfering with one of their unique metabolic pathways. Antibiotic therapy is highly successful if it is carried out in the recommended manner. Otherwise, as discussed in the Health reading on page 345, resistant strains can develop. Unfortunately, there are no vaccines for STDs caused by bacteria; therefore, preventing transmission becomes all the more important. Abstinence or monogamous relations (always the same partner) with someone who is free of an STD will prevent transmission. Otherwise, the use of a condom with a vaginal spermicide containing nonoxynol-9 and avoidance of oral/genital contact is recommended.

Bacteria lack a membrane-bound nucleus and other membranous organelles found in eukaryotic cells, but they are capable of an independent existence. Most are free-living, but a few cause human diseases that can often be cured by antibiotic therapy.

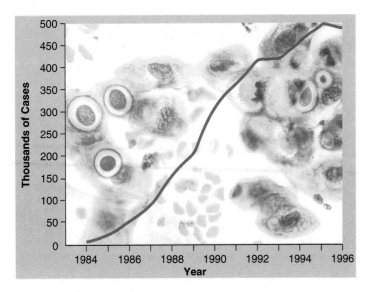

Figure 16.9 **Chlamydial infection.**
A graph depicting the incidence of new cases of chlamydia in the United States from 1984 to 1996 is superimposed on a photomicrograph of a cell containing different stages of the organism.

Chlamydia

Chlamydia is named for the tiny bacterium that causes it (*Chlamydia trachomatis*). For years, chlamydiae were considered to be more closely related to viruses than to bacteria, but today it is known that these organisms are cellular. Even so, they are obligate parasites due to their inability to produce ATP molecules. After chlamydia enters a cell by endocytosis, the life cycle occurs inside the endocytic vacuole, which eventually bursts and liberates many new infective chlamydiae.

New chlamydial infections are more numerous than any other sexually transmitted disease. From 1984 through 1996, reported incidences of chlamydia increased dramatically, from less than 50,000 cases per year to about 500,000 cases. Some estimate that the actual incidence could be as high as 6 million new cases per year. For every reported case in men, more than five cases are detected in women (Fig. 16.9). This is mainly due to increased detection of asymptomatic infections through screening. The low rates in men suggest that many of the sex partners of women with chlamydia are not diagnosed or reported.

Symptoms

Chlamydial infections of the lower reproductive tract usually are mild or asymptomatic, especially in women. About 8–21 days after infection, men may experience a mild burning sensation on urination and a mucoid discharge. Women may have a vaginal discharge along with the symptoms of a urinary tract infection. Unfortunately, a physician mistakenly may diagnose it as a gonorrheal or urinary infection and prescribe the wrong type of antibiotic, or the person may never seek medical help. In these instances, there is a particular risk of the infection spreading from the cervix to the oviducts so that **pelvic inflammatory disease (PID)** results. This very painful condition can result in a blockage of the oviducts, with the possibility of sterility or ectopic pregnancy (Fig. 16.10).

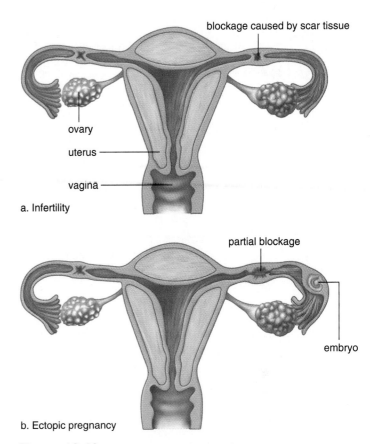

a. Infertility

b. Ectopic pregnancy

Figure 16.10 **Pelvic inflammatory disease.**
Following a chlamydial infection, the oviducts can be **(a)** completely blocked by scar tissue so that infertility results or **(b)** partially blocked so that fertilization occurs but the embryo is unable to pass to the uterus. The growing embryo can cause the oviduct to burst.

Some believe that chlamydial infections increase the possibility of premature and stillborn births. If a newborn comes in contact with chlamydia during delivery, pneumonia or inflammation of the eyes can result. Erythromycin eyedrops at birth prevent this occurrence. If a chlamydial infection is detected in a pregnant woman, treatment with erythromycin should be used during pregnancy.

Diagnosis and Treatment

New and faster laboratory tests are now available for detection of a chlamydial infection. Their expense sometimes prevents public clinics from using them, however. These criteria could help physicians decide which women should be tested: no more than 24 years old; having had a new sex partner within the preceding two months; having a cervical discharge; bleeding during parts of the vaginal exam; and using a nonbarrier method of contraception. A chlamydial infection is cured with the antibiotics tetracycline, doxycycline, and azithromycin; therefore, treatment should begin immediately.

PID and sterility are possible effects of a chlamydial infection in women. This condition may accompany a gonorrheal infection, discussed next.

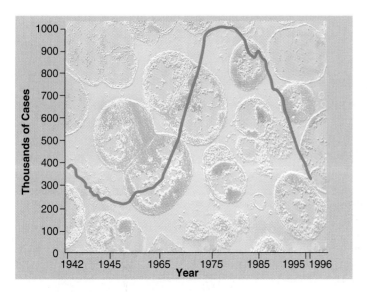

Figure 16.11   Gonorrhea.
A graph depicting the incidence of new cases of gonorrhea in the United States from 1942 to 1996 is superimposed on a photomicrograph of a urethral discharge from an infected male. Gonorrheal bacteria (*Neisseria gonorrhoeae*) occur in pairs; for this reason, they are called diplococci.

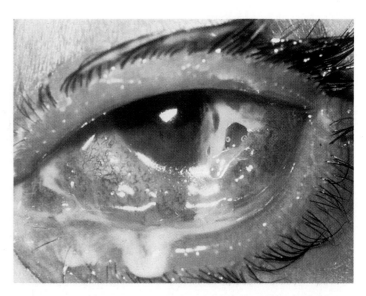

Figure 16.12   Secondary sites for a gonorrheal infection.
Gonorrhea infection of the eye is possible whenever the bacterium comes in contact with the eyes. This can happen when the newborn passes through the birth canal. Manual transfer from the genitals to the eyes is also possible.

Gonorrhea

Gonorrhea is caused by the bacterium *Neisseria gonorrhoeae,* which is a diplococcus, meaning that generally there are two spherical cells in close proximity. The incidence of gonorrhea has been declining since an all-time high in 1978 (Fig. 16.11). However, it can be noted that gonorrhea rates among African Americans is 28 times greater than the rate among Caucasians. Also, women using the birth-control pill have a greater risk of gonorrhea because hormonal contraceptives cause the genital tract to be more receptive to pathogens.

Symptoms

The diagnosis of gonorrhea in men is not difficult as long as they display typical symptoms (as many as 20% of men may be asymptomatic). The patient complains of pain on urination and has a milky urethral discharge three to five days after contact. In women, the bacteria may first settle within the vagina or near the cervix, from which they may spread to the oviducts. Unfortunately, the majority of women are asymptomatic until they develop severe pains in the abdominal region due to *PID (pelvic inflammatory disease)* (Fig. 16.10). PID from gonorrhea is especially apt to occur in women using an IUD (intrauterine device) as a birth-control measure.

PID from either a chlamydial or gonorrheal infection affects about one million women a year in the United States. The total number of ectopic pregnancies in the United States in 1992 was estimated at about 20 cases per 1,000 pregnancies, and many of these were most likely due to PID.

Gonorrhea proctitis is an infection of the anus, with symptoms including anal pain, and blood or pus in the feces. Oral/genital contact can cause infection of the mouth, throat, and the tonsils. Gonorrhea can spread to internal parts of the body, causing heart damage or arthritis. If, by chance, the person touches infected genitals and then his or her eyes, a severe eye infection can result (Fig. 16.12).

Eye infection leading to blindness can occur as a baby passes through the birth canal. Because of this, all newborn infants receive erythromycin eyedrops as a protective measure.

Transmission and Treatment

The chances of getting a gonorrheal infection are good. Women have 50–60% risk, while men have a 20% risk of contracting the disease after even a single exposure to an infected partner. Therefore, the preventive measures listed in the introduction on page 339 should be followed.

Blood tests for gonorrhea are being developed, but in the meantime, it is necessary to diagnose the condition by microscopically examining the discharge of men or by growing a culture of the bacterium from either the male or the female to positively identify the organism. Because there is no blood test, it is very difficult to recognize asymptomatic carriers, who are capable of passing on the condition without realizing it. If the infection is diagnosed, gonorrhea can usually be cured using the antibiotics penicillin or tetracycline; however, resistance to antibiotic therapy is becoming more common. In 1994, resistance was noted in as many as 40% of strains tested.

Gonorrheal infections are presently on the decline; however, resistance to antibiotic therapy is becoming increasingly prevalent.

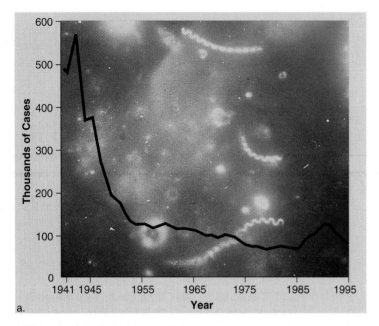

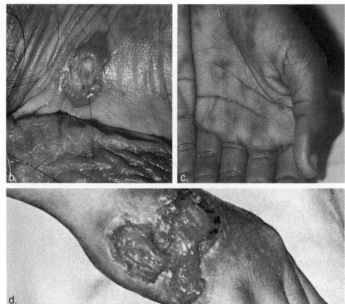

Figure 16.13 **Syphilis.**

a. A graph depicting the incidence of total cases of syphilis in the United States from 1941 to 1995. **b.** First stage of syphilis is a chancre where the bacterium enters the body. **c.** Second stage is a body rash that occurs even on the palms of the hands and soles of the feet. **d.** In the tertiary stage, gummas may appear on skin or internal organs.

Syphilis

Syphilis is caused by a bacterium called *Treponema pallidum,* an actively motile, corkscrewlike organism that is classified as a spirochete. The number of new cases of syphilis in 1995 were the fewest reported in the United States since 1977 (Fig. 6.13).

Syphilis has three stages, which can be separated by latent stages in which the bacteria are not multiplying. During the primary stage, a hard chancre (ulcerated sore with hard edges) indicates the site of infection. The chancre can go unnoticed, especially since it usually heals spontaneously, leaving little scarring. During the secondary stage, proof that bacteria have invaded and spread throughout the body is evident when the individual breaks out in a rash. Curiously, the rash does not itch and is seen even on the palms of the hands and the soles of the feet. There can be hair loss and infectious gray patches on the mucous membranes, including the mouth. These symptoms disappear of their own accord.

Not all cases of secondary syphilis go on to the tertiary stage. Some spontaneously resolve the infection, and some do not progress beyond the secondary stage. During a tertiary stage, which lasts until the patient dies, syphilis may affect the cardiovascular system, and weakened arterial walls (aneurysms) are seen, particularly in the aorta. In other instances, the disease may affect the nervous system. The patient may become mentally impaired, blind, walk with a shuffle, or show signs of insanity. Gummas, large destructive ulcers, may develop on the skin or within the internal organs. Congenital syphilis is caused by syphilitic bacteria

crossing the placenta. The child is stillborn, or born blind, with many other possible anatomical malformations.

Diagnosis and Treatment

Diagnosis of syphilis can be made by blood tests or by microscopic examination of fluids from lesions. One blood test is based on the presence of reagin, an antibody that appears during the course of the disease. Currently, the most used test is Rapid Plasma Reagin (RPR), which is used for screening large numbers of test serums. Because the blood tests can give a false positive, they are followed up by microscopic detection. Dark-field microscopic examination of fluids from lesions can detect the living organism. Alternately, test serum is added to a freeze-dried *T. pallidum* on a slide. Any antibodies present that react to *T. pallidum* are detected by fluorescent-labeled antibodies to human gamma globulin. The organism will now fluoresce when examined under a fluorescence microscope.

Syphilis is a devastating disease. Control of syphilis depends on prompt and adequate treatment of all new cases; therefore, it is crucial for all sexual contacts to be traced so that they can be treated. The cure for all stages of syphilis is some form of penicillin. To prevent transmission, the general instructions given in the introduction (see p. 339) should be faithfully followed.

Syphilis is a devastating sexually transmitted disease, which is curable by antibiotic therapy. The chance of it developing resistance is always a threat.

16.3 Other Sexually Transmitted Diseases

Bacteria are organisms classified in the kingdom Monera. Three other kingdoms also contain organisms that cause sexually transmitted diseases. They are kingdoms Protista, Fungi, and Animalia.

Protozoa, in the kingdom Protista, are unicellular organisms that are said to be animal-like because their single cell functions as does an animal cell. Most types of protozoa have their own mode of locomotion. There are some that move by extensions of the cytoplasm, called pseudopods; some that move by cilia; and some that have flagella. Protozoa are often found in an aquatic environment such as freshwater ponds, and the ocean simply teems with them. All protozoa require an outside source of nutrients, but only the parasitic ones take this nourishment from a host. Parasitic protozoa cause malaria, amoebic dysentery, and giardiasis, among other diseases. Also, the protozoan called *Trichomonas vaginalis*, as discussed on this page, is transmitted sexually and causes one form of vaginitis in women.

Most *fungi* are nonpathogenic decomposers, as are bacteria, but a few are parasitic. Usually, the body of a fungus is made up of filaments called hyphae, but yeasts are an exception since they are single cells. Almost everyone is familiar with yeasts. Certain types are used to make bread rise and to produce wine, beer, and whiskey because of their ability to ferment sugar. Antibiotic therapy and taking the birth-control pill can disrupt the normal balance of flora that live on mucous membranes, and the result can be a yeast infection. *Candida albicans* is a yeast that can cause diaper rash, thrush (a mouth infection), and a form of vaginitis in women.

Although we tend to think of organisms in the kingdom Animalia as being macroscopic, actually some are quite small, even microscopic. An animal is a motile, multicellular organism with specialized tissues. Since animals require an outside source of nutrients, some are parasitic. Various types of worms, such as pinworms, tapeworms, and hookworms, are well-known parasites that inhabit the body of other animals. Lice are insects that can infect the hair of humans. The head louse is well known for infecting the hair of schoolchildren; the crab louse, as discussed on this page, is sexually transmitted.

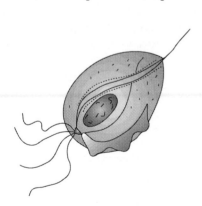

Figure 16.14 *Trichomonas vaginalis.*
Vaginitis is often caused by *Trichomonas vaginalis*, a pear-shaped protozoan that uses flagella to move about.

Vaginitis

Women very often have *vaginitis*, or infection of the vagina, which is caused by the flagellated protozoan *Trichomonas vaginalis* (Fig. 16.14) or the yeast *Candida albicans*. The protozoan infection causes a frothy white or yellow foul-smelling vaginal discharge accompanied by itching, and the yeast infection causes a thick, white, curdy vaginal discharge, also accompanied by itching. *Trichomonas* is most often acquired through sexual intercourse, and an asymptomatic partner is usually the reservoir of infection. *Candida albicans,* however, is a normal organism found in the vagina; its growth simply increases beyond normal under certain circumstances. For example, women taking the birth-control pill or who have been on antibiotic therapy are sometimes prone to yeast infections.

Pubic Lice (Crabs)

The parasitic crab louse *Phthirus pubis* (an insect) takes its name from its resemblance to a small crab (Fig. 16.15). Crabs, as it is commonly called, can be contracted by direct contact with an infested person or by contact with his or her clothing or bedding. The females lay their eggs around the base of the hair, and these eggs hatch within a few days to produce a larger number of animals that suck blood from their host and cause severe itching, particularly at night. The pubic hair, underarm hair, and even the eyebrows can be infected.

In contrast to most types of sexually transmitted diseases, self-diagnosis and self-treatment that does not require shaving is possible. Usually an infested person will find and identify the adult lice and/or their numerous eggs on or near pubic hairs. Medications such as lindane are applied to the infested area, and all sexual partners need to be treated also. Undergarments, sheets, and night clothing should be washed by machine in hot water.

Aside from viruses and bacteria, a protozoan and yeast can cause vaginitis, and an animal is the cause of a condition called crabs.

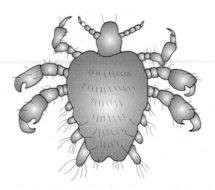

Figure 16.15 *Phthirus pubis.*
This parasitic crab louse infects the pubic hair of humans.

Bioethical Issue

Carriers of disease are persons who do not appear to be ill but who can pass on an infectious disease. Carriers of sexually transmitted diseases can pass on an infection to their partners, and even sometimes to people who have never had sexual relations with them. Hepatitis C and HIV are transmitted by blood to blood contact, as when drug abusers share needles.

The only way society can protect itself is to identify carriers and remove them from areas or activities where transmission of the pathogen is most likely. Sometimes it's difficult to identify activities that might pass on a pathogen like HIV. A few people believe that they have acquired HIV from their dentists, and while this is generally believed to be unlikely, medical personnel are still required to identify themselves when they are carriers of HIV. Magic Johnson is a famous basketball star who is HIV positive. When he made it known that he was a carrier, some of his teammates refused to play basketball with him and he had no choice but to retire. Later, Johnson did play again. There are other sports in which transmission of HIV is more likely. The Centers for Disease Control did a statistical study to try to figure the odds of acquiring HIV from another football player. They figured that the odds were 1 in 85 million.

The odds might be higher for boxing, a bloody sport. When two brothers, one of whom had AIDS, got into a vicious fight, the infected brother repeatedly bashed his head against his brother's. Both men bled profusely and soon after, the previously uninfected brother tested positive for the virus. The possibility of transmission of HIV in the boxing ring has caused several states to require boxers to undergo routine HIV testing. If they are HIV positive, they can't fight.

Questions

1. Should all people who are HIV positive identify themselves? Why or why not? By what method would they identify themselves at school, at work, and other places?
2. What type of contact would you have with an HIV positive individual? Would you play on the same team? Would you have sexual relations with an HIV positive individual if a condom was used? Discuss.
3. Is it ethical to make an HIV positive individual feel stigmatized? Discuss.

Summarizing the Concepts

16.1 Viral in Origin

Human beings are subject to sexually transmitted diseases (STDs). Some of these are viral in origin. Viruses are tiny particles that always have an outer capsid of protein and an inner core of nucleic acid that may be DNA or RNA. Viruses reproduce within living cells; upon entry, uncoating occurs; viral DNA is replicated and capsid proteins are made. Following assembly, viruses bud from the cell. Retroviruses are RNA viruses that have a DNA stage. Latency occurs when viral DNA integrates into the host chromosome, and latency can lead to cancer.

HIV, the virus that causes AIDS, is a retrovirus that can integrate itself into the DNA of the host cell. When active, HIV eventually causes a collapse of the immune system. Then, opportunistic diseases occur and lead to the death of the individual. The number of AIDS cases has steadily increased and is now spreading through the heterosexual population.

Genital warts are caused by human papillomavirus (HPV), which is a cuboidal DNA virus that reproduces in the nuclei of skin cells. Genital warts is a disease characterized by warts on the penis and foreskin in men and near the vaginal opening in women. Genital herpes is caused by herpes simplex virus: type 1 usually causes cold sores and fever blisters, while type 2 often causes genital herpes. Genital herpes is a disease that causes painful blisters on the genitals.

Hepatitis B, which is spread in the same manner as AIDS, can lead to liver failure. There is a vaccine for hepatitis B, and everyone is encouraged to be inoculated.

Medications have been developed to control AIDS and genital herpes, but there is no cure, and no vaccines are available. Of all the viral STDs, a vaccine for only hepatitis B exists at the present time.

16.2 Bacterial in Origin

Some STDs are bacterial in origin. Bacteria are cells that lack a nucleus and the membranous organelles typical of animal cells. Bacteria reproduce by binary fission and may form endospores. Most bacteria are aerobic decomposers that perform useful services for humans, but a few cause disease. Antibiotics are available to cure bacterial STDs unless a resistant strain is involved.

Chlamydia is caused by a tiny bacterium of the same name. These tiny bacteria develop inside phagocytic vacuoles that eventually burst and liberate infective chlamydiae. Chlamydia can be asymptomatic but can produce symptoms of a urinary tract infection. Both chlamydia and gonorrhea can result in PID, leading to sterility and ectopic pregnancy. Gonorrhea is caused by the bacterium *Neisseria gonorrhoeae*, a diplococcus. Gonorrhea may not cause symptoms, particularly in women, but in men there may be painful urination and a thick, greenish-yellow discharge. Syphilis is caused by a bacterium called *Treponema pallidum*, an actively motile, corkscrewlike organism. Syphilis is a systemic disease that should be cured in its early stages before deterioration of the nervous system and cardiovascular system possibly takes place.

STDs seriously affect one's health, and all possible steps should be taken to prevent their spread. Abstinence, a monogamous relationship, or use of a condom with a vaginal spermicide that contains nonoxynol-9 can help prevent their transmission.

16.3 Other Sexually Transmitted Diseases

Other STDs are caused by a protozoan (vaginitis), a fungus (yeast infection), or insect (crab lice infestation).

Studying the Concepts

1. Describe the structure and life cycle of a DNA virus. 340
2. Describe the cause and symptoms of an HIV infection. 342–43
3. Among which groups of society is AIDS now increasing most rapidly? How might transmission be prevented? 342
4. Give the cause and symptoms for genital warts, genital herpes, and a heptitis B infection. 343–45
5. Discuss the treatment for STDs caused by viruses. How might these viruses be prevented from spreading? 345
6. State the three shapes of bacteria, and describe the structure of a prokaryotic cell. 346

7. Describe the symptoms and results of a chlamydial infection and gonorrhea in men and in women. What is PID, and how does it affect reproduction? 348–49
8. Describe the three stages of syphilis. 350
9. How does the newborn acquire an infection of AIDS, herpes, chlamydia, gonorrhea, or syphilis? What effects do these infections have on infants? 343, 344, 348, 349, 350
10. Describe the symptoms of vaginitis and pubic lice. 351

Testing Your Knowledge of the Concepts

In questions 1–4, match the disease to the descriptions.
a. HIV
b. genital warts
c. genital herpes
d. syphilis
e. gonorrhea

_____ 1. Often appears with chlamydia, and both conditions can cause PID.

_____ 2. Multidrug therapy is successful but a resistant strain has now appeared.

_____ 3. Painful ulcers can lead to passage from mother to child.

_____ 4. Has a tertiary stage in which cardiovascular and nervous system involvement appears.

In questions 5–7, indicate whether the statement is true (T) or false (F).

_____ 5. Intravenous drug users who share needles cannot get an HIV infection, because the disease is sexually transmitted.

_____ 6. There is a vaccine available for hepatitis B, a disease which can lead to liver failure.

_____ 7. Penicillin therapy will cure both gonorrhea and syphilis.

In questions 8–18, fill in the blanks.

8. Bacteria reproduce asexually by _____.
9. All viruses have an inner core of _____ and a capsid of _____.
10. AIDS is caused by a type of RNA virus known as a _____.
11. Herpes simplex virus type 1 causes _____, and type 2 causes _____.
12. Bacterial cells have one of three shapes: _____, _____, and _____.
13. Women are often asymptomatic for a chlamydial infection until they develop _____.
14. The use of a _____ can help reduce the risk of acquiring an STD.

15. In the tertiary stage of syphilis, there may be large sores called _____.
16. These three sexually transmitted diseases are usually curable by antibiotic therapy: _____, _____, and _____.
17. Women who take the birth-control pill are more likely to acquire vaginitis due to a _____ infection.
18. Viruses can only reproduce in a _____ host cell.
19. Compose a sentence for each box to explain the life cycle of a DNA virus.

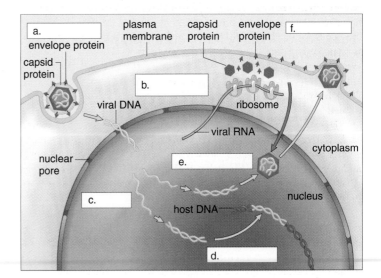

Applying Your Knowledge to the Concepts

These questions pertain to sexually transmitted diseases.

1. In the United States, the total number of STDs of bacterial origin remains about the same each year, but the total number of STDs of viral origin increases each year. Explain the reason for this.

2. Some animal viruses are said to have developed disguises that make it more difficult for the host immune system to recognize them as foreign objects. How is this done?

3. How may the advent of the birth-control pill have resulted in an increase in STDs?

4. If a young female has contracted genital warts, why should she be sure to have a yearly Pap test?

Understanding the Terms

acquired immunodeficiency
   syndrome (AIDS) 342
antibiotic 347
bacteria 346
binary fission 346
chlamydia 348
endospore 346
genital herpes 344
genital warts 343
gonorrhea 349

hepatitis 345
host 340
opportunistic infection 342
pathogen 340
pelvic inflammatory
   disease (PID) 348
retrovirus 341
syphilis 350
virus 340

Match the terms to these definitions:

a. _____ Inflammation of the liver.

b _____ Nonliving, obligate, intracellular parasite consisting of an outer capsid and an inner core of nucleic acid.

c. _____ Disease state of the reproductive organs usually caused by a chlamydial infection or gonorrhea.

d. _____ Organism a parasite lives on or in, to obtain nutrients.

e. _____ Medicine that specifically interferes with bacterial metabolism and in that way cures humans of a bacterial disease.

Applying Technology to the Concepts

Your study of sexually transmitted diseases is supported by these available technologies:

Exploring the Internet

The Mader Home Page provides resources and tools as you study this chapter.

http://www.mhhe.com/biosci/genbio/mader

Dynamic Human 2.0 CD-ROM
Lymphatic System → Clinical Concepts
           → HIV-AIDS

HealthQuest CD-ROM
4 Communicable Diseases

AIDS Supplement

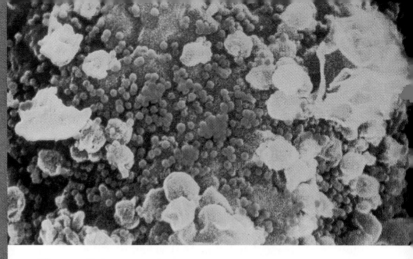

Figure S.1 **HIV (green) budding from an infected T lymphocyte.**

Supplement Concepts

Figure S.1 shows HIV (human immunodeficiency viruses) budding from a T lymphocyte, their primary host. No wonder the immune system falters in a person with an HIV infection. The very cells that orchestrate the immune response are under viral attack. As a society, we are bearing a tremendous expense to treat those infected with multiple drug therapy. Will there instead be a vaccine sometime soon? If so, the immune system would be primed to control the infection before it established itself. There might be something about the virus—mode of transmission or the course of the disease—that will make any type of vaccine ineffective. The burden, it seems, is on the individual. We all must come to realize the importance of our T lymphocytes and take measures to protect them from possible destruction by HIV. The various ways to prevent infection discussed on page 362 should be followed faithfully.

An HIV infection leads to AIDS (acquired immunodeficiency syndrome). The full name of AIDS can be explained in this way: *acquired* means that the condition is caught rather than inherited; *immune deficiency* means that the virus attacks the immune system so there is greater susceptibility to certain infections and cancer; and *syndrome* means that some fairly typical opportunistic infections and cancers usually occur in the infected person. These are the conditions that cause AIDS patients to die.

An HIV infection precedes the occurrence of AIDS by several years. At first, the body is able to deal effectively with the infection, and only later do symptoms arise that permit a diagnosis of AIDS. Therefore, the number of persons diagnosed as having AIDS is much smaller than those that are actually infected with HIV. In the United States, about two million are believed to be infected with the virus, but only about 650,000 individuals have been diagnosed as having AIDS. The U.S. Centers for Disease Control and Prevention (CDC) is now using a new definition for AIDS that allows more persons to be counted as having AIDS. The new definition includes persons with severely impaired immune systems but who have not necessarily come down with a life-threatening disease.

S.I Origin and Scope of the AIDS Pandemic

It's generally accepted that HIV originated in Africa and then spread to the United States and Europe by way of the Caribbean. Just recently, HIV was found in a preserved 1959 blood sample taken from a man who lived in an African country now called the Democratic Republic of the Congo. Even before this discovery, scientists speculated that an immunosuppressive virus may have evolved into HIV during the late 1950s. Perhaps originally the virus was found only in monkeys—there are monkeys in Africa infected with immunosuppressive viruses that could have mutated to HIV after humans ate monkey meat.

British scientists have been able to show that AIDS came to their country perhaps as early as 1959. They examined the preserved tissues of a Manchester seaman who died that year and concluded that he most likely died of AIDS. Similarly, it's thought that HIV entered the United States on numerous occasions as early as the 1950s. But the first documented case is a 15-year-old male who died in Missouri in 1969 with skin lesions now known to be characteristic of an AIDS-related cancer. Doctors froze some of his tissues because they could not identify the cause of death. Researchers also want to test the preserved tissue samples of a 49-year-old Haitian who died in New York in 1959 of the type of pneumonia now known to be AIDS related.

Throughout the 1960s, it was the custom in the United States to list leukemia as the cause of death in immunodeficient patients. Most likely some of these people actually died of AIDS. Since HIV is not extremely infectious, it took several decades for the number of AIDS cases to increase to the point that AIDS became recognizable as a specific and separate disease. The name AIDS was coined in 1982, and HIV was found to be the cause of AIDS in 1983-84.

AIDS most likely originated in the 1950s, but it wasn't until 1983 that HIV was recognized as the cause of AIDS.

Prevalence of AIDS Today

Currently, it's estimated that 6 million people a year, or 16,000 persons a day, come down with AIDS worldwide. Over 2 million people a year die of AIDS, including 460,000 children. To date, as many as 40 million people may have contracted HIV, and almost 12 million have died. AIDS is a pandemic because the disease is prevalent in the entire human population around the globe. Even so, the disease is concentrated in certain countries.

Since AIDS originated in Africa, it is not surprising that most infected people live in Africa (66%), but new infections are now occurring at the fastest rate in Southeast Asia and the Indian subcontinent, areas which now account for 20% of all those infected. The explosion of infections in Southeast Asia and India is attributed to a commercial sex industry. A study conducted in northern Thailand estimated that of newly infected military recruits, most have a history of sexual contact with female prostitutes.

Just as AIDS is not distributed equally around the globe, so the distribution of AIDS in any particular country is also likely to be uneven. In the United States, AIDS is concentrated in certain large cities such as New York City, Los Angeles, Chicago, and Washington, D.C., but AIDS is now spreading to smaller cities and rural areas as well. Similarly, AIDS was at one time largely restricted to the male homosexual community in the United States. Now, however, the number of new HIV infections due to homosexual contact has leveled off and the number due to heterosexual contact is on the rise.

In 1988, Caucasian males accounted for 60% of AIDS cases and black and Hispanic for only 39%. Today, the situation is reversing; 38% of new cases are occurring among Caucasians and 61% are occurring in blacks and Hispanics. One of every 50 African-American males may now be infected with HIV. Hispanics represent only about 15% of all AIDS cases but again have a higher percentage rate of new cases today than do Caucasians. Formerly, too, AIDS was largely restricted to males, but now new infections in females has the greater rate of increase.

There is one more statistic to call to your attention. It used to be that new cases of AIDS were largely found in persons older than 40 years. Now, one-quarter of all new AIDS cases in the United States occur among people younger than 40 years, and the incidence is still rising (Fig. S.2). All teenagers and young adults should be aware that their age group will soon account for most of new HIV infections. It takes several years for this infection to develop into AIDS. As of 1992, AIDS had become the leading cause of death for U.S. men aged 25–44 years and the third leading cause of death among women in this age group.

Certain large cities and minorities have a higher prevalence of AIDS. Young people have an increasing incidence of AIDS compared to older adults in which the incidence has leveled off.

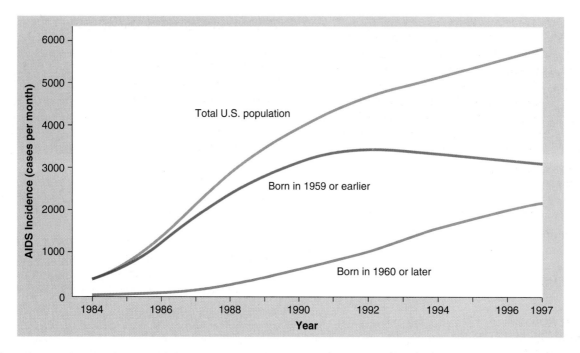

Figure S.2 **Monthly AIDS incidence per age.**
The rate of AIDS incidence is slowing down in the overall U.S. population and for individuals born in or before 1959. The AIDS incidence for individuals born after 1960 continues to rise. Today, teenagers and young adults are very much at risk for newly acquired HIV infections which do not become AIDS until about ten years later.

At-Risk Behaviors

Table S.1 lists the most frequent ways by which HIV is transmitted. HIV is transmitted by sexual contact with an infected person, including vaginal or rectal intercourse and oral/genital contact. Also, needle-sharing among intravenous drug users is high-risk behavior. A less common mode of transmission (and now rare in countries where blood is screened for HIV) is through transfusions of infected blood or blood-clotting factors. Babies born to HIV-infected women may become infected before or during birth, or through breast-feeding after birth.

As mentioned, HIV first spread through the homosexual community in the United States, and in this country male-to-male sexual contact still accounts for the largest percentage of new AIDS cases in the United States. But the *rate* of HIV infections is now rising faster among heterosexuals and intravenous drug users than homosexuals. Currently only 15% of all people with AIDS are women, but they account for 20% of all newly diagnosed cases. Today, new infections are most often seen in minority women. The rise in the incidence of AIDS among women of reproductive age is paralleled by a rise in the incidence of AIDS in children younger than 13. The chance of an infected mother passing HIV to her unborn child is thought to be about 15–25%.

Behaviors that help prevent transmission are discussed later in this supplement. These behaviors should be followed by everyone in order to prevent the spread of AIDS.

New HIV infections are increasing faster among heterosexuals than homosexuals; women now account for nearly 20% of new AIDS cases.

Table S.1	Transmission of HIV-1
Possible Routes	
Homosexual and heterosexual contact	
Intravenous drug use	
Transfusion (unlikely in U.S.)	
Crossing placenta during pregnancy; breast-feeding	
Increasing the Risk	
Promiscuous behavior (large number of partners, sex with a prostitute)	
Drug abuse with needle sharing	
Presence of another sexually transmitted disease	

S.2 Phases of an HIV Infection

Although it is convenient to speak of the AIDS virus, there are at least two viruses that can cause AIDS. HIV-1 is common in the United States and most other countries of the world. But HIV-2 is usually the cause of an infection in West Africa. An HIV-2 infection appears to be less virulent than an HIV-1 infection.

The description below pertains to an HIV-1 infection. The HIV-1 virus primarily infects helper T lymphocytes that are also called CD4 T lymphocytes, or simply CD4 T cells. These cells display a molecule called CD4 on their surface. Helper T cells, you will recall, ordinarily stimulate B lymphocytes to produce antibodies and cytotoxic T cells to destroy cells that are infected with a virus. The Centers for Disease Control and Prevention now recognize three categories of an HIV-1 infection. AIDS is the final stage of an infection.

Category A: Acute Phase

A normal CD4 T cell count is at least 800 cells per cubic millimeter of blood. This first phase of an HIV infection is characterized by a CD4 T cell count of 500 per mm^3 or greater (Fig. S.3). This count is sufficient for the immune system to function normally.

Today investigators are able not only to track the blood level of CD4 T cells, but they can also monitor the viral load. The viral load is the number of HIV particles that are in the blood. At the start of an HIV-1 infection, the virus is replicating ferociously and the killing of CD4 T cells is evident because the blood level of these cells drops dramatically. For a few weeks, however, people don't usually have any symptoms at all. Then, a few (1–2%) do have mononucleosis-like symptoms that may include fever, chills, aches, swollen lymph nodes, and an itchy rash. These symptoms disappear and there are no other symptoms for quite some time. Even so, the person is highly infectious despite a negative HIV blood text. The HIV blood test commonly used at clinics is not yet positive because it tests for the presence of antibodies and not for the presence of HIV itself. This means that HIV can still be transmitted even when the HIV blood test is negative.

After a period of time, the body responds to the infection by increased activity of immune cells and the HIV blood test becomes positive. During this phase the number of CD4 T cells is greater than the viral load (Fig. S.3). But most investigators believe that a great unseen battle is going on. The body is staying ahead of the hordes of viruses entering the blood by producing as many as one to two billion new helper T lymphocytes each day. This is called the "kitchen sink model" for CD4 loss. The sink's faucet (production of new CD4 T cells) and the sink's drain (destruction of CD4 T cells) are wide open. As long as the body's production of new CD4 T cells is able to keep pace with the destruction of these cells by HIV and by cytotoxic T cells, the person has a healthy immune system that can deal with the infection.

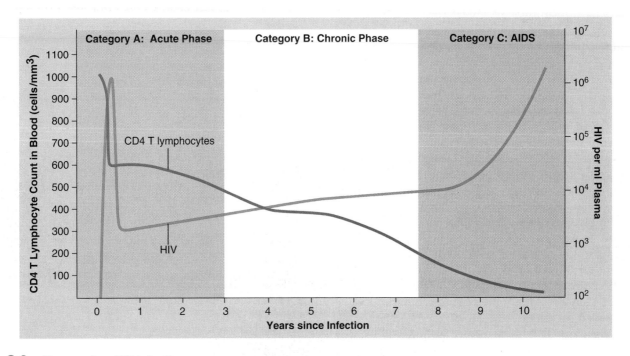

Figure S.3 **Stages of an HIV infection.**
In category A individuals, the number of HIV in plasma rises upon infection and then falls. The number of CD4 T lymphocytes falls, but stays above 400 per mm^3. In category B individuals, the number of HIV in plasma is slowly rising and the number of T lymphocytes is decreasing. In category C individuals, the number of HIV in plasma rises dramatically as the number of T lymphocytes falls below 200 per mm^3.

Category B: Chronic Infection

Several months to several years after infection, an untreated individual will probably progress to category B. During this stage, the CD4 T cell count is 200 to 499 per mm³ of blood and, most likely, symptoms will begin to appear. There will be swollen lymph nodes in the neck, armpits, or groin that persist for three months or more. Other symptoms that indicate category B are severe fatigue not related to exercise or drug use; unexplained persistent or recurrent fevers, often with night sweats; persistent cough not associated with smoking, a cold, or the flu; and persistent diarrhea. Also possible are signs of nervous system impairment, including loss of memory, inability to think clearly, loss of judgment, and/or depression.

When the individual develops non-life-threatening but recurrent infections, it is a signal that full-blown AIDS will occur shortly. One possible infection is thrush, a *Candida albicans* infection that is identified by the presence of white spots and ulcers on the tongue and inside the mouth. The fungus may also spread to the vagina, resulting in a chronic infection there. Another frequent infection is herpes simplex, with painful and persistent sores on the skin surrounding the anus, the genital area, and/or the mouth.

Category C: AIDS

When a person has AIDS, the CD4 T cell count has fallen below 200 per mm³ and the lymph nodes have degenerated. The patient is extremely thin and weak due to persistent diarrhea and coughing, and will most likely develop one of the opportunistic infections. An **opportunistic infection** is one that has the *opportunity* to occur only because the immune system is severely weakened. Persons with AIDS die from one or more of these diseases and not from the HIV infection itself:

- *Pneumocystis carinii* pneumonia. The *lungs* become useless as they fill with fluid and debris due to an infection with a protozoan.
- *Mycobacterium tuberculosis*. This bacterial infection, usually of the *lungs*, is seen more often as an infection of lymph nodes and other organs in patients with AIDS.
- Toxoplasmic encephalitis is caused by a one-cell parasite that ordinarily lives in cats and other animals. Many persons harbor a latent infection in the *brain* or muscle, but in AIDS patients, the infection leads to loss of brain cells, seizures, and weakness.
- Kaposi's sarcoma is an unusual *cancer* of blood vessels, which gives rise to reddish purple, coin-size spots and lesions on the skin.
- Invasive cervical cancer. This *cancer* of the cervix spreads to nearby tissues. This condition was added to the list when AIDS became more common in women.

Although there are newly developed drugs to deal with opportunistic diseases, most AIDS patients are repeatedly hospitalized due to weight loss, constant fatigue, and multiple infections (Fig. S.4). Death usually follows in 2–4 years.

During the acute phase of an HIV infection, the person is highly infectious although there may be no symptoms; during the chronic phase, there may be swollen lymph nodes and various infections; during the last phase, which is called AIDS, the patient usually succumbs to an opportunistic infection.

a. AIDS patient, Tom Moran, July 1987

b. AIDS patient, Tom Moran, early January 1988

c. AIDS patient, Tom Moran, late January 1988

Figure S.4   The course of an AIDS infection.
These photos show the effect of an HIV infection in one individual who progressed through all the stages of AIDS.

S.3 Treatment for HIV

Until a few years ago, an HIV infection almost invariably led to AIDS and an early death. The medical profession was able to treat the opportunistic infections that stemmed from immune failure, but had no drugs for controlling HIV itself. But since late 1995, scientists have gained a much better understanding of the structure of HIV and its life cycle. Now therapy is available that successfully controls HIV replication and keeps patients in the chronic phase of infection so that they do not develop AIDS.

HIV Structure and Life Cycle

HIV are *retroviruses,* meaning that their genetic material consists of RNA instead of DNA. The HIV particle has an envelope acquired when it buds from an infected cell. Inside the HIV capsid are two strands of RNA and three enzymes of interest: reverse transcriptase, integrase, and protease (Fig. S.5).

For HIV-1 to enter a CD4 T cell or any other type of immune cell, an envelope protein, called gp120 (gp for glyco-

protein and 120 for its weight), must first bind with a CD4 receptor in the host-cell plasma membrane ①. Then a change in shape of the gp120 molecule allows it to also bind to a co-receptor. Ordinarily, a CD4 receptor is a binding site for various signaling molecules, and the co-receptor is a binding site for a signaling molecule called a chemokine. Chemokines are cytokines that attract immune cells to the site of an infection.

After binding occurs, the HIV virus fuses with the plasma membrane so that the RNA strands plus the enzymes mentioned enter the host cell ②. Now HIV uses its enzyme called reverse transcriptase to make a DNA copy, called cDNA, of its genetic material ③. This single DNA strand replicates ④ and the viral enzyme integrase splices the double-stranded DNA into a hostchromosome ⑤. The term **HIV provirus** refers to viral DNA integrated into host DNA. HIV is usually transmitted to another person by means of T lymphocytes that contain a provirus. Also, proviruses serve as a reservoir for HIV during drug treatment. Even if drug therapy results in an undetectable viral load, investigators know that there are still proviruses inside infected lymphocytes.

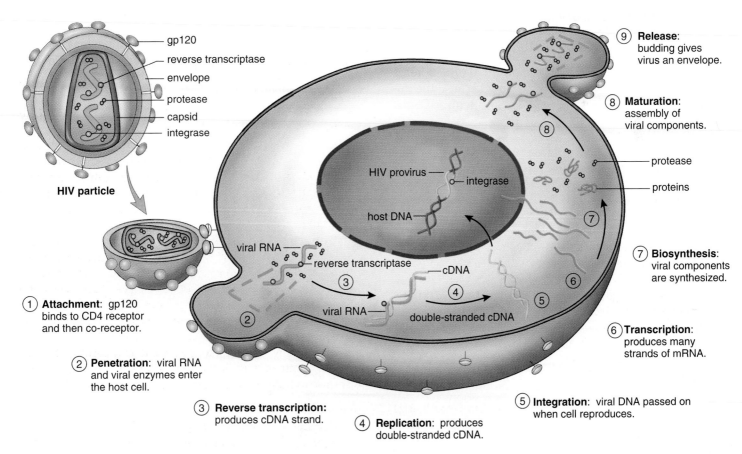

Figure S.5   Reproduction of HIV-1.
HIV-1 is a retrovirus that utilizes reverse transcription to produce cDNA (DNA copy of RNA genes). cDNA integrates into the cell's chromosomes before it reproduces and buds from the cell.

When an infected T cell is activated, perhaps by a new and different infection, the provirus becomes active and directs the production of more viral RNA ⑥. Each strand of viral RNA brings about synthesis of protein, many of which are capsid proteins ⑦. The viral enzyme protease cleaves viral proteins so that they are a size suitable for viral assembly. Assembly of RNA strands and capsids now occurs ⑧. Many viruses then bud from the host cell before it dies ⑨.

The life cycle of an HIV virus includes transmission to a new host. Body secretions, such as semen of an infected male, contain proviruses inside CD4 T lymphocytes. When this semen is discharged into the vagina, rectum, or mouth, infected CD4 T cells migrate through the organ's lining and enter the body. The receptive partner in anal-rectal intercourse appears to be most at risk because the lining of the rectum is a thin, single-cell layer. CD4 macrophages present in tissues are believed to be the first infected when proviruses enter the body. When these macrophages move to the lymph nodes, HIV begins to infect CD4 T cells. HIV can hide out in local lymph nodes for some time; but eventually, the lymph nodes degenerate, and large numbers of HIV enter the general bloodstream. Now the viral load begins to increase; and when it exceeds the CD4 T cell count, the individual progresses to the final stage of an HIV infection.

HIV-1 is a retrovirus that infects immune cells, such as helper T lymphocytes, carrying a CD4 receptor and a co-receptor for a chemokine. HIV is transmitted as a provirus inside infected CD4 cells.

Drug Therapy

There is no cure for AIDS, but a treatment called highly active antiretroviral therapy (HAART) is usually able to stop HIV replication to such an extent the viral load becomes undetectable. Even so, investigators know that about one million viruses are still present, including those that exist only as inactive proviruses.

HAART utilizes a combination of two types of drugs: nucleotide analogs and protease inhibitors. The well-publicized drug called AZT and several others are nucleotide analogs. Analogs are chemically altered forms, so that when reverse transcriptase chooses them instead of a normal nucleotide, reverse transcription stops and viral DNA is not produced. Protease inhibitors block the action of the viral enzyme called protease, which is required for normal viral assembly. When HIV protease is blocked, the resulting virus lacks the capacity to cause infection.

A common multidrug therapy consists of two analogs such as AZT and 3TC plus a protease inhibitor. The use of two analogs at the same time seems to retain the sensitivity of HIV to both of them. Otherwise, the virus will probably mutate to a resistant strain. It is important to realize that these drugs are very expensive, cause side effects such as diarrhea, neuropathy (painful or numb feet), hepatitis, and possibly diabetes, and that the regimen of pill taking throughout the day is very demanding. If the drugs are not taken as prescribed or if therapy is stopped, resistance may occur. Unfortunately, an HIV strain resistant to all known drugs has been reported—persons who become infected with this strain have no drug therapy available to them.

The sooner drug therapy begins after infection, the better the chances that the immune system will not be destroyed by HIV. And medication must be continued indefinitely. Investigators have found that when HAART is discontinued, the virus rebounds. It may also be possible to help the immune system counter the virus by injecting the patient with a stimulatory cytokine like interferon or interleukin-2. A novel idea is to administer antibodies that will bind to the CD4 receptor and/or chemokines that will bind to the co-receptor. If these molecules bind before HIV does, then HIV will not be able to enter CD4 cells. Alternately, the patients could be given a vaccine.

Drug therapy during the last few weeks of pregnancy along with birth by cesarean section seems to prevent the transmission of HIV from mother to child. Treatment with AZT alone during the last few weeks of pregnancy cuts transmission of HIV to an offspring by half.

Vaccines

There is a general consensus that control of the AIDS epidemic will not occur until a vaccine is developed. Yet an effective vaccine is not in sight.

Some investigators are trying to develop a vaccine in the traditional way. Traditionally, vaccines are made by weakening a pathogen so that it will not cause disease when it is injected into the body. Since the antigens are intact, however, it will cause an immune response including both the production of antibodies by B cells and the stimulation of cytotoxic T cells. Thus far, this approach has not been successful with regard to HIV.

Others are working on subunit vaccines that utilize just a single HIV protein, such as gp120, as the vaccine. Recently, researchers were heartened to find that an injection of DNA for gp120 alone can act as a vaccine. After entering cells, the DNA directs protein synthesis so that viral protein appears in the plasma membrane where it serves as an antigen to stimulate the immune system.

Early trials have demonstrated that a combination of various vaccines may be the best strategy to bring about a response in both B lymphocytes and cytotoxic T cells.

Combination drug therapy has met with encouraging success against an HIV infection. Immunotherapy and the possibility of a vaccine are also being pursued.

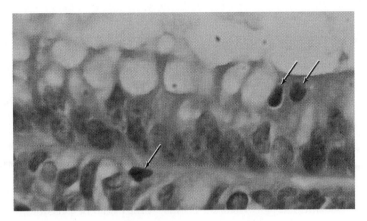

Figure S.6   **Transmission by way of the rectum.**
Arrows show lymphocytes moving from the lumen of the rectum through the lining of the rectum to enter the body.

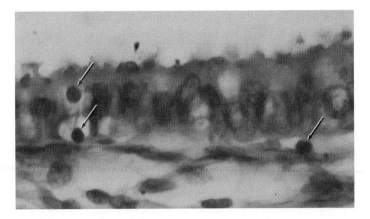

Figure S.7   **Transmission by way of the uterus.**
Arrows show lymphocytes moving through from the uterine cavity through the lining of the uterus to enter the body.

S.4   Preventing Transmission of HIV

Sexual behaviors and drug-related activities are the major means by which AIDS is transmitted in the United States (see Table S.1). HIV is spread by passing virus-infected CD4 cells found in body secretions or in blood from one person to another (Figs. S.6 and S.7).

Sexual activities transmit HIV. Therefore,

1. Abstain from sexual intercourse or develop a long-term monogamous (always the same partner) sexual relationship with a partner who is free of HIV.
2. Refrain from multiple sex partners or having relations with someone who has multiple sex partners. If you have sex with two other people and each of these has sex with two people and so forth, the number of people who are relating is quite large.
3. Remember that the prevalence of AIDS is presently higher among homosexuals and bisexuals than among those that are heterosexual. Also, having relations with an intravenous drug user is risky because the behavior of this group risks AIDS. Be aware that anyone who already has another sexually transmitted disease is more susceptible to an HIV infection.
4. Avoid anal-rectal intercourse (in which the penis is inserted into the rectum) because this behavior increases the risk of infection. The lining of the rectum is thin, and infected CD4 T cells can easily enter the body here. Also, the rectum is supplied with many blood vessels, and insertion of the penis into the rectum is likely to cause tearing and bleeding that facilitate the entrance of HIV (Fig. S.6). The vaginal lining is thick and difficult to penetrate, but the lining of the uterus is only one cell thick and does allow CD4 T cells to enter (Fig. S.7). Uncircumcised males are more likely than circumcised males to become infected because vaginal secretions can remain under the foreskin for a long period of time.

5. Practice safe sex. If you do not know for certain that your partner has been free of HIV for the past five years, always use a latex condom during sexual intercourse. Be sure to follow the directions supplied by the manufacturer. Use of a water-based spermicide containing nonoxynol-9 in addition to the condom can offer further protection because nonoxynol-9 immobilizes the virus and virus-infected lymphocytes.
6. Avoid fellatio (kissing and insertion of the penis into a partner's mouth) and cunnilingus (kissing and insertion of the tongue into the vagina) because they may be a means of transmission. The mouth and gums often have cuts and sores that facilitate the entrance of infected CD4 T cells.
7. Practice penile, vaginal, oral, and hand cleanliness. Be aware that hormonal contraceptives make the female genital tract receptive to the transmission of HIV and other sexually transmitted diseases.

Drug use transmits HIV. Therefore,

1. Stop, if necessary, or do not start the habit of injecting drugs into your veins.
2. If you are a drug user and cannot stop your behavior, then always use a new sterile needle for injection or one that has been cleaned in bleach.
3. Aside from intravenous drug use, do not use alcohol or any drugs in a way that may prevent you from being able to control your behavior.

At present, there is no way to stop the AIDS epidemic. But these behaviors will help prevent the spread of AIDS and decrease the projected number of new cases.

Chapter 17

bober

Development and Aging

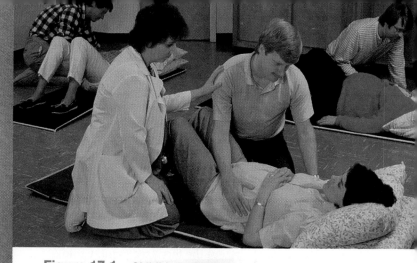

Figure 17.1 **Childbirth classes.**
At childbirth classes, both expectant parents learn how to facilitate the birthing process.

Chapter Concepts

"At last," thought Mary Jean as she left the doctor's office. "I'm going to have a baby." Bill was sure to be overjoyed and so would the prospective grandparents. Bill and Mary Jean had been trying to have a baby for three years with no success. Recently, they had been talking to their doctor about artificial insemination, in vitro fertilization, and other methods of getting pregnant. It was a wonderful surprise and a relief to find that the pregnancy had happened naturally.

Now there are many decisions for Mary Jean. "Where should I have the baby—at a birthing center or hospital? Should I breast- or bottle-feed? How much weight can I gain?" She would discuss all these concerns with Bill and have them sign up soon for childbirth classes (Fig. 17.1). She wanted Bill to be with her every step of the way.

Thinking back over her Human Biology course, Mary Jean knew that the embryo had implanted itself by now in the uterine wall. At two months it would have a human appearance and become a fetus floating within a watery sac. While there, the fetus would be completely dependent upon the placenta, a special structure through which nourishment and oxygen are supplied and wastes are removed. But only at eight months or more would her child have an excellent chance of becoming an independently functioning human being. She was glad to have a doctor who would oversee her health while she waited for that all important day.

This chapter is about the events and the processes that occur as a single cell becomes a complex organism. The reproductive systems were discussed in a previous

chapter. A male continually produces sperm in the testes, a female releases one egg a month from the ovaries. The release of the egg, called ovulation, usually occurs on about day 14 of an ovarian cycle that lasts 28 days. The hormones produced by the ovary control the uterine cycle, which consists of menstruation, a proliferative phase during which the endometrium (uterine lining) thickens due to the increased production of glands and blood vessels, and a secretory phase during which the endometrium secretes nutrient substances to sustain a developing embryo. These two phases prepare the uterus to accept the fertilized egg (zygote).

This time for Mary Jean and Bill, everything went perfectly. The 200 to 600 million sperm released by Bill during sexual intercourse were ejaculated against the opening of the cervix at the far end of the vagina. The semigelatinous seminal fluid protected the sperm from the acid of the vagina for several minutes, and many managed to enter the uterus. Whether or not sperm enter the uterus depends in part on the consistency of the cervical mucus. Three to four days prior to ovulation and on the day of ovulation, the mucus is watery, and the sperm can penetrate it easily. During the other days of the uterine cycle, the mucus is thicker and has a sticky consistency, and the sperm can rarely penetrate it. Some high-speed sperm reach the oviducts in only 30 minutes, but generally, the journey from vagina to oviduct takes several hours. Many sperm get lost along the way, and only several hundred sperm ever reach the oviducts.

17.1 Fertilization

Fertilization (Fig. 17.2) normally occurs in the upper third of the oviduct, where the sperm encounter an egg if ovulation has occurred. The sperm must swim against the downward current created by the ciliary action of the epithelium lining the oviducts. However, it is believed that uterine and oviduct contractions help transport the sperm and that prostaglandins within seminal fluid promote these contractions.

Fertilization occurs after a sperm and egg have interacted. A sperm has three distinct parts: a head, a middle piece, and a tail. The tail is a flagellum, which allows the sperm to swim toward the egg, and the middle piece contains energy-producing mitochondria. The head contains a haploid nucleus capped by a membrane-bound acrosome containing enzymes that allow the sperm to penetrate the egg.

Once sperm have made their way past the layers of nutrient cells around the egg, several mechanisms assure fertilization takes place in a species-specific manner (Fig. 17.2). The egg has a plasma membrane, a vitelline envelope, and a jelly coat. The acrosomal enzymes digest a hole in the jelly coat as the acrosome extrudes a filament that attaches to receptors located in the vitelline envelope. This is a lock-and-

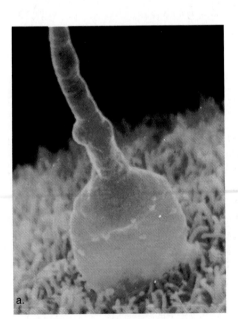

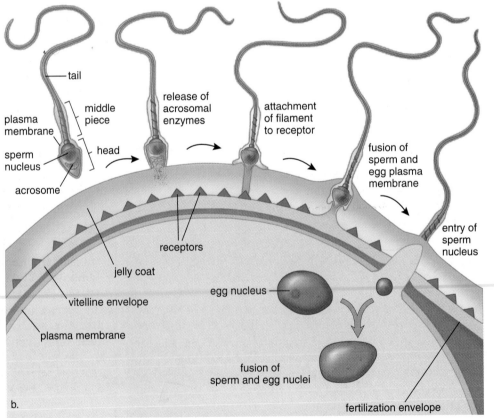

Figure 17.2 **Fertilization.**
a. During fertilization, a single sperm enters the egg. **b.** A head of a sperm has a membrane-bound acrosome filled with enzymes. When released, these enzymes digest away the jelly coat around the egg, and the acrosome extrudes a filament that attaches to a receptor on the vitelline membrane. Now the sperm nucleus enters and fuses with the egg nucleus, and the resulting zygote begins to divide. The vitelline membrane becomes the fertilization membrane, which prohibits any more sperm from entering the egg.

TesT

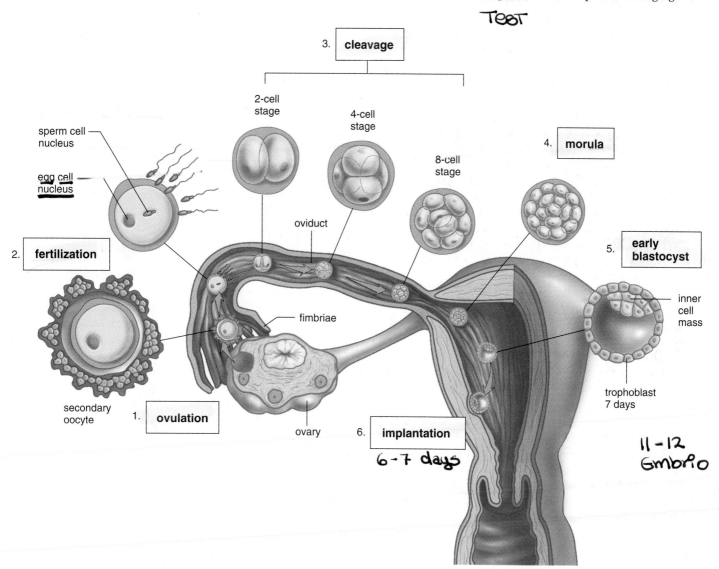

3. **cleavage**

2-cell stage

sperm cell nucleus

egg cell nucleus

2. **fertilization**

secondary oocyte

1. **ovulation**

4-cell stage

8-cell stage

oviduct

fimbriae

ovary

4. **morula**

5. **early blastocyst**

inner cell mass

trophoblast 7 days

6. **implantation**

6-7 days

11-12 Embrio

Figure 17.3 Human development before implantation.
Structures and events proceed counterclockwise. At ovulation, the secondary oocyte leaves the ovary. A single sperm penetrates the zona pellucida, and fertilization occurs in the oviduct. As the zygote moves along the oviduct, it undergoes cleavage to produce an embryo that implants itself in the uterine lining.

key reaction that is species-specific. Then, the egg plasma membrane and sperm plasma membrane fuse, allowing the sperm nucleus to enter.

A **zygote** is present when the sperm nucleus fuses with the egg nucleus. Following fusion, the egg plasma membrane and the vitelline envelope undergo changes that prevent the entrance of any other sperm. The vitelline envelope is now called the fertilization membrane.

Following fertilization, the zygote begins dividing and becomes an **embryo.** The developing embryo travels very slowly down the oviduct to the uterus, where after two to three days it implants itself in the prepared uterine lining (Fig. 17.3). Upon **implantation,** the woman is pregnant. If implantation takes place, the uterine lining is maintained, because the membrane surrounding the embryo begins to produce a hormone called HCG (human chorionic gonadotropin)

that prevents degeneration of the corpus luteum. Pregnancy tests, which are readily available in hospitals, clinics, and now even drug and grocery stores, are based on the fact that HCG is present in the blood and urine of a pregnant woman. There is also an early detection test that can be done in a doctor's office; the results are available within the hour. The signs that often prompt a woman to have a pregnancy test are cessation of menstruation, increased frequency of urination, morning sickness, and increase in the size and fullness of the breasts, as well as darkening of the areolae, a pigmented area surrounding the nipples.

Following fertilization of an egg by a sperm, the zygote begins developing, and the embryo implants itself in the uterine lining. Pregnancy has now taken place.

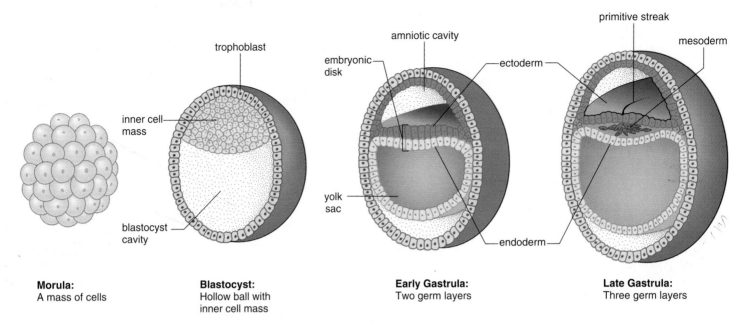

Figure 17.4   Early developmental stages in cross section.
Cleavage results in the inner cell mass. Morphogenesis occurs as cells rearrange themselves, and differentiation is exemplified first by the formation of three different germ layers and then by the formation of organs such as the neural tube.

17.2 Human Development before Birth

This section considers the major processes and events that take place from the time of fertilization to the time of birth.

Processes of Development

Embryonic development includes these processes:

Cleavage   Immediately after fertilization, the zygote begins to divide so that at first there are 2, then 4, 8, 16, and 32 cells, and so forth. Increase in size does not accompany these divisions.

Morphogenesis   *Morphogenesis* refers to the shaping of the embryo and is first evident when certain cells are seen to move, or migrate, in relation to other cells. By these movements, the embryo begins to assume various shapes.

Differentiation   *Differentiation* occurs as cells take on a specific structure and function. Nerve cells have long processes that conduct nerve impulses, and muscle cells contain contractile elements.

Growth   During embryonic development, cell division is accompanied by an increase in size of the daughter cells, and *growth* (in the true sense of the term) takes place.

These processes can be observed in the early developmental stages, which humans share with all animals (Fig. 17.4).

Morula

Cleavage is a process that occurs during the first stage of development. During cleavage, cell division without growth results in a mass of tiny cells. The cells are uniform in size because the cytoplasm has been equally divided among them. This solid mass of cells is called a *morula*, which means a bunch of berries.

Blastula

Morphogenesis begins as the cells of the morula position themselves to create a cavity. All animal blastulas have an empty cavity, but since the human blastula is called a **blastocyst,** the cavity is called the blastocyst cavity. The inner cell mass is at one end of the blastocyst.

Gastrula

In humans, a space called the amniotic cavity appears above the inner cell mass. The inner cell mass becomes an embryonic disk composed of two layers: the upper layer is called ectoderm, and the lower layer is called the endoderm.

Now, the bilayered disk elongates to form a primitive streak, at the midline region of the embryo. Morphogenesis continues as some of the upper cells within the primitive streak invaginate and spread out between the ectoderm and endoderm. The invaginating cells are the mesoderm layer.

Differentiation is in progress because **ectoderm, mesoderm,** and **endoderm** are the germ layers that give rise to all the other tissues and organs of the body (Table 17.1). Gastrulation is the movement of cells that results in a *gastrula,* an embryo composed of three germ layers.

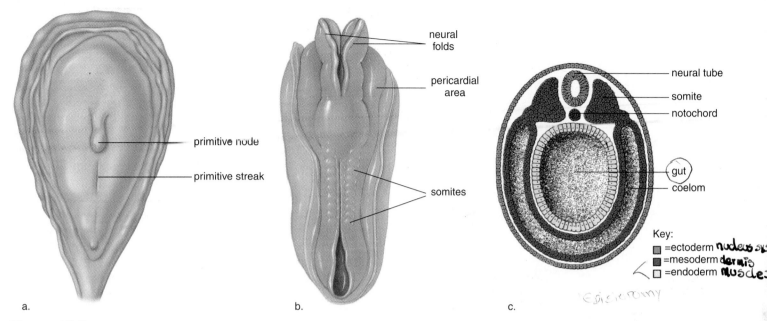

Figure 17.5 Primitive streak and neurula.
a. At 16 days, the primitive node marks the extent of the primitive streak where invagination results in a three-layered embryo. **b.** At 21 days, the neural tube is forming along the midline of the body. The pericardial area contains the primitive heart, and the somites give rise to the muscles and the vertebrae, which replace the notochord. **c.** Generalized cross section of a human neurula. Each of the germ layers can be associated with the later development of particular organs as listed in Table 17.1.

Neurula

Newly formed mesoderm cells that lie along the main axis become the notochord, a dorsal supporting rod. (In humans, the notochord is later replaced by the vertebral column.) The nervous system develops from ectoderm located just above the notochord. A neural plate thickens into neural folds that fuse, forming a neural tube. The neural tube develops into the spinal cord and the brain.

Neurulation involves **induction,** a process by which one tissue influences the development of another tissue. Experiments have shown that the nervous system does not form unless there is a notochord present. Today, investigators believe that induction explains the process of differentiation. Induction requires direct contact or the production of a chemical by one tissue that most likely activates certain genes in the cells of the other tissue. These genes then direct how differentiation is to occur.

Midline mesoderm not contributing to the formation of the notochord becomes two longitudinal masses of tissue. From these blocklike portions of mesoderm, called somites, the muscles of the body and the vertebrae of the spine develop. The coelom, an embryonic body cavity that forms at this time, is completely lined by mesoderm. In humans, the coelom becomes the thoracic and abdominal cavities. Figure 17.5 shows an intact human neurula and a generalized cross section of the embryo indicating the location of the three germ layers.

Table 17.1	Germ Layers and Organ Development	
Ectoderm	**Mesoderm**	**Endoderm**
Skin epidermis, including hair, nails, and sweat glands	All muscles	Lining of digestive tract, trachea, bronchi, lungs, gallbladder, and urethra
Nervous system, including brain, spinal cord, ganglia, and nerves	Dermis of skin	Liver
Retina, lens, and cornea of eye	All connective tissue, including bone, cartilage, and blood	Pancreas
Inner ear	Blood vessels	Thyroid, parathyroid, and thymus glands
Lining of nose, mouth, and anus	Kidneys	Urinary bladder
Tooth enamel	Reproductive organs	

Development involves certain processes that can be related to particular stages of early development.

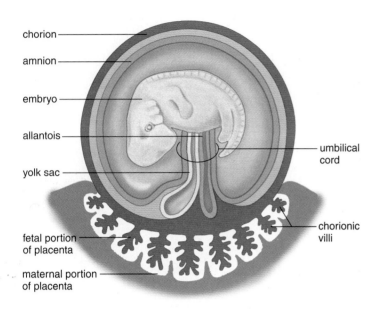

Figure 17.6   **The extraembryonic membranes.**
The chorion and amnion surround the embryo. The two other extraembryonic membranes, the yolk sac and allantois, contribute to the umbilical cord.

Extraembryonic Membranes

One of the major events in early development is the establishment of the extraembryonic membranes (Fig. 17.6). The term **extraembryonic membranes** is apt because these membranes extend out beyond the embryo. One of the membranes, the **amnion,** provides a fluid environment for the developing embryo and fetus. It is a remarkable fact that all animals, even land-dwelling humans, develop in a watery medium. Amniocentesis is a process by which amniotic fluid and the fetal cells floating in it are withdrawn for examination. One authority describes the functions of amniotic fluid in this way:

> The colorless amniotic fluid by which the fetus is
> surrounded serves many purposes. It prevents the walls
> of the uterus from cramping the fetus and allows it
> unhampered growth and movement. It encompasses the
> fetus with a fluid of constant temperature which is a mar-
> velous insulator against cold and heat. Above all, it acts
> as an excellent shock absorber. A blow on the mother's
> abdomen merely jolts the fetus, and it floats away.[1]

The **yolk sac** is another extraembryonic membrane. Yolk is a nutrient material utilized by other animal embryos—the yellow of a chick's egg is yolk. However, in humans, the yolk sac contains no yolk and is the first site of red blood cell formation. Part of this membrane becomes incorporated into the umbilical cord. Another extraembryonic membrane, the **allantois,** contributes to the circulatory system: its blood vessels become umbilical blood vessels that transport fetal blood to and from the placenta. The **chorion,** the outer extraembryonic membrane, becomes part of the **placenta** (Fig. 17.7), where the fetal blood exchanges gases, nutrients, and wastes with maternal blood.

[1]A. F. Guttmacher. *Pregnancy, Birth and Family Planning.* (New York: New American Library, 1974), p. 74.

Fetal Circulation

Fetal circulation involves the placenta which begins formation once the embryo is implanted fully. The placenta has a fetal side contributed by the chorion and a maternal side consisting of uterine tissues. Notice in Figure 17.7 how projections called **chorionic villi** meet maternal blood vessels—the blood of the mother and the fetus never mix since exchange always takes place across plasma membranes. Carbon dioxide and other wastes move from the fetal side to the maternal side, and nutrients and oxygen move from the maternal side to the fetal side of the placenta by diffusion. The **umbilical cord** stretches between the placenta and the fetus. Although it may seem that the umbilical cord travels from the placenta to the intestine, actually it simply contains the blood vessels taking fetal blood to and from the placenta. The umbilical cord is the lifeline of the fetus because it contains the umbilical arteries and vein, which transport waste molecules (carbon dioxide and urea) to the placenta for disposal and take oxygen and nutrient molecules from the placenta to the rest of the fetal circulatory system.

By the tenth week, the placenta is formed fully and begins to produce progesterone and estrogen. These hormones have two effects, due to their negative feedback effect on the mother's hypothalamus and anterior pituitary. They prevent any new follicles from maturing, and they maintain the lining of the uterus—the corpus luteum is not needed now. There is usually no menstruation during pregnancy. Harmful chemicals in the mother's blood can cross the placenta, and this is of particular concern during the embryonic period, when various structures are first forming. Each organ or part seems to have a sensitive period during which a substance can alter its normal function. The reading on pages 370–71 concerns the origination of birth defects and explains ways to detect genetic defects before birth.

Path of Fetal Blood

Blood within the fetal aorta travels to the various branches, including the iliac arteries, which connect to the *umbilical arteries* leading to the placenta (Fig. 17.7). Exchange between maternal blood and fetal blood takes place here. The *umbilical vein* carries blood rich in nutrients and oxygen away from the placenta to the fetus. The umbilical vein enters the liver and then joins the *venous duct,* which merges with the inferior vena cava, a vessel that returns blood to the heart.

It is interesting to note that the umbilical arteries and vein run alongside one another in the umbilical cord, which is cut at birth, leaving only the umbilicus (navel).

Features of the fetal heart can be related to the fact that the fetus does not use its lungs for gas exchange. Much of the blood entering the right atrium is shunted into the left atrium through the *oval opening* (foramen ovale) between the two atria. Also, any blood that does enter the right ventricle and is pumped into the pulmonary trunk is shunted into the aorta by way of the *arterial duct* (ductus arteriosus).

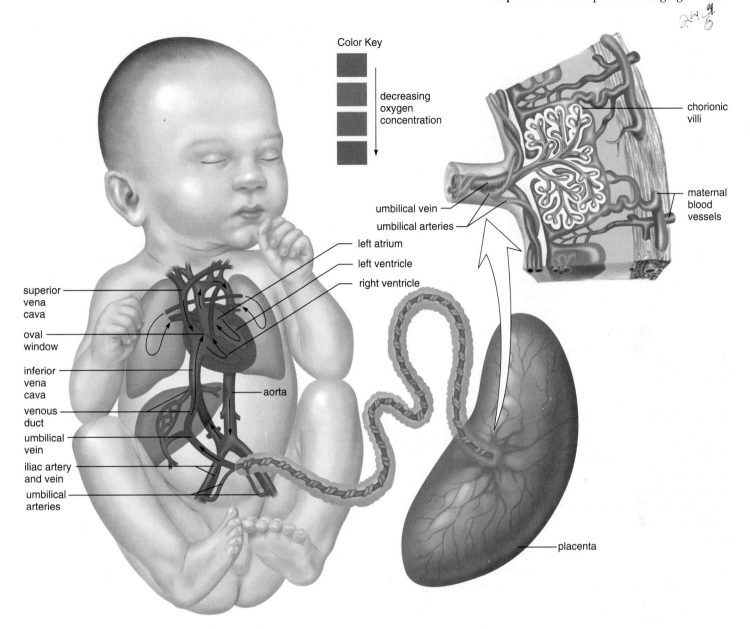

Figure 17.7   Fetal circulation and the placenta.
The lungs are not functional in the fetus, and the blood passes directly from the right atrium to the left atrium or from the right ventricle to the aorta. The umbilical arteries take fetal blood to the placenta where exchange of molecules between fetal and maternal blood takes place across the walls of the chorionic villi. Oxygen and nutrient molecules diffuse into the fetal blood, and carbon dioxide and urea diffuse from fetal blood. The umbilical vein returns blood from the placenta to the fetus.

The most common of all cardiac defects in the newborn is the persistence of the oval opening. With the tying of the cord and the expansion of the lungs, blood enters the lungs in quantity. Return of this blood to the left side of the heart usually causes a flap to cover the opening. Incomplete closure occurs in nearly one out of four individuals, but even so, passage of the blood from the right atrium to the left atrium rarely occurs because either the opening is small or it closes when the atria contract. In a small number of cases, the passage of impure blood from the right side to the left side of the heart is sufficient to cause a "blue baby." Such a condition now can be corrected by open heart surgery.

The arterial duct closes because endothelial cells divide and block off the duct. Remains of the arterial duct and parts of the umbilical arteries and vein later are transformed into connective tissue.

Fetal circulation shunts blood away from the lungs, toward and away from the placenta within the umbilical blood vessels located within the umbilical cord. Exchange of substances between fetal blood and maternal blood takes place at the placenta, which forms from the chorion, an extraembryonic membrane, and uterine tissue.

Preventing Birth Defects

It is believed that at least 1 in 16 newborns has a birth defect, either minor or serious, and the actual percentage may be even higher. It is estimated that only 20% of all birth defects are due to heredity. Those that are hereditary can sometimes be detected before birth. Amniocentesis allows the fetus to be tested for abnormalities of development; chorionic villi sampling allows the embryo to be tested; and a new method has been developed for screening eggs to be used for in vitro fertilization (Fig. 17A).

It is recommended that all females take everyday precautions to protect any future and/or presently developing embryos and fetuses from defects. Proper nutrition is a must (deficiency in folic acid causes neural tube defects). X-ray diagnostic therapy should be avoided during pregnancy because X rays cause mutations in the developing embryo or fetus. Children born to women who received X-ray treatment are apt to have birth defects and/or to develop leukemia later. Toxic chemicals, such as pesticides and many organic industrial chemicals, which are also mutagenic, can cross the placenta. Cigarette smoke not only contains carbon monoxide but also other fetotoxic chemicals. Babies born to smokers are often underweight and subject to convulsions.

Pregnant Rh⁻ women should receive an Rh immunoglobulin injection to prevent the production of Rh antibodies. These antibodies can cause nervous system and heart defects.

Sometimes, birth defects are caused by pathogens. Females can be immunized before the childbearing years for rubella (German measles), which in particular causes birth defects such as deafness. Unfortunately, immunization for sexually transmitted diseases is not possible. The AIDS virus can cross the placenta, and over 1,500 babies who contracted AIDS while in their mother's womb are now mentally retarded. When a mother has herpes, gonorrhea, or chlamydia, newborns can become infected as they pass through the birth canal. Blindness and other physical and mental defects may develop. Birth by cesarean section could prevent these occurrences.

Pregnant women should not take any type of drug except with a doctor's prescription. Certainly illegal drugs, such as marijuana, cocaine, and heroin, should be completely avoided. "Cocaine babies" now make up 60% of drug-affected babies. Severe fluctuations in blood pressure that are produced by the use of cocaine temporarily deprive the developing brain of oxygen. Cocaine babies have visual problems, lack coordination, and are mentally retarded. The drugs aspirin, caffeine (present in coffee, tea, and cola), and alcohol should be severely limited. It is not unusual for babies of drug addicts and alcoholics to display withdrawal symptoms and to have various abnormalities. Babies born to women who drink while pregnant are apt to have fetal alcohol syndrome (FAS). These babies have decreased weight, height, and head size, with malformation of the head and face. Mental retardation is common in FAS infants.

Medications can also cause problems. When the synthetic hormone DES was given to pregnant women to prevent miscarriage, their daughters showed various abnormalities of the reproductive organs and an increased tendency toward cervical cancer. Other sex hormones, including birth-control pills, can possibly cause abnormal fetal development, including abnormalities of the sex organs. The tranquilizer thalidomide is well known for having caused deformities of the arms and legs in children born to women who took the drug. Therefore, a woman has to be very careful about taking medications while pregnant.

Now that physicians and laypeople are aware of the various ways in which birth defects can be prevented, it is hoped that the incidence of birth defects will decrease in the future.

Figure 17A Three methods for genetic-defect testing before birth.

a. Amniocentesis is usually performed from the 15th to the 17th week of pregnancy. A long needle is passed through the abdominal wall to withdraw a small amount of amniotic fluid, along with fetal cells. Since there are only a few cells in the amniotic fluid, testing may be delayed as long as four weeks until cell culture produces enough cells for testing purposes. About 40 tests are available for different defects.

b. Chorionic villi sampling is usually performed from the 5th to the 12th week of pregnancy. The doctor inserts a long, thin tube through the vagina into the uterus. With the help of ultrasound, which gives a picture of the uterine contents, the tube is placed between the lining of the uterus and the chorion. Then a sampling of the chorionic villi cells is obtained by suction. Chromosome analysis and biochemical tests for genetic defects can be done immediately on these cells.

c. Screening eggs for genetic defects is a new technique. Preovulatory eggs are removed by aspiration after a laparoscope (optical telescope) is inserted into the abdominal cavity through a small incision in the region of the navel. The first polar body is tested. If the woman is heterozygous (Aa) and the defective gene (a) is found in the polar body, then the egg must have received the normal gene (A). Normal eggs then undergo in vitro fertilization and are placed in the prepared uterus. At present, only one in ten attempts results in a birth, but it is known ahead of time that the child will be normal for the genetic traits tested.

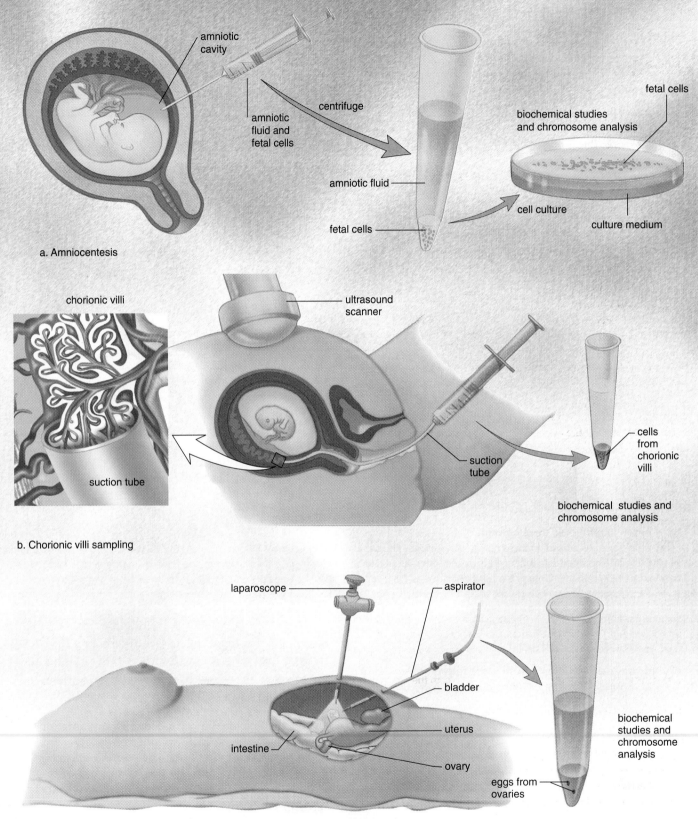

a. Amniocentesis

amniotic cavity

amniotic fluid and fetal cells

centrifuge

amniotic fluid

fetal cells

cell culture

biochemical studies and chromosome analysis

fetal cells

culture medium

b. Chorionic villi sampling

chorionic villi

ultrasound scanner

suction tube

suction tube

cells from chorionic villi

biochemical studies and chromosome analysis

c. Obtaining eggs for screening

laparoscope

aspirator

bladder

uterus

intestine

ovary

biochemical studies and chromosome analysis

eggs from ovaries

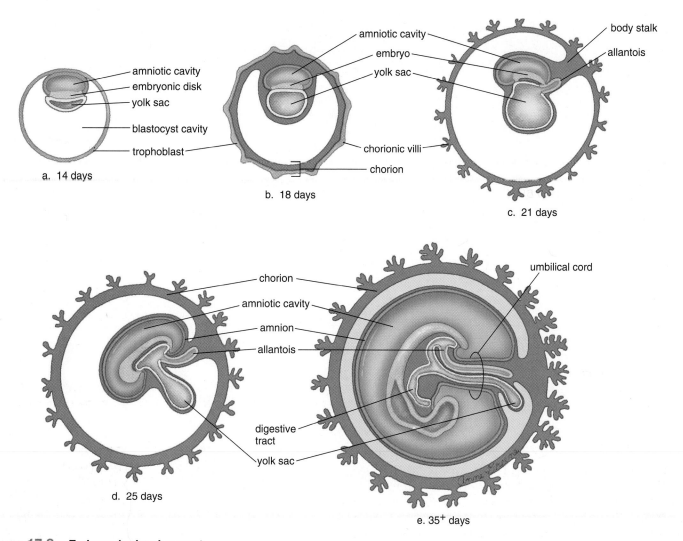

Figure 17.8   Embryonic development.
a. At first, no organs are present in the embryo, only tissues. The amniotic cavity is above the embryo, and the yolk sac is below. **b.** The chorion is developing villi, so important to exchange between mother and child. **c.** The allantois and yolk sac are two more extraembryonic membranes. **d.** These extraembryonic membranes are positioned inside the body stalk as it becomes the umbilical cord. **e.** At 28 days, the embryo has a head region and a tail region. The umbilical cord takes blood vessels between the embryo and the chorion (placenta).

Embryonic Development

Embryonic development lasts from fertilization to the end of the second month. All major organs develop during this time.

First Month

Immediately after fertilization, the zygote divides repeatedly as it passes down the oviduct to the uterus. The resulting morula becomes the hollow blastocyst with an inner cell mass to one side. The blastocyst is bounded by a layer of cells that becomes the chorion (see Fig. 17.3). The early appearance of the chorion emphasizes the complete dependence of the developing embryo on this extraembryonic membrane. The inner cell mass is the embryo. Once it has arrived in the uterus on the fourth or fifth day after fertilization, it spends about two or three days, and then the blastocyst begins to implant itself in the uterine lining (see Fig. 17.3).

By the end of the second week, implantation is completed. The ever-growing number of cells becomes a gastrula with three germ layers. The amniotic cavity is seen above the embryo, and the yolk sac is below (Fig. 17.8a). During the third week, another extraembryonic membrane, the allantois, makes its appearance briefly, but later it and the yolk sac become part of the umbilical cord as it forms (Fig. 17.8c–d). Organs are already developed, including the spinal cord and heart.

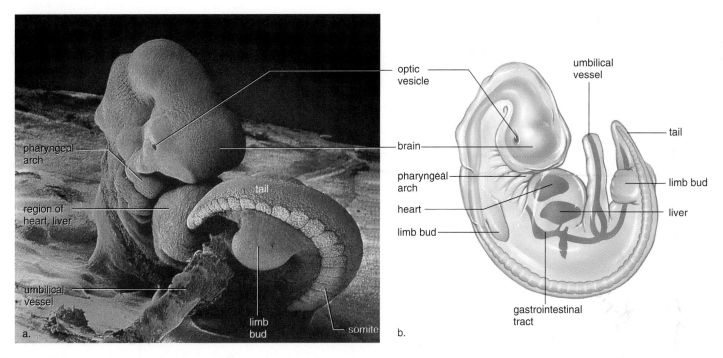

Figure 17.9 **Human embryo at the beginning of the fifth week.**
a. Scanning electron micrograph. **b.** The embryo is curled so that the head touches the heart. The organs of the gastrointestinal tract are forming. The bones in the tail will regress and become those of the coccyx. The arms and legs will develop from the bulges that are called limb buds.

By the end of the first month, the placenta is forming. The embryo has a nonhuman appearance largely due to the presence of a tail, but also because the arms and legs, which begin as limb buds, resemble paddles. The head is much larger than the rest of the embryo, and the whole embryo bends under its weight (Fig. 17.9). The eyes, ears, and nose are just appearing. The enlarged heart beats and the bulging liver takes over the production of blood cells for blood, which will carry nutrients to the developing organs and wastes from the developing organs.

Second Month

At the end of two months, the embryo's tail has disappeared, and the arms and legs are more developed, with fingers and toes apparent (Fig. 17.10). The head is very large, the nose is flat, the eyes are far apart, and the ears are distinctively present. Internally, all major organs have appeared. Embryonic development is now finished; Table 17.2 outlines the main events.

At the end of the embryonic period, all organ systems are established, and there is a mature and functioning placenta. The embryo is only about 38 mm (1½ inches) long.

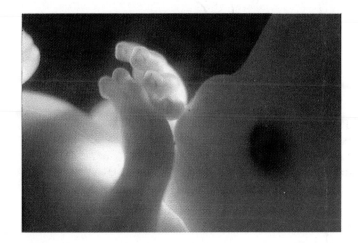

Figure 17.10 **Two-month-old embryo.**
The embryonic period is over, and from now on a more human appearance takes shape.

Table 17.2 also outlines the events for the mother. When first pregnant, the mother may experience nausea and vomiting, loss of appetite, and fatigue. Other changes are swelling and tenderness of the breasts, increased urination, and irregular bowel movements. Some women, however, report increased energy levels and a general sense of well-being during this time.

Table 17.2   Human Development

Time	Events for Mother	Events for Baby
Embryonic Development		
First week	Ovulation occurs.	Fertilization occurs. Cell division begins and continues. Chorion appears.
Second week	Symptoms of early pregnancy (nausea, breast swelling and tenderness, fatigue) are present.	Implantation. Amnion and yolk sac appear. Embryo has tissues. Placenta begins to form.
Third week	First missed menstruation. Blood pregnancy test is positive.	Nervous system begins development. Allantois and blood vessels are present. Placenta is well formed. *Heart*
Fourth week	Urine pregnancy test is positive.	Limb buds form. Heart is noticeable and beating. Nervous system is prominent. Embryo has tail. Other systems form.
Fifth week	Uterus is the size of hen's egg. Frequent need to urinate due to pressure of growing uterus on bladder.	Embryo is curved. Head is large. Limb buds show divisions. Nose, eyes, and ears are noticeable.
Sixth week	Uterus is the size of an orange.	Fingers and toes are present. Cartilaginous skeleton.
Two months	Uterus can be felt above the pubic bone.	All systems are developing. Bone is replacing cartilage. Refinement of facial features. 38 mm (1½ inches). *Placenta develops*
Fetal Development		*because white cromosome*
Third month	Uterus is the size of a grapefruit.	Possible to distinguish sex. Fingernails appear.
Fourth month	Fetal movement is felt by those who have been pregnant before.	Skeleton visible. Hair begins to appear. 150 mm (6 inches), 170 g (6 oz). *1½ lb*
Fifth month	Fetal movement is felt by those who have not been pregnant before. Uterus reaches up to level of umbilicus and pregnancy is obvious.	Protective cheesy coating begins to be deposited. Heartbeat can be heard.
Sixth month	Doctor can tell where baby's head, back, and limbs are. Breasts have enlarged, nipples and areolae are darkly pigmented, and colostrum is produced.	Body is covered with fine hair. Skin is wrinkled and reddish.
Seventh month	Uterus reaches halfway between umbilicus and rib cage.	Testes descend into scrotum. Eyes are open. 300 mm (12 inches), 1,350 g (3 lb).
Eighth month	Weight gain is averaging about a pound a week. Difficulty in standing and walking because center of gravity is thrown forward.	Body hair begins to disappear. Subcutaneous fat begins to be deposited.
Ninth month	Uterus is up to rib cage, causing shortness of breath and heartburn. Sleeping becomes difficult.	Ready for birth, 530 mm (20½ inches), 3,400 g (7½ lb).

(Handwritten annotations: "Reflex action." next to Sixth week/Two months; "Heartbeat Fetus" and "baby can be born & survive" bracketing Fourth–Seventh months)

Fetal Development

Fetal development extends from the third to the ninth month as shown in Table 17.2. The fetus has a human appearance, but refinements are still taking place.

Third and Fourth Months

At the beginning of the third month (Fig. 17.11), head growth begins to slow down as the rest of the body increases in length. Epidermal refinements, such as eyelashes, eyebrows, hair on the head, fingernails, and nipples, appear.

Cartilage is replaced by bone as ossification centers appear in the bones. The skull has six large fontanels (membranous areas or soft spots), which permit a certain amount of flexibility as the head passes through the birth canal and allow rapid growth of the brain during infancy. The fontanels usually close by 2 years of age.

Sometime during the third month, it is possible to distinguish males from females. Once the testes differentiate, they produce androgens, the male sex hormones. The androgens, especially testosterone, stimulate the growth and differentiation of the male external genitals. In the absence of androgens, female genitals form. The ovaries do not produce estrogen because there is plenty of it circulating in the mother's bloodstream.

At this time, the testes or ovaries are located within the abdominal cavity. Later, in the last trimester of male fetal development, the testes descend into the scrotal sacs of the scrotum. Sometimes the testes fail to descend, in which case an operation can be performed to place them in their proper location.

Figure 17.11 **The three- to four-month-old fetus looks human.**
Face, hands, and fingers are well defined.

By the end of the fourth month, the fetus is less than 150 mm (6 in) in length and weighs a little more than 170 grams (6 oz).

During the third and fourth months, the skeleton is becoming ossified. The sex of the individual is now distinguishable.

Fifth through Seventh Months

During the fifth through seventh months (Fig. 17.12), the mother begins to feel fetal movement. At first, there is only a fluttering sensation, but as the legs grow and develop, kicks and jabs are felt. The fetal heartbeat is loud enough to be heard when a physician applies a stethoscope to the mother's abdomen. The fetus is in the fetal position with the head bent down and in contact with the flexed knees.

The wrinkled skin is covered by a fine down called **lanugo.** The lanugo is coated with a white, greasy, cheese-like substance called **vernix caseosa,** which is believed to protect the delicate skin from the amniotic fluid. During these months, the eyelids open fully.

At the end of this period, the fetus is almost 300 mm (12 in), and the weight has increased to almost 1,350 grams (3 lb). It is possible that if born now, the baby will survive; however, the lungs lack surfactant, which reduces surface tension within the alveoli (air sacs). Babies born without sur-factant risk respiratory distress syndrome or collapsed lungs.

Eighth and Ninth Months

As the end of development approaches, the fetus usually ro-tates so that the head is pointed toward the cervix. However, if the fetus does not turn, then the likelihood of a breech birth

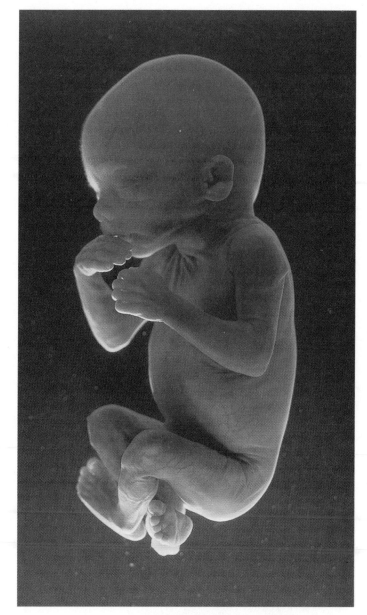

Figure 17.12 **A five- to seven-month-old fetus.**

(rump first) may prescribe a cesarean section. It is very difficult for the cervix to expand enough to accommodate this form of birth, and asphyxiation of the baby is more likely to occur.

At the end of nine months, the fetus is about 530 mm (20½ in) long and weighs about 3,400 grams (7½ lb) (Fig. 17.13). Weight gain is due largely to an accumulation of fat beneath the skin. Full-term babies have a better chance of survival.

From the fifth to the ninth month, the fetus continues to grow and to gain weight. Babies born after six or seven months may survive, but full-term babies have a better chance of survival.

Table 17.2 continues with the effects of pregnancy on the mother.

17.3 Birth

Investigation and experimentation show that when the fetal brain is sufficiently mature, the hypothalamus causes the pituitary to stimulate the adrenal cortex so that androgens are released into the bloodstream. The placenta utilizes androgens as a precursor for estrogen, a molecule that stimulates the local production of prostaglandins and oxytocin. All three of these molecules cause the uterus to contract and expel the fetus.

The uterus has contractions throughout pregnancy. At first, these are light, lasting about 20–30 seconds and occurring every 15–20 minutes. Near the end of pregnancy, the contractions may become stronger and more frequent so that the woman may think that she is in labor. However, the onset of true labor is marked by uterine contractions that occur regularly every 15–20 minutes and last for 40 seconds or more.

Stage I

Prior to or at the first stage of **parturition,** which includes labor and birth of the fetus, there can be a "bloody show" caused by expulsion of a mucous plug from the cervical canal. This plug prevents bacteria and sperm from entering the uterus during pregnancy.

Uterine contractions during the first stage of labor occur in such a way that the cervical canal slowly disappears as the lower part of the uterus is pulled upward toward the baby's head (Fig. 17.13*b*). This process is called *effacement,* or "taking up the cervix." With further contractions, the baby's head acts as a wedge to assist cervical dilation. If it has not occurred already, the amniotic membrane is apt to rupture during this stage, releasing the amniotic fluid, which leaks out the vagina (sometimes referred to as breaking water). The first stage of labor ends once the cervix is dilated completely.

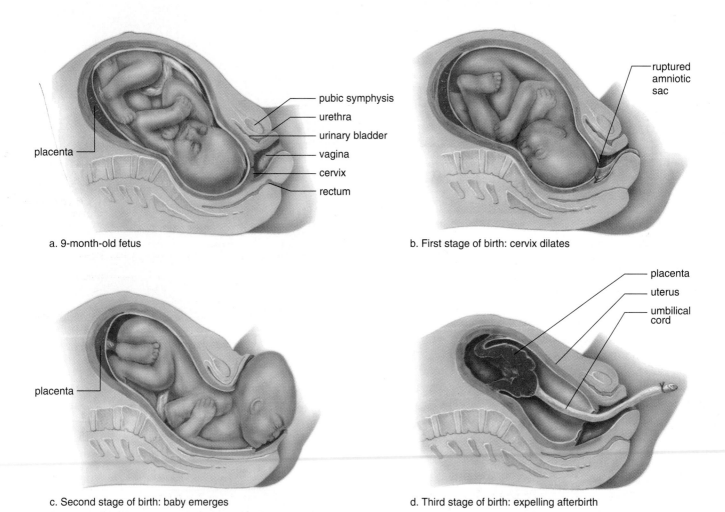

a. 9-month-old fetus

b. First stage of birth: cervix dilates

c. Second stage of birth: baby emerges

d. Third stage of birth: expelling afterbirth

Figure 17.13   Three stages of parturition.
a. Position of fetus just before birth begins. **b.** Dilation of cervix. **c.** Birth of baby. **d.** Expulsion of afterbirth.

Stage 2

During the second stage of parturition, the uterine contractions occur every 1–2 minutes and last about one minute each. They are accompanied by a desire to push, or bear down. As the baby's head gradually descends into the vagina, the desire to push becomes greater. When the baby's head reaches the exterior, it turns so that the back of the head is uppermost (Fig. 17.13c). Since the vaginal orifice may not expand enough to allow passage of the head without tearing, an **episiotomy** often is performed. This incision, which enlarges the opening, is sewn together later and heals more perfectly than a tear. As soon as the head is delivered, the baby's shoulders rotate so that the baby faces either to the right or the left. At this time, the physician may hold the head and guide it downward, while one shoulder and then the other emerges. The rest of the baby follows easily.

Once the baby is breathing normally, the umbilical cord is cut and tied, severing the child from the placenta. The stump of the cord shrivels and leaves a scar, which is the umbilicus.

Stage 3

The placenta, or *afterbirth,* is delivered during the third stage of labor (Fig. 17.13d). About 15 minutes after delivery of the baby, uterine muscular contractions shrink the uterus and dislodge the placenta. The placenta then is expelled into the vagina. As soon as the placenta and its membranes are delivered, the third stage of labor is complete.

During the first stage of birth, the cervix dilates; during the second stage, the child is born; and during the third stage, the afterbirth is expelled.

Female Breast and Lactation

A female breast contains 15 to 25 lobules, each with its own milk duct, which begins at the nipple and divides into numerous other ducts that end in blind sacs called *alveoli* (Fig. 17.14).

During pregnancy, the breasts enlarge as the ducts and alveoli increase in number and size. The same hormones that affect the mother's breast can also affect the child's. Some newborns, including males, even secrete a small amount of milk for a few days.

Usually, there is no production of milk during pregnancy. The hormone *prolactin* is needed for lactation to begin, and the production of this hormone is suppressed because of the feedback control that the increased amount of estrogen and progesterone during pregnancy has on the pituitary. Once the baby is delivered, however, the pituitary begins secreting prolactin. It takes a couple of days for milk production to begin, and, in the meantime, the breasts produce **colostrum,** a thin, yellow, milky fluid rich in protein, including antibodies.

The continued production of milk requires a suckling child. When a breast is suckled, the nerve endings in the areola are stimulated, and a nerve impulse travels along neural pathways from the nipples to the hypothalamus, which directs the pituitary gland to release the hormone *oxytocin.* When this hormone arrives at the breast, it causes contraction of the lobules so that milk flows into the ducts (called milk letdown), where it may be drawn out of the nipple by the suckling child. Some women choose to breast-feed and some do not.

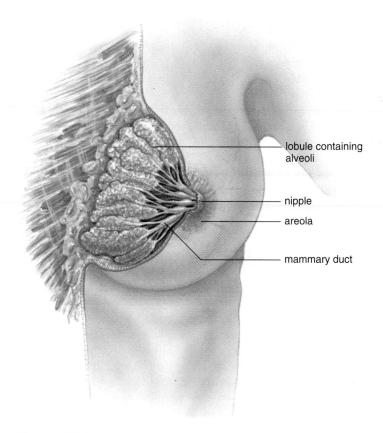

Figure 17.14   Female breast anatomy.
The female breast contains lobules consisting of ducts and alveoli. The alveoli are lined by milk-producing cells in the lactating (milk-producing) breast.

17.4 Human Development after Birth

Development does not cease once birth has occurred but continues throughout the stages of life: infancy, childhood, adolescence, and adulthood. **Aging** encompasses these progressive changes that contribute to an increased risk of infirmity, disease, and death (Fig. 17.15).

Today, there is great interest in **gerontology,** the study of aging, because there are now more older individuals in our society than ever before, and the number is expected to rise dramatically. In the next half-century, those over age 75 will rise from the present 13 million to 34–45 million, and those over age 80 will rise from 3 million to 6 million individuals. The human life span is judged to be a maximum of 110–115 years. The present goal of gerontology is not necessarily to increase the life span, but to increase the health span, the number of years that an individual enjoys the full functions of all body parts and processes.

Theories of Aging

There are many theories about what causes aging. Three of these are considered here.

Genetic in Origin

Several lines of evidence indicate that aging has a genetic basis. (1) The number of times a cell divides is species-specific. The maximum number of times human cells divide is around 50. Perhaps as we grow older, more and more cells are unable to divide, and instead, they undergo degenerative changes and die. (2) Some cell lines may become nonfunctional long before the maximum number of divisions has occurred. Whenever DNA replicates, mutations can occur, and this can lead to the production of nonfunctional proteins. Eventually, the number of inadequately functioning cells can build up, which contributes to the aging process. (3) The children of long-lived parents tend to live longer than those of short-lived parents. Recent work suggests that when an animal produces fewer free radicals, it lives longer. Free radicals are unstable molecules that carry an extra electron. In order to stabilize themselves, free radicals donate an electron to another molecule like DNA or proteins (e.g., enzymes) or lipids found in plasma membranes. Eventually these molecules are unable to function, and the cell is destroyed. There are genes that code for antioxidant enzymes that detoxify free radicals. This research suggests that animals with particular forms of these genes—and therefore more efficient antioxidant enzymes—live longer.

Whole-Body Process

A decline in the hormonal system can affect many different organs of the body. For example, type II diabetes is common in older individuals. The pancreas makes insulin, but the cells lack the receptors that enable them to respond.

Figure 17.15   Aging.
Aging is a slow process during which the body undergoes changes that eventually bring about death, even if no marked disease or disorder is present. Medical science is trying to extend the human life span and the health span, the length of time the body functions normally.

Menopause in women occurs for a similar reason. There is plenty of follicle-stimulating hormone in the bloodstream, but the ovaries do not respond. Perhaps aging results from the loss of hormonal activities and a decline in the functions they control.

The immune system, too, no longer performs as it once did, and this can affect the body as a whole. The thymus gland gradually decreases in size, and eventually most of it is replaced by fat and connective tissue. The incidence of cancer increases among the elderly, which may signify that the immune system is no longer functioning as it should. This idea is substantiated, too, by the increased incidence of autoimmune diseases in older individuals.

It is possible, though, that aging is not due to the failure of a particular system that can affect the body as a whole, but to a specific type of tissue change that affects all organs and even the genes. It has been noticed for some time that proteins—such as collagen fibers which are present in many support tissues—become increasingly cross-linked as peo-

ple age. Undoubtedly, this cross-linking contributes to the stiffening and the loss of elasticity characteristic of aging tendons and ligaments. It may also account for the inability of such organs as the blood vessels, the heart, and the lungs to function as they once did. Some researchers have now found that glucose has the tendency to attach to any type of protein, which is the first step in a cross-linking process. They are presently experimenting with drugs that can prevent cross-linking.

Extrinsic Factors

The current data about the effects of aging are often based on comparisons of the characteristics of the elderly to younger age groups; but perhaps today's elderly were not as aware when they were younger of the importance of, for example, diet and exercise to general health. It is possible, then, that much of what we attribute to aging is instead due to years of poor health habits.

Consider, for example, osteoporosis. This condition is associated with a progressive decline in bone density in both males and females so that fractures are more likely to occur after only minimal trauma. Osteoporosis is common in the elderly—by age 65, one-third of women will have vertebral fractures, and by age 81, one-third of women and one-sixth of men will have suffered a hip fracture. While there is no denying that a decline in bone mass occurs as a result of aging, certain extrinsic factors are also important. The occurrence of osteoporosis itself is associated with cigarette smoking, heavy alcohol intake, and perhaps inadequate calcium intake. Not only is it possible to eliminate these negative factors by personal choice, it is also possible to add a positive factor. A moderate exercise program has been found to slow down the progressive loss of bone mass.

Even more important, a proper diet that includes at least five servings of fruits and vegetables a day and a sensible exercise program will most likely help eliminate cardiovascular disease, the leading cause of death today. Experts no longer believe that the cardiovascular system necessarily suffers a large decrease in functioning ability with age. Persons 65 years of age and older can have well-functioning hearts and open coronary arteries if their health habits are good and they continue to exercise regularly.

Effect of Age on Body Systems

Data about how aging affects body systems should be accepted with reservations.

Skin

As aging occurs, skin becomes thinner and less elastic because the number of elastic fibers decreases and the collagen fibers undergo cross-linking, as discussed previously. Also, there is less adipose tissue in the subcutaneous layer; therefore, older people are more likely to feel cold. The loss of thickness partially accounts for sagging and wrinkling of the skin.

Homeostatic adjustment to heat is also limited because there are fewer sweat glands for sweating to occur. There are fewer hair follicles, so the hair on the scalp and the extremities thins out. The number of oil (sebaceous) glands is reduced, and the skin tends to crack. Older people also experience a decrease in the number of melanocytes, making hair gray and skin pale. In contrast, some of the remaining pigment cells are larger, and pigmented blotches appear in skin.

Processing and Transporting

Cardiovascular disorders are the leading cause of death among the elderly. The heart shrinks because there is a reduction in cardiac muscle cell size. This leads to loss of cardiac muscle strength and reduced cardiac output. Still, it is observed that the heart, in the absence of disease, is able to meet the demands of increased activity. It can increase its rate to double or triple the amount of blood pumped each minute even though the maximum possible output declines.

Because the middle layer of arteries contains elastic fibers, which most likely are subject to cross-linking, the arteries become more rigid with time, and their size is further reduced by plaque, a buildup of fatty material. Therefore, blood pressure readings gradually rise. Such changes are common in individuals living in Western industrialized countries but not in agricultural societies. A low cholesterol and saturated fatty acid diet has been suggested as a way to control degenerative changes in the cardiovascular system.

There is reduced blood flow to the liver, and this organ does not metabolize drugs as efficiently as before. This means that as a person gets older, less medication is needed to maintain the same level in the bloodstream.

Cardiovascular problems often are accompanied by respiratory disorders, and vice versa. Growing inelasticity of lung tissue means that ventilation is reduced. Because we rarely use the entire vital capacity, these effects are not noticed unless there is increased demand for oxygen.

There is also reduced blood supply to the kidneys. The kidneys become smaller and less efficient at filtering wastes. Salt and water balance are difficult to maintain, and the elderly dehydrate faster than young people. Difficulties involving urination include incontinence (lack of bladder control) and the inability to urinate. In men, the prostate gland may enlarge and reduce the diameter of the urethra, making urination so difficult that surgery is often needed.

The loss of teeth, which is frequently seen in elderly people, is more apt to be the result of long-term neglect than aging. The digestive tract loses tone, and secretion of saliva and gastric juice is reduced, but there is no indication of reduced absorption. Therefore, an adequate diet, rather than vitamin and mineral supplements, is recommended. There are common complaints of constipation, increased amount of gas, and heartburn, and gastritis, ulcers, and cancer can also occur.

Integration and Coordination

It is often mentioned that while most tissues of the body regularly replace their cells, some at a faster rate than others, the brain and the muscles ordinarily do not. However, contrary to previous opinion, recent studies show that few neural cells of the cerebral cortex are lost during the normal aging process. This means that cognitive skills remain unchanged even though there is characteristically a loss in short-term memory. Although the elderly learn more slowly than the young, they can acquire and remember new material. It is noted that when more time is given for the subject to respond, age differences in learning decrease.

Neurons are extremely sensitive to oxygen deficiency, and if neuron death does occur, it may be due not to aging itself but to reduced blood flow in narrowed blood vessels. Specific disorders, such as depression, Parkinson disease, and Alzheimer disease, are sometimes seen, but they are not common. Reaction time, however, does slow, and more stimulation is needed for hearing, taste, and smell receptors to function as before. After age 50, there is a gradual reduction in the ability to hear tones at higher frequencies, and this can make it difficult to identify individual voices and to understand conversation in a group. The lens of the eye does not accommodate as well and also may develop a cataract. Glaucoma, the buildup of pressure due to increased fluid, is more likely to develop because of a reduction in the size of the anterior cavity of the eye.

Loss of skeletal muscle mass is not uncommon, but it can be controlled by a regular exercise program. There is a reduced capacity to do heavy labor, but routine physical work should be no problem. A decrease in the strength of the respiratory muscles and inflexibility of the rib cage contribute to the inability of the lungs to expand as before, and reduced muscularity of the urinary bladder contributes to difficulties with urination.

As noted before, aging is accompanied by a decline in bone density. Osteoporosis, characterized by a loss of calcium and mineral from bone, is not uncommon, but there is evidence that proper health habits can prevent its occurrence. Arthritis, which causes pain upon movement of the joint, is also seen.

Weight gain occurs because the basal metabolism decreases and inactivity increases. Muscle mass is replaced by stored fat and retained water.

The Reproductive System

Females undergo menopause, and thereafter the level of female sex hormones in blood falls markedly. The uterus and the cervix are reduced in size, and there is a thinning of the walls of the oviducts and the vagina. The external genitals become less pronounced. In males, the level of androgens falls gradually over the age span of 50–90, but sperm production continues until death.

Figure 17.16 **Remaining active.**
The aim of gerontology is to allow the elderly to enjoy living.

It is of interest that as a group, females live longer than males. Although their health habits may be better, it is also possible that the female sex hormone estrogen offers women some protection against cardiovascular disorders when they are younger. Males suffer a marked increase in heart disease in their forties, but an increase is not noted in females until after menopause, after which women lead men in the incidence of stroke. Men are still more likely than women to have a heart attack, however.

Aging Well

We have listed many adverse effects due to aging, but it is important to emphasize that while such effects are seen, they are not a necessary occurrence (Fig. 17.16). We must discover any extrinsic factors that precipitate these adverse effects and guard against them. Just as it is wise to make the proper preparations to remain financially independent when older, it is also wise to realize that biologically successful old age begins with the health habits developed when we are younger.

Bioethical Issue

The fetus is subject to harm by maternal use of medicines and drugs of abuse including nicotine and alcohol. Also, various sexually transmitted diseases, notably an HIV infection, can be passed on to the fetus by way of the placenta. Women need to be aware of the need to protect their unborn child from harm. Indeed their behavior should be protective if they are sexually active even if they are using a recognized form of birth control. Harm can occur before a woman realizes she is pregnant!

Because we are now aware of the need for maternal responsibility before a child is born, there has been a growing acceptance of prosecuting women when a newborn has a condition such as fetal alcohol syndrome. This condition could only have been caused by the drinking habits of the mother. Employers have also become aware that they might be subject to prosecution. To protect themselves, Johnson

Controls, a U.S. battery manufacturer developed a fetal protection policy. No woman who could bear a child was offered a job that might expose her to toxins that could negatively affect the development of her baby. To get such a job, a woman had to show that she had been sterilized or was otherwise incapable of having children. In 1991, the U.S. Supreme Court declared this policy unconstitutional on the basis of sexual discrimination. The decision was hailed as a victory for women, but was it? The decision was written in such a way that women alone, and not an employer, is responsible for any harm done to the fetus by workplace toxins.

Some have noted that prosecuting women for causing prenatal harm can itself have a detrimental effect. The women may tend to avoid prenatal treatment, thereby increasing the risk to their children. Or they may opt for an abortion in

order to avoid the possibility of prosecution. The women feel they are in a no-win situation. If they have a child that has been harmed due to their behavior, they are bad mothers or if they abort, they are also bad mothers.

Questions

1. Do you believe women should be prosecuted if their child is born with a preventable condition? Why or why not?
2. Is the woman or physician responsible when a woman of child-bearing age takes a prescribed medication that does harm to the unborn? Is the employer or the woman responsible when a workplace toxin does harm to an unborn?
3. Should sexually active women who can bear a child be expected to avoid substances or situations that could possibly harm an unborn even if they are using birth control? Why or why not?

Summarizing the Concepts

17.1 Fertilization

Only one sperm head actually enters the egg, and this sperm's nucleus fuses with the egg nucleus. The zygote begins to develop into an embryo, which travels down the oviduct and embeds itself in the uterine lining. Cells surrounding the embryo produce HCG, and the presence of this hormone indicates that the female is pregnant.

17.2 Human Development before Birth

The processes of development (cleavage, morphogenesis, differentiation, and growth) occur during the early developmental stages, which consist of the morula, blastocyst, gastrula, and neurula. The extraembryonic membranes, including the placenta, are special features of human development. Fetal lungs do not operate, and the fetal circulation takes blood to the placenta, where exchange takes place.

Human development consists of embryonic (first two months) and fetal (third to ninth month) development. During the embryonic period, the extraembryonic membranes appear and serve important functions: the embryo acquires organ systems. During the fetal period, there is a refinement of these systems.

During the third and fourth months, it is obvious that the skeleton is becoming ossified. The sex of the individual is now distinguishable. From the fifth to the ninth months, the fetus continues to grow and to

gain weight. Babies born after six or seven months may survive, but full-term babies have a better chance of survival.

17.3 Birth

Birth, or parturition, has three phases. During the first stage, the cervix dilates to allow passage of the baby's head and body. The amnion usually bursts sometime during this stage. During the second stage, the baby is born and the umbilical cord is cut. As the baby takes his or her first breath, anatomical changes convert fetal circulation to adult circulation. During the third stage, the placenta is delivered.

Milk is not produced during pregnancy because of hormonal suppression, but once the child is born, milk production begins. Prolactin promotes the production of milk, and oxytocin allows milk letdown.

17.4 Human Development after Birth

Development after birth consists of infancy, childhood, adolescence, and adulthood. Young adults are at their prime, and then the aging process begins. Aging encompasses progressive changes from about age 20 on that contribute to an increased risk of infirmity, disease, and death. Perhaps aging is genetic in origin, perhaps it is due to a change that affects the whole body, or perhaps it is due to extrinsic factors.

Studying the Concepts

1. Describe the process of fertilization and the events immediately following it. 364–65
2. What is the basis of the pregnancy test? 365
3. What are the processes of development? Relate these processes to the early development stages. 366–67
4. Name the four extraembryonic membranes, and give a function for each. 368
5. Describe the structure and function of the placenta. 368
6. During which period of pregnancy does a woman have to be the most careful about the intake of medications and other drugs? 368
7. Describe the vascular components of the umbilical cord and how they relate to fetal circulation. 368
8. Specifically, what events normally occur during the first, second, third, and fourth weeks of development? 372–73, 374 What events normally happen during the second through the ninth months? 373–75
9. In general, describe the physical changes in the mother during pregnancy. 374
10. What are the three stages of birth? Describe the events of each stage. 376–77
11. Describe the suckling reflex. 377
12. Discuss three theories of aging. What are the major changes in body systems that have been observed as adults age? 378–80

Testing Your Knowledge of the Concepts

In questions 1–4, match the stage of development to the descriptions below.
a. morula
b. blastula
c. gastrula
d. neurula

_____ 1. Invagination of cells along the primitive streak occurs.
_____ 2. Notochord induces formation of neural tube.
_____ 3. Morphogenesis begins as a cavity forms.
_____ 4. Cleavage has resulted in a ball of cells.

In questions 5–7, indicate whether the statement is true (T) or false (F).

_____ 5. All major organs form during embryonic development.
_____ 6. The umbilical vein carries blood rich in nutrients and oxygen to the fetus.
_____ 7. In most deliveries, the head appears before the rest of the body.

In questions 8–15, fill in the blanks.

8. The _____ membranes include the chorion, the _____, the yolk sac, and the allantois.
9. Once the embryo arrives at the uterus, it begins to _____ itself in the uterine lining.
10. Fertilization occurs when the _____ nucleus fuses with the _____.
11. The zygote divides as it passes down a uterine tube. This process is called _____.
12. When cells take on a specific structure and function, _____ has occurred.
13. Fetal development begins at the end of the _____ month.
14. During development, the nutrient needs of the developing embryo (fetus) are served by the _____.
15. The hormone _____ is required for milk letdown during the suckling reflex.

16. Label this diagram.

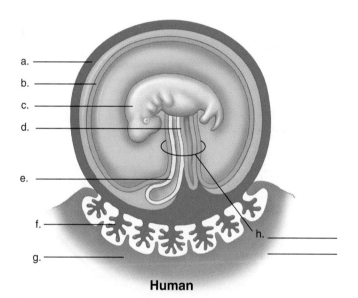

a. _____
b. _____
c. _____
d. _____
e. _____
f. _____
g. _____
h. _____

Human

Applying Your Knowledge to the Concepts

These questions pertain to development and aging.

1. Fertilization of an egg that has a chance of becoming an embryo normally takes place in the upper one-third of the oviduct. Explain the benefit for this.
2. Fetal hemoglobin is different than adult hemoglobin. It has a greater affinity for oxygen. What advantage is there to this difference?
3. Osteoporosis is very common in older people, particularly women. What factors can you control that will decrease your chances of suffering from osteoporosis?
4. Rarely more than one sperm enters an egg. If any egg were fertilized by more than one sperm, what would be the probable result?

Understanding the Terms

aging 378 ✓
allantois 368 *umbilical vessels*
amnion 368 *fluid*
blastocyst 366 *cell mass*
chorion 368 *placenta*
chorionic villi 368
colostrum 377 *— milk*
ectoderm 366 *— nervous system* *outer*
embryo 365 *fetus*
endoderm 366 *—Inner, muscle*
episiotomy 377 *surgery*
extraembryonic
   membranes 368

fertilization 364 *creates zygote*
gerontology 378
implantation 365 *attachment*
induction 367
lanugo 375 *hair*
mesoderm 366 *middle*
parturition 376
placenta 368
umbilical cord 368
vernix caseosa 375
yolk sac 368 *blood cell formation*
zygote 365

Match the terms to these definitions:

a. *Lanugo*   Short, fine hair that is present during the later portion of fetal development.
b. *Parturition* Process and the developmental stages by which a cell becomes specialized for a particular function.
c. *Amnion* Extraembryonic membrane that forms a fluid-filled sac around the embryo.
d. *Fertilization* Union of a sperm nucleus and an egg nucleus, which creates a zygote with the diploid number of chromosomes.
e. *Endoderm* Inner germ layer that lines the archenteron/gut of the gastrula; it becomes the lining of the digestive and respiratory tracts and associated organs.

Applying Technology to the Concepts

Your study of development and aging is supported by these available technologies:

Essential Study Partner CD-ROM
Animals → Development
Visit the Mader web site for related ESP activities.

Exploring the Internet
The Mader Home Page provides resources and tools as you study this chapter.

http://www.mhhe.com/biosci/genbio/mader

HealthQuest CD-ROM
1 Stress Management & Mental Health
→ Gallery → Suicide Rates
5 Cardiovascular Health → Gallery
→ Leading Causes of Death

Further Readings for Part 5

Alcamo, I. E. 1997. *AIDS: The biological basis.* 2d ed. Dubuque, Iowa: Wm. C. Brown Publishers. This easily understood book focuses on the biological basis of AIDS.

Alexander, N. J. March/April 1996. Barriers to sexually transmitted diseases. *Scientific American Science & Medicine* 3(2):32. Article discusses the effectiveness of certain contraceptives in protecting women against STDs.

Black, P. H. November/December 1995. Psychoneuroimmunology: Brain and immunity. *Scientific American Science & Medicine* 2(6):16. Article discusses the role of stress in susceptibility to infections, and cancer and HIV progression.

Caldwell, J. C., and Caldwell, P. March 1996. The African AIDS epidemic. *Scientific American* 273(3):62. Article discusses a factor most likely responsible for causing the high rate of AIDS transmission in Africa.

Cox, F. D. 1996. *The AIDS booklet.* 4th ed. Dubuque, Iowa: Brown & Benchmark Publishers. This easy to read, informative booklet covers the transmission, prevention, and treatment of AIDS.

Crooks, R., and Baur, K. 1996. *Our sexuality.* 6th ed. Redwood City, Calif.: Benjamin/Cummings Publishing. Introduction to the biological, psychosocial, behavioral, and cultural aspects of sexuality.

Curiel, T. September/October 1997. Gene therapy: AIDS-related malignancies. *Science & Medicine* 4(5):4. The field of AIDS-related gene therapies is advancing.

Duan, L., and Pomerantz, R. J. May/June 1996. Intracellular antibodies for HIV-1 gene therapy. *Science & Medicine* 3(3):24. Article discusses the cloning of synthetic antibody fragments that can inhibit the function of viral proteins.

Duke, R. C., et al. December 1996. Cell suicide in health and disease. *Scientific American* 275(6):80. Failures in the processes of cellular self-destruction may give rise to cancer, AIDS, Alzheimer disease, and some genetic diseases.

Dusenbery, D. B. 1996. *Life at small scale: The behavior of microbes.* New York: Scientific American Library. This easy-to-read, well-illustrated text describes how microbes respond to the physical demands of their environment.

Ingber, D. E. January 1998. The architecture of life. *Scientific American* 278(1):48. Simple mechanical rules may govern development, tissue organization, and cellular movement.

MacDonald, P. C., and Casey, M. L. March/April 1996. Preterm birth. *Scientific American Science & Medicine* 3(2):42. Article discusses the role of oxytocin, prostaglandins, and infections in the initiation of human labor.

Mader, S. S. 1990. *Human reproductive biology.* 2d ed. Dubuque, Iowa: Wm. C. Brown Publishers. An introductory text covering human reproduction in a clear, easily understood manner.

Markowitz, M. H. June 1998. A new dawn in AIDS treatments. *Discover* 19(6):S-6. A new combination therapy greatly reduces viral replication.

Newman, J. December 1995. How breast milk protects newborns. *Scientific American* 273(6):76. Human milk contains special antibodies that boost the newborn's immune system.

Nowak, M. A., and McMichael, A. J. August 1995. How HIV defeats the immune system. *Scientific American* 273(2):58. Full-blown AIDS results when the proliferating virus finally overwhelms the body's defenses, a process that may take years.

Nusslein-Volhard, C. August 1996. Gradients that organize embryo development. *Scientific American* 275(2):54. Nobel Prize-winning researcher describes how chemical gradients of substances called morphogens give an evolving embryo its shape.

O'Brien, S. J., and Dean, M. September 1997. In search of AIDS-resistance genes. *Scientific American* 277(3):44. Some genes deter the AIDS virus; studying these genes may lead to prevention and treatment strategies.

Packer, C. July/August 1998. Why menopause? *Natural History* 107(6):24. Article addresses reasons why menopause occurs so early in life, compared to other aging processes.

Pennisi, E. 17 July 1998. Genome reveals wiles and weak points of syphilis. *Science* 281(5375):324. Gene sequencing of the spirochete that causes syphilis may finally make it possible to culture spirochetes for study.

Prescott, L. M., et al. 1996. *Microbiology.* 3d ed. Dubuque, Iowa: Wm. C. Brown Publishers. This introductory text covers all major areas of microbiology.

Ricklefs, R. E., and Finch, C. E. 1995. *Aging: A natural history.* New York: Scientific American Library. This text emphasizes the nature of aging and the mechanisms of physiological deterioration.

Ross, I. K. 1995. *Aging of cells, humans and societies.* Dubuque, Iowa: Wm. C. Brown Publishers. Presents current concepts on aging.

Science. June 19, 1998. AIDS research news. The issue contains an entire section on AIDS research, vaccine progress, treatment successes and failures.

Science & Medicine. Special Report 5(2):36. March/April 1998. Dynamics of HIV infection. HIV-1 may use a variety of coreceptors to gain entry into cells.

Scientific American. July 1998. Defeating AIDS: What will it take? 279(1):81. Nine separate articles address AIDS problems and issues.

Stolley, P. D., and Lasky, T. 1995. *Investigating disease patterns: The science of epidemiology.* New York: Scientific American Library. The process of epidemiology and its contribution to the understanding of disease is covered in this interesting, easy-to-read book.

Tortora, G. J., et al. 1995. *Microbiology: An introduction.* Redwood City, Calif.: Benjamin/Cummings Publishing. This introductory microbiology text presents the diversity of microbial life and the roles of microbes in nature and in our daily lives.

Van De Graaff, K. M. 1995. *Survey of infectious and parasitic diseases.* Dubuque, Iowa: Wm. C. Brown Publishers.

Wallace, D. C. August 1997. Mitochondrial DNA in aging and disease. *Scientific American* 277(2):40. Genes in mitochondria have been linked to certain diseases, and could be important in age-related disorders.

Weindruch, R. January 1996. Caloric restriction and aging. *Scientific American* 274(1):64. Consuming fewer calories may increase longevity.

Weiss, R. November 1997. Aging—new answers to old questions. *National Geographic* 192(5):2. The mechanics of human aging are studied.

Zhong, G., and Brunham, R. C. September/October 1998. Chlamydial resistance to host defense. *Science & Medicine* 5(5):38. Chlamydia may produce anti-apoptosis factors.

Human beings practice sexual reproduction, which requires gamete production. Gametes carry half the total number and various combinations of chromosomes and genes. It is sometimes possible to determine the chances of an offspring receiving a particular parental gene and, therefore, inheriting a genetic disorder.

Genes, now known to be constructed of DNA, control not only the metabolism of the cell but also, ultimately, the characteristics of the individual. The step-by-step procedure by which DNA specifies protein synthesis has been discovered. Biotechnology is a new and burgeoning field that permits the extraction of DNA from one organism and its insertion in a different organism for a purpose useful to human beings.

Cancer is a cellular disease brought on by DNA mutations that transform a normal cell into a cancer cell. Cancer-causing genes are derived from normal genes, which keep cell division under control. Because environmental factors play a major role in causing or promoting cancer, it may be possible to reduce its incidence. Knowledge about the detection and treatment of cancer is improving daily.

Chapter 18

Chromosomal Inheritance

Chapter Concepts

18.1 **Chromosomal Inheritance**
- Normally, humans inherit 22 pairs of autosomes and one pair of sex chromosomes for a total of 46 chromosomes. 386
- Normally, males have the sex chromosomes XY and females have the sex chromosomes XX. 386
- Abnormalities arise when humans inherit an abnormal number or type of chromosome. 387

18.2 **Human Life Cycle**
- The human life cycle involves two types of cell divisions: mitosis and meiosis. 390

18.3 **Mitosis**
- Mitosis, cell division in which the chromosomal number remains constant, is involved in growth and repair. 391

18.4 **Meiosis**
- Meiosis, cell division in which the chromosomal number is reduced by one-half, is involved in gamete production. 394

Mary was a very attractive young woman who liked to play ice hockey. Her high school gym teacher took an interest in her ability and gave her extra coaching. She hoped that one day Mary would play on an Olympic team.

But something was wrong. Mary was sixteen and still not menstruating. Her parents decided to have her undergo a series of medical tests. Much to the surprise of everyone, Mary had an X and Y chromosome in the nucleus of her cells. She was a chromosomal male.

The doctor explained to Mary and her parents that Mary had testicular feminization syndrome. She has internal testes that produce testosterone but her cells won't respond to it. Her genitals are like those of a female and she has well-developed breasts. However, she will never be able to have children.

Mary will be able to go on and play hockey in the Olympics but she has to always carry a letter explaining her condition. Otherwise she will be disqualified because of her sex chromosomes.

This chapter describes various syndromes that occur when people inherit an abnormal chromosomal number. Chromosomal abnormalities are apt to occur during meiosis, the type of cell division needed for gamete production.

18.1   Chromosomal Inheritance

In a nondividing cell, the nucleus contains indistinct and diffuse *chromatin*, but in a dividing cell, chromatin becomes the short and thick *chromosomes*. A cell may be photographed just prior to division so that a picture of the chromosomes is obtained. The picture may be entered into a computer and the chromosomes electronically arranged by pairs (Fig. 18.1).

Individual chromosomes are recognized by their size, location of the centromere (a constriction), and characteristic banding due to staining. The resulting display of pairs of chromosomes is called a **karyotype.** Although both males and females have 23 pairs of chromosomes, one of these pairs is of unequal length in males. The larger chromosome of this pair is the **X chromosome** and the smaller is the **Y chromosome.** Females have two X chromosomes in their karyotype; males have one X and one Y. The X chromosome and Y chromosome are called the **sex chromosomes** because they contain the genes that determine sex.

The other chromosomes, known as **autosomes,** include all of the pairs of chromosomes except the X and Y chromosomes. Each pair of autosomes in the human karyotype is numbered.

1. Blood is centrifuged to separate out blood cells.

2. Only white blood cells are transferred and treated to stop cell division.

3. Sample is fixed, stained, and spread on a microscope slide.

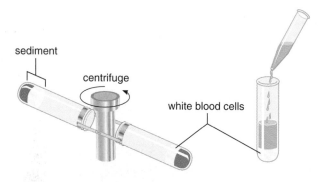

4. Slide is examined microscopically, and the chromosomes are photographed. Computer arranges the chromosomes into pairs.

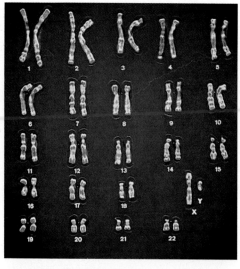

5. Karyotype: Chromosomes are paired by size, centromere location, and banding patterns.

Figure 18.1   Human karyotype preparation.
As illustrated here, the stain used can result in chromosomes with a banded appearance. The bands help researchers identify and analyze the chromosomes.

Down Syndrome

A **syndrome** is a group of characteristics or symptoms that appear together and tend to indicate the presence of a particular disorder. *Down syndrome* (Fig. 18.2) is easily recognized by these characteristics: short stature; an eyelid fold; stubby fingers; a wide gap between the first and second toes; a large, fissured tongue; a round head; a palm crease, the so-called simian line; and, unfortunately, mental retardation, which can sometimes be severe.

Down syndrome is also called trisomy 21 because the individual usually has three copies of chromosome 21. In most instances, the egg had two copies instead of one of this chromosome. (In 23% of the cases studied, however, the sperm had the extra chromosome 21.) The chances of a woman having a Down syndrome child increase rapidly with age, starting at about age 40. The frequency of Down syndrome is 1 in 800 births for mothers under 40 years of age and 1 in 80 for mothers over 40 years of age.

Although an older woman is more likely to have a Down syndrome child, most babies with Down syndrome are born to women younger than age 40 because this is the age group having the most babies. Amniocentesis (removing fluid and cells from the amniotic sac surrounding the fetus) followed by karyotyping can detect a Down syndrome child. However, young women are not usually encouraged to undergo this procedure because the risk of complications from the procedure is greater than the risk of having a Down syndrome child. It has been observed that a woman carrying a Down syndrome child sometimes has a lower than usual amount of a substance called α-fetoprotein (AFP) in her blood. These results should certainly be followed up with amniocentesis and karyotyping.

It is known that the genes that cause Down syndrome are located on the bottom third of chromosome 21 (Fig. 18.2*b*), and extensive investigative work has been directed toward discovering the specific genes responsible for the characteristics of the syndrome. Thus far, investigators have discovered several genes that may account for various conditions seen in persons with Down syndrome. For example, they have located genes most likely responsible for the increased tendency toward leukemia, cataracts, accelerated rate of aging, and mental retardation. The gene for mental retardation, dubbed the *Gart* gene, causes an increased level of purines in the blood, a finding associated with mental retardation. One day it might be possible to control the expression of the *Gart* gene even before birth so that at least this symptom of Down syndrome does not appear.

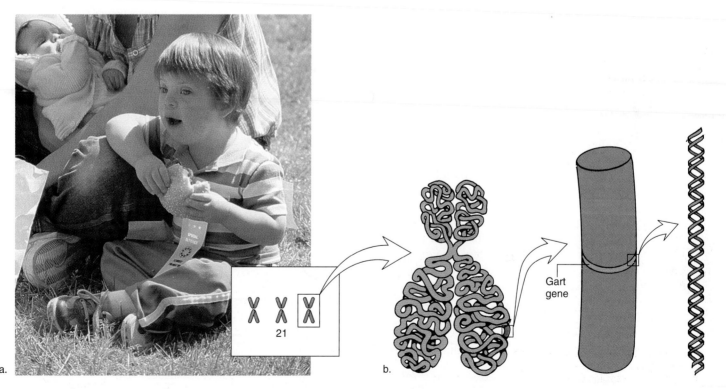

a.

b.

Figure 18.2 **Down syndrome.**
a. Common characteristics of the syndrome include a wide, rounded face and a fold of the upper eyelids. Mental retardation, along with an enlarged tongue, makes it difficult for a person with Down syndrome to speak distinctly. **b.** Karyotype of an individual with Down syndrome shows an extra chromosome 21. More sophisticated technologies allow investigators to pinpoint the location of specific genes associated with the syndrome. An extra copy of the *Gart* gene, which leads to a high level of purines, may account for the mental retardation seen in persons with Down syndrome.

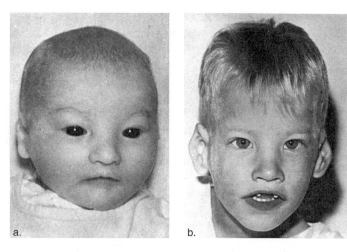

Figure 18.3   Cri du chat syndrome.
a. An infant and **(b)** an older child with this syndrome.

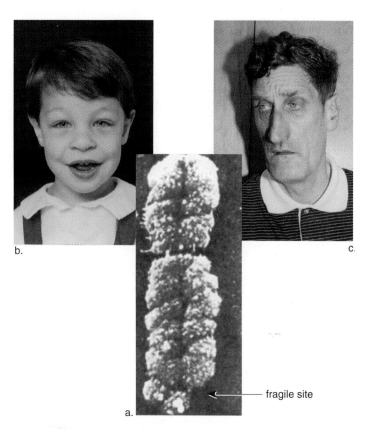

fragile site

Figure 18.4   Fragile X syndrome.
a. An arrow points out the fragile site of this fragile X chromosome.
b. A young person with the syndrome appears normal but **c.** with age, the elongated face has a prominent jaw, and the ears noticeably protrude.

Cri du Chat Syndrome

A chromosomal deletion is responsible for *cri du chat* (cat's cry) *syndrome,* which has a frequency of 1 in 50,000 live births (Fig. 18.3). An infant with this syndrome has a moon face, small head, and a cry that sounds like the meow of a cat because of a malformed larynx. An older child has an eyelid fold and misshapen ears that are placed low on the head. Severe mental retardation becomes evident as the child matures. A karyotype shows that a portion of one chromosome 5 is missing (deleted), while the other chromosome 5 is normal, as are all other chromosomes.

Fragile X Syndrome

Males outnumber females by about 25% in institutions for the mentally retarded. In some of these males, the X chromosome is nearly broken, leaving the tip hanging by a flimsy thread. These males are said to have *fragile X syndrome* (Fig. 18.4).

Fragile X syndrome occurs in one in 1,000 male births and one in 2,500 female births. As children, fragile X syndrome individuals appear to be normal except they may be hyperactive or autistic. Their speech is delayed in development and is often repetitive in nature. As adults, they are short in stature with a long face. The jaw is prominent, and there are big, usually protruding ears (Fig. 18.4*b* and *c*). Males also have large testicles. Stubby hands, lax joints, and a heart defect may also be seen. Symptoms, including mental retardation, are not as severe in females.

The inheritance of an abnormal number of chromosomes or a defective chromosome can result in a syndrome with particular symptoms.

Sex Chromosomal Inheritance

The sex chromosomes in humans are called X and Y. Since women are XX, an egg always bears an X, but since males are XY, a sperm can bear an X or a Y. Therefore, the sex of the newborn child is determined by the father. If a Y-bearing sperm fertilizes the egg, then the XY combination results in a male. On the other hand, if an X-bearing sperm fertilizes the egg, the XX combination results in a female. All factors being equal, there is a 50% chance of having a girl or a boy.

♀ \ ♂	X	Y
X	XX	XY

However, for reasons that are not clear, more males than females are conceived, but from then on, the death rate among males is higher than for females. By age 85, there are twice as many females as males.

Too Many/Too Few Sex Chromosomes

Turner syndrome occurs in one in 6,000 births. The individual is XO, with one sex chromosome, an X; the O signifies the absence of a second sex chromosome. These females are short, have a broad chest, and webbed neck. The ovaries, oviducts, and uterus are very small and nonfunctional. Turner females do not undergo puberty or menstruate, and there is a lack of breast development (Fig. 18.5*a*). They are usually of normal intelligence and can lead fairly normal lives, but they are infertile even if they receive hormone supplements.

Klinefelter syndrome occurs in one in 1,500 births. These males with two or more X chromosomes in addition to a Y chromosome are sterile. The testes and prostate gland are underdeveloped, and there is no facial hair. Also, there may be some breast development (Fig. 18.5*b*). Affected individuals have large hands and feet and very long arms and legs. They are usually slow to learn but not mentally retarded unless they inherit more than two X chromosomes.

The *triplo-X syndrome* occurs in one in 1,500 births. A triplo-X female has three or more X chromosomes. It might be supposed that the XXX female is especially feminine, but this is not the case. Although in some cases there is a tendency toward learning disabilities, most triplo-X females have no apparent physical abnormalities except that they may have menstrual irregularities, including early onset of menopause.

Jacob syndrome occurs in one in 1,000 births. These XYY males are usually taller than average, suffer from persistent acne, and tend to have speech and reading problems. At one time, it was suggested that these men were likely to be criminally aggressive, but it has since been shown that the incidence of such behavior among them may be no greater than among XY males.

Individuals sometimes are born with the sex chromosomes XO (Turner syndrome), XXY (Klinefelter syndrome), XXX (triplo-X syndrome), and XYY (Jacob syndrome). No matter how many X chromosomes there are, an individual with a Y chromosome is usually a male.

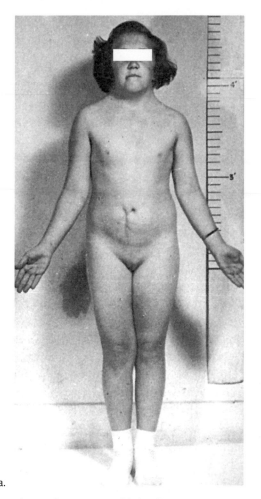
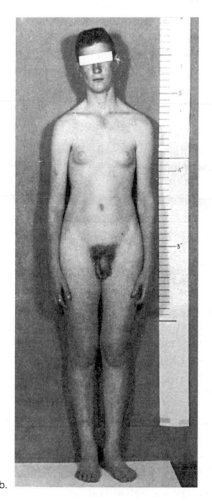

a.
b.

Figure 18.5   Abnormal sex chromosomal inheritance.
a. Female with Turner (XO) syndrome, which includes a web neck, short stature, and immature sexual features. **b.** A male with Klinefelter (XXY) syndrome, which is marked by small testes and development of the breasts in some cases.

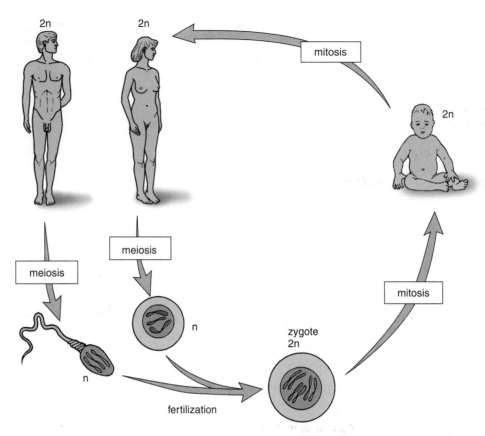

Figure 18.6   Life cycle of humans.
Meiosis in males is a part of sperm production, and meiosis in females is a part of egg production. When a haploid sperm fertilizes a haploid egg, the zygote is diploid. The zygote undergoes mitosis as it develops into a newborn child. Mitosis continues after birth until the individual reaches maturity, and then the life cycle begins again.

Table 18.1	Mitosis Versus Meiosis		
Location	**Cell Division**	**Description**	**Result**
Somatic (body) cells	Mitosis	2n (diploid) → 2n (diploid)	Growth and repair
Sex organs	Meiosis	2n (diploid) → n (haploid)	Gamete production

18.2 Human Life Cycle

The human life cycle involves growth and sexual reproduction (Fig. 18.6). During growth, cells divide by a process called *mitosis*, which ensures that each and every cell has a complete number of chromosomes. Sexual reproduction requires the production of sex cells, which have half the number of chromosomes. A type of cell division called *meiosis* reduces the chromosomal number by one-half.

Meiosis occurs in the sex organs. In males, it produces the cells that become sperm; in females, it produces the cells that become eggs. The sperm and the egg are the sex cells, or **gametes.** Gametes contain half the number of chromosomes compared to **somatic** (body) **cells**—one chromosome from each of the pairs of chromosomes. This is called the **haploid (n)** number of chromosomes; the haploid number of chromosomes in humans is 23.

A new individual comes into existence when a sperm fertilizes an egg. The resulting zygote has the **diploid (2n)** number of chromosomes. Each parent contributes one chromosome to each of the pairs of chromosomes present. As the individual develops, *mitosis* occurs and each somatic (body) cell has the diploid number of chromosomes. In humans, the diploid number is 46 because there are 23 pairs of chromosomes.

Table 18.1 summarizes the major differences between mitosis and meiosis in multicellular animals.

The life cycle of humans requires two types of cell division: mitosis and meiosis.

18.3 Mitosis

Mitosis is cell division that produces *two daughter cells, each with the same number and kinds of chromosomes as the parent cell, the cell that divides.*[1] Therefore, following mitotic cell division, the parent cell and the daughter cells are genetically identical. Mitosis occurs as part of the cell cycle.

Cell Cycle

The **cell cycle** consists of mitosis and interphase. The cell divides, and then it enters interphase before dividing again. Therefore, **interphase** is the interval of time between cell divisions. The length of time required for the entire cell cycle varies according to the organism and even the type of cell within the organism, but 18–24 hours is typical for animal cells. Mitosis lasts less than an hour to slightly more than 2 hours; for the rest of the time, the cell is in interphase.

It used to be said that interphase was a resting stage, but we now know that this is not the case. The organelles are metabolically active and are carrying on their normal functions. If the cell is going to divide, *DNA replication* occurs. During replication, DNA is copied and each chromosome becomes duplicated. A duplicated chromosome is composed of two sister chromatids. *Sister chromatids* are genetically identical—they contain the same genes. Also, organelles, including the *centrioles*, duplicate. A nondividing cell has one pair of centrioles, but in a cell that is going to divide, this pair duplicates, and there are two pairs of centrioles outside the nucleus.

The cell cycle includes mitosis and interphase. During interphase, DNA replication results in each chromosome having sister chromatids. The organelles, including centrioles, also duplicate during interphase.

Overview of Mitosis: 2n → 2n

During interphase, the chromosomes are indistinct chromatin, but when mitosis is going to occur, chromatin becomes condensed, and the chromosomes become visible. Before mitosis begins, the parental cell is 2n, and the sister chromatids are held together in a region called the **centromere.** At the completion of mitosis, each chromosome consists of a single **chromatid.**

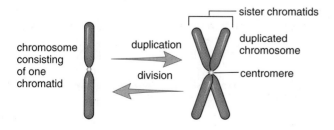

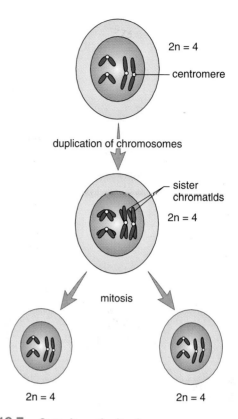

Figure 18.7 Overview of mitosis.
The blue chromosomes were inherited from one parent and the red chromosomes were inherited from the other parent.

Figure 18.7 gives an overview of mitosis; for simplicity, only four chromosomes are depicted. (In determining the number of chromosomes, it is necessary to count only the number of independent centromeres.) During mitosis, the centromeres divide, the sister chromatids separate, and one of each kind goes into each daughter cell. Therefore, each daughter cell gets a complete set of chromosomes and is 2n. (Following separation, each chromatid is called a chromosome.) Since each daughter cell receives the same number and kinds of chromosomes as the parent cell, each is genetically identical to each other and to the parent cell.

Mitosis occurs in humans when tissues grow or when repair occurs. Following fertilization, the zygote begins to divide mitotically, and mitosis continues during development and the life span of the individual. In the adult, tissues differ as to their ability to divide; nervous and muscle tissue cells seem to lose the ability to divide. Epidermal cells, which line the respiratory tract and the digestive tract and form the outer layer of the skin, divide continuously. Stem cells in the red bone marrow divide to produce millions of blood cells every day. Whenever repair takes place, as when a broken bone is mended, mitosis has occurred.

Following mitosis, each of two daughter cells has the same number and kinds of chromosomes as the parental cell.

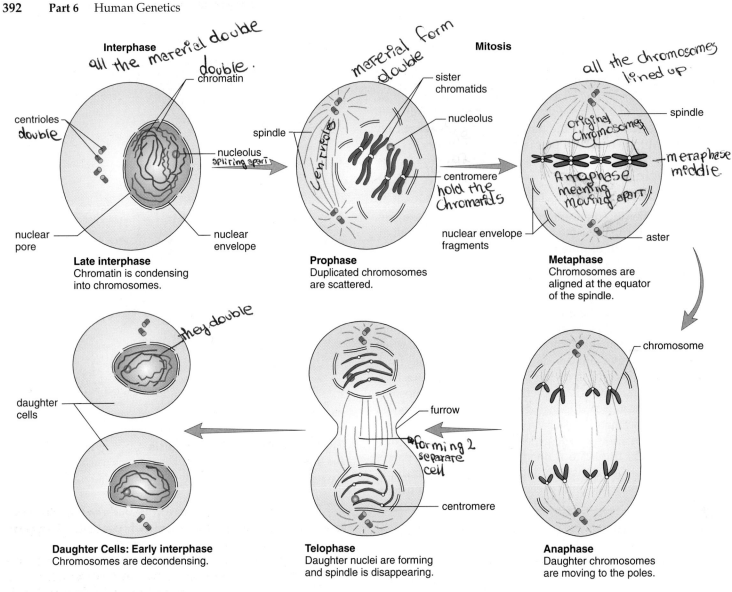

Figure 18.8   Interphase and mitosis.
The blue chromosomes were inherited from one parent and the red chromosomes were inherited from the other parent.

Stages of Mitosis

As an aid in describing the events of mitosis, the process is divided into four phases: prophase, metaphase, anaphase, and telophase (Fig. 18.8). Although the stages of mitosis are depicted as if they were separate, they are actually continuous and flow from one stage to another with no noticeable interruption.

Prophase

The events of **prophase** indicate that cell division is about to occur. The two pairs of centrioles outside the nucleus begin moving away from each other toward opposite ends of the nucleus. Spindle fibers appear between the separating centriole pairs, the nuclear envelope begins to fragment, and the nucleolus begins to disappear.

The chromosomes are now visible. Each is composed of sister chromatids held together at a centromere. Spindle fibers attach to the centromeres as the chromosomes continue to shorten and to thicken. At prophase, chromosomes are randomly placed in the nucleus and have not yet aligned at the equator of the spindle.

Structure of the Spindle   At the end of prophase, a cell has a fully formed spindle. A **spindle** has poles, asters, and fibers. The **asters** are arrays of short microtubules that radiate from the poles, and the fibers are bundles of microtubules that stretch between the poles. Microtubule organizing centers (MTOC) are associated with the centrioles at the poles. These centers organize microtubules when the cell is not dividing—it is likely they also organize the spindle. Centrioles may assist in this function, but their location at the poles of a spindle could be simply to ensure that each daughter cell receives a pair of centrioles.

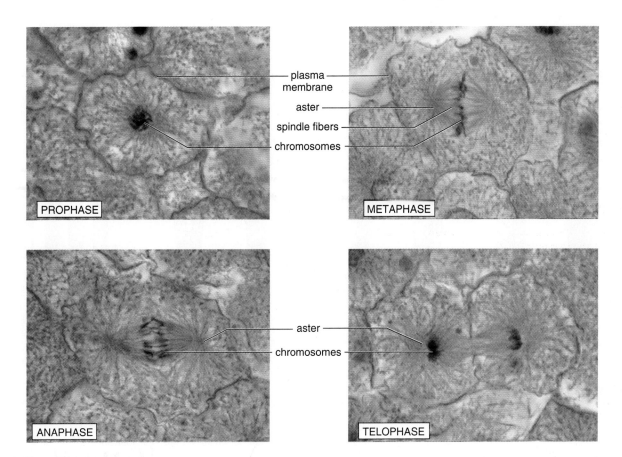

Figure 18.9   **Micrographs of mitosis occurring in a whitefish embryo.**

Metaphase

During **metaphase,** the nuclear envelope is fragmented, and the spindle occupies the region formerly occupied by the nucleus. The chromosomes are now at the *equator* (center) of the spindle. Metaphase is characterized by a fully formed spindle, with the chromosomes, each having two sister chromatids, aligned at the equator (Fig. 18.9). At the close of metaphase, the centromeres uniting the chromatids split.

Anaphase

At the start of **anaphase,** the sister chromatids separate. *Once separated, the chromatids are called chromosomes.* Separation of the sister chromatids ensures that each cell receives a copy of each type of chromosome and thereby has a full complement of genes. During anaphase, the daughter chromosomes move to the poles of the spindle. Anaphase is characterized by the diploid number of chromosomes moving toward each pole.

Function of the Spindle   The spindle brings about chromosomal movement. Two types of spindle fibers are involved in the movement of chromosomes during anaphase. One type extends from the poles to the equator of the spindle; there they overlap. As mitosis proceeds, these fibers increase in length, and this helps push the chromosomes apart. The chromosomes themselves are attached to other spindle fibers that simply extend from their centromeres to the poles. These fibers get shorter and shorter as the chromosomes move toward the poles, and eventually disappear. These fibers pull the chromosomes apart.

Spindle fibers, as stated earlier, are composed of microtubules. Microtubules can assemble and disassemble by the addition or subtraction of tubulin (protein) subunits. This is what enables spindle fibers to lengthen and shorten and what ultimately causes the movement of the chromosomes.

Telophase

Telophase begins when the chromosomes arrive at the poles. During telophase, the chromosomes become indistinct chromatin again. The spindle disappears as nucleoli appear, and nuclear envelope components reassemble in each cell. Telophase is characterized by the presence of two daughter nuclei.

In animal cells, a slight indentation called a **cleavage furrow** passes around the circumference of the cell. Actin filaments form a contractile ring, and as the ring gets smaller and smaller, the cleavage furrow pinches the cell in half. As a result, each cell becomes enclosed by its own plasma membrane.

Following mitosis, each daughter cell is 2n. When the sister chromatids separate during anaphase, each newly forming cell receives the same number and kinds of chromosomes as the parental cell.

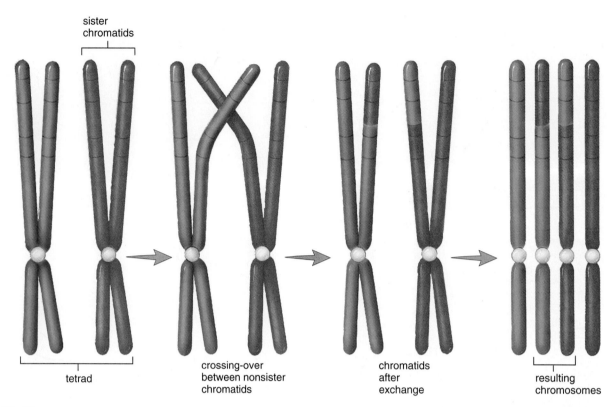

sister
chromatids

tetrad

crossing-over
between nonsister
chromatids

chromatids
after
exchange

resulting
chromosomes

Figure 18.10   Crossing-over.
When homologous chromosomes are in synapsis, the nonsister chromatids exchange genetic material. The illustration shows only one crossover per chromosome pair, but the average is slightly more than two per chromosome pair in humans. Following crossing-over, there is a different combination of genes on each chromatid.

18.4 Meiosis

Meiosis, which requires two cell divisions, results in *four daughter cells, each having one of each kind of chromosome and therefore half the number of chromosomes as the parent cell.*[2] The parent cell has the 2n number of chromosomes, while the daughter cells have the n number of chromosomes. Therefore, meiosis is often called reduction division. Following meiotic cell division, the daughter cells are not genetically identical, and neither is identical to the parent cell.

Overview of Meiosis: 2n → n

Meiosis results in four daughter cells because it consists of two divisions called meiosis I and meiosis II. Before meiosis I begins, each chromosome has duplicated and is composed of two sister chromatids. The parental cell is 2n. Recall that when a cell is 2n, the chromosomes occur in pairs. For example, the 46 chromosomes of humans occur in 23 pairs of chromosomes. These pairs are called **homologous chromosomes.**

During meiosis I, the homologous chromosomes of each pair come together and line up side by side due to a means of attraction still unknown. This so-called synapsis results in a **tetrad,** an association of four chromatids that stay in close proximity until they separate. During synapsis, nonsister chromatids may exchange genetic material. The exchange of genetic material between chromatids is called **crossing-over.** Crossing-over recombines the genes of the parental cell without the loss or gain of genetic material (Fig. 18.10).

Following synapsis during meiosis I, the homologous chromosomes of each pair separate. This separation means that one chromosome from each homologous pair will be found in each daughter cell. There are no restrictions as to which chromosome goes to each daughter cell, and therefore, all possible combinations of chromosomes occur within the daughter cells.

Notice that following meiosis I, the daughter cells have half the number of chromosomes and the chromosomes are still duplicated (Fig. 18.11). Again, counting the number of centromeres tells the number of chromosomes in each daughter cell.

During meiosis I, homologous chromosomes separate, and the daughter cells receive one of each pair. The daughter cells are not genetically identical. The chromosomes are still duplicated.

---

[2]The term *meiosis* technically refers only to nuclear division, but for convenience, it is used here to refer to the division of the entire cell.

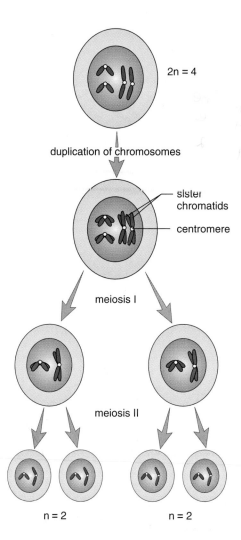

2n = 4

duplication of chromosomes

sister
chromatids

centromere

meiosis I

meiosis II

n = 2   n = 2

Figure 18.11   Overview of meiosis.
Following duplication of chromosomes, the parent cell undergoes
two divisions, meiosis I and meiosis II. During meiosis I, homologous
chromosomes separate, and during meiosis II, chromatids separate.
The final daughter cells are haploid. (The blue chromosomes were
inherited from one parent, and the red chromosomes were inherited
from the other parent.)

When meiosis II begins, the chromosomes are still duplicated. Therefore, no duplication of chromosomes is
needed between meiosis I and meiosis II. The chromosomes
are **dyads** because each one is composed of two sister
chromatids. During meiosis II, the sister chromatids separate in each of the cells from meiosis I. Each of the resulting
four daughter cells has the haploid number of chromosomes
(Fig. 18.11).

During meiosis II, the sister chromatids separate,
and the resulting four daughter cells are each
haploid.

The Importance of Meiosis

Because of meiosis, the chromosomal number stays constant in each generation of humans. The *gametes* are the
sperm in males and eggs in females. In humans, meiosis occurs in the testes and ovaries during the production of the
gametes. Meiosis is not complete in egg production until

fertilization occurs. When a haploid sperm fertilizes a haploid egg, the new individual has the diploid number of
chromosomes. There are three ways in which the new individual is assured a different combination of genes than either parent:

1. Crossing-over recombines the genes on the sister
   chromatids of homologous pairs.
2. Following meiosis, each gamete has a different
   combination of chromosomes.
3. Upon fertilization, recombination of chromosomes
   occurs.

Following meiosis, each of four daughter cells has
the haploid number of chromosomes, whereas the
parental cell was diploid. Meiosis takes place in
the testes and ovaries during the production of the
gametes. It ensures that the chromosomal number
stays constant in each generation and that the
new individual has a different combination of
genes than either parent.

Stages of Meiosis

The same four stages in mitosis—prophase, metaphase, anaphase, and telophase—occur during both meiosis I and meiosis II.

The First Division

The stages of meiosis I are diagrammed in Figure 18.12*a*. During *prophase I*, the spindle appears while the nuclear envelope fragments and the nucleolus disappears. The homologous chromosomes, each having two sister chromatids, undergo **synapsis,** forming tetrads. Crossing-over occurs now, but for simplicity, this event has been omitted from Figure 18.12. In *metaphase I*, tetrads line up at the equator of the spindle. During *anaphase I*, homologous chromosomes of each pair separate and move to opposite poles of the spindle. During *telophase I,* nucleoli appear and nuclear envelopes form as the spindle disappears. In certain species, the plasma membrane furrows to give two cells, and in others, the second division begins without benefit of complete furrowing. Regardless, each daughter cell contains only one chromosome from each homologous pair. The chromosomes are dyads, and each has two sister chromatids. No replication of DNA occurs during a period of time called interkinesis.

The Second Division

The stages of meiosis II for an animal cell are diagrammed in Figure 18.12*b*. At the beginning of *prophase II*, a spindle appears while the nuclear envelope disassembles and the nucleolus disappears. Dyads (one dyad from each pair of homologous chromosomes) are present, and each attaches to the spindle independently. During *metaphase II,* the dyads are lined up at the equator. At the close of metaphase, the centromeres split. During *anaphase II*, the sister chromatids of each dyad separate and move toward the poles. Each pole receives the same number of chromosomes. In *telophase II,* the spindle disappears as nuclear envelopes form. The plasma membrane furrows to give two complete cells, each of which has the haploid, or n, number of chromosomes. Since each cell from meiosis I undergoes meiosis II, there are four daughter cells altogether.

Meiosis involves two cell divisions. During meiosis I, tetrads form and crossing-over occurs. Homologous chromosomes separate, and each daughter cell receives a pair of sister chromatids. During meiosis II, separation of chromatids in daughter cells from meiosis I results in four daughter cells, each with the haploid number of chromosomes.

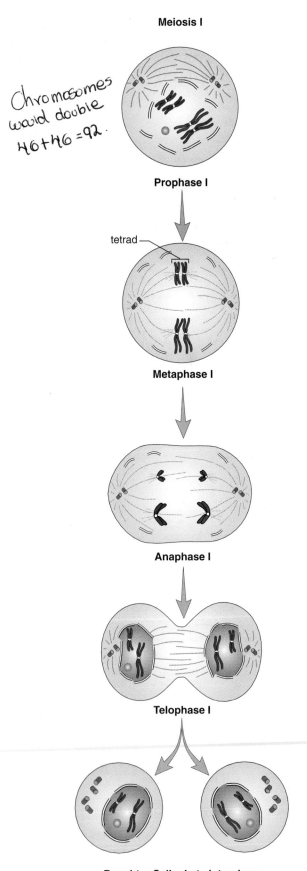

Meiosis I

Chromosomes would double 46+46 = 92.

Prophase I

tetrad

Metaphase I

Anaphase I

Telophase I

a.   **Daughter Cells: Late interphase**

Meiosis II

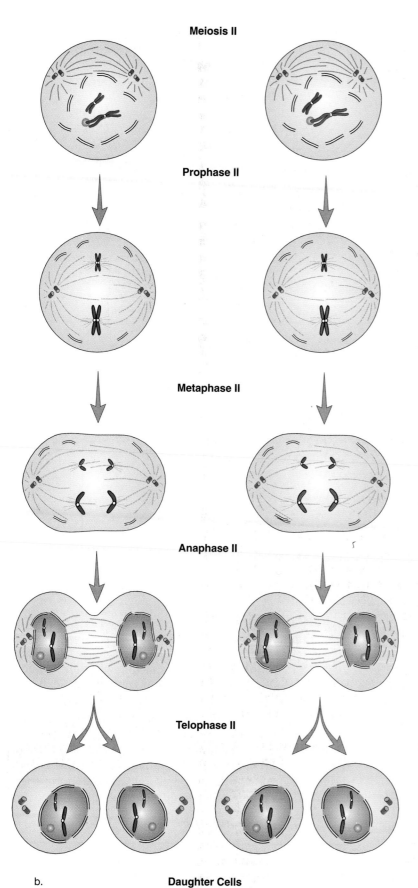

Prophase II

Metaphase II

Anaphase II

Telophase II

b.   **Daughter Cells**

Figure 18.12   Meiosis I and meiosis II.
a. During meiosis I, homologous chromosomes undergo synapsis and then separate so that each daughter cell has only one chromosome from each original homologous pair. For simplicity's sake, the results of crossing-over have not been depicted. Notice that each daughter cell is haploid and each chromosome still has two chromatids. **b.** During meiosis II, sister chromatids separate. Each daughter cell is haploid, and each chromosome consists of one chromatid. (The blue chromosomes were inherited from one parent, and the red chromosomes were inherited from the other parent.)

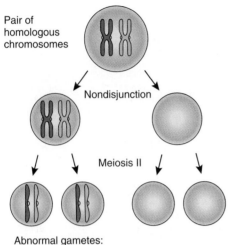

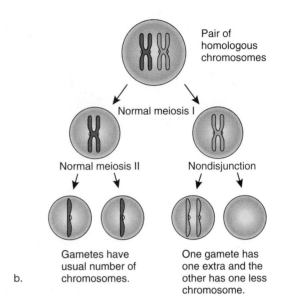

Syndrome	Sex	Chromosomes	Frequency	
			Abortuses	*Births*
Down	M or F	Trisomy 21	1/40	1/800
Patau	M or F	Trisomy 13	1/33	1/15,000
Edward	M or F	Trisomy 18	1/200	1/6,000
Turner	F	XO	1/18	1/6,000
Triplo-X	F	XXX (or XXXX)	0	1/1,500
Klinefelter	M	XXY (or XXXY)	0	1/1,500
Jacob	M	XYY	?	1/1,000

c.

Figure 18.13   Nondisjunction of autosomes during meiosis.
a. Nondisjunction can occur during meiosis I if homologous chromosomes fail to separate and **(b)** during meiosis II if the sister chromatids fail to separate completely. In either case, certain abnormal gametes carry an extra chromosome (n + 1) or lack a chromosome (n − 1). **c.** Frequency of syndromes.

Nondisjunction

Abnormal chromosomal constitutions (Fig. 18.13) can be due to nondisjunction. **Nondisjunction** is the failure of homologous chromosomes or sister chromatids to separate during the formation of gametes. Nondisjunction can occur during meiosis I if both members of a homologous pair of chromosomes go into the same daughter cell, or during meiosis II when the sister chromatids fail to separate and both daughter chromosomes go into the same gamete.

Figure 18.13*a* shows nondisjunction during meiosis I. The homologous chromosomes are duplicated in the parental cell. The daughter cells are abnormal because the daughter cell on the left received both members of the homologous pair and the daughter cell on the right received neither. Following meiosis II, all four daughter cells are abnormal. If the homologous pair was for chromosome 21, the first two daughter cells could result in Down syndrome after

fertilization. If the chromosomes were X chromosomes, Klinefelter syndrome could result. If they were Y chromosomes, XYY syndrome could result following fertilization. The two daughter cells on the right could result in Turner syndrome after fertilization (YO zygotes die off).

In Figure 18.13*b*, nondisjunction occurs during meiosis II. Meiosis I occurred normally and each daughter cell has one of the members of the homologous pair. During meiosis II, the sister chromatids fail to separate and the same gamete receives both daughter chromosomes. The abnormal gametes can result in the same syndromes discussed in the preceding paragraph.

Nondisjunction results in gametes with an abnormal number of chromosomes and, following fertilization, the syndromes listed in Figure 18.13.

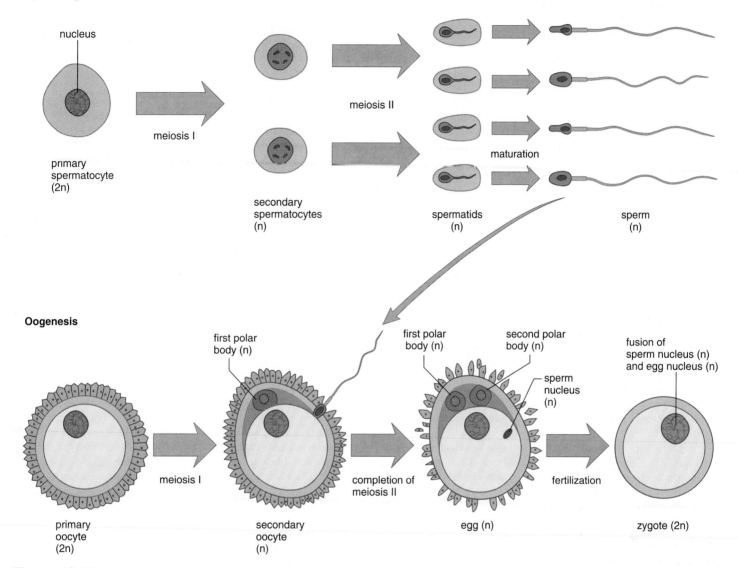

Figure 18.14   Spermatogenesis and oogenesis.
Spermatogenesis produces four viable sperm, whereas oogenesis produces one egg and two polar bodies. Notice that oogenesis does not go to completion unless the secondary oocyte is fertilized. In humans, both sperm and egg have 23 chromosomes each; therefore, following fertilization, the zygote has 46 chromosomes.

Spermatogenesis and Oogenesis

Spermatogenesis and oogenesis occur in the sex organs—the testes in males and the ovaries in females. During **spermatogenesis,** sperm are produced, and during **oogenesis,** eggs are produced. The gametes appear differently in the two sexes (Fig. 18.14), and meiosis is different, too. The process of meiosis in males always results in four cells that become sperm. Meiosis in females produces only one egg. Meiosis I results in one large cell called a secondary oocyte and one polar body. After meiosis II, there is one egg and two polar bodies. The **polar bodies** are a way to discard unnecessary chromosomes while retaining much of the cytoplasm in the egg. The cytoplasm serves as a source of nutrients for the developing embryo.

Spermatogenesis, once started, continues to completion, and mature sperm result. In contrast, oogenesis does not necessarily go to completion. Only if a sperm fertilizes the secondary oocyte does it undergo meiosis II and become an egg. Regardless of this complication, however, both the sperm and the egg contribute the haploid number of chromosomes to the zygote (fertilized egg). In humans, each contributes 23 chromosomes.

Spermatogenesis, which occurs in the testes of males, produces sperm. Oogenesis, which occurs in the ovaries of females, produces eggs. Meiosis is a part of spermatogenesis and oogenesis; therefore, both sperm and egg are haploid.

Bioethical Issue

Cloning is making exact multiple copies of DNA or a cell or an organism. The first two procedures have been around for some time. Through biotechnology, bacteria produce cloned copies of human DNA. When a single bacterium reproduces asexually on a petri dish, a colony results. Each member of the colony is a clone of the original cell. Now for the first time in our history, it is possible to produce a clone of a vertebrate. No sperm and egg are required. The DNA of an adult cell is placed in an egg that undergoes development to become an exact copy of the organism that donated the DNA. Some people fear that billionaires and celebrities will hasten to make multiple copies of themselves. Oth-

ers feel that this is unlikely. Rather, they see possibilities of a different type of cloning.

Suppose it were possible to use the DNA of a burn victim to produce embryonic cells that are cajoled to become skin cells. These cells could be used to provide grafts of new skin. Would this be a proper use of cloning in humans?

Or suppose parents want to produce a child free of a genetic disease. Scientists produce a zygote through in vitro fertilization, and then they clone the zygote to produce any number of cells. Genetic engineering to correct the defect doesn't work on all the cells—only a few. They implant just those few in the uterus where

development continues to term. Would this be a proper use of cloning in humans?

What if science progressed to producing children with increased intelligence or athletic prowess in the same way? Would this be an acceptable use of cloning in humans?

Questions

1. Presently, research in the cloning of humans is banned. Should it be? Why or why not?
2. Under what circumstances might cloning in humans be acceptable? Explain.
3. Is cloning to produce improved breeds of farm animals acceptable? Why or why not?

Summarizing the Concepts

18.1 Chromosomal Inheritance

A human karyotype ordinarily shows 22 homologous pairs of autosomes and one pair of sex chromosomes. The sex pair is an X and a Y chromosome in males and two X chromosomes in females.

Abnormalities sometimes occur. The major inherited autosomal abnormality is Down syndrome, in which the individual inherits three copies of chromosome 21. Cri du chat occurs when part of chromosome 5 is deleted. Examples of abnormal sex chromosomal inheritance are a fragile X chromosome and abnormal chromosomal numbers: Turner syndrome (XO), Klinefelter syndrome (XXY), triplo-X syndrome (XXX) , and Jacob syndrome (XYY).

18.2 Human Life Cycle

The life cycle of higher organisms requires two types of cell divisions: mitosis and meiosis. Mitosis is responsible for growth and repair, while meiosis is required for gamete production.

18.3 Mitosis

Mitosis assures that all cells in the body have the diploid number and the same kinds of chromosomes. The cell cycle includes mitosis and an additional stage termed interphase. During interphase, DNA replication causes each chromosome to have sister chromatids.

Mitosis has the following phases: prophase—at first the chromosomes have no particular arrangement but later the chromosomes are attached to spindle fibers; metaphase, when the chromosomes are aligned at the equator; anaphase, when the sister chromatids separate, becoming daughter chromosomes that move toward the poles; and telophase, when new nuclear envelopes form around the daughter chromosomes. The cytoplasm is partitioned by furrowing in human cells.

18.4 Meiosis

Meiosis involves two cell divisions. During meiosis I, the homologous chromosomes (following crossing-over between nonsister chromatids) separate, and during meiosis II, the sister chromatids separate. The result is four cells with the haploid number of chromosomes in single copy. Meiosis is a part of gamete formation in humans. Nondisjunction during meiosis leads to gametes with an abnormal number of chromosomes and the syndromes discussed in this chapter.

Spermatogenesis in males usually produces four viable sperm, while oogenesis in females produces one egg and two polar bodies. Oogenesis does not go on to completion unless a sperm fertilizes the developing egg.

Among sexually reproducing organisms, such as humans, each individual is genetically different because

1. Crossing-over recombines the genes on the sister chromatids of homologous pairs.

2. Following meiosis, each gamete has a different combination of chromosomes.

3. Upon fertilization, recombination of chromosomes occurs.

Studying the Concepts

1. Describe the normal karyotype of a human being. What is the difference between a male and a female karyotype? 386

2. Describe Down syndrome, the most common autosome abnormality in humans. What is cri du chat syndrome? 387–88

3. Name a chromosomal abnormality other than an abnormal number of sex chromosomes. List four sex chromosomal abnormalities that have to do with an abnormal number of sex chromosomes. 388–89

4. Draw a diagram describing the human life cycle. What is mitosis? What is meiosis? 390

5. Describe the stages of mitosis, including in your description the terms centrioles, nucleolus, spindle, and furrowing. 392–93

6. How do the terms diploid (2n) and haploid (n) pertain to meiosis? 394

7. Describe the stages of meiosis I, including the terms tetrad and dyad in your description. 396

8. Compare the stages of meiosis II to a mitotic division. 396

9. What is the importance of mitosis and meiosis in the life cycle of humans? 390, 395

10. What is nondisjunction, and how does it occur? 398

11. How does spermatogenesis in males compare to oogenesis in females? 399

Testing Your Knowledge of the Concepts

In questions 1–5, match the syndrome to chromosomal makeup.
a. Down syndrome
b. Turner syndrome
c. Klinefelter syndrome
d. cri du chat syndrome
e. Jacob
_____ 1. deletion in chromosome 5
_____ 2. extra chromosome 21
_____ 3. X0
_____ 4. XXY
_____ 5. autosomal abnormality

In questions 6–11, indicate whether the statement is true (T) or false (F).
_____ 6. Normally a person has 46 chromosomes in their karyotype.
_____ 7. Nondisjunction can occur during meiosis I or meiosis II.
_____ 8. During meiosis I chromatids separate, and during meiosis II the members of a homologous pair separate.
_____ 9. The chromosomes are aligned at the equator of the spindle during anaphase.
_____ 10. Half of your chromosomes were inherited from your father and half were inherited from your mother.
_____ 11. Homologous chromosomes pair during prophase of mitosis.

In questions 12–16, fill in the blanks.
12. A karyotype shows an individual's _____.
13. The sex chromosomes of a male are called _____ and _____.
14. If the parent cell has 24 chromosomes, the daughter cells following mitosis will have _____ chromosomes.
15. Meiosis in males is a part of _____, and meiosis in females is a part of _____.
16. Oogenesis will go to completion unless _____ occurs.
17. Which of these drawings represents metaphase I?

 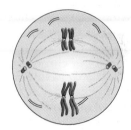

18. How do you know your choice represents metaphase I?

Applying Your Knowledge to the Concepts

These questions pertain to chromosomal inheritance.
1. What is the reason that Turner, Klinefelter, and triplo-X syndromes can all occur as a result of nondisjunction in either the sperm or egg, but XYY can occur only as a result of nondisjunction in the sperm?
2. What are four changes that occur during prophase that are reversed during telophase?

3. Turner syndrome and Klinefelter syndrome are referred to as 2n − 1 and 2n + 1, respectively. What is the meaning of this terminology or symbol?
4. Normally, the two cells that result from the first mitotic division of the zygote stay together. If, instead, the two cells separate and each develops on its own, what is the result?

Understanding the Terms

anaphase 393
aster 392
autosome 386
cell cycle 391
centromere 391
chromatid 391
cleavage furrow 393
crossing-over 394
diploid (2n) 390
dyad 395
gamete 390
haploid (n) 390
homologous chromosome 394
interphase 391
karyotype 386
meiosis 394

metaphase 393
mitosis 391
nondisjunction 398
oogenesis 399
polar body 399
prophase 392
sex chromosome 386
somatic cell 390
spermatogenesis 399
spindle 392
synapsis 396
syndrome 387
telophase 393
tetrad 394
X chromosome 386
Y chromosome 386

Match the terms to these definitions:
a. _____ One of the two identical parts of a chromosome following replication of DNA.
b. _____ Repeating sequence of events in eukaryotic cells consisting of interphase, when growth and DNA synthesis occurs, and mitosis, when cell division occurs.
c. _____ Structure consisting of fibers, poles, and asters (if animal cell) that brings about the movement of chromosomes during cell division.
d. _____ Nonfunctioning daughter cell that has little cytoplasm and is formed during oogenesis.
e. _____ Half the diploid number; the number of chromosomes in the gametes.

Applying Technology to the Concepts

Your study of chromosomal inheritance is supported by these available technologies:

Essential Study Partner CD-ROM
Genetics → Cell Division
Visit the Mader web site for related ESP activities.

Exploring the Internet
The Mader Home Page provides resources and tools as you study this chapter.

http://www.mhhe.com/biosci/genbio/mader

Dynamic Human 2.0 CD-ROM
Human Body → Explorations → Cell Cycle

Life Science Animations 3D Video
10 Mitosis
11 Meiosis
12 Crossing-over

Chapter *19*

Genes and Medical Genetics

Figure 19.1 **Inheritance.**
We know that physical traits pass from parent to child. What about behavioral traits? Do the likes and dislikes of parents influence their children?

Chapter Concepts

Intelligent. Homosexual. Loving. Aggressive. Overweight. Do any of these characteristics describe you? If so, you can probably place some of the responsibility on your genes.

Biologists have long linked physical **traits,** such as facial characteristics, to the genes. Now, after deciphering the human body's vast network of genes, they have extended the influence of genes to our most intimate traits. It's possible the findings could lead to new treatments for various ills. For example, some biologists are studying newfound genes involved in obesity in order to create drugs that fight fat. Others are engineering lab mice without specific genes for aggression, watching the animals to see how they act.

Even though dozens of new behavior-related genes are sure to surface in the future, we may never know just how genetics and environment combine to make us who we are. Social scientists argue that behavioral traits may be controlled to a degree by genes, but childrearing, peer groups, and other social conditions also shape the personality. The classic "nature" versus "nurture" debate continues despite the conclusion by many that it might be a 50/50 situation.

Just what are the genes that control our physical features and, at least to a degree, our behavioral features? This chapter shows that it is sometimes helpful to think of

genes as being units of chromosomes, particularly when we want to calculate the chances of passing on a genetic disorder. But we have already mentioned that genes are actually DNA molecules that specify protein synthesis. We are now beginning to decipher how this function can directly affect the structure of the cell and the organism.

19.1 Genotype and Phenotype

Genotype refers to the genes of the individual. Alternate forms of a gene having the same position (locus) on a pair of chromosomes and affecting the same trait are called **alleles.** It is customary to designate an allele by a letter, which rep-

resents the specific characteristic it controls; a **dominant allele** is assigned an uppercase (capital) letter, while a **recessive allele** is given the same letter, lowercase. In humans, for example, unattached (free) earlobes are dominant over attached earlobes, so a suitable key would be *E* for unattached earlobes and *e* for attached earlobes.

Since autosomal alleles occur in pairs, the individual normally has two alleles for a trait. Just as one of each pair of chromosomes is inherited from each parent, so too is one of each pair of alleles inherited from each parent. Figure 19.2 shows three possible fertilizations and the resulting genetic makeup of the zygote and, therefore, the individual. In the first instance, the chromosome of both the sperm and egg carries an *E.* Consequently, the zygote and subsequent individual have the alleles *EE,* which may be called a *homozygous* (pure) *dominant* genotype. A person with genotype *EE* obviously has unattached earlobes. The physical appearance of the individual, in this case unattached earlobes, is called the **phenotype.**

In the second fertilization, the zygote received two recessive alleles (*ee*), and the genotype is called *homozygous* (pure) *recessive.* An individual with this genotype has the recessive phenotype, which is attached earlobes. In the third fertilization, the resulting individual has the alleles *Ee,* which is called a *heterozygous* genotype. A heterozygote shows the dominant characteristic; therefore, the phenotype of this individual is unattached earlobes.

These examples show that a dominant allele contributed from only one parent can bring about a particular phenotype. A recessive allele must be received from both parents to bring about the recessive phenotype.

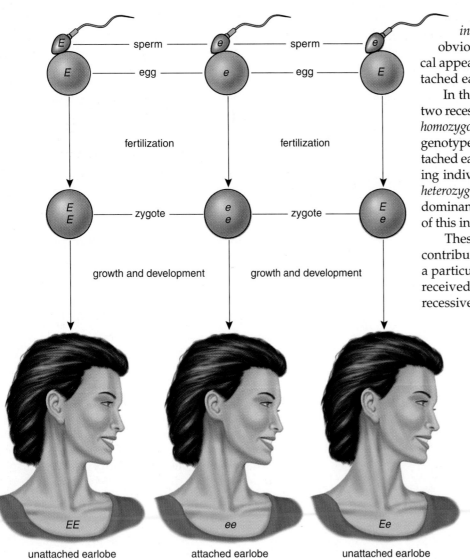

| The genotype, whether homozygous dominant (*EE*), homozygous recessive (*ee*), or heterozygous (*Ee*), tells what genes a person carries. The phenotype—for example, attached or free earlobes—tells what the person looks like. |

Figure 19.2   Genetic inheritance.
Individuals inherit a minimum of two alleles for every characteristic of their anatomy and physiology. The inheritance of a single dominant allele (*E*) causes an individual to have unattached earlobes; two recessive alleles (*ee*) cause an individual to have attached earlobes. Notice that each individual receives one allele from the father (by way of a sperm) and one allele from the mother (by way of an egg).

19.2 Dominant/Recessive Traits

The alleles designated by *E* and *e* are on part of the chromosomes. An individual has two alleles for each trait because a chromosome pair carries alleles for the same traits. How many alleles for each trait will be in the gametes? One, because chromosome pairs separate during meiosis I.

Forming the Gametes

During gametogenesis, the chromosome number is reduced. Whereas the individual has 46 chromosomes, a gamete has only 23 chromosomes. (If this did not happen, each new generation of individuals would have twice the number of chromosomes as their parents.) Reduction of the chromosome number occurs when the pairs of chromosomes separate as meiosis occurs. Since the alleles are on the chromosomes, they also separate during meiosis, and therefore, the gametes carry only one allele for each trait. If an individual carried the alleles *EE*, all the gametes would carry an *E* since that is the only choice. Similarly, if an individual carried the alleles *ee*, all the gametes would carry an *e*. What if an individual were *Ee*? Figure 19.3 shows that half of the gametes would carry an *E* and half would carry an *e*. Figure 19.4 gives the genotypes for certain other traits in humans, and you can practice deciding what alleles the gametes would carry for these genotypes.

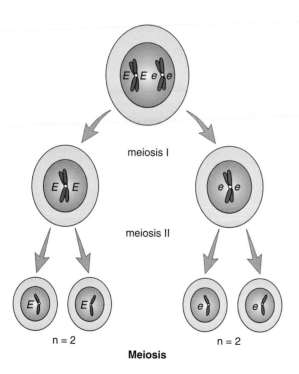

Figure 19.3 Gametogenesis.
Because the pairs of chromosomes separate during meiosis, which occurs during gametogenesis, the gametes have only one allele for each trait.

a. Widow's peak: *WW* or *Ww*   b. Straight hairline: *ww*

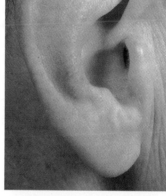 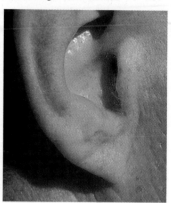

c. Unattached earlobes: *EE* or *Ee*   d. Attached earlobes: *ee*

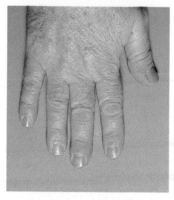 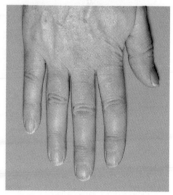

e. Short fingers: *SS* or *Ss*   f. Long fingers: *ss*

 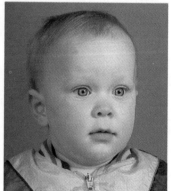

g. Freckles: *FF* or *Ff*   h. No freckles: *ff*

Figure 19.4 Common inherited characteristics in human beings.
The notations indicate which characteristics are dominant and which are recessive.

Figuring the Odds

Many times parents would like to know the chances of a child having a certain genotype and, therefore, a certain phenotype. If one of the parents is homozygous dominant (*EE*), the chances of their having a child with unattached earlobes is 100%, because this parent has only a dominant allele (*E*) to pass on to the offspring. On the other hand, if both parents are homozygous recessive (*ee*), there is a 100% chance that each of their children will have attached earlobes. However, if both parents are heterozygous, then what are the chances that their child will have unattached or attached earlobes? To solve a problem of this type, it is customary first to indicate the genotype of the parents and their possible gametes.

Genotypes:	*Ee*	*Ee*
Gametes:	*E* and *e*	*E* and *e*

Second, a **Punnett square** is used to determine the phenotypic ratio among the offspring when all possible sperm are given an equal chance to fertilize all possible eggs (Fig. 19.5). The possible sperm are lined up along one side of the square, and the possible eggs are lined up along the other side of the square (or vice versa). The ratio among the offspring in this case is 3:1 (three children with unattached earlobes to one with attached earlobes). This means that there is a 3/4 chance (75%) for each child to have unattached earlobes and a 1/4 chance (25%) for each child to have attached earlobes.

Another cross of particular interest is that between a heterozygous individual (*Ee*) and a pure recessive (*ee*). In this case, the Punnett square shows that the ratio among the offspring is 1:1, and the chance of the dominant or recessive phenotype is 1/2, or 50% (Fig. 19.6).

Comparing the two crosses (Fig. 19.5 and Fig. 19.6), each child has a 75% chance of having the dominant phenotype if the two parents are heterozygous and a 50% chance if one parent is heterozygous and the other is recessive. Each child has a 25% chance of having the recessive phenotype if the parents are heterozygous and a 50% chance if one parent is heterozygous and the other is homozygous recessive.

If both parents are heterozygous, each child has a 25% chance of exhibiting the recessive phenotype. If one parent is heterozygous and the other is homozygous recessive, each child has a 50% chance of exhibiting the recessive phenotype.

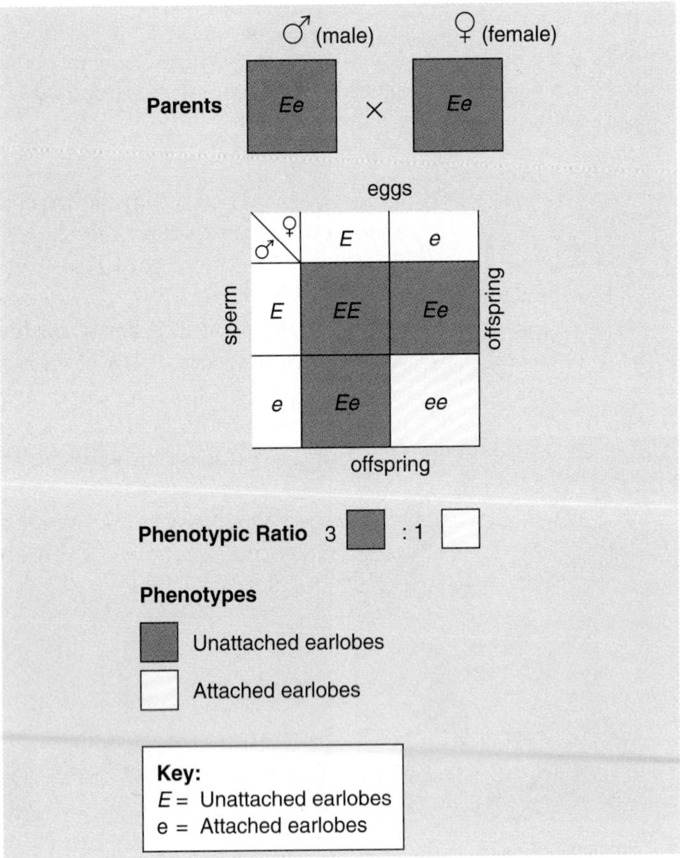

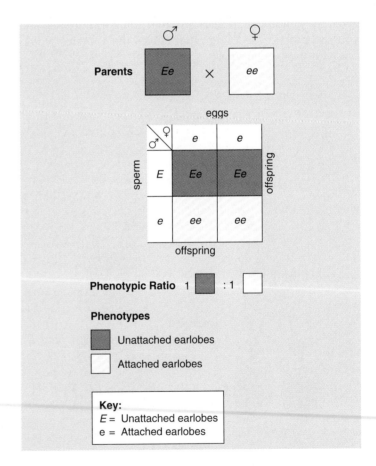

Figure 19.5 Heterozygous-by-heterozygous cross.
When the parents are heterozygous, each child has a 75% chance of having the dominant phenotype and a 25% chance of having the recessive phenotype.

Figure 19.6 Heterozygous-by-homozygous recessive cross.
When one parent is heterozygous and the other recessive, each child has a 50% chance of having the dominant phenotype and a 50% chance of having the recessive phenotype.

Dominant Disorders

Of the many autosomal (non-sex-linked) dominant disorders, we will discuss only two.

Neurofibromatosis

Neurofibromatosis, sometimes called von Recklinghausen[1] disease, is one of the most common genetic disorders. It affects roughly one in 3,000 people, including an estimated 100,000 in the United States. It is seen equally in every racial and ethnic group throughout the world.

At birth or later, the affected individual may have six or more large, tan spots (known as cafe-au-lait) on the skin. Such spots may increase in size and number and may get darker. Small benign tumors (lumps) called neurofibromas may occur under the skin or in various organs. Neurofibromas are made up of nerve cells and other cell types.

The expression of neurofibromatosis varies. In most cases, symptoms are mild, and patients live a normal life. In some cases, however, the effects are severe. Skeletal deformities, including a large head, are seen, and eye and ear tumors can lead to blindness and hearing loss. Many children with neurofibromatosis have learning disabilities and are hyperactive.

In 1990, researchers isolated the gene for neurofibromatosis, which was known to be on chromosome 17. By analyzing the DNA (deoxyribonucleic acid), they determined that the gene was huge and actually included three even smaller genes. This was only the second time that nested genes have been found in humans. The gene for neurofibromatosis is a tumor-suppressor gene active in controlling cell division. When it mutates, a benign tumor develops.

Huntington Disease

One in 20,000 persons in the United States has *Huntington disease*, a neurological disorder that leads to progressive degeneration of brain cells, which in turn causes severe muscle spasms and personality disorders (Fig. 19.7). Most people appear normal until they are of middle age and have already had children who might also be stricken. Occasionally, the first signs of the disease are seen in these children when they are teenagers or even younger. There is no effective treatment, and death comes ten to fifteen years after the onset of symptoms.

Several years ago, researchers found that the gene for Huntington disease was located on chromosome 4. A test was developed for the presence of the gene, but few want to know if they have inherited the gene because as yet there is no treatment for Huntington disease. After the gene was isolated in 1993, an analysis revealed that it contains many repeats of the base triplet CAG (cytosine, adenine, and

Figure 19.7   Huntington disease.
Persons with this condition gradually lose psychomotor control of the body. At first there are only minor disturbances, but the symptoms become worse over time.

guanine). Normal persons have 11 to 34 copies of the triplet, but affected persons tend to have 42 to more than 120 copies. The more repeats present, the earlier the onset of Huntington disease and the more severe the symptoms. It also appears that persons most at risk have inherited the disorder from their fathers. The latter observation is consistent with a new hypothesis called **genomic imprinting.** The genes are imprinted differently during formation of sperm and egg, and therefore, the sex of the parent passing on the disorder becomes important.

It is now known that there are a number of other genetic diseases whose severity and time of onset vary according to the number of triplet repeats present within the gene. Moreover, the genes for these disorders are also subject to genomic imprinting.

There are many autosomal dominant disorders in humans. Among these are neurofibromatosis and Huntington disease.

---

[1]Although neurofibromatosis is commonly associated with Joseph Merrick, the severely deformed nineteenth-century Londoner depicted in *The Elephant Man*, ressearchers believe Merrick actually suffered from a much rarer disorder called Proteus syndrome.

Recessive Disorders

Of the many autosomal recessive disorders, we will discuss only three.

Tay-Sachs Disease

Tay-Sachs disease is a well-known genetic disease that usually occurs among Jewish people in the United States, most of whom are of central and eastern European descent. At first, it is not apparent that a baby has Tay-Sachs disease. However, development begins to slow down between four months and eight months of age, and neurological impairment and psychomotor difficulties then become apparent. The child gradually becomes blind and helpless, develops uncontrollable seizures, and eventually becomes paralyzed. There is no treatment or cure for Tay-Sachs disease, and most affected individuals die by the age of three or four.

Tay-Sachs disease results from a lack of the enzyme hexosaminidase A (Hex A) and the subsequent storage of its substrate, a fatty substance known as glycosphingolipid, in lysosomes. Although more and more lysosomes build up in many body cells, the primary sites of storage are the cells of the brain, which account for the onset and the progressive deterioration of psychomotor functions.

Persons heterozygous for Tay-Sachs have about half the level of Hex A activity found in normal individuals. Prenatal diagnosis of the disease also is possible following either amniocentesis or chorionic villi sampling.

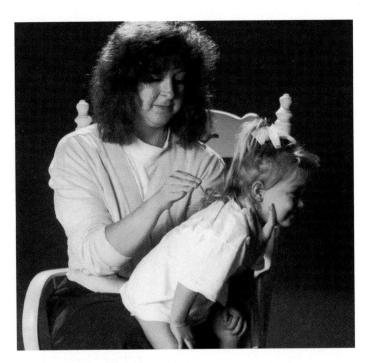

Figure 19.8   Cystic fibrosis.
The mucus in the lungs of a child with cystic fibrosis should be periodically loosened by clapping the back. A new treatment destroys the cells that tend to build up in the lungs. The white blood cells are destroyed by one drug, and another drug does away with the DNA the cells leave behind. Judicious use of antibiotics controls pulmonary infection, and aggressive use of other drugs thins mucous secretions.

Cystic Fibrosis

Cystic fibrosis is the most common lethal genetic disease among Caucasians in the United States. About one in 20 Caucasians is a carrier, and about one in 2,500 births has the disorder. In these children, the mucus in the bronchial tubes and pancreatic ducts is particularly thick and viscous, interfering with the function of the lungs and pancreas. To ease breathing, the thick mucus in the lungs has to be manually loosened periodically (Fig. 19.8), but still the lungs become infected frequently. The clogged pancreatic ducts prevent digestive enzymes from reaching the small intestine, and to improve digestion, patients take digestive enzymes mixed with applesauce before every meal.

In the past few years, much progress has been made in our understanding of cystic fibrosis, and new treatments have raised the average life expectancy to 28 years of age. Research has demonstrated that chloride ions (Cl^-) fail to pass through plasma membrane channel proteins in these patients. Ordinarily, after chloride ions have passed through the membrane, water follows. It is thought that lack of water is the cause of abnormally thick mucus in bronchial tubes and pancreatic ducts. The cystic fibrosis gene, which is located on chromosome 7, has been isolated, and attempts have been made to insert it into nasal epithelium, so far with little success. Genetic testing for the gene in adult carriers and in fetuses is possible; if present, couples have to decide whether to risk having a child with the condition or whether abortion is an option.

Phenylketonuria (PKU)

Phenylketonuria (PKU) occurs once in 5,000 births, so it is not as frequent as the disorders previously discussed. However, it is the most commonly inherited metabolic disorder to affect nervous system development. First cousins who marry are more apt to have a PKU child.

Affected individuals lack an enzyme that is needed for the normal metabolism of the amino acid phenylalanine, and an abnormal breakdown product, phenylketone, accumulates in the urine. The PKU gene is located on chromosome 12, and a prenatal DNA test can determine the presence of this allele. Years ago, the urine of newborns was tested at home for phenylketone in order to detect PKU. Presently, newborns are routinely tested in the hospital for elevated levels of phenylalanine in the blood. If elevated levels are detected, newborns are placed on a diet low in phenylalanine, which must be continued until the brain is

fully developed, around age seven, or else severe mental retardation develops.

There are many autosomal recessive disorders in humans. Among these are Tay-Sachs disease, cystic fibrosis, and phenylketonuria (PKU).

Pedigree Charts

When a genetic disorder is autosomal dominant, an individual with the alleles *AA* or *Aa* will have the disorder. When a genetic disorder is recessive, only individuals with the alleles *aa* will have the disorder. Genetic counselors often construct pedigree charts to determine whether a condition is dominant or recessive. A pedigree chart shows the pattern of inheritance for a particular condition. Consider these two possible patterns of inheritance:

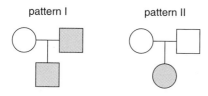

In both patterns, males are designated by squares and females by circles. Shaded circles and squares are affected individuals. A line between a square and a circle represents a union. A vertical line going downward leads, in these patterns, to a single child. (If there are more children, they are placed off a horizontal line.) Which pattern of inheritance do you suppose represents an autosomal dominant characteristic, and which represents an autosomal recessive characteristic?

In pattern I, the child is affected, as is one of the parents. When a disorder is dominant, an affected child usually has at least one affected parent. Of the two patterns, this one shows a dominant pattern of inheritance. Figure 19.9 shows a typical pedigree chart for a dominant disorder. Other ways to recognize an autosomal dominant pattern of inheritance are also given.

In pattern II, the child is affected, but neither parent is; this can happen if the condition is recessive and the parents are *Aa*. Notice that the parents are carriers because they appear to be normal but are capable of having a child with a genetic disorder. Figure 19.10 shows a typical pedigree chart for a recessive genetic disorder. Other ways to recognize an autosomal recessive pattern of inheritance are also given in the figure.

It is important to realize that "chance has no memory"; therefore, each child born to heterozygous parents has a 25% chance of having the disorder. In other words, it is possible that if a heterozygous couple has four children, each child might have the condition.

Dominant and recessive alleles have different patterns of inheritance.

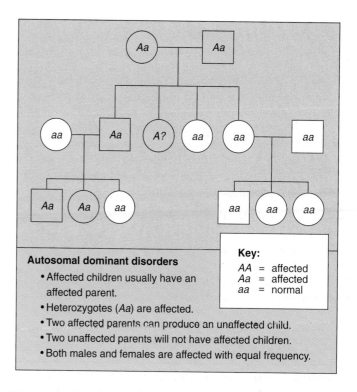

Autosomal dominant disorders

- Affected children usually have an affected parent.
- Heterozygotes (*Aa*) are affected.
- Two affected parents can produce an unaffected child.
- Two unaffected parents will not have affected children.
- Both males and females are affected with equal frequency.

Key:
AA = affected
Aa = affected
aa = normal

Figure 19.9   Autosomal dominant pedigree chart.
The list gives ways to recognize an autosomal dominant disorder.

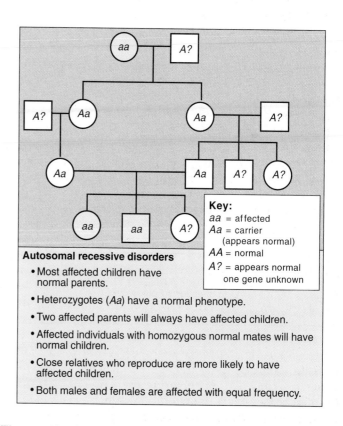

Autosomal recessive disorders

- Most affected children have normal parents.
- Heterozygotes (*Aa*) have a normal phenotype.
- Two affected parents will always have affected children.
- Affected individuals with homozygous normal mates will have normal children.
- Close relatives who reproduce are more likely to have affected children.
- Both males and females are affected with equal frequency.

Key:
aa = affected
Aa = carrier
     (appears normal)
AA = normal
A? = appears normal
     one gene unknown

Figure 19.10   Autosomal recessive pedigree chart.
Only those affected with the recessive genetic disorder are shaded. The list gives ways to recognize an autosomal recessive disorder.

Genetic Counseling

Now that potential parents are becoming aware that many illnesses are caused by faulty genes, more couples are seeking genetic counseling. The counselor studies the backgrounds of the couple and tries to determine if any immediate ancestor may have had a genetic disorder. A pedigree chart may be constructed. Then the counselor studies the couple. As much as possible, laboratory tests are performed on all persons involved.

Tests are now available for a large number of genetic diseases. For example, chromosomal tests are available for cystic fibrosis, neurofibromatosis, and Huntington disease. Blood tests can identify carriers of thalassemia and sickle-cell disease. By measuring enzyme levels in blood, tears, or skin cells, carriers of enzyme defects can also be identified for certain inborn metabolic errors, such as Tay-Sachs disease. From this information, the counselor can sometimes predict the chances of a child having the disorder.

Whenever the woman is pregnant, chorionic villi sampling can be done early, and amniocentesis can be done later in the pregnancy. These procedures, which were illustrated in chapter 17, allow the testing of embryonic and fetal cells, respectively, to determine if the unborn child has a genetic disorder. If so, treatment may be available even before birth, or parents may decide whether or not to end the pregnancy (Table 19A).

Table 19A Test and Treatment for Some Human Genetic Disorders

Name	Description	Chromosome	Incidence Among Newborns	Status
Autosomal Recessive Disorders				
Cystic fibrosis	Mucus in the lungs and digestive tract is thick and viscous, making breathing and digestion difficult	7	One in 2,500 Caucasians	Allele located; chromosome test now available;* treatment being investigated
Tay-Sachs disease	Neurological impairment and psychomotor difficulties develop early, followed by blindness and uncontrollable seizures before death occurs, usually before age 5	15	One in 3,600 eastern European Jews	Biochemical test now available*
Phenylketonuria	Inability to metabolize phenylalanine, and if a special diet is not begun, mental retardation develops	12	One in 5,000 Caucasians	Biochemical test now available; treatment available
Autosomal Dominant Disorders				
Neurofibromatosis	Benign tumors occur under the skin or deeper	17	One in 3,000	Allele located; chromosome test now available*
Huntington disease	Minor disturbances in balance and coordination develop in middle age and progress toward severe neurological disturbances leading to death	4	One in 20,000	Allele located; chromosome test now available*
Incomplete Dominance				
Sickle-cell disease	Poor circulation, anemia, internal hemorrhaging, due to sickle-shaped red blood cells	11	One in 500 African Americans	Chromosome test now available*
X-Linked Recessive				
Hemophilia A	Propensity for bleeding, often internally, due to the lack of a blood clotting factor	X	One in 15,000 male births	Treatment available
Duchenne muscular dystrophy	Muscle weakness develops early and progressively intensifies until death occurs, usually before age 20	X	One in 5,000 male births	Allele located; biochemical tests of muscle tissue available; treatment being investigated

*Prenatal testing is done.

19.3 Polygenic Traits

Polygenic inheritance occurs when one trait is governed by two or more sets of alleles, and the individual has a copy of all allelic pairs. Each dominant allele has a quantitative effect on the phenotype, and these effects are additive. The result is a continuous variation of phenotypes, resulting in a distribution of these phenotypes that resembles a bell-shaped curve. The more genes involved, the more continuous the variation and the distribution of the phenotypes. Also, environmental effects cause many intervening phenotypes; in the case of height, differences in nutrition assure a bell-shaped curve (Fig. 19.11).

Skin Color

Just how many pairs of alleles control skin color is not known, but a range in colors can be explained on the basis of two pairs. When a very dark person reproduces with a very light person, the children have medium-brown skin; when two people with medium-brown skin reproduce with one another, the children range in skin color from very dark to very light. This can be explained by assuming that skin color is controlled by two pairs of alleles and that each capital letter contributes to the color of the skin:

Phenotype	Genotypes
Very dark	*AABB*
Dark	*AABb* or *AaBB*
Medium brown	*AaBb* or *AAbb* or *aaBB*
Light	*Aabb* or *aaBb*
Very light	*aabb*

Notice again that there is a range in phenotypes and that there are several possible phenotypes in between the two extremes. Therefore, the distribution of these phenotypes is expected to follow a bell-shaped curve—few people have the extreme phenotypes, and most people have the phenotype that lies in the middle between the extremes.

Polygenic Disorders

Many human disorders, such as cleft lip and/or palate, clubfoot, congenital dislocations of the hip, hypertension, diabetes, schizophrenia, and even allergies and cancers, are most likely controlled by polygenes and are subject to environmental influences. Therefore, many investigators are in the process of considering the nature versus nurture question; that is, what percentage of the trait is controlled by genes and what percentage is controlled by the environment? Thus far, it has not been possible to come to precise, generally accepted percentages for any particular trait.

In recent years, reports have surfaced that all sorts of behavioral traits such as alcoholism, phobias, and even suicide can be associated with particular genes. No doubt behavioral traits are to a degree controlled by genes, but again, it is impossible at this time to determine to what degree. And very few scientists would support the idea that these traits are predetermined by our genes.

Many human traits most likely controlled by polygenes are subject to environmental influences. The frequency of the phenotypes of such traits follows a bell-shaped curve.

a.

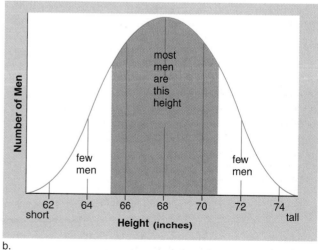

b.

Figure 19.11 **Polygenic inheritance.**
When you record the heights of a large group of young men **(a)** the values follow a bell-shaped curve **(b)**. Such a continuous distribution is due to control of a trait by several sets of alleles. Environmental effects are also involved.

19.4 Multiple Allelic Traits

When a trait is controlled by **multiple alleles,** the gene exists in several allelic forms. But each person usually has only two of the possible alleles. A person's blood type is determined by multiple alleles.

ABO Blood Types

Three alleles for the same gene control the inheritance of ABO blood types. These alleles determine the presence or absence of antigenic glycoproteins on the red blood cells.

A = A antigen on red blood cells
B = B antigen on red blood cells
O = Neither A nor B antigen on red blood cells

Alleles *A* and *B* are dominant over *O*, and are fully expressed in the presence of the other. This is called codominance. Therefore,

Phenotype	Possible Genotype
A	*AA, AO*
B	*BB, BO*
AB	*AB*
O	*OO*

An examination of possible matings between different blood types sometimes produces surprising results; for example,

Genotypes of parents: *AO* × *BO*
Possible genotypes of children: *AB, OO, AO, BO*

Therefore, from this particular mating, every possible phenotype (types AB, O, A, B blood) is possible.

Blood typing can sometimes aid in paternity suits. However, a blood test of a supposed father can only suggest that he might be the father, not that he definitely is the father. For example, it is possible, but not definite, that a man with type A blood (having genotype *AO*) is the father of a child with type O blood. On the other hand, a blood test sometimes can definitely prove that a man is not the father. For example, a man with type AB blood cannot possibly be the father of a child with type O blood. Therefore, blood tests can be used in legal cases only to exclude a man from possible paternity.

Inheritance by multiple alleles occurs when a gene exists in more than two allelic forms. However, each individual usually inherits only two alleles for these genes.

As a point of interest, the Rh factor is inherited separately from A, B, AB, or O blood types. When you are Rh positive, there is a particular antigen on the red blood cells, and when you are Rh negative, it is absent. It can be assumed that the inheritance of this antigen is controlled by a single allelic pair in which simple dominance prevails: the Rh-positive allele is dominant over the Rh-negative allele.

19.5 Incompletely Dominant Traits

The field of human genetics also has examples of codominance and incomplete dominance. **Codominance** occurs when alleles are equally expressed in a heterozygote. We have already mentioned that the multiple alleles controlling blood type are codominant. An individual with the genotype *AB* has type AB blood. Recall that skin color is controlled by polygenes in which all dominants (capital letters) add equally to the phenotype. For this reason, it is possible to observe a range of skin colors from very dark to very light.

Incomplete dominance is exhibited when the heterozygote has an intermediate phenotype between that of either homozygote. For example, incomplete dominance pertains to the inheritance of curly versus straight hair in Caucasians. When a curly-haired Caucasian reproduces with a straight-haired Caucasian, their children will have wavy hair. When two wavy-haired persons reproduce, the expected phenotypic ratio among the offspring is 1:2:1; that is, one curly-haired child to two with wavy hair to one with straight hair (Fig. 19.12).

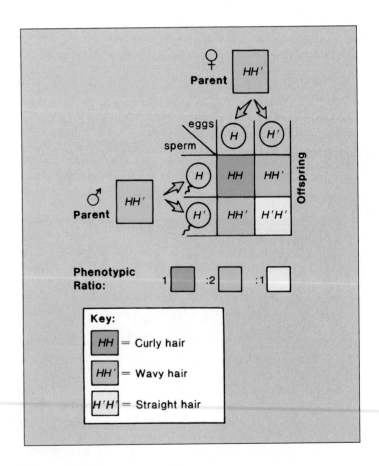

Figure 19.12   Incomplete dominance.
Among Caucasians, neither straight nor curly hair is dominant. When two wavy-haired individuals reproduce, each offspring has a 25% chance of having either straight or curly hair and a 50% chance of having wavy hair, the intermediate phenotype.

Sickle-Cell Disease

Sickle-cell disease is an example of a human disorder that is controlled by incompletely dominant alleles. Individuals with the $Hb^A Hb^A$ genotype are normal, those with the $Hb^S Hb^S$ genotype have sickle-cell disease, and those with the $Hb^A Hb^S$ genotype have the sickle-cell trait. Two individuals with sickle-cell trait can produce children with all three phenotypes, as indicated in Figure 19.13.

In persons with sickle-cell disease, the red blood cells aren't biconcave disks like normal red blood cells; they are irregular. In fact, many are sickle shaped. The defect is caused by an abnormal hemoglobin (Hb^S) that accumulates inside the cells. Normal hemoglobin (Hb^A) differs from Hb^S by one amino acid in the protein globin. Although globin is 146 amino acids long, just one change in the sixth position results in sickle-shaped cells. Because sickle-shaped cells can't pass along narrow capillary passageways like disk-shaped cells, they clog the vessels and break down. This is why persons with sickle-cell disease suffer from poor circulation, anemia, and poor resistance to infection. Internal hemorrhaging leads to further complications, such as jaundice, episodic pain of the abdomen and joints, and damage to internal organs. Persons with sickle-cell trait do not usually have any sickle-shaped cells unless they experience dehydration or mild oxygen deprivation. Still, at present, most investigators believe that no restrictions on physical activity are needed for persons with the sickle-cell trait.

Among regions of malaria-infested Africa, infants with sickle-cell disease die, but infants with sickle-cell trait have a better chance of survival than the normal homozygote. The malaria parasite normally reproduces inside red blood cells. But a red blood cell of a sickle-cell trait infant becomes sickle-shaped if it becomes infected with a malaria-causing parasite. As the cell becomes sickle-shaped and it loses potassium, the parasite dies. The protection afforded by the sickle-cell trait keeps the allele prevalent in populations exposed to malaria. As many as 60% of the population in malaria-infested regions of Africa have the allele. In the United States, about 10% of the African-American population carries the allele.

In a recent study, 22 children around the world were given bone marrow transplants from healthy siblings, and 16 were completely cured of sickle-cell disease. Optimism over these results is dampened by the knowledge that only 6% of patients met the medical criteria for receiving treatment and the drugs used in the procedure result in infertility. Also, if unsuccessful, health could worsen instead of improve, and there is a 10% risk of death from the treatment.

Innovative therapies are still being explored. For example, persons with sickle-cell disease produce normal fetal hemoglobin during development, and drugs that turn on the genes for fetal hemoglobin in adults are being developed. Mice have been genetically engineered to produce sickled red blood cells in order to test new antisickling drugs and various genetic therapies.

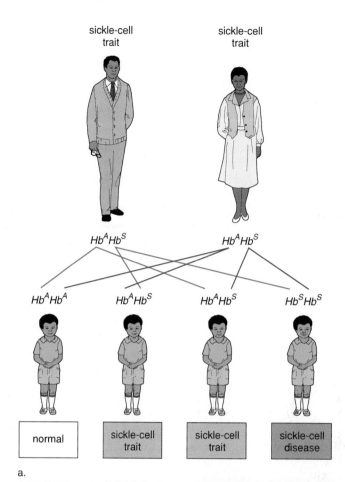

a.

b.

Figure 19.13   **Inheritance of sickle-cell disease.**
a. In this example, both parents have the sickle-cell trait. Therefore, each child has a 25% chance of having sickle-cell disease or of being perfectly normal and a 50% chance of having the sickle-cell trait. **b.** Sickled cells. Individuals with sickle-cell disease have sickled red blood cells that tend to clump, as illustrated here.

Sickle-cell disease is an inherited, lifelong disorder that is being investigated on many fronts.

19.6 Sex-Linked Traits

The sex chromosomes contain genes just as the autosomal chromosomes do. Some of these genes determine whether the individual is a male or a female. Investigators have found that the Y chromosome has an *SRY* gene (sex-determining region Y gene). When this gene is lacking from the Y chromosome, the individual is a female even though the chromosomal inheritance is XY.

Traits controlled by alleles on the **sex chromosomes** are said to be **sex-linked;** an allele that is only on the X chromosome is **X-linked,** and an allele that is only on the Y chromosome is Y-linked. Most sex-linked alleles are on the X chromosome, and the Y chromosome is blank for these. Very few alleles have been found on the Y chromosome, as you might predict, since it is much smaller than the X chromosome.

The X chromosomes carry many genes unrelated to the sex of the individual, and we will look at a few of these in depth. It would be logical to suppose that a sex-linked trait is passed from father to son or from mother to daughter, but this is not the case. A male always receives a sex-linked condition from his mother, from whom he inherited an X chromosome. The Y chromosome from the father does not carry an allele for the trait. Usually the trait is recessive; therefore, a female must receive two alleles, one from each parent, before she has the condition.

X-Linked Alleles

When considering X-linked traits, the allele on the X chromosome is shown as a letter attached to the X chromosome. For example, the key for red-green color blindness is as follows:

X^B = normal vision
X^b = color blindness

The possible genotypes in both males and females are as follows:

$X^B X^B$	=	female who has normal color vision
$X^B X^b$	=	carrier female who has normal color vision
$X^b X^b$	=	female who is color blind
$X^B Y$	=	male who has normal vision
$X^b Y$	=	male who is color blind

Carriers are individuals that appear normal but can pass on an allele for a genetic disorder. Note that the second genotype is a carrier female because although a female with this genotype appears normal, she is capable of passing on an allele for color blindness. Color-blind females are rare because they must receive the allele from both parents; color-blind males are more common since they need only one recessive allele to be color blind. The allele for color blindness has to be inherited from their mother because it is on the X chromosome; males only inherit the Y chromosome from their father.

Now, let us consider a mating between a man with normal vision and a heterozygous woman (Fig. 19.14). What are

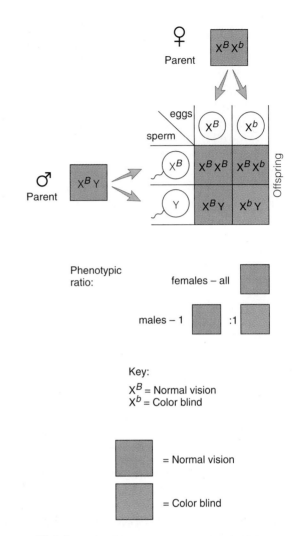

Figure 19.14   Cross involving an X-linked allele.
The male parent is normal, but the female parent is a carrier; an allele for color blindness is located on one of her X chromosomes. Therefore, each son stands a 50% chance of being color blind. The daughters will appear to be normal but each one stands a 50% chance of being a carrier.

the chances of this couple having a color-blind daughter? A color-blind son? All daughters will have normal color vision because they all receive an X^B from their father. The sons, however, have a 50% chance of being color blind, depending on whether they receive an X^B or an X^b from their mother. The inheritance of a Y chromosome from their father cannot offset the inheritance of an X^b from their mother. Notice in Figure 19.14, the phenotypic results for sex-linked problems are given separately for males and females.

The X chromosome carries alleles that are not on the Y chromosome. Therefore, a recessive allele on the X chromosome is expressed in males.

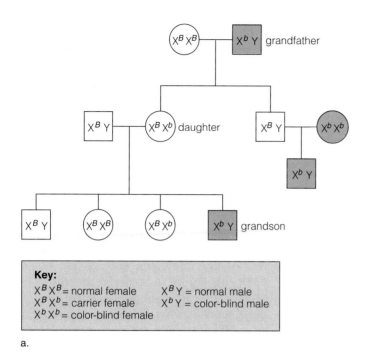

X-linked Recessive Disorders

- More males than females are affected.
- An affected son can have parents who have the normal phenotype.
- For a female to have the characteristic, her father must also have it. Her mother must have it or be a carrier.
- The characteristic often skips a generation from the grandfather to the grandson.
- If a woman has the characteristic, all of her sons will have it.

b.

Among 205 catalogued X-linked recessive disorders are:
- Agammaglobulinemia—lack of immunity to infections
- Color blindness—inability to distinguish certain colors
- Hemophilia—defect in blood-clotting mechanisms
- Muscular dystrophy (some forms)—progressive wasting of muscles
- Spinal ataxia (some forms)—spinal cord degeneration

c.

Figure 19.15   X-linked recessive pedigree chart.
The list gives ways of recognizing an X-linked recessive disorder.

X-Linked Disorders

Figure 19.15 gives a pedigree chart for an X-linked recessive condition. It also lists ways to recognize this pattern of inheritance. X-linked conditions can be dominant or recessive, but most known are recessive. More males than females have the trait because recessive alleles on the X chromosome are always expressed in males since the Y chromosome does not have a corresponding allele. If a male has an X-linked recessive condition, his daughters are often carriers; therefore, the condition passes from grandfather to grandson. Females who have the condition inherited the allele from both their mother and their father; and all the sons of such a female will have the condition. Three well-known X-linked recessive disorders are color blindness, muscular dystrophy, and hemophilia.

Color Blindness

In humans, there are three different classes of cone cells, the receptors for color vision in the retina of the eyes. Only one pigment protein is present in each type of cone cell; there are blue-sensitive, red-sensitive, and green-sensitive cone cells. The gene for the blue-sensitive protein is autosomal, but the genes for the red- and green-sensitive proteins are on the X chromosome. About 8% of Caucasian men have red-green color blindness. Most of these see brighter greens as tans, olive greens as browns, and reds as reddish-browns. A few cannot tell reds from greens at all. They see only yellows, blues, blacks, whites, and grays. Opticians have special charts by which they detect those who are color blind.

Muscular Dystrophy

Muscular dystrophy, as the name implies, is characterized by a wasting away of the muscles. The most common form, *Duchenne muscular dystrophy*, is X-linked and occurs in about one out of every 3,600 male births. Symptoms, such as waddling gait, toe walking, frequent falls, and difficulty in rising, may appear as soon as the child starts to walk. Muscle weakness intensifies until the individual is confined to a wheelchair. Death usually occurs by age 20; therefore, affected males are rarely fathers. The recessive allele remains in the population by passage from carrier mother to carrier daughter.

Recently, the gene for muscular dystrophy was isolated, and it was discovered that the absence of a protein now called dystrophin is the cause of the disorder. Much investigative work determined that dystrophin is involved in the release of calcium from the sarcoplasmic reticulum in muscle fibers. The lack of dystrophin causes calcium to leak into the cell, which promotes the action of an enzyme that dissolves muscle fibers. When the body attempts to repair the tissue, fibrous tissue forms, and this cuts off the blood supply so that more and more cells die.

A test is now available to detect carriers for Duchenne muscular dystrophy. Also, various treatments are being attempted. Immature muscle cells can be injected into muscles, and for every 100,000 cells injected, dystrophin production occurs in 30–40% of muscle fibers. The gene for dystrophin has been inserted into the thigh muscle cells of mice, and about 1% of these cells then produced dystrophin.

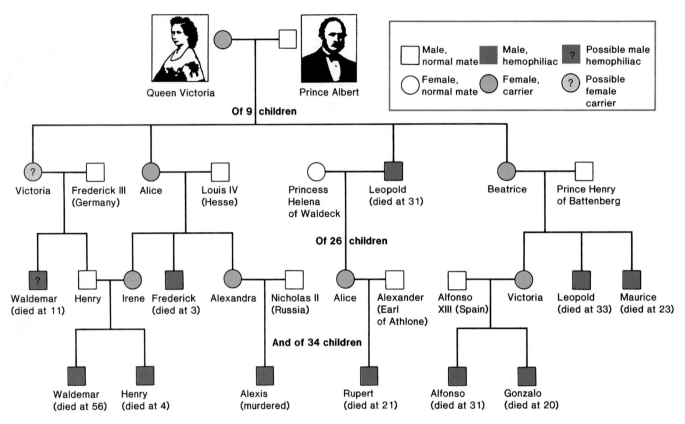

Figure 19.16   A simplified pedigree showing the X-linked inheritance of hemophilia in European royal families.
Because Queen Victoria was a carrier, each of her sons had a 50% chance of having the disease, and each of her daughters had a 50% chance of being a carrier. This pedigree shows only the affected individuals. Many others are unaffected, such as the members of the present British royal family.

Hemophilia

About one in 10,000 males is a hemophiliac. There are two common types of hemophilia: hemophilia A is due to the absence or minimal presence of a clotting factor known as factor IX, and hemophilia B is due to the absence of clotting factor VIII.

Hemophilia is called the bleeder's disease because the affected person's blood does not clot. Although hemophiliacs bleed externally after an injury, they also suffer from internal bleeding, particularly around joints. Hemorrhages can be checked with transfusions of fresh blood (or plasma) or concentrates of the clotting protein. Unfortunately, some hemophiliacs have contracted AIDS after receiving blood or using a blood concentrate, but donors are now screened more closely, and donated blood is now tested for HIV. Also, factor VIII is now available as a biotechnology product.

At the turn of the century, hemophilia was prevalent among the royal families of Europe, and all of the affected males could trace their ancestry to Queen Victoria of England. Figure 19.16 shows that of Queen Victoria's 26 grandchildren, five grandsons had hemophilia and four granddaughters were carriers. Because none of Queen Victoria's forebears or relatives were affected, it seems that the faulty allele she carried arose by mutation either in Victoria or in one of her parents. Her carrier daughters, Alice and Beatrice, introduced the gene into the ruling houses of Russia and Spain, respectively. Alexis, the last heir to the Russian throne before the Russian Revolution, was a hemophiliac. There are no hemophiliacs in the present British royal family because Victoria's eldest son, King Edward VII, did not receive the gene and therefore could not pass it on to any of his descendants.

Certain traits that have nothing to do with the gender of the individual are controlled by genes on the X chromosomes. Males have only one X chromosome, and therefore, X-linked recessive alleles are expressed.

Sex-Influenced Traits

Not all traits we associate with the gender of the individual are sex-linked traits. Some are simply **sex-influenced traits;** that is, the phenotype is determined by autosomal genes that are expressed differently in males and females.

It is possible that the sex hormones determine whether these genes are expressed or not. Pattern baldness (Fig. 19.17) is thought to be influenced by the male sex hormone testosterone because males who take the hormone to increase masculinity begin to lose their hair. A more detailed explanation has been suggested by some investigators. It has been reasoned that due to the effect of hormones, men require only one allele for baldness in order for the condition to appear, whereas women require two alleles. In other words, the allele for baldness acts as a dominant in men but as a recessive in women. This means that men who have a bald father and a mother with hair have a 50% chance at best and a 100% chance at worst of going bald. Women who have a bald father and a mother with hair have no chance at best and a 50% chance at worst of going bald.

Another sex-influenced trait of interest is the length of the index finger. In females, an index finger longer than the fourth finger (ring finger) seems to be dominant. In males, an index finger longer than the fourth finger seems to be recessive.

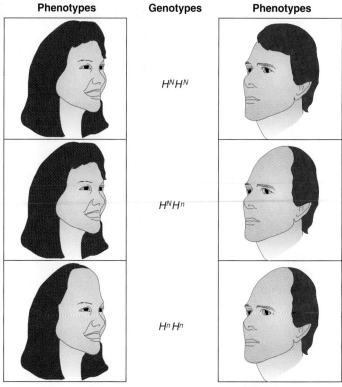

Phenotypes   **Genotypes**   **Phenotypes**

$H^N H^N$

$H^N H^n$

$H^n H^n$

H^N = Normal hair growth
H^n = Pattern baldness

Figure 19.17   Pattern baldness, a sex-influenced characteristic.

Due to hormonal influences, the presence of only one gene for baldness causes the condition in men, whereas the condition does not occur in women unless they possess both genes for baldness.

Bioethical Issue

If the results were absolutely private, would you want to know if you have a genetic disease whose symptoms may not appear for several decades? If the disease were curable, it might be a good idea. Testing before symptoms appear encourages increased vigilance for the onset of breast and colon cancers. If gene therapy is the only way out, you might be first in line if such therapy is developed in the future.

Huntington disease is a particularly frightening genetic disorder, especially if you have seen a loved one suffer the mental and musculoskeletal deterioration that finally results in death. Although there is no cure for Huntington disease, you would be able to warn your children if you test positive, or it might help you decide whether to have a child. If you have this dominant allele, you will develop the condition and your children have a 50%

chance of developing the condition. Children usually begin to have symptoms earlier than their parents did.

Jason Brandt of the Johns Hopkins University School of Medicine has developed a program of pretest and posttest counseling for those who decide they want to be tested for Huntington disease. So far, about 200 have been tested, and almost twice that number have dropped out of pretest counseling, signifying no doubt they really didn't want to be tested.

Brandt refuses to perform the test on persons he feels could not psychologically handle the results if they were positive. Surprisingly, he finds that people rarely make major life changes as a result of knowing they have the allele. Among 63 who tested positive, ten got married to persons who knew they would come down with Huntington disease, and ten

others went on to have more children. One patient made a radical change after she knew she was negative for Huntington disease. She divorced her present husband, remarried, had another child and took up a career as a physical therapist.

Questions

1. Is it a person's right to be tested for genetic diseases regardless or do you think Brandt should forgo testing until he considers the person ready to receive bad news?

2. Is it ethical to discourage people with a dominant allele for Huntington disease from having children? Why or why not?

3. If gene therapy for Huntington disease became available should it be made available to adults? children? embryos? Why or why not?

Summarizing the Concepts

19.1 Genotype and Phenotype
It is customary to use letters to represent the genotype of individuals. Homozygous dominant (two capital letters) and heterozygous (a capital letter and a lowercase letter) exhibit the dominant phenotype. Homozygous recessive (two lowercase letters) exhibit the recessive phenotype.

19.2 Dominant/Recessive Traits
Whereas the individual has two alleles (copies of a gene) for each trait, the gametes have only one allele for each trait. A Punnett square can help determine the phenotype ratio among offspring because all possible sperm types of a particular male are given an equal chance to fertilize all possible egg types of a particular female. When a heterozygous individual reproduces with another heterozygote, there is a 75% chance the offspring will have the dominant phenotype and a 25% chance the child will have the recessive phenotype. When a heterozygous individual reproduces with a pure recessive, the offspring has a 50% chance of either phenotype.

19.3 Polygenic Traits
Polygenic traits are controlled by more than one pair of alleles, and the individual inherits each allelic pair. Skin color illustrates polygenic inheritance.

19.4 Multiple Allelic Traits
When a trait is controlled by multiple alleles, each person has one pair of all possible alleles. ABO blood types illustrate inheritance by multiple alleles.

19.5 Incompletely Dominant Traits
Determination of inheritance is sometimes complicated by codominance (e.g., blood type) and by incomplete dominance (e.g., hair curl among Caucasians).

19.6 Sex-Linked Traits
Sex-linked genes are located on the sex chromosomes, and most of these alleles are X-linked because they are on the X chromosome. Since the Y chromosome is blank for X-linked alleles, males are more apt to exhibit the recessive phenotype, which they inherited from their mother. If a female does have the phenotype, then her father must also have it, and her mother must be a carrier or have the trait. Usually, X-linked traits skip a generation and go from maternal grandfather to grandson by way of a carrier daughter.

Sex-influenced traits are controlled by an allele that acts as if it were dominant in one sex but recessive in the other. Most likely the activity of such genes is influenced by the sex hormones.

Studying the Concepts

1. What is the difference between the genotype and the phenotype of an individual? For which phenotype are there two possible genotypes?   404
2. Why do the gametes have only one allele for each trait?   405
3. What is the chance of a child with the dominant phenotype from these crosses?   406

   $AA = AA$
   $Aa = AA$
   $Aa = Aa$
   $aa = aa$

4. From which of the crosses in question 3 can there be an offspring with the recessive phenotype? Explain.   406
5. Give examples of genetic disorders caused by inheritance of a single dominant allele and disorders caused by inheritance of two recessive alleles.   407–8
6. List ways to recognize an autosomal dominant genetic disorder and an autosomal recessive genetic disorder when examining a pedigree chart.   409
7. Give examples of these patterns of inheritance: polygenic inheritance, multiple alleles, and incomplete dominance.   411–12
8. Give examples of genetic disorders caused by the patterns of inheritance in question 7.   411–13
9. If a trait is on an autosome, how would you designate a homozygous dominant female? If a trait is on an X chromosome, how would you designate a homozygous dominant female?   414
10. List ways to recognize an X-linked recessive disorder when examining a pedigree chart. Why do males exhibit such disorders more often than females?   415
11. Give examples of genetic disorders caused by X-linked recessive alleles.   415–16

Applying Your Knowledge to the Concepts

These questions pertain to genes and medical genetics.
1. You sometimes hear the expression "The child gets his eyes from his grandmother." Is there any scientific basis for this statement? Explain.
2. Some genetic disorders are more prevalent in people of a particular ethnic group. Present a possible explanation for this.
3. It is said that there are no carriers for Huntington disease. What is the justification for this statement?
4. A woman with type AB blood gives birth to a type B child. She identifies a man with type O blood as the father. He claims on the basis of his blood type that he could not be the father. As a well-informed judge, how would you rule in this case? Explain.

Testing Your Knowledge of the Concepts

In questions 1–4, match the answers to the genetics problem.

a. 0%
b. 25%
c. 50%
d. 100%

_____ 1. A woman heterozygous for polydactyly (6 fingers and toes) reproduces with a man without the condition. Polydactyly is dominant; what are the chances a child will have the condition?

_____ 2. Parents who do not have Tay-Sachs disease produce a child who has Tay-Sachs disease (recessive). What are the chances the next child will have Tay-Sachs?

_____ 3. One parent has sickle-cell disease (incompletely dominant), and the other is perfectly normal. What are the chances a child will have sickle-cell trait?

_____ 4. Black hair is dominant over blond hair. What are the chances a homozygous black-haired woman who reproduces with a blond-haired man will have brown-haired children?

5. Identify each of these genetic disorders.
   a. Mucus in lungs and digestive tract is thick and viscous.
   b. Neurological impairment and psychomotor difficulties develop early.
   c. Benign tumors occur under the skin or deeper.
   d. Minor disturbances in balance and coordination develop in middle age and progress toward severe mental disturbances.

In questions 6–8, indicate whether the statement is true (T) or false (F).

_____ 6. In a pedigree chart, it is observed that the trait skips a generation and passes from grandfather to grandson. The trait must be sex-linked dominant.

_____ 7. For a girl to have hemophilia (X-linked recessive), both of her parents must have it.

_____ 8. The genotype of a female who has a color-blind father but a homozygous normal mother is $X_B X_b$.

In questions 9 and 10, fill in the blanks.

9. Mary has a widow's peak, and John has a straight hairline. This would be a description of their _____.

10. The genotype of a color-blind man with attached earlobes (recessive) is _____.

11. Determine if the characteristic possessed by the shaded males and females in the pedigree chart below is autosomal dominant, autosomal recessive, or X-linked recessive:

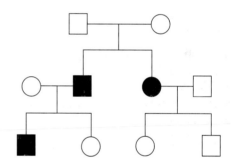

Understanding the Terms

allele 404
carrier 414
codominance 412
dominant allele 404
genomic imprinting 407
genotype 404
incomplete dominance 412
multiple alleles 412
phenotype 404

polygenic inheritance 411
Punnett square 406
recessive allele 404
sex chromosome 414
sex-influenced trait 417
sex-linked 414
trait 403
X-linked 414

Match the terms to these definitions:

a. _____ Chromosome responsible for the development of characteristics associated with gender; an X or Y chromosome.

b. _____ Genes of any individual for (a) particular trait(s).

c. _____ Gridlike device used to calculate the expected results of simple genetic crosses.

d. _____ Individual that appears normal but is capable of transmitting an allele for a genetic disorder.

e. _____ Pattern of inheritance in which more than one allelic pair controls a trait; each dominant allele has a quantitative effect on the phenotype.

Applying Technology to the Concepts

Your study of genes and medical genetics is supported by these available technologies:

Essential Study Partner CD-ROM

Genetics → **Mendelian Genetics**
         → **Chromosomes**

Visit the Mader web site for related ESP activities.

Exploring the Internet

The Mader Home Page provides resources and tools as you study this chapter.

http://www.mhhe.com/biosci/genbio/mader

Chapter 20

DNA and Biotechnology

Chapter Concepts

20.1 DNA and RNA Structure and Function

- DNA is the genetic material, and therefore, its structure and function constitute the molecular basis of inheritance. 422
- When DNA replicates, two exact copies result. RNA structure is similar to, but also different from, DNA. RNA occurs in three forms, each with a specific function. 424

20.2 Gene Expression

- Proteins are composed of amino acids and function, in particular, as enzymes and as structural elements of membranes and organelles in cells. 426
- DNA's genetic information codes for the sequence of amino acids in a protein during protein synthesis. 426
- There are various levels of genetic control in human cells. 431

20.3 Biotechnology

- Recombinant DNA technology is the basis for biotechnology, an industry that produces many products. 432
- Modern-day biotechnology is expected to revolutionize medical care and agriculture. It also permits gene therapy and the mapping of human chromosomes. 434

Figure 20.1   Cindy Cutshall.
Cindy was sick and couldn't play baseball until she underwent gene therapy. She doesn't like to talk about the procedure. She says, "It's too weird."

Cindy Cutshall was born with a rare immune disorder called SCID (severe combined immunodeficiency). Her white blood cells were ineffective because they couldn't produce a critical enzyme known as ADA. She was in and out of the hospital with infections, and her parents feared she would probably die young.

And so, in 1993, the Cutshalls agreed to let Cindy undergo gene therapy, sponsored by the National Institutes of Health (NIH), the country's top biomedical funding agency. Gene therapy is based on the idea that it is possible to supply a patient with a missing or defective gene. NIH researchers took white blood stem cells from Cindy, added the gene that specifies ADA, and injected the improved cells back into her. Stem cells were chosen because they reside in the bone marrow where they continually produce more white blood cells. Combined with drug treatment, the therapy apparently worked. Cindy is well and her white blood cells contain functioning ADA genes even today.

Like Cindy, over 600 patients have now received some form of gene therapy in treatment for AIDS, cancer, cystic fibrosis, and other diseases. In many cases the patient's cells have not received an adequate number of working genes to make a difference. But once efficient methods of delivery have been developed, gene therapy may one day improve the lives of millions.

The gene that Cindy received was a specific piece of **DNA (deoxyribonucleic acid)** taken from a human chromosome. This piece of DNA is responsible for the production of ADA, the enzyme that Cindy needed in her cells. In this chapter, we will learn that any particular gene is just one section of one of the chromosomes, and that a **gene** codes for a specific sequence of amino acids in a protein. Gene therapy is only one aspect of the burgeoning field of biotechnology.

20.1 DNA and RNA Structure and Function

DNA, the genetic material, is found principally in the chromosomes, which are located in the nucleus of a cell. (Small amounts of DNA are also found in mitochondria.)

Figure 20.2 shows the levels of organization of chromosomes. While the metaphase chromosome is a very compact

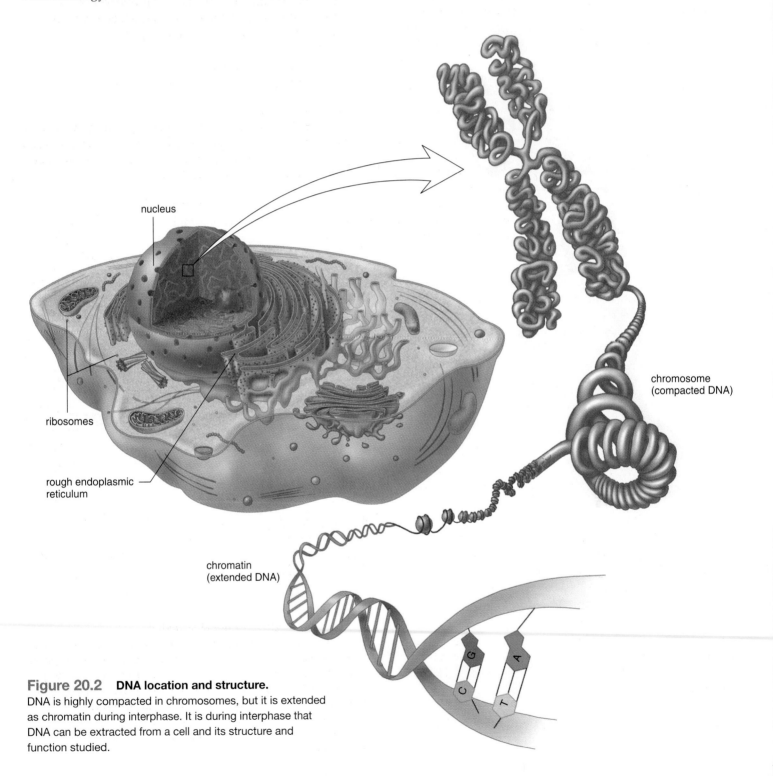

Figure 20.2   DNA location and structure.
DNA is highly compacted in chromosomes, but it is extended as chromatin during interphase. It is during interphase that DNA can be extracted from a cell and its structure and function studied.

structure, chromatin is fine threads in which the DNA is extended. The DNA molecule is a double-stranded helix in which the strands are held together by hydrogen bonding. The structure of DNA permits replication of the molecule prior to cell division.

DNA Structure and Replication

DNA is a type of nucleic acid, and like all nucleic acids, it is formed by the sequential joining of molecules called nucleotides. Nucleotides, in turn, are composed of three smaller molecules—a *phosphate*, a *sugar*, and a *nitrogen-containing base*. The sugar in DNA is deoxyribose, which accounts for its name, *deoxyribonucleic acid*. The four nucleotides that make up the DNA molecule have the following bases: **adenine (A), guanine (G), cytosine (C)**, and **thymine (T)** (Fig. 20.3).

When nucleotides join, the sugar and the phosphate molecules become the backbone of a strand, and the bases project to the side. In DNA, there are two strands of nucleotides; consequently, DNA is double stranded. Weak hydrogen bonds between the bases hold the strands together. Each base is bonded to another particular base, called **complementary base pairing**—adenine (A) is always paired with thymine

(T), and cytosine (C) is always paired with guanine (G), and vice versa. The dotted lines in Figure 20.4a represent the hydrogen bonds between the bases. The structure of DNA is said to resemble a ladder; the sugar-phosphate backbone makes up the sides of the ladder, and the paired bases are the rungs. The ladder structure of DNA twists to form a spiral staircase called a *double helix* (Fig. 20.4b).

DNA is a double helix with a sugar-phosphate backbone on the outside and paired bases on the inside. Complementary base pairing occurs: adenine (A) pairs with thymine (T), and guanine (G) pairs with cytosine (C).

Replication of DNA

Between cell divisions, chromosomes must duplicate before cell division can occur again. **Replication** of DNA occurs as a part of chromosome duplication. Replication has been found to require the following steps:

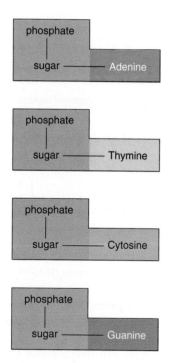

Figure 20.3   DNA nucleotides.
DNA contains four different kinds of nucleotides, molecules that in turn contain a phosphate, a sugar, and a base. The base of a DNA nucleotide can be either adenine (A), thymine (T), cytosine (C), or guanine (G).

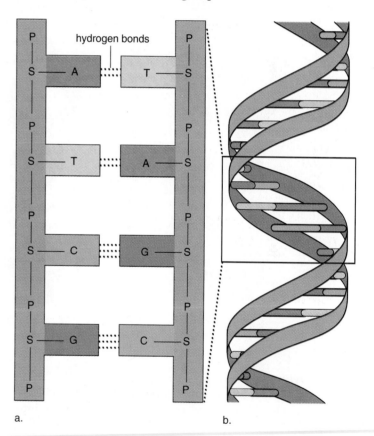

a.   b.

Figure 20.4   DNA is double stranded.
a. When the DNA nucleotides join, they form two strands so that the structure of DNA resembles a ladder. The phosphate (P) and sugar (S) molecules make up the sides of the ladder, and the bases make up the rungs of the ladder. Each base is weakly bonded (dotted lines) to its complementary base (T is bonded to A and vice versa; G is bonded to C and vice versa). **b.** The DNA ladder twists to give a double helix. Each chromatid of a duplicated chromosome contains one double helix.

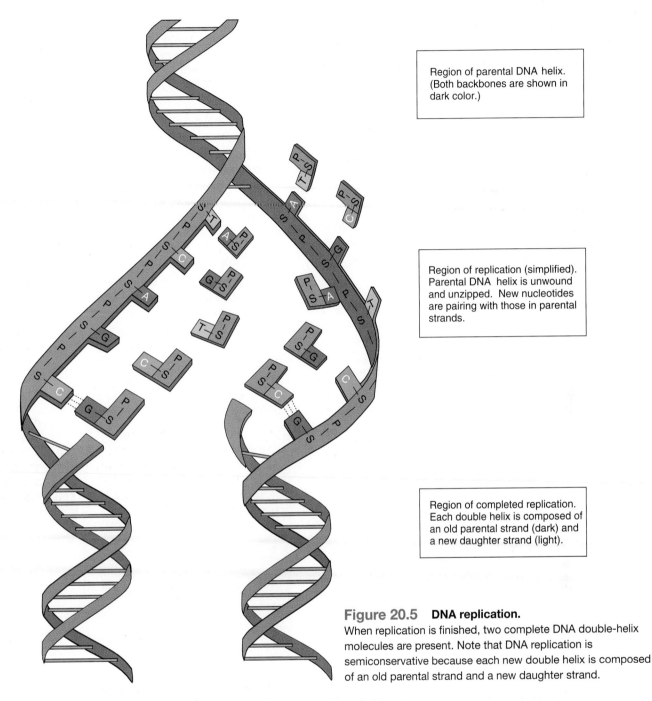

Region of parental DNA helix. (Both backbones are shown in dark color.)

Region of replication (simplified). Parental DNA helix is unwound and unzipped. New nucleotides are pairing with those in parental strands.

Region of completed replication. Each double helix is composed of an old parental strand (dark) and a new daughter strand (light).

Figure 20.5   DNA replication.
When replication is finished, two complete DNA double-helix molecules are present. Note that DNA replication is semiconservative because each new double helix is composed of an old parental strand and a new daughter strand.

1. The hydrogen bonds between the two strands of DNA break as enzymes unwind and "unzip" the molecule.
2. New nucleotides, always present in the nucleus, fit into place beside each old (parental) strand by the process of complementary base pairing.
3. These new nucleotides become joined by an enzyme called *DNA polymerase.*
4. When the process is finished, two complete DNA molecules are present, identical to each other and to the original molecule.

Each new double helix is composed of an old (parental) and a new (daughter) strand (Fig. 20.5). Because each strand of DNA serves as a **template,** or mold, for the production of a complementary strand, DNA replication is called semiconservative. Rarely, a replication error occurs—the new sequence of the bases is not exactly like that of a parental strand. But if an error does occur, the cell has repair enzymes that usually fix them. Replication errors that persist are a source of **mutations,** a permanent change in a gene causing a change in the phenotype.

DNA replication results in two double helixes. DNA unwinds and unzips, and new (daughter) strands, each complementary to an old (parental) strand, form.

The Structure and Function of RNA

RNA (ribonucleic acid) is made up of nucleotides containing the sugar ribose. This sugar accounts for the scientific name of this polynucleotide. The four nucleotides that make up the RNA molecule have the following bases: adenine (A), guanine (G), cytosine (C), and **uracil (U)**. (Fig. 20.6a). Notice that in RNA, the base uracil replaces the base thymine.

RNA, unlike DNA, is single stranded (Fig. 20.6b), but the single RNA strand sometimes doubles back on itself. Similarities and differences between these two nucleic acid molecules are listed in Table 20.1.

In general, RNA is a helper to DNA, allowing protein synthesis to occur according to the blueprint that DNA provides. There are three types of RNA, each with a specific function in protein synthesis.

Ribosomal RNA

Ribosomal RNA (rRNA) forms off a DNA template in the nucleolus of a nucleus. Ribosomal RNA joins with proteins made in the cytoplasm to form the subunits of ribosomes. The subunits leave the nucleus and come together in the cytoplasm when polypeptide synthesis is about to begin. Polypeptides are synthesized at the ribosomes, which in low-power electron micrographs look like granules arranged along the endoplasmic reticulum, a system of tubules and saccules within the cytoplasm. Some ribosomes appear free in the cytoplasm or in clusters called polyribosomes.

Messenger RNA

Messenger RNA (mRNA) forms off a DNA template in the nucleus and then carries its genetic information from the chromosome to the ribosomes in the cytoplasm where protein synthesis occurs. Messenger RNA occurs as a linear molecule.

Transfer RNA

Transfer RNA (tRNA) also forms off a DNA template in the nucleus. Appropriate to its name, tRNA transfers amino acids to the ribosomes, where the amino acids are joined, forming a protein. There are 20 different types of amino acids in proteins; therefore, there has to be at least this number of tRNAs functioning in the cell. Each type of tRNA is designed to carry only one type of amino acid.

RNA is a polynucleotide that functions during protein synthesis in various ways: there are three types of RNA, each with a specific role.

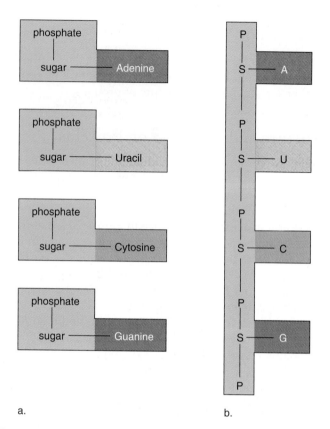

a.   b.

Figure 20.6   RNA structure.
a. The four nucleotides in RNA each have a phosphate (P) molecule; the sugar (S) is ribose; and the base may be either adenine (A), uracil (U), cytosine (C), or guanine (G). **b.** RNA is single stranded. The sugar and phosphate molecules join to form a single backbone, and the bases project to the side.

Table 20.1	DNA-RNA Similarities and Differences
DNA-RNA Similarities	
Both are nucleic acids	
Both are composed of nucleotides	
Both have a sugar-phosphate backbone	
Both have four different type bases	
DNA-RNA Differences	

DNA	*RNA*
Found in nucleus	Found in nucleus and cytoplasm
The genetic material	Helper to DNA
Sugar is deoxyribose	Sugar is ribose
Bases are A, T, C, G	Bases are A, U, C, G
Double stranded	Single stranded
Is transcribed (to give mRNA)	Is translated (to give proteins)

20.2 Gene Expression

DNA provides the cell with a blueprint for the synthesis of proteins in a cell. Before discussing mechanics of gene expression, let's review the structure of proteins.

Structure and Function of Proteins

Proteins are composed of individual units called *amino acids.* Twenty different amino acids are commonly found in proteins, which are synthesized at the ribosomes in the cytoplasm of cells. Proteins differ because the number and order of their amino acids differ. Notice in Figure 20.7 that the sequence of amino acids in a portion of one protein can differ completely from the sequence of amino acids in a portion of another protein. The unique sequence of amino acids in a protein leads to it having a particular shape, and shape of a protein helps determine its function.

Proteins are found in all parts of the body; some are structural proteins and some are enzymes. The protein hemoglobin is responsible for the red color of red blood cells. The antibodies, as well as albumin, are well-known plasma proteins. Muscle cells contain the proteins actin and myosin, which give them substance and their ability to contract.

Enzymes are organic catalysts that speed reactions in cells. The reactions in cells form metabolic or chemical pathways. A pathway can be represented as follows:

$$\begin{array}{cccc} E_A & E_B & E_C & E_D \\ A \rightarrow B \rightarrow C \rightarrow D \rightarrow E \end{array}$$

In this pathway, the letters are molecules, and the notations over the arrows are enzymes: molecule A becomes molecule B, and enzyme E_A speeds the reaction; molecule B becomes molecule C, and enzyme E_B speeds the reaction, and so forth. Notice that each reaction in the pathway has its own enzyme: enzyme E_A can only convert A to B, enzyme E_B can only convert B to C, and so forth. For this reason, enzymes are said to be *specific.*

The DNA Code

The occurrence of inherited metabolic disorders first suggested that genes are responsible for the metabolic workings of a cell. In phenylketonuria (PKU), mental retardation is caused by the inability to convert phenylalanine to tyrosine. In albinism, there is no natural pigment in the skin because tyrosine cannot be converted to melanin. Each condition is caused by the deficiency of a particular enzyme. Even in the early 1900s, these conditions were called inborn errors of metabolism.

After many laboratory experiments, it became clear that a gene is a segment of DNA that codes for a protein. If a person inherits a faulty gene, the code is abnormal, and a particular enzyme does not function as it should. We now know that DNA contains a triplet code—every three bases (*triplet*) represent (*codes*) one amino acid (Table 20.2).

The genetic code is essentially universal. The same codons stand for the same amino acids in most organisms from bacteria to humans. This illustrates the remarkable biochemical unity of living things and suggests that all living things have a common evolutionary ancestor.

Transcription

Gene expression, which results in a protein product, requires two steps, called transcription and translation. It is helpful to recall that transcription is often used to signify making a copy of certain information, while translation means to put this information into another language.

In the same manner that DNA serves as a template for the production of itself, it also serves as a template for the production of RNA. **Transcription** is the synthesis of an RNA molecule off a DNA template. Although all three types of RNA are transcribed off a DNA template, just now we are interested in the transcription of *messenger RNA* (*mRNA*). During transcription, a segment of DNA helix unwinds and unzips, and complementary RNA nucleotides pair with the DNA nucleotides of one strand: G pairs with

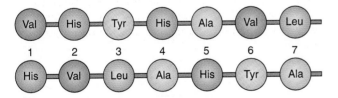

Figure 20.7   Structure of proteins.
Proteins differ by the sequence of their amino acids; the top row shows one possible sequence, and the bottom row shows another possible sequence.

Table 20.2	Some DNA Codes and RNA Codons		
DNA Code	mRNA Codon	tRNA Anticodon	Amino Acid
TTT	AAA	UUU	Lysine
TGG	ACC	UGG	Threonine
CCG	GGC	CCG	Glycine
CAT	GUA	CAU	Valine

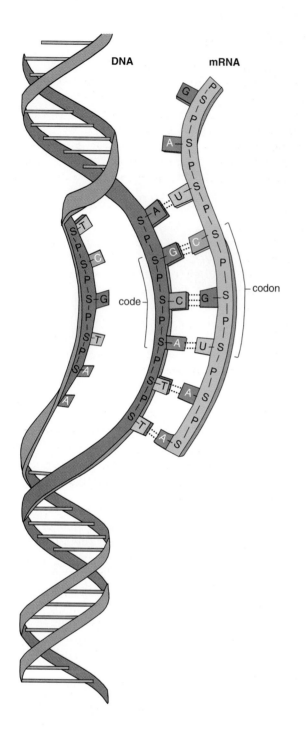

C, U pairs with A, and A pairs with T in DNA. An enzyme called **RNA polymerase** joins the nucleotides together, and the mRNA that results has a sequence of bases complementary to those of a gene. While DNA contains a **triplet code** in which every three bases stand for one amino acid, mRNA contains **codons,** each of which is made up of three bases that also stand for the same amino acid (Table 20.2).

In Figure 20.8, mRNA has formed and is ready to move away from DNA and make its way to the cytoplasm. Referring again to the structure of the cell, recall that DNA is found in the nucleus while protein synthesis occurs at the ribosomes, which are located free in the cytoplasm or on the endoplasmic reticulum (ER) (see Fig. 20.2). (Proteins produced by ribosomes free in the cytoplasm stay in the cell. Those produced by ribosomes attached to the ER are for transport outside the cell.)

Processing of mRNA

Most genes in humans are interrupted by segments of DNA that are not part of the gene. These portions are called introns because they are intragene segments. The other portions of the gene are called exons because they are ultimately expressed. They result in a protein product.

When DNA is transcribed, the mRNA contains bases that are complementary to both exons and introns, but before the mRNA exits the nucleus, it is *processed*. During processing, the nucleotides complementary to the introns are spliced out by ribozymes. **Ribozymes** are organic catalysts composed of RNA and not protein. There has been much speculation about the role of introns. It is possible that they allow crossing-over within a gene during meiosis. It is also possible that introns divide a gene into domains that can be joined in different combinations to give different protein products in different cells.

In human cells, processing occurs in the nucleus. After the mRNA strand is processed, it passes from the cell nucleus into the cytoplasm. There it becomes associated with the ribosomes.

Following transcription, mRNA has a sequence of bases complementary to one of the DNA strands. It contains codons and moves into the cytoplasm where it becomes associated with the ribosomes.

Figure 20.8   Transcription.
During transcription, mRNA is formed when nucleotides complementary to the sequence of bases in a portion of DNA (i.e., a gene) join. Note that DNA contains a triplet code (sequences of three bases) and that mRNA contains codons (sequences of three complementary bases). *Top*—mRNA transcript is ready to move into the cytoplasm. *Middle*—transcription has occurred, and mRNA nucleotides have joined together. *Bottom*—rest of DNA molecule.

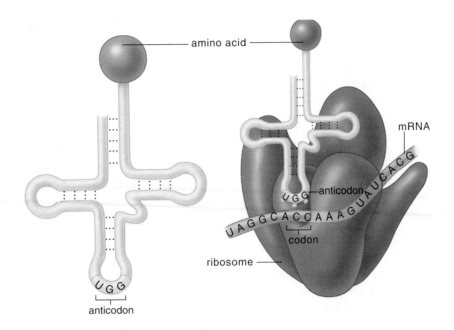

a. tRNA–amino acid   **b. tRNA–amino acid at ribosome**

Figure 20.9   Anticodon-codon base pairing.
a. tRNA molecules have an amino acid attached to one end and an anticodon at the other end. **b.** The anticodon of a tRNA molecule is complementary to a codon. The pairing between codon and anticodon at a ribosome ensures that the sequence of amino acids in a polypeptide is that directed originally by DNA. If the codon is ACC, the anticodon is UGG, and the amino acid is threonine.

Translation

Translation is the synthesis of a polypeptide under the direction of an mRNA molecule. During translation, *transfer RNA (tRNA)* molecules bring amino acids to the ribosomes. Usually, there is more than one tRNA molecule for each of the 20 amino acids found in proteins. The amino acid binds to one end of the molecule (Fig. 20.9a). Therefore, the entire complex is designated as tRNA–amino acid.

At the other end of each tRNA, there is a specific **anticodon,** a group of three bases that is complementary to an mRNA codon (Fig. 20.9b). The tRNA molecules come to the ribosome where each anticodon pairs with a codon in the order directed by the sequence of the mRNA codons. In this way, the order of codons in mRNA brings about a particular order of amino acids in a protein.

If the codon is ACC, what is the anticodon, and what amino acid will be attached to the tRNA molecule? Inspection of Table 20.2 (p. 426) allows us to determine this:

Codon	Anticodon	Amino Acid
ACC	UGG	Threonine

Polypeptide synthesis requires three steps: initiation, elongation, and termination.

1. During *initiation,* mRNA binds to the smaller of the two ribosomal subunits; then the larger subunit joins the smaller one.
2. During *elongation,* the polypeptide lengthens one amino acid at a time (Fig. 20.10). A ribosome is large enough to accommodate two tRNA molecules: the incoming tRNA molecule and the outgoing tRNA molecule. The incoming tRNA–amino acid complex receives the peptide from the outgoing tRNA. The ribosome then moves laterally so that the next mRNA codon is available to receive an incoming tRNA–amino acid complex. In this manner, the peptide grows, and the linear structure of a polypeptide comes about. (The particular shape of a polypeptide comes about after termination, as the amino acids interact with one another.)
3. Then *termination* of synthesis occurs at a codon that means stop and does not code for an amino acid. The ribosome dissociates into its two subunits and falls off the mRNA molecule.

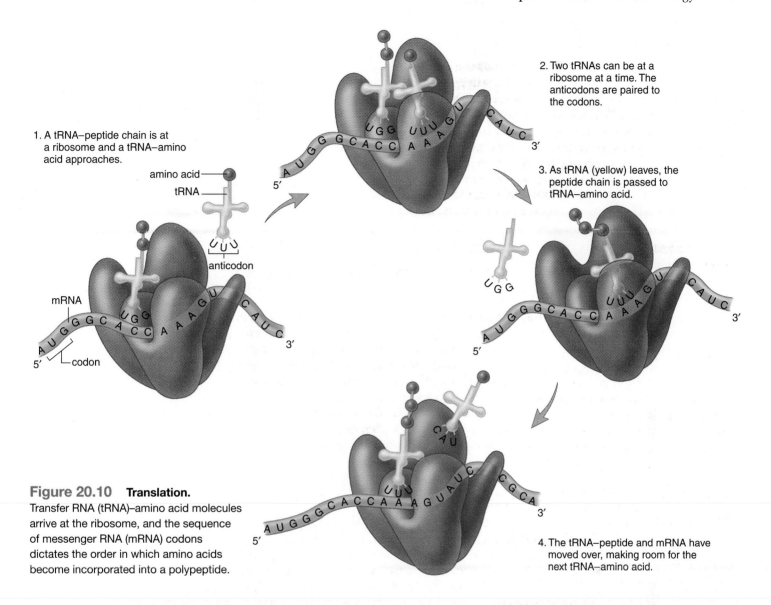

1. A tRNA–peptide chain is at a ribosome and a tRNA–amino acid approaches.

amino acid
tRNA
anticodon
mRNA
codon

2. Two tRNAs can be at a ribosome at a time. The anticodons are paired to the codons.

3. As tRNA (yellow) leaves, the peptide chain is passed to tRNA–amino acid.

4. The tRNA–peptide and mRNA have moved over, making room for the next tRNA–amino acid.

Figure 20.10 Translation.
Transfer RNA (tRNA)–amino acid molecules arrive at the ribosome, and the sequence of messenger RNA (mRNA) codons dictates the order in which amino acids become incorporated into a polypeptide.

As soon as the initial portion of mRNA has been translated by one ribosome, and the ribosome has begun to move down the mRNA, another ribosome attaches to the mRNA. Therefore, several ribosomes, collectively called a **polyribosome,** can move along one mRNA at a time. And several polypeptides of the same type can be synthesized using one mRNA molecule (Fig. 20.11).

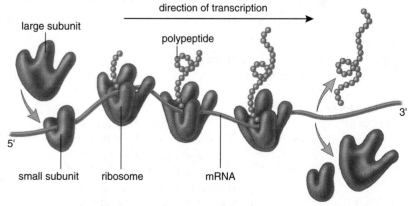

direction of transcription
large subunit
polypeptide
small subunit
ribosome
mRNA

Figure 20.11 Polyribosome structure.
Several ribosomes, collectively called a polyribosome, move along a messenger RNA (mRNA) molecule at one time. They function independently of one another; therefore, several polypeptides can be made at the same time.

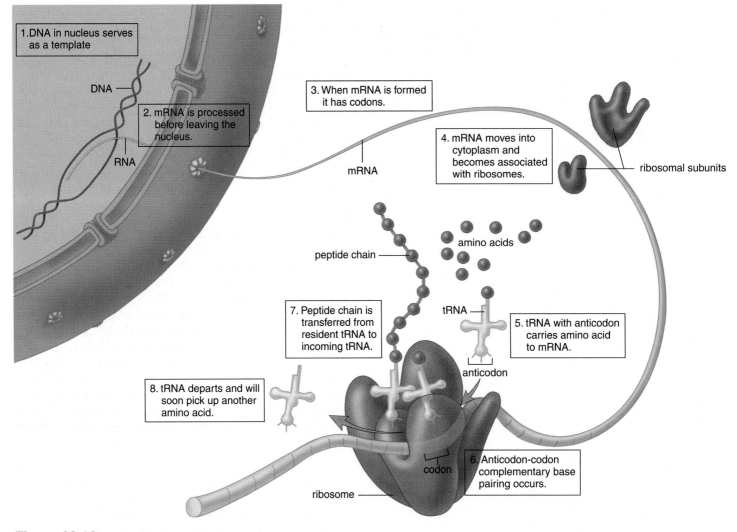

Figure 20.12   **Summary of gene expression.**

Let's Review Gene Expression

The following list, along with Table 20.3 and Figure 20.12, provides a brief summary of the events involved in gene expression that results in a protein product.

1. DNA in the nucleus contains a *triplet code.* Each group of three bases stands for a specific amino acid.
2. During transcription, a segment of a DNA strand serves as a template for the formation of mRNA. The bases in mRNA are complementary to those in DNA; every three bases is a *codon* for a certain amino acid.
3. Messenger RNA (mRNA) is processed before it leaves the nucleus, during which time the introns are removed.
4. Messenger RNA (mRNA) carries a sequence of codons to the *ribosomes,* which are composed of rRNA and proteins.
5. Transfer RNA (tRNA) molecules, each of which is bonded to a particular amino acid, have anticodons that pair complementarily to the codons in mRNA.

Table 20.3	Participants in Gene Expression	
Molecule	**Special Significance**	**Definition**
DNA	Triplet code	Sequence of bases in threes
mRNA	Codon	Complementary sequence of bases in threes
tRNA	Anticodon	Sequence of three bases complementary to codon
Amino acids	Building blocks	Transported to ribosomes by tRNAs
Protein	Enzymes and structural proteins	Amino acids joined in a predetermined order

6. During translation, tRNA molecules and their attached amino acids arrive at the ribosomes, and the linear sequence of codons of the mRNA determines the order in which the amino acids become incorporated into a protein.

The Control of Gene Expression

All cells receive a copy of all genes; however, cells differ as to which genes are being actively expressed. Muscle cells, for example, have a different set of genes that are turned on in the nucleus and proteins that are active in the cytoplasm than do nerve cells. In eukaryotic cells a variety of mechanisms regulates gene expression from transcription to protein activity. These mechanisms can be grouped under four primary levels of control, two of which pertain to the nucleus and two of which pertain to the cytoplasm.

1. *Transcriptional control:* In the nucleus, a number of mechanisms serve to control which genes are transcribed and/or the rate at which transcription of the genes occurs. These include the organization of chromatin and the use of transcription factors that initiate transcription, the first step in gene expression.
2. *Posttranscriptional control:* Posttranscriptional control occurs in the nucleus after DNA is transcribed and mRNA is formed. How mRNA is processed before it leaves the nucleus and also the speed with which mature mRNA leaves the nucleus can affect the amount of gene expression.
3. *Translational control:* Translational control occurs in the cytoplasm after mRNA leaves the nucleus and before there is a protein product. The life expectancy of mRNA molecules (how long they exist in the cytoplasm) can vary, as can their ability to bind ribosomes. It is also possible that some mRNAs may need additional changes before they are translated at all.
4. *Posttranslational control:* Posttranslational control, which also occurs in the cytoplasm, occurs after protein synthesis. The polypeptide product may have to undergo additional changes before it is biologically functional. Also, a functional enzyme is subject to

feedback control—the binding of an enzyme's product can change its shape so that it is no longer able to carry out its reaction.

Activated Chromatin

For a gene to be transcribed in human cells, the chromosome in that region must first decondense. The chromosomes within the developing egg cells of many vertebrates are called lampbrush chromosomes because they have many loops that appear to be bristles (Fig. 20.13). Here mRNA is being synthesized in great quantity; then protein synthesis can be carried out, despite rapid cell division during development.

Transcription Factors

In human cells, **transcription factors** are DNA-binding proteins. Every cell contains many different types of transcription factors, and a specific combination is believed to regulate the activity of any particular gene. After the right combination of transcription factors binds to DNA, an RNA polymerase attaches to DNA and begins the process of transcription.

As cells mature, they become specialized. Specialization is determined by which genes are active, and therefore, perhaps, by which transcription factors are present in that cell. Signals received from inside and outside the cell could turn on or off genes that code for certain transcription factors. For example, the gene for fetal hemoglobin ordinarily gets turned off as a newborn matures—one possible treatment for sickle-cell disease is to turn this gene on again.

Control of gene expression occurs at four levels. The first level, transcriptional control, involves organization of chromatin, and the use of transcription factors that bind to DNA.

Figure 20.13 **Lampbrush chromosomes.**
These chromosomes, seen in amphibian egg cells, give evidence that when mRNA is being synthesized, chromosomes decondense. Each chromosome has many loops extending from its axis (white). Many mRNA transcripts are being made off these DNA loops (red).

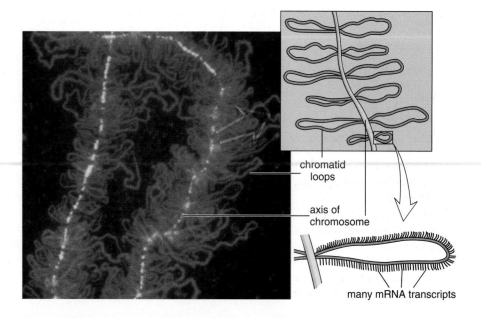

chromatid loops

axis of chromosome

many mRNA transcripts

20.3 Biotechnology 🔘

Genetic engineering is the use of technology to alter the genome of viruses, bacteria, and other cells for medical or industrial purposes. Biotechnology includes genetic engineering and other techniques that make use of natural biological systems to produce a product or to achieve an end desired by human beings. Genetic engineering has the capability of altering the genotype of unicellular organisms and the genotype of plants and animals, including ourselves.

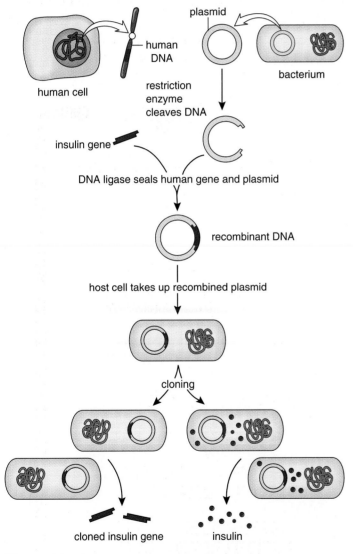

Figure 20.14   Cloning of a human gene.
Human DNA and plasmid DNA are cleaved by a specific type of restriction enzyme and spliced together by the enzyme DNA ligase. Gene cloning is achieved when a host cell takes up the recombinant plasmid, and as the cell reproduces, the plasmid is replicated. Multiple copies of the gene are now available to an investigator. If the insulin gene functions normally as expected, insulin may also be retrieved.

The Cloning of a Gene

The **cloning** of a gene produces many identical copies. Recombinant DNA technology is used when a very large quantity of the gene is required. The use of the polymerase chain reaction (PCR) creates fewer copies within a laboratory test tube.

Recombinant DNA Technology

Recombinant DNA (rDNA) contains DNA (deoxyribonucleic acid) from two or more different sources. To make rDNA, a technician often begins by selecting a **vector,** the means by which rDNA is introduced into a host cell. One common type of vector is a plasmid. **Plasmids** are small accessory rings of DNA. The ring is not part of the bacterial chromosome and can be replicated independently. Plasmids were discovered by investigators studying the reproduction of the intestinal bacterium *Escherichia coli* (*E. coli*).

Two enzymes are needed to introduce foreign DNA into vector DNA (Fig. 20.14). The first enzyme, called a **restriction enzyme,** cleaves plasmid DNA, and the second, called **DNA ligase,** seals foreign DNA into the opening created by the restriction enzyme.

Hundreds of restriction enzymes occur naturally in bacteria, where they stop viral reproduction by cutting up viral DNA. They are called restriction enzymes because they *restrict* the growth of viruses, but they are also called *molecular scissors* because each one cuts DNA at a specific cleavage site. For example, the restriction enzyme called *Eco*RI always cuts double-stranded DNA in this manner when it has this sequence of bases:

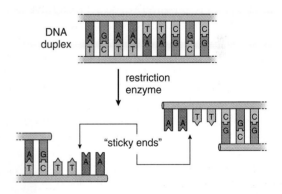

Notice there is now a gap into which a piece of foreign DNA can be placed if it ends in bases complementary to those exposed by the restriction enzyme. To assure this, it is only necessary to cleave the foreign DNA with the same type of restriction enzyme. The single-stranded, but complementary, ends of the two DNA molecules are called "sticky ends" because they can bind a piece of foreign DNA by complementary base pairing. Sticky ends facilitate the insertion of foreign DNA into vector DNA.

The second enzyme needed for preparation of rDNA, DNA ligase, is a cellular enzyme that seals any breaks in a

DNA molecule. Genetic engineers use this enzyme to seal the foreign piece of DNA into the vector. DNA splicing is now complete; an rDNA molecule has been prepared.

Bacterial cells take up recombinant plasmids, especially if they are treated to make them more permeable. Thereafter, if the inserted foreign gene is replicated and actively expressed, the investigator can recover either the cloned gene or a protein product (Fig. 20.14).

In order for gene expression to occur in a bacterium, the gene has to be accompanied by the proper regulatory regions. Also, the gene should not contain introns because bacterial cells do not have the necessary enzymes to process primary messenger RNA (mRNA). However, it's possible to make a mammalian gene that lacks introns. The enzyme called reverse transcriptase can be used to make a DNA copy of mature mRNA. This DNA molecule, called *complementary DNA (cDNA)*, does not contain introns. Alternatively, it is possible to manufacture small genes in the laboratory. A machine called a DNA synthesizer joins together the correct sequence of nucleotides, and this resulting gene also lacks introns.

The Polymerase Chain Reaction

The **polymerase chain reaction (PCR)** can create millions of copies of a single gene, or any specific piece of DNA, in a test tube. PCR is very specific—the targeted DNA sequence can be less than one part in a million of the total DNA sample! This means that a single gene among all the human genes can be amplified (copied) using PCR.

Before carrying out PCR, double-stranded DNA is denatured into single strands by heating. *Primers,* sequences of about 20 bases that are complementary to the bases on one side of the "target DNA," are then added. The primers are needed because DNA polymerase does not start the replication process; it only continues or extends the process. After the primers bind by complementary base pairing to the single-stranded DNA, DNA polymerase (the enzyme that carries out DNA replication) copies the target DNA (Fig. 20.15).

PCR has been in use for several years, and now almost every laboratory has automated PCR machines to carry out the procedure. Automation became possible after a temperature-insensitive (thermostable) DNA polymerase was extracted from the bacterium *Thermus aquaticus,* which lives in hot springs. The enzyme can withstand the high temperature used to separate double-stranded DNA; therefore, replication need not be interrupted by the need to add more enzyme.

Analyzing DNA Segments

Following PCR, the DNA can be subjected to **DNA fingerprinting.** When the DNA of a person is treated with restriction enzymes, the result is a unique collection of different-sized fragments. During a process called gel electrophoresis, the fragments can be separated according to their charge/size ratios, and the result is a pattern of distinc-

tive bands. If two DNA patterns match, then there is a high probability that the DNA came from the same person. Because of PCR, DNA from a single sperm is enough to identify a suspected rapist. DNA was used to successfully identify a teenage murder victim from eight-year-old remains. Skeletal DNA was compared to the DNA of the victim's parents. A DNA profile resembles that of one's parents because it is inherited.

DNA profiling following PCR has many other applications. When the DNA matches that of a virus, mutated gene, or oncogene, then we know that a viral infection, genetic disorder, or cancer is present. Mitochondrial DNA sequences in modern living populations have been used to determine the evolutionary history of human populations.

Recombinant DNA technology and the polymerase chain reaction are two ways to clone a gene.

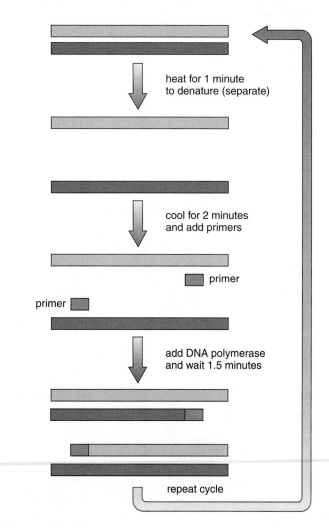

Figure 20.15   Polymerase chain reaction (PCR).
PCR is performed in a laboratory test tube. Primers (pink), which are DNA sequences complementary to the bases at one end of target DNA, are necessary for DNA polymerase to make a copy of each DNA strand.

Biotechnology Products

Today, bacteria, plants, and animals are genetically engineered to produce biotechnology products. Organisms that have had a foreign gene inserted into them are called **transgenic organisms.**

From Bacteria

Recombinant DNA technology is used to produce bacteria that reproduce in large vats called bioreactors. If the foreign gene is replicated and actively expressed, a large amount of protein product can be obtained. Biotechnology products produced by bacteria, such as insulin, human growth hormone, t-PA (tissue plasminogen activator), and hepatitis B vaccine, are now on the market (Fig. 20.16).

Transgenic bacteria have been produced to promote the health of plants. For example, bacteria that normally live on plants and encourage the formation of ice crystals have been changed from frost-plus to frost-minus bacteria. Also, a bacterium that normally colonizes the roots of corn plants has now been endowed with genes (from another bacterium) that code for an insect toxin. The toxin protects the roots from insects.

Bacteria can be selected for their ability to degrade a particular substance, and then this ability can be enhanced by genetic engineering. For instance, naturally occurring bacteria that eat oil can be genetically engineered to do an even better job of cleaning up beaches after oil spills (Fig. 20.17). Industry has found that bacteria can be used as biofilters to prevent airborne chemical pollutants from being vented into the air. They can also remove sulfur from coal before it is burned and help clean up toxic waste dumps. One such strain was given genes that allowed it to clean up levels of toxins that would have killed other strains. Further, these bacteria were given "suicide" genes that caused them to self-destruct when the job had been accomplished.

Organic chemicals are often synthesized by having catalysts act on precursor molecules or by using bacteria to carry out the synthesis. Today, it is possible to go one step further and to manipulate the genes that code for these enzymes. For instance, biochemists discovered a strain of bacteria that is especially good at producing phenylalanine, an organic chemical needed to make aspartame, the dipeptide sweetener better known as NutraSweet. They isolated, altered, and formed a vector for the appropriate genes so that various bacteria could be genetically engineered to produce phenylalanine.

Many major mining companies already use bacteria to obtain various metals. Genetic engineering may enhance the ability of bacteria to extract copper, uranium, and gold from low-grade sources. Some mining companies are testing genetically engineered organisms that have improved bioleaching capabilities.

Bacteria are being genetically altered to perform all sorts of tasks, not only in the factory, but also in the environment.

Figure 20.16   Biotechnology products.
Products like clotting factor VIII, which is administered to hemophiliacs, can be made by transgenic bacteria, plants, or animals. After being processed and packaged, it is sold as a commercial product.

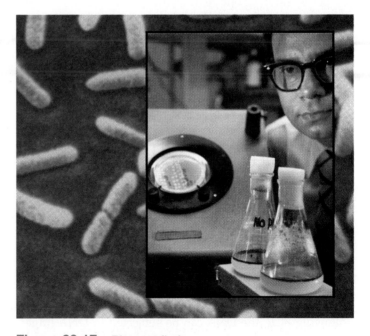

Figure 20.17   Bioremediation.
Bacteria capable of decomposing oil have been engineered and patented by the investigator, Dr. Chakrabarty. In the inset, the flask toward the rear contains oil and no bacteria; the flask toward the front contains the bacteria and is almost clear of oil.

From Plants

Techniques have been developed to introduce foreign genes into immature plant embryos, or into plant cells that have had the cell wall removed and are called protoplasts. It is possible to treat protoplasts with an electric current while they are suspended in a liquid containing foreign DNA. The electric current makes tiny, self-sealing holes in the plasma membrane through which genetic material can enter. Then a protoplast will develop into a complete plant.

Foreign genes transferred to cotton, corn, and potato strains have made these plants resistant to pests because their cells now produce an insect toxin. Similarly, soybeans have been made resistant to a common herbicide. Some corn and cotton plants are both pest and herbicide resistant. These and other genetically engineered crops are now reaching the marketplace.

Plants are also being engineered to produce human proteins, such as hormones, clotting factors, and antibodies, in their seeds. One type of antibody made by corn can deliver radioisotopes to tumor cells, and another made by soybeans can be used as treatment for genital herpes. A weed called mouse-eared cress has been engineered to produce a biodegradable plastic (polyhydroxybutyrate, or PHB) in cell granules.

Genetically engineered crops are now reaching the market, and medicines made by genetically engineered plants will soon be used to treat cancer and other types of diseases.

Ecology Focus

Biotechnology: Friend or Foe?

In a hungry world you would expect herbicide-resistant crops to be greeted with enthusiasm—genetically engineered wheat and corn offer the possibility of a more bountiful harvest and the feeding of many more people. There are some, however, who see dangers lurking in the use of biotechnology to develop new and different strains of plants and animals. And these doomsayers are not just anybody—they are ecologists.

First, we have to consider that herbicide-resistant crops will allow farmers to use more herbicide than usual in order to kill off weeds. Then, too, suppose this new form of wheat is better able to compete in the wild. Certainly we know of plants that have become pests when transported to a new environment: prickly pear cactus took over many acres of Australia; an ornamental tree, the melaleuca, has invaded and is drying up many of the swamps in Florida. Such plants spread because they are able to overrun the native plants of an area. Perhaps genetically engineered plants will also spread beyond their intended areas and be out of control. Or worse, suppose herbicide-resistant wheat were to hybridize with a weed, making the weed also resistant and able to take over other agricultural fields. As more and different herbicides are used to kill off the weed, the environment would be degraded. And similar concerns pertain to any transgenic organism, whether a bacterium, animal, or plant.

In the past, humans have been quick to believe that a new advance was the answer to a particular problem. The pesticide DDT was going to kill off mosquitos, making malaria a disease of the past. And this worked until mosquitos became resistant. Today, we know that DDT accumulates in the tissues of humans, possibly contributing to all manner of health problems, from reduced immunity to reproductive infertility. When antibiotics were first introduced, it was hoped a disease like tuberculosis would be licked forever. Resistant strains of tuberculosis have now evolved to threaten us all.

More and more transgenic varieties have been developed and are being tested in agricultural fields. A few of these have a weedy relative in the wild with which they could hybridize. Agricultural officials point out, however, that genetically engineered plants have been growing in fields since 1994, and although hybridization with weedy relatives may have occurred, a "superweed" has not emerged. Still, say botanists, it could happen in the future. Laboratory studies in Denmark showed that a hybrid of transgenic oilseed rape and field mustard did resist herbicides and was able to produce highly fertile pollen. Ecologists maintain it may be only a matter of time before a "superweed" does appear in the wild.

From Animals

Techniques have been developed to insert genes into the eggs of animals. It is possible to microinject foreign genes into eggs by hand, but another method uses vortex mixing. The eggs are placed in an agitator with DNA and silicon-carbide needles, and the needles make tiny holes through which the DNA can enter. When these eggs are fertilized, the resulting offspring are transgenic animals. Using this technique, many types of animal eggs have taken up the gene for bovine growth hormone (bGH). The procedure has been used to produce larger fishes, cows, pigs, rabbits, and sheep. Genetically engineered fishes are now being kept in ponds that offer no escape to the wild because there is much concern that they will upset or destroy natural ecosystems.

Gene pharming, the use of transgenic farm animals to produce pharmaceuticals, is being pursued by a number of firms. Genes that code for therapeutic and diagnostic proteins are incorporated into the animal's DNA, and the proteins appear in the animal's milk. There are plans to produce drugs for the treatment of cystic fibrosis, cancer, blood diseases, and other disorders. Antithrombin III, for preventing blood clots during surgery, is currently being produced by a herd of goats, and clinical trials have begun. Figure 20.18*b, c* outlines the procedure for producing transgenic mammals: DNA containing the gene of interest is injected into donor eggs. Following in vitro fertilization, the zygotes are placed in host females where they develop. After female offspring mature, the product is secreted in their milk.

USDA scientists have been able to genetically engineer mice to produce human growth hormone in their urine instead of in milk. They expect to be able to use the same technique on larger animals. Urine is a preferable vehicle for a biotechnology product than milk because all animals in a herd urinate—only females produce milk; animals start to urinate at birth—females don't produce milk until maturity; and it's easier to extract proteins from urine than from milk.

Xenotransplantation

Scientists have begun the process of genetically engineering animals to serve as organ donors for humans who need a transplant. **Xenotransplantation** is the use of animal organs instead of human organs in transplant patients. We now have the ability to transplant kidneys, heart, liver, pancreas, lung, and other organs for two reasons. First, solutions have been developed that preserve donor organs for several hours, and second, rejection of transplanted organs can be prevented by immunosuppressive drugs. Unfortunately, however, there are not enough human donors to go round. Fifty thousand Americans needed transplants in 1996, but only 20,000 patients got them. As many as 4,000 died that year while waiting for an organ.

It's no wonder, then, that scientists are suggesting that we should get organs from a source other than another human. You might think that apes, such as the chimpanzee or man. You might think that apes, such as the chimpanzee or the baboon might be a scientifically suitable species for this purpose. But apes are slow breeders and probably cannot be counted on to supply all the organs needed. Anyway, many people might object to using apes for this purpose. In contrast, animal husbandry has long included the raising of pigs as a meat source and pigs are prolific. A female pig can become pregnant at six months and can have two litters a year, each averaging about ten offspring.

Ordinarily, humans would violently reject transplanted pig organs. Genetic engineering, however, can make these organs less antigenic. Scientists have produced a strain of pigs whose organs would most likely, even today, survive for a few months in humans. They could be used to keep a patient alive until a human organ was available. The ultimate goal is to make pig organs as widely accepted by humans as type O blood. A person with type O blood is called a universal donor because the red blood cells carry no ABO antigens.

As xenotransplantation draws near, other concerns have been raised. Some experts fear that animals might be infected with viruses, akin to Ebola virus or the virus that causes "mad cow" disease. After infecting a transplant patient, these viruses might spread into the general populace and begin an epidemic. Scientists believe that HIV was spread to humans from monkeys when humans ate monkey meat. Those in favor of using pigs for xenotransplantation point out that pigs have been around humans for centuries without infecting them with any serious diseases.

Cloning of Animals

Imagine that an animal has been genetically altered to produce a biotechnology product or to serve as an organ donor. What would be the best possible way to get identical copies of this animal? If cloning of the animal was possible, you could get many exact copies of this animal. Asexual reproduction through cloning would be the preferred procedure to use. Cloning is a form of asexual reproduction because it requires only the genes of that one animal. For many years it was believed that adult vertebrate animals could not be cloned. Although each cell contains a copy of all the genes, certain genes are turned off in mature specialized cells. Different genes are expressed in muscle cells, which contract, compared to nerve cells, which conduct nerve impulses, and to glandular cells, which secrete. Cloning of an adult vertebrate would require that all genes of an adult cell be turned on again if development is to proceed normally. It has long been thought this would be impossible.

But in 1997, scientists at the Raslin Institute in Scotland announced that they achieved this feat and had produced a cloned sheep called Dolly. In 1998, genetically altered calves were cloned in the United States using the same method. An alternate method, used at the University of Hawaii for the cloning of mice, was so successful that

a.

Figure 20.18   Genetically engineered animals.

a. This goat is genetically engineered to produce antithrombin III, which is secreted in her milk. This researcher and many others are involved in the project.
b. The procedure to produce a transgenic animal.
c. The procedure to clone a transgenic animal.

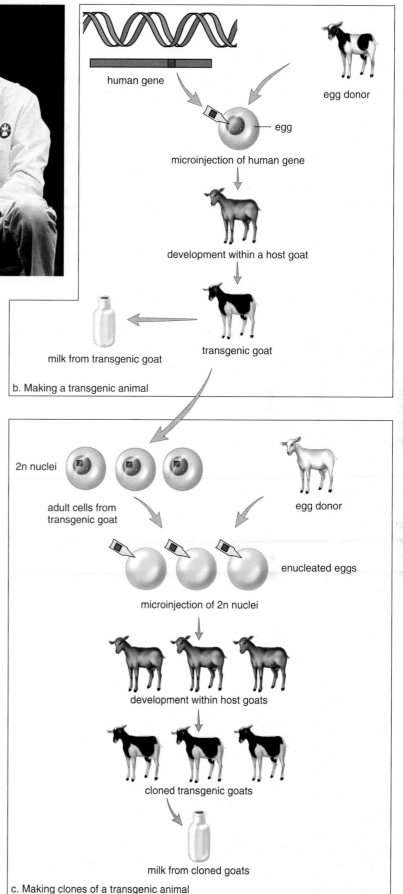

clones of clones were produced. Figure 20.18*c* suggests if enucleated eggs have been injected with 2n nuclei of adult cells, they can be coaxed to begin development. The offspring have the genotype and phenotype of the adult that donated the nuclei; therefore, the adult has been cloned. In the procedure that produced cloned mice, the 2n nuclei were taken from cumulus cells. Cumulus cells are those that cling to an egg after ovulation occurs. A specially prepared chemical bath was used to stimulate the eggs to divide and begin development. Now that scientists have a method to clone mammals, this procedure will undoubtedly be used routinely. Some are even beginning to think about the cloning of humans despite the objections of many and a presidential order that the procedure is not to be developed in the United States.

Transgenic animals, which secrete a biotechnology product in their milk, and pigs, whose organs can be used for xenotransplantation, have been produced. Procedures have been developed to allow the cloning of these animals.

The Human Genome Project

The Human Genome Project is a massive effort originally funded by the U.S. government and now increasingly by U.S. pharmaceutical companies to map the human chromosomes. Many nonprofit and for profit biochemical laboratories about the world are now involved in the project which has two primary goals. The first goal is to construct a genetic map of the human genome. The aim is to show the sequence of genes along the length of each type chromosome, such as depicted for the X chromosome in Figure 20.19. If the estimate of 1,000+ humans genes is correct, each chromosome on average would contain about 50 alleles.

The map for each chromosome is presently incomplete, and in many instances scientists rely on the placement of enzyme restriction sites that differ between individuals. These sites eventually allow scientists to pinpoint disease-causing genes because a particular restriction site and a defective gene are often inherited together. For example, it is known that persons with Huntington disease have a unique site where a restriction enzyme cuts DNA. The test for Huntington disease relies on this difference from the normal.

The genetic map of a chromosome can be used not only to detect defective genes, but possibly also to tailor treatments to the individual. Only certain hypertension patients benefit from a low-salt diet, and it would be useful to know which patients these are. Myriad Genetics, a genome company, has developed a test for a mutant angiotensinogen gene because they want to see if patients with this mutation are the ones that benefit from a low-salt diet. Several other mutant genes have also been correlated with specific drug treatments (Table 20.4). One day the medicine you take might carry a label that it is effective only in persons with genotype #101!

The second goal is to construct a base sequence map. There are three billion base pairs in the human genome, and it's estimated it would take an encyclopedia of 200 volumes, each with 1,000 pages, to list all of these. Yet, this goal of the Human Genome Project is expected to be reached by the year 2004, if not earlier.

The methodology, thus far, has been to first chop up the genome into small pieces, each just 1,000 to 2,000 base pairs long. PCR instruments copy the pieces many times and then an automatic DNA sequencer determines the order of the base pairs. You need many DNA copies because of the way the sequencer works. A computer program later strings the sequenced pieces together in the correct order by looking for base sequence overlaps between them. Instrumentation has gradually improved, and recently one scientist, J. Craig Venter, has founded a company which he says will sequence the entire genome in three years.

Venter plans on using what is called a whole-genome shotgun sequencing method. He plans on working with the entire human genome at once. Each overlapping frag-

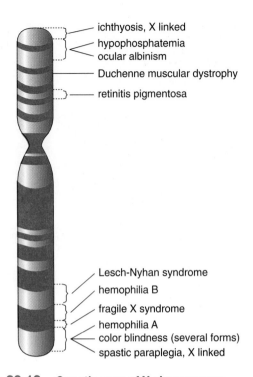

Figure 20.19 **Genetic map of X chromosome.**
The human X chromosome has been partially mapped, and this is the order of some of the genes now known to be on this chromosome.

ment will be about 5,000 bases long, and he will sequence the ends of each fragment using only powerful new sequencing machines. Again, a computer program will string the fragments together by looking for overlapping regions. Why is Venter going off on his own, instead of participating in the worldwide effort by many laboratories and scientists to sequence the human genome? His backers expect to market the whole-genome database to subscribers, and to patent rare but pharmacologically interesting genes. Private enterprise is giving new impetus to the field now called genomics.

Knowing the base sequence of normal genes may make it possible one day to treat certain human ills by administering normal genes and/or their protein products to those who suffer from a genetic disease.

Table 20.4	Customizing Drug Treatments	
Mutant Gene for	**Disease**	**Treatment**
Apolipoprotein E	Alzheimer	Experimental Glaxo Wellcome drug
Cytochrome P-450	Cancer	Amonafide
Chloride gate	Cystic fibrosis	Pulmozyme
Dopamine receptor D4	Schizophrenia	Clozapine
Angiotensinogen	Hypertension	Low-salt diet

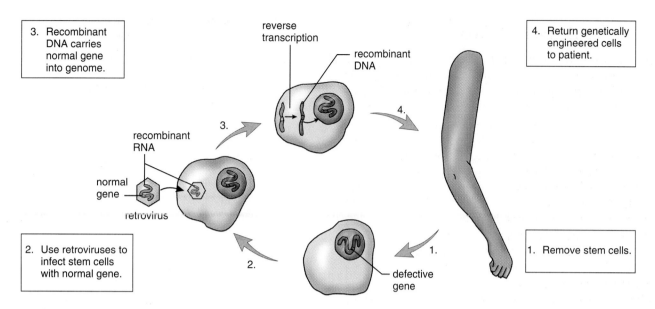

3. Recombinant DNA carries normal gene into genome.

reverse transcription

recombinant DNA

4. Return genetically engineered cells to patient.

recombinant RNA

normal gene

retrovirus

2. Use retroviruses to infect stem cells with normal gene.

1. Remove stem cells.

defective gene

Figure 20.20 Ex vivo gene therapy in humans.
Bone marrow stem cells are withdrawn from the body, a virus is used to insert a normal gene into them, and they are returned to the body.

Gene Therapy

Gene therapy gives a patient a normal gene to make up for a faulty gene. Gene therapy also includes the use of genes to treat various human illnesses. There are ex vivo (outside the living organism) and in vivo (inside the living organism) methods of gene therapy.

Ex Vivo Methods

Cindy Cutshall, discussed previously (p. 421), and another young girl with severe combined immunodeficiency syndrome (SCID), underwent ex vivo gene therapy several years ago. These girls lacked an enzyme that is involved in the maturation of T and B cells, and therefore, they were subject to life-threatening infections. Bone marrow stem cells were removed from their blood and infected with a retrovirus that carried a normal gene for the enzyme (Fig. 20.20). Then the cells were returned to the girls. Genetically engineered stem cells are preferred because they produce other cells with the same genes.

Among the 100-plus gene therapy trials, gene therapy is being used for treatment of familial hypercholesterolemia, a condition that develops when liver cells lack a receptor for removing cholesterol from the blood. The high levels of blood cholesterol make the patient subject to fatal heart attacks at a young age. In a newly developed procedure, a small portion of the liver is surgically excised and infected with a retrovirus containing a normal gene for the receptor. Chemotherapy in cancer patients often kills off healthy cells as well as cancer cells. In one ex vivo clinical trial, bone marrow stem cells from 30 women with late-stage ovarian cancer were infected with a virus carrying a gene to make them more tolerant of chemotherapy.

Once the bone marrow stem cells were protected, it was possible to increase the level of chemotherapy to kill the cancer cells.

In Vivo Methods

Other gene therapy procedures use viruses, laboratory-grown cells, or even synthetic carriers to introduce genes directly into patients. If in vivo therapy is used, no cells are removed from the patient. For example, liposomes—microscopic vesicles that spontaneously form when lipoproteins are put into a solution—have been coated with the gene to cure cystic fibrosis and then sprayed into patients' nostrils. Retroviruses carrying an anti-cancer gene have been injected directly into cancerous tumors with the hope that the anti-cancer genes will cure the cancer cells. In one trial no cures were reported, but the lung tumors shrank in three patients and stopped growing in three others.

Perhaps it will be possible also to use in vivo therapy to cure hemophilia, diabetes, Parkinson disease, or AIDS. To treat hemophilia, patients could get regular doses of cells that contain normal clotting-factor genes. Or such cells could be placed in *organoids,* artificial organs that can be implanted in the abdominal cavity. To cure Parkinson disease, dopamine-producing cells could be grafted directly into the brain. In a recent and surprising move, some researchers are investigating the possibility of directly correcting the base sequence of patients with a genetic disorder.

The Human Genome Project produces information useful to gene therapists. Researchers are envisioning all sorts of ways to cure human genetic disorders as well as many other types of illnesses.

Bioethical Issue

Somatic gene therapy attempts to treat or prevent human illnesses. Some day, for example, it may be possible to give children who have cystic fibrosis, Huntington disease, or any other genetic disorder a normal gene to make up for the inheritance of a faulty gene. And gene therapy is even more likely for treatment of diseases like cancer, AIDS, and heart disease. Germ line gene therapy is the term that is now being used to mean the use of gene therapy solely to improve the traits of an individual. It would be the genetic equivalent of procedures like body building, liposuction, or hair transplants. If genetic interference occurred earlier—that is, on the eggs, sperm, or embryo, it's possible that it would indeed affect the germ line—that is, all the future descendants of the individual.

How might germ line gene therapy become routine? Consider this scenario. Presently, a gene for VEGF (vascular endothelial growth factor), a protein produced by cells to grow new blood vessels, is being used to treat atherosclerosis. Improved arterial circulation in the legs of some recent patients did away with the threat of possible amputation. This same gene is now being considered for the treatment of blocked coronary arteries. There may be instances, though, in which healthy people want to grow new blood vessels for enhancement purposes. Runners might want to improve their circulation in order to win races, and parents might think that increased circulation to the brain might increase the intelligence of their children. Bioethicist Eric Juengst of Case Western thinks that as a society, there

is nothing we can now do to prevent us from crossing the line between the use of gene therapy for therapeutic purposes and its use for enhancement reasons.

Questions

1. Do you approve of gene therapy to treat or prevent human illnesses? after they occur? or possibly before they occur, if genetic testing shows the likelihood of their development? Why or why not?

2. Do you approve of gene therapy to enhance the characteristics of an individual? after a condition like baldness occurs? or possibly before it occurs if genetic testing shows the likelihood of it occurring? Why or why not?

3. Do you approve of somatic gene therapy of embryos? germ line gene therapy of embryos? Explain.

Summarizing the Concepts

20.1 DNA and RNA Structure and Function

DNA is a double helix composed of two nucleic acid strands that are held together by weak hydrogen bonds between the bases: A is bonded to T, and C is bonded to G. During replication, the DNA strands unzip, and then a new complementary strand forms opposite to each old strand. This results in two identical DNA molecules.

RNA is a single-stranded nucleic acid in which A pairs with U (uracil), while G pairs with C. Both DNA and RNA are involved in gene expression.

20.2 Gene Expression

Proteins differ from one another by the sequence of their amino acids. DNA has a code that specifies this sequence. Gene expression requires transcription and translation. During transcription, the DNA code (triplet of three bases) is passed to an mRNA that then contains codons. Introns are removed from mRNA during mRNA processing. During translation, tRNA molecules bind to their amino acids, and then their anticodons pair with the mRNA codons. In the end, each protein has a sequence of amino acids according to the blueprint provided by the sequence of nucleotides in DNA.

Control of gene expression can occur at four levels in a human cell: at the time of transcription; after transcription and during mRNA processing; during translation; and after translation—before, at, or after protein synthesis. It is known that chromatin has to be extended for transcription to occur and that there are transcription factors that control the activity of genes in human cells.

20.3 Biotechnology

Two methods are currently available for making copies of DNA. Recombinant DNA contains DNA from two different sources. Human DNA can be inserted into a plasmid, which is taken up by bacteria.

When the plasmid replicates, the gene is cloned and its protein is produced. The polymerase chain reaction (PCR) uses the enzyme DNA polymerase to make multiple copies of target DNA. PCR has many applications. Following PCR, it is possible to determine if the DNA of an infectious organism or any particular sequence of DNA is present. The DNA can be subjected to DNA profiling to see if it matches DNA from another source.

Transgenic organisms are now routinely made. Crops resistant to pests and herbicides are commercially available. Transgenic animals have been given various genes, in particular the one for bovine growth hormone (bGH). Animals have also been genetically engineered to secrete a protein product in their milk. Pigs have been genetically altered to serve as a source of organs for transplant patients. Cloning of animals is now possible.

The Human Genome Project has two goals. The first goal is to construct a genetic map which would show the sequence of the genes on each chromosome. Although knowledge of the location of genes is expanding, the genetic map still contains many RFLPs (restriction fragment length polymorphisms) instead as the first step toward finding more genes. The second goal of the project is to sequence the bases. To do this, researchers rely on PCR and automatic sequencing instruments. As the sequencers have improved, so has the speed with which DNA is sequenced. The hope is that knowing the location and sequence of bases in a gene will promote the possibility of gene therapy.

Human gene therapy is undergoing clinical trials. Ex vivo therapy involves withdrawing cells from the patient, inserting a functioning gene, usually via a retrovirus, and then returning the treated cells to the patient. Many investigators are trying to develop in vivo therapy in which viruses, laboratory-grown cells, or synthetic carriers will be used to carry healthy genes directly into the patient.

Studying the Concepts

1. Describe the structure of DNA and how this structure contributes to the ease of DNA replication. 423–24
2. Describe the structure of RNA and compare it to the structure of DNA. 425
3. Name and discuss the role of three different types of RNA. 425
4. Describe the structure and function of a protein and the manner in which DNA codes for a particular protein. 426
5. Describe the process of transcription and the three steps of translation. If the code is TTT; CAT; TGG; CCG, what are the codons, and what is the sequence of amino acids? 426–29
6. What are the four levels of genetic control in human cells? Describe two means by which transcription is regulated. 431
7. Describe precisely how you might clone a gene using recombinant DNA technology. When might you use a PCR instrument? 432–33
8. Naturally occurring bacteria have been genetically engineered to perform what services? 434
9. What types of genetically engineered plants are now available and/or are expected in the near future? 435
10. What types of animals have been genetically engineered and for what purposes? Discuss the advantages of cloning transgenic animals. 436
11. What are the two primary goals of the Human Genome Project? What are the possible benefits of the project for human society? 438
12. How is gene therapy in humans currently being done? 439

Testing Your Knowledge of the Concepts

In questions 1–4, match the nucleotide to its function.
a. DNA
b. mRNA
c. tRNA
d. rRNA

_____ 1. Joins with proteins to form subunits of a ribosome.
_____ 2. Contains codons which determine the sequence of amino acids in a polypeptide.
_____ 3. Contains a code and serves as a template for the production of RNA.
_____ 4. Brings amino acids to the ribosomes during the process of transcription.

In questions 5–7, indicate whether the statement is true (T) or false (F).

_____ 5. The two types of enzymes needed to make a recombinant DNA plasmid are a restriction enzyme and DNA ligase.
_____ 6. Replication of DNA is semiconservative, meaning that DNA contains a code and mRNA contains codons.
_____ 7. If the DNA code is ATC, GGG, CGC, then mRNA is TAG, CCC, GCG.

In questions 8–12, fill in the blanks.
8. The current ex vivo gene therapy clinical trials use a(n) _____ as a vector to insert healthy genes into the patient's cells.
9. The polymerase chain reaction makes many _____ of a short DNA segment.
10. The DNA code is a _____ code, meaning that every three bases stands for an _____.
11. The sequence of mRNA codons dictates the sequence of amino acids in a protein. This step in protein synthesis is called _____.
12. Bacteria, plants, and animals that have been genetically engineered are called _____ organisms.
13. This is a segment of a DNA molecule. (Remember that only the template strand is transcribed.) What are (a) the RNA codons and (b) the tRNA anticodons?

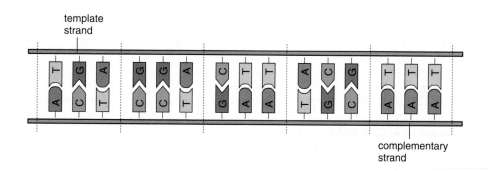

template
strand

complementary
strand

Applying Your Knowledge to the Concepts

These questions pertain to DNA and biotechnology.

1. The backbone of DNA strands contains atoms joined by covalent bonds, whereas the bases between the two strands are hydrogen bonded. Based upon your knowledge of bonds, what is the significance of the type of bonding existing in a DNA molecule?

2. If you knew the amino acid sequence for insulin, how would you go about making a gene that codes for insulin?

3. A change in a sequence of DNA occurs so that the mRNA codon reads AUC rather than AUU. Both of these code for the amino acid isoleucine. Argue that this is not a mutation.

4. The DNA code is redundant—as many as six different codons can code for an amino acid. What benefit arises because the code is redundant?

Understanding the Terms

adenine (A) 423
anticodon 428
cloning 432
codon 427
complementary base
    pairing 423
cytosine (C) 423
DNA (deoxyribonucleic
    acid) 422
DNA fingerprinting 433
DNA ligase 432
gene 422
gene therapy 439
genetic engineering 432
guanine (G) 423
messenger RNA (mRNA) 425
mutation 424
plasmid 432
polymerase chain
    reaction (PCR) 433

polyribosome 429
recombinant DNA (rDNA)
    432
replication 423
restriction enzyme 432
ribosomal RNA (rRNA) 425
ribozyme 427
RNA (ribonucleic acid) 425
RNA polymerase 427
template 424
thymine (T) 423
transfer RNA (tRNA) 425
transcription 426
transcription factor 431
transgenic organism 434
translation 428
triplet code 427
uracil (U) 425
vector 432
xenotransplantation 436

Match the terms to these definitions:

a. _____ Cluster of ribosomes attached to the same mRNA molecule; each ribosome is producing a copy of the same polypeptide.

b. _____ Free-living organism in the environment that has had a foreign gene inserted into it.

c. _____ Using fragment lengths resulting from restriction enzyme cleavage to identify particular individuals; also called DNA fingerprinting.

d. _____ Process resulting in the production of a strand of RNA that is complementary to a segment of DNA.

e. _____ One of four organic bases in the nucleotides composing the structure of DNA and RNA.

Applying Technology to the Concepts

Your study of DNA and biotechnology is supported by these available technologies:

Essential Study Partner CD-ROM

Genetics → DNA

→ Protein Synthesis

→ Recombinant DNA

Visit the Mader web site for related ESP activities.

Life Science Animations 3D Video

13 Structure of DNA
14 DNA Replication
16 DNA Mutations
18 Transcription
19 Translation
20 Polyribosomes

Exploring the Internet

The Mader Home Page provides resources and tools as you study this chapter.

http://www.mhhe.com/biosci/genbio/mader

Chapter 21

Cancer

Chapter Concepts

Figure 21.1   Sunbathing.
Sunbathing can lead to skin cancer. Melanoma is a particularly malignant skin cancer.

If there's one thing 25-year-old Emily loves, it's lying in the sun. "Baking on the beach" is what she calls it (Fig. 21.1). A little baby oil, a radio, some sunglasses. She does it for hours. Unfortunately, while Emily rubs in the oil, she doesn't notice the small mole on the back of her right arm. Dark and curvy, the little mole looks like a splotch of spilled ink. Doctors call it melanoma.

If Emily is lucky, she or her physician will see the mole before the cancer cells underneath break away and spread. If no one notices the mole, the cells will likely travel through Emily's circulatory system, making their way to her lungs, liver, brain, and ovaries. When that happens, it will be too late.

About 32,000 cases of melanoma are diagnosed each year. One in five persons diagnosed dies within five years. For most people, it's easy to avoid the disease. Most important, don't "bake on the beach and don't visit tanning machines." Protective clothing is a must and wearing sunscreen may also help.

The melanoma that developed on Emily's right arm went through certain phases. During *initiation,* a single cell underwent a mutation that caused it to begin to divide repeatedly. Then, *promotion* occurred as other factors encouraged a tumor to develop, and the tumor cells continued to divide. As they divided, other mutations occurred. Finally, *progression* took place as more mutations gave one cell a selective advantage over the other cells. This process can be repeated several times, until there is a cell that has the ability to invade surrounding tissues. This is called *metastasis.*

21.1 Cancer Cells

Cancer is a genetic disease requiring a series of mutations, each propelling cells toward the development of a **tumor,** an abnormal mass of cells. **Carcinogenesis,** the development of cancer, is a gradual, stepwise process, and it may be decades before a person notices any sign or symptom of a tumor. Although tumor cells share common characteristics, each type of cancer has its own particular sequence of mutations that has resulted in uncontrolled growth. It is common to designate cancers according to the organ of origin. Appendix E gives pertinent information about cancers at specific sites.

Characteristics of Cancer Cells

Cancer cells share characteristics that distinguish them from normal cells (Table 21.1).

Cancer Cells Lack Differentiation

Most cells are specialized; they have a specific form and function that suits them to the role they play in the body. Cancer cells are nonspecialized and do not contribute to the functioning of a body part. A cancer cell does not look like a differentiated epithelial, muscular, nervous, or connective tissue cell; instead, it looks distinctly abnormal. Normal cells can undergo the cell cycle for about 50 times, and then they die. Cancer cells can enter the cell cycle repeatedly, and in this way, they are potentially immortal.

Cancer Cells Have Abnormal Nuclei

The nuclei of cancer cells are enlarged, and there may be an abnormal number of chromosomes. The chromosomes have mutated; some parts may be duplicated and some may be deleted, for example. In addition, *gene amplification* (extra copies of specific genes) is seen much more frequently than in normal cells. Ordinarily cells with damaged DNA undergo **apoptosis,** or programmed cell death. Cancer cells fail to undergo apoptosis involving a series of enzymatic reactions that lead to the death of the cell.

Cancer Cells Form Tumors

Normal cells anchor themselves to a substratum and/or adhere to their neighbors. They exhibit *contact inhibition—* when they come in contact with a neighbor, they stop dividing. In culture, normal cells form a single layer that covers the bottom of a petri dish. Cancer cells have lost all restraint; they pile on top of one another to grow in multiple layers.

Normal cells do not grow and divide unless they are stimulated to do so by a growth factor. Cancer cells have a reduced need for a growth factor, such as epidermal growth factor, in order to grow and divide. Conversely, cancer cells no longer respond to inhibitory growth factors such as transforming growth factor beta (TGF-β) from their neighbors. Their growth, termed a *neoplasia* contains cells that are disorganized, a condition termed *anaplasia.* During carcinogenesis, the most aggressive cell becomes the dominant cell of the tumor (Fig. 21.2).

Table 21.1	Characteristics of Normal Cells Versus Cancer Cells	
Characteristic	**Normal Cells**	**Cancer Cells**
Differentiation	Yes	No
Nuclei	Normal	Abnormal
Tumor Formation	Controlled	Uncontrolled
Contact inhibition	Yes	No
Growth factors	Required	Not required
Angiogenesis	No	Yes
Metastasis	No	Yes

Cancer Cells Induce Angiogenesis

Angiogenesis is the formation of new blood vessels. To grow larger than, say, a pea, a tumor must have a well-developed capillary network to bring it nutrients and oxygen. Tumors secrete angiogenic proteins that stimulate nearby blood vessels to branch and send capillaries to the tumor. This blood flow not only nourishes the tumor, it also enables it to grow larger and to metastasize. Some modes of cancer treatment are aimed at preventing angiogenesis from occurring.

Cancer Cells Metastasize

A *benign tumor* is a disorganized, usually encapsulated, mass that does not invade adjacent tissue. Cancer in situ is a tumor located in its place of origin, before there has been any invasion of normal tissue. Malignancy is present when **metastasis** establishes new tumors distant from the primary tumor. To metastasize, cancer cells must make their way across a basement membrane and into a blood vessel or lymphatic vessel. Cancer cells produce proteinase enzymes that degrade the membrane and allow them to invade underlying tissues. Cancer cells tend to be motile because they have a disorganized internal cytoskeleton and lack intact actin filament bundles. After traveling through the blood or lymph, cancer cells may start tumors elsewhere in the body.

The patient's prognosis (probable outcome) is dependent on the degree to which the cancer has progressed: (1) whether the tumor has invaded surrounding tissues; (2) if so, whether lymph nodes are involved; and (3) whether there are metastatic tumors in distant parts of the body. With each progressive step, the prognosis becomes less favorable.

Cancer cells are nonspecialized, have abnormal chromosomes, and divide uncontrollably. Because they are not constrained by their neighbors, they form a tumor. Then they metastasize, forming new tumors wherever they relocate.

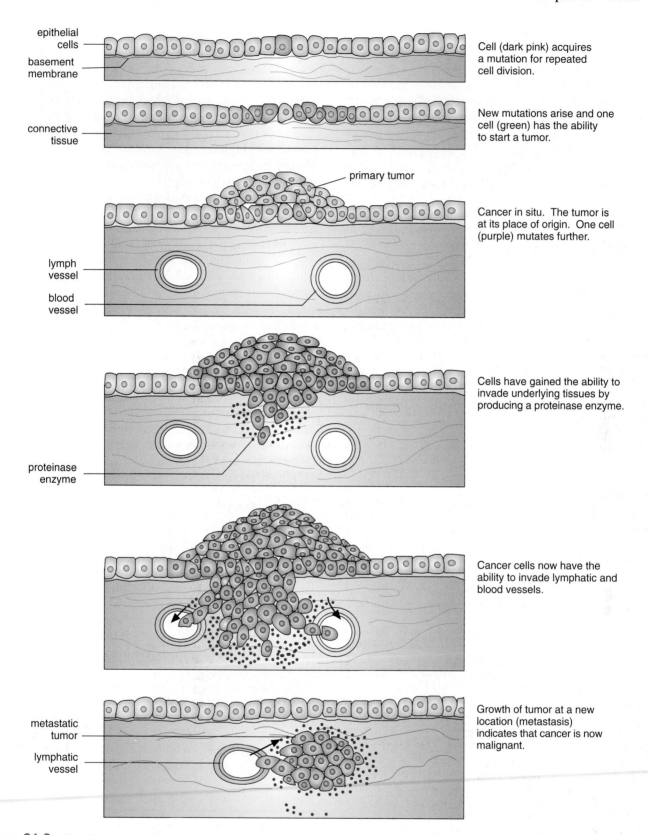

epithelial cells

basement membrane

Cell (dark pink) acquires a mutation for repeated cell division.

connective tissue

New mutations arise and one cell (green) has the ability to start a tumor.

primary tumor

Cancer in situ. The tumor is at its place of origin. One cell (purple) mutates further.

lymph vessel

blood vessel

Cells have gained the ability to invade underlying tissues by producing a proteinase enzyme.

proteinase enzyme

Cancer cells now have the ability to invade lymphatic and blood vessels.

metastatic tumor

lymphatic vessel

Growth of tumor at a new location (metastasis) indicates that cancer is now malignant.

Figure 21.2 Carcinogenesis.
The development of cancer requires a series of mutations leading first to a localized tumor and then metastatic tumors. With each successive step toward cancer, the most genetically altered and aggressive cell becomes the dominant type of the tumor.

21.2 Origin of Cancer

Mutations in at least four classes of genes are associated with the development of cancer. (1) The nucleus has a DNA repair system that corrects mutations occurring when DNA replicates. There are enzymes that read a newly forming DNA strand, making sure it has the same sequence of bases as the template strand. Any mistakes are then corrected by DNA repair enzymes. Mutations that arise in the genes encoding these various enzymes can contribute to carcinogenesis. (2) When a gene is being actively transcribed, the chromosome must decondense in a particular area. Mutations in genes that code for proteins involved in regulating the structure of chromatin help send cells down the road to cancer. (3) DNA segments called **telomeres** occur at the ends of chromosomes and protect them from damage. Telomeres usually get shorter every time chromosomes replicate prior to cell division. Once the cell has divided about 50 times, the shortened telomeres signal the cell to enter senescence and stop dividing. In cancer cells, an enzyme called telomerase is active and keeps telomeres at a constant so that they keep on dividing over and over again. (4) **Proto-oncogenes** code for proteins that stimulate the cell cycle, and **tumor-suppressor genes** code for proteins that inhibit the cell cycle. When cancer develops, normal regulation of the cell cycle no longer occurs due to mutations in these two types of genes.

Regulation of the Cell Cycle

The cell cycle is a series of stages that occur in this sequence: G_1 stage—organelles begin to double in number; S stage—replication of DNA occurs and duplication of chromosomes occurs; G_2 stage—synthesis of proteins occurs that prepares the cell for mitosis; M stage—mitosis occurs. The passage of a cell from G_1 to the S stage is tightly regulated, as is the passage of a cell from G_2 to mitosis.

Both oncogenes and tumor-suppressor genes are a part of a *regulatory pathway* that involves extracellular *growth factors*, plasma membrane *growth factor receptors*, various *signaling proteins* within the cytoplasm, and a number of *genes* within the nucleus. In the stimulatory pathway, a growth factor released by a neighboring cell is received by a plasma membrane receptor, and this sets in motion a whole series of enzymatic reactions that ends when proteins that can trigger cell division enter the nucleus (Fig. 21.3). In the inhibitory pathway a growth-inhibitory factor released by a neighboring cell is received by a plasma membrane receptor, and this sets in motion a whole series of enzymatic reactions that ends when proteins that inhibit cell division enter the nucleus. The balance between stimulatory signals and inhibitory signals determines whether proto-oncogenes are active or tumor-suppressor genes are active.

Oncogenes

Proto-oncogenes are so called because a mutation can cause them to become **oncogenes** (cancer-causing genes). An oncogene may code for a faulty receptor in the stimulatory pathway. A faulty receptor may be able to start the stimulatory process even when no growth factor is present! Or an oncogene may produce an abnormal protein product or else abnormally high levels of a normal product that stimulates the cell cycle to begin or go to completion. In either case, uncontrolled growth threatens.

Researchers have identified perhaps one hundred oncogenes that can cause increased growth and lead to tumors. The oncogenes most frequently involved in human cancers belong to the *ras* gene family. An alteration of only a single nucleotide pair is sufficient to convert a normally functioning *ras* proto-oncogene to an oncogene. The *ras*K oncogene is found in about 25% of lung cancers, 50% of colon cancers, and 90% of pancreatic cancers. The *ras*N oncogene is associated with **leukemias** (cancer of blood-forming cells) and lymphomas (cancers of lymphoid tissue), and both *ras* oncogenes are frequently found in thyroid cancers.

Tumor-Suppressor Genes

When a tumor-suppressor gene undergoes a mutation, inhibitory proteins fail to be active and the regulatory balance shifts in favor of cell cycle stimulation. Researchers have identified about a half-dozen tumor-suppressor genes. The *RB* tumor-suppressor gene was discovered when the inherited condition retinoblastoma was being studied. If a child receives only one normal *RB* gene and that gene mutates, eye tumors develop in the retina by the age of three. The *RB* gene has now been found to malfunction in cancers of the breast, prostate, and bladder, among others. Loss of the *RB* gene through chromosome deletion is particularly frequent in a type of lung cancer called small cell lung carcinoma. How the RB protein fits into the inhibitory pathway is known. When a particular growth-inhibitory factor attaches to a receptor, the RB protein is activated. An active RB protein turns off the expression of a proto-oncogene, whose product initiates cell division.

Another major tumor-suppressor gene is called *p53,* a gene that is more frequently mutated in human cancers than any other known gene. It has been found that the p53 protein acts as a transcription factor and as such is involved in turning on the expression of genes whose products are cell cycle inhibitors. *p53* can also stimulate apoptosis, programmed cell death.

Each cell contains regulatory pathways involving proto-oncogenes and tumor-suppressor genes that code for components active in the pathways.

Visual Focus

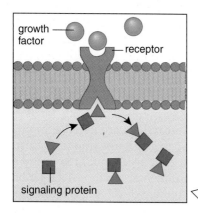

growth factor — receptor

signaling protein

Heredity

Organic chemicals

Radiation

Viruses

Oncogene

These agents can bring about the activation of oncogenes and the inactivation of tumor-suppressor genes.

The reception of a growth factor activates signaling proteins in a stimulatory pathway that extends to the nucleus.

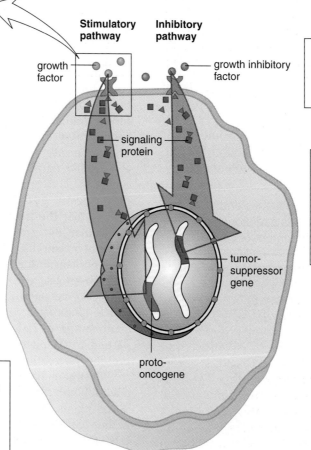

Stimulatory pathway   **Inhibitory pathway**

growth factor

signaling protein

growth inhibitory factor

The reception of a growth-inhibitory factor activates signaling proteins in an inhibitory pathway that extends to the nucleus.

2. Water is the only inorganic liquid that exists in nature and is the solvent in which most substances must be dissolved before cells cna use, or eliminate them.

tumor-suppressor gene

proto-oncogene

Proto-oncogenes code for a growth factor, a receptor, or a signaling protein in a regulatory pathway within the cell. If a proto-oncogene becomes an oncogene, the end result can be active cell division.

Tumor-suppressor genes code for a protein in a signaling regulatory pathway. If a tumor-suppressor gene is inactivated, the end result can be active cell division.

Figure 21.3   Origin of cancer.
Two types of regulatory pathways extend from the plasma membrane to the nucleus. In the stimulatory pathway, plasma membrane receptors receive growth-stimulatory factors. Then, proteins within the cytoplasm and proto-oncogenes within the nucleus stimulate the cell cycle. In the inhibitory pathway, plasma membrane receptors receive growth-inhibitory factors. Then, proteins within the cytoplasm and tumor-suppressor genes within the nucleus inhibit the cell cycle from occurring. Whether cell division occurs or not depends on the balance of stimulatory and inhibitory signals received. Hereditary and environmental factors cause mutations of proto-oncogenes and tumor-suppressor genes. These mutations can cause uncontrolled growth and a tumor.

Apoptosis

Most, if not all, cells contain apoptotic enzymes that can bring about their own self-destruction. These enzymes are known as ICE-like proteases because they structurally resemble interleukin-1 converting enzyme (ICE). Ordinarily, apoptotic enzymes are inactive and cause no harm, but when they are activated, they cleave the structural components of the cell, including its genetic material. Apoptosis is an important aspect of development. For example, when the webbing between the fingers is removed as the hand develops, apoptotic enzymes are involved. Apoptosis in the mature individual is a way to regulate cell population size and constituency. As you may recall, T lymphocytes are continually produced in red bone marrow and mature in the thymus. Any T lymphocytes that bear receptors capable of recognizing the body's own cells undergo apoptosis.

Recently, it has been suggested that the mutation of a proto-oncogene to an oncogene or the mutation of a tumor-suppressor gene ordinarily brings about apoptosis of the cell (Fig. 21.4). This is regarded as a safeguard to prevent the development of tumors in the body. If DNA is damaged in any way, the p53 protein inhibits the cell cycle, and this gives cellular enzymes the opportunity to repair the damage. But if DNA damage should persist, the *p53* gene goes on to bring about apoptosis of the cell. No wonder many types of tumors contain cells that lack an active *p53* gene. Some cancer cells have a *p53* gene but make a large amount of bcl-2, a protein which binds to and inactivates the p53 protein.

Apoptosis of cells that contain oncogenes and mutated tumor-suppressor genes is a reasonable way for the body to prevent the development of cancer.

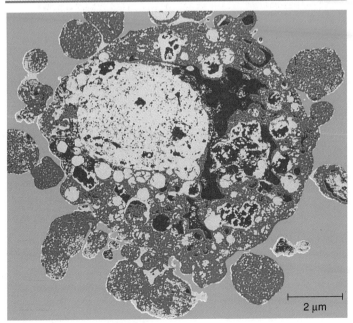

Figure 21.4   Apoptosis.
This cell is undergoing apoptosis, programmed cell death. Failure of a precancer cell to undergo apoptosis is another cause of cancer aside from mutations that lead to a tumor.

21.3 Causes and Prevention of Cancer

A **mutagen** is an agent that increases the chances of a mutation, while a **carcinogen** is an environmental agent that can contribute to the development of cancer. Carcinogens are often mutagenic.

Carcinogens

Among the best-known mutagenic carcinogens are (1) radiation, (2) organic chemicals (e.g., tobacco smoke, foods, pollutants), and (3) viruses. Researchers are now beginning to study the possibility that specific groups due to predisposing genetic traits, ethnicity, age, or sex might be more susceptible to particular environmental carcinogens than other groups. This knowledge will help develop guidelines to increase protection of those groups most at risk of particular environmental carcinogens.

Radiation

Some forms of radiation are carcinogenic. Ultraviolet radiation in sunlight and tanning lamps is most likely responsible for the dramatic increases seen in skin cancer the past several years. Today there are at least six cases of skin cancer for every one case of lung cancer. Nonmelanoma skin cancers are usually curable through surgery, but melanoma skin cancer tends to spread and is responsible for 1–2% of total cancer deaths in the United States. Many researchers now believe that the frequency of sunburns during childhood rather than the cumulative exposure to sunlight is the key factor in melanoma.

Another natural source of radiation is radon gas. In the very rare house with an extremely high level of exposure, the risk of developing lung cancer is thought to be equivalent to smoking a pack of cigarettes a day. Therefore, the combination of radon gas and smoking cigarettes can be particularly dangerous. Radon levels can be lowered by improving the ventilation of a building.

Most of us have heard about the damaging effects of the nuclear bomb explosions or accidental emissions from nuclear power plants. For example, more cancer deaths are expected in the vicinity of the Chernobyl Power Station (in the former U.S.S.R.), which suffered a terrible accident in 1986. Usually, however, diagnostic X rays account for most of our exposure to artificial sources of radiation. The benefits of these procedures can far outweigh the possible risk, but it is still wise to avoid any X-ray procedures that are not medically warranted.

Despite much publicity, scientists have not been able to show a clear relationship between cancer and radiation from electric power lines, household appliances, and cellular telephones.

Organic Chemicals

Tobacco smoke, foods, and pollutants all contain organic chemicals that can be carcinogenic.

Prevention of Cancer

There is clear evidence that the risk of certain types of cancer can be reduced by adopting protective behaviors and the right diet.

Protective Behaviors

These behaviors help prevent cancer:

Don't smoke Cigarette smoking accounts for about 30% of all cancer deaths. Smoking is responsible for 90% of lung cancer cases among men and 79% among women—about 87% altogether. Those who smoke two or more packs of cigarettes a day have lung cancer mortality rates 15 to 25 times greater than nonsmokers. Smokeless tobacco (chewing tobacco or snuff) increases the risk of cancers of the mouth, larynx, throat, and esophagus.

Don't sunbathe Almost all cases of basal-cell and squamous-cell skin cancers are considered to be sun related. Further, sun exposure is a major factor in the development of melanoma, and the incidence of this cancer increases for those living near the equator.

Avoid alcohol Cancers of the mouth, throat, esophagus, larynx, and liver occur more frequently among heavy drinkers, especially when accompanied by tobacco use (cigarettes or chewing tobacco).

Avoid radiation Excessive exposure to ionizing radiation can increase cancer risk. Even though most medical and dental X rays are adjusted to deliver the lowest dose possible, unnecessary X rays should be avoided. Excessive radon exposure in homes increases the risk of lung cancer, especially in cigarette smokers. It is best to test your home and take the proper remedial actions.

Be tested for cancer Do the shower check for breast cancer or testicular cancer. Have other exams done regularly by a physician.

Be aware of occupational hazards Exposure to several different industrial agents (nickel, chromate, asbestos, vinyl chloride, etc.) and/or radiation increases the risk of various cancers. Risk from asbestos is greatly increased when combined with cigarette smoking.

Be aware of hormone therapy Estrogen therapy to control menopausal symptoms increases the risk of endometrial cancer. However, including progesterone in estrogen replacement therapy helps to minimize this risk.

The Right Diet

Statistical studies have suggested that persons who follow certain dietary guidelines are less likely to have cancer. The following dietary guidelines greatly reduce your risk of developing cancer:

Avoid obesity The risk of cancer (especially colon, breast, and uterine cancers) is 55% greater among obese women and 33% greater among obese men, compared to people of normal weight.

Lower total fat intake A high-fat intake has been linked to development of colon, prostate, and possibly breast cancers.

Eat plenty of high-fiber foods These include whole-grain cereals, fruits, and vegetables. Studies have indicated that a high-fiber diet protects against colon cancer, a frequent cause of cancer deaths. It is worth noting that foods high in fiber also tend to be low in fat!

Increase consumption of foods that are rich in vitamins A and C Beta-carotene, a precursor of vitamin A, is found in dark green, leafy vegetables; carrots; and various fruits. Vitamin C is present in citrus fruits. These vitamins are called antioxidants because in cells they prevent the formation of free radicals (organic ions that have an unpaired electron) that can possibly damage DNA. Vitamin C also prevents the conversion of nitrates and nitrites into carcinogenic nitrosamines in the digestive tract.

Reduce consumption of salt-cured, smoked, or nitrite-cured foods Salt-cured or pickled foods may increase the risk of stomach and esophageal cancers. Smoked foods, like ham and sausage, contain chemical carcinogens similar to those in tobacco smoke. Nitrites are sometimes added to processed meats (e.g., hot dogs and cold cuts) and other foods to protect them from spoilage; as mentioned previously, nitrites are converted to nitrosamines in the digestive tract.

Include vegetables from the cabbage family in the diet The cabbage family includes cabbage, broccoli, brussels sprouts, kohlrabi, and cauliflower. These vegetables may reduce the risk of gastrointestinal and respiratory tract cancers.

Be moderate in the consumption of alcohol People who drink and smoke are at an unusually high risk for cancers of the mouth, larynx, and esophagus.

Tobacco Smoke   Tobacco smoke contains a number of organic chemicals that are known carcinogens, and it is estimated that smoking is also implicated in the development of cancers of the mouth, larynx, bladder, kidney, and pancreas. The greater the number of cigarettes smoked per day, the earlier the habit starts, and a high tar content all contribute to the possibility of cancer. When smoking is combined with drinking alcohol, the risk of these cancers increases even more (Fig. 21.5).

Passive smoking, or inhalation of someone else's tobacco smoke, is also dangerous and probably causes a few thousand deaths each year.

Foods   Statistically, studies suggest that a diet rich in saturated fats and low in fiber rivals tobacco in causing cancer, especially colon, rectal, and prostate cancer. Obesity seems to increase the risk of colon, kidney, and gallbladder cancers.

Animal testing of certain food additives, such as red dye #2 and the synthetic sweetener saccharin, has shown that these substances are cancer-causing in very high doses. In humans, salty foods and fats appear to make a significant contribution to cancer.

The good news is that eating vegetables and fruits is protective against the development of cancer. Antioxidants in these foods neutralize free radicals that are released during metabolism and can damage a cell's DNA.

Pollutants   Industrial chemicals, such as benzene and carbon tetrachloride, and industrial materials, such as vinyl chloride and asbestos fibers, are also associated with the development of cancer. Pesticides and herbicides are dangerous not only to pets and plants but also to our own health because they contain organic chemicals that can cause mutations. A panel of experts assembled by the National Academy of Sciences' Institute of Medicine has found conclusive evidence that exposure to dioxin, a contaminant of the herbicide Agent Orange used during the Vietnam War, can be linked to cancers of lymphoid tissues and those of muscles and certain connective tissues. Similarly, another study has found that these and other cancers were found more often in those exposed to dioxin accidently released after an explosion in a chemical plant over Seveso, Italy, on July 10, 1976, than in the rest of the population.

Viruses   At least three DNA viruses have been linked to human cancers: hepatitis B virus to liver cancer, human papillomavirus to cancer of the cervix, and Epstein-Barr virus to Burkitt's lymphoma (a cancer of the lymphoid tissues) and nasopharyngeal cancer.

In China, almost all persons have been infected with the hepatitis B virus, and this correlates with the high incidence of liver cancer in that country. For a long time, circumstances suggested that cervical cancer was a sexually transmitted disease, and now human papillomaviruses are routinely isolated from cervical cancers. Burkitt's lymphoma occurs frequently in Africa, where virtually all children are infected with the Epstein-Barr virus. In China, the Epstein-Barr virus is isolated in nearly all nasopharyngeal cancer specimens. It's believed

Figure 21.5   Tobacco smoke and alcohol.
Smoking cigarettes, especially when combined with drinking of alcohol, is associated with cancer of the lungs, mouth, larynx, kidney, bladder, pancreas, and many other cancers.

that environmental factors must be involved in determining the final effect of an Epstein-Barr viral infection because the virus is associated with different diseases in different countries (in the United States, this virus causes mononucleosis).

RNA-containing retroviruses, in particular, are known to cause cancers in animals. In humans, the retrovirus HTLV-1 (human T-cell lymphotropic virus, type 1) has been shown to cause adult T-cell leukemia. This disease occurs frequently in parts of Japan, the Caribbean, and Africa, particularly in those regions where people are known to be infected with the virus.

Heredity

Particular types of cancer seem to run in families. For instance, the risk of developing breast, lung, and colon cancers increases two- to threefold when first-degree relatives have had these cancers. Investigators have pinpointed the location of a gene called *BRCA1* (Breast Cancer gene #1), which was first identified in a large family whose female members are prone to breast cancer.

Certain childhood cancers seem to be due to the inheritance of a dominant gene. Retinoblastoma is an eye tumor that usually develops by age three; Wilm's tumor is characterized by numerous tumors in both kidneys. In adults, several family syndromes (e.g., Li-Fraumeni cancer family syndrome, Lynch cancer family syndrome, and Warthin cancer family syndrome) are known. Those who inherit a dominant allele develop tumors in various parts of the body.

Specific mutations cause cancer. Carcinogens (e.g., tobacco smoke and fats), heredity, and immunodeficiency all play a role in the development of cancer.

21.4 Diagnosis and Treatment

Due to improved means of detection and treatment, there were 5,760 fewer deaths due to cancer in 1997 in the United States than in 1996. This seems like a small number until you realize that the population grew by 2.5 million and the average age continued to increase. The death rate fell from one in four persons to one in five persons dying from cancer. Researchers are optimistic that the rate of decline will accelerate and that each passing year will have far better statistics than the year before.

Diagnosis of Cancer

Diagnosis of cancer before metastasis is difficult although treatment at this stage is usually more successful. The American Cancer Society, Inc., publicizes seven warning signals that spell out the word CAUTION and that everyone should be aware of:

C   hange in bowel or bladder habits;
A   sore that does not heal;
U   nusual bleeding or discharge;
T   hickening or lump in breast or elsewhere;
I   ndigestion or difficulty in swallowing;
O   bvious change in wart or mole;
N   agging cough or hoarseness.

Unfortunately, some of these symptoms are not obvious until cancer has progressed to one of its later stages.

Routine Screening Tests

The aim of medicine is to develop tests for cancer that are relatively easy to do, cost little, and are fairly accurate. So far, only the Pap smear for cervical cancer fulfills these three requirements. A physician merely takes a sample of cells from the cervix, which are examined microscopically for signs of abnormality. Regular Pap smears are credited with preventing over 90% of deaths from cervical cancer.

Breast cancer is not as easily detected, but three procedures are recommended. Every woman should do a monthly breast self-examination (see the Health reading on page 452). During an annual physical examination, recommended especially for women above age 40, a physician does this same procedure. While helpful, an examination may not detect lumps before metastasis has already taken place. The third recommended procedure, *mammography,* which is an X-ray study of the breast, is expected to do this (Fig. 21.6). However, mammograms do not show all cancers, and new tumors may develop during the interval between mammograms. The objective is that a mammogram will reveal a lump that is too small to be felt and at a time when the cancer is still highly curable.

Screening for colon cancer is also dependent upon three types of testing. A digital rectal examination performed by a physician is actually of limited value because only a portion of the rectum can be reached by finger. With flexible

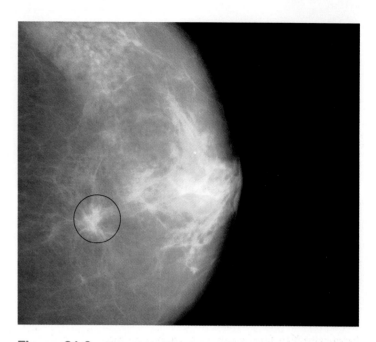

Figure 21.6   Mammogram.
X-ray image of the breast can find tumors too small to be palpable.

sigmoidoscopy, the second procedure, a much larger portion of the colon can be examined by using a thin, pliable, lighted tube. Finally, a stool blood test (fecal occult blood test) consists of examining a stool sample to detect any hidden blood. The sample is smeared on a slide, and a chemical is added that changes color in the presence of hemoglobin. This procedure is based on the supposition that a cancerous polyp bleeds, although some polyps do not bleed, and bleeding is not always due to a polyp. Therefore, the percentage of false negatives and false positives is high. All positive tests are followed up by a colonoscopy, an examination of the entire colon, or by X ray after a barium enema. The colonoscope can detect polyps that are destroyed by laser as soon as they are detected.

Other tests in routine use are blood tests to detect leukemia and urinalysis for the diagnosis of bladder cancer. Newer tests under consideration are tumor marker tests and tests for oncogenes.

Tumor Marker Tests

Blood tests for tumor antigens/antibodies are called *tumor marker tests.* They are possible because tumors release substances that provoke an antibody response in the body. For example, if an individual has already had colon cancer, it is possible to use the presence of an antigen called CEA (for carcinoembryonic antigen) to detect any relapses. When the CEA level rises, additional tumor growth has occurred.

There are also tumor marker tests that can be used as an adjunct procedure to detect cancer in the first place. They are not reliable enough to count on solely, but in conjunction with physical examination and ultrasound (see following), they are considered useful. There is a prostate-specific

Shower Check for Cancer

The American Cancer Society urges women to do a breast self-exam and men to do a testicle self-exam every month. Breast cancer and testicular cancer are far more curable if found early, and we must all take on the responsibility of checking for one or the other.

Breast Self-Exam for Women

1. Check your breasts for any lumps, knots, or changes about one week after your period.
2. Place your right hand behind your head. Press firmly with the pads of your fingers (Fig. 21A). Move your *left* hand over your *right* breast in a circle. Also check the armpit.
3. Now place your left hand behind your head and check your *left* breast with your *right* hand in the same manner as before. Also check the armpit.
4. Check your breasts while standing in front of a mirror right after you do your shower check. First, put your hands on your hips and then raise your arms above your head (Fig. 21B). Look for any changes in the way your breasts look; dimpling of the skin, changes in the nipple, or redness or swelling.

5. If you find any changes during your shower or mirror check, see your doctor right away.

You should know that the best check for breast cancer is a mammogram. When your doctor checks your breasts, ask about this. See Appendix E, which gives the warning signals for breast cancer.

Testicle Self-Exam for Men

1. Check your testicles once a month.
2. Roll each testicle between your thumb and finger as shown in Figure 21C. Feel for hard lumps or bumps.
3. If you notice a change or have aches or lumps, tell your doctor right away so he or she can recommend proper treatment.

Cancer of the testicles can be cured if you find it early. You should also know that prostate cancer is the most common cancer in men. Men over age 50 should have an annual health checkup that includes a prostate examination. See Appendix E, which gives the warning signs for prostate cancer.

Information provided by the American Cancer Society. Used by permission.

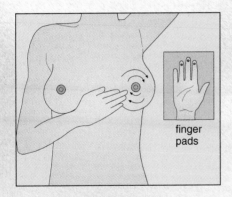

Shower check for breast cancer.

Figure 21A **Shower check for breast cancer.**

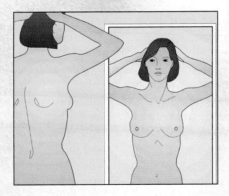

Mirror check for breast cancer.

Figure 21B **Mirror check for breast cancer.**

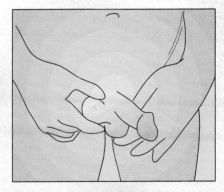

Shower check for testicular cancer.

Figure 21C **Shower check for testicular cancer.**

antigen (PSA) test for prostate cancer, a CA-125 test for ovarian cancer, and an alpha-fetoprotein (AFP) test for liver tumors, for example.

Tests for Cancer Genes
Identification of the genetic mutations that drive carcinogenesis is providing a variety of DNA tests that are expected to become routine. Tests are available for the presence of

mutations that may raise the risk of colon, breast, and thyroid cancers and melanoma.

Researchers believe that the presence of the *ras* gene can be used to indicate the possibility of colon and bladder cancer. They have found that when a person has a tumor that contains a *ras* oncogene, this same oncogene can be detected in stool and urine samples. Genetic testing for inherited breast cancer may also be possible. Researchers have been able to

develop a genetic test for *BRCA1* (Breast Cancer gene #1) mentioned earlier. Those who test positive for inheritance of the gene can choose either to have prophylactic surgery or to be frequently examined for signs of breast cancer. Mutations in the *RET* gene seem to signify thyroid cancer, and mutations in the *p16* gene appear to be linked to melanoma. Genetic markers can also be used to determine if there are cancer cells remaining after the tumor has been removed. In one study, 50% of the patients had tumor cells in apparently "clean" margins or in lymph nodes believed to be free of them.

In addition to specific mutations in oncogenes and tumor-suppressor genes, the testing of microsatellites can yield useful information. Microsatellites are small (micro) DNA regions that have di-, tri-, or tetranucleotide repeats. People differ according to the number of base repeats they have. Typically a sample of normal DNA (from lymphocytes) and clinical DNA (from a bodily fluid such as urine) are compared. A difference in the repeat pattern signifies a chromosomal deletion such as occurs when cancer is present. Clinical trials to detect bladder tumors have been so successful that a larger trial involving many institutions is now under way.

Another new genetic marker is the presence of telomerase, the enzyme that keeps telomeres a constant length in cancer cells. Telomerase is an enzyme whose structure contains a small amount of RNA. The RNA serves as a template for DNA bases in telomeres.

Confirming the Diagnosis

There are ways to confirm a diagnosis of cancer without major surgery. Needle biopsies allow removal of a few cells for examination, and sophisticated techniques such as laparoscopy permit a viewing of body parts. Computerized axial tomography (or CAT scan) uses computer analysis of scanning X-ray images to create cross-sectional pictures that portray a tumor's size and location. Magnetic resonance imaging (MRI) is another type of imaging technique that depends on computer analysis. MRI is particularly useful for analysis of tumors in tissues surrounded by bone, such as tumors of the brain or spinal cord. A radioactive scan obtained after a radioactive isotope is administered can reveal any abnormal isotope accumulation due to a tumor. During ultrasound, echoes of high-frequency sound waves directed at a part of the body are used to reveal the size, shape, and location of tissue masses. Ultrasound can confirm tumors of the stomach, prostate, pancreas, kidney, uterus, and ovary.

There are standard procedures to detect specific cancers; for example, the Pap smear for cervical cancer, mammograms for breast cancer, and the stool blood test for colon cancer. Tumor marker tests and genetic marker tests are new ways to test for cancer. Biopsy and imaging are used to confirm the diagnosis of cancer.

Treatment of Cancer

Certain standard methods of therapy have been used for quite some time. Other methods of therapy are in clinical trials and, if they prove to be successful, will become more generally available in the future.

Standard Methods of Therapy

Surgery, radiation, and chemotherapy are the standard methods of therapy (Fig. 21.7). Surgery alone is sufficient for cancer in situ. But because there is always the danger that some cancer cells were left behind, surgery is often preceded by and/or followed by radiation therapy.

Radiation   Radiation is mutagenic, and dividing cells such as cancer cells are more susceptible to its effects. The theory is that radiation will cause cancer cells to mutate and undergo apoptosis. Powerful X rays or gamma rays can be administered by using an externally applied beam or, in some instances, by implanting tiny radioactive sources into the patient's body. Cancer of the cervix and larynx, early stages of prostate cancer, and Hodgkin's disease are often treated with radiation therapy only.

Although X rays and gamma rays are the mainstays of radiation therapy, protons and neutrons also work well. Proton beams can be aimed at the tumor like a rifle bullet hitting the bull's-eye of a target.

Chemotherapy   Chemotherapy is a way to catch cancer cells that have spread throughout the body. Most chemotherapeutic drugs kill cells by damaging their DNA or interfering with DNA synthesis. The hope is that all cancer cells will be killed while leaving untouched enough normal cells to allow the body to keep functioning. Whenever possible, chemotherapy is specifically designed for the particular cancer. For example, in Allen's lymphoma/leukemia, it is known that a small portion of a chromosome 9 is missing, and, therefore, DNA metabolism differs in the cancerous cells compared to normal cells. Specific chemotherapy for this cancer provides the patient with a drug designed to exploit this metabolic difference and destroy the cancerous cells.

One drug, taxol, extracted from bark of the Pacific yew tree, was found to be particularly effective against advanced ovarian cancers as well as breast, head, and neck tumors. Taxol interferes with microtubules needed for cell division. Now chemists have synthesized a family of related drugs, called taxoids, which may be more powerful with less side effects than the original.

Certain types of cancer, such as leukemias, lymphomas, and testicular cancer, are now successfully treated by combination chemotherapy alone. There is an 80% survival rate of children with childhood leukemia. Hodgkin's disease, a lymphoma, once killed two out of three patients. Now, combination therapy of four different

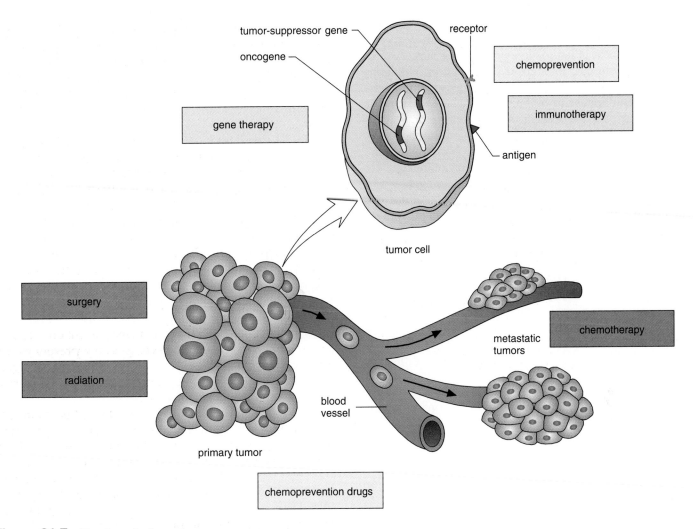

Figure 21.7   Treatment of cancer.
In standard use (dark blue), surgery and radiation are used to rid the body of a localized tumor. Chemotherapy is more effective and used when the cancer has metastasized. Future therapies (light blue) include immunotherapy, gene therapy and chemoprevention. Immunotherapy zeros in on plasma membrane receptors and antigens as a way to identify cancer cells. Gene therapy corrects the genotype of cancer cells, and chemoprevention prevents metastasis and angiogenesis. The ultimate feat would be to prevent the occurrence of cancer in the first place.

drugs can wipe out the disease in a matter of months in three out of four patients, even when the cancer is not diagnosed immediately. In other cancers—most notably breast and colon cancer—chemotherapy can reduce the chance of recurrence after surgery has removed all detectable traces of the disease.

Chemotherapy sometimes fails because cancer cells become resistant to one or several chemotherapeutic drugs. When cancer cells become resistant to combinations of drugs, it is called multidrug resistance. This occurs because all the drugs are capable of interacting with a plasma membrane carrier that pumps them out of the cell. Researchers are testing drugs known to poison the pump in an effort to restore efficacy of the drugs.

Some physicians have found that they get better results when chemotherapy is used first, followed by surgery or radiation. This is called induced chemotherapy. Another pos-

sibility is to use combinations of drugs with nonoverlapping patterns of toxicity because cancer cells can't become resistant to many different types at once.

Bone Marrow Transplants   The red bone marrow contains large populations of dividing cells; therefore, red bone marrow is particularly prone to destruction by chemotherapeutic drugs. In bone marrow autotransplantation, a patient's stem cells are harvested and stored before chemotherapy begins. Quite high doses of radiation or chemotherapeutic drugs are then given within a relatively short period of time. This prevents multidrug resistance from occurring, and the treatment is more likely to catch each and every cancer cell. Then, the stored stem cells, which are needed to produce blood cells, are returned to the patient by injection. They automatically make their way to bony cavities where they regenerate red bone marrow.

Future Methods of Therapy

Due to our molecular understanding of cancer cells, more therapies are being investigated than ever before. New treatments are likely to be a combination of these approaches in order to combat cancer from many directions.

Immunotherapy   Immunotherapy is the use of the body's immune system to promote the health of the body. Figure 21.7 shows that antigens are present on cancer cells, and this suggests that immunotherapy could possibly be successful in the treatment of cancer. In the case of cancer, vaccines prepare the patient's own immune system to react against cancer cells that bear specific antigens. On the other hand, the administration of antibodies is a form of passive immunity because the patient has not made the antibodies that bind to the cell.

Only now has immunotherapy begun to show some promise in the treatment of cancer. Monoclonal antibodies are antibodies of the same type because they are produced by the same plasma cell. Monoclonal antibodies can be designed to zero in on plasma membrane receptors of cancer cells, but alone they are not effective at killing cancer cells. To increase the killing power of monoclonal antibodies, they are linked to radioactive isotopes or chemotherapeutic drugs.

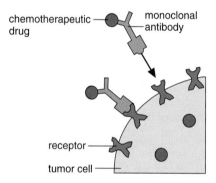

Herceptin is a monoclonal antibody that binds to a growth factor receptor found on the surface of about 30% of breast cancer cells. Because this monoclonal antibody can be combined with a chemotherapy agent, it delivers a double punch. In a recent trial, tumors shrank in 49% of the women receiving treatment. Bexxar is a monoclonal antibody that is attached to radioactive iodine. In lymphoma patients, Bexxar seeks out a target protein, called cd20, on the surface of affected lymphocytes.

Scientists have been able to isolate tumor antigens that can serve as vaccines. One vaccine called Melacine, which mobilizes the immune system against melanoma, is heading toward possible FDA approval because it has fewer serious side effects compared to chemotherapy.

Gene Therapy   With greater understanding of genes and carcinogenesis, it is not unreasonable to think that gene therapy can help cure human cancers.

Recently a retrovirus carrying a normal *p53* gene was injected directly into tumor cells of patients with lung cancer.

The tumors shrank in three patients and stopped growing in the other three. Researchers believe that *p53* expression is only needed for 24 hours to trigger programmed cell death. And the *p53* gene seems to trigger cell death only in cancer cells—elevating the *p53* level in a normal cell doesn't do any harm, possibly because apoptosis requires extensive DNA damage.

Some investigators prefer working with adenoviruses rather than retroviruses. Ordinarily when adenoviruses infect a cell they first produce a protein that inactivates *p53*. In a cleverly designed procedure, investigators genetically engineered an adenovirus that lacks the gene for this protein. Now, the adenovirus can infect and kill only cells that lack a *p53* gene. Which cells are those? Tumor cells, of course. Another plus to this procedure is that the injected adenovirus need infect only a small number of cancer cells because it will spread through the cancer, killing tumor cells as it goes. This genetically engineered virus is now in clinical trials.

Chemoprevention   The aim of chemoprevention is to prevent the spread of cancer or even possibly prevent cancer from occurring in the first place. If metastasis and angiogenesis can be controlled, then cancer might become a disease that you live with rather than die from.

Drugs that *prevent metastasis* are a distinct possibility. In the early stages of colon cancer a mutation in the *Apc* tumor-suppressor gene is accompanied by a marked increase in an enzyme known as COX-2. An inhibitor of this enzyme known as MF-tricyclic has been shown to be effective in preventing metastasis in laboratory animals and may soon undergo clinical trials to see if it is effective against the development of colon cancer.

Investigators have found that cells having a high level of a protein called nm23 (nonmetastatic 23) do not metastasize. Someday this protein might be used as a drug. In the meantime, human trials of the drug CAI (carboxyamide aminoimidazoles) are under way because this drug prevents metastasis in some, as yet, unknown way.

When cancer cells derived from epithelial cells metastasize, they produce proteinase enzymes that allow them to make their way through the basement membrane and the wall of capillaries. Surprisingly, cancer cells themselves produce an inhibitor or proteinase enzymes. Proteinase action occurs only if the number of enzyme molecules is greater than the number of TIMP (tissue inhibitor metalloproteinase) molecules. This means that TIMPs or drugs that act like them may offer an approach to prevent metastasis (Fig. 21.8).

Reversing angiogenesis is a proposed therapy that is now well under way. Angiogenesis occurs when tumor cells cause nearby blood vessels to branch and create a capillary network that brings them nutrients and growth factors they need to keep on expanding. Blood appearing between menstrual periods or in the urine, stool, or sputum is a danger signal that angiogenesis may have occurred in the uterus, bladder, colon, or lung, respectively. Antiangiogenic drugs confine and reduce tumors by breaking up the network of

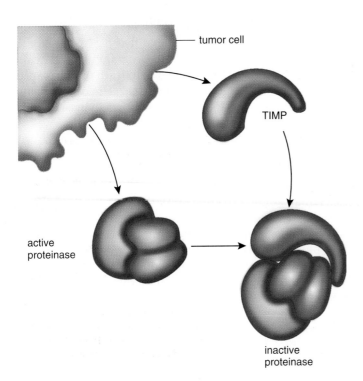

Figure 21.8 Chemoprevention.
Cancer cells produce proteinase enzymes that allow them to metastasize and TIMP, a proteinase inhibitor. Perhaps it would be possible to isolate, produce, and administer the TIMP to cancer patients. In this way metastasis would be prevented.

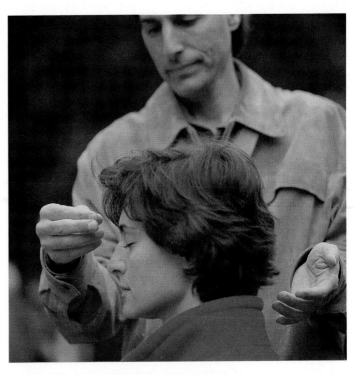

Figure 21.9 Complementary therapy.
Some cancer patients turn to complementary therapy for assistance. One form of complementary therapy is the transfer of energy from one person to another in order to restore the recipient to good health.

new capillaries in the vicinity of a tumor. A number of antiangiogenic compounds are currently being tested in clinical trials. Two highly effective drugs, called angiostatin and endostatin, have been shown to inhibit angiogenesis in laboratory animals and are expected to do the same in humans.

Chemoprevention can also include *promoting differentiation.* In promyelocytic leukemia, cells have too many retinoic acid (a cousin of vitamin A) receptors. Strangely enough, the treatment is the administration of retinoic acid because this leads to differentiation and the concomitant cessation of cell growth.

The ultimate in chemoprevention is to *prevent carcinogenesis.* The hormone estrogen sometimes promotes breast cancer. Tamoxifen is a drug that binds to estrogen receptors and thereby prevents estrogen from binding. For 25 years, tamoxifen has been used to limit breast cancer recurrence in women already treated for the disease. In a recent trial to see if tamoxifen could prevent breast cancer from ever occurring, twice as many women receiving a placebo developed breast cancer compared to the group that received tamoxifen. Unfortunately, however, tamoxifen seemed to increase risk of uterine cancer. Still researchers and others are cheered by the fact that breast cancer prevention appears to be a possibility. Another drug called raloxifene will now be tested to see if it will prevent breast cancer without increasing the risk of uterine cancer.

New screening methods have led to an increased detection of early prostate cancer. Similar to the effect that estrogen has on breast cancer, the hormone testosterone seems to increase the possibility of prostate cancer. A large clinical trial that involves the use of a testosterone inhibitor is now under way and the results are expected in about five years.

Complementary Therapies Many patients are interested in therapies outside the mainstream of conventional medicine. The list of possible alternative therapies includes tai chi, homeopathy, biofeedback, acupuncture, and exotic foods. Although there have been no studies to show which of these might be beneficial, many patients still want to avail themselves of therapies that might better be called complementary therapies. Due to the insistence of patients, institutions are beginning to make these therapies available (Fig. 21.9). The goal is to develop a way of working with cancer patients that integrates both conventional and alternative therapies.

Surgery followed by radiation and/or chemotherapy is the standard method of treating cancer. Many new therapies are undergoing clinical trials. The hope is that combination therapy can allow people to live with cancer rather than die from cancer.

Bioethical Issue

In the 1950s and 1960s, the U.S. government performed atomic bomb tests in Nevada. Radioactive iodine released into the air increased the risk of thyroid cancer for all U.S. citizens, and especially those who lived in Nevada, western states north of the test site, and states east of the site. People who were children at the time have an increased risk due to their small size at the time, and the fact that they probably drank milk from cows that fed on contaminated grass. It can take several decades for thyroid cancer to develop after exposure to radiation.

Physicians for Social Responsibility (PSR) are concerned because it took the government more than thirty years to decide that on the average people received 2 rads (radiation absorbed doses) of radiation, and some children may have received levels above 120 rads. The Agency for Toxic Substances and Disease Registry sets a dose of 10 rads as the threshold for entry into a monitoring program.

Still, the National Cancer Institute, who released its report in two stages during 1997 and 1998 only after media pressure, does not believe that a monitoring program is necessary. Their argument is that there is no evidence that early detection of thyroid cancer will reduce mortality. Even so, the Physicians for Social Responsibility believe that monitoring should be offered, and further, a major research effort should begin on how to identify and treat people with thyroid cancer. A spokesperson for the group said, "Nuclear testing has exposed millions of American people to dangerous radioactivity without their knowledge or consent, and then the information about the exposure was withheld from them by federal officials. The Department of Health and Human Services should provide clear direction to citizens, physicians, and public health officials about [the] next steps [to be taken]."

Questions

1. Suppose the federal government didn't realize at the time that citizens would be exposed to radioactivity from nuclear bomb testing. Should it still be held responsible for cases of thyroid cancer that develop in the states mentioned? Why or why not?

2. Do you think the federal government should institute a monitoring program even though the bomb tests only *increased* the risk of thyroid cancer? Why or why not?

3. If you can prove you were a child living in one of the states most exposed to radioactivity, should the government pay your medical expenses if you develop thyroid cancer? Why or why not?

Summarizing the Concepts

21.1 Cancer Cells

Cancer cells have characteristics that distinguish them from normal cells. They are nondifferentiated and enter the cell cycle repeatedly. They have abnormal nuclei with many chromosomal irregularities. They form tumors because they do not exhibit contact inhibition. They induce angiogenesis and cause nearby blood vessels to form a capillary network that services the tumor. They metastasize and form tumors in other parts of the body.

21.2 Origin of Cancer

Each cell contains two types of regulatory pathways. The cell surface receptors for growth factors, signaling proteins, and proto-oncogenes in the stimulatory network promote the cell cycle and cause the cell to divide. The cell surface receptors for growth-inhibitory factors, the signaling proteins, and tumor-suppressor genes in the inhibitory pathway repress the cell cycle and cause the cell to stop dividing. Whether the cell divides or not depends on which pathway is most active at the time.

When proto-oncogenes mutate, becoming oncogenes, and tumor-suppressor genes mutate, the regulatory pathways no longer function as they should, and uncontrolled growth results. Usually, cells that bear damaged DNA undergo apoptosis. Apoptosis fails to take place in cancer cells.

21.3 Causes and Prevention of Cancer

Certain environmental factors are carcinogens. They bring about mutations that lead to cancer. Ultraviolet radiation is a well-known carcinogen for melanoma. Carcinogenic organic chemicals include those in tobacco smoke, foods, and pollutants. Tobacco smoke and diet are believed to account for 60% of all cancer deaths. Industrial chemicals including pesticides and herbicides are carcinogenic. Certain viruses such as hepatitis B, human papillomaviruses, and Epstein-Barr virus cause certain specific cancers. Cancers that run in families are most likely due to the inheritance of mutated genes that contribute to carcinogenesis.

21.4 Diagnosis and Treatment

There are procedures to detect specific cancers; for example, the Pap smear for cervical cancer, mammograms for breast cancer, and the stool blood test for colon cancer, to name a few. But new ways, such as tumor marker tests and oncogene tests, are being developed. Biopsy and imaging are used to confirm the diagnosis of cancer.

Surgery followed by radiation and/or chemotherapy is the standard method of treating cancer. Chemotherapy involving bone marrow transplants has now become fairly routine. Future methods of therapy include immunotherapy, gene therapy, chemoprevention, and complementary therapies. Immunotherapy involves the use of monoclonal antibodies that can search out and find cancer cells. Gene therapy relies on viruses which can carry normal genes into cells. Chemoprevention is an up and coming field in which drugs are being developed to prevent metastasis, reverse angiogenesis, promote differentiation, and prevent carcinogenesis.

Studying the Concepts

1. Why is cancer called a genetic disease? 444
2. List and discuss five characteristics of cancer cells that distinguish them from normal cells. 444
3. What are oncogenes and tumor-suppressor genes? What role do they play in the regulatory pathways that control cell division and involve cell surface receptors and signaling proteins? 446
4. What is apoptosis, and what role does apoptosis play in carcinogenesis? 448
5. Name three types of carcinogens, and give examples of each type. What role does heredity play in the development of cancer? 448–50
6. What are the standard ways to detect cervical cancer, breast cancer, and colon cancer? 451–53

7. Describe and give examples of tumor marker tests and oncogene tests. 451–53
8. What are the standard methods of treatment for cancer? Explain why a bone marrow transplant is sometimes used in conjunction with chemotherapy. 453–55
9. What two immunotherapy approaches have met with some success recently? 455
10. Describe two investigative experiments utilizing gene therapy. Why would you expect them to be successful? 455
11. Chemoprevention involves what four types of drugs? Why were investigators cheered by the results of a clinical trial in which women took tamoxifen? 455–56

Testing Your Knowledge of the Concepts

In questions 1–4, match the treatment to the description.
a. immunotherapy
b. gene therapy
c. chemoprevention
d. radiation

_____ 1. Inject a virus carrying a *p53* gene directly into a tumor.

_____ 2. Administer angiostatin to prevent angiogenesis.

_____ 3. Prepare monoclonal antibodies that will search and find a patient's cancer cells.

_____ 4. Use proton beams that can be aimed directly at a tumor.

In questions 5–7, indicate whether the statement is true (T) or false (F).

_____ 5. Without angiogenesis, cancer cells are not able to survive.

_____ 6. The tumor-suppressor gene *p53* ordinarily induces apoptosis, an action that prevents carcinogenesis.

_____ 7. Since standard chemotherapy works so well, there is no need to develop new methods of therapy for cancer.

In questions 8–17, fill in the blanks.

8. The _____ virus is a carcinogen for cancer of the cervix.
9. Autotransplants of bone marrow stem cells permit a much higher dosage of _____ than otherwise.
10. To prevent cancer, you should avoid _____, which are environmental agents associated with the development of cancer.
11. _____ contains many organic chemical carcinogens and is associated with one-third of all cancers.
12. Cancer cells _____; they travel to distant body parts and start new tumors.
13. The mutation of _____ and _____ genes leads to uncontrolled growth and cancer.

14. The *RB* gene is a _____; it normally keeps *c-myc* from being an oncogene.
15. Cancer cells are sensitive to radiation therapy and chemotherapy because they are constantly _____.
16. Autotransplants of bone marrow permit a much higher dosage of _____ than otherwise.
17. _____ includes the use of monoclonal antibodies to carry a chemotherapeutic drug.
18. (a–g) Identify these portions of a cancer cell and associated structures. h. Which portion(s) would be used to make a vaccine against cancer? i. Which portion(s) would be the site of action for a chemoprevention drug? j. Which portion(s) could be involved in a regulatory pathway? k. Which portion(s) would be a likely target for radiation therapy? l. Which portion(s) would be a likely target for chemotherapy?

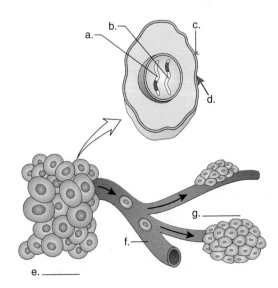

Applying Your Knowledge to the Concepts

These questions pertain to cancer.

1. What is the reason that a person undergoing chemotherapy often loses her/his hair?

2. One is advised to eat plenty of high-fiber foods as a prevention against colon cancer. What is the reason that high-fiber foods have this effect?

3. Estrogen therapy to control menopausal symptoms increases the risk of endometrial cancer. What is the explanation for this?

4. Ideally, a cancer drug or therapy should have what specific characteristics? Why don't the standard methods of cancer therapy meet these requirements?

Understanding the Terms

angiogenesis 444
apoptosis 444
cancer 444
carcinogen 448
carcinogenesis 444
chemotherapy 453
immunotherapy 455
leukemia 446

metastasis 444
mutagen 448
oncogene 446
proto-oncogene 446
telomere 446
tumor 444
tumor-suppressor gene 446

Match the terms to these definitions.

a. _____ Use of any immune system component such as antibodies, cytotoxic T cells, or lymphokines to promote the health of the body, such as curing cancer.

b. _____ Environmental agent that contributes to the development of cancer.

c. _____ Normal gene involved in cell growth and differentiation that becomes an oncogene through mutation.

d. _____ Formation of new blood vessels such as a capillary network.

e. _____ Spread of cancer from the place of origin throughout the body caused by the ability of cancer cells to migrate and invade tissues.

Applying Technology to the Concepts

Your study of cancer is supported by these available technologies:

Essential Study Partner CD-ROM
Visit the Mader web site for related ESP activities.

Exploring the Internet
The Mader Home Page provides resources and tools as you study this chapter.

http://www.mhhe.com/biosci/genbio/mader

Dynamic Human 2.0 CD-ROM
Lymphatic System → Clinical Concepts → Cancer
Reproductive System → Clinical Concepts → Breast Cancer

HealthQuest CD-ROM
6 Cancer
7 Tobacco → Gallery → Tumor Initiation

Further Readings for Part 6

Berns, M. W. April 1998. Laser scissors and tweezers. *Scientific American* 278(62):4. New laser techniques allow manipulation of chromosomes and other structures inside cells.

Blaser, M. J. February 1996. The bacteria behind ulcers. *Scientific American* 274(2):104. Acid-loving pathogens are linked to stomach ulcers and stomach cancer.

Borek, C. November/December 1997. Antioxidants and cancer. *Science & Medicine* 4(6):52. The importance of supplemental antioxidant vitamins depends on factors such as diet and lifestyle.

Duke, R. C., et al. December 1996. Cell suicide in health and disease. *Scientific American* 275(6):80. Failures in the processes of cellular self-destruction may give rise to cancer, AIDS, Alzheimer disease, and some genetic diseases.

Galili, U. September/October 1998. Anti-Gal antibody prevents xenotransplantation. *Science & Medicine* 5(5):28. Prevention of interaction of the anti-Gal antibody with pig cells is necessary to the progress of xenotransplantation.

Garnick, M. B., and Fair, W. R. December 1998. Combating prostate cancer. *Scientific American* 279(6):74. Article details the recent developments in diagnosis and treatment of prostate cancer.

Glausiusz, J. May 1998. The great gene escape. *Discover* 19(5):90. Genes from genetically engineered plants can escape from crops into the wild, causing resistance in wild plants.

Goldberg, J. April 1998. A head full of hope. *Discover* 19(4):70. Article discusses a new gene therapy for killing brain cancer cells.

Greider, C. W., and Blackburn, E. H. February 1996. Telomeres, telomerase, and cancer. *Scientific American* 274(2):92. The enzyme telomerase rebuilds the chromosomes of tumor cells; this enzyme is being researched as a target for anticancer treatments.

Haseltine, W. A. March 1997. Discovering genes for new medicines. *Scientific American* 276(3):92. New medical products being developed are a result of recent genetic analyses of the human genome.

Johnson, G. B. 1996. *How scientists think.* Dubuque, Iowa: Wm. C. Brown Publishers. Presents the rationale behind 21 important experiments in genetics and molecular biology that became the foundation for today's research.

Kher, U. January 1998. A man-made chromosome. *Discover* 18(1):40. Researchers announce a promising new gene carrier, a human artificial chromosome, for use in gene therapy.

Lasic, D. D. May/June 1996. Liposomes. *Science & Medicine* 3(3):34. Liposomes can be used to deliver drugs or genes for gene therapy.

Leffell, D. J., and Brash, D. E. July 1996. Sunlight and skin cancer. *Scientific American* 275(1):52. Discusses the sequence of changes that may occur in skin cells after exposure to UV rays.

Miller, R. V. January 1998. Bacterial gene swapping in nature. *Scientific American* 278(1):66. The study of the process of DNA exchange between bacteria can help limit the risks of releasing genetically engineered microbes into the environment.

Nicolaou, K. C., et al. June 1996. Taxoids: New weapons against cancer. *Scientific American* 274(6):94. Chemists are synthesizing a family of drugs related to taxol for the treatment of cancer.

Nielson, P. E. September/October 1998. Peptide nucleic acids. *Science & Medicine* 5(5):48. Peptide nucleic acids are synthetic molecules that mimic DNA and can substitute for DNA in gene therapy applications.

Nurse, P., et al. October 1998. Understanding the cell cycle. *Nature Medicine* 4(1):1103. The medical relevance of cell-cycle research is discussed.

O'Brochta, D. A., and Atkinson, P. W. December 1998. Building a better bug. *Scientific American* 279(6):90. Transgenic insect technology could decrease pesticide use, and prevent certain infectious diseases. Article discusses the production of a transgenic insect.

Paolella, P. 1998. *Introduction to molecular biology.* Dubuque, Iowa: WCB/McGraw-Hill. For the undergraduate science major, this is an introductory text which explores the processes and mechanisms of gene function and control.

Pennisi, E. 13 November 1998. Training viruses to attack cancers. *Science* 282(5392):1244. Certain viruses can replicate in and kill cancer cells, but leave normal tissue intact.

Plomerin, R., and DeFries, J. C. May 1998. The genetics of cognitive abilities and disabilities. *Scientific American* 278(5):62. Genes involved in cognitive abilities and disabilities, including dyslexia, are being sought.

Plunkett, M. J., and Ellman, J. A. April 1997. Combinatorial chemistry and new drugs. *Scientific American* 276(4):68. Combinatorial chemistry is a process that allows chemists to produce millions of molecules quickly.

Pool, R. May 1998. Saviors. *Discover* 19(5):52. Genetic engineering may make animal organs compatible for human transplants.

Ronald, P. C. November 1997. Making rice disease-resistant. *Scientific American* 277(5):100. Genetic engineering is being used to protect rice from disease.

Russell, P. J. 1996. *Genetics.* 4th ed. New York: HarperCollins College Publishers. This easy-to-read text emphasizes an inquiry-based approach to genetics; explores many research experiments that led to important genetic advances.

Sapolsky, R. October 1997. A gene for nothing. *Discover* 18(10):40. Article discusses why genes don't determine behavior.

Scientific American editors. June 1997. Special report: Making gene therapy work. 276(6):95. Researchers discuss obstacles to gene therapy.

Scientific American Special Issue. September 1996. What you need to know about cancer. 275(3). The entire issue is devoted to the causes, prevention, and early detection of cancer, and cancer therapies—conventional and future.

Shcherbak, Y. M. April 1996. Ten years of the Chernobyl Era. *Scientific American* 274(4):44. Article discusses the medical aftermath of the accident.

Stix, G. October 1997. Growing a new field. *Scientific American* 277(4):15. Tissue engineers try to grow organs in the laboratory.

Tamarin, R. 1996. Principles of genetics. 5th ed. Dubuque, Iowa: Wm. C. Brown Publishers. This text emphasizes the major areas of genetics and genetic research.

Van Noorden, C. J. F., et al. March/April 1998. Metastasis. *American Scientist* 86(2):130. The mechanisms by which cancer cells metastasize are discussed.

Velander, W. H., et al. January 1997. Transgenic livestock as drug factories. *Scientific American* 276(1):70. Pigs, cows, sheep, and other farm animals can be bred to produce large amounts of medicinal proteins in their milk.

Weaver, R. F., and Hedrick, P. W. 1997. *Basic genetics.* 3d ed. Dubuque, Iowa: Wm. C. Brown Publishers. Basic concepts are presented in a concise, easy-to-understand manner. There is an emphasis on problem solving in both the text and chapter end matter.

Wikonkal, N. M., and Brash, D. E. September/October 1998. Squamous cell carcinoma. *Science & Medicine* 5(5):18. Mutations of tumor-suppressor gene *p53* are commonly found in squamous cell carcinomas.

Wills, C. January 1998. A sheep in sheep's clothing? *Discover* 18(1):22. Some pros and cons of cloning are discussed.

Wilmut, I. December 1998. Cloning for medicine. *Scientific American* 279(6):58. Cloning holds many benefits for the advancement of medical science and animal husbandry.

Wolffe, A. P. November/December 1995. Genetic effects of DNA packaging. *Scientific American Science & Medicine* 2(6):68. The regulation of DNA coiling in the chromosomes adds to the properties of the genes involved in several genetic diseases.

Zimmer, C. January 1998. Hidden unity. *Discover* 18(1):46. Studies suggest the same basic gene may be involved in the development of certain features in both vertebrates and invertebrates.

PART

VII

Human Evolution and Ecology

Evidence for the theory of evolution is drawn from many areas of biology. Charles Darwin was the first to present extensive evidence, and he suggested that those organisms best suited to a particular environment are the ones that survive and reproduce most successfully. Evolution causes life to have a history, and it is possible to trace the ancestry of humans even from the first cell or cells.

The world's diverse forms of life live within ecosystems, where energy flows and chemicals cycle. Humans have greatly modified the ecosystems, and worldwide, the human population keeps increasing in size so that an ever-greater amount of energy and raw materials are needed each year. Since 1850, the human population has expanded so rapidly that some doubt there will be sufficient energy and food to permit the same degree of growth in the future. The human-impacted ecosystem depends on natural ecosystems not only because they absorb pollutants but also because natural ecosystems are inherently stable. Every possible step should be taken to protect natural ecosystems to help ensure the continuance of the human species.

Chapter 22

Evolution

Chapter Concepts

22.1 **Evidence for Evolution**
- The fossil record, comparative biochemistry and anatomy, and biogeography all provide evidence for evolution. 462

22.2 **The Evolutionary Process**
- Charles Darwin formulated a mechanism for an evolutionary process that results in adaptation to the environment. 465

22.3 **Organic Evolution**
- Organic evolution accounts for the diversity of life from the first cell(s) to all other living things, including humans. 466

22.4 **Modern Humans Evolve**
- Humans are primates, and many of their physical traits are the result of their ancestors' adaptations to living in trees. 469
- It is possible to trace the evolution of humans using a series of fossils found in Africa, Europe, and Asia. 469
- All human races are classified as *Homo sapiens sapiens*. 472

Sam sat down to write a paper on the evolution and diversity of life. "This is an overwhelming task" he thought. "Why did I choose this topic anyway?" Then he remembered why. His instructor had said that perhaps if we understood the past, it would help us know how best to manage the environment so that life can continue to exist in the future.

In his paper, Sam wrote that bacteria alone had existed on earth from the evolution of life to 1.5 billion years ago. Unicellular organisms remained in the oceans and increased in complexity until multicellular organisms evolved about 650 million years ago. Plants invaded the land environment about 400 million years ago, and they were followed by fungi, invertebrates, and finally vertebrates. The Age of Reptiles lasted from 300 million to 65 million years ago, when the dinosaurs died out. Mammals had their origins some 150 million years ago, but it was not until the dinosaurs had vanished that mammals became abundant. Direct ancestors to humans (hominids) did not appear until about 3 million years ago, which is very recent, compared to how long life has been evolving. If the history of the earth is measured using a 24-hour timescale that starts at midnight, humans do not appear until one half minute before the next midnight (Fig. 22.1).

22.1 Evidence for Evolution

The fossil record, comparative anatomy and biochemistry, and biogeography all support the theory of evolution.

Fossils and the History of Life

Fossils are the remains or evidence of some organism that lived long ago. Nearly all fossils are of organisms that are now extinct (there are no living forms), and some fossils are older than others. Today, it is possible to date fossils by using isotopes. The oldest fossils of bacteria suggest that life began 3.5 billion years ago. Thereafter, fossils get more and more complex. Our knowledge of the history of life is based primarily on the fossil record.

In some instances, it is even possible to trace a line of descent over vast amounts of time. Researchers have now traced the modern-day horse to an animal that was about the size of a dog with four toes on each front foot and three toes on each hind foot. When grasslands replaced the forest home of this animal, it was millions of years before a larger

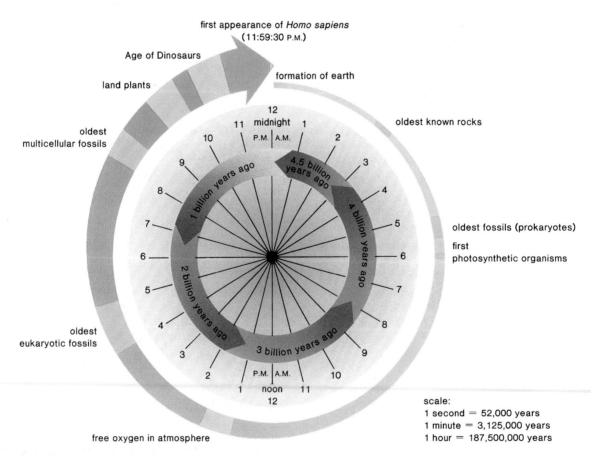

Figure 22.1   History of earth.
History of earth measured on a 24-hour timescale (yellow) compared to actual years (red). A very large portion of life's history is devoted to the evolution of unicellular organisms. (Prokaryotes do not have a nucleus; eukaryotes do have a true nucleus.) The first multicellular organisms do not appear until just after 8 P.M., and humans are not on the scene until less than a minute before midnight.

size provided the strength needed for combat; a larger skull made room for a larger brain; elongated legs ending in hooves provided greater speed to escape enemies; and durable, grinding teeth enabled the animal to feed efficiently on grasses.

The fossil record broadly traces the history of life and, more specifically, allows us to study the history of particular groups.

Comparative Anatomical Evidence

Diverse organisms sometimes share anatomical similarities. Vertebrate forelimbs are used for flight (birds and bats), orientation during swimming (whales and seals), running (horses), climbing (arboreal lizards), or swinging from tree branches (monkeys). Yet all vertebrate forelimbs contain the same sets of bones organized in similar ways, despite their dissimilar functions (Fig. 22.2). The most plausible explanation for this unity is that the basic forelimb plan originated with a common ancestor, and then the plan was modified in the succeeding groups as each continued along its own evolutionary pathway. Structures that are similar because they were inherited from a common ancestor are called **homologous structures.**

The unity of plan shared by vertebrates extends to their embryological development (Fig. 22.3). At some time during development, all vertebrates have a supporting dorsal rod, called a notochord, and exhibit paired pharyngeal pouches. In fishes and amphibian larvae, these pouches develop into functioning gills. In humans, the pouches are modified into such structures as the middle ear, thymus, or parathyroid glands. Why do pharyngeal pouches appear and then later undergo modification? The most likely explanation is that fishes are ancestral to other vertebrate groups.

Vestigial structures are anatomical features that are fully developed in one group of organisms but are reduced and may or may not have a function in similar groups. Some bird species (e.g., ostrich), for example, have wings that are greatly reduced, and they do not fly. Similarly, snakes do not use limbs for locomotion and yet some have remnants of a pelvic girdle and legs. Humans have a tailbone but no tail. Vestigial structures are thought to occur because organisms inherit their anatomy from their ancestors; they are traces of an organism's evolutionary history.

Organisms share a unity of plan when they are closely related because of common descent. This is substantiated by comparative anatomy and embryological development.

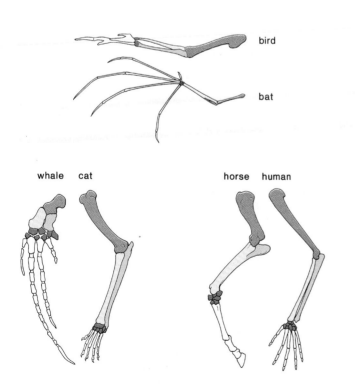

Figure 22.2 Vertebrate forelimbs.
The same bones are present (they are color-coded) in all vertebrates (animals with a backbone) but designed for different functions. The unity of plan is evidence of a common ancestor.

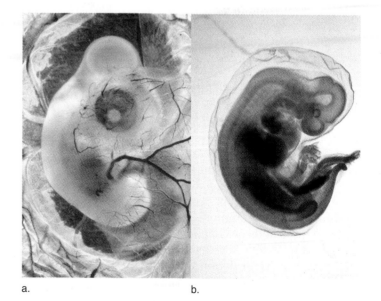

Figure 22.3 Early developmental stage of chick and pig.
a. Chick embryo. **b.** Pig embryo. At these comparable early developmental stages, the two have many features in common, although eventually they are completely different animals. This is evidence they evolved from a common ancestor.

Comparative Biochemical Evidence

Almost all living organisms use the same basic biochemical molecules, including DNA, ATP, and many identical or nearly identical enzymes. Further, almost all organisms utilize the same nuclear DNA triplet code and the same 20 amino acids in their proteins. Organisms even share the same introns and hypervariable regions. There is obviously no functional reason why these elements need be so similar. But their similarity can be explained by descent from a common ancestor.

Further, the degree of similarity in amino acid sequences of proteins and the degree of similarity in DNA base sequences is consistent with data regarding the anatomical similarities and, therefore, the relatedness of organisms.

> All organisms have certain biochemical molecules in common. The degree of similarity between DNA base sequences and amino acid sequences is thought to indicate the degree of relatedness between organisms.

Biogeographical Evidence

Biogeography is the study of the distribution of plants and animals throughout the world. Such distributions are consistent with the hypothesis that related forms evolved in one locale and then spread out into other accessible regions. Why are there no rabbits in South America even though the environment is quite suitable to them? Most likely because rabbits originated someplace else, and they had no means to reach South America.

The islands of the world have many unique animals and plants found no place else, even when the soil and climate are the same as other places. Why are there so many species of finches on the Galápagos Islands located off the coast of Ecuador? Since they are not on the mainland, the reasonable explanation is that all these types of finches are descended from a common ancestor that came to the islands by chance.

Physical factors, such as the location of continents, often determine where a population can spread. Both cacti and euphorbia are plants adapted similarly to a hot, dry environment—they both are succulent, spiny, flowering plants. Why do cacti grow in North American deserts

and euphorbia grow in African deserts when each would do well on the other continent? It seems obvious that they just happened to evolve on their respective continents.

At one time in the history of the earth, South America, Antarctica, and Australia were all connected. Marsupials (pouched mammals) arose at this time and today are found in both South America and Australia. When Australia separated and drifted away, the marsupials diversified into many different forms suited to various environments (Fig. 22.4). They were free to do so because there were no placental mammals in Australia. In other regions, such as South America where there are placental mammals, marsupials are not as diverse.

> The distribution of organisms on the earth is explainable by assuming that related forms evolved in one locale, where they then diversified and/or spread out into other accessible areas.

Evolution is one of the great unifying theories of biology. In science, the word theory is reserved for those conceptual schemes that are supported by a large number of observations and have not yet been found lacking. The theory of evolution has the same status in biology that the germ theory of disease has in medicine.

Australian native cat, *Kasyurus*, a carnivore of forests

Kangaroo, *Macropus*, a herbivore of plains and forests

Sugar glider, *Petaurista*, a tree dweller

Figure 22.4   Three marsupials from Australia.
Each type of marsupial is adapted to a different way of life. All of the marsupials of Australia presumably evolved from a common ancestor that entered Australia some 60 million years ago.

22.2 The Evolutionary Process

Based on much evidence he collected, Charles Darwin formulated a theory of **natural selection** around 1860 to explain the evolutionary process. The following are critical to understanding natural selection.

Existence of Variations

Individual members of a population vary in physical characteristics. Such variations can be passed on from generation to generation. Darwin was never able to determine the cause of variations or how they are passed on. Today, we realize that genes determine the appearance of an organism and that mutations can cause new variations to arise.

Struggle for Existence

Darwin knew that a socioeconomist, Thomas Malthus, had stressed the reproductive potential of human beings. He proposed that death and famine were inevitable because the human population tended to increase faster than the supply of food. Darwin applied this concept to all organisms and saw that members of plant or animal populations must compete with one another for available resources. Darwin calculated the reproductive potential of elephants. Assuming a life span of about 100 years and a breeding span of from 30–90 years, a single female will probably bear no fewer than six young. If all these young survived and continued to reproduce at the same rate, after only 750 years, the descendants of a single pair of elephants would number about 19 million! Such reproductive potential necessitates a *struggle for existence;* only certain members of a population survive and reproduce those characteristics that give them a competitive advantage.

Survival of the Fittest

Darwin noted that in *artificial selection* humans choose which plants or animals will reproduce. This selection process brings out certain traits. For instance, there are many varieties of dogs, each of which was derived from the wild wolf. In a similar way, several varieties of vegetables can be traced to a single type.

In contrast to artificial selection, *natural selection* occurs because certain members of a population happen to have a variation that makes them more suited to the environment. For example, any variation that increases the speed of a hoofed animal will help it escape predators and live longer; a variation that reduces water loss will help a desert plant survive; and one that increases the sense of smell will help a coyote find its prey. Therefore, we would expect organisms with these traits to live longer and, consequently, reproduce to a greater extent.

Adaptation to the Environment

An **adaptation** is a trait that helps an organism be more suited to its environment. We can especially recognize an adaptation when unrelated organisms living in a particular environment display similar characteristics (Fig. 22.5).

Natural selection results in the adaptation of populations to their specific environments. Because of differential reproduction generation after generation, adaptive traits are more and more common in each succeeding generation.

The following listing summarizes the theory of evolution as developed by Darwin.

1. There are inheritable variations among the members of a population.

2. Many more individuals are produced each generation than can survive and reproduce.

3. Individuals with adaptive characteristics are more likely to be selected to reproduce by the environment.

4. Gradually, over long periods of time, a population can become well adapted to a particular environment.

5. The end result of organic evolution is many different species, each adapted to specific environments.

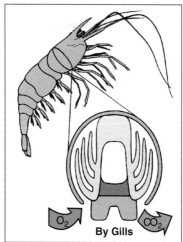

Figure 22.5 Adaptation to aquatic environment.
Distantly related animals living in the same environment have similar adaptations. Both crayfishes (arthropods) and fishes (chordates) breathe by gills that are finely divided and vascularized outgrowths of the body.

22.3 Organic Evolution

Organic evolution began with the evolution of the first form of life, that is, the first cell or cells (Fig. 22.6).

Origin of Life

The sun and the planets probably formed from aggregates of dust particles and debris about 4.6 billion years ago. Intense heat produced by gravitational energy and radioactivity caused the earth to become stratified into several layers. Heavier atoms of iron and nickel became the molten liquid core, and dense silicate minerals became the semiliquid mantle. Upwellings of volcanic lava produced the first crust. The size of the earth is such that its gravitational field is strong enough to have an atmosphere. The earth's primitive atmosphere was not the same as today's atmosphere. It is thought that the primitive atmosphere was produced by outgassing from the interior, particularly by volcanic action. In that case, the primitive atmosphere would have consisted mostly of water vapor (H_2O), nitrogen (N_2), and carbon dioxide (CO_2), with only small amounts of hydrogen (H_2) and carbon monoxide (CO). The primitive atmosphere had little, if any, free oxygen.

Small Organic Molecules

At first the earth was so hot that water was present only as a vapor that formed dense, thick clouds. Then as the earth cooled, water vapor condensed to liquid water, and rain began to fall. It rained in such enormous quantity over hundreds of millions of years that the oceans of the world were produced. The atmospheric gases, dissolved in rain, were carried down into newly forming oceans. The energy sources on the primitive earth included heat from volcanoes and meteorites, radioactivity from isotopes in the earth's crust, powerful electric discharges in lightning, and solar radiation, especially ultraviolet radiation. In the presence of such abundant energy, the primitive gases reacted with one another and produced small organic compounds. With the accumulation of these small organic compounds, the oceans became a warm, organic soup containing a variety of organic molecules.

Macromolecules

The newly formed organic molecules likely joined to produce still larger molecules and then macromolecules. There are two hypotheses of interest concerning this stage in the origin of life. One is the RNA-first hypothesis, which suggests that only the macromolecule RNA (ribonucleic acid) was needed at this time to progress toward formation of the first cell or cells. This hypothesis was formulated after the discovery that RNA can sometimes be both a substrate and an enzyme. Such RNA molecules are called ribozymes. It would seem, then, that RNA could have carried out the processes of life commonly associated today with DNA (deoxyribonucleic acid, the genetic material) and proteins (enzymes). Some viruses today have RNA genes; therefore, the first genes could have been RNA. And the first enzymes also could have been RNA molecules, since we now know that ribozymes exist. Those who support this hypothesis say that it was an "RNA world" some 4 billion years ago.

Another hypothesis is termed the protein-first hypothesis. Sidney Fox has shown that amino acids polymerize abiotically when exposed to dry heat. He suggests that amino acids collected in shallow puddles along the rocky shore and the heat of the sun caused them to form proteinoids, small polypeptides that have some catalytic properties. When proteinoids are returned to water, they form microspheres, structures composed only of protein that have many properties of a cell. It is possible that the first polypeptides had enzymatic properties, and some proved to be more capable than others. Those that led to the first cell or cells had a selective advantage. This hypothesis assumes that DNA genes came after protein enzymes arose. After all, it is protein enzymes that are needed for DNA replication.

The Protocell

Before the first true cell arose, there would have been a protocell, a structure that had a lipid-protein membrane and carried on energy metabolism (Fig. 22.6). Fox has shown that if lipids are made available to microspheres, lipids tend to become associated with microspheres, producing a lipid-protein membrane.

Eventually, a semipermeable-type boundary may have formed about the droplet. In a liquid environment, phospholipid molecules automatically form droplets called liposomes. Perhaps the first membrane formed in this manner. In that case, the protocell could have contained only RNA, which functioned as both genetic material and enzymes.

The protocell would have had to carry on nutrition so that it could grow. Nutrition was no problem because the protocell existed in the ocean, which at that time contained small organic molecules that could have served as food. Therefore, the protocell likely was a heterotroph, an organism that takes in preformed food. Notice that this suggests that heterotrophs preceded autotrophs, organisms that make their own food.

The True Cell

A true cell is a membrane-bound structure that can carry on protein synthesis needed to produce the enzymes that allow DNA to replicate. The central dogma of genetics states that DNA directs protein synthesis and that there is a flow of information from DNA → RNA → protein. It is possible that this sequence developed in stages.

According to the RNA-first hypothesis, RNA would

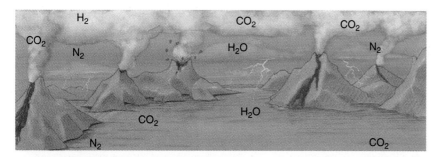

a. The primitive atmosphere contained gases, including water vapor, that escaped from volcanoes; as the water vapor cooled, some gases were washed into the oceans by rain.

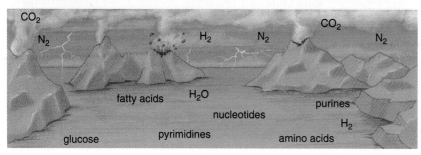

b. The availability of energy from volcanic eruption and lightning allowed gases to form simple organic molecules.

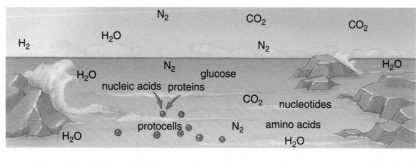

c. Simple organic molecules could have joined to form proteins and nucleic acids, which became incorporated into membrane-bounded spheres. The spheres became the first cells, called protocells.

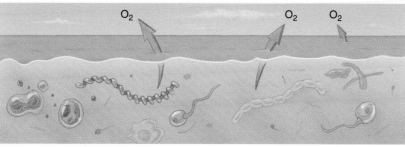

d. Eventually, various types of organisms evolved. Some of these were oxygen-producing photosynthesizers. The presence of oxygen in the atmosphere was necessary for aerobic cellular respiration to evolve.

Figure 22.6   **A model for the origin of life.**

have been the first to evolve, and the first true cell would have had RNA genes. These genes would have directed and enzymatically carried out protein synthesis. Viruses with RNA genes have a protein enzyme called reverse transcriptase that uses RNA as a template to form DNA. Perhaps with time, reverse transcription occurred within the protocell, and this is how DNA genes arose. Once there were DNA genes, then protein synthesis would have been carried out in the manner dictated by the central dogma of genetics.

According to the protein-first hypothesis, proteins, or at least polypeptides, were the first of the three (i.e., DNA, RNA, and protein) to arise. Only after the protocell devel-

oped sophisticated enzymes did it have the ability to synthesize DNA and RNA from small molecules provided by the ocean. Researchers point out that because a nucleic acid is a very complicated molecule, the likelihood that RNA arose de novo (on its own) is minimal. It seems more likely that enzymes were needed to guide the synthesis of nucleotides and then nucleic acids.

Once the protocells acquired genes that could replicate, they became cells capable of reproducing, and evolution began.

Evolution and Classification of Living Things

As mentioned, the fossil record and molecular data allow us to trace the evolution of life, including particular groups of organisms, since life began. These same data are used to classify organisms. **Taxonomy** is that part of biology dedicated to naming, describing, classifying, and relating species. A species is a group of similarly constructed organisms capable of interbreeding and producing fertile offspring. Table 22.1 shows you how taxonomists classify human beings into different categories. As we move from genus to kingdom, more and more different types of species are included in each successive category. Only human beings are in the genus *Homo,* but many different types of

animals are in the animal kingdom. Notice that in the example given, species within the same genus share very specific characteristics, but those that are in the same kingdom have only general characteristics in common.

Taxonomists give each species a scientific name in Latin. The scientific name is a binomial (*bi* means *two, nomen* means *name*). For example, the name for humans is *Homo sapiens.* The first word is the genus, and the second word is a specific epithet for that species. (Note that both words are in italics but only the genus is capitalized.) Scientific names are universally used by biologists so as to avoid confusion. Common names tend to overlap and often are in the language of a particular country.

Taxonomy is an attempt to make sense out of the bewildering variety of life on earth. Species are classified according to their presumed evolutionary relationship; those placed in the same genus are the most closely related, and those placed in separate kingdoms are the most distantly related. As more is known about evolutionary relationships between species, taxonomy changes. Presently, many biologists recognize five kingdoms, which they believe to be related as shown in Figure 22.7. Such a diagram, called an evolutionary tree, portrays the evolutionary history of a group of organisms—in this case, all living things.

Table 22.1	Classification Categories	
Categories	**For Humans**	**Description**
Kingdom	Animalia	Multicellular, moves, ingests food
Phylum	Chordata	Dorsal supporting rod and nerve cord
Class	Mammalia	Hair, mammary glands
Order	Primates	Adapted to climb trees
Family	Hominidae	Adapted to walk erect
Genus	*Homo*	Large brain, tool use
Species	*sapiens*	

Taxonomists classify living things into categories according to their evolutionary relationships and use evolutionary trees to depict these relationships.

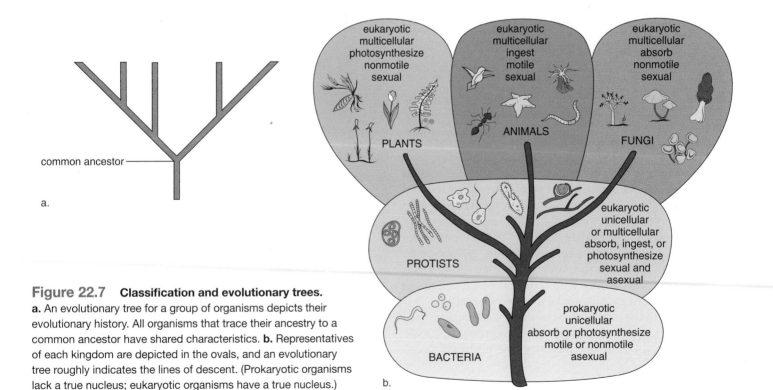

Figure 22.7   Classification and evolutionary trees.
a. An evolutionary tree for a group of organisms depicts their evolutionary history. All organisms that trace their ancestry to a common ancestor have shared characteristics. **b.** Representatives of each kingdom are depicted in the ovals, and an evolutionary tree roughly indicates the lines of descent. (Prokaryotic organisms lack a true nucleus; eukaryotic organisms have a true nucleus.)

22.4 Modern Humans Evolve

Humans are **primates** (order primates), which are placental mammals adapted to living in trees.

Primates

The limbs of primates are mobile, as are the hands, because the thumb (and in nonhuman primates, the big toe as well) is opposable; that is, the thumb can touch each of the other fingers. In primates, the snout is shortened considerably, allowing the eyes to move to the front of the head. The stereoscopic vision (or depth perception) that results permits primates to make accurate judgments about the distance and position of adjoining tree limbs. Gestation is lengthy in a primate, allowing time for good forebrain development. One birth at a time is the norm in primates; it is difficult to care for several offspring while moving from limb to limb. The juvenile period of dependency is extended, and there is an emphasis on learned behavior and complex social interactions.

Among primates, humans are most closely related to the apes. There are four types of modern apes: gibbons, orangutans, gorillas, and chimpanzees. Gibbons, the smallest of the apes, have extremely long arms, which are specialized for swinging between tree limbs. The orangutan is a large ape but nevertheless spends a great deal of time in trees. In contrast, the gorilla, the largest of the apes, spends most of its time on the ground. Chimpanzees, which are at home both in the trees and on the ground, are the most humanlike of the apes in appearance.

Humans can be distinguished from modern apes by locomotion and posture, dental features, and other characteristics (Fig. 22.8).

These characteristics especially distinguish primates from other mammals: opposable thumb (and in some cases, big toe); expanded forebrain; emphasis on learned behaviors; single birth; extended period of parental care.

Hominids

Modern humans are **hominids** (family Hominidae) as are their immediate ancestors. The hominid line of descent begins with the **australopithecines,** which evolved and diversified in eastern Africa from about 4.4 MYA (millions of years ago) to almost 1 MYA.

The most famous of these fossils, dated about 3.4 MYA, is scientifically called *Australopithecus afarensis* but is better known by its field name, Lucy. (The name derives from the Beatles' song "Lucy in the Sky with Diamonds.") Although her brain was quite small, the shapes and relative proportions of her limbs indicate that Lucy walked upright. There are even footprints that show how Lucy walked. *A. afarensis* males, which were discovered later, are much larger than the females. Taking into account their smaller body size, the relative brain size is about one-third of ours, and the jaw is heavy. The cheek teeth are enormous, and in males, large canine teeth project forward. *A. afarensis* had descendants, and there may have been as many as ten species of hominids about 2 MYA in Africa, some of which were quite robust. One of these descendents is classified in the genus *Homo*.

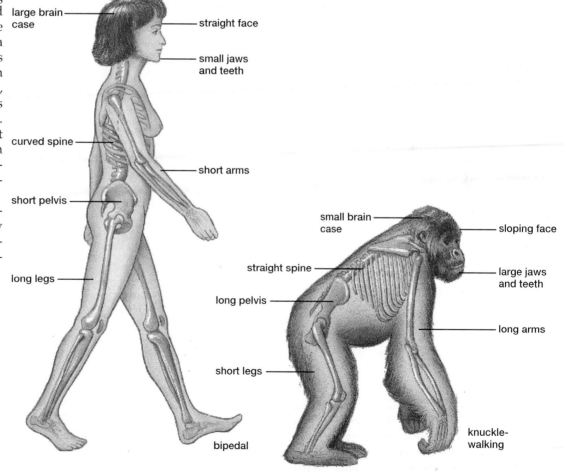

Labels — Modern Human: large brain case, straight face, small jaws and teeth, curved spine, short arms, short pelvis, long legs, bipedal

Labels — Modern Ape: small brain case, sloping face, straight spine, large jaws and teeth, long pelvis, long arms, short legs, knuckle-walking

Modern Human   **Modern Ape**

Figure 22.8 **Comparison of human skeletal features with those of a gorilla.**

Homo habilis

Homo habilis is dated as early as 2 MYA (Fig. 22.9). The name means handy man, and this hominid is often accompanied by stone tools. Why is this fossil classified within our own genus? *H. habilis* was small—about the size of Lucy—but the brain at 700 cc is about 45% larger.

The stone tools made by *H. habilis* are called Oldowan tools because they were first identified as tools at a place called Olduvai Gorge. Oldowan tools are simple and look rather clumsy (Fig. 22.10), but perhaps stone flakes were also used. The flakes would have been sharp and able to scrape away hide and cut tendons to easily remove meat from a carcass. Perhaps a division of labor arose, with certain individuals serving as hunters and others as gatherers. Speech would have facilitated their cooperative efforts, and later they most likely shared their food and ate together. In this way, society and culture could have begun. Prior to the development of culture, adaptation to the environment necessitated a biological change. The acquisition of culture provided an additional way by which adaptation was possible. And the possession of culture by *H. habilis* may have hastened the extinction of the australopithecines.

H. habilis warrants classification as a *Homo* because of brain size, posture, and dentition. Circumstantial evidence suggests the use of tools and also the development of culture.

Homo erectus

Homo erectus is the name assigned to hominid fossils found in Africa, Asia, and Europe and dated between 1.9 and 0.5 MYA. Compared to *H. habilis*, *H. erectus* had a larger brain (about 1,000 cc), more pronounced brow ridges, a flatter face, and a nose that projects like ours. These hominids not only stood erect, they most likely had a striding gait like ours.

It is believed that *H. erectus* first appeared in Africa and then migrated into Asia and Europe. Such an extensive population movement is a first in the history of humankind and a tribute to the intellectual and physical skills of the species. *H. erectus* was also the first hominid to use fire and fashion more advanced tools, called Acheulean tools after a site in France (Fig. 22.10). There are heavy teardrop-shaped axes and cleavers as well as

flakes, which were probably used for cutting and scraping. Some believe that *H. erectus* was a systematic hunter and brought kills to the same site over and over again. In one location, there are over 40,000 bones and 2,647 stones. These sites could have been "home bases" where social interaction occurred and a prolonged childhood allowed time for much learning. Perhaps a language evolved and a culture more like our own developed.

H. erectus, which evolved from *H. habilis*, had a striding gait, made well-fashioned tools (perhaps for hunting), and could control fire. This hominid migrated into Europe and Asia from Africa about 1 MYA.

Modern Humans

The out-of-Africa hypothesis proposes that *H. sapiens* became fully modern only in Africa, and thereafter migrated to Europe and Asia about 100,000 years before present. Modern humans may have interbred to a degree with popula-

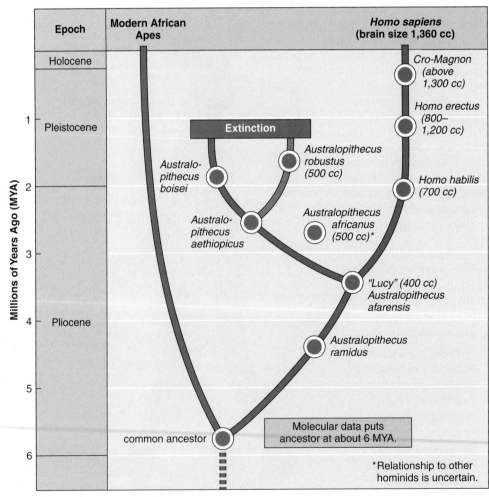

Figure 22.9   Evolutionary tree for hominids.
This tree suggests that African apes and hominids had a common ancestor about 6 million years ago (MYA). Each group has been on their own evolutionary pathway since that time. *Australopithecus afarensis* (Lucy) is a common ancestor for all the other australopithecines and for the human line of descent.

tions of *H. erectus* in Europe and Asia, but in effect, the modern humans supplanted them.

A brain capacity larger than 1,000 cc allows a species to be classified as *Homo sapiens.* The **Neanderthals** (*H. sapiens neanderthalensis*) take their name from Germany's Neander Valley, where one of the first Neanderthal skeletons, dated some 200,000 years ago, was discovered. The Neanderthals had massive brow ridges, and the nose, the jaws, and the teeth protruded far forward. The forehead was low and sloping, and the lower jaw sloped back without a chin.

The Neanderthals are thought to be an archaic *H. sapiens,* and most likely not in the main line of *Homo* descent. Surprisingly, however, the Neanderthal brain was, on the average, slightly larger than that of modern humans (1,400 cc compared to 1,360 cc in most modern humans). The Neanderthals were heavily muscled, especially in the shoulders and the neck. The bones of the limbs were shorter and thicker than those of modern humans. It is hypothesized that a larger brain than that of modern humans was required to control the extra musculature. They lived in Eurasia during the last Ice Age, and their sturdy build could have helped conserve heat.

The Neanderthals give evidence of being culturally advanced. They most likely successfully hunted bears, woolly mammoths, rhinoceroses, reindeer, and other contemporary animals. They used and could control fire, and they even buried their dead with flowers and tools, indicating that they may have had a religion.

In keeping with the out-of-Africa hypothesis, it is increasingly believed that modern humans, by custom called **Cro-Magnon** after a fossil location in France, entered Eurasia 100,000 years ago or even earlier. Cro-Magnons (*H. sapiens sapiens*) had a thoroughly modern appearance. They made advanced stone tools, called Aurignacian tools (Fig. 22.10), and were such accomplished hunters that they may have caused the extinction of many larger mammals, such as the giant sloth, the mammoth, the saber-toothed tiger, and the giant ox.

Cro-Magnons hunted cooperatively and most likely lived in small groups, with the men hunting by day while the women remained at home with the children. The Cro-Magnon culture included art. They sculpted small figurines out of reindeer bones and antlers. They also painted beautiful drawings of animals on cave walls in Spain and France.

The main line of hominid descent is now believed to include *A. afarensis, H. habilis, H. erectus,* and Cro-Magnon.

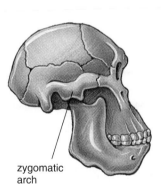

zygomatic arch

Australopithecus
low forehead
projecting face
large brow ridge
small brain case
large zygomatic arch

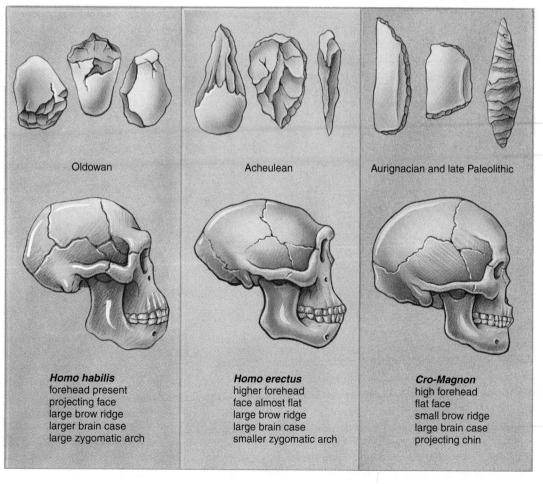

Oldowan   Acheulean   Aurignacian and late Paleolithic

Homo habilis
forehead present
projecting face
large brow ridge
larger brain case
large zygomatic arch

Homo erectus
higher forehead
face almost flat
large brow ridge
large brain case
smaller zygomatic arch

Cro-Magnon
high forehead
flat face
small brow ridge
large brain case
projecting chin

Figure 22.10   Comparative hominid skull anatomy and tools.
Oldowan tools are crude; Acheulean tools are better made and more varied; Aurignacian tools are well designed for their specific purposes.

We Are One Species

Human beings are diverse, but even so, we are all classified as *H. sapiens sapiens*. This is consistent with the biological definition of species because it is possible for all types of humans to interbreed and to bear fertile offspring. While it may appear that there are various "races," molecular data show that the DNA base sequence varies as much between individuals of the same ethnicity as between individuals of different ethnicity.

It is generally accepted that the phenotype is adapted to the climate of a region. Although it might seem as if dark skin is a protection against the hot rays of the sun, it has been suggested that it is actually a protection against ultraviolet ray absorption. Dark-skinned persons living in southern regions and light-skinned persons living in northern regions absorb the same amount of radiation. (Some absorption is required for vitamin D production.) Other features that correlate with skin color, such as hair type and eye color, may simply be side effects of genes that control skin color.

Differences in body shape represent adaptations to temperature. A squat body with short limbs and a short nose retains more heat than an elongated body with long limbs and a long nose. Also, almond-shaped eyes, a flat nose and forehead, and broad cheeks are believed to be adaptations to the last Ice Age.

While it always has seemed to some that physical differences warrant assigning humans to different "races," this contention is not borne out by the molecular data mentioned previously.

Bioethical Issue

Evolution is a scientific theory. So is the cell theory, which says that all organisms are composed of cells, and so is the atomic theory that says all matter is composed of atoms. Yet, no one argues that schools should teach alternatives to the cell theory or the subatomic theory. Confusion reigns over the use of the expression, "the theory of evolution." But the term *theory* in science is reserved for those ideas that scientists have found to be all encompassing because they are based on data collected in a number of different fields.

No wonder most scientists in our country are dismayed when state legislatures or school boards rule that teachers must put forward a variety of "theories" on the origin of life, including one that runs contrary to the mass of data that supports the theory of evolution. An institute in California called the Institute for Creation Research advocates that students be taught an "intelligent-design theory" which says that DNA could never have arisen without the involvement of an "intelligent agent," and that gaps in the fossil record mean that species arose fully developed with no antecedents.

Since our country forbids the mingling of church and state—no purely religious ideas can be taught in the schools—the advocates for an "intelligent-design theory" are careful to never mention the Bible nor any strictly religious ideas (i.e., God created the world in seven days). Still, teachers who have a solid scientific background do not feel comfortable teaching an "intelligent-design theory" because it does not meet the test of a scientific theory. Science is based on hypotheses that have been tested by observation and/or experimentation. A scientific theory has stood the test of time—no hypotheses have been supported by observation and/or experimentation that run contrary to the theory. On the contrary, the theory of evolution is supported by data collected in such wide-ranging fields as development, anatomy, geology, biochemistry, and so forth.

The polls consistently show that nearly half of all Americans prefer to believe the Old Testament account of how God created the world in seven days. That, of course, is their right, but should schools be required to teach an "intelligent-design theory" that traces its roots back to the Old Testament, and is not supported by observation and experimentation?

Questions

1. Should teachers be required to teach an "intelligent-design theory" of the origin of life in schools? Why or why not?
2. Should schools rightly teach that science is based on data collected by the testing of hypotheses by observation and experimentation? Why or why not?
3. Should schools be required to show that the "intelligent-design theory" does not meet the test of being scientific? Why or why not?

Summarizing the Concepts

22.1 Evidence for Evolution

The theory of evolution explains the history and diversity of life. Evidence for evolution can be taken from the fossil record, comparative biochemistry and anatomy, and biogeography.

22.2 The Evolutionary Process

Darwin not only presented evidence in support of organic evolution, he showed that evolution was guided by natural selection. Due to reproductive potential, there is a struggle for existence among members of the same species. Those members that possess variations more suited to the environment will most likely have more offspring than other members. Because of this natural selection process, there is a gradual change in species composition, which leads to adaptation to the environment.

22.3 Organic Evolution

Life came into being with the first cell(s), and thereafter diversification accounts for the evolution of all other life forms. The classification of organisms parallels their presumed evolutionary history.

22.4 Modern Humans Evolve

Humans are mammalian primates, as are apes. Humans differ from apes in regard to mode of locomotion, shape of jaw, and brain size.

The first hominid (humans and immediate ancestors) was *A. afarensis*, which could walk erect but had a small brain. *H. habilis* made tools, but *H. erectus* was the first fossil to have a brain size of more than 1,000 cc. *H. erectus* migrated from Africa into Europe and Asia. They used fire and may have been big-game hunters.

Most likely, *H. sapiens* evolved in Africa but then migrated to Europe and Asia, where this species supplanted the previous humans living there. One of the species was *H. sapiens neanderthalensis*. The Neanderthals did not have the physical traits of modern humans, but they did have culture.

Studying the Concepts

1. Show that the fossil record, comparative biochemistry and anatomy, and biogeography all give evidence of evolution. 462–64
2. What are the five aspects of Darwin's theory of evolution? 465
3. Describe the events that led to the origin of the first cell(s). 466–67
4. What are the major categories of classification? What is the goal of modern taxonomists? 468
5. Name several primate characteristics still retained by humans. How do modern humans differ from modern apes? 469
6. Draw a hominid evolutionary tree. 470
7. How do the australopithecines, *Homo erectus*, and Cro-Magnon differ anatomically from one another and from apes? 469–71
8. Which humans were tool users? Walked erect? Used fire? Drew pictures? 469–71
9. Discuss the reason(s) we know that all humans belong to the same species and why the concept of "race" is unnecessary. 472

Testing Your Knowledge of the Concepts

In questions 1–4, match the evolutionary evidence to the description:

a. biogeography
b. fossil record
c. comparative biochemistry
d. comparative anatomy

_____ 1. Species change over time.
_____ 2. Forms of life are variously distributed.
_____ 3. A group of related species has a unity of plan.
_____ 4. The same types of molecules are found in all living things.

In questions 5–15, indicate whether the statement is true (T) or false (F).

_____ 5. Evolutionary success is judged by reproductive success, or the number of offspring.
_____ 6. The end result of natural selection is adaptation to the environment.
_____ 7. An unusually dry wind is believed to have produced the first cell(s).
_____ 8. The forelimbs of vertebrates have the same bones but are designed for different functions.
_____ 9. Because the early developmental stages of a chick and pig are so different it is easy to tell one embryo from the other.
_____ 10. The very many different kinds of marsupials in Australia are descended from a common ancestor.
_____ 11. Survival of the fittest means that one type of animal is able to kill another type outright.

_____ 12. The protocell must have been able to make its own food through photosynthesis.
_____ 13. The phylum classification category contains organisms that are more closely alike than the order classification category.
_____ 14. *Homo habilis* is named for his ability to make stone tools.
_____ 15. Despite differences in appearance, all ethnic groups of human beings belong to one species.

In questions 16 and 17, fill in the blanks.

16. The australopithecines could probably walk _____, but they had a _____ brain.
17. The two varieties of *Homo sapiens* from the fossil record are _____ and _____.
18. Complete this table to describe the classification of humans:

Category	Name	Examples
Kingdom	_____	_____
Phylum	_____	_____
Class	_____	_____
Order	_____	_____
Family	_____	_____
Genus	_____	_____
Species	_____	_____

Applying Your Knowledge to the Concepts

These questions pertain to the evolution of organisms.

1. On the basis of Darwin's theory of evolution, how would you explain the fact that certain insects, flies in particular, have become resistant to DDT?

2. Organisms can be classified as anaerobic or aerobic depending on whether they do not require oxygen or do require oxygen for respiration. Yet, even aerobic organisms start the respiratory process with anaerobic reactions. On an evolutionary basis, explain this.

3. Many individuals have great difficulty with the concept that humans evolved from apes. They do not like to be that closely associated with gorillas or chimpanzees. What information could you provide that might ease their concern?

4. Few people have a complete set of fully erupted wisdom teeth. What does this suggest about the evolution of the human jaw and what do you predict will be the evolutionary consequences?

Understanding the Terms

adaptation 465
australopithecine 469
Cro-Magnon 471
fossil 462
hominid 469
Homo erectus 470
Homo habilis 470

homologous structure 463
natural selection 465
Neanderthal 471
primate 469
taxonomy 467
vestigial structure 463

Match the terms to these definitions:

a. _____ Process by which populations become adapted to their environment.

b. _____ Modification in structure, function, or behavior that increases likelihood of surviving and reproducing.

c. _____ Science of naming, classifying, and showing relationships of organisms.

d. _____ Animal that belongs to the order Primates; the order of mammals that includes monkeys, apes, and humans.

e. _____ Any remains of an organism that have been preserved in the earth's crust.

Applying Technology to the Concepts

Your study of evolution is supported by these available technologies:

Essential Study Partner CD-ROM

Evolution

Visit the Mader web site for related ESP activities.

Exploring the Internet

The Mader Home Page provides resources and tools as you study this chapter.

http://www.mhhe.com/biosci/genbio/mader

Chapter 23

Ecosystems

Chapter Concepts

Figure 23.1   Piñon trees.
Piñon trees are a population of organisms in an ecosystem that also includes deer mice, which feed on piñon nuts.

When people first began dying in the Southwest of an unknown cause in 1993, no one knew to blame the warm, wet weather. Or the fat little deer mice scampering about. And absolutely no one would have pointed a finger at a new strain of a Korean hemorrhagic fever. But these unlikely factors combined to launch one of the decade's most deadly emerging diseases: hantavirus.

First found in Korea some 40 years ago, hantavirus—which is deadly in humans—can be carried by common deer mice. Normally, that's not a problem. But in 1993, deer mice in New Mexico and nearby states experienced a population boom. The boom stemmed from unusually heavy spring rains, which nourished trees carrying piñon nuts—a favorite food of the deer mouse (Fig. 23.1). This slight change in the weather was enough to spur a tenfold increase in the deer mouse population. Suddenly, the mice were everywhere—inside garages, in the backyard, on Indian reservations in the region. With the mice came hantavirus, carried in rodent feces and urine.

This example shows that human beings are indeed a part of natural **ecosystems,** which contain populations of organisms such as deer mice, piñon trees, and yes, humans. In ecosystems, populations interact among themselves and the physical environment. The saying goes that in an ecosystem everything is connected to

everything else: warm, wet weather led to plentiful piñon nuts, which led to a deer mouse explosion, and finally, illness in humans. This chapter examines the interworkings of ecosystems and how they have been impacted by human beings. While an example like this might make it seem as if we should further restrict natural ecosystems, this is not the case. The case will be made that they need to be preserved, albeit managed, if the human species is to continue.

Thankfully, the illness outbreak caused by the hantavirus ended. Rodent control efforts, disease surveillance programs, and research on a hantavirus vaccine should prevent similar outbreaks in the future.

Table 23.1	Ecological Terms
Term	**Definition**
Ecology	Study of the interactions of organisms with each other and with the physical environment
Population	All the members of the same species that inhabit a particular area
Community	All the populations that are found in a particular area
Ecosystem	A community and its physical environment; nonliving (abiotic) and living (biotic) components
Biosphere	All the communities on earth whose members exist in air, water, and on land

23.1   The Nature of Ecosystems

When the earth was formed, the outer crust was covered by ocean and barren land. Over time, aquatic organisms filled the seas and terrestrial organisms colonized the land so that eventually there were many complex communities of living things. A **community** is made up of all the **populations** in a particular area, such as a forest or pond. When we study a community, we are considering only the populations of organisms that make up that community, but when we study an *ecosystem*, we are concerned with the community and its physical environment (Table 23.1).

Succession

To the human eye, it seems as if the communities stay pretty much the same from year to year. Actually, one of the most outstanding characteristics of natural communities is their changing nature. The dynamic nature of communities is dramatically illustrated by succession. **Succession** is a sequential change in the relative dominance of species within a community. Most likely you have observed that an abandoned field slowly changes to be more like the surrounding natural area. Although the final result will be determined by the species that migrate there, it is possible to suggest the interim stages of succession.

Primary succession on land begins on bare rock. At first the rock is subjected to weathering by wind and rain, followed by the invasion of lichens and mosses (Fig. 23.2). They cause the buildup of soil, permitting low-lying grasses and then larger plants to take over. Depending on the area, pine trees and then broadleaf trees will eventually take root. Secondary succession, which begins with an abandoned field, will also go through a herbaceous stage before shrubs and then trees possibly appear. When ecologists first observed succession, they called the final stage of succession the *climax community*. They felt that each particular area had its own climax community. For example, in the United States, a deciduous forest is typical of the Northeast, a prairie is natural to the Midwest, and a semidesert covers the Southwest (Fig. 23.3). Today, we realize that many factors determine what particular species are found within a community, and it is not predetermined from the start.

The dynamic nature of communities is witnessed by their changing nature when succession occurs. Climax communities are threatened by disturbances.

a.

b.

c.

d.

Figure 23.2   Example of primary succession.
a. Lichens can grow on bare rock. **b.** Grasses take hold. **c.** Larger, nonwoody plants spread throughout the area. **d.** Trees become established.

a. Southwest desert

c. Coastal Pacific Northwest rain forest

b. California chaparral

d. Midwest prairie

Figure 23.3   Biological communities of the United States.
a. A desert gets limited rainfall a year. Because daytime temperatures are high, most animals are active at night when it is cooler. **b.** The California chaparral recovers quickly after a fire. **c.** A coastal rain forest of the Pacific Northwest receives a plentiful amount of rain. **d.** Midwest prairies contain grasses and rich soil built up by the remains of grasses.

Biotic Components of an Ecosystem

An ecosystem possesses both nonliving (abiotic) and living (biotic) components. The abiotic components include resources, such as sunlight and inorganic nutrients, and conditions, such as type of soil, water availability, prevailing temperature, and amount of wind (Fig. 23.4).

The biotic components of an ecosystem are the various populations of organisms. Each population in an ecosystem has a **habitat** and a niche. The habitat of an organism is its place of residence, that is, where it can be found, such as under a log or at the bottom of the pond. The **niche** of an organism is its profession or total role in the community. A description of an organism's niche includes its interactions with the physical environment and with the other organisms in the community. For example, the trees in Figure 23.4a take solar energy, carbon dioxide, and water from the abiotic environment but have many interactions with other organisms, such as woodpeckers (Fig. 23.4b). Woodpeckers feed on parasitic grubs from a tree, which also provides a habitat for the woodpecker's young. All of these behaviors are aspects of the woodpecker's niche, as is apparent from reviewing the listing given in the figure.

The populations are often categorized in this way:

Producers are autotrophic organisms with the capability of carrying on photosynthesis and making food (organic nutrients) for themselves (and indirectly for the other populations as well). In terrestrial ecosystems, the producers are predominantly green plants, while in freshwater and marine ecosystems, the dominant producers are various species of algae.

Consumers are heterotrophic organisms that use preformed food. It is possible to distinguish four types of consumers, depending on their food source. **Herbivores** feed directly on green plants; they are termed primary consumers. **Carnivores** feed only on other animals and are thus secondary, or tertiary, consumers. **Omnivores** feed on both plants and animals. Therefore, a caterpillar feeding on a leaf is a herbivore; a green heron feeding on a fish is a carnivore; a human being eating both leafy green vegetables and beef is an omnivore. **Decomposers,** a fourth type of consumer, feed on detritus. **Detritus** is the remains of plants and animals following their death and fragmentation by soil organisms. The bacteria and fungi of decay are important detritus feeders, but so are other soil organisms, such as earthworms and various small arthropods. The importance of the latter can be demonstrated by placing leaf litter in bags with mesh too fine to allow soil animals to enter; the leaf litter does not decompose well, even though bacteria and fungi are present. Small soil organisms precondition the detritus so that bacteria and fungi can break it down to inorganic matter that producers can use again.

Energy Flow and Chemical Cycling

When we diagram all the biotic components of an ecosystem, as in Figure 23.5, it is possible to illustrate that ecosystems are characterized by two fundamental phenomena: energy flow and chemical cycling. *Energy flow* begins when producers absorb solar energy, and chemical cycling begins when producers take in inorganic nutrients from the physical environment. Thereafter, producers make food for themselves and indirectly for the other populations of the

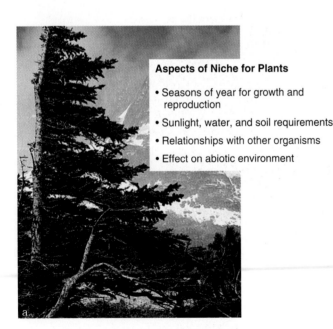

Aspects of Niche for Plants

- Seasons of year for growth and reproduction
- Sunlight, water, and soil requirements
- Relationships with other organisms
- Effect on abiotic environment

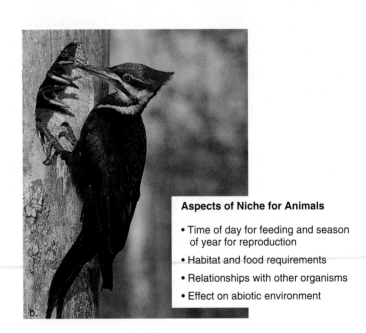

Aspects of Niche for Animals

- Time of day for feeding and season of year for reproduction
- Habitat and food requirements
- Relationships with other organisms
- Effect on abiotic environment

Figure 23.4 Biotic components of an ecosystem.
The biotic components of an ecosystem are influenced by the abiotic components as in **(a),** where the force of the wind has affected tree growth. Each biotic component has a niche whose aspects are listed for plants **(a)** and animals **(b).**

ecosystem. Energy flow occurs because all the energy content of organic food is eventually converted to heat, which dissipates in the environment. Therefore, most ecosystems cannot exist without a continual supply of solar energy. *Chemicals cycle;* that is, the original inorganic nutrients are returned to the producers within and between ecosystems.

Only a portion of the food made by autotrophs is passed on to heterotrophs because plants use organic molecules to fuel their own cellular respiration. Only about 55% of the food made by producers is available to heterotrophs. Similarly, only a small percentage of food taken in by heterotrophs is available to higher level consumers. Figure 23.6 shows why. A certain amount of the food eaten by a herbivore is never digested and is eliminated as feces. Metabolic wastes are excreted as urine. Of the assimilated energy, a large portion is utilized during cellular respiration and thereafter becomes heat. Only the remaining food which is converted into increased body weight (for additional offspring) becomes available to carnivores.

The elimination of feces and urine by a heterotroph, and indeed the death of all organisms, does not mean that substances are lost to an ecosystem. They represent the food made available to decomposers. Since decomposers can be food for other heterotrophs of an ecosystem, the situation can get a bit complicated. Still, we can conceive that all the solar energy that enters an ecosystem eventually becomes heat. And this is consistent with the observation that ecosystems are dependent on a continual supply of solar energy.

Energy flow in an ecosystem is a consequence of two fundamental laws of thermodynamics. The first law states that energy cannot be created or destroyed, and the second law states that when energy is transformed from one form to another, there is always a loss of some usable energy as heat. This means that as primary consumers feed on producers, and secondary consumers feed on primary consumers, as shown in Figure 23.6, some energy is lost to the environment as heat. Eventually, all the energy that entered the system is dissipated.

Since energy cannot be created or completely transformed from one form to another, ecosystems are unable to function unless there is a constant energy input. This input usually comes from the sun, the ultimate source of energy for our planet. It can't be said that the sun always supplies energy for an ecosystem because there are well-known exceptions. A unique community of organisms existing around deep-sea hypothermal vents in the ocean rely on chemosynthetic bacteria rather than photosynthetic algae and plants to produce food. Scalding-hot jets of water spew out of cracks in the earth's crust and supply chemosynthetic bacteria with hydrogen sulfide, a molecule they can oxidize to supply energy for synthesis of organic molecules. These bacteria that exist freely or mutualistically within the tissues of organisms are the start of food chains for a community that includes huge tube worms and clams.

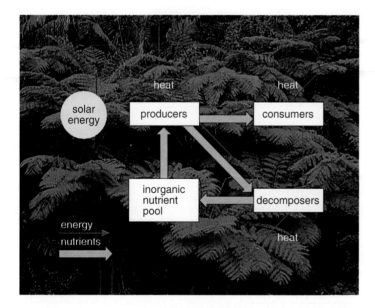

Figure 23.5   **Nature of an ecosystem.**
Chemicals cycle but energy flows through an ecosystem. All the energy derived from the sun eventually dissipates as heat as energy transformations repeatedly occur.

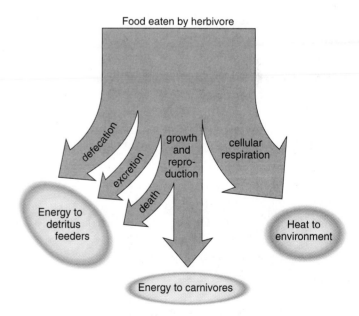

Figure 23.6   **Energy balances.**
Only about 10% of the food energy taken in by a herbivore is passed on to carnivores. A large portion goes to detritus feeders in the ways indicated, and another large portion is used for cellular respiration.

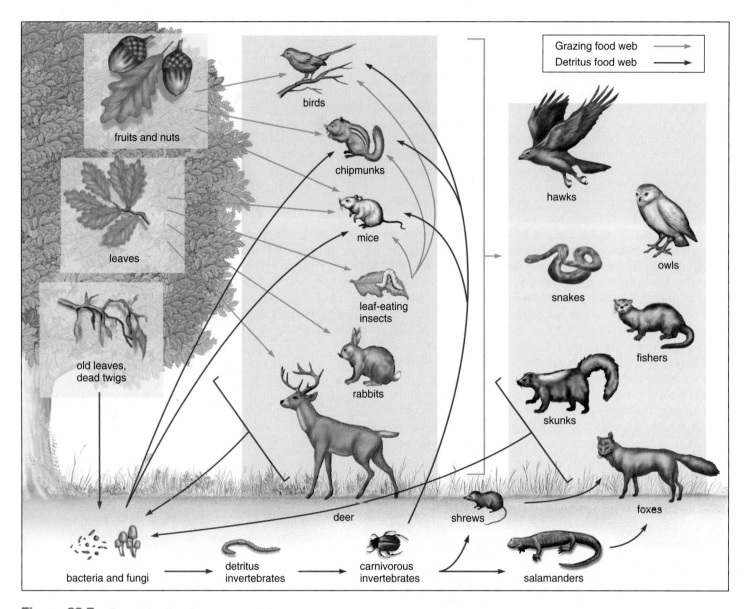

Figure 23.7   Forest food webs.
Two linked food webs are shown for a forest ecosystem: a grazing food web and a detrital food web.

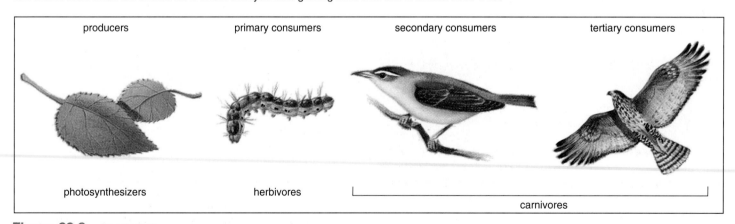

Figure 23.8   Food chain.
Trace this grazing food chain in the grazing food web depicted in Figure 23.7.

Food Webs and Trophic Levels

The principles we have been discussing can now be applied to an actual example—a forest in New Hampshire. In this forest, the producers include sugar maple, beech, and yellow birch trees. The complicated feeding relationships that exist in natural ecosystems are called food webs. A **food web** shows how organisms acquire their food. For example, Figure 23.7 shows that insects in the form of caterpillars feed on leaves, while mice, rabbits, and deer feed on leaf tissue at or near the ground. Birds, chipmunks, and mice feed on fruits and nuts, but they are in fact omnivores because they also feed on caterpillars. These herbivores and omnivores all provide nutrients for a number of different carnivores. This portion of the diagram is called a **grazing food web** because it begins with aboveground plant material.

The lower half of Figure 23.7 is devoted to the **detrital food web.** The bacteria and fungi of decay can be food for other larger decomposers, which feed on organic matter in the soil. Because some of these, like shrews and salamanders, become food for aboveground carnivores, the detrital and the grazing food webs are connected.

We naturally tend to think that aboveground vegetation like trees are the largest storage form of organic matter and energy, but this is not necessarily the case. In this particular forest, the organic matter lying on the forest floor and mixed into the soil contains much more energy than does the leaf matter of living trees. The soil contains over twice as much energy as the forest floor. Therefore, more energy in a forest may be funneling through the detrital food web than through the grazing food web.

Trophic Levels

You can see that Figure 23.7 would allow us to link organisms one to another in a straight line manner, according to who eats whom. Such diagrams are called **food chains** (Fig. 23.8). For example, in the grazing food web we can find this **grazing food chain:**

leaves → caterpillars → tree birds → hawks

And in the detrital food web we could find this **detrital food chain:**

dead organic matter → soil microbes → earthworms → etc.

A **trophic level** is all the organisms that feed at a particular link in a food chain. In the grazing food web, going from left to right, the trees are primary producers (first trophic level), the first series of animals are primary consumers (second trophic level), and the next group of animals are secondary consumers (third trophic level) and so forth.

Ecological Pyramids

Ecologists portray the energy relationships between trophic levels in the form of **ecological pyramids,** diagrams whose building blocks designate the various trophic levels (Fig. 23.9). (We need to keep in mind that sometimes organisms don't fit into one trophic level. For example, chipmunks feed on fruits and nuts, but they also feed on leaf-eating insects.)

A *pyramid of numbers* simply tells how many organisms there are at each trophic level. It's easy to see that a pyramid of numbers could be completely misleading. For example, in Figure 23.7 you would expect each tree to contain numerous caterpillars; therefore there would be more herbivores than autotrophs! The problem, of course, has to do with size. Autotrophs can be tiny, like microscopic algae, or they can be big like beech trees; similarly, herbivores can be small like caterpillars, or they can be large like elephants.

Pyramids of biomass eliminate size as a factor since biomass is the number of organisms multiplied by their weight. You would certainly expect the biomass of producers to be greater than the biomass of the herbivores, and that of the herbivores to be greater than the carnivores. In some aquatic ecosystems such as lakes and open seas, where algae are the only producers, the herbivores may have a greater biomass

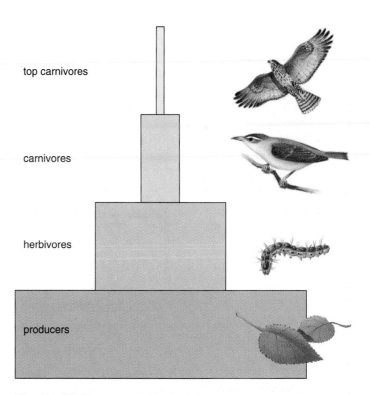

Figure 23.9   Ecological pyramid.

An ecological pyramid shows the relationship between either the number of organisms, the biomass, or the amount of energy theoretically available at each trophic level.

than the producers when you take their measurements. Why? The reason is that over time, the algae reproduce rapidly, but they are also consumed at a high rate. Pyramids, like this one, that have more herbivores than producers are called inverted pyramids:

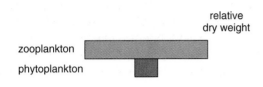

There are ecological *pyramids of energy* also, and they generally have the appearance of Figure 23.9. Ecologists are now beginning to rethink the usefulness of utilizing pyramids to describe energy relationships. One problem is what to do with the decomposers, which are rarely included in pyramids, and yet a large portion of energy becomes detritus in many ecosystems.

There is a rule of 10% with regard to biomass (or energy) pyramids. It says that, in general, the amount of biomass (or energy) from one level to the next is reduced by a magnitude of 10. Thus, if an average of 1,000 kg of plant material is consumed by herbivores, about 100 kg is converted to herbivore tissue, 10 kg to first-level carnivores, and 1 kg to second-level carnivores. The rule of 10% suggests that few carnivores can be supported in a food web. This is consistent with the observation that each food chain has from three to four links, rarely five.

23.2 Global Biogeochemical Cycles

All organisms require a variety of organic and inorganic nutrients. Carbon dioxide and water are necessary for photosynthesis. Nitrogen is a component of all the structural and functional proteins and nucleic acids that sustain living tissues. Phosphorus is essential for ATP and nucleotide production. In contrast to energy, inorganic nutrients are used over and over again by autotrophs.

Since the pathways by which chemicals circulate through ecosystems involve both living (biosphere) and nonliving (geological) components, they are known as **biogeochemical cycles.** For each element, chemical cycling may involve (1) a reservoir—a source normally unavailable to producers, such as fossilized remains, rocks, and deep-sea sediments; (2) an exchange pool—a source from which organisms do generally take elements, such as the atmosphere or soil; and (3) the biotic community—through which chemicals move along food chains, perhaps never entering a pool (Fig. 23.10).

There are two general categories of biogeochemical cycles. In a *gaseous cycle,* exemplified by the carbon and nitrogen cycles, the element returns to and is withdrawn from the atmosphere as a gas. In the *sedimentary cycle,* exemplified by the phosphorus cycle, the chemical is absorbed from the sediment by plant roots, passed to heterotrophs, and is eventually returned to the soil by decomposers, usually in the same general area.

The diagrams on the next few pages make it clear that nutrients can flow between terrestrial and aquatic ecosystems. In the nitrogen and phosphorus cycles, these nutrients run off from a terrestrial to an aquatic ecosystem and in that way enrich aquatic ecosystems. Decaying organic material in aquatic ecosystems can be a source of nutrients for intertidal inhabitants like fiddler crabs. Seabirds feed on fish but deposit guano (droppings) on land, and in that way phosphorus from the water is deposited on land. It would seem that anything put into the environment in one ecosystem could find its way to another ecosystem. Scientists find the soot from urban areas and pesticides from agricultural fields in the snow and animals of the Arctic.

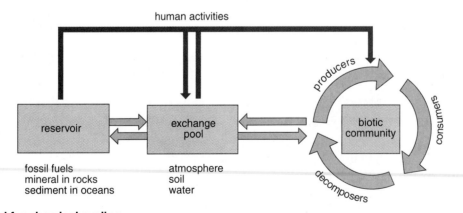

Figure 23.10   Model for chemical cycling.
Nutrients cycle between these components of ecosystems: Reservoirs such as fossil fuels, minerals in rocks, and sediments in oceans are normally unavailable sources, but pools such as those in the atmosphere, soil, and water are available sources of chemicals for the biotic community. Human activities remove chemical nutrients from reservoirs and pools and make them available to the biotic community, and the result can be pollution.

The Water Cycle

The **water (hydrologic) cycle** is described in Figure 23.11. Fresh water is distilled from salt water. The sun's rays cause fresh water to evaporate from seawater, and the salts are left behind. Vaporized fresh water rises into the atmosphere, cools, and falls as rain over the oceans and the land.

Water evaporates from land and from plants (evaporation from plants is called transpiration). It also evaporates from bodies of fresh water, but since land lies above sea level, gravity eventually returns all fresh water to the sea. In the meantime, water is contained within standing waters (lakes and ponds), flowing water (streams and rivers), and groundwater.

When rain falls, some of the water sinks or percolates into the ground and saturates the earth to a certain level. The top of the saturation zone is called the groundwater table, or simply, the water table. Sometimes groundwater is also located in **aquifers,** rock layers that contain water and will release it in appreciable quantities to wells or springs. Aquifers are recharged when rainfall and melted snow percolate into the soil. In some parts of the country, especially arid areas and southern Florida, withdrawals from aquifers exceed any possibility of recharge. This is called "groundwater mining." In these locations the groundwater is dropping, and residents may run out of groundwater, at least for irrigation purposes, within a few short years. Fresh water, which makes up only about 3% of the world's supply of water, is called a renewable resource because a new supply is always being produced. But it is possible to run out of fresh water when the available supply is not adequate and/or is polluted so that it is not usable.

In the water cycle, fresh water evaporates from the bodies of water. Water that falls on land enters the ground, surface waters, or aquifers. Water ultimately returns to the ocean—even the quantity that remains in aquifers for some time.

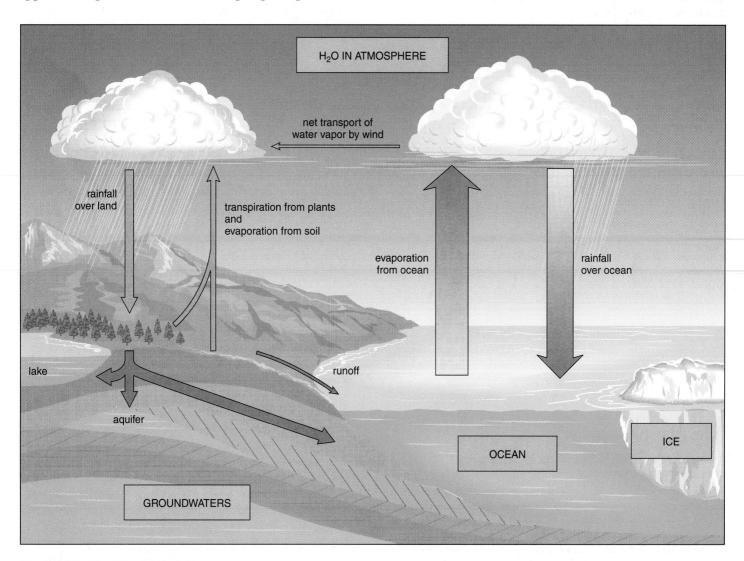

Figure 23.11   The water cycle.
Evaporation from the ocean exceeds rainfall, so there is a net movement of water vapor onto land where rainfall results in surface water and groundwater that flow back to the sea. On land, transpiration by plants contributes to evaporation. The width of the arrows indicates the relative quantities of water involved.

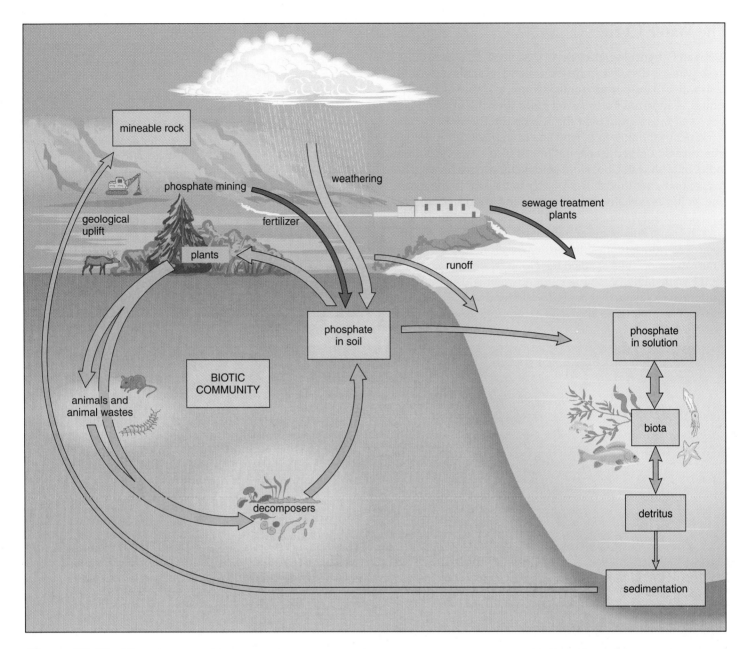

Figure 23.12   Phosphorus cycle.
Weathering of rocks releases phosphate, which enters producers and then cycles through organisms. Phosphorus which becomes a part of oceanic sediments is lost to the system for many years. The biotic community plus the reservoir and exchange pool are identified by an orange box. The width of the arrows signifies the various transfer rates. Phosphate mining is done by humans for the production of fertilizer. Fertilizer runoff and discharges from sewage treatment plants enrich bodies of water.

The Phosphorus Cycle

On land, the weathering of rocks makes phosphate ions (PO_4^{3-} and HPO_4^{2-}) available to plants, which take phosphate up from the soil (Fig. 23.12). Some of this phosphate runs off into aquatic ecosystems, where algae acquire phosphate from the water before it becomes trapped in sediments. Phosphate in sediments becomes available only when a geological upheaval exposes sedimentary rocks to weathering once more. Phosphorus does not enter the atmosphere; therefore, the phosphorus cycle is called a sedimentary cycle.

Producers use phosphate in a variety of molecules, including phospholipids, ATP, and the nucleotides that become a part of DNA and RNA. Animals eat producers and incorporate some of the phosphate into teeth, bones, and shells that take many years to decompose. Death and decay of all organisms and also decomposition of animal wastes do, however, make phosphate ions available to producers once again. Phosphate is usually a limiting inorganic nutrient in most ecosystems because the available amount has already been taken up by organisms.

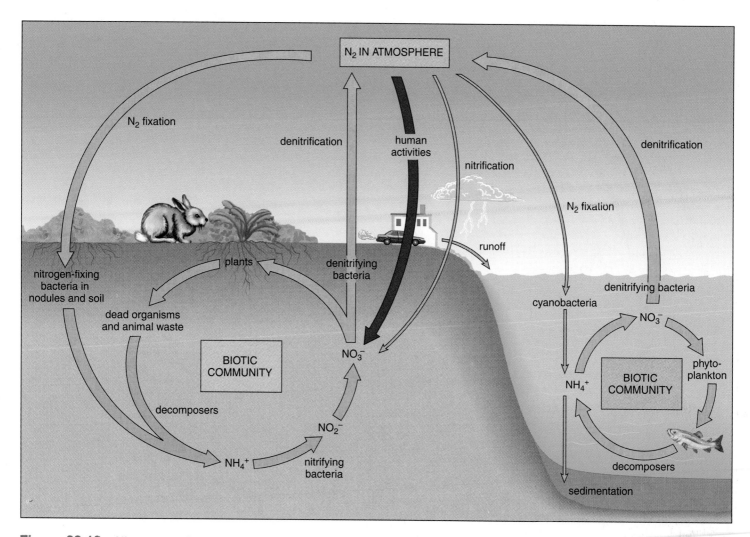

Figure 23.13   Nitrogen cycle.
The biotic community on land and in the sea, plus the reservoir and exchange pool, are identified by an orange box. The width of the arrows signifies the various transfer rates. Nitrogen fixation by bacteria in nodules and soil just about balances that returned to the atmosphere by denitrification. The production of fertilizer by humans removes much atmospheric nitrogen and this eventually enriches bodies of water due to runoff.

Humans Alter Transfer Rates

Human beings boost the supply of phosphate by mining phosphate ores for fertilizer production, animal feed supplements, and detergents. Animal wastes from livestock feedlots, fertilizers from cropland, and the discharge of untreated and treated municipal sewage all add excess phosphate to nearby waters. Then, when algae grow in excess, a condition called an algal bloom occurs.

In the phosphorus cycle, weathering makes phosphate available to producers, followed by consumers. Death and decay of all organisms make phosphate available to producers once again. Phosphate trapped in sedimentary rock becomes available only following a geological upheaval.

The Nitrogen Cycle

Nitrogen is an abundant element in the atmosphere. Nitrogen gas (N_2) makes up about 78% of the atmosphere by volume, yet nitrogen deficiency commonly limits plant growth. Plants cannot make use of nitrogen gas, and therefore, they depend on various types of bacteria that are able to take up nitrogen gas from the atmosphere (Fig. 23.13).

Nitrogen Gas Becomes Fixed

Nitrogen fixation occurs when nitrogen gas (N_2) is reduced and added to organic compounds, which can be utilized by plants. Some cyanobacteria in aquatic ecosystems and some free-living bacteria in soil are able to reduce nitrogen gas to ammonium (NH_4^+). Other nitrogen-fixing bacteria live in nodules on the roots of legumes (Fig. 23.13). They make reduced nitrogen and organic compounds available to the host plant.

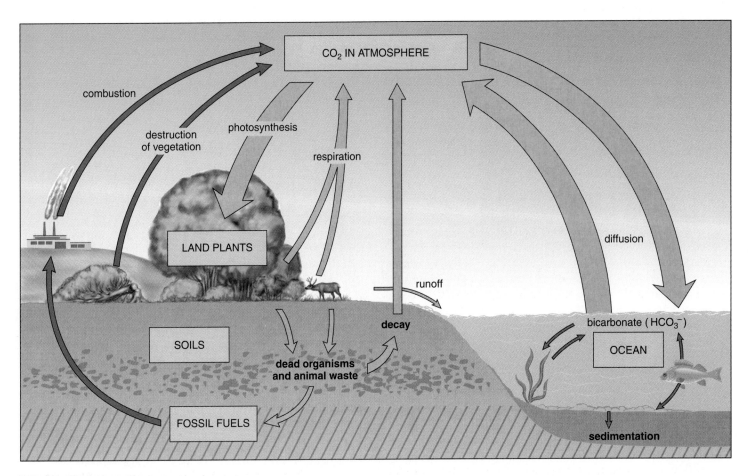

Figure 23.14   Carbon cycle.
The transfer rate of carbon dioxide into the atmosphere due to respiration just about matches the rate due to withdrawal by plants, as indicated by the width of the arrows. The burning of fossil fuels and trees has been adding carbon dioxide to the atmosphere. The exchange pool and/or reservoirs are identified by an orange box. The width of the arrows signifies the various transfer rates.

Plants take up both NH_4^+ and nitrate (NO_3^-) from the soil. After plants take up NO_3^- from the soil, it is enzymatically reduced to NH_4^+ and is used to produce amino acids and nucleic acids.

Nitrogen Gas Becomes Nitrates

Nitrification is the production of nitrates. Nitrogen gas (N_2) is converted to nitrate (NO_3^-) in the atmosphere when cosmic radiation, meteor trails, and lightning provide the high energy needed for nitrogen to react with oxygen. Also, humans make a most significant contribution to the nitrogen cycle when they convert nitrogen gas to ammonium and urea for use in fertilizers.

Ammonium (NH_4^+) in the soil is converted to nitrate by certain soil bacteria in a two-step process. First, nitrite-producing bacteria convert ammonium to nitrite (NO_2^-), and then nitrate-producing bacteria convert nitrite to nitrate. These two groups of bacteria are called the nitrifying bacteria.

Notice the subcycle in the nitrogen cycle that involves only ammonium, nitrites, and nitrates. This subcycle does not depend on the presence of nitrogen gas at all (see Fig. 23.13).

Denitrification is the conversion of nitrate to nitrous oxide and nitrogen gas, which enters the atmosphere. There are denitrifying bacteria that act under anaerobic conditions in both aquatic and terrestrial ecosystems. In the nitrogen cycle, denitrification counterbalanced nitrogen fixation until humans started making fertilizer, which requires nitrogen fixation. Now, excess nitrates run off into bodies of fresh water and cause water pollution.

In the nitrogen cycle, nitrogen-fixing bacteria (in nodules and in the soil) reduce nitrogen gas; nitrifying bacteria convert ammonium to nitrate; denitrifying bacteria convert nitrate back to nitrogen gas.

The Carbon Cycle

The relationship between photosynthesis and aerobic cellular respiration should be kept in mind when discussing the carbon cycle. Recall that this equation in the forward direction represents aerobic cellular respiration, and in the reverse direction, for simplicity's sake, is used to represent photosynthesis.

$$C_6H_{12}O_6 + 6O_2 \underset{\text{photosynthesis}}{\overset{\text{aerobic cellular respiration}}{\rightleftharpoons}} 6CO_2 + 6H_2O$$

The equation tells us that aerobic cellular respiration releases carbon dioxide, the molecule needed for photosynthesis, and photosynthesis releases oxygen, the molecule needed for aerobic respiration. Animals are dependent on green organisms, not only to produce organic food and energy, but also for a supply of oxygen. However, since producers both photosynthesize and respire, they can function independently of the animal world.

In the carbon cycle, organisms in both terrestrial and aquatic ecosystems exchange carbon dioxide with the atmosphere (Fig. 23.14). On land, plants take up carbon dioxide from the air, and through photosynthesis, they incorporate carbon into food that is used by autotrophs and heterotrophs alike. When organisms respire, a portion of this carbon is returned to the atmosphere as carbon dioxide.

In aquatic ecosystems, the exchange of carbon dioxide with the atmosphere is indirect. Carbon dioxide from the air combines with water to produce bicarbonate ion (HCO_3^-), a source of carbon for algae that produce food for themselves and for heterotrophs. Similarly, when aquatic organisms respire, the carbon dioxide they give off becomes bicarbonate ion. The amount of bicarbonate in the water is in equilibrium with the amount of carbon dioxide in the air.

Reservoirs Hold Carbon

Living and dead organisms contain organic carbon and serve as one of the reservoirs for the carbon cycle. The world's biota, particularly trees, contains 800 billion tons of organic carbon, and an additional 1,000–3,000 billion metric tons are estimated to be held in the remains of plants and animals in the soil. Before decomposition can occur, some of these remains are subjected to physical processes that transform them into coal, oil, and natural gas. We call these materials the **fossil fuels.** Most of the fossil fuels were formed during the Carboniferous period, 286–360 million years ago, when an exceptionally large amount of organic matter was buried before decomposing. Another reservoir is the inorganic carbonate that accumulates in limestone and in calcium carbonate shells. Many marine organisms have calcium carbonate shells that remain in bottom sediments long after the organisms have died. Geological forces change these sediments into limestone.

Figure 23.15   Greenhouse effect.
Much carbon dioxide which enters the atmosphere comes from coal-burning power plants. Carbon dioxide is largely responsible for the greenhouse effect and possible global warming.

Humans Alter Transfer Rates

The activities of human beings have increased the amount of carbon dioxide and other gases in the atmosphere. Data from monitoring stations record an increase of 20 ppm (parts per million) in carbon dioxide in only 22 years. (This is equivalent to 42 billion metric tons of carbon.) This buildup is primarily attributed to the burning of fossil fuels and wood (Fig. 23.15). The oil-well fires that were set during the Persian Gulf War of 1991 added carbon dioxide to the atmosphere, but the ongoing burning of tropical rain forests is of even more concern. When we do away with forests, we reduce a reservoir that takes up excess carbon dioxide. At this time, the oceans are believed to be taking up most of the excess carbon dioxide; the burning of fossil fuels in the last 22 years has probably released 78 billion metric tons of carbon, yet the atmosphere registers an increase of "only" 42 billion metric tons.

There is much concern that an increased amount of carbon dioxide (and other gases) in the atmosphere is causing a global warming. These gases allow the sun's rays to pass through, but they absorb and reradiate heat back to the earth, a phenomenon called the greenhouse effect.

In the carbon cycle, photosynthesis removes, but respiration returns, carbon dioxide to the atmosphere. Forests and dead organisms are carbon reservoirs, as is the ocean.

23.3 Human-Impacted Ecosystems

Mature natural ecosystems are stable in the sense that they perpetuate themselves and require little, if any, additional materials each year. The many and varied populations are held in dynamic balance by the interactions between species, such as competition and predation. The amount of energy that flows through and the amount of matter that cycles is appropriate to support these populations. **Pollution,** defined as any undesirable change in the environment that can be harmful to humans and other life, does not normally occur. Human-impacted ecosystems, however, are quite different, as is shown in Figure 23.16.

Human-impacted ecosystems essentially have two added parts: the *country,* where agriculture and animal husbandry are found, and the *city,* where most people live and where industry is carried on. This representation of human-impacted ecosystems, although simplified, allows us to see that these systems require two major inputs: *fuel energy* and *raw materials* (e.g., metals, wood, synthetic materials). The use of these necessarily results in *waste* and *pollution* as outputs.

The Country

Modern U.S. agriculture produces exceptionally high yields per acre, but this bounty is dependent on a combination of the five variables given here.

1. **Planting of a few genetic varieties.** The majority of farmers specialize in growing only one variety of a crop. Wheat farmers plant the same type of wheat, and corn

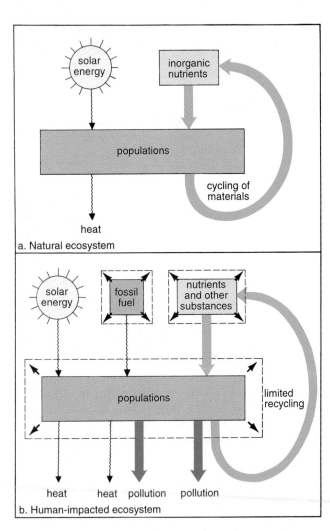

Figure 23.16   The natural ecosystem compared to the human-impacted ecosystem.

a. In natural ecosystems, the inputs and outputs are minimal. **b.** In the human-impacted ecosystem, the ever-increasing size of the human population and the concomitant use of more fossil fuels and other supplements cause an ever-greater amount of wastes, including pollutants.

Figure 23.17   The country.

Crops are tended with heavy farming equipment that operates on fossil fuel. High yields are dependent upon a generous supply of fertilizers, pesticides, herbicides, and water.

farmers plant the same type of corn (Fig. 23.17). This so-called monoculture agriculture is subject to attack by a single type of parasite. For example, a single parasitic mold reduced the 1970 corn crop by 15%, and the results could have been much worse because 80% of the nation's corn acreage was susceptible.

2. **Heavy use of fertilizers, pesticides, and herbicides.** Fertilizer production requires a large energy input, and fertilizer runoff contributes to water pollution. Pesticides reduce soil fertility because they kill off beneficial soil organisms as well as pests, and some pesticides, like alar, have been accused of increasing the long-term risk of cancer, particularly in children. Herbicides, especially those containing the contaminant dioxin, have been charged with causing adverse reproductive effects and cancer.

3. **Generous irrigation.** River waters sometimes are redirected for the purpose of irrigation, in which case "used water" returns to the river carrying a heavy concentration of salt. The salt content of the Rio Grande River in the Southwest is so high that the government has built a treatment plant to remove the salt. Water also is sometimes taken from aquifers (underground rivers), whose water content can be so reduced that it becomes too expensive to pump out more water. Farmers in Texas already are facing this situation.

4. **Excessive fuel consumption.** Energy is consumed on the farm for many purposes. Irrigation pumps already have been mentioned, but large farming machines also are used to spread fertilizers, pesticides, and herbicides, and to sow and harvest the crops. It is not incorrect to suggest that modern farming methods transform fossil fuel energy into food energy. Supplemental fossil fuel energy also contributes to animal husbandry yields. At least 50% of all cattle are kept in feedlots, where they are fed grain. Chickens are raised in a completely artificial environment, where the climate is controlled and each bird has its own cage to which food is delivered on a conveyor belt. Animals raised under these conditions often have antibiotics and hormones added to their feed to increase yield.

5. **Loss of land quality.** Evaporation of excess water on irrigated lands can result in a residue of salt. This process, termed salinization, makes the land unsuitable for the growth of crops. Between 25% and 35% of the irrigated western croplands are thought to have excessive salinity. Soil erosion is also a serious problem. It is said that we are mining the soil because farmers are not taking measures to prevent the loss of topsoil. The Department of Agriculture estimates that erosion is causing a steady drop in the productivity of land equivalent to the loss of 1.25 million acres per year. Even more fertilizers, pesticides, and energy supplements will be required to maintain present-day yield.

Figure 23.18   The city.
The city is dependent upon the country to supply it with food and other resources. The larger the city, the more resources required to support it.

The City

The city (Fig. 23.18) is dependent on the country to meet its needs. For example, each person in the city requires several acres of land for food production. Overcrowding in cities does not mean that less land is needed; each person still requires a certain amount of land to ensure survival. Unfortunately, however, as the population increases, the suburbs and the cities tend to encroach on agricultural areas and rangeland.

The city houses workers for both commercial businesses and industrial plants. Solar and other renewable types of energy rarely are used; cities currently rely mainly on fossil fuel in the form of coal and oil. The city does not conserve resources. An office building with continuously burning lights and windows that cannot be opened shows how energy is wasted. Another example is people who drive cars long distances instead of carpooling or taking public transportation and who drive short distances instead of walking or bicycling. Also, materials are not recycled, and products are designed for rapid replacement.

The burning of fossil fuels for transportation, commercial needs, and industrial processes causes air and water pollution. This pollution is compounded by the chemical and solid waste pollution that results from the manufacture of many products. Consider that any product used by the average consumer (house, car, washing machine) causes pollution and waste, both during its production and when it is discarded. Humans themselves produce much sewage that is discharged into bodies of water, often after only minimal treatment.

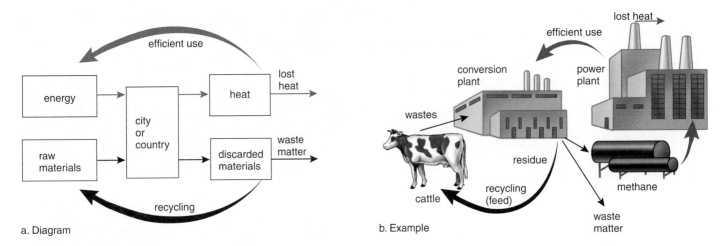

a. Diagram

b. Example

Figure 23.19   Recycling.

a. In order to cut down on the amount of lost heat and waste matter, heat could be used more efficiently, and discarded materials could be recycled. **b.** For example, instead of allowing cattle waste to enter a water supply, it could be sent to a conversion plant that produces methane gas. (The remaining residue could be converted into feed for cattle.) Excess heat, which arises from the burning of methane gas to produce electricity, could be cycled back to the conversion plant.

The Solution

In human-impacted ecosystems, fuel combustion by-products, sewage, fertilizers, pesticides, and solid wastes all are added to the environment in the hope that natural cycles will cleanse the biosphere of these pollutants. But we have exploited natural ecosystems to the extent that the environment is overloaded.

More and more natural ecosystems are impacted because an ever-increasing number of people want to maintain a standard of living that requires many goods and services. But we can call a halt to this spiraling process if we achieve zero population growth and if we conserve energy and raw materials. Conservation can be achieved in three ways: (1) wise use of only what is actually needed; (2) recycling of nonfuel minerals, such as iron, copper, lead, and aluminum; and (3) use of renewable energy resources and development of more efficient ways to utilize all forms of energy (Fig. 23.19a).

Some farmers have adopted organic farming methods and do not apply fertilizers, pesticides, or herbicides to their crops. They use cultivation of row crops to control weeds, crop rotation to combat major pests, and the growth of legumes to supply nitrogen fertility to the soil. Some farmers use natural predators and parasites instead of pesticides to control insects. Most of these farmers switched farming methods because they were concerned about the health of their families and livestock and had found that the chemicals were sometimes ineffective. A study of about 40 farms showed that organic farming, for the most part, was just as profitable as conventional farming. Crop yields were lower, but so were operating costs. Organic farms required about two-fifths as much fossil energy to produce one dollar's worth of crop. The method of plowing and the utilization of crop rotation resulted in one-third less soil erosion. The researchers concluded it would be well to determine how far farmers can move in the direction of reduced agricultural chemical use and still maintain the quality of the product. They noted that a modest application of fertilizer would have improved the protein content of the crop.

As another example, consider a plant that was built in Lamar, Colorado, which produces methane from feedlot animals' wastes (Fig. 23.19b). The methane is burned in the city's electrical power plant, and the heat given off is used to incubate the anaerobic digestion process that produces the methane. In addition, a protein feed supplement is produced from the residue of the digestion process. This system represents a cyclical use of material and an efficient use of energy similar to that found in nature. Many other such processes for achieving this end have been, and could be, devised.

It is important to consider that ecosystems should be developed and managed with care so that their integral nature is preserved. This is often called "working with nature" rather than against nature. This principle is explored in the ecology reading on the next page.

Ecology Focus

Preservation of the Everglades

Originally, the Everglades encompassed the whole of southern Florida from Lake Okeechobee down to Florida Bay (Fig. 23A). Now, largely in the Everglades National Park alone do we find the vast saw-grass prairie, interrupted occasionally by a cypress dome or hardwood tree island. Within these islands, both temperate and tropical evergreen trees grow among the dense and tangled vegetation. Mangrove, or salt-tolerant trees, are found along sloughs (creeks) and at the shoreline. Only the roots of the red mangrove can tolerate the sea constantly. The prop roots of this tree protect over 40 different types of juvenile fishes as they grow to maturity. During the wet season, from May to November, animals are dispersed throughout the region, but in the dry season, from December to April, they congregate wherever pools of water are found. Alligators are famous for making "gator holes" where water collects, and fish, shrimps, crabs, birds, and a host of living things survive until the rains come again. Almost everyone is captivated by the birds that find the ready supply of fish they need for daily existence at these holes. The large and beautiful herons, egrets, roseate spoonbill, and anhinga are awesome. These birds once numbered in the millions; now they number only in the thousands. Why is this?

At the turn of the century, settlers began to drain the land just south of Lake Okeechobee to grow crops on the soil enriched by partially decomposed saw grass. The large dike that now rings the lake prevents the water from taking its usual course: over the banks of Lake Okeechobee and slowly southward. In times of flooding, water can be shunted through the St. Lucie Canal to the Atlantic Ocean or through the canalized Caloosahatchee River to the Gulf of Mexico. In times of drought, water is contained not only in the lake but also in three so-called conservation areas established to the south of the lake. Water must be conserved for the irrigation of the farmland and to recharge the Biscayne aquifer (underground river), which supplies drinking water for the cities on the east coast of Florida. Containing and moving the water from place to place has required the construction of over 2,250 kilometers of canals, 125 water control stations, and 18 large pumping stations. Now the Everglades National Park re-

ceives water only when it is discharged artificially from a conservation area. This disruption of the natural flow of water has affected the reproduction pattern of the birds, which is attuned to the natural wet-dry season turnover.

It took considerable human effort and a huge financial investment to control nature and to establish the Everglades Agricultural Area. Has this attempt to bend nature to human will been worthwhile? The area does, in fact, produce more sugar than Hawaii and a large proportion of the vegetables consumed in the United States each winter. But this has not been without a price. The rich soil, built up over thousands of years, is disappearing and most likely will be unable to sustain conventional agriculture after the year 2000. It has been suggested that at that time we might use the Everglades Agricultural Area for the growth of aquatic plants. Perhaps it should have been decided in the beginning to work with nature by growing aquatic plants instead of conventional plants. Then all the canals and pumping stations would have been unnecessary, the water would still flow from Lake Okeechobee to the Everglades as it had for eons, and the birds today would still number in the millions.

Figure 23A   The Everglades.
The Everglades once extended from Lake Okeechobee south to Florida Bay. Now it encompasses only three conservation areas and the Everglades National Park.

Bioethical Issue

Peter Jutro, a scientist working for the U.S. Environmental Protection Agency, wants to research native traditions for clues on how to preserve ecosystems. He has run into opposition from the indigenous groups because they mistrust conservationists. Take, as an example, the fact that the Kuna people in Panama refused to renew the lease for a Smithsonian Institution's study of reef ecology for the past 21 years. Much of the trouble seems to have come from the failure of scientists to explain their program to local communities. In a meeting of the Kuna congress, the scientists were accused by the Kuna of "stealing their knowledge, stealing their reefs, stealing their sand." Local people find it hard to see a difference between a scientific study and commercial ventures which exploit their areas for minerals, timber, and other resources.

Laura Snook, a forester from Duke University, researches the growing habits of mahogany trees in Mexico. She came to the conclusion that because the trees grow slowly, logging shouldn't be done too fast. Most local foresters resented her findings, but she was able to establish a good relationship with two women foresters who were open to her ideas. Now local people are changing the way they replant mahogany trees so that the resource will be there for some time to come. The point is that conservationists working with people from different cultural backgrounds probably need to communicate their goals more clearly and involve local people in project planning. Perhaps they should learn to accept slower timescales and styles of decision-making different from our own.

Questions

1. Should U.S. scientists be studying ecosystems in such far-flung places as Panama, Alaska, Mexico, and the Amazon? Why or why not?
2. Should we insert political correctness into negotiations with local groups in order to bring about conservation? Why or why not?
3. How far should a scientist go to establish communication with local groups in order to preserve the environment in other countries? Explain.

Summarizing the Concepts

23.1 The Nature of Ecosystems

The process of succession from either bare rock or disturbed land results in a climax community. An ecosystem is a community of organisms plus the physical environment. Each population in an ecosystem has a habitat and niche. Some populations are producers and some are consumers. Producers are autotrophs that produce their own food. Consumers are heterotrophs that take in preformed food. Consumers may be herbivores, carnivores, omnivores, or decomposers.

Energy flows through an ecosystem. Photosynthesizers transform solar energy into food (organic nutrients) for themselves and all heterotrophs. As herbivores feed on plants for algae, and carnivores feed on herbivores, some energy is converted to heat. Feces, urine, and dead bodies become food for decomposers. Eventually all the solar energy that enters an ecosystem is converted to heat, and thus ecosystems require a continual supply of solar energy.

Chemicals are not lost from the biosphere as is energy. They recycle within and between ecosystems. Decomposers return some proportion of inorganic nutrients to autotrophs, and other portions are imported or exported between ecosystems in global cycles.

Ecosystems contain food webs, and a diagram of a food web shows how the various organisms are connected by eating relationships. Grazing food chains begin with vegetation that is fed on by a herbivore, which becomes food for a carnivore, and so forth. In detrital food chains, a decomposer acts on organic material in the soil, and when it is fed on by a carnivore, the two food webs are joined. A trophic level is all the organisms that feed at a particular link in food chains.

Ecological pyramids show trophic levels stacked one on the other like building blocks. Generally they show that biomass and energy content decrease from one trophic level to the next. Most pyramids pertain to grazing food webs and largely ignore the detrital food web portion of an ecosystem.

23.2 Global Biogeochemical Cycles

Biogeochemical cycles contain reservoirs, components of ecosystems like fossil fuels, sediments, and rocks that contain elements available on a limited basis to living things. Pools are components of ecosystems like the atmosphere, soil, and water—which are ready sources of nutrients for living things. Nutrients cycle among the members of the biotic component of an ecosystem.

In the water cycle, evaporation over the ocean is not compensated by rainfall. Evaporation from terrestrial ecosystems includes transpiration from plants. Rainfall over land results in bodies of fresh water plus groundwater, including aquifers. Eventually all water returns to the oceans.

In the phosphorus cycle, the reservoir is sediments at the bottom of the ocean. Only upheavals make sedimentary rock available on land to plants.

In the nitrogen cycle, the reservoir is the atmosphere, and nitrogen (N_2) gas must be converted to nitrate (NO_3^-) for use by producers. Nitrogen-fixing bacteria, particularly in root nodules, make organic nitrogen available to plants. Other bacteria active in the nitrogen cycle are the nitrifying bacteria, which convert ammonium to nitrate, and the denitrifying bacteria, which reconvert nitrate to N_2.

In the carbon cycle, the reservoir is organic matter, calcium carbonate shells, and limestone. The exchange pool is the atmosphere: photosynthesis removes carbon dioxide, and respiration and combustion add carbon dioxide.

23.3 Human-Impacted Ecosystems

If undisturbed, mature, natural ecosystems remain in dynamic balance and need the same amount of energy each year. Additional nutrient inputs are minimal because chemicals cycle within and between ecosystems. In human-impacted ecosystems, the human population size constantly increases. In the country, farmers plant only certain high-yield varieties of plants, which require such supplements as fertilizers, pesticides, and water. In the city, the populace is wasteful of energy and materials. Therefore, there is much pollution. We and future generations must find ways to conserve energy, use excess heat, recycle materials, and reduce pollution.

Studying the Concepts

1. What is succession, and how does it result in a climax community? 476

2. Distinguish between abiotic and biotic components of an ecosystem. What are the aspects of niche for a plant? An animal? 478

3. Name four different types of consumers found in natural ecosystems. 478

4. Tell why energy must flow but chemicals can cycle in an ecosystem. 478–79

5. Describe two types of food webs and two types of food chains typically found in terrestrial ecosystems. Which of these typically moves more energy through an ecosystem? 481

6. What is a trophic level? An ecological pyramid? 481–82

7. Give examples of reservoirs and pools in biogeochemical cycles. Which of these is less accessible to biotic communities? 482

8. Draw a diagram to illustrate the water, phosphorus, nitrogen, and carbon biogeochemical cycles, and include in your diagram the manner in which humans disturb the equilibrium of these cycles. 483–87

9. Compare and contrast the basic characteristics of a natural ecosystem to the characteristics of the human-impacted ecosystem. 488–89

10. Suggest ways by which it would be possible to make the human-impacted ecosystem more cyclical so pollution might be better controlled. 490

Testing Your Knowledge of the Concepts

In questions 1–4, match the population to the description.

a. producer
b. consumer
c. decomposer
d. herbivore

_____ 1. Heterotroph that feeds on plant material.

_____ 2. Autotroph that manufactures organic nutrients.

_____ 3. Any type of heterotroph that feeds on plant material or feeds on other animals.

_____ 4. Heterotroph that breaks down detritus as a source of nutrients.

In questions 5–7, indicate whether the statement is true (T) or false (F).

_____ 5. As organic nutrients are converted to inorganic nutrients, energy flow occurs in an ecosystem.

_____ 6. Human-impacted ecosystems are characterized by an ever-increasing use of material and energy.

_____ 7. Trophic levels decrease in biomass and energy content as indicated by an ecological pyramid.

In questions 8–9, fill in the blanks.

8. During the process of denitrification, nitrate is converted to

_____ .

9. In the carbon cycle, when living organisms _____ , carbon dioxide (CO_2) is returned to the exchange pool.

10. Label this diagram of an ecosystem.

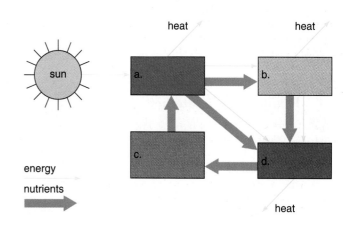

Applying Your Knowledge to the Concepts

These questions pertain to ecosystems.

1. A group of concerned citizens found out that a plant growing wild in a rather large natural area was being used as a stimulant by some of the high-school students. They proposed a systematic elimination of all these plants. The local Audubon group opposed this action, saying that it could affect the entire ecosystem. Explain the reasoning behind the Audubon's statement.

2. Pigs are fed corn. Use this information to suggest why pork costs much more per pound than corn.

3. Indicate the reasons that replacing a forest or grassland area with a factory and parking lot increases the amount of carbon dioxide in the atmosphere.

4. What is the reason(s) that introduction of an exotic species into an ecosystem often results in the new species overrunning or displacing the native species?

Understanding the Terms

aquifer 483
biogeochemical cycle 482
biosphere 476
carnivore 478
community 476
consumer 478
decomposer 478
detrital food web 481
detritus 478
ecological pyramid 481
ecology 476
ecosystem 475
food chain 481
food web 481

fossil fuel 487
grazing food chain 481
grazing food web 481
habitat 478
herbivore 478
niche 478
nitrogen fixation 485
omnivore 478
pollution 488
population 476
producer 478
succession 476
trophic level 481
water (hydrologic) cycle 483

Match the terms to these definitions.

a. _____ Animal that feeds on both plants and animals.

b. _____ Complete set of food links between populations in a community.

c. _____ Series of ecological stages by which the community in a particular area gradually changes until a community persists.

d. _____ Pictorial graph representing the biomass, organism number, or energy content of each trophic level in a food web, from the producer to the final consumer population.

e. _____ Nonliving organic matter.

Applying Technology to the Concepts

Your study of ecosystems is supported by these available technologies:

Essential Study Partner CD-ROM

Ecology → Ecosystems

Visit the Mader web site for related ESP activities.

Exploring the Internet

The Mader Home Page provides resources and tools as you study this chapter.

http://www.mhhe.com/biosci/genbio/mader

Life Science Animations 3D Video

42 Nutrient Cycling

Chapter 24

Population Concerns

Chapter Concepts

Figure 24.1   The American family.
Should young couples think about how their children will contribute to population statistics and environmental stress before having a family?

Linda L. is worried. At age 40, she expected to have made it. She expected to be successful. Comfortable. Peaceful. Instead, Linda—like millions of other Americans—worries about two major things: her kids, who will soon be entering high school, and her parents, who may soon need live-in help.

Linda is caught in the aftermath of the baby boom, a period of time in the 1950s when the American birthrate skyrocketed (Fig. 24.1). When the so-called baby boomers were in their 20s, they were young and vibrant, and they added to American productivity. Today, however, the boomers are aging. Their children worry that elderly boomers will drain resources—not to mention national attention—from younger adults.

Suddenly, Linda fears that nursing homes, Alzheimer disease, and other topics involving the aged will swamp political agendas and public issues in the near future. Meanwhile, adults her age often find themselves taking care of both their children and parents at the same time.

Linda's predicament shows that population statistics should interest all of us because they can impact our lives. By international standards, the U.S. population age shift is not devastating. Far worse population concerns have long haunted some countries. In overpopulated China, for example, couples are punished for conceiving more than one

child. Other less-developed countries of the world are experiencing marked population increases that lead to environmental consequences for us all. A reduction in the size of the tropical rain forests is expected to reduce biodiversity and contribute to possible global warming.

24.1 Human Population Growth

The human population growth curve is a **J**-shaped growth curve (Fig. 24.2). In the beginning, growth of the human population was relatively slow, but as more reproducing individuals were added, growth increased, until the curve began to slope steeply upward. It is apparent from the position of 1995 on the growth curve in Figure 24.2 that growth is quite rapid now. The world population increases at least the equivalent of a medium-sized city (200,000) every day and the equivalent of a country the size of Germany (80 million) every year. These startling figures are a reflection of the fact that a very large world population is undergoing exponential growth.

Exponential Growth

Mathematically speaking, **exponential growth,** or geometric increase, occurs in the same manner as compound interest; that is, the percentage increase is added to the principal before the next increase is calculated. With regard to populations, the percentage increase is termed the **growth rate,** which is applied per year. Because of exponential growth, each year's increase will be greater than the previous year's increase since the population size has increased in the meantime. In fact, the world's population grows by a larger amount each year even when the growth rate decreases slightly. The increase in size is dramatically large because the world population is very large.

The Growth Rate

The growth rate of a population is determined by considering the difference between the number of persons born per year (birthrate, or natality) and the number of persons who

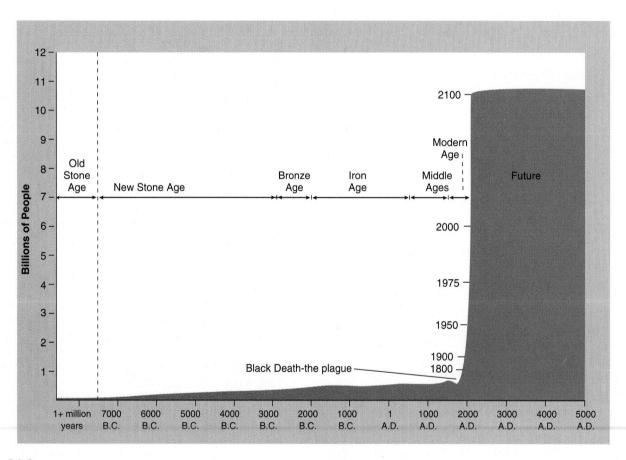

Figure 24.2   **Growth curve for human population.**
The human population is now undergoing rapid exponential growth. Since the growth rate is declining, it is predicted that the population size may level off at 8 billion or increase to as much as 11 billion by 2100, depending on the speed with which the growth rate declines.
Source: Population Reference Bureau and United Nations, World Population Projections to 2100 (1998).

die per year (death rate, or mortality). It is customary to record these rates per 1,000 persons. For example, Canada at the present time has a birthrate of 13 per 1,000 per year, but it has a death rate of 6 per 1,000 per year. This means that Canada's population growth, or simply its growth rate, is

$$\frac{13 - 6}{1,000} = \frac{7}{1,000} = 0.007 \times 100 = 0.7\%$$

Notice that while birthrate and death rate are expressed in terms of 1,000 persons, the growth rate is expressed per 100 persons, or as a percentage.

After 1750, the world population growth rate steadily increased, until it peaked at 2% in 1965. It has fallen since then to 1.5%. Yet, there is an ever-greater increase in the world population each year because of exponential growth. The explosive potential of the present world population can be appreciated by considering the doubling time.

The Doubling Time

The **doubling time** (*d*)—the length of time it takes for the population size to double—can be calculated by dividing 70 (the demographic constant) by the growth rate (*gr*):

$$d = 70/gr$$

d = Doubling time

gr = Growth rate

70 = Demographic constant

If the present world growth rate of 1.5% continues, the world population will double in 47 years.

$$d = 70/1.5 = 47 \text{ years}$$

This means that in 47 years, the world will need double the amount of food, jobs, water, energy, and so on to maintain the same standard of living.

It is of grave concern to many that the amount of time needed to add each additional billion persons to the world population has taken less and less time. The first billion didn't occur until 1800; the second billion arrived in 1930; the third billion in 1960; and today there are 5.8 billion. Only if the growth rate continues to decline can there be zero population growth when births equal deaths and the population size remains steady. Figure 24.2 shows the population may level off at 8, 10.5, or 14.2 billion, depending on the speed with which the growth rate declines.

The Carrying Capacity

The growth curve for many nonhuman populations is **S**-shaped—the population tends to level off at a certain size. Figure 24.3 illustrates an example based on actual data for the growth of a fruit fly population reared in a culture bottle. In the beginning, the numbers were small, and population growth was minimal. Then the flies began to

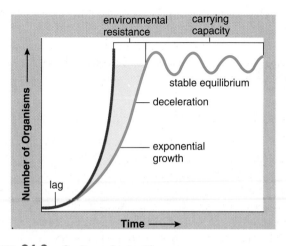

Figure 24.3   **S-shaped growth curve.**
In nature, populations initially grow exponentially; but instead of following a J-shaped growth curve, they follow an S-shaped curve because of environmental resistance. Population size enters a steady state when the carrying capacity is reached.

multiply rapidly. Notice that the curve began to rise dramatically, just as the human population curve does now. At this time, a population is demonstrating its biotic potential. **Biotic potential** is the maximum growth rate under ideal conditions. It usually is not demonstrated for long because of an opposing force called environmental resistance. **Environmental resistance** includes all the factors that cause early death of organisms and therefore prevent the population from producing as many offspring as it might otherwise. In the fruit fly example, we can speculate that environmental resistance included the limiting factors of food and space. Fruit fly wastes also may have limited the population size. When environmental resistance sets in, biotic potential is overcome, and the slope of the growth curve begins to decline. This is the inflection point of the curve.

The eventual size of any population represents a compromise between biotic potential and environmental resistance. This compromise occurs at the carrying capacity of the environment. The **carrying capacity** is the maximum population that the environment can support for an indefinite period. The carrying capacity of the earth for humans has not been determined. Some authorities think the earth is potentially capable of supporting 50–100 billion people. Others think we already have more humans than the earth can adequately support.

Exponential growth is usually curtailed by environmental resistance. Whether this will hold true for the human population remains to be seen.

a.

b.

Figure 24.4   More-developed countries (MDCs) versus less-developed countries (LDCs).
a. In the MDCs, most people enjoy a high standard of living. **b.** In the LDCs, the majority of people are poor and have few amenities.

The More-Developed Versus Less-Developed Countries

The countries of the world today can be divided into two groups. The more-developed countries (MDCs), typified by countries in North America and Europe, are those in which population growth is under control and the people enjoy a good standard of living. The less-developed countries (LDCs), typified by a few countries in Latin America, Africa, and Asia, are those in which population growth is out of control and the majority of people live in poverty (Fig. 24.4). (Sometimes the term *third-world countries* is used to mean the less-developed countries. This term was introduced by those who thought of the United States and Europe as the first world and the former USSR as the second world.)

Presently, population growth in the *more-developed countries* (MDCs) is minimal, while population growth in the *less-developed countries* (LDCs) is increasing rapidly. The MDCs doubled their populations between 1850 and 1950. This was largely due to a decline in the death rate, the development of modern medicine, and improved socioeconomic conditions. The decline in the death rate was followed shortly thereafter by a decline in the birthrate, so that populations in the MDCs experienced only modest growth between 1950 and 1975. This sequence of events (i.e., decreased death rate followed by decreased birthrate) is termed a **demographic transition.**

The growth rate for the MDCs is now about 0.1%. The populations of a few of the MDCs—Italy, Denmark, Hungary, Sweden—are not growing or are actually decreasing in size. In contrast, there is no leveling off and no end in sight to U.S. population growth. Although the United States has a growth rate of 0.6%, many people immigrate to the United States each year. In addition, a baby boom between 1947 and 1964 means that a large number of women are still of reproductive age.

Although the death rate began to decline steeply in the LDCs following World War II with the importation of modern medicine from the MDCs, the birthrate remained high. The growth rate of the LDCs peaked at 2.5% between 1960 and 1965. Since that time, a demographic transition has begun: the decline in the death rate has slowed and the birthrate has fallen. A growth rate of 1.8% is expected by the end of the century. Still, because of exponential growth, discussed previously, the population of the LDCs may explode from 4.6 billion today to 10.2 billion in 2100. Most of this growth will occur in Africa, Asia, and Latin America.

Comparing Age Structure

The LDCs are experiencing a population momentum because they have more women entering the reproductive years than older women leaving them. Populations have

three age groups: dependency, reproductive, and postre-productive. One way of characterizing population is by these age groups. This is best visualized when the proportion of individuals in each group is plotted on a bar graph, thereby producing an age-structure diagram (Fig. 24.5).

Laypeople are sometimes under the impression that if each couple has two children, zero population growth will take place immediately. However, **replacement reproduction,** as it is called, will still cause most countries today to continue growing due to the age structure of the population. If there are more young women entering the reproductive years than there are older women leaving them, then replacement reproduction will still give a positive growth rate.

Many MDCs have a stabilized age-structure diagram, but most LDCs have a youthful profile—a large proportion of the population is younger than the age of 15. Since there are so many young women entering the reproductive years, the population will still expand greatly, even after replacement reproduction is attained. The more quickly replacement reproduction is achieved, however, the sooner zero population growth will result.

Age-structure diagrams can be used to predict future population growth.

Possible Solutions

It will be almost impossible to reduce poverty and bring about sustainable development to all of Africa, Asia, and Latin America if more than 6 billion people are added to this area within the next century. However, ways to greatly reduce the expected increase have been suggested.

1. Establish and/or strengthen family planning programs. A decline in growth rate is seen in countries with good family planning programs supported by community leaders. Presently, 25% of women in sub-Saharan Africa say they would like to delay or stop childbearing, yet they are not practicing birth control; likewise, 15% of women in Asia and Latin America have an unmet need of birth control. It is estimated that accessibility to family planning could lower the expected population increase of LDCs to 8.3 billion instead of 10.2 billion by 2100.

2. Use social progress to reduce the desire for large families. Many couples in the LDCs presently desire as many as four to six children. But providing available education, raising the status of women, and reducing child mortality are desirable social improvements that could cause them to think differently. It is estimated that such improvements could further reduce the expected population increase in the LDCs to 7.3 billion by the year 2100.

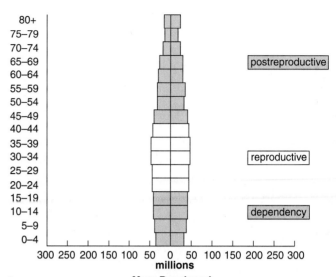

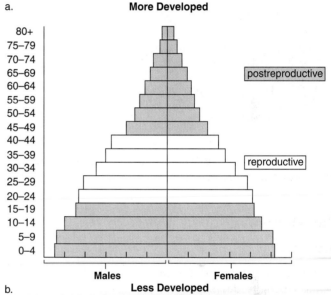

Figure 24.5 **Age-structure diagrams (1998).**
The diagrams illustrate that **(a)** the populations of MDCs are approaching stabilization, whereas **(b)** the populations of LDCs will expand rapidly due to the shape of their age-structure diagram.
Source: United Nations Population Division, 1998.

3. Delay the onset of childbearing. A delay in the onset of childbearing and wider spacing of births could reduce the present population growth rate. If childbearing begins five years later than usual, a further reduction of 1.2 billion births might be possible, so that 6.1 rather 10.2 billion persons would result in the LDCs by the year 2100.

Although the less-developed countries are now undergoing demographic transition and the growth rate is declining, they are still expanding dramatically because of exponential growth.

24.2 The Human Population and Pollution

As the human population increases in size, more and more ecosystems are impacted by humans. Ecosystems are then no longer able to process and rid the biosphere of wastes which accumulated and are called pollutants. Pollutants are substances added to the environment, particularly by human activities, that lead to undesirable effects for living things, including humans. Human activities and pollutants are having detrimental effects in the air, in the water, and on land.

Global Climate Change

The earth's climate has fluctuated in the past. We all know that ice ages have occurred in the history of the earth. Presently we are enjoying a moderate temperature that the earth has not seen since about 130,000 years ago. But many are concerned that the global climate will continue to warm and at a rate ten times faster than any time in the past. Let's examine the reasons why global warming is expected.

In 1850, atmospheric carbon dioxide was about 280 parts per million (ppm) and today it is about 350 ppm. This increase is largely due to the burning of fossil fuels and the burning and clearing of forests to make way for farmland and pasture. The oceans are currently taking up about one-half of the carbon dioxide emitted, or else the increase would be much higher than this stated amount. The emission of other gases due to human activities is also taking place. The amount of methane given off by oil and gas wells, rice paddies, and all sorts of organisms including domesticated cows is increasing by about 1% a year. Altogether the following pollutants are expected to contribute to global warming:

Gas	From
Carbon dioxide (CO_2)	Fossil fuel and wood burning
Nitrous oxide (N_2O)	Fertilizer use and animal wastes
Methane (CH_4)	Biogas (bacterial decomposition, particularly in the guts of animals, in sediments, and in flooded rice paddies)
Chlorofluorocarbons (CFCs)	Freon, a refrigerant
Halons (halocarbons, $C_xF_xBr_x$)	Fire extinguishers
Ozone	Photochemical smog in troposphere

These gases are known as greenhouse gases because just like the panes of a greenhouse they allow solar radiation to pass through but hinder the escape of infrared heat back into space. Figure 24.6*a* shows the earth's radiation and energy balances. One thing to be learned from this diagram is that water vapor is a greenhouse gas: clouds also reradiate heat back to earth. If the earth's temperature rises due to the **greenhouse effect,** more water will evaporate, forming more clouds, setting up a positive feedback effect that could exacerbate global warming.

The greenhouse gases differ in their ability to absorb specific wavelengths of infrared radiation. Atmospheric carbon dioxide may already be absorbing as much infrared radiation as it can. This means that adding a molecule of one of the other greenhouse gases will have more of an effect than adding a molecule of carbon dioxide. The greenhouse effectiveness of methane is now about 25 times that of carbon dioxide. Nitrous oxide is about 200 times more effective than carbon dioxide.

Today data collected around the world show a steady rise in the concentration of the various greenhouse gases. These data are used to generate computer models that predict the earth may warm to temperatures never before experienced by living things. The global climate has already warmed about 0.6°C since the industrial revolution. Computer models are unable to consider all possible variables, but the earth's temperature may rise 1.5°–4.5°C by 2060 if greenhouse emissions continue at the current rates (Fig. 24.6*b*).

Predicted Consequences

Global warming will bring about climate changes, which computer models attempt to forecast. It is predicted that as the oceans warm, temperatures in the polar regions will rise to a greater degree than other regions. Glaciers would melt, and sea levels will rise, not only due to this melting but also because water expands as it warms. Water evaporation will increase, and most likely there will be increased precipitation along the coasts and dryer conditions inland. The occurrence of droughts will reduce agricultural yields and also cause trees to die off. Expansion of forests into Arctic areas will most likely not offset the loss of forests in the temperate zones. Coastal agricultural lands such as the deltas of Bangladesh, India, and China would be inundated, and billions will have to be spent to keep coastal cities, like New York, Boston, Miami, and Galveston in the United States, from disappearing into the sea.

Carbon dioxide, methane, and other gases known as the greenhouse gases impede the escape of infrared radiation from the surface of the earth. Ever greater concentrations of these gases are expected to lead to a global warming that will have drastic consequences on the hydrosphere and lithosphere.

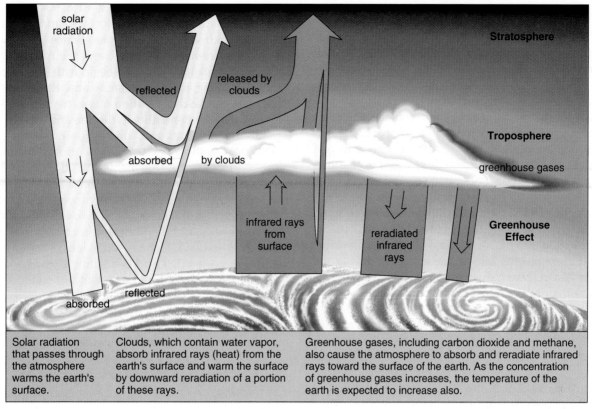

| Solar radiation that passes through the atmosphere warms the earth's surface. | Clouds, which contain water vapor, absorb infrared rays (heat) from the earth's surface and warm the surface by downward reradiation of a portion of these rays. | Greenhouse gases, including carbon dioxide and methane, also cause the atmosphere to absorb and reradiate infrared rays toward the surface of the earth. As the concentration of greenhouse gases increases, the temperature of the earth is expected to increase also. |

a.

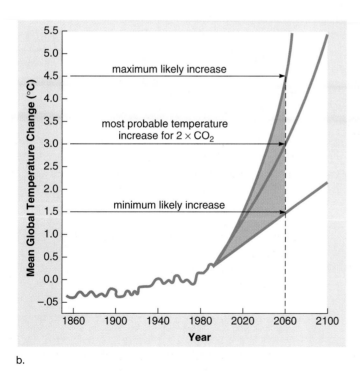

b.

Figure 24.6 Global climate change.

a. Earth's radiation and energy balances showing that the greenhouse gases contribute to global surface warming. **b.** Mean global temperature change is expected to rise. Because of uncertainties concerning the increase in atmospheric greenhouse gases and in computer modeling, maximum and minimum temperature rises are also given.

Global Chemical Climate

Pure water has a pH of 7 because dissociation of water releases an equal number of H^+ and OH^- ions. However, the small amount of carbon dioxide in the atmosphere (0.03%) combines with water to produce a weak solution of carbonic acid and an increased number of H^+ ions. Therefore, rain normally has a pH of about 5.6 rather than 7.0. But near urban areas, rain has pH levels nearer 4.0 than 5.6. Cloud and fog droplets are almost always even more acidic than rain. In some fogs, the pH of the droplets has been found to be as low as 1.7—close to that of battery acid. It is no wonder that vegetation and other materials exposed to such fogs deteriorate rapidly.

Our changing global chemical climate is due to human activities. The coal and oil routinely burned by power plants emit sulfur dioxide into the air. Kuwait oil has a high sulfur content, and therefore the oil well fires started during the Persian Gulf War released much sulfur dioxide into the atmosphere. Automobile exhaust routinely puts nitrogen oxides in the air. Both sulfur dioxide and nitrogen oxides are converted to acids when they combine with water vapor in the atmosphere. These acids return to earth as either wet deposition (acid rain or snow) or dry deposition (sulfate and nitrate salts).

The sulfur dioxide and nitrogen oxides are emitted in one locale, but the deposition occurs in another, across state and national boundaries. Increased deposition of acids has drastically affected forests and lakes in northern Europe, Canada, and northeastern United States (Fig. 24.7) because their soils are naturally acidic and their surface waters are only mildly alkaline (basic) to begin with. The forests in these areas are dying, and their waters cannot support normal fish populations. **Acid deposition** causes trees to weaken and increases their susceptibility to disease and insects. It also kills small invertebrates and decomposers so that the entire ecosystem is threatened. The new global chemical climate also reduces agricultural yields and corrodes marble, metal, and stonework, an effect that is noticeable in cities.

Acid deposition causes heavy metals to be present in drinking water. It leaches aluminum from the soils into surrounding waters and also dissolves copper from pipes and lead from solder, which is used to join pipes. A statistical study suggested that acid deposition is implicated in the increased incidence of lung cancer and possibly colon cancer in United States residents of the East Coast.

Acid deposition, which is caused by the emission of sulfur dioxide and nitrogen oxides into the atmosphere, is harmful to ecosystems and human health.

a.

b.

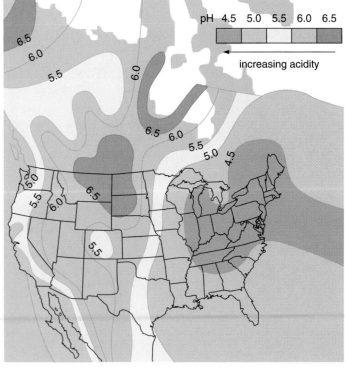

c.

Figure 24.7   Acid deposition.
a. Many forests in higher elevations of northeastern North America and Europe are dying due to acid deposition. **b.** Air pollution due to emissions from factories and fossil fuel burning is the major cause of acid deposition. **c.** The numbers in this map of the United States are average pH recordings of deposition. Winds carry acid pollutants toward the northeast from other parts of the country.

Photochemical Smog

Another aspect of the changing chemical climate is increased concentrations of ozone at ground level, resulting from reactions between nitrogen oxides, hydrocarbons, and sunlight. **Photochemical smog** contains two air pollutants—nitrogen oxides (NO_x) and hydrocarbons (HC)—that react with one another in the presence of sunlight to produce ozone (O_3) and **PAN (peroxyacetylnitrate).** Both nitrogen oxides and hydrocarbons come from fossil fuel combustion, but additional hydrocarbons come from various other sources as well, including paint solvents and pesticides.

Ozone and PAN are commonly referred to as oxidants. Breathing ozone affects the respiratory and nervous systems, resulting in respiratory distress, headache, and exhaustion. These symptoms are particularly apt to appear in young people; therefore, in Los Angeles, where ozone levels are often high, schoolchildren must remain inside the school building whenever the ozone level reaches 0.24 ppm (parts per million by weight). Ozone is especially damaging to plants, resulting in leaf mottling and reduced growth.

Carbon monoxide (CO) is another gas that comes from the burning of fossil fuels in the industrial Northern Hemisphere. High levels of carbon monoxide increase the formation of ozone. Carbon monoxide also combines preferentially with hemoglobin and thereby prevents hemoglobin from carrying oxygen. Breathing large quantities of automobile exhaust can even result in death because of this effect. Of late, it has been discovered that the amount of carbon monoxide over the Southern Hemisphere—from the burning of tropical forests—is equal to that over the Northern Hemisphere.

Normally, warm air near the ground is able to escape into the atmosphere. Sometimes, however, air pollutants, including smog and soot, are trapped near the earth due to a thermal inversion. During a **thermal inversion,** there is cold air at ground level beneath a layer of warm stagnant air above. This often occurs at sunset, but turbulence usually mixes these layers during the day. Some areas surrounded by hills are particularly susceptible to the effects of a temperature inversion because the air tends to stagnate, and there is little turbulent mixing (Fig. 24.8).

Photochemical smog contains ozone and PAN, which are sometimes trapped near ground level due to a thermal inversion.

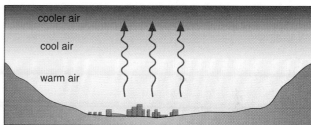

a. Normal pattern

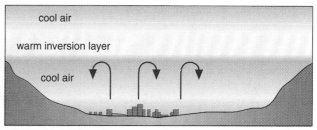

b. Thermal inversion

Figure 24.8   **Thermal inversion.**
a. Normally, pollutants escape into the atmosphere when warm air rises. **b.** During a thermal inversion, a layer of warm air (warm inversion layer) overlies and traps pollutants in cool air below. **c.** Los Angeles is particularly susceptible to thermal inversions, and this accounts for why this city is the "air pollution capital" of the United States.

Stratospheric Ozone Depletion

The earth's atmosphere is divided into layers (see Fig. 24.6). The troposphere envelops us as we go about our day-to-day lives. Ozone in the troposphere is a pollutant, but in the stratosphere, some 50 kilometers above the earth, ozone (O_3) forms a layer, called the *ozone shield*, that absorbs most of the wavelengths of harmful ultraviolet (UV) radiation so that they do not strike the earth. Life on land is threatened if the ozone shield is reduced. UV radiation impairs crop and tree growth and also kills plankton (microscopic plant and animal life) that sustain oceanic life. Without an adequate ozone shield, food sources and health are threatened. UV radiation causes mutations that can lead to skin cancer and can make the lenses of the eyes develop cataracts. It also is believed to adversely affect the immune system and the ability to resist infectious diseases.

Depletion of the ozone layer within the stratosphere in recent years is, therefore, of serious concern. It became apparent in the 1980s that some worldwide depletion of ozone had occurred, and that by the 1990s there was a severe depletion of some 40–50% above the Antarctic every spring (Fig. 24.9). Severe depletions of the ozone layer are commonly called **"ozone holes."** Detection devices now tell us that there is an ozone hole above the Arctic as well, and ozone holes could also develop within northern and southern latitudes, where many people live. Whether or not these holes develop depends on prevailing winds, weather conditions, and the type of particles in the atmosphere. A United Nations Environment Program report predicts a 26% rise in cataracts and nonmelanoma skin cancers for every 10% drop in the ozone level. A 26% increase translates into 1.75 million additional cases of cataracts and 300,000 more cases of skin cancer every year, worldwide.

The cause of ozone depletion can be traced to chlorine atoms (Cl) that are released in the troposphere but rise into the stratosphere. Chlorine atoms combine with ozone and strip away the oxygen atoms one by one. One atom of chlorine can destroy up to 100,000 molecules of ozone before settling to the earth's surface as chloride many years later. These chlorine atoms come from the breakdown of **chlorofluorocarbons (CFCs),** chemicals much in use by humans from 1955 to 1990. The best-known CFC is Freon, a heat transfer agent still found in refrigerators and air conditioners today. CFCs were used as cleaning agents and during the production of Styrofoam found in coffee cups, egg cartons, insulation, and paddings. Their use as a propellant in spray cans has been outlawed in the United States and several other countries but not in western Europe. Although most countries of the world have agreed to stop using CFCs by the year 2000, CFCs already in the atmosphere will be there for over a hundred years before they are spent.

Air pollutants are involved in causing four major environmental effects: global warming, ozone shield destruction, acid deposition, and photochemical smog. Each pollutant may be involved in more than one of these.

Figure 24.9 **Stratospheric ozone.** These satellite observations show that the amount of ozone over the Antarctic (the South Pole) between October 1979 and October 1990 fell by more than 50%. Green represents an average amount of ozone, blue less, and purple still less.

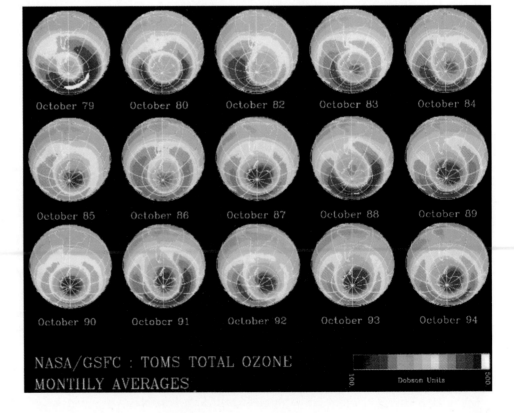

Surface Waters, Aquifers, and Oceans

Fresh water is used for domestic purposes including drinking water, crop irrigation, industrial uses, and energy production. Surface water from rivers, lakes, and underground rivers called **aquifers** are used to meet these needs. The water in aquifers is a vast natural resource, but to ensure a continual supply withdrawals cannot exceed deposits. Countries worldwide are withdrawing water from aquifers. In this country, the farmers of the Midwest withdraw water from aquifers up to 50 times faster than nature replaces it. China, with a population of one billion, is also mining its water to meet the needs of its people, despite the estimate that its aquifers can sustain only 650 million people. In the United States, the government still heavily subsidizes water so that the incentive to use water carefully and efficiently is lacking.

Pollution of Surface Waters

Besides excessive use, pollution of surface water, groundwater, and the oceans is another reason why we are running out of fresh water. Solid wastes include not only household trash but also sewage sludge, agricultural residues, mining refuse, and industrial wastes. Every year, the U.S. population discards billions of tons of solid wastes, some on land and some in fresh and marine waters. Point sources are sources of pollution that are easily identifiable, and nonpoint sources are those caused by runoff from the land (Fig. 24.10).

The United States spends $9 billion a year on cleanup but only $200 million yearly to prevent contamination. It would be best to place the emphasis on preventing contaminants from entering the environment rather than cleaning up pollution. Likewise, recycling can save industry money, as witnessed by the 3M Corporation, which reported savings of $1.2 billion by recycling waste and preventing pollution.

Sewage   Sewage treatment can help degrade organic wastes, which otherwise can cause oxygen depletion in lakes and rivers. As the oxygen level decreases, the diversity of life is greatly reduced. Also, human feces can contain pathogenic microorganisms that cause cholera, typhoid fever, and dysentery. In less-developed countries, where the population is growing and where sewage treatment is practically nonexistent, many children die each year from these diseases. Typically, sewage treatment plants use bacteria to break down organic matter to inorganic nutrients, like nitrates and phosphates, which then enter surface waters. These types of nutrients, which also can enter waters by fertilizer runoff and soil erosion, lead to *cultural eutrophication,* an acceleration of the natural process by which bodies of water fill in and disappear. First, the nutrients cause overgrowth of algae. Then, when the algae die, oxygen is used up by the decomposers, and the water's capacity to support life is reduced. Massive fish kills are sometimes the result of cultural eutrophication.

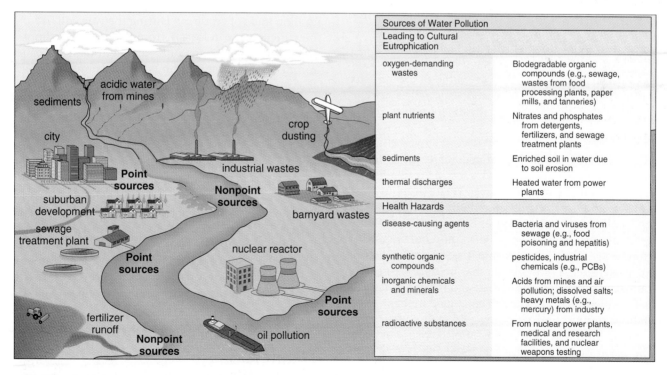

Figure 24.10   **Sources of surface water pollution.**
Many bodies of water are dying due to the introduction of pollutants from point sources, which are easily identifiable, and nonpoint sources, which cannot be specifically identified.

Agricultural and Industrial Wastes   In areas of intensive animal husbandry, or where there are many septic tanks, ammonium ion (NH_4^+) released from animal and human waste is converted by soil bacteria to soluble nitrate, which moves down through the soil (percolates) into underground water supplies. Between 5% and 10% of all wells examined in the United States have nitrate levels higher than the recommended maximum.

Industrial wastes can include heavy metals and organochlorides, such as those in some pesticides. These materials are not degraded readily under natural conditions nor in conventional sewage treatment plants. Sometimes, they accumulate in the mud of deltas and estuaries of highly polluted rivers and cause environmental problems if they are disturbed. Industrial pollution is being addressed in many industrialized countries but usually has low priority in less-developed countries. These wastes enter bodies of water and are subject to **biological magnification** (Fig. 24.11). Decomposers are unable to break down these wastes. They enter and remain in the body because they are not excreted. Therefore, they become more concentrated as they pass along a food chain. Notice in Figure 24.11 that the dots representing DDT become more concentrated as they pass from producer to tertiary consumer. Biological magnification is most apt to occur in aquatic food chains, since there are more links in aquatic food chains than there are in terrestrial food chains. Humans are the final consumers in both types of food chains, and in some areas, human milk contains detectable amounts of DDT and PCBs, which are organochlorides.

Industry also pollutes aquifers. Previously, industry ran wastewater into a pit, from which pollutants could seep into the ground. Wastewater and chemical wastes were also injected into deep wells, from which pollutants constantly discharged. Both of these customs have been or are in the process of being phased out. It is very difficult for industry to find other ways to dispose of wastes, especially since citizens do not wish to live near waste treatment plants.

Pollution of Oceans

Coastal regions are not only the immediate receptors for local pollutants, they are also the final receptors for pollutants carried by rivers that empty at a coast. Waste dumping also occurs at sea, and ocean currents sometimes transport both trash and pollutants back to shore. Examples are the non-biodegradable plastic bottles, pellets, and containers that now commonly litter beaches and the oceans' surfaces. Some of these, such as the plastic that holds a six-pack of cans, cause the death of birds, fishes, and marine mammals that mistake them for food and get entangled in them.

Offshore mining and shipping add pollutants to the oceans. Some 5 million metric tons of oil a year—or more than one gram per 100 square meters of the oceans' surfaces—end up in the oceans. Large oil spills kill plankton, fish fry, and shellfishes, as well as birds and marine mammals. The largest tanker spill in U.S. territorial waters occurred on March 24, 1989, when the tanker *Exxon Valdez*

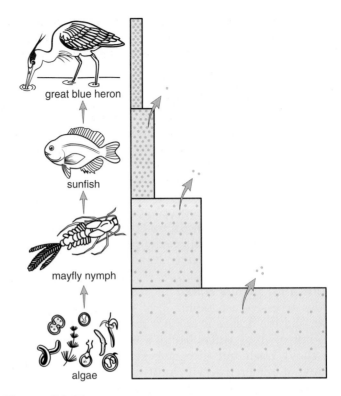

Figure 24.11   **Biological magnification.**
A poison (dots) such as DDT, which is minimally excreted (arrows) becomes maximally concentrated as it passes along a food chain due to the ever smaller biomass of each higher trophic level.

struck a reef in Alaska's Prince William Sound and leaked 44 million liters of crude oil. During the war with Iraq, 120 million liters were released from damaged onshore storage tanks into the Persian Gulf—an event that was called environmental terrorism. Although petroleum is biodegradable, the process takes a long time because the environs do not ordinarily contain a large bacterial population for degrading petroleum. Once the oil washes up onto beaches, it takes many hours of work and millions of dollars to clean it up.

In the last 50 years, we have polluted the seas and exploited their resources to the point that many species are at the brink of extinction. Fisheries once rich and diverse, such as George's Bank off the coast of New England, are in severe decline. Haddock was once the most abundant species in this fishery, but now it accounts for less than 2% of the total catch. Cod and bluefin tuna have suffered a 90% reduction in population size. In warm, tropical regions, many areas of coral reefs are now overgrown with algae because the fish that normally keep the algae under control have been killed off.

Soil Erosion, Desertification, and Deforestation

Whereas 20% of the world's population lived in cities in 1950, it is predicted that 60% will live in cities by the year 2000. Near cities, the development of new housing areas has

led to urban sprawl. Such areas tend to take over agricultural land or simply further degrade the land in the area. The land has been degraded in many ways. Here, we will discuss some of the greatest concerns.

Soil Erosion and Desertification

In agricultural areas, wind and rain carry away about 25 billion tons of topsoil yearly, worldwide. If this rate of loss continues, the earth will lose practically all of its topsoil by the middle of the next century. Soil erosion causes a loss of productivity that is compensated by increased use of fertilizers, pesticides, and fossil fuel energy. One answer to the problem of erosion is to adopt soil conservation measures such as are employed by many farmers in the United States. For example, farmers can use strip-cropping and contour farming to control soil erosion.

Desertification is the transformation of marginal lands to desert conditions because of overgrazing and overfarming (Fig. 24.12). Desertification has been particularly evident along the southern edge of the Sahara Desert in Africa, where it is estimated that 240,000 square miles of once-productive grazing land has become desert in the last 50 years. However, desertification also occurs in this country. The U.S. Bureau of Land Management, which opens up federal lands for grazing, reports that much of the rangeland it manages is in poor or bad condition, with much of its topsoil gone and with greatly reduced ability to support forage plants.

Humans use land in various ways. Farms, towns, and cities are located on land. Agricultural land quality is threatened by soil erosion, which can lead to desertification.

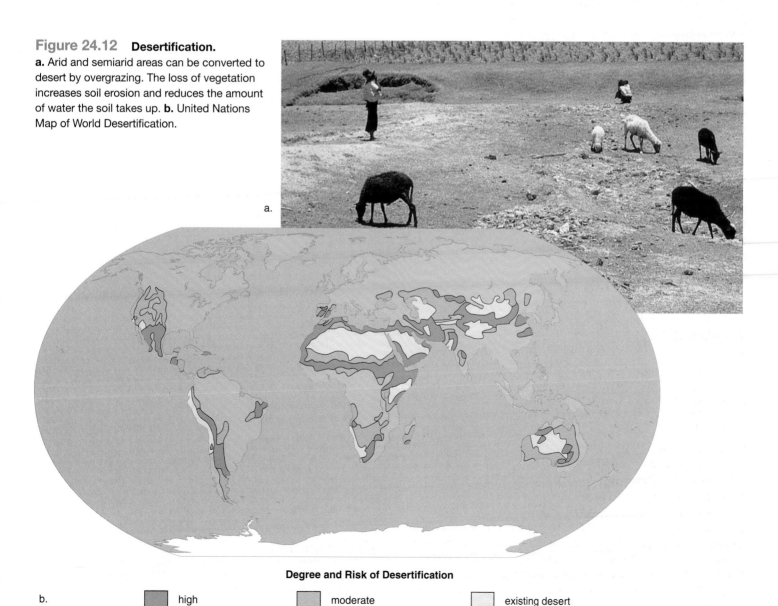

Figure 24.12   Desertification.
a. Arid and semiarid areas can be converted to desert by overgrazing. The loss of vegetation increases soil erosion and reduces the amount of water the soil takes up. **b.** United Nations Map of World Desertification.

a.

Degree and Risk of Desertification

b.   high   moderate   existing desert

Deforestation

In the northern hemisphere, vast stands of trees have been felled and turned into paper and wood products like posts and particleboard. Thousands of miles of new logging roads have been built to facilitate removal of trees from areas that have been clear cut. The animals that live in these forests—moose, porcupine, lynx, and snowshoe hare—have been displaced. Songbirds who migrate there during the summer no longer find shelter there. Although the logging companies are to replant, conservationists wonder if companies have the necessary expertise or if wildlife can sustain themselves in the meantime.

Tropical rain forests are home to more wildlife than temperate forests. For example, temperate forests across the entire United States contain about 400 tree species. In the rain forest, a typical ten-hectare area holds as many as 750 types of trees. The fresh waters of South America are inhabited by an estimated 5,000 fish species; on the eastern slopes of the Andes, there are 80 or more species of frogs and toads; and in Ecuador, there are more than 1,200 species of birds—roughly twice as many as those inhabiting all of the United States and Canada. Therefore, a very serious side effect of deforestation in tropical countries is a loss of biological diversity.

A National Academy of Sciences study estimated that a million species of plants and animals are in danger of disappearing within 20 years as a result of **deforestation** in tropical countries. Many of these life-forms have never been studied, and yet they may be useful sources of food or medicines.

Logging of tropical forests occurs because industrialized nations prefer furniture made from costly tropical woods and because people want to farm the land (Fig. 24.13). In Brazil, the government allows citizens to own any land they clear in the Amazon forest (along the Amazon River). When they arrive, the people practice slash-and-burn agriculture, in which trees are cut down and burned to provide nutrients and space to raise crops. Unfortunately, the fertility of the land is sufficient to sustain agriculture for only a few years. Once the cleared land is incapable of sustaining crops, the farmer moves on to another part of the rain forest to slash and burn again. In the meantime, cattle ranchers move in. Cattle ranchers are the greatest beneficiaries of deforestation, and increased ranching is therefore another reason for tropical rain forest destruction. A newly begun pig-iron industry in Brazil also indirectly results in further exploitation of the rain forest. The pig iron must be processed before it is exported, and smelting the pig iron requires the use of charcoal (burnt wood).

There is much concern worldwide about the loss of biological diversity due to the destruction of tropical rain forests.

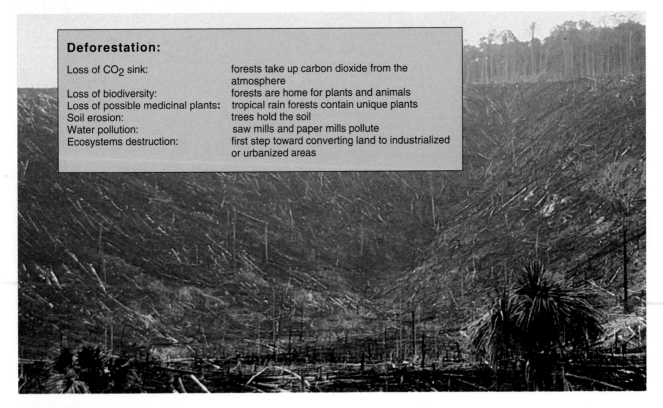

Deforestation:

Loss of CO$_2$ sink:	forests take up carbon dioxide from the atmosphere
Loss of biodiversity:	forests are home for plants and animals
Loss of possible medicinal plants:	tropical rain forests contain unique plants
Soil erosion:	trees hold the soil
Water pollution:	saw mills and paper mills pollute
Ecosystems destruction:	first step toward converting land to industrialized or urbanized areas

Figure 24.13   Deforestation.
Forest destruction leads to the detrimental effects listed.

24.3 The Human Population and Biodiversity

One level of biodiversity is understood by everyone because it includes all the different species of bacteria, protists, fungi, plants, and animals on earth. Genetic variation is a finer level of biodiversity, which helps maintain reproductive vitality and assists adaptation to new environments. And finally, different species live in widely different types of communities making up the biosphere. Therefore, biodiversity is considered to have these three levels: genetic diversity, species diversity, and community diversity.

Extinction

There have been several mass extinctions in the history of the earth followed by periods of recovery. Recovery from the last mass extinction resulted in the highest level of biodiversity the earth has known. Human activities that reduce biodiversity may have begun about 30,000 years ago, when Cro-Magnon may have killed off many large animals such as the giant sloth, the mammoth, the saber-toothed tiger, and the giant ox. *Hunting* by individuals for food and sport and commercial hunting are collectively one of the major causes for an estimated extinction of 15,000 to 30,000 species a year (Fig. 24.14). Sharks

are disappearing from the oceans due to overexploitation, and we certainly know that many fish stocks are being depleted by overfishing. Similarly, on land humans hunt and collect organisms for both pleasure and profit. Wealthy sportsmen now fly into otherwise unreachable Arctic regions to hunt polar bear from the air. Spotted cats (tigers, cheetahs, leopards, jaguars, etc.) are endangered because customers will pay more for furs as these animals become more and more rare. As with sharks, some species have been hurried toward extinction because only certain parts are desired. Rhino horn powder is used as an aphrodisiac and as a cure for snake bites, kidney disease, and other illnesses. At least a thousand black rhinos a year are killed illegally on preserves in Kenya—only to acquire the horn. Likewise elephants are killed only to acquire their tusks. Cacti are collected from the wild for sale in other countries and, on one expedition, dealers removed all of the native cacti from an island off the Baja California coast. The removal of the cacti threatens the lives of the animals that depend upon them for food and shelter.

Another major cause of extinction is *habitat destruction* outright or by fragmentation into small pieces that cannot support the same species richness as before. By the year 2010, very little undisturbed rain forest will exist outside of national parks and other relatively small protected areas. Animal husbandry, agricultural fields, towns and cities, and also human activities like mining and the building of dams destroy ecosystems.

Commercial hunting: A trawl net needs to be towed through the water for only a few minutes before it is full.

Illegal hunting: In Africa, black rhinoceroses are illegally killed in preserved areas only for their horns.

Destruction of ecosystems: Strip mining for coal destroys habitats and pollutes the surroundings.

Introduction of new species: The brown tree snake *(Boiga irregularis)* has devastated endemic bird populations after being introduced into many Pacific islands.

Pollution: Solid wastes are obvious signs of pollution in ecosystems; pesticides cannot be seen but are deadly to wildlife.

Pollution: Pesticides kill bees, and declining populations of bees threaten the plants that depend on them for pollination.

Figure 24.14   Extinction.
Human activities cause extinction in many ways. Some of them are illustrated here.

The accidental or purposeful *introduction of new species* into an ecosystem can cause the extinction of endemic species. The brown tree snake somehow slipped into Guam from southwestern Pacific islands in the late 1940s. Since then, it has wiped out 9 of 11 native bird species, leaving the forests eerily quiet. The carp, an Asian fish that can tolerate polluted waters, is now more prevalent than our own native fishes in certain waters.

Pollution also takes it toll. As we have discussed, the global climate may change so rapidly that many species will be unable to adjust their ranges and may become extinct. Pollution can kill species even if biological communities appear to be undisturbed. Biological magnification of pesticides has caused the abundance of predatory birds to decrease, and acid deposition is implicated in a worldwide decline in amphibian populations.

Human activities are on the verge of causing a massive extinction of species and the loss of ecosystems throughout the biosphere. The loss of biodiversity will most likely be detrimental to humans since they depend on the natural environment for raw materials, food, medicines, and other goods and services.

Conservation Biology

As we have discussed, scientists have only begun to formulate theories about the complex interactions within the biosphere and, therefore, they do not yet know how best to counter the rapid and worldwide decline in biodiversity. *Conservation biology* is a relatively new scientific discipline that brings together people and knowledge from many differ-ent fields to attempt to solve the biodiversity crisis. Conservation biology attempts to understand the effects of human activities on species, communities, and ecosystems, and develop practical approaches to preventing the extinctions of species and the destruction of ecosystems. In the past, ecologists have preferred to study the workings of ecosystems not tainted by human activities, and wildlife managers have been concerned with managing a small number of species for the marketplace and for recreation. Therefore, neither endeavor has addressed the possibility of preserving entire biological communities, although humans are active in the area. Conservation biologists must be able to draw from scientific research and experience in the field to develop a management program that will preserve an ecosystem.

Conservation biology interfaces with many disciplines, including ecological ethics. Most conservation biologists believe that biological diversity is a good thing, and that each species has a value all its own, regardless of its direct material value to humans. This viewpoint runs counter to the idea that the value of a species depends on the goods and services it provides or could potentially provide to humans. Still, many conservationists are willing to work with governmental agencies that promote the concept of **sustainability:** that it is possible for development to meet economic needs while protecting the environment for future generations. Some economists argue that as per capita income increases, environmental degradation first increases, and then begins to decrease as people become affluent enough to begin to protect the environment.

Conservation biology is the scientific study of biodiversity, leading to the preservation of species and the management of ecosystems for sustainable human welfare.

Bioethical Issue

The many types of animals in a coral reef form a complex community that is admired by both snorkelers and scuba divers. The various types of fish and shellfish in a coral reef are sources of food for millions of people. Like a tropical forest, coral reefs are most likely sources of medicines yet to be discovered. And a reef serves as a storm barrier that protects the shoreline and provides a safe harbor for ships.

Reefs around the globe are being destroyed. Tons of soil from deforested tracts of land bring nutrients that stimulate the growth of all kinds of algae. This has contributed to population explosion of the crown-of-thorn starfish that are devouring Australia's 1,200 mile-long Great Barrier Reef. Reefs are also being damaged by pollutants that seep into the sea from facto-ries, farm fields, and sewers. Stress, combined with unusually warm seawater, has caused the corals to expel their symbiotic colorful algae, which carry on photosynthesis and help sustain them. So-called coral bleaching has been noticed in reefs of the Pacific Ocean and Caribbean. Might worldwide global warming also contribute to coral bleaching and death?

Marine scientist Eduardo Gomez estimates that 90% of coral reefs of the Philippines are dead or deteriorating due to pollution, but especially due to overfishing. The methods are sinister, including the use of dynamite to kill the fish, making it easier to scoop them up, use of cyanide to stun the fish to capture them alive, and using satellite navigation systems to home in on areas where mature fish are spawning to reproduce. If all large herbivores are killed off, seaweed overgrows and kills the coral.

Paleobiologist Jeremy Jackson of the Smithsonian Tropical Research Institute near Panama City wonders if he is doing enough to warn the public that reefs around the world are in danger. He estimates that we may lose 60% of all coral reefs by the year 2050.

Questions

1. Do you think it would be possible to make the public care about the loss of coral reefs? Explain.
2. When and under what circumstances do dire predictions help preserve the environment?
3. Considering what is causing the loss of coral reefs, would it be possible to save them? How?

Summarizing the Concepts

24.1 Human Population Growth

The human population is expanding exponentially, and it is unknown when growth will level off. Presently, each year exhibits a large increase, and the doubling time is now about 47 years. Populations have a biotic potential for increase in size. Biotic potential is normally held in check by environmental resistance, thereby producing an **S**-shaped growth curve, leveling off at the carrying capacity of the environment.

The MDCs underwent a demographic transition between 1950 and 1975, but the LDCs are just now undergoing demographic transition. Due particularly to increases in Africa, Asia, and Latin America, an explosion in LDC's populations from 4.3 billion to 10.2 billion is expected. Support for family planning, human development, and delayed childbearing could help prevent such a large increase.

Most LDCs have a youthful age profile—a large proportion of the population is younger than age 15. Since there are so many young women entering the reproductive years, the population will still expand greatly, even after replacement reproduction is attained. The more quickly replacement reproduction is achieved, however, the sooner zero population growth will result.

24.2 The Human Population and Pollution

An increasing human population is causing air, water, and land pollution.

Like the panes of a greenhouse, carbon dioxide, nitrous oxide, methane, and CFCs allow the sun's rays to pass through but impede the release of infrared wavelengths. It is predicted that a buildup in these "greenhouse gases" will lead to a global warming. The effects of global warming could be a rise in sea level and a change in climate patterns. An effect on agriculture could follow.

Global chemical climate changes have already occurred. Sulfur dioxide and nitrogen oxide react with water vapor to form acids that contribute to acid deposition. Acid deposition is killing lakes and forests and also corrodes marble, metal and stonework.

Hydrocarbons and nitrogen oxides react to form smog, which contains ozone and PAN. These oxidants are harmful to animal and plant life.

Ozone shield destruction is particularly associated with CFCs. CFCs rise into the stratosphere and release chlorine. Chlorine causes ozone to break down. Since ozone prevents harmful ultraviolet radiation from reaching the surface of the earth, reduction in the amount of ozone will lead to skin cancer and decreased productivity of the oceans.

Surface waters, aquifers, and oceans are all being affected by human activities. Water in aquifers is being withdrawn up to 50 times faster than it can be replaced. Pollution of underground water supplies—as well as of lakes and oceans— is also a serious problem.

Solid wastes, including hazardous wastes, are deposited on land. The latter, including metals, organochlorides, and nuclear wastes, may contaminate water supplies The oceans are the final recipients of wastes deposited in rivers and along the coasts.

Soil erosion reduces the quality of land and leads to desertification. Desertification is the transformation of marginal lands to desert conditions because of overgrazing and overfarming.

Temperate forests and tropical rain forests in Southeast Asia and Oceania, Central and South America, and Africa are being cut to provide wood for domestic use and export. Slash-and-burn agriculture also reduces tropical rain forests. The loss of biological diversity due to the destruction of tropical rain forests will be immense. Many of these threatened organisms could possibly be of benefit to humans if we had time to study and domesticate them.

24.3 The Human Population and Biodiversity

Human activities have brought about a biodiversity crisis. Individuals and commercial hunting, habitat destruction or fragmentation, and introduction of new species and pollution are all major causes of species extinction.

Conservation biology is a new discipline that pulls together information from a number of biological fields to determine how best to manage ecosystems for the benefit of all species including human beings.

Studying the Concepts

1. Draw a growth curve to represent exponential growth, and explain why a curve representing population growth usually levels off. 496

2. Calculate the growth rate and the doubling time for a population in which the birthrate is 20 per 1,000 and the death rate is 2 per 1,000. 497

3. Distinguish between MDCs and LDCs. Include a reference to age-structure diagrams. 498–99

4. Explain why the population of LDCs is expected to increase tremendously. What steps could be taken to prevent this from occurring? 498–99

5. How and why is the global climate expected to change, and what are the predicted consequences of this change? 500–1

6. What causes acid deposition, and what are its effects? 502

7. How does photochemical smog develop, and what is thermal inversion? 503

8. Of what benefit is the ozone shield? What pollutant in particular should be associated with stratospheric ozone depletion, and what are the consequences of this depletion? 504

9. What are several ways in which surface waters, aquifers, and oceans can be polluted? What is biological magnification? 505–6

10. Explain how soil erosion and desertification are related. 507

11. What are the primary ecological concerns associated with the destruction of rain forests? 508

12. Explain the primary causes of the biodiversity crisis and the goals of conservation biology. 509–10

Testing Your Knowledge of the Concepts

In questions 1–4, match the molecule to an environmental problem below.
a. sulfur dioxide
b. hydrocarbons
c. CFCs
d. carbon dioxide
_____ 1. photochemical smog
_____ 2. global warming
_____ 3. ozone shield destruction
_____ 4. acid deposition

In questions 5–17, indicate whether the statement is true (T) or false (F).
_____ 5. After a country has undergone the demographic transition, the death rate and the birthrate are both high.
_____ 6. Pesticides and radioactive wastes are both subject to biological magnification.
_____ 7. If global warming occurs, it is predicted that rising waters will threaten many coastal cities.
_____ 8. The human population growth curve has now leveled off since the LDCs are no longer growing.
_____ 9. The carrying capacity is the maximum population the environment can support indefinitely.
_____ 10. The MDCs but not the LDCs have a pyramid shaped age-structure diagram.
_____ 11. A society in which people earn more and have a better lifestyle tends to experience less population growth.
_____ 12. Carbon dioxide from the burning of fossil fuels is one of the major greenhouse gases.
_____ 13. Acid deposition affects just those areas in which there are factories and power plants.

_____ 14. Thermal inversions make the tropics colder and the Arctic warmer.
_____ 15. Ozone depletion in the stratosphere can be traced to magnesium ions that are released into the troposphere.
_____ 16. Nutrients released by sewage treatment plants have a polluting effect.
_____ 17. When tropical forests are destroyed, more carbon dioxide (CO_2) than before is added to the atmosphere.

In questions 18 and 19, fill in the blanks.
18. Photochemical smog contains ozone and PAN, which are sometimes trapped near ground level due to a _____.
19. If the age-structure diagram has a pyramid shape, there are more women _____ the reproductive years than older women leaving them.
20. Label this **S**-shaped growth curve.

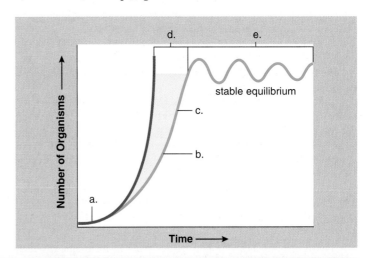

Applying Your Knowledge to the Concepts

These questions pertain to population concerns.
1. What are the two factors, for decreasing the growth rate? Explain.
2. How long would it take to stabilize the world's population if the growth rate is reduced to zero? Explain.
3. Humans, as well as other animals, have been dumping their wastes into the environment for thousands of years. What is the reason(s) that this appears to be such a problem today?

4. Some individuals believe that the carrying capacity of the earth is between 50–100 billion people; others believe that the present population of 5.5–6 billion people already exceeds the earth's carrying capacity. How is it possible for so-called "experts" to arrive at such different numbers?

Understanding the Terms

acid deposition 502
aquifer 505
biological magnification 506
biotic potential 497
carrying capacity 497
chlorofluorocarbons (CFCs) 504
deforestation 508
demographic transition 498
desertification 507
doubling time 497

environmental resistance 497
exponential growth 496
greenhouse effect 500
growth rate 496
ozone hole 504
PAN (peroxyacetylnitrate) 503
photochemical smog 503
replacement reproduction 499
sustainability 510
thermal inversion 503

Match the terms to these definitions.
a. _____ Water-bearing stratum of permeable rock that constitutes an underground reservoir.
b. _____ The yearly percentage of increase or decrease in the size of a population.
c. _____ Transformation of marginal lands to desert conditions.
d. _____ Largest number of organisms of a particular species that can be maintained indefinitely in an ecosystem.
e. _____ Number of years it takes for a population to double in size.

Applying Technology to the Concepts

Your study of population concerns is supported by these available technologies:

Essential Study Partner CD-ROM
Ecology → **Human Impact**
Visit the Mader web site for related ESP activities.

Exploring the Internet
The Mader Home Page provides resources and tools as you study this chapter.

http://www.mhhe.com/biosci/genbio/mader

Further Readings for Part 7

Agnew, N., and Demas, M. September 1998. Preserving the Laetoli footprints. *Scientific American* 279(3):44. This article recaps the discovery of hominid footprints in East Africa, and explains how they were reburied in order to preserve them.

Allen, J. L., editor. 1997. Annual editions: *Environment 97/98.* Guilford, Conn: Dushkin/McGraw-Hill. This volume is a collection of articles pertaining to the problems and issues of the environment and environmental quality.

Balick, M. J., and Cox, P. A. 1996. *Plants, people, and culture: The science of ethnobotany.* New York: Scientific American Library. This interesting, well-illustrated book discusses the medicinal and cultural uses of plants, and the importance of rain forest conservation.

Baskin, Y. 1997. *The work of nature: How the diversity of life sustains us.* Washington, D.C.: Island Press. This book shows the value of goods and services provided by intact systems.

Begon, M., et al. 1996. *Population ecology: A unified study of animals and plants.* 3d ed. Oxford: Blackwell Scientific Publications. The present state of population ecology is described in this introductory-level text.

Béland, P. May 1996. The beluga whales of the St. Lawrence River. *Scientific American* 274(5):74. Pollution and hydroelectric projects take their toll on the Beluga whale.

Bioscience. September 1998. Flooding: Natural and managed disturbances. 48(9):677. This special issue is devoted to flooding.

Borman, F. H., and Likens, G. E. 1994. *Pattern and process in a forested ecosystem.* New York: Springer-Verlag. Disturbance, development, and the steady state based on the Hubbard Brook Ecosystem Study is presented.

Botzler, R. G., and Armstrong, S. J. 1998. *Environmental ethics: Divergence and convergence.* 2d ed. Boston: McGraw-Hill, Inc. This text for upper-division undergraduate students is a comprehensive and balanced introduction to the field of environmental ethics.

Castillon, D. A. 1996. *Conservation of natural resources: A resource management approach.* 2d ed. Dubuque, Iowa: Brown & Benchmark Publishers. Written for beginning students, this edition contains many new examples of successful environmental programs.

Collier, M. P., et al. January 1997. Experimental flooding in Grand Canyon. *Scientific American* 276(1):82. Periodic human-manufactured floods may improve the canyon environment.

Cox, G. 1997. *Conservation ecology.* 2d ed. Dubuque, Iowa: Wm. C. Brown Publishers. Discusses the nature of the biosphere, the threats to its integrity, and ecologically sound responses.

Cunningham, W. P., and Saigo, B. W. 1997. *Environmental science: A global concern.* Dubuque, Iowa: Wm. C. Brown Publishers.

Provides scientific principles plus insights into the social, political, and economic systems impacting the environment.

Daily, G. C., editor. 1997. Nature's services: *Societal dependence on natural ecosystems.* Washington, D.C.: Island Press. This book shows the value of goods and services provided by intact systems.

de Steiguer, J. E. 1997. *The age of environmentalism.* Intended as supplemental reading for undergraduate and graduate environmental studies courses, this book explores the writings of Rachel Carson and other early environmental scholars.

Duxbury, A. C., and Duxbury, A. B. 1997. *An introduction to the world's oceans.* 5th ed. Dubuque, Iowa: Wm. C. Brown Publishers. This is an introductory oceanographic text that studies the scientific processes that govern the oceans and the earth.

Futuyma, D. J. 1995. *Science on trial: The case for evolution.* Sunderland, Mass.: Sinauer Associates. Presents the evidence for evolution and the operation of the evolutionary process versus that proposed by the doctrine of creation.

Goldfarb, T. D., editor. 1997. *Sources: Notable selections in environmental studies.* Guilford, Conn.: Dushkin Publishing Group/Brown & Benchmark Publishers. This source book is a collection of essays, excerpts, and journal articles that have shaped the world's understanding of the environment.

Goldfarb, T. D., editor. 1997. *Taking sides: Clashing views on controversial environmental issues.* 7th ed. Guilford, Conn.: Dushkin/McGraw-Hill. This text is a collection of essays, presenting pro and con views on selected environmental controversies.

Gore, R. July 1997. The dawn of humans. *National Geographic* 192(1):96. Scientists are finding early human remains from one million years ago.

Gore, R. September 1997. The dawn of humans. *National Geographic* 192(3):92. A footprint dated about 117,000 years ago was discovered in southern Africa.

Hammond, A. L., et al., editors. 1996. *World resources 1996-97: The urban environment.* New York: Oxford University Press. This report provides accurate information on urban environmental management.

Hedin, L. O., and Likens, G. E. December 1996. Atmospheric dust and acid rain. *Scientific American* 274(6):88. Although there has been a reduction in acidic air pollutants, acid rain continues to be a problem.

Hollister, C. D., and Nadis, S. January 1998. Burial of radioactive waste under the seabed. *Scientific American* 278(1):60. Radioactive wastes might be safely disposed of in mudflats found on the deep ocean floor.

Holloway, M. October 1997. Field and stream: A new way to identify the inhabitants of an ecosystem. *Scientific American* 277(4):24. A new method developed in Alaska determines ecosystem makeup.

Johanson, D. C. March 1996. Face-to-face with Lucy's family. *National Geographic* 189(3):96. New fossils from Ethiopia provide more information about human evolution.

Leakey, M., and Walker, A. June 1997. Early hominid fossils from Africa. *Scientific American* 276(6):74. An arm bone unearthed in 1965 recently proved the existence of a new species of *Australopithecus*, showing ancestral humans existed 4 million years ago.

Levin, H. L. 1997. *The earth through time.* 5th ed. Philadelphia: Saunders College Publishing. Provides an in-depth study of evolutionary and geological eras and events.

Lewin, R. 1997. Patterns in evolution: The new molecular view. New York: *Scientific American* Library. This easy-to-read book explores how genetic information is providing insight into evolutionary events.

Miller, G. T. 1996. *Living in the environment.* 9th ed. Belmont, Calif.: Wadsworth Publishers. Designed for use in an introductory course on environmental science, this book discusses how the environment is being used and abused, and what individuals can do to protect the environment.

Mitchell, J. G. February 1996. Our polluted runoff. *National Geographic* 189(2):106. Eighty percent of U.S. water pollution is due to land runoff not resulting from municipal or industrial sources.

Odum, E. 1997. *A bridge between science and society.* 3d ed. Sunderland, Mass.: Sinauer Associates. Introduces the principles of modern ecology as they relate to threats to the biosphere.

Pinter, N., and Brandon, M. T. April 1997. How erosion builds mountains. *Scientific American* 276(4):74. The building of mountains depends on the destructive power of water and wind, volcanic eruptions, and seismic plate collisions.

Plucknett, D. L., and Winkelmann, D. L. September 1995. Technology for sustainable agriculture. *Scientific American* 273(3):182. The practice of sustainable agriculture, increasing productivity while protecting the environment, will be difficult to achieve in developing countries.

Rice, R. E., et al. April 1997. Can sustainable management save tropical forests? *Scientific American* 276(4):44. The strategy of replacing trees harvested for lumber in the rain forests often fails.

Scientific American Quarterly. Fall 1998. The oceans. *Scientific American* 9(3). This issue's articles discuss the origins of earth's water, polar ice cap melting, weather, pollution and legal issues, aquaculture, mineral mining, and marine diversity.

Sharpe, G. W., et al. 1995. *Introduction to forest and renewable resources.* 6th ed. New York: McGraw-Hill, Inc. This text for students of forestry presents the changing policies and practices in the conservation and management of forests and other renewable resources.

Smith, R. 1996. *Ecology and field biology.* 5th ed. New York: Harper & Row. Presents a balanced introduction to ecology—plant and animal, theoretical and applied, physiological and behavioral, and population and ecosystem.

Stiling, P. D. 1996. *Ecology: Theories and applications.* 2d ed. Upper Saddle River, New Jersey: Prentice-Hall. This text for ecology majors stresses the role of evolution and behavior in ecology, and uses scientific studies to illustrate plant and animal distribution.

Strickberger, M. 1995. *Evolution.* 2d ed. Boston: Jones and Bartlett Publishers. Presents the basics of evolutionary theories.

Tattersall, I. April 1997. Out of Africa Again . . . and Again? *Scientific American* 276(4):60. Hominids may have migrated out of Africa several times, with each emigration sending a different species.

Wenke, R. 1996. *Patterns in prehistory: Humankind's first three million years.* 4th ed. New York: Oxford University Press. Provides a comprehensive review of world prehistory.

Appendix A

Answer Key

This appendix contains the answers to the Testing Your Knowledge of the Concepts, Applying Your Knowledge to the Concepts, and Understanding the Terms questions, which appear at the end of each chapter.

Introduction

Testing Your Knowledge of the Concepts

1. d; **2.** e; **3.** a; **4.** c; **5.** T; **6.** F; **7.** T; **8.** copy; **9.** Society; **10. a.** Investigator notices that when a dye is present bacteria live despite exposure to sunlight. **b.** Dye protects bacteria against death by UV light. **c.** One hundred plates containing bacteria and dye and another one hundred plates containing only bacteria are exposed to UV light. The bacteria on both plates die. **d.** Dye does not protect bacteria against death by UV light.

Applying Your Knowledge to the Concepts

1. When one undergoes a medical examination, homeostasis is being checked. Urine and blood tests plus various other tests are done.

2. Humans have much more control over their environment than other animals. Therefore our population is growing rapidly with an accompanying depletion of resources and production of pollutants.

3. The biochemistry of bacteria and humans is quite similar; therefore, if a chemical affects the biochemistry of bacteria, it will probably affect ours also.

4. Malfunction of digestive system: nutrients would enter blood; circulatory system: nutrients would not be transported to and wastes would not be transported from cells; immune system: no protection from disease; respiratory system: oxygen would not enter blood; nervous system and endocrine systems: coordination of systems would be absent; musculoskeletal system: lack of ability to move away from danger.

Understanding the Terms:

a. scientific theory; **b.** hypothesis; **c.** energy; **d.** adaptation; **e.** homeostasis

Chapter 1

Testing Your Knowledge of the Concepts

1. b; **2.** c; **3.** a; **4.** d; **5.** F; **6.** F; **7.** T; **8.** glycerol, fatty acids; **9.** pentose, phosphate, base; **10. a.** monomers; **b.** condensation; **c.** polymer; **d.** hydrolysis

Applying Your Knowledge to the Concepts

1. In a covalent bond, two atoms are sharing the same electrons. In an ionic bond and a hydrogen bond, there is simply an attraction between the two ions.

2. As indicated, carbon dioxide combines with water to form carbonic acid, and this makes the carbonated drink acidic; once the bottle is opened, the carbon dioxide escapes and the pH increases.

3. Carbon can form bonds with as many as four different atoms; therefore, it serves well as the core or skeleton for organic molecules.

Hydrogen and oxygen, the other common atoms of all organic molecules, can only form one and two bonds, respectively.

4. A protein can have four levels of structure; only the first level involves the sequence of amino acids. Obviously levels two to four must be affected by heating. Heating changes the shape of the enzyme and makes it ineffective.

Understanding the Terms:

a. protein; **b.** enzyme; **c.** cellulose; **d.** neutron; **e.** acid

Chapter 2

Testing Your Knowledge of the Concepts

1. c; **2.** a; **3.** d; **4.** b; **5.** T; **6.** F; **7.** T; **8.** electron transport system, cristae; **9.** 2, 36; **10. a.** nucleus—DNA specifies; **b.** nucleolus—RNA helps; **c.** rough endoplasmic reticulum—produces; **d.** smooth endoplasmic reticulum—modifies; **e.** Golgi apparatus—further modifies; **11. a.** glycolysis; **b.** ATP; **c.** transition reaction; **d.** CO_2; **e.** Krebs cycle; **f.** CO_2; **g.** ATP; **h.** electron transport system; **i.** ATP; **j.** $\frac{1}{2} O_2$; **k.** H_2O

Applying Your Knowledge to the Concepts

1. Glycoprotein. These molecules serve to identify differences in individuals, such as blood type.

2. Lysosome. The lysosome contains hydrolytic digestive enzymes.

3. A microtubule contains tubulin proteins, not actin proteins.

4. In the case of human beings, the rest of the chemical energy is tied up in the two lactate molecules still remaining following fermentation.

Understanding the Terms

a. glycolysis; **b.** diffusion; **c.** active site; **d.** Golgi apparatus; **e.** endoplasmic reticulum

Chapter 3

Testing Your Knowledge of the Concepts

1. c; **2.** b; **3.** a; **4.** d; **5.** T; **6.** F; **7.** F; **8.** Homeostasis; **9.** not striated (smooth); **10. a.** columnar epithelium, lining of intestine (digestive tract), protection and absorption; **b.** cardiac muscle, wall of heart, pumps blood; **c.** compact bone, skeleton, support and protection.

Applying Your Knowledge to the Concepts

1. Epithelial tissue. Capillaries are composed of squamous epithelial tissue. Alveoli of the lungs have squamous epithelial tissue. Glomerular capsules in the kidney are composed of squamous epithelial tissue. Intestinal mucosa has an outer layer of columnar epithelium.

2. Muscular activity generates heat energy. If you exercise, this activity generates the heat energy that helps keep the body temperature normal. If you stand around, the homeostatic mechanisms of the body stimulate muscles to contract weakly and rapidly (shivering), which generates heat energy and helps keep the body temperature normal.

3. Cartilage gives more flexibility during the birth process. And it allows easier growth in the skeleton of the newborn.

4. Smooth muscle functions without conscious input from the nervous system; it is found in many of the systems that carry on "housekeeping" duties that do not require a sudden response to a stimulus. Skeletal muscle is under conscious control and generally permits rapid response to a stimulus that has been monitored by the nervous system.

Understanding the Terms
a. ligament; b. epidermis; c. striated; d. homeostasis; e. spongy bone

Chapter 4

Testing Your Knowledge of the Concepts
1. d; 2. b; 3. c; 4. a; 5. T; 6. F; 7. F; 8. bile, emulsifies; 9. acid, basic; 10. a. salivary glands; b. esophagus; c. stomach; d. liver; e. gallbladder; f. pancreas; g. small intestine; h. large intestine; i. sugar and amino acids; j. lipids; k. water

Applying Your Knowledge to the Concepts
1. The folds, villi, and microvilli produce this additional surface area, which provides more membrane through which nutrients may pass.

2. The liver stores glucose as glycogen and breaks down the glycogen to glucose as the need arises. It also converts amino acids to glucose molecules.

3. They cannot store bile. Therefore, they should eat less fatty food and eat fats in smaller amounts.

4. A certain amount of fat is necessary in the diet; one of the fatty acids is necessary for the production of phospholipids, an important component of plasma membranes of cells. Just make sure to reduce the amount of fat in the diet to 30% or somewhat less.

Understanding the Terms
a. duodenum; b. sphincter; c. lipase; d. defecation; e. gallbladder

Chapter 5

Testing Your Knowledge of the Concepts
1. b, phagocytize; 2. a, hemoglobin; 3. d, blood clotting; 4. a; 5. F; 6. T; 7. T; 8. antibodies; 9. no; 10. a. blood pressure; b. osmotic pressure; c. blood pressure; d. osmotic pressure

Applying Your Knowledge to the Concepts
1. The air of Los Angeles is highly polluted from the large number of automobiles and industries. Most air pollutants adversely affect the respiratory system.

2. Both these groups of animals are very active, and thus, a rich supply of oxygen is an advantage. The lack of a nucleus in the red blood cell allows more space for hemoglobin, and the red blood cell can carry more oxygen.

3. Women lose a small amount of blood during menstruation; therefore, the iron lost in the hemoglobin in the red blood cells must be replaced. Loss of blood due to surgery or an accident also requires a larger amount of iron intake to permit the manufacture of additional red blood cells.

4. Although white blood cells are larger than red blood cells, they are very flexible and can move through capillary walls in a manner often referred to as an amoeboid movement.

Understanding the Terms
a. hemoglobin; b. agglutination; c. plasma; d. lymph; e. prothrombin

Chapter 6

Testing Your Knowledge of the Concepts
1. b; 2. a; 3. c; 4. b; 5. F; 6. T; 7. T; 8. fat, cholesterol; 9. stroke; 10. a. aorta; b. left pulmonary arteries; c. pulmonary trunk; d. left pulmonary veins; e. left atrium; f. semilunar valves; g. atrioventricular (mitral) valve; h. left ventricle; i. septum; j. inferior vena cava; k. right ventricle; l. chordae tendineae; m. atrioventricular (tricuspid) valve; n. right atrium; o. right pulmonary veins; p. right pulmonary arteries; q. superior vena cava. See also Figure 6.5, page 129.

Applying Your Knowledge to the Concepts
1. The additional weight requires increased units in the cardiovascular system; thus, the heart must exert more pressure to deliver the blood. This puts added tension on the arteries and also additional work on the heart. Also, some of the fat collects around the heart, which restricts its pumping action.

2. A high enough level of drugs should be administered to prevent the immune system from rejecting the tissue; at the same time, if the immune system is depressed too much, the patient is unable to combat disease, producing organisms that enter the body. Transplant patients are very susceptible to the common cold and similar types of viruses.

3. The atria serve largely as collecting chambers and, as indicated by their thin walls, do not pump blood with much pressure. The expanding of the ventricles causes a negative pressure, which will result in the blood, under greater pressure in the atria, entering the ventricles without the atria pumping.

4. This indicates that the atria are receiving the electrical impulse from the pacemaker, but that there is some sort of blockage between the pacemaker and the AV node.

Understanding the Terms
a. diastole; b. vena cava; c. fibrinogen; d. hemoglobin; e. systemic circuit

Chapter 7

Testing Your Knowledge of the Concepts
1. b; 2. c; 3. c, d; 4. d; 5. F; 6. T; 7. T; 8. tissue fluid, subclavian; 9. filter; 10. thymus; 11. the complement system; 12. plasma, memory; 13. antibody; 14. cytokines; 15. APC; 16. vaccines (or antigens); 17. monoclonal; 18. memory, memory; 19. a. antigen-binding site; b. light chain; c. heavy chain. d. variable region; e. constant region; f. spherical. See also Figure 7.6, page 151.

Applying Your Knowledge to the Concepts
1. The lymph node serves as a filter for foreign material, including bacteria. The reaction of the lymph node tissue with this foreign material results in the swelling and consequently, pain from the node.

2. Immunological memory is dependent upon the presence of memory B cells. In the case of certain types of memory B cells, they decline in number so that they will not be able to respond rapidly to an antigen (disease-producing organism). A booster shot will stimulate the production of more memory B cells.

3. Immunosuppressive drugs prevent immune responses in general, and therefore, the body is more susceptible to all pathogens.

4. Histamine causes capillaries to dilate and become more permeable, resulting in the so-called "runny" nose and stuffiness that often accompany a cold. It also constricts air passageways, causing some difficulty in breathing. Antihistamine counteracts these effects of histamine.

Understanding the Terms

a. vaccine; **b.** lymph; **c.** antigen; **d.** apoptosis; **e.** B lymphocyte

Chapter 8

Testing Your Knowledge of the Concepts

1. b; **2.** a; **3.** c; **4.** d; **5.** F; **6.** T; **7.** F; **8.** expanded; **9.** hemoglobin; **10. a.** nasal cavity; **b.** nostril; **c.** pharynx; **d.** epiglottis; **e.** glottis; **f.** larynx; **g.** trachea; **h.** bronchus; **i.** bronchiole. See also Figure 8.2, page 166.

Applying Your Knowledge to the Concepts

1. The nose contains a filtering system of hairs and cilia that remove foreign particles so that they do not enter the lungs. Also, the surface area of nasal cavities, with their convoluted walls, warms the air before it enters the lungs.

2. No. The buildup of carbon dioxide will stimulate the inspiration center in the brain and force the child to breathe.

3. The active part of the body will be releasing more CO_2 and heat; thus, the affinity of hemoglobin for oxygen will decrease, and more oxygen will be released in the tissue fluid surrounding the active cells. (Also, there will be a dilation of the blood vessels in that area.)

4. Both the respiratory center and the cardiac center are very sensitive to the CO_2 level in the blood. If CO_2 falls below a certain level, breathing ceases until CO_2 increases to the critical level.

Understanding the Terms

a. pharynx; **b.** diaphragm; **c.** bicarbonate ion; **d.** expiration; **e.** alveolus

Chapter 9

Testing Your Knowledge of the Concepts

1. d; **2.** b; **3.** a; **4.** c; **5.** F; **6.** T; **7.** T; **8.** salt; **9.** Urea; **10. a.** glomerulus; **b.** efferent arteriole; **c.** afferent arteriole; **d.** proximal convoluted tubule; **e.** loop of the nephron; **f.** descending limb; **g.** ascending limb; **h.** peritubular capillary network; **i.** distal convoluted tubule; **j.** renal vein; **k.** renal artery; **l.** collecting duct.

Applying Your Knowledge to the Concepts

1. The bladder lies in front of the uterus. As the uterus enlarges during pregnancy, it places pressure on the bladder, which results in the desire to urinate more often.

2. Alcohol inhibits the secretion of ADH, and this results in a dilute urine being secreted. This lowers the fluid content of the body and results in a dry mouth and feeling of thirst.

3. Load your body with sugar prior to being tested, thereby overloading the reabsorptive power of the kidneys. Then, excess sugar is excreted, and urinalysis gives a positive test for diabetes.

4. All material moving through the membrane in an artificial kidney must move from a higher concentration to a lower concentration by passive transport. In the human kidney, some materials are moved against the concentration gradient by active transport.

Understanding the Terms

a. loop of the nephron; **b.** tubular reabsorption; **c.** aldosterone; **d.** renal cortex; **e.** peritubular capillary network

Chapter 10

Testing Your Knowledge of the Concepts

1. c; **2.** g; **3.** a; **4.** d; **5.** f; **6.** e; **7.** e; **8.** a; **9.** b; **10.** c; **11.** f; **12.** d; **13.** F; **14.** T; **15.** osteon; **16.** spongy; **17.** osteoclast; **18.** sinuses; **19.** pelvic girdle, rib cage, **20. a.** coxal bone; **b.** femur, **c.** patella; **d.** tibia; **e.** fibula; **f.** metatarsals; **g.** phalanges; **h.** tarsals

Applying Your Knowledge to the Concepts

1. Osteoclasts cause absorption of the minerals in the bone, and if there is an insufficient amount of calcium in the blood, the osteoblasts cannot repair the bone.

2. Either the spinal cord or the spinal nerves were being compressed by the vertebrae or by damaged disks.

3. The knee is a synovial joint that is structured to be mobile in only one plane. Tackling from the side puts pressure on the joint in a manner to which the joint cannot respond, and consequently, the ligaments and tendons can suffer severe damage.

4. Jane's long bones are increasing in length, resulting in an increase in her height. Jane's grandmother is probably experiencing a decrease in the mass of her intervertebral disks which could result in a decrease in her height. She may also suffer from osteoporosis, which can result in her spinal cord shrinking and development of a hunchback.

Understanding the Terms

a. synovial joint; **b.** axial skeleton; **c.** menisci; **d.** fontanel; **e.** pectoral girdle

Chapter 11

Testing Your Knowledge of the Concepts

1. d; **2.** a; **3.** b; **4.** c; **5.** T; **6.** F; **7.** T; **8.** creatine phosphate; **9.** ACh; **10. a.** T tubule; **b.** sarcoplasmic reticulum; **c.** myofibril; **d.** Z line; **e.** sarcomere; **f.** sarcolemma of muscle fiber

Applying Your Knowledge to the Concepts

1. Isometric because muscles do not shorten and put stress on the joints.

2. For some reason, the neuromuscular junction is no longer working properly. Some think the junction acts like a fuse, protecting the muscle from being damaged by overstimulation.

3. The liver has no need of glucose like the muscles do. Having a ready supply of glucose means that the muscles can respond quickly to get the organism away from danger.

4. The aerobic or isotonic exercises, such as running, swimming, bicycling, etc.

Understanding the Terms

a. sarcomere; **b.** insertion; **c.** tetanus; **d.** oxygen debt; **e.** motor unit

Chapter 12

Testing Your Knowledge of the Concepts

1. d; **2.** a; **3.** b; **4.** c; **5.** F; **6.** T; **7.** F; **8.** synaptic cleft; **9.** sensory, motor; **10. a.** sensory neuron; **b.** interneuron; **c.** motor neuron; **d.** receptor; **e.** cell body; **f.** dendrites; **g.** axon; **h.** nucleus of Schwann cell; **i.** node of Ranvier; **j.** effector

Testing Your Knowledge of the Concepts

1. Anger and an excited state fall under "fight or flight." Although this complex response readies the body for action, it inhibits the digestive

tract. One aspect—blood is shunted from the skin and visceral organs to the skeletal muscles.

2. It causes paralysis, and ultimately, atrophy of the muscles innervated by the destroyed nerves. Muscles that don't receive stimuli from nerves atrophy.

3. The advantage lies in the rapid response to a stimulus. Many reflex actions are protective in nature. The disadvantage lies in no variation in the response to the stimulus.

4. Alcohol affects the higher levels of the brain first; it affects lower levels as the blood concentration increases. Its effects move from cerebrum (centers for conscious thought and speech) to cerebellum (centers for control of muscle activity) to the medulla (centers for control of breathing and heart rate—involuntary activities).

Understanding the Terms

a. nerve impulse; **b.** cerebral hemisphere; **c.** integration; **d.** motor neuron; **e.** reticular formation

Chapter 13

Testing Your Knowledge of the Concepts

1. b; **2.** a; **3.** d; **4.** c; **5.** F; **6.** F; **7.** T; **8.** vestibule, semicircular canals; **9.** left; **10. a.** retina; **b.** choroid; **c.** sclera; **d.** optic nerve; **e.** fovea centralis; **f.** ciliary body; **g.** lens; **h.** iris; **i.** pupil; **j.** cornea;

Testing Your Knowledge of the Concepts

1. A transducer converts one form of energy to another. Receptors in living organisms convert a variety of energies to electrical energy, which is then used to convey a message to the central nervous system.

2. Vitamin A, the precursor of retinal found in carrots, can help night blindness, which occurs when rods are not functioning as they should. Vitamin A cannot help red-green color blindness, an inherited condition that affects the cones.

3. The eye can respond to stimuli other than light. (It has a low threshold to light, however.)

4. The eye helps in positioning the body by allowing us to line up on some visual object.

Understanding the Terms:

a. cochlea; **b.** chemoreceptor; **c.** rhodopsin; **d.** sensation; **e.** spiral organ

Chapter 14

Testing Your Knowledge of the Concepts

1. f; **2.** a; **3.** e; **4.** b; **5.** d; **6.** c; **7.** F; **8.** T; **9.** T; **10.** T; **11.** atrial natriuretic hormone; **12.** negative; **13.** testosterone, estrogens, progesterone; **14. a.** inhibits; **b.** inhibits; **c.** releasing hormone; **d.** stimulating hormone; **e.** target gland hormone

Applying Your Knowledge to the Concepts

1. Epinephrine is a nonsteroid hormone and has its effect on existing enzymes; whereas testosterone is a steroid hormone and has its effect on the genetic material, resulting in the production of more enzymes.

2. Although the anterior pituitary secretes hormones that control other endocrine glands, it is the hypothalamus that controls the anterior pituitary's secretion of these hormones.

3. Thyroxin increases metabolism and thus resulted in the burning of extra calories, rather than them being stored as fat.

4. Melatonin is secreted by the pineal gland in larger amounts in lower light intensity, at night, and thus prepares a person for sleep. On an overseas trip, the hormone is taken about an hour before one wants to sleep.

Understanding the Terms

a. thyroid gland; **b.** diabetes mellitus; **c.** adrenocorticotropic hormone (ACTH); **d.** nonsteroid hormone; **e.** oxytocin

Chapter 15

Testing Your Knowledge of the Concepts

1. c; **2.** d; **3.** a; **4.** b; **5.** F; **6.** T; **7.** T; **8.** blood; **9.** testosterone; **10.** seminal vesicles; **11.** testosterone; **12.** vagina; **13.** follicle, endometrial; **14.** estrogens, progesterones; **15.** human chorionic gonadotropin (HCG); **16.** laboratory glassware; **17. a.** seminal vesicle; **b.** ejaculatory duct; **c.** prostate gland; **d.** bulbourethral gland; **e.** anus; **f.** vas deferens; **g.** epididymis; **h.** testis; **i.** scrotum; **j.** foreskin; **k.** glans penis; **l.** penis; **m.** urethra; **n.** vas deferens; **o.** urinary bladder

Applying Your Knowledge to the Concepts

1. The female system opens into the abdominal cavity by way of the upper opening of the oviducts. The male system has the testes at the upper end of the vas deferens and thus does not open into the abdominal cavity.

2. Follicle-stimulating hormone (FSH), a hormone from the anterior pituitary, stimulates the development of a follicle in the ovary. The follicle produces estrogen that "feeds" back on the anterior pituitary and inhibits FSH release.

3. The acidic secretion in the vagina is a barrier to disease-producing organisms. The male prostate gland secretes an alkaline solution that helps neutralize the vaginal environment.

4. Women sometimes fail to ovulate because of low body weight. And although normal menstrual function results from interactions between the CNS, hypothalamus, anterior pituitary, ovaries, and target tissues, the ovaries are primarily responsible for controlling the changes that occur during the cycle. Therefore, failure to ovulate throws off the cycle. The low amount of fat tissue in these women may be important. Estrogen is synthesized in fat tissue, which may be an important source of this hormone.

Understanding the Terms

a. vagina; **b.** endometrium; **c.** prostate gland; **d.** follicle-stimulating hormone; **e.** vulva.

Chapter 16

Testing Your Knowledge of the Concepts

1. e; **2.** a; **3.** c; **4.** d; **5.** F; **6.** T; **7.** T; **8.** binary fission; **9.** nucleic acid, protein; **10.** retrovirus; **11.** coldsores, genital herpes; **12.** rod, spherical, spiral; **13.** PID; **14.** condom; **15.** gummas; **16.** gonorrhea, chlamydia, syphilis; **17.** yeast; **18.** living; **19. a.** Virus enters cell by endocytosis; uncoating occurs; **b.** Viral RNA is produced and travels to ribosomes where viral proteins are made. **c.** Copies of viral DNA are made. **d.** Viral DNA may integrate into host DNA (latency). **e.** Viral DNA and capsid proteins are assembled into new viruses; **f.** Virus exits by budding.

Applying Your Knowledge to the Concepts

1. STDs of bacterial origin can be treated and cured; whereas STDs of viral origin cannot be cured but, at present, only inhibited in their multiplication.

2. Some animal viruses, including HIV, cloak the host. These coatings

make it more difficult for the immune system to recognize the viruses as foreign.

3. The advent of the birth control pill makes it unnecessary to use other means of contraception, particularly the condom. Thus, STDs are more readily transmitted during sexual activity.

4. Pap tests detect the presence of cervical cancer, and there is a high correlation between the incidence of genital warts and cervical cancer.

Understanding the Terms

a. hepatitis; **b.** virus; **c.** pelvic inflammatory disease (PID); **d.** host; **e.** antibiotic

Chapter 17

Testing Your Knowledge of the Concepts

1. c; **2.** d; **3.** b; **4.** a; **5.** T; **6.** T; **7.** T; **8.** extraembryonic, amnion; **9.** implant; **10.** sperm, egg; **11.** cleavage; **12.** differentiation; **13.** second; **14.** placenta; **15.** oxytocin; **16. a.** chorion; **b.** amnion; **c.** embryo; **d.** allantois; **e.** yolk sac; **f.** fetal portion of placenta; **g.** maternal portion of placenta; **h.** chorionic villi

Applying Your Knowledge to the Concepts

1. The fertilized egg has time to develop to a stage so that it can implant itself in the uterine wall. If fertilization occurred in the uterus, the chances of implantation would be greatly reduced.

2. In the case of a low oxygen supply in the mother's blood, the fetus will be able to "pull" the oxygen from the mother's hemoglobin. This is an insurance for the well-being of the fetus.

3. Don't smoke, or drink alcohol to excess. Do get plenty of calcium in your diet and exercise regularly.

4. This condition is called polyspermy and prevents normal development of the zygote; therefore, the zygote would pass out with the menstrual flow or be absorbed.

Understanding the Terms

a. lanugo; **b.** differentiation; **c.** amnion; **d.** fertilization; **e.** endoderm

Chapter 18

Testing Your Knowledge of the Concepts

1. d; **2.** a; **3.** b; **4.** c; **5.** a; **6.** T; **7.** T; **8.** F; **9.** F; **10.** T; **11.** F; **12.** chromosomes; **13.** X, Y; **14.** 24; **15.** spermatogenesis; oogenesis; **16.** fertilization; **17.** Diagram to the right; **18.** Homologous pairs are at equator.

Applying Your Knowledge to the Concepts

1. Only nondisjunction of the sperm can result in a cell with two Ys (YY).

2. Chromatin becomes more dense and forms chromosomes, the nucleolus disappears, spindle fibers form, and the nuclear membrane fragments.

3. Turner syndrome is the result of having only one X chromosome; therefore, the cell contains one less than a complete 2n number. Klinefelter syndrome has 2 X plus 1 Y chromosome, and thus has a complete set, 2n, plus the extra X chromosome.

4. If the two cells separate and both develop, the result is identical twins.

Understanding the Terms

a. chromatid; **b.** cell cycle; **c.** spindle; **d.** polar body; **e.** haploid (n)

Chapter 19

Testing Your Knowledge of the Concepts

1. c; **2.** b; **3.** d; **4.** a; **5. a.** cystic fibrosis; **b.** Tay-Sachs disease; **c.** neuro fibromatosis; **d.** Huntington disease; **6.** F; **7.** F; **8.** T; **9.** phenotype; **10.** eeX^bY **11.** autosomal recessive

Applying Your Knowledge to the Concepts

1. Both of the parents might be carriers for a trait which could have been expressed in their grandmothers. The grandmothers would have been homozygous recessive.

2. Gene mutation may have occurred in a specific area of the world where a particular ethnic group lived, and there may have been little genetic exchange with outside groups.

3. A carrier is an individual who carries a hidden faulty gene; Huntington disease is caused by a dominant gene and thus is not hidden.

4. The blood type in this case neither proves nor disproves the man is the father. The mother could have passed the *B* gene to her offspring, and the suspected father would have passed an *O* gene, with the child then being genotype *BO*, type B.

Understanding the Terms

a. sex chromosome; **b.** genotype; **c.** Punnett square; **d.** carrier; **e.** polygenic inheritance

Chapter 20

Testing Your Knowledge of the Concepts

1. d; **2.** b; **3.** a; **4.** c; **5.** T; **6.** F; **7.** F; **8.** virus; **9.** copies; **10.** triplet, amino acids; **11.** Translation; **12.** transgenic; **13. a.** ACU´CCU´GGA´UGC´AAA; **b.** GA´GGA´CUU´ACG´UUU

Applying Your Knowledge to the Concepts

1. Covalent bonds are stronger and tend to hold the backbone together. Hydrogen bonds are much weaker and lend themselves to breaking to enable replication or transcription.

2. From the amino acid sequence, determine the mRNA base sequence using a table like Table 20.2; from the mRNA base sequence, determine the DNA base sequence; use a machine called a DNA synthesizer to make the gene.

3. If one defines a mutation as a change in genetic material resulting in a different phenotypic expression, then this is not a mutation.

4. Because the code is redundant, a change in DNA base sequence does not necessarily change the sequence of amino acids in a protein.

Understanding the Terms

a. polyribosome; **b.** transgenic organism; **c.** DNA profiling; **d.** transcription; **e.** adenine

Chapter 21

Testing Your Knowledge of the Concepts

1. b; **2.** c; **3.** a; **4.** d; **5.** T; **6.** T; **7.** F; **8.** human papilloma; **9.** chemotherapy; **10.** carcinogens; **11.** cigarette smoke; **12.** metastasize; **13.** proto-oncogenes, tumor-suppressor; **14.** tumor-suppressor gene; **15.** dividing; **16.** chemotherapy; **17.** immunotherapy; **18. a.** oncogene; **b.** tumor-suppressor gene; **c.** receptor; **d.** antigen; **e.** primary tumor; **f.** blood vessel; **g.** metastatic tumor; **h.** antigen (d); **i.** receptor, primary tumor (c, e); **j.** receptor, oncogene, tumor-suppressor gene (c, a, b); **k.** tumor (e); **l.** metastatic tumors (g).

Applying Your Knowledge to the Concepts

1. The chemicals are particularly hard on rapidly dividing cells. The hair bulb cells divide relatively rapidly in producing the hair shaft.

2. It is thought that feces remaining in the colon for lengthy periods results in the production of toxic materials which can serve as carcinogens. High fiber materials stimulate the movement of feces from the colon.

3. Estrogen stimulates growth of endometrial tissue; it would tend to amplify a cancer developing as cancer is an uncontrolled growth of cells.

4. A cancer drug should (1) medicate cancer cells without doing any harm to normal cells, and (2) do such a thorough job that the cancer will not come back. The standard therapies of surgery, radiation, and chemotherapy do not meet these criteria. Surgery which requires the removal of organs or portions of organs must necessarily also remove some normal cells along with the tumor. Radiation and chemotherapy bring about destruction of predominantly cancer cells but also some normal cells. And none of these methods guarantees that all cancer cells have been eradicated, and therefore the cancer may come back.

Understanding the Terms

a. immunotherapy; **b.** carcinogen; **c.** proto-oncogene; **d.** angiogenesis; **e.** metastasis

Chapter 22
Testing Your Knowledge of the Concepts

1. b; **2.** a; **3.** d; **4.** c; **5.** T; **6.** T; **7.** F; **8.** T; **9.** F; **10.** T; **11.** F; **12.** F; **13.** F; **14.** T; **15.** T; **16.** erect, small; **17.** Neanderthal, Cro-Magnon;

18.

Name	Example
Animalia	Multicellular, moves, ingests food
Chordata	Animals with dorsal supporting rod and nerve cord
Mammalia	Animals with hair, mammary glands
Primates	Apes, humans
Hominidae	hominids, *A. afarensis*
Homo	*H. habilis*, *H. erectus*
H. sapiens	Neanderthal, Cro-Magnon

See also Table 22.1, page 468.

Applying Your Knowledge to the Concepts

1. The DDT killed all those flies that were not resistant to it; the genetic material of the flies that were not killed enabled them to live in the presence of DDT. These flies lived to reproduce and thus a strain of flies resistant to DDT developed.

2. It is thought that the primitive earth contained little, if any, gaseous oxygen. The first cells had to carry on anaerobic respiration. As aerobic organisms evolved from the more primitive cells, they retained the enzymes for the reactions of anaerobic respiration, and aerobic respiration is built on top of these reactions.

3. Both present day humans and present day great apes have evolved a great deal from their primitive ancestors; therefore, both evolved from a distant relative.

4. A shorter jaw with fewer or no wisdom teeth must be adaptive today. If so, the number of people with fewer or no wisdom teeth will continue to increase.

Understanding the Terms

a. natural selection; **b.** adaptation; **c.** taxonomy; **d.** primate; **e.** fossil

Chapter 23
Testing Your Knowledge of the Concepts

1. d; **2.** a; **3.** b; **4.** c; **5.** T; **6.** T; **7.** T; **8.** nitrous oxide and nitrogen gas; **9.** respire; **10. a.** producers; **b.** consumers; **c.** inorganic nutrient pool; **d.** decomposers

Applying Your Knowledge to the Concepts

1. All plants and animals in an ecosystem are related either directly or indirectly to one another. If one member of a food web is removed, it will affect the other members of the web, for example.

2. Pigs are primary consumers and as such they trap only about 10% of the energy of corn. In addition, there is the cost of growing both corn and pigs and processing both for marketing.

3. Removing the forest or grassland results in a smaller amount of carbon dioxide being removed from the atmosphere. Building a factory and parking lot will result in more fossil fuels being consumed in the building and operation of the factory and thus more carbon dioxide being released into the atmosphere.

4. In many cases there are no natural predators to keep the new species in check, and they reproduce in large numbers, taking over the food supply of other animals.

Understanding the Terms

a. omnivore; **b.** food web; **c.** succession; **d.** ecological pyramid; **e.** detritus

Chapter 24
Testing Your Knowledge of the Concepts

1. b; **2.** d; **3.** c; **4.** a; **5.** F; **6.** T; **7.** T; **8.** F; **9.** T; **10.** F; **11.** T; **12.** T; **13.** F; **14.** F; **15.** F; **16.** T; **17.** T; **18.** thermal inversion; **19.** entering; **20. a.** lag; **b.** exponential growth; **c.** deceleration; **d.** environmental resistance; **e.** carrying capacity

Applying Your Knowledge to the Concepts

1. The death rate must increase or the birthrate decrease. Either will reduce the ratio of births to deaths.

2. The population will never be reduced unless the growth rate is reduced to below zero. It will continue to grow for about one generation and then will level off at a larger population than now exists.

3. The human population is increasing rapidly, producing more wastes; and a significant amount of the wastes are non-biodegradable.

4. Some experts believe that it will be possible to solve various problems like food supply and waste disposal, particularly if we all assume a less affluent lifestyle. They believe that much more food can be produced by better farming methods, and that the chief problem of food supply stems from inadequate distribution of food. They believe we could recycle more and develop methods so that humans would produce less waste.

Understanding the Terms

a. aquifer; **b.** growth rate; **c.** desertification; **d.** carrying capacity; **e.** doubling time

Appendix B

Metric System

Unit and Abbreviation	Metric Equivalent	Approximate English-to-Metric Equivalents	Units of Temperature
Length			
nanometer (nm)	$= 10^{-9}$ m		
micrometer (µm)	$= 10^{-6}$ m		
millimeter (mm)	$= 0.001 \ (10^{-3})$ m		
centimeter (cm)	$= 0.01 \ (10^{-2})$ m	1 inch = 2.54 cm 1 foot = 30.5 cm	
meter (m)	$= 100 \ (10^{2})$ cm $= 1,000$ mm	1 foot = 0.30 m 1 yard = 0.91 m	
kilometer (km)	$= 1,000 \ (10^{3})$ m	1 mi = 1.6 km	
Weight (mass)			
nanogram (ng)	$= 10^{-9}$ g		
microgram (µg)	$= 10^{-6}$ g		
milligram (mg)	$= 10^{-3}$ g		
gram (g)	$= 1,000$ mg	1 ounce = 28.3 g 1 pound = 454 g	
kilogram (kg)	$= 1,000 \ (10^{3})$ g	= 0.45 kg	
metric ton (t)	$= 1,000$ kg	1 ton = 0.91 t	
Volume			
microliter (µl)	$= 10^{-6}$ l $(10^{-3}$ ml$)$		
milliliter (ml)	$= 10^{-3}$ liter $= 1$ cm^3 (cc) $= 1,000$ mm^3	1 tsp = 5 ml 1 fl oz = 30 ml	
liter (l)	$= 1,000$ ml	1 pint = 0.47 liter 1 quart = 0.95 liter 1 gallon = 3.79 liter	
kiloliter (kl)	$= 1,000$ liter		

Thermometer scale (F° / C°):

F° values: 230, 220, 212°→210, 200, 190, 180, 170, 160°→160, 150, 140, 134°/131°→130, 120, 105.8°→110, 98.6°→100, 90, 80, 68.6°→70, 60, 50, 40, 32°→30, 20, 10, 0, −10, −20, −30, −40

C° values: 110, 100—100°, 90, 80, 70—71°, 60, 57°, 50, 40—41°, 37°, 30, 20—20.3°, 10, 0—0°, −10, −20, −30, −40

°C	°F	
100	212	Water boils at standard temperature and pressure
71	160	Flash pasteurization of milk
57	134	Highest recorded temperature in the United States, Death Valley, July 10, 1913
41	105.8	Average body temperature of a marathon runner in hot weather
37	98.6	Human body temperature
20.3	68.6	Human survival is still possible at this temperature
0	32.0	Water freezes at standard temperature and pressure

To convert temperature scales:

$$°C = \frac{5(°F - 32)}{9}$$

$$°F = \frac{9°C}{5} + 32$$

Appendix C

Periodic Table of the Elements

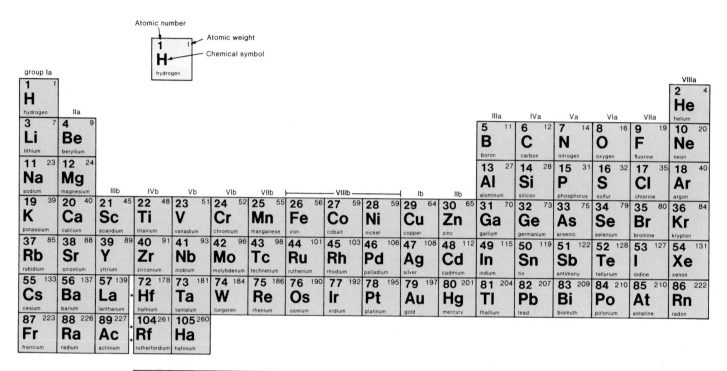

Appendix D

Drugs of Abuse

	Drugs	Often Prescribed Brand Names	Medical Uses	Potential Physical Dependence	Potential Psychological Dependence	Tolerance
Narcotics	Opium	Dover's Powder, Paregoric	Analgesic, antidiarrheal	High	High	Yes
	Morphine	Morphine	Analgesic	High	High	Yes
	Codeine	Codeine	Analgesic, antitussive	Moderate	Moderate	Yes
	Heroin	None	None	High	High	Yes
	Meperidine (Pethidine)	Demerol, Pethadol	Analgesic	High	High	Yes
	Methadone	Dolophine, Methadone, Methadose	Analgesic, heroin substitute	High	High	Yes
	Other Narcotics	Dilaudid, Leritine, Numorphan, Percodan	Analgesic, antidiarrheal, antitussive	High	High	Yes
Depressants	Chloral Hydrate	Noctec, Somnos	Hypnotic	Moderate	Moderate	Probable
	Barbiturates	Amytal, Butisol, Nembutal, Phenobarbitol, Seconal, Tuinal	Anesthetic, anticonvulsant, sedation, sleep	High	High	Yes
	Glutethimide	Doriden	Sedation, sleep	High	High	Yes
	Methaqualone	Optimil, Parest, Quaalude, Somnafac, Sopor	Sedation, sleep	High	High	Yes
	Tranquilizers	Equanil, Librium, Miltown, Serax, Tranxene, Valium	Antianxiety, muscle relaxant, sedation	Moderate	Moderate	Yes
	Other Depressants	Clonopin, Dalmane, Dormate, Noludar, Placydil, Valmid	Antianxiety, sedation, sleep	Possible	Possible	Yes
Stimulants	Cocaine*	Cocaine	Local anesthetic	Possible	High	Yes
	Amphetamines	Benzedrine, Biphetamine, Desoxyn, Dexedrine	Hyperkinesis, narcolepsy, weight control	Possible	High	Yes
	Phenmetrazine	Preludin	Weight control	Possible	High	Yes
	Methylphenidate	Ritalin	Hyperkinesis	Possible	High	Yes
	Other Stimulants	Bacarate, Cylert, Didrex, Ionamin, Plegine, Pondimin, Pro-Sate, Sanorex, Voranil	Weight control	Possible	Possible	Yes
Hallucinogens	LSD	None	None	None	Degree unknown	Yes
	Mescaline	None	None	None	Degree unknown	Yes
	Psilocybin-Psilocyn	None	None	None	Degree unknown	Yes
	MDA	None	None	None	Degree unknown	Yes
	PCP†	Sernylan	Veterinary anesthetic	None	Degree unknown	Yes
	Other Hallucinogens	None	None	None	Degree unknown	Yes
Cannabis	Marijuana, Hashish, Hashish Oil	None	Glaucoma	Degree unknown	Moderate	Yes

Source: Drugs of Abuse, produced by the Affairs in Cooperation with the Office of Public Science and Technology.

*Designated a narcotic under the Controlled Substances Act.

†Designated a depressant under the Controlled Substances Act.

Duration of Effects (in hours)	Usual Methods of Administration	Possible Effects	Effects of Overdose	Withdrawal Syndrome
3–6	Oral, smoked	Euphoria, drowsiness, respiratory depression, constricted pupils, nausea	Slow and shallow breathing, clammy skin, convulsions, coma, possible death	Watery eyes, runny nose, yawning, appetite loss, irritability, tremors, panic, chills
3–6	Injected, smoked			
3–6	Oral, injected			
3–6	Injected, sniffed			
3–6	Oral, injected			
12–24	Oral, injected			
3–6	Oral, injected			
5–8	Oral			
1–16	Oral, injected	Slurred speech, disorientation, drunken behavior without odor of alcohol	Shallow respiration, cold and clammy skin, dilated pupils, weak and rapid pulse, coma, possible death	Anxiety, insomnia, tremors, delirium, convulsions, possible death
4–8	Oral			
4–8	Oral			
4–8	Oral			
4–8	Oral			
2	Injected, sniffed	Increased alertness, excitation, euphoria, dilated pupils, increased pulse rate and blood pressure, insomnia, loss of appetite	Agitation, increased body temperature, hallucinations, convulsions, possible death	Apathy, long periods of sleep, irritability, depression, disorientation
2–4	Oral, injected			
2–4	Oral			
2–4	Oral			
2–4	Oral			
Variable	Oral	Illusions and hallucinations (with the exception of MDA), poor perception of time and distance	Longer, more intense "trip" episodes, psychosis, possible death	Withdrawal syndrome not reported
Variable	Oral, injected			
Variable	Oral			
Variable	Oral, injected, sniffed			
Variable	Oral, injected, smoked			
Variable	Oral, injected, sniffed			
2–4	Oral, smoked	Euphoria, relaxed inhibitions, increased appetite, disoriented behavior	Fatigue, paranoia, possible psychosis	Insomnia, hyperactivity, and decreased appetite reported in a limited number of individuals

Appendix E

Selected Types of Cancer

Cancers are classified according to the type of tissue from which they arise. Carcinomas, the most common type, are cancers of epithelial tissues; sarcomas are cancers arising in muscle or connective tissue (especially bone or cartilage); leukemias are cancers of the blood; and lymphomas are cancers of lymphoid tissue. The chance of developing cancer in a particular tissue shows a positive correlation to the rate of cell division; new blood cells arise at a rate of 2,500,000 cells per second, and epithelial cells also reproduce at a high rate.

Five-year survival rates for the most common cancers are given in Table E.

Breast Cancer

Signs and Symptoms: Pre-clinical radiographic signs seen on a mammogram. Breast changes, such as a lump, thickening, swelling, dimpling, skin irritation, distortion, retraction, scaliness, pain, tenderness of the nipple, or nipple discharge.

Risk Factors: The risk of breast cancer increases with age. The risk is higher in the woman who has a personal or family history of breast cancer; some forms of benign breast disease; early onset of menstruation; late menopause; lengthy exposure to cyclic estrogen; never having children or having the first live birth at a later age; and higher education and socioeconomic status. International variability in breast cancer incidence rates correlates with variations in diet, especially fat intake, although a causal role for dietary factors has not been firmly established. Additional factors that may be associated with increased breast cancer risk and that are currently under study include pesticide and other chemical exposures, alcohol consumption, induced abortion, and physical inactivity. A majority of women will have one or more risk factors for breast cancer. However, most risks are at such a low level that they only partly explain the high frequency of the disease in the population. To date, knowledge about risk factors has not translated into practical ways to prevent breast cancer. Since women may not be able to alter their personal risk factors, the best opportunity at present for reducing mortality is through early detection.

Early Detection: The American Cancer Society recommends that asymptomatic women aged 40 to 49 should have a screening mammogram every one to two years; and women aged 50 and over should have a mammogram every year. In addition, a clinical breast exam is recommended every three years for women 20 to 40, and every year for women over 40. The American Cancer Society also recommends monthly breast self-exam as a routine good health habit for women 20 years or older. Most breast lumps are not cancer, but only a physician can make a diagnosis.

Mammography is recognized as a valuable diagnostic technique for women who have findings suggestive of breast cancer. When a suspicious area is identified on a mammogram, or when a woman has a suspicious lump, mammography can help determine if there are other lesions too small to be felt in the same or opposite breast. Since a small percentage of breast cancers may not be seen on a mammogram, all suspicious lumps should be biopsied for a definitive diagnosis, even when current or recent mammography findings are described as normal.

Treatment: Taking into account the medical situation and the patient's preferences, treatment may involve lumpectomy (local removal of the tumor), mastectomy (surgical removal of the breast), radiation therapy, chemotherapy, or hormone therapy. Often, two or more methods are used in combination. Patients should discuss possible options for the best management of their breast cancer with their physicians.

In recent years, new techniques have made breast reconstruction possible after mastectomy, and the cosmetic results usually are good. Reconstruction has become an important part of treatment and rehabilitation. Bone marrow transplantation is another new type of treatment that is under study.

Lung Cancer

Signs and Symptoms: Persistent cough, sputum streaked with blood, chest pain, recurring pneumonia or bronchitis.

Risk Factors: Cigarette smoking is by far the most important risk factor in the development of lung cancer. Other factors include exposure to certain industrial substances, such as arsenic; certain organic chemicals and asbestos, particularly for persons who smoke; radiation exposure from occupational, medical, and environmental sources; air pollution; tuberculosis; radon exposure, especially in cigarette smokers; and environmental tobacco smoke in nonsmokers.

Early Detection: Because symptoms often don't appear until the disease is advanced, early detection is difficult. In smokers who stop smoking when precancerous changes are found, damaged lung tissue often returns to normal. Smokers who persist in smoking may form abnormal cell growth patterns that lead to cancer. Chest X ray, analysis of the types of cells contained in sputum, and fiberoptic examination of the bronchial passages assist diagnosis.

Treatment: Determined by the type and stage of the cancer. Options include surgery, radiation therapy, and

Table E	Five-Year Survival Rates by Stage at Diagnosis*			
Site	All Stages %	Local %	Regional %	Distant %
Oral	53	81	42	18
Colon	62	93	67	8
Rectum	60	88	55	5
Pancreas	4	15	5	2
Lung	14	49	18	2
Skin (melanoma)	88	95	61	16
Female Breast	84	97	76	21
Cervix	69	91	49	9
Uterus	84	96	66	27
Ovary	46	93	55	25
Prostate	89	100	94	31
Bladder	81	94	49	6
Kidney	59	88	60	9

*Adjusted for normal life expectancy. This chart is based on cases diagnosed from 1986 to 1993, followed through 1994.

Source: NCI surveillance, Epidemiology, and End Results Program, 1997.

chemotherapy. For many localized cancers, surgery is usually the treatment of choice. Because the disease has usually spread by the time it is discovered, radiation therapy and chemotherapy are often needed in combination with surgery as the treatment of choice; on this regimen, a large percentage of patients experience remission, which in some cases is long-lasting.

Prostate Cancer

Signs and Symptoms: Weak or interrupted urine flow; inability to urinate, or difficulty starting or stopping the urine flow; the need to urinate frequently, especially at night; blood in the urine; pain or burning on urination; continuing pain in lower back, pelvis, or upper thighs. Most of these symptoms are nonspecific and may be similar to those caused by benign conditions such as infection or prostate enlargement.

Risk Factors: The incidence of prostate cancer increases with age; over 80% of all prostate cancers are diagnosed in men over age 65. African Americans have the highest prostate cancer incidence rates in the world; the disease is common in North America and Northwestern Europe and is rare in the Near East, Africa, and South America. There may be some familial tendency, but it is unclear whether this is due to genetic or environmental factors. International studies suggest that dietary fat may also be a factor.

Early Detection: Every man aged 40 and over should have a digital rectal exam as part of his regular annual physical checkup. In addition, the American Cancer Society recommends that men aged 50 and over have an annual prostate-specific antigen blood test. If either result is suspicious, further evaluation in the form of transrectal ultrasound should be performed.

Treatment: Surgery, radiation, and/or hormones and anticancer drugs are treatment options. Hormone treatment and anticancer drugs may control prostate cancer for long periods by shrinking the size of the tumor, thus relieving pain. Careful observation without immediate active treatment may be appropriate for individuals with low-grade and/or early-stage tumors.

Colon and Rectal Cancer

Signs and Symptoms: Rectal bleeding, blood in the stool, a change in bowel habits.

Risk Factors: Personal or family history of colorectal cancer or polyps, and inflammatory bowel disease have been associated with increased colorectal cancer risk. Other possible risk factors include physical inactivity and high-fat and/or low-fiber diet. Recent studies have suggested that estrogen replacement therapy and nonsteroidal anti-inflammatory drugs such as aspirin may reduce colorectal cancer risk.

Early Detection: Digital rectal exam, stool blood test, and sigmoidoscopy are recommended by the American Cancer Society to detect colon or rectal cancer in asymptomatic patients. These tests offer the best opportunity for the diagnosis and removal of polyps, and hence the prevention of these cancers.

Digital rectal exam is performed by a physician during an office visit. The American Cancer Society recommends that this exam be performed annually after age 40.

The stool blood test is a simple method to test feces for hidden blood. The specimen is obtained by the patient at home and returned to the physician's office, a hospital, or a clinic for analysis. The Society recommends annual testing after age 50.

In sigmoidoscopy, the physician uses a hollow, lighted tube or a fiberoptic sigmoidoscope to inspect the rectum and lower colon. The American Cancer Society recommends sigmoidoscopy every three to five years after age 50.

If any of these tests reveal possible problems, more extensive studies, such as colonoscopy (exam of the entire colon) and barium enema (an X-ray procedure in which the intestines are viewed), may be needed.

Treatment: Surgery, at times combined with radiation, is the most effective method of treating colorectal cancer. The role of chemotherapy and immunologic agents may be beneficial in postoperative patients with cancerous lymph nodes.

Colostomy (creation of an abdominal opening for elimination of body wastes) is seldom needed for colon cancer and is infrequently required for rectal cancer. The American Cancer Society has a patient assistance program for those who do have permanent colostomies.

Bladder Cancer

Signs and Symptoms: Blood in the urine. Usually associated with increased frequency of urination.

Risk Factors: Smoking is the greatest risk factor in bladder cancer, with smokers experiencing twice the risk of non-smokers. Smoking is estimated to be responsible for approximately 47% of the bladder cancer deaths among men and 37% among women. People living in urban areas and workers exposed to dye, rubber, or leather also are at higher risk. Chlorination in water has also been associated with a small increase in bladder cancer risk.

Early Detection: Bladder cancer is diagnosed by examination of the bladder wall with a cytoscope, a slender tube fitted with a lens and light that can be inserted into the tract through the urethra.

Treatment: Surgery, alone or in combination with other treatments, is used in over 90% of cases. Preoperative chemotherapy, alone or with radiation before cystectomy (bladder removal), has improved some treatment results.

Lymphoma

Signs and Symptoms: Enlarged lymph nodes, itching, fever, night sweats, anemia, and weight loss. Fever can come and go in periods of several days or weeks.

Risk Factors: Risk factors are largely unknown, but in part involve reduced immune function and exposure to certain infectious agents. Persons with organ transplants are at higher risk due to altered immune function. Human immunodeficiency virus (HIV) and human T lymphocyte leukemia/lymphoma virus-I (HTLV-I) are associated with increased risk of non-Hodgkin lymphoma. Burkitt lymphoma in Africa is partly caused by the Epstein-Barr herpes virus. Other possible risk factors include occupational exposures to herbicides and perhaps other chemicals.

Treatment: Hodgkin disease: chemotherapy and radiotherapy are useful for most patients. Non-Hodgkin lymphoma: early stage, localized lymph node disease can be treated with radiotherapy. Patients with later stage disease often benefit from the addition of chemotherapy. New programs using highly specific monoclonal antibodies directed at lymphoma cells and improved techniques in bone marrow preservation are under investigation in selected patients who relapse after standard treatment.

Cervical Cancer

Signs and Symptoms: Abnormal uterine bleeding or spotting; abnormal vaginal discharge. Pain and systemic symptoms are late manifestations of the disease.

Risk Factors: First intercourse at an early age, multiple sexual partners, or partners who have had multiple sexual partners are at increased risk of developing the disease. Other risk factors include genital warts and cigarette smoking.

Early Detection: The Pap test is a simple procedure that can be performed at appropriate intervals by health care professionals as part of a pelvic exam. A small sample of cells is swabbed from the cervix, transferred to a slide, and examined under a microscope. This test should be performed annually with a pelvic exam in women who are, or have been, sexually active or who have reached age 18 years. After three or more consecutive annual exams with normal findings, the Pap test may be performed less frequently at the discretion of the physician.

Treatment: Cervical cancers generally are treated by surgery or radiation, or by a combination of the two. In precancerous (in situ) stages, changes in the cervix may be treated by cryotherapy (the destruction of cells by extreme cold), by electrocoagulation (the destruction of tissue through intense heat by electric current), or by local surgery.

Skin Cancer

Signs and Symptoms: Any change in the skin, especially a change in the size or color of a mole or other darkly pigmented growth or spot. Scaliness, oozing, bleeding, or change in the appearance of a bump or nodule, the spread of pigmentation beyond its border, a change in sensation, itchiness, tenderness, or pain.

Risk Factors: Excessive exposure to ultraviolet radiation; fair complexion; occupational exposure to coal tar, pitch, creosote, arsenic compounds, or radium; family history.

Protective clothing and sunscreen should be worn. The sun's ultraviolet rays are strongest between 10 A.M. and 3 P.M.; exposure at these times should be avoided. Sunscreen

comes in various strengths, ranging from those facilitating gradual tanning to those that allow practically no tanning. Because of the possible link between severe sunburns in childhood and greatly increased risk of melanoma in later life, children, in particular, should be protected from the sun.

Early Detection: Early detection is critical. Recognition of changes in skin growths or the appearance of new growths is the best way to find early skin cancer. Adults should practice skin self-exam once a month, and suspicious lesions should be evaluated promptly by a physician. Basal and squamous cell skin cancers often take the form of a pale, waxlike, pearly nodule, or a red, scaly, sharply outlined patch. A sudden or progressive change in a mole's appearance should be checked by a physician. Melanomas often start as small, molelike growths that increase in size, change color, become ulcerated, and bleed easily from a slight injury. A simple **ABCD** rule outlines the warning signals of melanoma: **A** is for asymmetry. One-half of the mole does not match the other half. **B** is for border irregularity. The edges are ragged, notched, or blurred. **C** is for color. The pigmentation is not uniform. **D** is for diameter greater than 6 mm. Any sudden or progressive increase in size should be of special concern.

Treatment: There are five methods of treatment: surgery (used in 90% of cases), radiation therapy, electrodesiccation (tissue destruction by heat), cryosurgery (tissue destruction by freezing), and laser therapy for early skin cancer. For malignant melanoma, the primary growth must be adequately excised, and it may be necessary to remove nearby lymph nodes. Removal and microscopic examination of all suspicious moles is essential. Advanced cases of melanoma are treated according to the characteristics of the case.

Oral Cavity and Pharynx Cancer

Signs and Symptoms: A sore that bleeds easily and does not heal; a lump or thickening; a red or white patch that persists. Difficulty in chewing, swallowing, or moving tongue or jaws are often late symptoms.

Risk Factors: Cigarette, cigar, or pipe smoking; use of smokeless tobacco; excess use of alcohol.

Early Detection: Cancer can affect any part of the oral cavity, including the lip, tongue, mouth, and throat. Dentists and primary care physicians have the opportunity during regular checkups to see abnormal tissue changes and to detect cancer at an early, curable stage.

Treatment: Principal methods are radiation therapy and surgery. In advanced disease, chemotherapy is being studied as an adjunct to surgery.

Leukemia

Signs and Symptoms: Fatigue, paleness, weight loss, repeated infections, bruising easily, and nosebleeds or other hemorrhages. In children, these symptoms can appear sud-

denly. Chronic leukemia can progress slowly and with few symptoms.

Risk Factors: Leukemia strikes both sexes and all ages. Causes of most leukemias are unknown. Persons with Down syndrome and certain other genetic abnormalities have higher than usual incidence rate of leukemia. It has also been linked to excessive exposure to ionizing radiation and to certain chemicals such as benzene, a commercially used toxic liquid that is also present in lead-free gasoline. Certain forms of leukemia and lymphoma are caused by a retrovirus, HTLV-I (human T lymphocyte leukemia/lymphoma virus-I).

Early Detection: Because symptoms often resemble those of other, less serious conditions, leukemia can be difficult to diagnose early. When a physician does suspect leukemia, diagnosis can be made using blood tests and bone marrow biopsy.

Treatment: Chemotherapy is the most effective method of treating leukemia. Various anticancer drugs are used, either in combination or as single agents. Transfusions of blood components and antibiotics are used as supportive treatments. To eliminate hidden cells, therapy of the central nervous system has become standard treatment, especially in acute lymphocytic leukemia. Under appropriate conditions, bone marrow transplantation may be useful in the treatment of certain leukemias.

Pancreas Cancer

Signs and Symptoms: Cancer of the pancreas generally occurs without symptoms until it is in advanced stages and thus is considered a "silent" disease.

Risk Factors: Very little is known about what causes the disease or how to prevent it. Risk increases after age 50, with most cases occurring between ages 65 and 79. Smoking is a risk factor; incidence rates are more than twice as high for smokers as nonsmokers. Some studies have suggested associations with chronic pancreatitis, diabetes, or cirrhosis. In countries where the diet is high in fat, pancreatic cancer rates are higher.

Early Detection: At present, only surgical biopsy yields a certain diagnosis, and because of the "silent" course of the disease, the need for biopsy is likely to be obvious only after the disease has advanced. Researchers are focusing on ways to diagnose pancreatic cancer before symptoms occur. Ultrasound imaging and computerized tomography scans are under investigation.

Treatment: Surgery, radiation therapy, and anticancer drugs are treatment options, but they have had little influence on the outcome. Diagnosis is usually so late that none of these is used.

Ovarian Cancer

Signs and Symptoms: Ovarian cancer is often "silent," showing no obvious signs or symptoms until late in its

development. The most common sign is enlargement of the abdomen, which is caused by the accumulation of fluid. Rarely will there be abnormal vaginal bleeding. In women over 40, vague digestive disturbances (stomach discomfort, gas, distention) that persist and cannot be explained by any other cause may indicate the need for a thorough evaluation for ovarian cancer.

Risk Factors: Risk for ovarian cancer increases with age. Women who have never had children are more likely to develop ovarian cancer than those who have. Pregnancy and the use of oral contraceptives appear to be protective against ovarian cancer. Women who have had breast cancer or have a family history of ovarian cancer are at increased risk. With the exception of Japan, industrialized countries have the highest incidence rates.

Early Detection: Periodic, thorough pelvic examinations are important. The Pap test, useful in detecting cervical cancer, only rarely uncovers ovarian cancer. Transvaginal ultrasound and a tumor marker, CA 125, may assist diagnosis. Women over the age of 40 should have a cancer-related checkup every year.

Treatment: Surgery, radiation therapy, and drug therapy are treatment options. Surgery usually includes the removal of one or both ovaries (oophorectomy), the uterus (hysterectomy), and the Fallopian tubes (salpingectomy). In some very early tumors, only the involved ovary will be removed, especially in young women. In advanced disease, an attempt is made to remove all intraabdominal disease to enhance the effect of chemotherapy.

Cancer in Children

New Cases: An estimated 8,700 new cases in 1998. As a childhood disease, cancer is rare. Common sites include the blood and bone marrow, bone, lymph nodes, brain, nervous system, kidneys, and soft tissues.

Some of the main childhood cancers are as follows:

1. Leukemia.
2. Osteogenic sarcoma is a bone cancer that may cause no pain at first; swelling in the area of the tumor is often the first sign. Ewing's sarcoma is another type of cancer that arises in bone.
3. Neuroblastoma can appear anywhere but usually in the abdomen, where a swelling occurs.
4. Rhabdomyosarcoma, the most common soft tissue sarcoma, can occur in the head and neck area, genitourinary area, trunk, and extremities.
5. Brain cancers in early stages may cause headaches, blurred or double vision, dizziness, difficulty in walking or handling objects, and nausea.
6. Lymphomas and Hodgkin disease are cancers that involve the lymph nodes but also may invade bone marrow and other organs. They may cause swelling of lymph nodes in the neck, armpit, or groin. Other symptoms may include general weakness and fever.
7. Retinoblastoma, an eye cancer, usually occurs in children under age 4. When detected early, cure is possible with appropriate treatment.
8. Wilms' tumor, a kidney cancer, may be recognized by a swelling or lump in the abdomen.

Childhood cancers can be treated by a combination of therapies. Treatment is coordinated by a team of experts including oncologic physicians, pediatric nurses, social workers, psychologists, and others who assist children and their families.

Source: From the American Cancer Society, Inc. Cancer Facts & Figures—1998. Used with permission.

Appendix F

Acronyms

Acronym	Meaning	Acronym	Meaning
ACh	acetylcholine	IUD	intrauterine device
AChE	acetylcholinesterase	IVF	in vitro fertilization
ACTH	adrenocorticotropic hormone	LDC	less-developed country
AD	Alzheimer disease	LDL	low-density lipoprotein
ADH	antidiuretic hormone	LH	luteinizing hormone
ADP	adenosine diphosphate	MDC	more-developed country
AIDS	acquired immunodeficiency syndrome	MHC	major histocompatibility complex
ANH	atrial natriuretic hormone	MI	myocardial infarction
APC	antigen-presenting cell	MRI	magnetic resonance imaging
ATP	adenosine triphosphate	MSH	melanocyte-stimulating hormone
AV node	atrioventricular node	MYA	millions of years ago
BP	before present	NAD^+	nicotinamide adenine dinucleotide
CAM	crassulacean-acid metabolism	$NADP^+$	nicotinamide adenine dinucleotide phosphate
cAMP	cyclic adenosine monophosphate		
CAPD	continuous ambulatory peritoneal dialysis	NE	norepinephrine
CAT	computerized axial tomography	NGF	nerve growth factor
CCK	cholecystokinin	NGU	nongonococcal urethritis
cDNA	complementary deoxyribonucleic acid	NOX	nitrogen oxides
CEA	carcinoembryonic antigen	NPS	nail patella syndrome
CFC	chlorofluorocarbon	PAN	peroxyacetyl nitrate
cGMP	cyclic guanosine monophosphate	PCB	polychlorinated biphenyl
CNS	central nervous system	PCR	polymerase chain reaction
CoA	coenzyme A	PG	prostaglandin
CVA	cardiovascular accident	PGA	phosphoglycerate
DNA	deoxyribonucleic acid	PGAL	phosphoglyceraldehyde
DTP	diphtheria, tetanus, whooping cough	PGAP	diphosphoglycerate
ECG (or EKG)	electrocardiogram	PID	pelvic inflammatory disease
ER	endoplasmic reticulum (rough ER, smooth ER)	PNS	peripheral nervous system
		PRL	prolactin
FAD	flavin adenine dinucleotide	PSA	prostate-specific antigen
FAP	fixed action patterns	PS I	photosystem I
FAS	fetal alcohol syndrome	PS II	photosystem II
FSH	follicle-stimulating hormone	PTH	parathyroid hormone
GABA	gamma-aminobutyric acid	rDNA	recombinant deoxyribonucleic acid
GDGF	glial-derived growth factor	RFLP	restriction fragment length polymorphism
GEM	genetically engineered microbe	RNA	ribonucleic acid
GH	growth hormone	rRNA	ribosomal ribonucleic acid
GIFT	gamete intrafallopian transfer	RuBP	ribulose bisphosphate
GIP	gastric inhibitory peptide	SA node	sinoatrial node
GMP	guanosine monophosphate	SAD	seasonal affective disorder
GnRH	gonadotropic-releasing hormone	SCID	severe combined immunodeficiency syndrome
GTP	guanosine triphosphate		
HC	hydrocarbon	SPF	sun protection factor
HCG	human chorionic gonadotropin	STD	sexually transmitted disease
HDL	high-density lipoprotein	THC	tetrahydrocannabinol
HDN	hemolytic disease of the newborn	TIMP	tissue inhibitor metalloproteinase
HHb	hemoglobin	tPA	tissue plasminogen activator
HIV	human immunodeficiency virus	TSH	thyroid-stimulating hormone
HPV	human papillomavirus	UV	ultraviolet
ICSH	interstitial cell-stimulating hormone		

Glossary

A

acetylcholine (ACh) (uh-seet-ul-KOH-leen) Neurotransmitter active in both the peripheral and central nervous systems.   251

acetylcholinesterase (AChE) (uh-SEET-ul-koh-luh-nes-tuh-rays, -rayz) Enzyme that breaks down acetylcholine bound to postsynaptic receptors within a synapse.   251

acid Solution in which pH is less than 7; a substance that contributes or liberates hydrogen ions (protons) in a solution.   23

acid deposition Acid rain or snow due to the presence of sulfate and/or nitrate produced by commercial and industrial activities.   502

acquired immunodeficiency syndrome (AIDS) (im-yuh-noh-dih-FISH-un-see) Disease caused by a group of related retroviruses and transmitted via body fluids; characterized by failure of the immune system.   342

acromegaly (ak-roh-MEG-uh-lee) Condition resulting from an increase in growth hormone production after adult height has been achieved.   300

acrosome (AK-ruh-sohm) Covering on the tip of a sperm cell's nucleus that is believed to contain enzymes necessary for fertilization.   321

actin (AK-tin) One of two major proteins of muscle; makes up thin filaments in myofibrils of muscle fibers. See myosin.   231

action potential Polarity changes due to the movement of ions across the plasma membrane of an active neuron; nerve impulse.   248

active site Region on the surface of an enzyme where the substrate binds and where the reaction occurs.   54

active transport Transfer of a substance into or out of a cell from a region of lower concentration to a region of higher concentration by a process that requires a carrier and an expenditure of energy.   48

acute bronchitis (brahn-KY-tis, brahng-) Infection of the primary and secondary bronchi.   179

adaptation Fitness of an organism for its environment, including the process by which it becomes fit and is able to survive and to reproduce. Also, decrease in the excitability of a receptor in response to continuous constant-intensity stimulation.   465

Addison disease Condition resulting from a deficiency of adrenal cortex hormones; characterized by low blood glucose, weight loss, and weakness.   305

adenine (A) (AD-un-een) One of four nitrogen bases in nucleotides composing the structure of DNA and RNA.   423

adhesion junction Junction between cells in which the adjacent plasma membranes do not touch but are held together by intercellular filaments attached to buttonlike thickenings.   64

adipose tissue (AH-duh-pohs) Fat storing loose connective tissue.   64

ADP (adenosine diphosphate) (ah-DEN-ah-seen dy-FAHS-fayt) Nucleotide with two phosphate groups that can accept another phosphate group and become ATP.   35

adrenal cortex (uh-DREE-nul KOR-teks) Outer portion of the adrenal gland; secretes hormones such as mineralocorticoid aldosterone and glucocorticoid cortisol.   303

adrenal gland (uh-DREE-nul) Gland that lies atop a kidney; the adrenal medulla produces the hormones epinephrine and norepinephrine, and the adrenal cortex produces the glucocorticoid and mineralocorticoid hormones.   303

adrenal medulla (uh-DREE-nul muh-DUL-uh) Inner portion of the adrenal gland; secretes the hormones epinephrine and norepinephrine.   303

adrenocorticotropic hormone (ACTH) (uh-DREE-noh-kawrt-ih-koh-troh-pik) Hormone secreted by the anterior lobe of the pituitary gland that stimulates activity in the adrenal cortex.   298

aerobic cellular respiration Metabolic reactions that provide energy to cells by the step-by-step oxidation of substrates with the concomitant buildup of ATP molecules.   55

agglutination (uh-gloot-un-AY-shun) Clumping of cells, particularly in reference to red blood cells involved in an antigen-antibody reaction.   120, 278

aging Progressive changes over time, leading to loss of physiological function and eventual death.   378

agranular leukocyte White blood cell that does not contain distinctive granules.   115

albumin (al-BYOO-mun) Plasma protein of the blood having transport and osmotic functions.   118

aldosterone (al-DAHS-tuh-rohn) Hormone secreted by the adrenal cortex that regulates the sodium and potassium balance of the blood.   197, 304

allantois (uh-LAN-toh-is) Extraembryonic membrane that accumulates nitrogenous wastes in reptiles and birds and contributes to the formation of umbilical blood vessels in mammals, including humans.   368

allele (uh-LEEL) Alternative form of a gene located at a particular chromosome site (locus).   404

allergen (AL-ur-jun) Foreign substance capable of stimulating an allergic response.   158

allergy Immune response to substances that usually are not recognized as foreign.   158

alveolus (pl., alveoli) (al-VEE-uh-lus) Air sac of a lung.   169

amino acid Monomer of a protein; takes its name from the fact that it contains an amino group ($-NH_2$) and an acid group ($-COOH$).   31

amnion (AM-nee-ahn) Extraembryonic membrane of reptiles, birds, and mammals that forms an enclosing, fluid-filled sac.   368

ampulla (am-POOL-uh, -PUL-uh) Base of a semicircular canal in the inner ear.   289

amygdala (uh-MIG-duh-luh) Almond-shaped mass of gray matter in the anterior extremity of the temporal lobe.   262

anabolic steroid (a-nuh-BAHL-ik) Synthetic steroid that mimics the effect of testosterone.   309

anaphase Mitosis phase during which daughter chromosomes move toward poles of spindle.   393

androgen (AN-druh-jun) Male sex hormone (e.g., testosterone).   309

anemia (uh-NEE-mee-uh) Inefficiency in the oxygen-carrying ability of blood due to a shortage of hemoglobin.   113

aneurysm (AN-yuh-riz-um) Ballooning of a blood vessel.   138

angina pectoris (an-JY-nuh PEK-tuh-ris) Condition characterized by thoracic pain resulting from occluded coronary arteries; precedes a heart attack.   138

angiogenesis (an-jee-oh-JEN-uh-sis) Formation of new blood vessels; one mechanism by which cancer spreads.   444

angioplasty (AN-jee-uh-plas-tee) Surgical procedure for treating clogged arteries, in which a plastic tube is threaded through a major blood vessel toward the heart and then a balloon at the end of the tube is inflated, forcing open the vessel.   138

anorexia nervosa (a-nuh-REK-see-uh nur-VOH-suh) Eating disorder characterized by a morbid fear of gaining weight. Symptoms include distorted body image, absence of menstrual cycle for at least three months, excessive exercise, self-induced vomiting or misuse of laxatives.   105

anterior pituitary (pih-TOO-ih-tair-ee) Portion of the pituitary gland that produces six types of hormones and is controlled by hypothalamic-releasing and release-inhibiting hormones.   298

antibiotic Drug that interferes with a bacterial enzyme and in that way causes the death of a particular bacterium.   347

antibody (AN-tih-bahd-ee) Protein produced in response to the presence of an antigen; each antibody combines with a specific antigen.   150

antibody-mediated immunity Specific mechanism of defense in which plasma cells derived from B cells produce antibodies that combine with antigens.   151

anticodon (an-tih-KOH-dahn) "Triplet" of bases in tRNA that pairs with a complementary triplet (codon) in mRNA.   428

antidiuretic hormone (ADH) (an-tih-dy-uh-RET-ik) Hormone secreted by the posterior pituitary that promotes the reabsorption of water from the collecting ducts, which receive urine produced by nephrons within the kidneys.   196, 298

antigen (AN-tih-jun) Foreign substance, usually a protein or a polysaccharide, that stimulates the immune system to react, such as to produce antibodies.   150

antigen-presenting cell (APC) Cell that displays the antigen to the cells of the immune system so they can defend the body against that particular antigen.   154

anus Outlet of the digestive tube.   88

aorta (ay-OR-tuh) Major systemic artery that receives blood from the left ventricle.   129

aortic bodies Sensory receptors in the aortic arch sensitive to oxygen content, carbon dioxide content, and blood pH.   172

apoptosis (ap-ohp-TOH-sis, -ahp-) Programmed cell death involving a cascade of specific cellular events leading to death and destruction of the cell.   151, 444

appendicular skeleton (ap-un-DIK-yuh-lur) Portion of the skeleton forming the pectoral girdle and upper extremities; and the pelvic girdle and lower extremities.   214

appendix (uh-PEN-diks) Small, tubular appendage that extends outward from the cecum of the large intestine.   88

aqueous humor (AY-kwee-us, AK-wee-) Clear, watery fluid between the cornea and lens of the eye.   279

aquifer (AHK-wuh-fur) Underground reservoir that lies between two nonporous layers of rock.   483, 505

arteriole (ar-TEER-ee-ohl) Vessel that takes blood from an artery to capillaries.   126

artery Vessel that takes blood away from the heart to arterioles; characteristically possessing thick elastic and muscular walls.   126

arthritis Joint inflammation.   218

articular cartilage (ar-TIK-yuh-lur) Hyaline cartilaginous covering over the articulating surface of the bones of synovial joints.   207

association area Region of the cerebral cortex related to memory, reasoning, judgment, and emotional feelings.   261

aster Short microtubule that extends outward from a spindle pole in animal cells during cell division.   392

asthma (AZ-muh, as-) Condition in which bronchioles constrict and cause difficulty in breathing.   180

astigmatism (uh-STIG-muh-tiz-um) Blurred vision due to an irregular curvature of the cornea or the lens.   283

atherosclerosis (ath-uh-roh-skluh-ROH-sis) Condition in which fatty substances accumulate abnormally beneath the inner linings of the arteries.   137

atom Smallest unit of matter that cannot be divided by chemical means.   16

ATP (adenosine triphosphate) (uh-DEN-uh-seen try-FAHS-fayt) Nucleotide with three phosphate groups. The breakdown of ATP into ADP + ℗ makes energy available for energy-requiring processes in cells.   35

atrial natriuretic hormone (ANH) (AY-tree-ul nay-tree-yoo-RET-ik) Substance secreted by the atria of the heart that accelerates sodium excretion so that blood volume and blood pressure decreases.   197, 304

atrioventricular bundle (ay-tree-oh-ven-TRIK-yuh-lur) Group of specialized fibers that conduct impulses from the atrioventricular node to the ventricular muscle of the heart; A-V bundle.   130

atrioventricular valve Valve located between the atrium and the ventricle.   128

atrium (AY-tree-um) Chamber; particularly an upper receiving chamber of the heart lying above the ventricles; either the left atrium or the right atrium.   128

atrophy (AT-ruh-fee) Wasting away or decrease in size of an organ or tissue.   237

auditory tube Extension from the middle ear to the nasopharynx for equalization of air pressure on the eardrum.   177, 286

australopithecine (aw-stray-loh-PITH-uh-syn) One of several species of *Australopithecus*, a genus that contains the first generally recognized hominids.   469

autoimmune disease Disease that results when the immune system mistakenly attacks the body's own tissues.   160

autonomic system (awt-uh-NAHM-ik) Branch of the peripheral nervous system that has control over the internal organs; consists of the sympathetic and parasympathetic systems.   255

autosome (AW-tuh-sohm) Chromosome other than a sex chromosome.   386

AV (atrioventricular) node Small region of neuromuscular tissue that transmits impulses received from the SA node to the ventricular walls.   130

axial skeleton (AK-see-ul) Portion of the skeleton that supports and protects the organs of the head, the neck, and the trunk.   212

axon (AK-sahn) Fiber of a neuron that conducts nerve impulses away from the cell body.   246

axon bulb Small swelling at the tip of one of many endings of the axon.   251

B

B lymphocyte (LIM-fuh-syt) Lymphocyte that matures in the bone marrow and, when stimulated by the presence of a specific antigen, gives rise to antibody-producing plasma cells.   150

Bacteria One of three domains of life; prokaryotic cells other than archaea with unique genetic, biochemical, and physiological characteristics.   346

ball and socket joint Most freely movable type of joint (e.g., the shoulder or hip joint).   218

basal body Structure in the cell cytoplasm, located at the base of a cilium or flagellum.   54

basal nuclei (BAY-sul) Subcortical nuclei deep within the white matter that serve as relay stations for motor impulses and produce dopamine to help control skeletal muscle activities.   261

base Solution in which pH is greater than 7; a substance that contributes or liberates hydroxide ions (OH^-) in a solution; alkaline; opposite of acid. Also, a term commonly applied to one of the components of a nucleotide.   23

basement membrane Layer of nonliving material that anchors epithelial tissue to underlying connective tissue.   62

basophil (BAY-suh-fil) Leukocyte with a granular cytoplasm and that is able to be stained with a basic dye.   115, 148

bile (byl) Secretion of the liver that is temporarily stored in the gallbladder before being released into the small intestine, where it emulsifies fat.   87

binary fission Bacterial reproduction into two daughter cells without the utilization of a mitotic spindle.   346

biodiversity Total number of species, the variability of their genes, and the ecosystems in which they live.   5

biogeochemical cycle (by-oh-jee-oh-KEM-ih-kul) Circulating pathway of an element through the biotic and abiotic components of ecosystems within the biosphere.   482

biological magnification Process by which nonexcreted substances like DDT become more concentrated in organisms in the higher trophic levels of the food chain.   506

biosphere (BY-uh-sfeer) That portion of the surface of the earth (air, water, and land) where living things exist.   4, 476

biotic potential (by-AHT-ik) Maximum population growth rate under ideal conditions.   497

birth-control pill Oral contraception containing estrogen and progesterone.   330

blastocyst (BLAS-tuh-sist) Early stage of human embryonic development that consists of a hollow fluid-filled ball of cells.   366

blind spot Area of the eye containing no rods or cones and where the optic nerve passes through the retina.   281

blood Type of connective tissue in which cells are separated by a liquid called plasma.   66

blood pressure Force of blood pushing against the inside wall of an artery.   132

bone Connective tissue having a hard matrix of mineral salts deposited around protein fibers.   65

brain Enlarged superior portion of the central nervous system located in the cranial cavity of the skull.   258

brain stem Portion of the brain consisting of the medulla oblongata, pons, and midbrain.   259

bronchiole (BRAHNG-kee-ohl) Smaller air passages in the lungs that eventually terminate in alveoli.   169

bronchus (pl., bronchi) (BRAHNG-kus) One of two major divisions of the trachea leading to the lungs.   169

buffer Substance or compound that prevents large changes in the pH of a solution.   24

bulbourethral gland (bul-boh-yoo-REE-thrul) Either of two small structures located below the prostate gland in males; adds secretions to semen.   318

bulimia nervosa (byoo-LEE-mee-uh, -LIM-ee-, nur-VOH-suh) Eating disorder characterized by binge eating followed by purging by self-induced vomiting or use of a laxative.   104

bursa (BUR-suh) Saclike, fluid-filled structure, lined with synovial membrane, that occurs near a joint.   218

bursitis (bur-SY-tis) Inflammation of any of the friction-easing sacs called bursae within the knee joint.   218

C

calcitonin (kal-sih-TOH-nin) Hormone secreted by the thyroid gland that helps to regulate the blood calcium level.   302

calorie Amount of heat energy required to raise the temperature of water 1°C.   22

cancer Malignant tumor whose nondifferentiated cells exhibit loss of contact inhibition, uncontrolled growth, and the ability to invade tissues and metastasize.   444

capillary (KAP-uh-lair-ee) Microscopic vessel connecting arterioles to venules and through the thin walls of which substances either exit or enter blood.   126

carbaminohemoglobin (kar-buh-MEE-noh-hee-muh-gloh-bun) Hemoglobin carrying carbon dioxide.   174

carbohydrate Organic compound characterized by the presence of CH_2O groups; includes monosaccharides, disaccharides, and polysaccharides.   27

carbonic anhydrase (kar-BAHN-ik an-HY-drays, -drayz) Enzyme in red blood cells that speeds the formation of carbonic acid from water and carbon dioxide.   174

carcinogen (kar-SIN-uh-jun) Environmental agent that causes mutations leading to the development of cancer.   448

carcinogenesis (kar-suh-nuh-JEN-uh-sis) Development of cancer.   444

carcinoma (kar-suh-NOH-muh) Cancer arising in epithelial tissue.   62

cardiac cycle (KAR-dee-ak) Series of myocardial contractions that constitute a complete heartbeat.   130

cardiac muscle Striated, involuntary muscle tissue found only in the heart.   67

cardiovascular system (kar-dee-oh-VAS-kyuh-lur) Organ system consisting of the blood, heart, and a series of blood vessels that distribute blood under the pumping action of the heart.   70

carnivore (KAR-nuh-vor) Secondary or higher consumer in a food chain, which therefore eats other animals.   478

carotid bodies (kuh-RAHT-id) Structures located at the branching of the carotid arteries and that contain chemoreceptors sensitive to the hydrogen ion concentration but also the level of carbon dioxide and oxygen in blood.   172

carrier Individual that appears normal but is capable of transmitting an allele for a genetic disorder.   414

carrying capacity Largest number of organisms of a particular species that can be maintained indefinitely in an ecosystem.   497

cartilage (KAR-tul-ij, KART-lij) Connective tissue in which the cells lie within lacunae separated by a flexible matrix.   64, 206

cataract (KAT-uh-rakt) Opaqueness of the lens of the eye, making the lens incapable of transmitting light.   279

cecum (SEE-kum) Small pouch that lies below the entrance of the small intestine, and is the blind end of the large intestine.   88

cell Structural and functional unit of an organism; the smallest structure capable of performing all the functions necessary for life.   2, 42

cell body Portion of a neuron that contains a nucleus and from which the fibers extend.   246

cell cycle Repeating sequence of events in eukaryotic cells consisting of the phases of interphase, when growth and DNA synthesis occurs, and the stages of mitosis, when cell division occurs.   391

cell theory One of the major theories of biology which states that all organisms are made up of cells; cells are capable of self-reproduction and cells come only from pre-existing cells.   42

cell-mediated immunity Specific mechanism of defense in which T cells destroy antigen-bearing cells.   154

cellulose (SEL-yuh-lohs, -lohz) Polysaccharide composed of glucose molecules; the chief constituent of a plant's cell wall.   28

central nervous system (CNS) Portion of the nervous system consisting of the brain and spinal cord.   246

centriole (SEN-tree-ohl) Short, cylindrical organelle in animal cells that contains microtubules in a 9 + 0 pattern; present in a centrosome and associated with the formation of basal bodies.   53

centromere (SEN-truh-meer) Constricted region of a chromosome where sister chromatids are attached to one another and where the chromosome attaches to a spindle fiber.   391

cerebellum (ser-uh-BEL-um) Part of the brain located posterior to the medulla oblongata and pons that coordinates skeletal muscles to produce smooth, graceful motions.   259

cerebral cortex (suh-REE-brul, SER-uh-brul KOR-teks) Outer layer of cerebral hemispheres; receives sensory information and controls motor activities.   260

cerebral hemisphere One of the large, paired structures that together constitute the cerebrum of the brain. 259

cerebrospinal fluid (sair-uh-broh-SPY-nul, suh-REE-broh-) Fluid found in the ventricles of the brain, in the central canal of the spinal cord, and in association with the meninges. 256

cerebrum (SAIR-uh-brum, suh-REE-brum) Main part of the brain consisting of two large masses, or cerebral hemispheres; the largest part of the brain in mammals.   259

cervix (SUR-viks) Narrow end of the uterus, which leads into the vagina. 322

chemoreceptor (kee-moh-rih-SEP-tur) Sensory receptor that is sensitive to chemical stimulation—for example, sensory receptors for taste and smell. 272

chemotherapy Treatment of cancer with drugs.   453

chlamydia (kluh-MID-ee-uh) Sexually transmitted disease caused by the bacterium *Chlamydia trachomatis*; often causes painful urination and swelling of the testes in men; is usually symptomless in women but can cause inflammation of the cervix or uterine tubes.   348

chlorofluorocarbons (klor-oh-floor-oh-KAR-buns) Organic compounds containing carbon, chlorine, and fluorine atoms. CFCs such as Freon can deplete the ozone shield by releasing chlorine atoms in the upper atmosphere.   504

chordae tendineae (KOR-dee TEN-din-ee-ee) Tough bands of connective tissue that attach the papillary muscles to the atrioventricular valves within the heart.   128

chorion (KOR-ee-ahn) Extraembryonic membrane functioning for respiratory exchange in reptiles and birds; contributes to placenta formation in mammals.   368

chorionic villi (kor-ee-AHN-ik VIL-eye) Treelike extensions of the chorion of the embryo, projecting into the maternal tissues at the placenta.   368

choroid (KOR-oyd) Vascular, pigmented middle layer of the eyeball.   278

chromatid One of the two side-by-side replicas in a duplicated chromosome. 391

chromatin (KROH-muh-tin) Threadlike network in the nucleus that is made up of DNA and proteins.   49

chromosome (KROH-muh-som) Rodlike structure in the nucleus seen during cell division; contains the hereditary units, or genes.   49

chronic bronchitis Obstructive pulmonary disorder that tends to recur, marked by inflamed airways filled with mucus, and degenerative changes in the bronchi, including loss of cilia.   179

chyme (kym) Thick, semi-liquid food material that passes from the stomach to the small intestine.   86

ciliary body (SIL-ee-air-ee) Structure associated with the choroid layer that contains ciliary muscle, which controls the shape of the lens of the eye.   278

cilium (pl., cilia) (SIL-ee-um) Short, hairlike extension from a cell, occurring in large numbers and used for cell mobility.   54, 62

circadian rhythm (sur-KAY-dee-un) Regular physiological or behavioral event that occurs on an approximately 24-hour cycle.   309

cirrhosis (sih-ROH-sis) Chronic, irreversible injury to liver tissue; commonly caused by frequent alcohol consumption.   89

cleavage furrow Indentation that begins the process of cleavage, by which animal cells undergo cytokinesis. 393

clonal selection theory States that the antigen selects which lymphocyte will undergo clonal expansion and produce more lymphocytes bearing the same type of receptor.   151

cloning Production of identical copies; in genetic engineering, the production of many identical copies of a gene. 432

clotting Process of blood coagulation, usually when injury occurs.   116

cochlea (KOHK-lee-uh, KOH-klee-uh) Portion of the inner ear that

resembles a snail's shell and contains the spiral organ, the sense organ for hearing.   286

cochlear canal (KOH-klee-ur) Canal within the cochlea that bears small hair cells that function as hearing receptors.   287

cochlear nerve Either of two cranial nerves that carry nerve impulses from the organ of Corti to the brain, thereby contributing to the sense of hearing; auditory nerve.   287

codominance Pattern of inheritance in which both alleles of a gene are equally expressed.   412

codon (KOH-dahn) "Triplet" of bases in mRNA that directs the placement of a particular amino acid into a polypeptide.   427

coelom (SEE-lum) Embryonic body cavity lying between the digestive tract and body wall that is completely lined by mesoderm; in humans, the embryonic coelom becomes the thoracic and abdominal cavities.   69

coenzyme (koh-EN-zym) Nonprotein organic molecule that aids the action of the enzyme to which it is loosely bound.   55

collagen fiber (KAHL-uh-jun) White fiber in the matrix of connective tissue, giving flexibility and strength.   64

collecting duct Tube that receives urine from the distal convoluted tubules of several nephrons within a kidney. 193

colon (KOH-lun) Portion of the large intestine that extends from the cecum to the rectum.   88

colony-stimulating factor (CSF) Protein that stimulates differentiation and maturation of white blood cells.   115

color vision Ability to detect the color of an object, dependent on three kinds of cone cells.   280

colostrum (kuh-LAHS-trum) Thin, milky fluid rich in proteins, including antibodies, that is secreted by the mammary glands a few days prior to or after delivery before true milk is secreted.   377

columnar epithelium (kuh-LUM-nur ep-uh-THEE-lee-um) Tissue with cylindrical cells that lines hollow organs such as the digestive tract. 62

community Group of many different populations that interact with one another.   476

compact bone Hard bone consisting of osteons (Haversian systems) cemented together.   65, 206

complement system Group of plasma proteins that form a nonspecific defense mechanism, often by puncturing microbes; it complements the antigen-antibody reaction.   150

complementary base pairing Pairing of bases between nucleic acid strands; adenine pairs with either thymine (DNA) or uracil (RNA), and cytosine pairs with guanine.   423

conclusion Statement made following an experiment as to whether the results support or falsify the hypothesis.   8

condensation synthesis Chemical change resulting in the covalent bonding of two monomers with the accompanying loss of a water molecule.   27

condom Latex sheath used to cover the penis during sexual intercourse; used as a contraceptive and to minimize the risk of transmitting infection. 331

cone cell Bright-light sensory receptor in the retina of the eye that detects color and provides visual acuity.   280

congestive heart failure Inability of the heart to maintain adequate circulation, especially of the venous blood returned to it.   139

connective tissue Type of tissue characterized by cells separated by a matrix that often contains fibers.   64

constipation (kahn-stuh-PAY-shun) Infrequent, difficult defecation caused by insufficient water in the feces. Chronic constipation is associated with hemorrhoids.   89

consumer Organism that feeds on another organism in a food chain; primary consumers eat plants, and secondary (or higher) consumers eat animals.   478

contraceptive (kahn-truh-SEP-tiv) Medication or device used to reduce the chance of pregnancy.   330

control group In experimentation, a sample that goes through all the steps of an experiment except the one being tested; a standard against which results of an experiment are checked.   10

cornea (KOR-nee-uh) Transparent, anterior portion of the outer layer of the eyeball.   278

coronary artery (KOR-uh-nair-ee) Artery that supplies blood to the wall of the heart.   135

corpus luteum (KOR-pus LOOT-ee-um) Yellow body that forms in the ovary from a follicle that has discharged its egg; it secretes progesterone.   325

cortisol (KOR-tuh-sawl) Glucocorticoid secreted by the adrenal cortex that results in increased blood glucose. 304

covalent bond (coh-VAY-lent) Chemical bond between atoms that results from the sharing of a pair of electrons.   20

cranial nerve (KRAY-nee-ul) Nerve that arises from the brain.   252

creatine phosphate (KREE-uh-teen FAHS-fayt) Compound unique to muscles that contains a high-energy phosphate bond.   235

creatinine (kree-AH-tuhn-een) Nitrogenous waste, the end product of creatine phosphate metabolism. 189

cretinism (KREE-tun-iz-um) Condition resulting from improper development of the thyroid in an infant.   301

Cro-Magnon (kroh-MAG-nun) Common name for the first fossils to be accepted as representative of modern humans.   471

crossing-over Exchange of corresponding segments of genetic material between nonsister chromatids of homologous chromosomes during synapsis of meiosis I.   394

cuboidal epithelium (kyoo-BOYD-ul) Tissue with cube-shaped cells, lining hollow organs such as the kidney tubules.   62

Cushing syndrome (KOOSH-ing) Condition resulting from hypersecretion of glucocorticoids; characterized by thin arms and legs and a "moon face," and accompanied by high blood glucose and sodium levels.   407

cyclic AMP (SY-klik, SIH-klik) ATP-related compound that acts as the second messenger in peptide hormone transduction; it initiates activity of the metabolic machinery.   295

cytokine (SY-tuh-kyn) Type of protein secreted by a T lymphocyte that attacks viruses, virally infected cells, and cancer cells.   157

cytoplasm (SY-tuh-plaz-um) Semifluid medium between the nucleus and the plasma membrane and that contains the organelles.   44

cytosine (C) (SY-tuh-seen) One of four nitrogen bases in nucleotides composing the structure of DNA and RNA.   423

cytoskeleton Fibrous protein elements found throughout the cytoplasm that help maintain the shape of the cell, anchor the organelles, and allow the cell and its organelles to move.   44

cytotoxic T cell (sy-tuh-TAHK-sik) T lymphocyte that attacks and kills antigen-bearing cells.   154

D

data Facts that are derived from observations and experiments pertinent to the matter under study.   8

decomposer Organism, usually a bacterium or fungus, that breaks down organic matter into inorganic nutrients that can be recycled in the environment.   478

defecation (def-ih-KAY-shun) Discharge of feces from the rectum through the anus.   88

deforestation (dee-for-uh-STAY-shun) Removal of trees from a forest in a way that ever reduces the size of the forest.   508

delayed allergic response Allergic response initiated at the site of the allergen by sensitized T cells, involving macrophages and regulated by cytokines.   158

demographic transition Decline in the birthrate following a reduction in the death rate so that the population growth rate is lowered. 498

denaturation (dee-nay-chuh-RAY-shun) Loss of normal shape by an enzyme so that it no longer functions; caused by a less than optimal pH and temperature.   32

dendrite (DEN-dryt) Fiber of a neuron, typically branched, that conducts signals toward the cell body.   246

dense fibrous connective tissue Type of connective tissue containing many collagen fibers packed together, and found in tendons and ligaments, for example.   64

dental caries (KAR-eez) Commonly called a cavity; tooth decay that occurs when bacteria within the mouth metabolize sugar and give off acids that erode teeth.   83

dermis (DUR-mus) Inner layer of skin that lies beneath the epidermis.   73

desertification (dih-zurt-uh-fuh-KAY-shun) Transformation of marginal lands to desert conditions.   507

detrital food chain (dih-TRYT-ul) Straight-line linking of organisms in the detrital food web according to who eats whom.   481

detritus (dih-TRYT-us) Partially decomposed remains of plants and animals found in soil and on the beds of bodies of water.   478

diabetes mellitus (dy-uh-BEE-teez MEL-ih-tus, muh-ly-tus) Condition characterized by a high blood glucose level and the appearance of glucose in the urine, due to a deficiency of insulin production or glucose uptake by cells.   307

diaphragm (DY-uh-fram) Dome-shaped horizontal sheet of muscle and connective tissue; also, a birth-control device consisting of a soft rubber or latex cup that fits over the cervix. 172, 330

diarrhea (dy-uh-REE-uh) Frequent watery defecation, often caused by digestive infection or stress.   89

diastole (dy-AS-tuh-lee) Relaxation of a heart chamber.   130

diastolic pressure (dy-uh-STAHL-ik) Arterial blood pressure during the diastolic phase of the cardiac cycle. 132

diencephalon (dy-en-SEF-uh-lahn) Portion of the brain in the region of the third ventricle that includes the thalamus and hypothalamus. 259

diffusion (dih-FYOO-zhun) Movement of molecules from a region of higher concentration to a region of lower concentration. 47

digestive system Consists of the mouth, esophagus, stomach, small intestine, and large intestine (colon) along with associated organs: teeth, tongue, salivary glands, liver, gallbladder, and pancreas. Receives food and digests it into nutrient molecules. 70

diploid (DIP-loyd) $2n$ number of chromosomes; twice the number of chromosomes found in gametes. 390

disaccharide (dy-SAK-uh-ryd) Sugar that contains two units of a monosaccharide; e.g., maltose. 27

distal convoluted tubule (DIS-tul TOO-byool) Highly coiled region of a nephron, that is distant from the glomerular capsule, where tubular secretion takes place. 193

diuretic (dy-uh-RET-ik) Drug used to counteract hypertension by inhibiting Na$^+$ reabsorption so that less water is reabsorbed in the nephron. 197

DNA (deoxyribonucleic acid) Nucleic acid found in cells; the genetic material that specifies protein synthesis in cells. 34, 422

DNA fingerprinting Using DNA fragment lengths, resulting from restriction enzyme cleavage, to identify particular individuals. 433

DNA ligase (LY-gays) Enzyme that links DNA fragments; used during production of recombinant DNA to join foreign DNA to vector DNA. 432

dominant allele Hereditary factor that expresses itself in the phenotype when the genotype is heterozygous. 404

dorsal-root ganglion (GANG-glee-un) Mass of sensory neuron cell bodies located in the dorsal root of a spinal nerve. 252

doubling time Number of years it takes for a population to double in size. 497

drug abuse Taking a drug at dose level and under circumstances that increase the potential for harmful effect. 265

duodenum (doo-uh-DEE-num) First portion of the small intestine into which secretions from the liver and pancreas enter. 87

dynamic equilibrium Maintenance of balance when the head and body are suddenly moved or rotated. 289

E

ecological pyramid Pictorial graph representing biomass, organism number, or energy content of each trophic level in a food web—from the producers to the final consumer populations. 481

ecology (ih-KAHL-uh-jee) Study of the interaction of organisms with each other and with the physical environment. 476

ecosystem (EK-uh-sis-tum, EE-kuh-) Region in which populations interact with each other and with the physical environment. 4, 475

ectoderm (EK-tuh-durm) Outer germ layer of the embryonic gastrula; it gives rise to the nervous system and skin. 367

edema (ih-DEE-muh) Swelling due to tissue fluid accumulation in the intercellular spaces. 146

egg Nonflagellate female gamete; also referred to as ovum. 322

elastic cartilage Cartilage composed of elastic fibers, allowing greater flexibility. 65

elastic fiber Yellow fiber in the matrix of connective tissue, providing flexibility. 64

electrocardiogram (ECG or EKG) (ih-lek-troh-KAR-dee-uh-gram) Recording of the electrical activity associated with the heartbeat. 131

electron Subatomic particle that has almost no weight and carries a negative charge; orbits in a shell about the nucleus of an atom. 17

electron transport system Chain of electron carriers in the cristae of mitochondria and thylakoid membrane of chloroplasts. The electrons release energy that is used to establish a hydrogen ion gradient. This gradient is associated with the production of ATP molecules. 56

element Substance that cannot be broken down into substances with different properties; composed of only one type atom. 16

embolus (EM-buh-lus) Moving blood clot that is carried through the bloodstream. 137

embryo (EM-bree-oh) Stage of a multicellular organism that develops from a zygote and before it becomes free living; in seed plants the embryo is part of the seed. 365

emphysema (em-fih-SEE-muh) Lung impairment caused by deterioration of the bronchioles, which trap air in alveoli. 179

emulsification (ih-mul-suh-fuh-KAY-shun) Breaking up of fat globules into smaller droplets by the action of bile salts. 29

endocrine gland (EN-duh-krin) Ductless organ that secretes (a) hormone(s) into the bloodstream. 297

endocrine system One of the major systems involved in the coordination of body activities; uses hormones as chemical messengers secreted into the bloodstream. 70

endoderm (EN-duh-durm) Inner germ layer that lines the archenteron/gut of the gastrula; it becomes the lining of the digestive and respiratory tract and associated organs. 367

endometriosis (en-doh-mee-tree-OH-sus) Presence of uterine tissue outside the uterus, which can contribute to infertility; possibly the result of irregular menstrual flow. 333

endometrium (en-doh-MEE-tree-um) Lining of the uterus, which becomes thickened and vascular during the uterine cycle. 323

endoplasmic reticulum (ER) (en-duh-PLAZ-mik reh-TIK-yuh-lum) Membranous system of tubules, vesicles, and sacs in cells, sometimes having attached ribosomes. Rough ER has ribosomes; smooth ER does not. 50

endorphin (en-DOR-fun) Neuropeptide synthesized in the pituitary gland that suppresses pain. 267

endospore Bacterium that has shrunk its cell, rounded up within the former plasma membrane, and secreted a new and thicker cell wall in the face of unfavorable environmental conditions. 346

environmental resistance Sum total of factors in the environment that limit the numerical increase of a population in a particular region. 497

enzyme (EN-zym) Organic catalyst, usually a protein that speeds up a reaction in cells due to its particular shape. 31

eosinophil (ee-oh-SIN-oh-fill) White blood cell containing cytoplasmic granules that stain with acidic dye. 115

epidermis (ep-uh-DUR-mus) In animals, outer layer of skin, composed of stratified squamous epithelium; in plants, outer tissue of roots, leaves, and nonwoody stems composed of epidermal cells. 73

epididymis (ep-ih-DID-uh-mus) Coiled tubule next to the testes where sperm mature and may be stored for a short time. 318

epiglottis (ep-uh-GLAHT-us) Structure that covers the glottis and closes off the air tract during the process of swallowing. 84, 168

epinephrine (ep-uh-NEF-rin) Hormone secreted by the adrenal medulla in times of stress; also called adrenaline. 303

epiphyseal plate (ep-uh-FIZ-ee-ul) Cartilaginous layer within the epiphysis of a long bone that grows. 208

episiotomy (ih-pee-zee-AHT-uh-mee) Surgical procedure performed during childbirth in which the opening of the vagina is enlarged to avoid tearing. 377

episodic memory Capacity of brain to store and retrieve information with regard to persons and events and such. 262

epithelial tissue (ep-uh-THEE-lee-ul) Type of tissue that covers the external surface of the body and lines its cavities. 62

erythropoietin (ih-rith-roh-poy-EE-tin) Hormone, produced by the kidneys, that speeds the maturation of red blood. 113, 189

esophagus (ih-SAHF-uh-gus) Tube that transports food from the mouth to the stomach. 85

essential amino acids Amino acids required in the human diet because the body cannot make them. 97

estrogen (ES-truh-jun) Female sex hormone, which, along with progesterone, maintains the primary sex organs and stimulates development of the female secondary sex characteristics. 309, 327

evolution Changes that occur in the members of a species with the passage of time, often resulting in increased adaptation of organisms to the environment. 2

excretion Removal of metabolic wastes from the body. 187

exophthalmic goiter (ek-sahf-THAL-mik) Enlargement of the thyroid gland accompanied by an abnormal protrusion of the eyes. 301

experiment Artificial situation devised to test a hypothesis. 10

expiration (ek-spuh-RAY-shun) Act of expelling air from the lungs; exhalation. 166

expiratory reserve volume (ik-SPY-ruh-tor-ee) Volume of air that can be forcibly exhaled after normal exhalation. 170

exponential growth Growth, particularly of a population, in which the increase occurs in the same manner as compound interest. 496

external respiration Exchange of oxygen and carbon dioxide between alveoli and blood. 174

extraembryonic membrane (ek-struh-em-bree-AHN-ik) Membrane that is not a part of the embryo but is necessary to the continued existence and health of the embryo. 368

F

facilitated transport Passive transfer of a substance into or out of a cell along a concentration gradient by a process that requires a carrier. 48

fat Organic molecule that contains glycerol and fatty acids and is found in adipose tissue of vertebrates. 29

fatty acid Molecule that contains a hydrocarbon chain and ends with an acid group. 29

fermentation Anaerobic breakdown of carbohydrates that results in organic end products such as alcohol and lactic acid. 57

fertilization Union of a sperm nucleus and an egg nucleus, which creates the zygote with the diploid number of chromosomes. 364

fiber Plant material that is nondigestible. Insoluble fiber has a laxative effect; soluble fiber prevents bile acids and cholesterol from being absorbed. 96

fibrin (FY-brun) Insoluble protein threads formed from fibrinogen during blood clotting. 116

fibrinogen (fy-BRIN-uh-jun) Plasma protein that is converted into fibrin threads during blood clotting. 116

fibroblast (FY-bruh-blast) Cell type of loose and fibrous connective tissue with cells at some distance from one another and separated by a jellylike matrix containing collagen and elastic fibers. 64

fibrocartilage (fy-broh-KAR-tul-ij, -KART-lij) Cartilage with a matrix of strong collagenous fibers. 65

fight or flight Response in which the sympathetic division accelerates heartbeat and dilates bronchi to provide muscles with a ready supply of glucose and oxygen. 255

fimbria (FIM-bree-uh) Fingerlike extension from the oviduct near the ovary. 322

flagellum (pl., flagella) (fluh-JEL-um) Slender, long extension that propels a cell through a fluid medium. 54

focus Manner by which light rays are bent by the cornea and lens, creating an image on the retina. 279

follicle (FAHL-ih-kul) Structure in the ovary that produces the egg and, in particular, the female sex hormones, estrogen and progesterone. 325

follicle-stimulating hormone (FSH) Hormone secreted by the anterior pituitary gland that stimulates the development of an ovarian follicle in a female or the production of sperm in a male. 321

fontanel (fahn-tun-EL) Membranous region located between certain cranial bones in the skull of a fetus or infant. 212

food chain Portion of a food web. A sequence that describes the energy flow from one population to the next in an ecosystem. Grazing food chains begin with a producer, and detritus food chains begin with partially decomposed organic matter. 481

food web Complex pattern of interlocking and crisscrossing food chains. 481

foramen magnum (fuh-RAY-mun MAG-num) Opening in the occipital bone of the vertebrate skull through which the spinal cord passes. 212

formed element Constituent of blood that is either cellular (red blood cells and white blood cells) or at least cellular in origin (platelets). 111

fossil Any past evidence of an organism that has been preserved in the earth's crust. 462

fossil fuel Remains of once-living organisms that are burned to release energy, such as coal, oil, and natural gas. 487

fovea centralis (FOH-vee-uh sen-TRA-lis, sen-TRAY-lis) Region of the retina consisting of densely packed cones that is responsible for the greatest visual acuity. 279

functional group Cluster of atoms that always behaves in a certain way. 26

G

gallbladder Saclike organ, associated with the liver, that stores and concentrates bile. 91

gamete (GA-meet, guh-MEET) Haploid reproductive cell. Two gametes, most often an egg or a sperm, join in fertilization to form a zygote. 333, 390

ganglion (GANG-glee-un) Collection of neuron cell bodies within the peripheral nervous system. 252

gap junction Junction between cells formed by the joining of two adjacent plasma membranes; it lends strength and allows ions, sugars, and small molecules to pass between cells. 64

gastric gland Gland within the stomach wall that secretes gastric juice. 86

gene (jeen) Unit of heredity that codes for a polypeptide and is passed on to offspring. 422

gene therapy Use of genetically engineered cells or other biotechnology techniques to treat human genetic and other disorders. 439

genetic engineering Use of technology to alter the genome of a living cell for medical or industrial use; bioengineered. 432

genital herpes (JEN-ih-tul HUR-peez) Sexually transmitted disease caused

by herpes simplex virus and sometimes accompanied by painful ulcers on the genitals.   344

genital warts Raised growths on the genitals due to a sexually transmitted disease caused by human papilloma virus.   343

genomic imprinting (jih-NOH-mik, jee-NAHM-ik) Genetic inheritance that is different depending upon the sex of the parent passing it on.   407

genotype (JEE-nuh-typ) Genes of any individual for (a) particular trait(s).   404

gerontology (jer-un-TAHL-uh-jee) Study of aging.   378

gland Epithelial cell or group of epithelial cells that are specialized to secrete a substance.   62

glaucoma (glow-KOH-muh, glaw-KOH-muh) Increasing loss of field of vision, caused by blockage of the ducts that drain the aqueous humor, creating pressure buildup and nerve damage.   279

glomerular capsule (gluh-MAIR-yuh-lur) Double-walled cup that surrounds the glomerulus at the beginning of the nephron.   193

glomerular filtrate Filtered portion of blood contained within the glomerular capsule.   195

glomerular filtration Movement of small molecules from the glomerulus into the glomerular capsule due to the action of blood pressure.   195

glomerulus (gluh-MAIR-uh-lus, gloh-MAIR-yuh-lus) Cluster; for example, the cluster of capillaries surrounded by the glomerular capsule in a nephron, where glomerular filtration takes place.   192

glottis (GLAHT-us) Opening for airflow into the larynx.   84, 168

glucagon (GLOO-kuh-gahn) Hormone secreted by the pancreas which causes the liver to break down glycogen and raises the blood glucose level.   306

glucocorticoid (GLOO-koh-KOR-tih-koyd) Any one of a group of hormones secreted by the adrenal cortex that influences carbohydrate, fat, and protein metabolism.   303

glucose (GLOO-kohs) Six-carbon sugar that organisms degrade as a source of energy during cellular respiration.   27

glycogen (GLY-koh-jun) Storage polysaccharide, found in animals, that is composed of glucose molecules joined in a linear fashion but having numerous branches.   28

glycolysis (gly-KAHL-uh-sis) Metabolic pathway found in the cytoplasm that participates in aerobic cellular respiration and fermentation; it converts glucose to two molecules of pyruvate.   56

Golgi apparatus (GAHL-jee) Organelle, consisting of concentrically folded saccules, which functions in the packaging, storage, and distribution of cellular products.   51

gonad (GOH-nad) Organ that produces sex cells; the ovary, which produces eggs, and the testis, which produces sperm.   318

gonadotropic hormone (goh-nad-uh-TRAHP-ic, -TROH-pic) Substance secreted by anterior pituitary that regulates the activity of the ovaries and testes; principally, follicle-stimulating hormone (FSH) and luteinizing hormone (LH).   298

gonadotropin-releasing hormone (GnRH) Hormone secreted by the hypothalamus that stimulates the anterior pituitary to secrete the gonadotropin hormones.   321

gonorrhea (gahn-nuh-REE-uh) Sexually transmitted disease caused by the bacterium *Neisseria gonorrhoeae* that causes painful urination and swollen testes in men and is usually symptomless in women, but can cause inflammation of the cervix and uterine tubes.   349

granular leukocyte (GRAN-yuh-lur LOO-kuh-syt) White blood cell with prominent granules in the cytoplasm.   115

gray matter Nonmyelinated nerve fibers in the central nervous system.   256

grazing food chain Straight-line linking of organisms in the grazing food web according to who eats whom.   481

grazing food web In ecosystems, interconnected food chains that begin with an autotroph (producer) and continue through a series of heterotrophs (consumers) to a top carnivore.   481

greenhouse effect Reradiation of solar heat toward the earth, caused by gases in the atmosphere.   500

growth factor Chemical messenger that stimulates mitosis and differentiation of target cells that have receptors for it; important in such processes as fetal development, tissue maintenance and repair, and hemopoiesis; sometimes a contributing factor in cancer.   310

growth hormone (GH) Substance secreted by the anterior pituitary; it promotes cell division, protein synthesis, and bone growth.   298

growth rate Yearly percentage of increase or decrease in the size of a population.   496

guanine (G) (GWAH-neen) One of four nitrogen bases in nucleotides composing the structure of DNA and RNA.   423

H

hair cell Mechanoreceptor in the inner ear that lies between the basilar membrane and the tectorial membrane and triggers action potentials in fibers of the auditory nerve.   286

hair follicle Tubelike depression in the skin in which a hair develops.   73

haploid (HAP-loyd) n number of chromosomes; half the diploid number; the number characteristic of gametes which contain only one set of chromosomes.   390

hard palate (PAL-it) Bony, anterior portion of the roof of the mouth.   82

heart Muscular organ located in the thoracic cavity, responsible for maintenance of blood circulation.   128

heart attack Damage to the myocardium due to blocked circulation in the coronary arteries; also called myocardial infarction (MI).   138

heartburn Burning pain in the chest that occurs when part of the stomach contents escape into the esophagus.   85

helper T cell T lymphocyte that releases cytokines and stimulates certain other immune cells to perform their respective functions.   154

hemodialysis (he-moh-dy-AL-uh-sus) Separation of smaller molecules from larger ones within blood.   200

hemoglobin (HEE-muh-gloh-bun) Red iron-containing pigment in blood that combines with and transports oxygen.   111, 174

hemolysis (he-MAHL-uh-sus) Rupture of red blood cells accompanied by the release of hemoglobin.   47

hemophilia (he-moh-FIL-ee-uh) Genetic disorder in which the affected individual is subject to uncontrollable bleeding.   416

hemorrhoids (HEM-uh-royds, HEM-royds) Abnormally dilated blood vessels of the rectum.   139

hepatic portal system (hih-PAT-ik) Portal system that begins at the villi of the small intestine and ends at the liver.   135

hepatic portal vein Vein leading to the liver and formed by the merging blood vessels of the small intestine.   135

hepatitis (hep-uh-TY-tis) Inflammation of the liver. Viral hepatitis occurs in several forms.   91, 345

herbivore ((h)UR-buh-vor) Primary consumer in a food chain; a plant eater.   478

hexose (HEK-sohs) Six-carbon sugar.   27

hinge joint Type of joint characterized by a convex surface of one bone fitting

into a concave surface of another bone so that movement is confined to one place, such as in the knee or interphalangeal joint.   218

hippocampus (hip-uh-KAM-pus) Part of the cerebral cortex where memories form.   262

histamine (HIS-tuh-meen, -mun) Substance, produced by basophils and mast cells in connective tissue, that causes capillaries to dilate.   148

HLA (human leukocyte associated) antigen Protein in a plasma membrane that identifies the cell as belonging to a particular individual and acts as a self-antigen.   154

homeostasis (hoh-mee-oh-STAY-sis) Maintenance of the internal environment, such as temperature, blood pressure, and other body conditions, within narrow limits.   2, 70

hominid (HAHM-uh-nid) Member of the family Hominid containing humans and their direct ancestors who are known only by the fossil record.   469

Homo erectus (HOH-moh ih-REK-tus) Hominid who used fire and migrated out of Africa to Eurasia.   470

Homo habilis (HOH-moh HAB-uh-lus) Hominid of 2 million years ago who is believed to have used tools.   470

homologous chromosome (hoh-MAHL-uh-gus, huh-MAHL-uh-gus) Similarly constructed chromosomes with the same shape and that contain genes for the same traits; also called homologues.   394

homologous structure Structure that is similar in two or more species because of common ancestry.   463

hormone (HOR-mohn) Chemical messenger produced in low concentrations that has physiological and/or developmental effects, usually in another part of the organism.   88, 294

host Organism a parasite lives on or in to obtain nutrients.

human chorionic gonadotropin hormone (HCG) (kor-ee-AHN-ik, goh-nad-uh-TRAHP-in, -TROH-pin) Gonadotropin hormone produced by the chorion that functions to maintain the uterine lining.   327

hyaline cartilage (HY-uh-lin) Cartilage whose cells lie in lacunae separated by a white translucent matrix containing very fine collagen fibers.   64

hydrogen bond Weak attraction between a hydrogen atom carrying a partial positive charge and an atom of another molecule carrying a partial negative charge.   21

hydrolysis (hy-DRAHL-ih-sis) Splitting of a covalent bond by the addition of water.   27

hydrolytic enzyme (hy-druh-LIT-ik) Enzyme that catalyzes a reaction in which the substrate is broken down with the addition of water.   92

hydrophilic (hy-druh-FIL-ik) Type of molecule that interacts with water by dissolving in water and/or forming hydrogen bonds with water molecules.   21

hydrophobic (hy-druh-FOH-bik) Type of molecule that does not interact with water because it is nonpolar.   21

hypertension Elevated blood pressure, particularly the diastolic pressure.   136

hypertrophy (hy-PUR-truh-fee) Increase in muscle size following long-term exercise.   237

hypothalamic-inhibiting hormone (hy-poh-thuh-LAH-mik) Hormone produced by the hypothalamus that inhibits the secretions of the anterior pituitary.   298

hypothalamic-releasing hormone Hormone produced by the hypothalamus that stimulates the secretions of the anterior pituitary.   298

hypothalamus (hy-poh-THAL-uh-mus) Part of the brain located below the thalamus that helps regulate the internal environment of the body and produces releasing factors that control the anterior pituitary.   259, 298

hypothesis (hy-PAHTH-ih-sis) Statement that is capable of explaining present data and is used to predict the outcome of future experimentation.   8

I

immediate allergic response Allergic response that occurs within seconds of contact with an allergen, caused by the attachment of the allergen to IgE antibodies.   158

immune system Population of cells, including leukoctyes and macrophages, that occur in most organs of the body and protect against foreign organisms, some foreign chemicals, and cancerous or other aberrant host cells.   150

immunity Ability of the body to protect itself from foreign substances and cells, including infectious microbes.   148

immunization (im-yuh-nuh-ZAY-shun) Strategy for achieving artificial immunity to the effects of specific disease-causing agents.   156

immunoglobulin (Ig) (im-yuh-noh-GLAHB-yuh-lin, -yoo-lin) Globular plasma protein that functions as an antibody.   152

immunotherapy Treatment of cancer with cytokines.   455

implantation Attachment and penetration of the embryo into the lining of the uterus (endometrium).   331, 365

incomplete dominance Pattern of inheritance in which the offspring shows characteristics intermediate between two extreme parental characteristics—for example, a red and white flower producing pink offspring.   412

incus (ING-kus) Middle of three ossicles of the ear that serve to conduct vibrations from the tympanic membrane to the oval window of the inner ear.   286

induction Ability of a chemical or a tissue to influence the development of another tissue.   367

infant respiratory distress syndrome Condition in newborns, especially premature ones, in which the lungs collapse because of a lack of surfactant lining the alveoli.   169

inferior vena cava (VEE-nuh KAY-vuh) One of the largest veins in the systemic circuit, it collects blood from the lower body regions.   134

inflammatory reaction Tissue response to injury that is characterized by redness, swelling, pain, and heat.   148

inner ear Portion of the ear consisting of a vestibule, semicircular canals, and the cochlea where balance is maintained and sound is transmitted.   286

inorganic molecule Type of molecule that is not derived from a living organism.   26

insertion End of a muscle that is attached to a movable bone.   227

inspiration (in-spuh-RAY-shun) Act of taking air into the lungs; inhalation.   166

inspiratory reserve volume (in-SPY-ruh-tohr-ee) Volume of air that can be forcibly inhaled after normal inhalation.   170

insulin (IN-suh-lin) Hormone secreted by the pancreas that lowers the blood glucose level by promoting the uptake of glucose by cells and the conversion of glucose to glycogen by the liver and skeletal muscles.   306

integration Summing up of excitatory and inhibitory signals by a neuron.   251, 272

integumentary system (in-teg-yoo-MEN-tuh-ree, -MEN-tree) Skin and accessory organs.   70

intercalated disks (in-tur-kuh-LAY-tud) Specialized junctions that hold adjacent cardiac muscle cells together and that appear as dense bands at right angles to the muscle striations.   67

interferon (in-tur-FEER-ahn) Protein formed by a cell infected with a virus that can increase the resistance of other cells to the virus.   150

interleukin (in-tur-LOO-kun) Chemical substance produced by macrophages and T lymphocytes that function as metabolic regulators (cytokines) of themselves and of neighboring cells.   157

internal respiration Exchange of oxygen and carbon dioxide between blood and tissue fluid.   174

interneuron Neuron found within the central nervous system that takes nerve impulses from one portion of the system to another.   246

interphase Interval between successive cell divisions; during this time, the chromosomes are extended and DNA replication and growth are occurring.   391

interstitial cell (in-tur-STISH-ul) Hormone-secreting cell located between the seminiferous tubules of the testes.   321

intervertebral disk (in-tur-VUR-tuh-brul) Layer of cartilage located between adjacent vertebrae.   214

intrauterine device (IUD) (in-truh-YOO-tur-in) Birth-control device consisting of a small piece of molded plastic inserted into the uterus, and believed to alter the uterine environment so that fertilization does not occur.   330

ion (EYE-un, -ahn) Atom or group of atoms carrying a positive or negative charge.   19

ionic bond (eye-AHN-ik) Bond created by an attraction between oppositely charged ions.   19

iris (EYE-ris) Muscular ring that surrounds the pupil and regulates the passage of light through this opening.   278

isotope (EYE-suh-tohp) One of two or more atoms with the same atomic number that differ in the number of neutrons and therefore in weight.   17

J

jaundice (JAWN-dis) Yellowish tint to the skin caused by an abnormal amount of bilirubin (bile pigment) in the blood, indicating liver malfunction.   91

joint Union of two or more bones; an articulation.   206

juxtaglomerular apparatus (juk-stuh-gluh-MER-yuh-lur) Structure located in the walls of arterioles near the glomerulus that regulates renal blood flow.   197

K

karyotype (KAR-ee-uh-typ) Arrangement of all the chromosomes within a cell by pairs in a fixed order.   386

kidney Organ in the urinary system that produces and excretes urine.   188

kingdom One of the seven categories, or taxa, used by taxonomists to group species.   2

kinin (KY-nen) Chemical mediator, released by damaged tissue cells and mast cells, which causes the capillaries to dilate and become more permeable.   148

Krebs cycle (krebz) Cyclical metabolic pathway found in the matrix of mitochondria that participates in aerobic cellular respiration; breaks down acetyl groups to carbon dioxide and hydrogen. Also called the citric acid cycle because the reactions begin and end with citrate.   56

L

lacteal (LAK-tee-ul) Lymphatic vessel in a villus of the intestinal wall.   87

lactose intolerance (LAK-tohs) Inability to digest lactose because of an enzyme deficiency. Symptoms include diarrhea, gas, and cramps.   92

lacuna (luh-KOO-nuh, -KYOO-nuh) Small pit or hollow cavity, as in bone or cartilage, where a cell or cells are located.   64

lanugo (luh-NOO-goh) Short, fine hair that is present during the later portion of fetal development.   375

large intestine Last major portion of the digestive tract, extending from the small intestine to the anus and consisting of the cecum, the colon, the rectum, and the anal canal.   88

laryngitis (lar-un-JY-tis) Infection of the larynx with accompanying hoarseness.   177

larynx (LAR-ingks) Cartilaginous organ located between pharynx and trachea that contains the vocal cords; voice box.   168

lens Clear membranelike structure found in the eye behind the iris; brings objects into focus.   279

leptin Protein hormone produced by adipose tissue that acts on the hypothalamus to signal satiety.   309

leukemia (loo-KEE-mee-uh) Cancer of the blood-forming tissues leading to the overproduction of abnormal white blood cells.   115, 446

ligament (LIG-uh-munt) Dense fibrous connective tissue that joins bone to bone at a joint.   64, 218

limbic system (LIM-bik) Portion of the brain concerned with memory and emotions.   261

lipase (LY-pays, LY-payz) Fat-digesting enzyme secreted by the pancreas.   90, 92

lipid (LIP-id, LY-pid) Organic compound that is insoluble in water; notably fats, oils, and steroids.   29

liver Large organ in the abdominal cavity that has many functions, such as production of proteins and bile and detoxification of harmful substances.   90

long-term potentiation (LTP) (puh-ten-shee-AY-shun) Enhanced response at synapses within the hippocampus, likely essential to memory storage.   262

loop of the nephron (NEF-rahn) Portion of the nephron lying between the proximal convoluted tubule and the distal convoluted tubule that functions in water reabsorption.   193

loose fibrous connective tissue Tissue composed mainly of fibroblasts that are widely separated by a matrix containing collagen and elastic fibers and found beneath epithelium.   64

lumen (LOO-mun) Cavity inside any tubular structure, such as the lumen of the digestive tract.   85

lungs Paired, cone-shaped organs within the thoracic cavity, functioning in internal respiration and containing moist surfaces for gas exchange.   169

luteinizing hormone (LH) (LOO-tee-uh-ny-zing, LOO-tee-ny-zing) Hormone produced by the anterior pituitary gland that stimulates the development of the corpus luteum in females and the production of testosterone in males.   321

lymph (limf) Fluid, derived from tissue fluid, that is carried in lymphatic vessels.   119, 146

lymph nodes Mass of lymphoid tissue located along the course of a lymphatic vessel.   147

lymphatic system (lim-FAT-ik) Mammalian organ system consisting of lymphatic vessels and lymphoid organs.   70, 146

lymphocyte (LIM-fuh-syt) Specialized white blood cell; occurs in two forms—T lymphocyte and B lymphocyte.   115

lysosome (LY-suh-sohm) Membrane-bounded organelle containing digestive enzymes.   52

M

macrophage (MAK-ruh-fayj) Large phagocytic cell derived from a monocyte that ingests microbes and debris.   148

malleus (MAL-ee-us) First of three ossicles of the ear that serve to

conduct vibrations from the tympanic membrane to the oval window of the inner ear.   286

maltase (MAHL-tays, -tayz) Enzyme produced in small intestine that breaks down maltose to two glucose molecules.   92

mast cell Cell to which antibodies, formed in response to allergens, attach, bursting the cell and releasing allergy mediators, which cause symptoms.   148

mastoiditis (mas-toyd-EYE-tis) Inflammation of the mastoid sinuses of the skull.   212

matrix (MAY-triks) Unstructured semifluid substance that fills the space between cells in connective tissues or inside organelles.   64

matter Anything that takes up space and has weight.   16

mechanoreceptor (mek-uh-noh-rih-SEP-tur) Sensory receptor that is sensitive to mechanical stimulation, such as that from pressure, sound waves, and gravity.   272

medulla oblongata (muh-DUL-uh ahb-lawng-GAH-tuh) Part of the brain stem controlling heartbeat, blood pressure, breathing, and other vital functions. It also serves to connect the spinal cord to the brain.   259

medullary cavity (muh-DUL-uh-ree) Cavity within the diaphysis of a long bone containing marrow.   207

megakaryocyte (meg-uh-KAR-ee-oh-syt, -uh-syt) Large bone marrow cell that gives rise to blood platelets.   116

meiosis (my-OH-sis) Type of cell division that in animals occurs during the production of gametes and results in four daughter cells with the haploid number of chromosomes.   394

melanocyte-stimulating hormone (MSH) Substance that causes melanocytes to secrete melanin in lower vertebrates.   73, 298

melatonin (mel-uh-TOH-nun) Hormone, secreted by the pineal gland, that is involved in biorhythms.   309

memory Capacity of the brain to store and retrieve information about past sensations and perceptions; essential to learning.   262

meninges (sing., meninx) (muh-NIN-jeez) Protective membranous coverings about the central nervous system.   69, 256

menisci (sing., meniscus) (muh-NIS-ky, -kee, -sy) Cartilaginous wedges that separate the surfaces of bones in synovial joints.   218

menopause (MEN-uh-pawz) Termination of the ovarian and uterine cycles in older women.   328

menstruation (men-stroo-AY-shun) Loss of blood and tissue from the uterus at the end of a uterine cycle.   327

mesoderm (MEZ-uh-durm, MES-) Middle germ layer of embryonic gastrula; gives rise to the muscles, the connective tissue, and the circulatory system.   367

messenger RNA (mRNA) Ribonucleic acid whose sequence of codons specifies the sequence of amino acids during protein synthesis.   425

metabolism All of the chemical changes that occur within a cell.   54

metaphase Mitosis phase during which chromosomes are aligned at the metaphase plate (equator) of the mitotic spindle.   393

metastasis (muh-TAS-tuh-sis) Spread of cancer from the place of origin throughout the body; caused by the ability of cancer cells to migrate and invade tissues.   444

microtubule (-TOO-byool) Organelle composed of 13 rows of globular proteins; found in multiple units within other organelles, such as the centriole, cilia, flagella, as well as spindle fibers.   53

microvillus (-VIL-us) Cytoplasmic projection from epithelial cells, usually containing actin filaments; microvilli greatly increase the surface area of a tissue.   62

midbrain Most superior part of the brain stem; contains traits and reflex centers.   259

middle ear Portion of the ear consisting of the tympanic membrane, the oval and round windows, and the ossicles; where sound is amplified.   286

mineral Homogeneous inorganic substance; certain minerals are required for normal metabolic functioning of cells and must be in the diet.   102

mineralocorticoid (min-ur-uh-loh-KOR-tih-koyd) Hormones secreted by the adrenal cortex that regulate salt and water balance, leading to increases in blood volume and blood pressure.   303

mitochondrion (my-tuh-KAHN-dree-un) Membranous organelle in which aerobic cellular respiration produces the energy carrier ATP.   52

mitosis (my-TOH-sis) Type of cell division in which daughter cells receive the exact chromosome and genetic makeup of the parent cell; occurs during growth and repair.   391

molecule Like or different atoms joined by a bond; the unit of a compound.   19

monoclonal antibody Antibody of one type produced by a single plasma cell.   158

monocyte (MAHN-uh-syt) Type of agranular leukocyte that functions as a phagocyte.   115

monosaccharide (mahn-uh-SAK-uh-ryd) Simple sugar; a carbohydrate that cannot be decomposed by hydrolysis.   27

motor neuron Neuron that takes nerve impulses from the central nervous system to the effectors.   246

motor unit Motor neuron and all the muscle fibers it innervates.   234

mucous membrane (MYOO-kus) Membrane that lines a cavity or tube that opens to the outside of the body; also called mucosa.   69

multiple allele Pattern of inheritance in which there are more than two alleles for a particular trait, although each individual has only two of these alleles.   412

muscle twitch Contraction of a whole muscle in response to a stimulus that causes an action potential in one or more muscle fibers.   234

muscular (contractile) tissue Type of tissue that contains cells capable of contracting; skeletal muscles are attached to the skeleton, smooth muscle is found within walls of internal organs, and cardiac muscle makes up the heart.   67

mutagen (MYOO-tuh-jun) Agent, such as radiation or a chemical, that brings about a mutation in DNA.   448

mutation Permanent change in DNA that leads to change in the phenotype.   424

myelin sheath (MY-uh-lin) Schwann plasma membranes that cover long neuron fibers, giving them a white, glistening appearance.   247

myocardium (MY-oh-kar-dee-um) Cardiac muscle in the wall of the heart.   128

myofibril (my-uh-FY-brul) Contractile portion of muscle fibers.   231

myoglobin (MY-uh-gloh-bin) Pigmented compound in muscle tissue that stores oxygen.   237

myogram Recording of a muscular contraction.   234

myosin (MY-uh-sin) One of two major proteins of muscle; makes up thick filaments in myofibrils and is capable of breaking down ATP. See actin.   231

myxedema (mik-sih-DEE-muh) Condition resulting from a deficiency of thyroid hormone in an adult.   301

N

nasal cavity One of two canals in the nose separated by a septum.   167

nasopharynx (nay-zoh-FAR-ingks) Region of the pharynx associated with the nasal cavity.   84

natural killer (NK) cell Lymphocyte that causes an infected or cancerous cell to burst.   150

natural selection Process by which populations become adapted to their environment.   465

Neanderthal (nee-AN-dur-thahl, -tahl) Hominid with a sturdy build who lived during the last Ice Age in Europe and the Middle East; hunted large game and has left evidence of being culturally advanced.   471

negative feedback Self-regulatory mechanism that is activated by an imbalance and results in a fluctuation above and below a mean.   77

nephron (NEF-rahn) Anatomical and functional unit of the kidney; kidney tubule.   191

nerve Bundle of nerve fibers outside the central nervous system.   68, 252

nerve impulse Action potential (electrochemical change) traveling along a neuron.   248

nervous system Consists of the brain, spinal cord, and associated nerves.   70

nervous tissue Type of animal tissue; contains nerve cells (neurons), which conduct impulses, and neuroglial cells, which support, protect, and provide nutrients to neurons.   68

neuroglial cell (noo-RAHG-lee-ul, noo-ROHG-lee-ul) One of several types of cells found in nervous tissue; that supports, protects, and nourishes neurons.   68, 246

neuromuscular junction Point of communication between a nerve cell and a muscle fiber.   232

neuron (NOOR-ahn, NYOOR-) Nerve cell that characteristically has three parts: dendrites, cell body, and axon.   68, 246

neurotransmitter Chemical stored at the ends of axons that is responsible for transmission across a synapse.   251

neutron (NOO-trahn) Subatomic particle that has a weight of one atomic mass unit, carries no charge, and is found in the nucleus of an atom.   17

neutrophil (NOO-truh-fil) Granular leukocyte that is the most abundant of the white blood cells; first to respond to infection.   115

niche (nich) Role an organism plays in community, including its habitat and its interactions with other organisms.   78

nitrogen fixation Process whereby nitrogen gas is reduced and nitrogen is added to organic compounds.   485

node of Ranvier (RAHN-vee-ay) Gap in the myelin sheath around a nerve fiber.   247

nondisjunction Failure of homologous chromosomes or sister chromatids to separate during the formation of gametes.   398

nonsteroid hormone Type of hormone that is not a steroid, such as a peptide, that is received by a plasma membrane receptor and brings about a change in metabolic reactions within a cell.   295

norepinephrine (NE) (NOR-ep-uh-NEF-rin) Neurotransmitter active in the peripheral and central nervous systems; also a hormone secreted by the adrenal medulla in times of stress.   251, 303

nuclear envelope Double membrane that surrounds the nucleus and is continuous with the endoplasmic reticulum.   49

nuclear pore Opening in the nuclear envelope which permits the passage of proteins into the nucleus and ribosomal subunits out of the nucleus.   49

nuclei (NOO-klee-eye, NYOO-) In the reticular formation, masses of cell bodies in the CNS.   259

nucleolus (noo-KLEE-uh-lus) Organelle found inside the nucleus where rRNA is produced for ribosome formation.   49

nucleoplasm (NOO-klee-uh-plaz-um) Semifluid medium of the nucleus, containing chromatin.   49

nucleotide Monomer of a nucleic acid that forms when a nitrogen base, a pentose sugar, and a phosphate join.   34

nucleus (NOO-klee-us, NYOO-) Region of a eukaryotic cell, containing chromosomes, that controls the structure and function of the cell.   44

O

obesity (oh-BEE-sih-tee) Excess adipose tissue; exceeding ideal weight by more than 20%.   104

oil Substance, usually of plant origin and liquid at room temperature, formed when a glycerol molecule reacts with three fatty acid molecules.   29

oil gland Gland of the skin, associated with hair follicle, that secretes sebum; also called sebaceous gland.   73

olfactory cell (ahl-FAK-tuh-ree, -tree, ohl-) Neuron modified as a sensory receptor for the sense of smell.   277

omnivore (AHM-nuh-vor) Organism in a food chain that feeds on both plants and animals.   478

oncogene (AHNG-koh-jeen) Gene that contributes to the transformation of a normal cell into a cancerous cell.   446

oogenesis (oh-uh-JEN-uh-sis) Production of an egg in females by the process of meiosis and maturation.   322, 399

opportunistic infection Infection that has an opportunity to occur because the immune system has been weakened.   342

optic nerve Either of two cranial nerves that carry nerve impulses from the retina of the eye to the brain, thereby contributing to the sense of sight.   279

organ Combination of two or more different tissues performing a common function.   2

organ system Group of related organs working together.   2

organelle (or-guh-NEL) Specialized structure within cells (e.g., nucleus, mitochondria, and endoplasmic reticulum).   44

organic molecule Molecule that always contains carbon (C) and hydrogen (H); organic molecules are associated with living things.   26

origin End of a muscle that is attached to a relatively immovable bone.   227

oscilloscope (ah-SIL-uh-skohp, uh-sil-) Apparatus that records changes in voltage by graphing them on a screen; used to study the nerve impulse.   248

osmosis (ahz-MOH-sis, ahs-) Movement of water from an area of higher concentration of water to an area of lower concentration of water across a differentially permeable membrane.   47

ossicle (AHS-ih-kul) One of the small bones of the middle ear—malleus, incus, stapes.   286

ossification Formation of bone tissue.   208

osteoblast Bone-forming cell.   208

osteoclast (AHS-tee-uh-klast) Cell that breaks down and causes reabsorption of bone.   208

osteocyte (AHS-tee-uh-syt) Mature bone cell.   208

osteoporosis Condition in which bones break easily because calcium is removed from them faster than it is replaced.   210

otitis media (oh-TY-tis MEE-dee-uh) Bacterial infection of the middle ear, characterized by pain and possibly by a sense of fullness, hearing loss, vertigo, and fever.   177

otolith (OH-tuh-lith) Calcium carbonate granule associated with ciliated cells in the utricle and the saccule.   289

outer ear Portion of ear consisting of the pinna and auditory canal.   286

oval window Membrane-covered opening between the stapes and the inner ear.   286

ovarian cycle (oh-VAIR-ee-un) Monthly changes occurring in the ovary that determine the level of sex hormones in the blood.   325

ovary In animals, the female gonad, the organ that produces eggs, estrogen, and progesterone; in flowering plants, the base of the pistil that protects ovules and along with associated tissues becomes a fruit.   309, 322

oviduct (OH-vuh-dukt) Tube that transports eggs to the uterus; also called uterine tube. 322

ovulation (ahv-yuh-LAY-shun, ohv-) Discharge of a mature egg from the follicle within the ovary. 322

oxygen debt Oxygen that is needed to metabolize lactate, a compound that accumulates during vigorous exercise. 235

oxyhemoglobin (ahk-see-HEE-muh-gloh-bin) Compound formed when oxygen combines with hemoglobin. 174

oxytocin (ahk-sih-TOH-sin) Hormone released by the posterior pituitary that causes contraction of uterus and milk letdown. 298

ozone shield Formed from oxygen in the upper atmosphere, it protects the earth from ultraviolet radiation. 504

P

pacemaker See SA (sinoatrial) node. 130

pain receptor Sensory receptor that is sensitive to chemicals released by damaged tissues or excess stimuli of heat or pressure. 272

PAN (peroxyacetylnitrate) (puh-rahk-see-uh-SEE-tul) Type of chemical found in photochemical smog. 503

pancreas (PANG-kree-us, PAN-) Elongate, flattened organ in the abdominal cavity that secretes enzymes into the small intestine (exocrine function) and hormones into the blood (endocrine function). 90, 306

pancreatic amylase (pang-kree-AT-ik AM-uh-lays, -layz) Enzyme in the pancreas that digests starch. 90, 92

pancreatic islets (of Langerhans) Distinctive group of cells within the pancreas that secretes insulin and glucagon. 306

Pap test Analysis done on cervical cells for detection of cancer. 323

parasympathetic division (par-uh-sim-puh-THET-ik) That part of the autonomic system that usually promotes activities associated with a restful state. 255

parathyroid gland (par-uh-THY-royd) One of four glands embedded in the posterior surface of the thyroid gland; produces parathyroid hormone. 302

parathyroid hormone (PTH) Hormone secreted by the four parathyroid glands that increases the blood calcium level and decreases the blood phosphate level. 302

parturition (par-tyoo-RISH-un, par-chuh-) Processes that lead to and include birth and the expulsion of the afterbirth. 376

pathogen (PATH-uh-jun) Disease-causing agent. 62, 111, 148, 340

pectoral girdle (PEK-tur-ul) Portion of the skeleton that provides support and attachment for the arms. 216

pelvic girdle Portion of the skeleton to which the legs are attached. 217

pelvic inflammatory disease (PID) Disease state of the reproductive organs caused by a sexually transmitted organism. 348

penis External organ in males through which the urethra passes and that serves as the organ of sexual intercourse. 318

pentose (PEN-tohs, -tohz) Five-carbon sugar; deoxyribose is the pentose sugar found in DNA; ribose is a pentose sugar found in RNA. 27

pepsin (PEP-sin) Protein-digesting enzyme secreted by gastric glands. 86, 92

peptidase (PEP-tih-days, -dayz) Intestinal enzyme that breaks down short chains of amino acids to individual amino acids that are absorbed across the intestinal wall. 90, 92

peptide bond Covalent bond that joins two amino acids. 32

perception Mental awareness of sensory stimulation. 272

perforin (PUR-for-in) Molecule in cytotoxic T cells that perforates the plasma membrane of the target cell to allow water and salts to enter and swell the cell, causing it to burst. 154

pericardium (paier-ih-KAR-dee-um) Protective serous membrane that surrounds the heart. 128

periodontitis (paier-ee-oh-dahn-TY-tus) Inflammation of the gums. 83

periosteum (paier-ee-AHS-tee-um) Fibrous connective tissue covering the surface of bone. 207

peripheral nervous system (PNS) (puh-RIF-ur-ul) Nerves and ganglia that lie outside the central nervous system. 246

peristalsis (paier-ih-STAWL-sis) Rhythmic contraction that serves to move the contents along in tubular organs, such as the digestive tract. 85

peritonitis (paier-ih-tuh-NY-tis) Generalized infection of the lining of the abdominal cavity. 69, 88

peritubular capillary network (paier-ih-TOO-byuh-lur) Capillary network that surrounds a nephron and functions in reabsorption during urine formation. 192

pH scale Measure of the hydrogen ion concentration [H^+]; any pH below 7 is acidic and any pH above 7 is basic. 24

phagocytosis (fag-uh-sy-TOH-sis) Taking in of bacteria and/or debris by engulfing; cell eating (verb, phagocytize). 115

pharynx (FAR-ingks) Portion of the digestive tract between the mouth and the esophagus which serves as a passageway for food and also air on its way to the trachea. 84, 167

phenotype (FEE-nuh-typ) Outward appearance of an organism caused by the genotype and environmental influences. 404

pheromone (FER-oh-mohn) Chemical substance secreted by one organism that influences the behavior of another. 294

phlebitis (flih-BY-tis) Inflammation of a vein. 139

phospholipid (fahs-foh-LIP-id) Molecule having the same structure as a neutral fat except one bonded fatty acid is replaced by a group that contains phosphate; an important component of plasma membranes. 30

photochemical smog Air pollution that contains nitrogen oxides and hydrocarbons, which react to produce ozone and PAN (peroxyacetylnitrate). 503

photoreceptor Light-sensitive sensory receptor. 272

pineal gland (PIN-ee-ul, PY-nee-ul) Gland—either at the skin surface on the dorsal side of the head (fish, amphibians) or in the third ventricle of the brain, (mammals)—that produces melatonin. 309

pituitary dwarf (pih-TOO-ih-taier-ee, -TYOO-) Person of normal proportions but small stature, caused by inadequate growth hormone. 300

pituitary gland Small gland that lies just inferior to the hypothalamus; the anterior pituitary produces several hormones, some of which control other endocrine glands; the posterior pituitary stores and secretes oxytocin and antidiuretic hormone. 298

placenta (pluh-SEN-tuh) Structure that forms from the chorion and the uterine wall and allows the embryo, and then the fetus, to acquire nutrients and rid itself of wastes. 327, 368

plaque (plak) Accumulation of soft masses of fatty material, particularly cholesterol, beneath the inner linings of the arteries. 98

plasma (PLAZ-muh) Liquid portion of blood. 66, 118

plasma cell Cell derived from a B-cell lymphocyte that is specialized to mass-produce antibodies. 151

plasma membrane Membrane surrounding the cytoplasm that consists of a phospholipid bilayer with embedded proteins; functions to regulate the entrance and exit of molecules from cell. 44

plasmid (PLAZ-mid) Self-duplicating ring of accessory DNA in the cytoplasm of bacteria.   432

platelet (PLAYT-lit) Cell fragment that is necessary to blood clotting; also called a thrombocyte.   66, 116

pleural membrane (PLOOR-ul) Serous membrane that encloses the lungs.   69, 172

pleurisy (PLOOR-ih-see) Infection of the pleural membranes.   69

pneumonectomy (noo-muh-NEK-tuh-mee, nyoo-) Surgical removal of all or part of a lung.   180

pneumonia (noo-MOHN-yuh, nyoo-) Infection of the lungs that causes alveoli to fill with mucus and pus.   179

podocyte (PAHD-uh-syt) Epithelial cell of Bowman's capsule attached to the outer surface of the glomerular capillary basement membrane by cytoplasmic foot processes.   193

polar body Nonfunctioning daughter cell, formed during oogenesis, that has little cytoplasm.   399

pollution Detrimental alternation of normal constituents of air, land, and water due to human activities.   500

polygenic inheritance (pahl-ee-JEN-ik) Pattern of inheritance in which many allelic pairs control a trait; each dominant allele has a quantitative effect on the phenotype.   411

polymerase chain reaction (PCR) (PAHL-uh-muh-rays, -rayz) Technique that uses the enzyme DNA polymerase to produce millions of copies of a particular piece of DNA.   433

polyp (PAHL-ip) Small, abnormal growth that arises from the epithelial lining.   89

polypeptide Polymer of many amino acids linked by peptide bonds.   32

polyribosome (pahl-ih-RY-buh-sohm) Cluster of ribosomes attached to the same mRNA molecule; each ribosome is producing a copy of the same polypeptide.   50, 429

polysaccharide (pahl-ee-SAK-uh-ryd) Carbohydrate composed of many bonded glucose units—for example, glycogen.   28

pons (pahnz) Portion of the brain stem above the medulla oblongata and below the midbrain; assists the medulla oblongata in regulating the breathing rate.   259

population All the members of the same species that inhabit a particular area.   4, 476

positive feedback Mechanism of homeostatic response in which the output intensifies and increases the likelihood of response, instead of countering it and canceling it.   77

posterior pituitary Portion of the pituitary gland that stores and secretes oxytocin and antidiuretic hormone which are produced by the hypothalamus.   298

prefrontal area Association area in the frontal lobe that receives information from other association areas and uses it to reason and plan actions.   261

primary motor area Area in the frontal lobe where voluntary commands begin; each section controls a part of the body.   260

primary somatosensory area (soh-mat-uh-SENS-ree, -suh-ree) Area dorsal to the central sulcus where sensory information arrives from skin and skeletal muscles.   260

primate (PRY-mayt) Animal that belongs to the order Primates, the order of mammals that includes prosimians, monkeys, apes, and humans.   469

principle Theory that is generally accepted by an overwhelming number of scientists. Also called law.   8

producer Organism at the start of a food chain that makes its own food (e.g., green plants on land and algae in water).   478

product Substance that forms as a result of a reaction.   54

progesterone (proh-JES-tuh-rohn) Female sex hormone secreted by the corpus luteum of the ovary and by the placenta.   309, 327

prolactin (PRL) (proh-LAK-tin) Hormone secreted by the anterior pituitary that stimulates the production of milk from the mammary glands.   298

prophase (PROH-fayz) Mitosis phase during which chromatin condenses so that chromosomes appear.   392

proprioceptor (proh-pree-oh-SEP-tur) Sensory receptor that assists the brain in knowing the position of the limbs.   274

prostaglandin (PG) (prahs-tuh-GLAN-din) Hormone that has various and powerful local effects.   310

prostate gland (PRAHS-tayt) Gland located around the male urethra below the urinary bladder; adds secretions to semen.   318

protein Organic compound that is composed of either one or several polypeptides.   31

prothrombin (proh-THRAHM-bin) Plasma protein that is converted to thrombin during the steps of blood clotting.   116

prothrombin activator Enzyme that catalyzes the transformation of the precursor prothrombin to the active enzyme thrombin.   116

proto-oncogene (PROH-toh-AHNG-koh-jeen) Normal gene that can become an oncogene through mutation.   446

proton Subatomic particle found in the nucleus of an atom that has a weight of one atomic mass unit and carries a positive charge; a hydrogen ion.   17

proximal convoluted tubule Highly coiled region of a nephron near the glomerular capsule, where tubular reabsorption takes place.   193

pulmonary artery (POOL-muh-naier-ee, PUUL-) Blood vessel that takes blood away from the heart to the lungs.   129, 134

pulmonary circuit That part of the circulatory system that takes deoxygenated blood to and oxygenated blood away from the gas-exchanging surfaces in the lungs.   134

pulmonary embolism (EM-buh-liz-um) Blockage of a pulmonary artery by a blood clot that commonly originates in a vein of the lower legs.   139

pulmonary fibrosis (fy-BROH-sis) Accumulation of fibrous connective tissue in the lungs; caused by inhaling irritating particles, such as silica, coal dust, or asbestos.   179

pulmonary tuberculosis Tuberculosis of the lungs, caused by the tubercule bacillus.   179

pulmonary vein Blood vessel that takes blood to the heart from the lungs.   129, 134

pulse Vibration felt in arterial walls due to expansion of the aorta following ventricle contraction.   132

Punnett square (PUN-ut) Gridlike device used to calculate the expected results of simple genetic crosses.   406

pupil (PYOO-pul) Opening in the center of the iris of the eye.   278

Purkinje fibers (pur-KIN-jee) Specialized muscle fibers that conduct the cardiac impulse from AV bundle into the ventricular walls.   130

R

reactant (re-AK-tunt) Substance that participates in a reaction.   54

recessive allele (uh-LEEL) Hereditary factor that expresses itself in the phenotype only when the genotype is homozygous.   404

recombinant DNA (rDNA) DNA that contains genes from more than one source.   432

red blood cell (erythrocyte) Formed element that contains hemoglobin and carries oxygen from the lungs to the tissues.   66, 111

red bone marrow Blood cell-forming tissue located in the spaces within spongy bone.   148, 206

reduced hemoglobin (HEE-muh-gloh-bun) Hemoglobin that is carrying hydrogen ions.   174

referred pain Pain perceived as having come from a site other than that of its actual origin.   275

reflex action Automatic, involuntary response of an organism to a stimulus. 84, 253

refractory period (rih-FRAK-tuh-ree) Time following an action potential when a neuron is unable to conduct another nerve impulse. 248

renal artery (REE-nul) Originates from the aorta and delivers blood to the kidney. 188

renal cortex (REE-nul KOR-teks) Outer portion of the kidney that appears granular. 191

renal medulla (REE-nul muh-DUL-uh) Inner portion of the kidney that consists of renal pyramids. 191

renin (REN-in) Enzyme released by kidneys that leads to the secretion of aldosterone and a rise in blood pressure. 197, 304

replacement reproduction Population in which each person is replaced by only one child. 499

replication Making an exact copy, as in the duplication of DNA. 424

reproduce To make a copy similar to oneself, as when one-celled organisms divide or humans have children. 2

reproductive system Organ system specialized for the production of offspring. 70

residual volume Amount of air remaining in the lungs after a forceful expiration. 170

respiratory center Group of nerve cells in the medulla oblongata that send out nerve impulses on a rhythmic basis, resulting in inspiration. 172

respiratory system Consists of the lungs and tubes that bring oxygen into the lungs and take carbon dioxide out. 70

resting potential Polarity across the plasma membrane of a resting neuron due to an unequal distribution of ions. 248

restriction enzyme Bacterial enzyme that stops viral reproduction by cleaving viral DNA; used to cut DNA at specific points during production of recombinant DNA. 432

reticular fiber (rih-TIK-yuh-lur) Very thin collagen fibers in the matrix of connective tissue, highly branched and forming delicate supporting networks. 64

reticular formation Complex network of nerve fibers within the brain stem that arouses the cerebrum. 259

retina (RET-n-uh, RET-nuh) Innermost layer of the eyeball, which contains the rod cells and the cone cells. 279

retinal (RET-n-al, -awl) Light-absorbing molecule, which is a derivative of vitamin A and a component of rhodopsin. 280

retrovirus (REH-tro-vy-rus) Virus that contains only RNA and carries out RNA → cDNA transcription, called reverse transcription. 341

rhodopsin (roh-DAHP-sun) Visual pigment found in the rods whose activation by light energy leads to vision. 280

rib cage Top and sides of the thoracic cavity; contains ribs and intercostal muscles. 172

ribosomal RNA (rRNA) (ry-buh-SOH-mul) RNA occurring in ribosomes, which are the structures involved in protein synthesis. 425

ribosome (RY-buh-sohm) Minute particle that is attached to endoplasmic reticulum or occurs loose in the cytoplasm and is the site of protein synthesis. 50

ribozyme (RY-buh-zym) Enzyme that carries out mRNA processing. 427

RNA (ribonucleic acid) (ry-boh-noo-KLEE-ik) Nucleic acid found in cells that assists DNA in controlling protein synthesis. 34, 425

RNA polymerase (PAHL-uh-muh-rays) Enzyme that speeds the formation of RNA from a DNA template. 427

rod cell Photoreceptor in vertebrate eyes that responds to dim light. 280

round window Membrane-covered opening between the inner ear and the middle ear. 286

S

SA (sinoatrial) node (sy-noh-AY-tree-ul) Small region of neuromuscular tissue that initiates the heartbeat; also called the pacemaker. 130

saccule (SAK-yool) Saclike cavity in the vestibule of the inner ear; contains sensory receptors for static equilibrium. 289

salivary amylase (SAL-uh-vair-ee AM-uh-lays, -layz) Secreted from the salivary glands; the first enzyme to act on starch. 83, 92

salivary gland Gland associated with the oral cavity that secretes saliva. 83

sarcolemma (sar-kuh-LEM-uh) Membrane that surrounds striated muscle cells. 231

sarcomere (SAR-kuh-mir) Structural and functional unit of a myofibril; contains actin and myosin filaments. 231

sarcoplasmic reticulum (sar-kuh-PLAZ-mik rih-tik-yuh-lum) Smooth endoplasmic reticulum of skeletal muscle cells; surrounds the myofibrils and stores calcium ions needed for myosin to bind to actin and therefore muscle contraction. 231

saturated fatty acid Fatty acid molecule that lacks double bonds between the atoms of its carbon chain. 29

Schwann cell (shwahn) Cell that surrounds a fiber of a peripheral nerve and forms the neurilemmal sheath and myelin. 247

science Human endeavor that considers only what is observable by the senses or by instruments that extend the ability of the senses. 8

scientific method Process by which scientists formulate a hypothesis, gather data by observation and experimentation, and come to a conclusion. 8

scientific theory Concept that joins together well-supported and related hypotheses; a conceptual scheme supported by a broad range of observations, experiments, and data. 8

sclera (SKLEER-uh) White, fibrous, outer layer of the eyeball. 278

scrotum (SKROH-tum) Pouch of skin that encloses the testes. 318

seasonal affective disorder (SAD) Depression and desire for sleep caused by lack of light, especially during winter months, and related to melatonin level. 309

selectively permeable Having degrees of permeability; the cell is impermeable to some substances and allows others to pass through at varying rates. 47

semantic memory Capacity of the brain to store and retrieve information with regard to words or numbers and such. 262

semen (SEE-mun) Thick, whitish fluid consisting of sperm and secretions from several glands of the male reproductive tract. 318

semicircular canal (sem-ih-SUR-kyuh-lur) One of three tubular structures within the inner ear that contain sensory receptors responsible for the sense of dynamic equilibrium. 289

semilunar valve (sem-ee-LOO-nur) Valve resembling a half moon located between the ventricles and their attached vessels. 128

seminal vesicle (SEM-uh-nul) Convoluted, saclike structure attached to the vas deferens near the base of the urinary bladder in males; adds secretions to semen. 318

seminiferous tubule (sem-uh-NIF-ur-us) Highly coiled duct within the male testes that produces and transports sperm. 321

sensation Conscious awareness of a stimulus due to nerve impulse sent to the brain from a sensory receptor by way of sensory neurons. 272

sensory adaptation Phenomenon of a sensation becoming less noticeable once it has been recognized by constant repeated stimulation. 272

sensory neuron Neuron that takes nerve impulses to the central nervous

system and typically has a long dendrite and a short axon; afferent neuron.   246

sensory receptor Structure specialized to receive information from the environment and to generate nerve impulses.   272

serous membrane (SEER-us) Membrane that covers internal organs and lines cavities without an opening to the outside of the body; also called serosa.   69

serum (SEER-um) Light yellow liquid left after clotting of blood.   116

sex chromosome Chromosome responsible for the development of characteristics associated with gender; an X or Y chromosome.   386, 414

sex-influenced trait Autosomal trait that is expressed differently in the two sexes.   417

sex-linked trait Phenotype that is controlled by a gene located on a sex chromosome, usually the X chromosome, whose pattern of inheritance differs in males and females.   414

sickle-cell disease Genetic disorder in which the affected individual has sickle-shaped red blood cells that are subject to hemolysis.   413

simple goiter (GOY-tur) Condition in which an enlarged thyroid produces low levels of thyroxine.   301

sinus (SY-nus) Cavity or hollow space in an organ such as the skull.   212

sinusitis (sy-nuh-SY-tis) Infection of the sinuses, caused by blockage of the openings to the sinuses, and characterized by postnasal discharge and facial pain.   177

skeletal muscle Striated, voluntary muscle tissue that makes up skeletal muscles; also called striated muscle.   67

skill memory Capacity of the brain to store and retrieve information with regard to the ability to perform motor activities, such as riding a bike.   262

skin Outer covering of the body; can be considered an organ system; also called the integumentary system.   72

sliding filament theory Movement of actin in relation to myosin; accounts for muscle contraction.   231

small intestine Long, tubelike chamber of the digestive tract between the stomach and large intestine.   87

smooth (visceral) muscle Nonstriated, involuntary muscles found in the walls of internal organs.   67

soap Salt formed from a fatty acid and an inorganic base.   29

sodium-potassium pump Transport protein in the plasma membrane that moves sodium ions out of and potassium ions into animal cells;

important in nerve and muscle cells.   248

soft palate (PAL-it) Entirely muscular posterior portion of the roof of the mouth.   82

somatic cell (soh-MAT-ik) In animals, a body cell, excluding those that undergo meiosis and become a sperm or egg.   390

somatic sense Sense associated with the skin, muscles, joints, or internal organs.   273

somatic system That portion of the peripheral nervous system containing motor neurons that control skeletal muscles.   253

special sense Sense that involves sensory receptors associated with specialized sensory organs, such as the eyes and ears.   273

species Group of similarly constructed organisms capable of interbreeding and producing fertile offspring; organisms that share a common gene pool.   9, 565, 568

specific epithet (EP-uh-thet) In the binomial system of taxonomy, the second part of an organism's name, which may be descriptive.   568

sperm Male sex cell with three distinct parts at maturity: head, middle piece, and tail.   321, 423

spermatogenesis (spur-mat-uh-JEN-ih-sis) Production of sperm in males by the process of meiosis and maturation.   98, 321, 399, 423

sphincter (SFINGK-tur) Muscle that surrounds a tube and closes or opens the tube by contracting and relaxing.   85, 217

spinal cord Part of the central nervous system; the nerve cord that is continuous with the base of the brain and housed within the vetebral column.   256, 332

spinal nerve Nerve that arises from the spinal cord.   252, 328

spindle Structure consisting of fibers, poles, and asters (if animal cell) that brings about the movement of chromosomes during cell division.   88, 392

spiral organ Portion of inner ear that permits hearing and consists of hair cells located on the basilar membrane within the cochlea; also called organ of Corti.   287, 363

spleen Large, glandular organ located in the upper left region of the abdomen that stores and purifies blood.   147, 265

sponge Aquatic, largely marine invertebrate, with saclike body perforated by pores, and a sessile filter feeder; member of the phylum Porifera.   621

spongy bone Porous bone found at the ends of long bones where blood cells are formed.   65, 197, 206, 370

spongy mesophyll (MEZ-uh-fil, MES-) In a plant leaf, the layer of mesophyll containing loosely packed, irregularly spaced cells that increase the amount of surface area for gas exchange; along with palisade mesophyll, it is the site of most of photosynthesis.   163

sporangium (spuh-RAN-jee-um) Structure within which spores are produced.   592, 601

spore Haploid reproductive cell, sometimes resistant to unfavorable environmental conditions, that is capable of producing a new individual which is also hapoid.   180, 584, 601

sporophyte (SPOR-uh-fyt) In the life cycle of a plant, the diploid generation that produces spores by meiosis.   180, 601

spring overturn Mixing process that occurs in spring in stratified lakes whereby the oxygen-rich top waters mix with nutrient-rich bottom waters.   733

squamous epithelium (SKWAY-mus, SKWAH-) Tissue with flat cells, lining hollow organs such as the lungs and blood vessels.   62

stapes (STAY-peez) Last of three ossicles of the ear that serve to conduct vibrations from the tympanic membrane to the oval window of the inner ear.   286

starch Storage polysaccharide found in plants that is composed of glucose molecules joined in a linear fashion.   28

static equilibrium Maintenance of balance when the head and body are motionless.   289

stem cell Any undifferentiated cell that can divide and differentiate into more functionally specific cell types such as blood cells and germ cells.   113

steroid (STEER-oyd) Type of lipid molecule having four interlocking rings; examples are cholesterol, progesterone, and testosterone.   30

steroid hormone Chemical messenger that is lipid soluble and therefore passes through the plasma and nuclear envelope to bind with a receptor inside the nucleus; the complex turns on specific genes leading to the production of particular proteins.   295

stimulus Change in the internal or external environment that a sensory receptor can detect leading to nerve impulses in sensory neurons.   272

stomach Muscular sac that mixes food with gastric juices to form chyme, which enters the small intestine.   86

striated (STRY-ayt-ud) Having bands; cardiac and skeletal muscle are

striated with light and dark bands. 67

stroke Condition resulting when an arteriole in the brain bursts or becomes blocked by an embolism; also called cerebrovascular accident. 138

subcutaneous layer (sub-kyoo-TAY-nee-us) Tissue layer that lies just beneath the skin and contains adipose tissue. 73

substrate Reactant in a reaction controlled by an enzyme. 54

succession Sequential change in the relative dominance of species within a community; primary succession begins on bare rock, and secondary succession begins where soil already exists. 476

superior vena cava (VEE-nuh KAY-vuh) Vein that returns blood from the lower limbs and the greater part of the pelvic and abdominal organs to the right atrium. 134

sustainability Global way of life that can continue indefinitely, because the economic needs of all peoples are met while still protecting the environment. 510

sustentacular (Sertoli) cell (sus-tun-TAK-yuh-lur) Elongated cell in the wall of the seminiferous tubules to which spermatids are attached during spermatogenesis. 321

suture (SOO-chur) Type of immovable joint articulation found between bones of the skull. 218

sweat gland Skin gland that secretes a fluid substance for evaporate cooling; also called sudoriferous gland. 73

sympathetic division That part of the autonomic system that usually promotes activities associated with emergency (fight or flight) situations. 255

synapse (SIN-aps, si-NAPS) Region between two nerve cells where the nerve impulse is transmitted from one to the other, usually from axon to dendrite. 251

synapsis (sih-NAP-sis) Pairing of homologous chromosomes during prophase I of meiosis. 396

synaptic cleft (sih-NAP-tik) Small gap between presynaptic and postsynaptic membranes of a synapse. 251

syndrome Group of symptoms that appear together and tend to indicate the presence of a particular disorder. 387

synovial joint (sih-NOH-vee-ul) Freely movable joint. 218

synovial membrane Membrane that forms the inner lining of the capsule of a freely movable joint. 69

syphilis (SIF-uh-lis) Sexually transmitted disease caused by the bacterium

Treponema pallidum that causes a painless chancre on the penis or cervix; if untreated, can lead to cardiac and central nervous system disorders. 350

systemic circuit That part of the circulatory system that serves body parts other than the gas-exchanging surfaces in the lungs. 132

systole (SIS-tuh-lee) Contraction of a heart chamber. 130

systolic pressure (sis-TAHL-ik) Arterial blood pressure during the systolic phase of the cardiac cycle. 132

T

T lymphocyte (LIM-fuh-syt) Lymphocyte that matures in the thymus and exists in four varieties, one of which kills antigen-bearing cells outright. 150

T (transverse) tubule Membranous channel that extends inward from a muscle fiber membrane and passes through the fiber. 231

taste bud Sense organ containing the receptors associated with the sense of taste. 276

taxonomy (tak-SAHN-uh-mee) Science of naming and classifying organisms into various categories. 468

tectorial membrane (tek-TOR-ee-ul) Membrane in the spiral organ (organ of Corti) that lies above and makes contact with the receptor cells for hearing. 272

telomere Tip of end of a chromosome. 446

telophase (TEL-uh-fayz) Mitosis phase during which the diploid number of daughter chromosomes are located at each pole. 393

template (TEM-plit) Pattern that serves as a mold for the production of an oppositely shaped structure; one strand of DNA is a template for a complementary strand. 424

tendon (TEN-dun) Fibrous connective tissue that joins muscle to bone. 64, 206, 227

testes (sing., testis) (TES-teez) Male gonads, the organ that produces sperm and testosterone. 309, 318

testosterone (tes-TAHS-tuh-rohn) In mammals, major male sex hormone produced by interstitial cells in the testes; it stimulates development of primary sex organs and maintains secondary sexual characteristics in males. 309, 321

tetanus (TET-n-us) Sustained muscle contraction without relaxation. 234

tetany (TET-n-ee) Severe twitching caused by involuntary contraction of the skeletal muscles due to a calcium imbalance. 302

tetrad Four chromatids that result when homologous chromosomes pair during meiosis I. 394

thalamus (THAL-uh-mus) Part of the brain located in the lateral walls of the third ventricle that serves as the integrating center for sensory input; it plays a role in arousing the cerebral cortex. 259

thermal inversion Temperature inversion that traps cold air and its pollutants near the earth, with the warm air above it. 503

thermoreceptor Sensory receptor that is sensitive to changes in temperature. 272

threshold Level of potential at which an action potential or nerve impulse is produced. 248

thrombin (THRAHM-bin) Enzyme that converts fibrinogen to fibrin threads during blood clotting. 116

thromboembolism (thrahm-boh-EM-buh-liz-um) Obstruction of a blood vessel by a thrombus that has dislodged from the site of its formation. 137

thrombus (THRAHM-bus) Blood clot that remains in the blood vessel where it formed. 137

thymine (T) (THY-meen) One of four nitrogen bases in nucleotides composing the structure of DNA. 423

thymus gland Organ that lies in the neck and chest area and is absolutely necessary to the development of immunity. 148, 309

thyroid gland Organ that lies in the neck and produces several important hormones, including thyroxin and calcitonin. 301

thyroid-stimulating hormone (TSH) Substance produced by the anterior pituitary that causes the thyroid to secrete thyroxin. 298

thyroxine (thy-RAHK-sin) Hormone secreted from the thyroid gland that promotes growth and development; in general, it increases the metabolic rate in cells. 301

tidal volume Amount of air normally moved in the human body during an inspiration or expiration. 170

tight junction Junction between cells when adjacent plasma membrane proteins join to form an impermeable barrier. 64

tissue Group of similar cells combined to perform a common function. 2, 62

tissue fluid Solution that bathes and services every cell in the body; also called interstitial fluid. 118

tone Continuous, partial contraction of muscle. 234

tonicity (toh-NIS-ih-tee) Degree to which the concentration of solute versus solvent causes fluids to move into or out of cells. 47

tonsillectomy (tahn-suh-LEK-tuh-mee) Surgical removal of the tonsils. 177

tonsillitis Infection of the tonsils that causes inflammation, and can spread to the middle ears. 82, 177

tonsils Partially encapsulated lymph nodules located in the pharynx. 147, 177

trachea (TRAY-kee-uh) Air tube in insects that transports oxygen in air to the tissues; also the windpipe in terrestrial vertebrates that takes air to the lungs. 168

tracheostomy (tray-kee-AHS-tuh-mee) Creation of an artificial airway by incision of the trachea and insertion of a tube. 168

tract Bundle of neurons forming a transmission pathway through the brain and spinal cord. 256

trait Specific term for a distinguishing phenotypic feature studied in heredity. 403

transcription Process resulting in the production of a strand of mRNA that is complementary to a segment of DNA. 426

transcription factor Protein that binds to DNA and thereby affects the initiation of transcription. 431

transfer RNA (tRNA) Molecule of RNA that carries an amino acid to a ribosome engaged in the process of protein synthesis. 425

transgenic organism Free-living organisms in the environment that have had a foreign gene inserted into them. 434

translation Process by which the sequence of codons in mRNA dictates the sequence of amino acids in a polypeptide. 428

triglyceride (trih-GLIS-uh-ryd) Neutral fat composed of glycerol and three fatty acids. 29

triplet code Genetic code (mRNA, tRNA) in which sets of three bases call for specific amino acids in the formation of polypeptides. 427

trophic level Feeding level of one or more populations in a food web. 481

tropomyosin (trahp-uh-MY-uh-sin, trohp-) Protein that blocks muscle contraction until calcium ions are present. 233

troponin (TROH-puh-nin) Protein that functions with tropomyosin to block muscle contraction until calcium ions are present. 233

trypsin (TRIP-sin) Protein-digesting enzyme secreted by the pancreas. 90, 92

tubular reabsorption Movement of nutrient molecules, as opposed to waste molecules, from the contents of the nephron into blood at the proximal convoluted tubule. 195

tubular secretion Movement of certain molecules from blood into the distal convoluted tubule of a nephron so that they are added to urine. 195

tumor (TOO-mur) Cells derived from a single mutated cell that has repeatedly undergone cell division; benign tumors remain at the site of origin and malignant tumors metastasize. 444

tumor-suppressor gene Gene that suppresses the development of a tumor; the mutated form contributes to the development of cancer. 446

tympanic membrane (tim-PAN-ik) Located between the outer and middle ear and receives sound waves; the eardrum. 286

U

ulcer (UL-sur) Open sore in the lining of the stomach; frequently caused by bacterial infection. 86

umbilical cord Acts as a tube connecting the fetus to the placenta, through which blood vessels pass. 368

unsaturated fatty acid Fatty acid molecule that has one or more double bonds between the atoms of its carbon chain. 29

uracil (U) (YOOR-uh-sil) One of four nitrogen bases in nucleotides composing the structure of RNA. 425

urea (yoo-REE-uh) Primary nitrogenous waste of humans derived from amino acid breakdown. 189

ureter (YOOR-uh-tur) One of two tubes that take urine from the kidneys to the urinary bladder. 188

urethra (yoo-REE-thruh) Tube that takes urine from the bladder to outside. 188, 318

uric acid (YOOR-ik) Waste product of nucleotide metabolism. 189

urinary bladder Organ where urine is stored before being discharged by way of the urethra. 188

urinary system Consists of the kidneys and urinary bladder; rids body of nitrogenous wastes and helps regulate fluid level and chemical contents of blood. 70

uterine cycle (YOO-tur-in, -tuh-ryn) Monthly occurring changes in the characteristics of the uterine lining (endometrium). 327

uterus (YOO-tur-us) Organ located in the female pelvis where the fetus develops; the womb. 322

utricle (YOO-trih-kul) Saclike cavity in the vestibule of the inner ear that contains sensory receptors for static equilibrium. 289

V

vaccine Antigens prepared in such a way that they can promote active immunity without causing disease. 156

vagina Organ that leads from the uterus to the vestibule and serves as the birth canal and organ of sexual intercourse in females. 323

valve Membranous extension of a vessel or the heart wall that opens and closes, ensuring one-way flow. 127

variable Factor that can cause an observable change during the progress of the experiment. 10

varicose veins (VAR-ih-kohs, VAIR-) Irregular dilation of superficial veins, seen particularly in lower legs, due to weakened valves within the veins. 139

vas deferens (vas DEF-ur-unz, -uh-renz) Tube that leads from the epididymis to the urethra in males. 318

vector (VEK-tur) In genetic engineering, a means to transfer foreign genetic material into a cell—for example, a plasmid. 432

vein Vessel that takes blood to the heart from venules; characteristically having nonelastic walls; in a leaf, veins are vascular bundles. 126

vena cava (VEE-nuh KAY-vuh) Large systemic vein that returns blood to the right atrium of the heart; either the superior or inferior vena cava. 129

ventilation Breathing; the process of moving air into and out of the lungs. 172

ventricle (VEN-trih-kul) Cavity in an organ, such as a lower chamber of the heart; or the ventricles of the brain, which are interconnecting cavities that produce and serve as a reservoir for cerebrospinal fluid. 128, 256

venule (VEN-yool, VEEN-) Vessel that takes blood from capillaries to a vein. 127

vernix caseosa (VUR-niks kay-see-oh-suh) Cheeselike substance covering the skin of the fetus. 375

vertebral column (VUR-tuh-brul) Backbone of vertebrates through which the spinal cord passes. 218

vertebrate (VUR-tuh-brit, -brayt) Referring to an animal with a backbone composed of vertebrae. 2

vertigo (VUR-tih-goh) Dizziness and a sense of rotation. 289

vesicle (VES-ih-kul) Small, membranous sac that stores substances within a cell. 50

vestibule (VES-tuh-byool) Space or cavity at the entrance of a canal, such as the cavity that lies between the semicircular canals and the cochlea. 286

vestigial structure (veh-STIJ-ee-ul, -STIJ-ul) Underdeveloped structure that was functional in some ancestor but is no longer functional in a particular organism.   463

villus (pl., villi) (VIL-us) Fingerlike projection from the wall of the small intestine that functions in absorption.   887

virus Noncellular obligate parasite of living cells consisting of an outer capsid and an inner core of nucleic acid.   340

visual accommodation Additional focusing power provided by the lens, as the ciliary muscle controlling the shape of the lens either relaxes or contracts.   279

visual field Area of vision for each eye.   282

vital capacity Maximum amount of air moved in or out of the human body with each breathing cycle.   170

vitamin Essential requirement in the diet, needed in small amounts; often a part of a coenzyme.   100

vitreous humor (VIT-ree-us) Clear gelatinous material between the lens of the eye and the retina.   279

vocal cord Fold of tissue within the larynx; creates vocal sounds when it vibrates.   168

vulva External genitals of the female that surround the opening of the vagina.   323

W

water (hydrologic) cycle Interdependent and continuous circulation of water from the ocean, to the atmosphere, to the land, and back to the ocean.   483

white blood cell (leukocyte) Formed element of which there are several types, each having a specific function in protecting the body from invasion by foreign substances and organisms.   66, 115

white matter Myelinated nerve fibers in the central nervous system.   256

X

X chromosome Female sex chromosome that carries genes involved in sex determination; see Y chromosome.   386

xenotransplantation (zen-uh-trans-plan-TAY-shun) Use of animal organs, instead of human organs, in human transplant patients.   436

X-linked Gene located on the X-chromosome that does not control a sexual feature of the organism.   414

Y

Y chromosome Male sex chromosome that carries genes involved in sex determination; see X chromosome.   386

yellow bone marrow Fat storage tissue found in the cavities within certain bones.   207

yolk sac Extraembryonic membrane that encloses yolk in reptiles and birds; in placental mammals it is the first site of blood cell formation.   368

Z

zygote (ZY-goht) Diploid cell formed by the union of two gametes; the product of fertilization.   322, 365

Credits

Line Art and Readings

Chapter opening text prepared by Kathryn Sergeant Brown.

Chapter 1

Page 25, Ecology Focus   Sources: G. Tyler Miller, *Living in the Environment,* Wadsworth Publishing Company, Belmont, CA, 1992; and Lester R. Brown, et al., *State of the World,* W.W. Norton and Company, Inc., New York, NY, 1993.

Chapter 4

Figure 4.15   Source: Data from T. T. Shintani, *Eat More, Weigh Less™ Diet,* 1983.
Page 99, Health Focus   Reprinted by permission from page 374 of *Understanding Nutrition,* Fifth Edition, by W.N. Whitney, et al.; Copyright © 1990 by West Publishing Company. All rights reserved.
Figure 4.17   Reprinted by permission from *Nutrition: Concepts and Controversies,* Sixth Edition, by Frances Sienkiewicz Sizer and Eleanor Noss Whitney. Copyright © 1994 by West Publishing Company. All Rights Reserved.

Chapter 6

Figure 6.3 from Kent M. Van De Graaff and Stuart Ira Fox, *Concepts of Human Anatomy and Physiology,* 3rd edition, © 1992 The McGraw-Hill Companies, Inc. All Rights Reserved. Reprinted by permission.

Chapter 8

Health Focus, Page 181: The Most Often Asked Questions about Tobacco and Health—and The Answers From "The Most Often Asked Questions About Smoking Tobacco and Health and . . . The Answers," revised July 1993. © American Cancer Society, Inc., Atlanta, GA. Used with permission.

Chapter 11

Figure 11.4 from Kent Van De Graaff and Stuart Ira Fox, *Concepts of Human Anatomy and Physiology,* 4th edition. Copyright © 1995 The McGraw-Hill Companies, Inc. All Rights Reserved. Reprinted by permission.

Chapter 16

Figures 16.5, 16.6, 16.9, 16.11, 16.13, 16.13a Source: Data from Division of STD Prevention. Sexually Transmitted Disease Surveillance, 1996. U.S. Department of Health and Human Services, Public Health Service. Atlanta: Centers for Disease Control and Prevention, September 1997.

Chapter 17

Figure 17.5 from John W. Hole, Jr. *Human Anatomy and Physiology,* 6th edition, © 1993 The McGraw-Hill Companies, Inc. All Rights Reserved. Reprinted by permission.
Figure 17.13 from Kent Van De Graaff and Stuart Ira Fox, *Concepts of Human Anatomy and Physiology,* 4th edition. Copyright © 1995 The McGraw-Hill Companies, Inc. All Rights Reserved. Reprinted by permission.
Figure 17.14 from Kent Van De Graaff and Stuart Ira Fox, *Concepts of Human Anatomy and Physiology,* 4th edition. Copyright © 1995 The McGraw-Hill Companies, Inc. All Rights Reserved. Reprinted by permission.

Chapter 21

Figure 21.8 from Lance A. Liotta, "Cancer Cell Invasion and Metastasis," *Scientific American,* February 1992, Illustration by Dana Burns-Pizer, p. 62; modified figure depicting tissue inhibitors of metalloproteinase.

Photographs

Introduction

Figure I.1: © John Cunningham/Visuals Unlimited; I.A(Background): © Barbara von Hoffman/Tom Stack & Associates; I.A(Toucan): © Ed Reschke/Peter Arnold, Inc.; I.A(Morpho): © Kjell Sandved/Butterfly Alphabet; I.A(Jaguar): © BIOS (Seitre)/Peter Arnold, Inc.; I.A(Orchid): © Max & Bea Hunn/Visuals Unlimited; I.A(Frog): © Kevin Schafer & Martha Hill/Tom Stack & Associates; IB: © George Holton/Photo Researchers, Inc.

Chapter One

Figure 1.1: © The McGraw-Hill Companies, Inc./Jim Shaffer, photographer; 1.3: © Biomed Commun./Custom Medical Stock Photos; 1.4: © Charles M. Falco/Photo Researchers, Inc.; 1.8a: © Martin Dohrn/SPL/Photo Researchers, Inc.; 1.8b: © Comstock, Inc.; 1.8c: © Marty Cooper/Peter Arnold, Inc.; 1.Aa: © Ray Pfortner/Peter Arnold, Inc.; 1.Ab: © John Miller/Tony Stone Images; 1.Ac: © Frederica Georgia/Photo Researchers, Inc.; 1.13,1.14,1.15: © Dwight Kuhn; 1.18: © Don W. Fawcett/Photo Researchers, Inc.; 1.19: © Richard C. Johnson/Visuals Unlimited.

Chapter Two

Figure 2.1(Bikers): © Pascal Rondeau/Tony Stone Images; 2.1(muscle cells): © Prof. P. Motta, Dept. of Anatomy, Univ. La Sapienza, Rome/Photo Researchers, Inc.; 2.2(left): © David M. Phillips/Visuals Unlimited; 2.2(middle): © Robert Caughey/Visuals Unlimited; 2.2(right): Warren Rosenberg/BPS/Tony Stone Images; 2.3: © Alfred Paisieka/Photo Researchers, Inc.; 2.7(top): Courtesy E.G. Pollock; 2.7(bottom): Courtesy Ron Milligan/Scripps Research Institute; 2.8b: © Barry F. King/Biological Photo Service; 2.9(top): © R. Rodewald/Biological Photo Service; 2.9(bottom): © K.G. Murti/Visuals Unlimited; 2.10a: Courtesy Keith Porter; 2.11: Courtesy of Kent McDonald, University of California, Berkeley; 2.12: © David M. Phillips/Photo Researchers, Inc.

Chapter Three

Figure 3.1: © Julie Houck/Westlight; 3.2(all): © Ed Reschke/Peter Arnold, Inc.; 3.4a-d, 3.6a-c: © Ed Reschke; 3.Ba: © Ken Greer/Visuals Unlimited; 3.Bb: © Dr. P. Marazzi/SPL/Photo Researchers, Inc.; 3.Bc: © James Stevenson/SPL/Photo Researchers, Inc.

Chapter Four

Figure 4.4b: © Biophoto Associates/Photo Researchers, Inc.; 4.5b: © Ed Reschke/Peter Arnold, Inc.; 4.5c: © St. Bartholomew's Hospital/PL/Photo Researchers, Inc.; 4.6(left): Photo by Susumu Ito, from Charles Flickinger, *Medical Cellular Biology,* W.B. Saunders, 1979; 4.6(right): © Manfred Kage/Peter Arnold, Inc.; 4.14: © The McGraw-Hill Companies, Inc./Bob Coyle, photographer; 4A: © F. Hache/Photo Researchers, Inc.; 4.16a: © Biophoto Associates/Photo Researchers, Inc.; 4.16b: © Ken Greer/Visuals Unlimited; 4.16c: © Biophoto Associates/Photo Researchers, Inc.; 4.18: © Carl Centineo/Medichrome/The Stock Shop, Inc.; 4.19: © Donna Day/Tony Stone Images; 4.20: © Tony Freeman/PhotoEdit.

Chapter Five

Figure 5.1: Courtesy Camp Good Days and Special Times; 5.3a: © Lennart Nilsson, "Behold Man" Little Brown and Company, Boston; 5.7: © Manfred Kage/Peter Arnold, Inc.; 5.10(both): Courtesy of Stuart I. Fox.

Chapter Six

Figure 6.1: © Mark Harmel/Tony Stone Images; 6.2, 6.7b, 6.7c: © Ed Reschke; 6A: © Biophoto Associates/Photo Researchers, Inc.

Chapter Seven

Figure 7.1: © Damien Lovegrove/SPL/Photo Researchers, Inc.; 7.3(thymus, spleen): © Ed Reschke/Peter Arnold, Inc.; 7.3(lymph): © Fred E. Hossler/Visuals Unlimited; 7.3(bone marrow): © R. Valentine/Visuals Unlimited; 7.7b: Courtesy Dr. Arthur J. Olson, Scripps Institute; 7A: © John Nuhn; 7.8A: © Boehringer Ingelheim International/photo Lennart Nilsson; 7.10a: © Matt Meadows/Peter Arnold, Inc.; 7.11: © Estate of Ed Lettau/Peter Arnold, Inc.; 7B: © Martha Cooper/Peter Arnold, Inc.

Chapter Eight

Figure 8.1: © Seth Resnick/Stock Boston; 8.6: © SIU/Visuals Unlimited; 8.Ac: © Bill Aron/Photo Edit; 8.13a,b: © Martin Rotker/Martin Rotker Photography.

Chapter Nine

Figure 9.1: © Jeff Greenberg/Unicorn Stock Photos; 9.5(left): © David M. Phillips/Visuals Unlimited; 9.5(right, bottom): © 1966 Academic Press, from A.B. Maunsbach, "J. Ultrastruct. Res." 15:242-282.; 9.6a: © Gennaro/Photo Researchers, Inc.; 9.11: © SIU/Peter Arnold, Inc.

Chapter Ten

Figure 10.1(both): © Ed Reschke; 10.3b: © Junebug Clark/Photo Researchers, Inc.; 10.A: © Royce Bair/Unicorn Stock Photos; 10.6b: © The McGraw-Hill Companies, Inc./Joe DeGrandis, photographer; 10.11(fibrous): © Ed Reschke/Peter Arnold, Inc.; 10.11(rest): © Ed Reschke.

Chapter Eleven

Figure 11.1: © Jim McHugh/Outline; 11.2: Courtesy Dr. Hugh E. Huxley; 11.5: © Ed Reschke/Peter Arnold, Inc.; 11.6: © Ed Reschke; 11.7: © Victor B. Eichler; 11.10: © Tim Davis/Photo Researchers, Inc.; 11.11: © G.W. Willis/Biological Photo Service.

Chapter Twelve

Figure 12.1a: © Gerhard Gscheidle/Peter Arnold, Inc.; 12.3b: © M.B. Bunge/Biological Photo Service; 12.4: © Linda Bartlett; 12.5: Courtesy Dr. E.R. Lewis, University of California, Berkeley; 12.11c: © Manfred Kage/Peter Arnold, Inc.; 12.12b: © Colin Chumbley/Science Source/Photo Researchers, Inc.; 12.17: © Marcus Raichle/Peter Arnold, Inc.; 12.19: © Lawrence Migdale/Photo Researchers, Inc.

Chapter Thirteen

Figure 13.1: © Karen Holsinger Mullen/Unicorn Stock Photos; 13.5: © Omikron/SPL/Photo Researchers, Inc.; 13.9: © Lennart Nilsson, from "The Incredible Machine"; 13.10: © Biophoto Associates/Photo Researchers, Inc.; 13.A: © Robert S. Preston and Joseph E. Hawkins, Kresge Hearing Research Institute, University of Michigan; 13.14: © P. Motta/SPL/Photo Researchers, Inc.

Chapter Fourteen

Figure 14.1: © James Darell/Tony Stone Images; 14.6(left): © Bob Daemmrich/Stock Boston; 14.6(right): © Ewing Galloway, Inc.; 14.7: Courtesy Department Illustrations, Washington University School of Medicine. From Clinical Pathological Conference, "Acromegaly, Diabetes, Hypermetabolism, Proteinura and Heart Failure", American Journal of Medicine 20 (1956) 133; 14.8: © Biophoto Associates/Photo Researchers, Inc.; 14.9: © John Paul Kay/Peter Arnold, Inc.; 14.13a: © Custom Medical Stock Photos; 14.13b: © NMSB/Custom Medical Stock Photos; 14.14a,b: "Atlas of Pediatric Physical Diagnosis," Second Edition by Zitelli & Davis, 1992. Mosby-Wolfe Europe Limited, London, UK.

Chapter Fifteen

Figure 15.3b: © Biophoto Associates/Photo Researchers, Inc.; 15.7: © Ed Reschke/Peter Arnold, Inc.; 15.10: © Dr. Landrum B. Shettles; 15.12a: © The McGraw-Hill Companies, Inc./Bob Coyle, photographer; 15.12b: © The McGraw-Hill Companies, Inc./Vincent Ho, photographer; 15.12c,d: © The McGraw-Hill Companies, Inc./Bob Coyle, photographer; 15.12e : © Hank Morgan/Photo Researchers, Inc.; 15.12f: © The McGraw-Hill Companies, Inc./Bob Coyle, photographer; 15.13: © M. Long/Visuals Unlimited.

Chapter Sixteen

Figure 16.1: © Paul Barton/The Stock Market; 16.2b: © K.G. Murti/Visuals Unlimited; 16.5: © CDC/Peter Arnold, Inc.; 16.6: © Charles Lightdale/Photo Researchers, Inc.; 16.7a: © David M. Phillips/Visuals Unlimited; 16.7b: © David M. Phillips/Visuals Unlimited; 16.7c: © R.G. Kessel - C.Y. Shih/Visuals Unlimited; 16.9: © G.W. Willis/BPS/Tony Stone Images; 16.11: © CNRI/SPL/Photo Researchers, Inc.; 16.12: Courtesy of Dr. Ira Abrahamson; 16.13a: Courtesy of Centers for Disease Control; 16.13b,c: © Carroll Weiss/Camera M.D.; 16.13d: © Science VU/Visuals Unlimited.

Aids Supplement

S.1: © NIBSC/SPL/Photo Researchers, Inc.; S.4(all): © Nicholas Nixon; S.6, S.7: Courtesy Dr. Jack Shields, from Lymphology 22:62-66 (1989).

Chapter Seventeen

Figure 17.1: © Joseph Nettis/Stock Boston; 17.2a: © David M. Phillips/Photo Researchers, Inc.; 17.9a: Lennart Nilsson, "A Child is Born," Dell Publishing Company; 17.10: © Petit Format/Science Source/Photo Researchers, Inc.; 17.11: © Claude Edelmann, Petit Format et Guigorz from "First Days of Life"/Black Star; 17.12: © James Stevenson/SPL/Photo Researchers, Inc.; 17.15: © John Cunningham/Visuals Unlimited; 17.16: © Richard Hutchings/Photo Edit.

Chapter Eighteen

Figure 18.1(4): © CNRI/SPL/Photo Researchers, Inc.; 18.1(5): © CNRI/SPL/Photo Researchers, Inc.; 18.2: © Jill Cannefax/EKM-Nepenthe; 18.3a,b: Courtesy of G.H. Valentine; 18.4a: © David M. Phillips/Visuals Unlimited; 18.4b,c: From R. Simensen and R. Curtis Rogers, "Fragile X Syndrome," AMERICAN FAMILY PHYSICIAN 39(5):186, May 1989. © American Academy of Family Physicians; 18.5a,b: Photograph by Earl Plunkett. Courtesy of G.H. Valentine; 18.9(all): © Michael Abbey/Photo Researchers, Inc.

Chapter Nineteen

Figure 19.1: © Mark Gibson/Visuals Unlimited; 19.4a: © Superstock; 19.4b: © Michael Grecco/Stock Boston; 19.4c-f: © The McGraw-Hill Companies, Inc./Bob Coyle, photographer; 19.4g: Courtesy of Mary L. Drapeau; 19.4h: © The McGraw-Hill Companies, Inc./Bob Coyle, photographer; 19.7(both): © Steve Uzzell; 19.8: Courtesy of the Cystic Fibrosis Foundation; 19.11a: © Kevin Fleming/Corbis; 19.13b: © Bill Longcore/Photo Researchers, Inc.

Chapter Twenty

Figure 20.1: © Elizabeth Fulford/AP Photo 1994; 20.11: Courtesy Alexander Rich; 20.13: From M.B. Roth and J.G. Gall, "Cell Biology" 105:1047-1054, 1987. © Rockefeller University Press; 20.16: © Will & Deni McIntyre/Photo Researchers, Inc.; 20.17(both): Courtesy of General Electric Research and Development Center; 20.18a: Courtesy of Genzyme Corporation and Tufts University School of Medicine.

Chapter Twenty One

Figure 21.1: © Adam Smith/Westlight; 21.4: © Dr. Gopal Murti/SPL/Photo Researchers, Inc.; 21.5: © Seth Joel/SPL/Photo Researchers, Inc.; 21.6: © Breast Scanning Unit, Kings College Hospital, London/SPL/Photo Researchers, Inc.; 21.9: © Joel Gordon.

Chapter Twenty Two

Figure 22.3a,b: © Carolina Biological Supply/Phototake; 22.4(Kangaroo): © George Holton/Photo Researchers, Inc.; 22.4(Australian native cat): © Tom McHugh/Photo Researchers, Inc.; 22.4(Sugar glider): © John Sundance/Jacana/Photo Researchers, Inc.

Chapter Twenty Three

Figure 23.1: © Tom Willock/Photo Researchers, Inc.; 23.2a: © Stephen Krasemann/Peter Arnold, Inc.; 23.2b: © Richard Ferguson; 23.2c: © Mary Thatcher/Photo Researchers, Inc.; 23.2d: © Karlene Schwartz; 23.3a: © G.R. Roberts Photo Library; 23.3b: © Bruce Iverson; 23.3c: © Peter K. Ziminiski/Visuals Unlimited; 23.3d: © Kevin Magee/Tom Stack & Associates; 23.4a: © Kevin Schafer/Tom Stack & Associates; 23.4b: © Gregory K. Scott/Photo Researchers, Inc.; 23.5: © Barbara J. Miller/Biological Photo Service; 23.15: © RayPfortner/Peter Arnold, Inc.; 23.17: © David Cavagnaro/Peter Arnold, Inc.; 23.18: © Alex S. Maclean/Peter Arnold, Inc.; 23.A: © Glenn Van Nimwegen.

Chapter Twenty Four

Figure 24.1: © Mark E. Gibson/Visuals Unlimited; 24.4a: © Myrleen Ferguson/Photo Edit; 24.4b: © Wolfgang Kaehler; 24.7a: © John Shaw/Tom Stack & Associates; 24.7b: © Thomas Kitchin/Tom Stack & Associates; 24.8c: © Bill Aron/Photo Edit; 24.9: Courtesy Arlin J. Krueger/Goddard Space Flight Center/NASA; 24.12a: © John Eastcott/Yva Momatiuck/Image Works; 24.13: © G. Prance/Visuals Unlimited; 24.14(Shrimp trawlers): © Ulrike Welsch/Photo Researchers, Inc.; 24.14(Rhino): © James Hancock/Photo Researchers, Inc.; 24.14(Mining): © Barbara Pfeffer/Peter Arnold, Inc.; 24.14(Snake): © John Mitchell/Photo Researchers, Inc.; 24.14(Tires): © Inga Spence/Tom Stack & Associates; 24.14(Honey bee): © L. West/Photo Researchers, Inc.

Index

I-1